ORGANIC REACTIVE INTERMEDIATES

This is Volume 26 of
ORGANIC CHEMISTRY
A series of monographs
Editors: ALFRED T. BLOMQUIST and HARRY WASSERMAN

A complete list of the books in this series appears at the end of the volume.

ORGANIC REACTIVE INTERMEDIATES

EDITED BY

Samuel P. McManus

Department of Chemistry
University of Alabama in Huntsville
Huntsville, Alabama

ACADEMIC PRESS   New York and London   1973
A Subsidiary of Harcourt Brace Jovanovich, Publishers

COPYRIGHT © 1973, BY ACADEMIC PRESS, INC.
ALL RIGHTS RESERVED.
NO PART OF THIS PUBLICATION MAY BE REPRODUCED OR
TRANSMITTED IN ANY FORM OR BY ANY MEANS, ELECTRONIC
OR MECHANICAL, INCLUDING PHOTOCOPY, RECORDING, OR ANY
INFORMATION STORAGE AND RETRIEVAL SYSTEM, WITHOUT
PERMISSION IN WRITING FROM THE PUBLISHER.

ACADEMIC PRESS, INC.
111 Fifth Avenue, New York, New York 10003

United Kingdom Edition published by
ACADEMIC PRESS, INC. (LONDON) LTD.
24/28 Oval Road, London NW1

LIBRARY OF CONGRESS CATALOG CARD NUMBER: 72-84371

PRINTED IN THE UNITED STATES OF AMERICA

CONTENTS

List of Contributors   vii
Preface   ix

1. Free Radicals
Earl S. Huyser

I. Introduction	1
II. Formation of Free Radicals	10
III. Reactions of Free Radicals	17
IV. Reactivities of Free Radicals	50
V. Kinetic Aspects of Free-Radical Chain Reactions	57
References	61

2. Carbenes and Related Species
D. Bethell

I. Introduction	61
II. Electronic States and Structures of Carbenes	65
III. The Formation of Transient Carbenes and Carbenoids in Solution	77
IV. Mechanism and Reactivity in Carbene and Carbenoid Reactions	89
V. Related Species	116
VI. Conclusion	119
References	120

3. Nitrenes
R. A. Abramovitch

I. Introduction: Nomenclature and Historical Background	127
II. Intermediacy of Nitrenes in Reactions	129
III. Electronic Structure and Spectra	131
IV. Methods of Generation of Nitrenes	135
V. Reactions	157
VI. Nitrenium Ions	181
VII. Applications	184
References	188

4. Carbonium Ions
S. P. McManus and C. U. Pittman, Jr.

I. General Aspects and Historical Background	194
II. Methods of Investigating Carbonium Ions	202
III. Methods of Formation	248
IV. Factors Affecting the Stability of Carbonium Ions	258
V. Reactions of Carbonium Ions	287
VI. Bridged Carbonium Ions	302
VII. Related Species	321
References	324

5. Carbanions
E. M. Kaiser and D. W. Slocum

I. General Aspects	337
II. Methods of Investigation of Carbanions	339
III. Preparation of Carbanions	341
IV. Factors Affecting the Stability of Carbanions	342
V. Reactions of Carbanions	344
References	415

6. Radical Ions
Glen A. Russell and Robert K. Norris

I. Introduction	423
II. Formation of Radical Ions	426
III. Methods of Investigation of Radical Ions	431
IV. Reactions Involving Radical Ions	434
References	444

7. Arynes
Ellis K. Fields

I. General Aspects and Historical Background	449
II. Methods of Investigating Arynes	455
III. Methods of Formation of Arynes	463
IV. Factors Affecting the Formation and Stability of Cyclic Acetylenes and Arynes	472
V. Reactions of Arynes	480
VI. Hetarynes	499
VII. Conclusion	503
References	504

AUTHOR INDEX	509
SUBJECT INDEX	533

LIST OF CONTRIBUTORS

Numbers in parentheses indicate the pages on which the authors' contributions begin.

R. A. ABRAMOVITCH, Department of Chemistry, University of Alabama, University, Alabama (127)

D. BETHELL, Department of Organic Chemistry, University of Liverpool, Liverpool, England (61)

ELLIS K. FIELDS, Research and Development Department, Amoco Chemicals Corporation, Naperville, Illinois (449)

EARL S. HUYSER, Department of Chemistry, University of Kansas, Lawrence, Kansas (1)

E. M. KAISER, Department of Chemistry, University of Missouri-Columbia, Columbia, Missouri (337)

S. P. MCMANUS, Department of Chemistry, University of Alabama in Huntsville, Huntsville, Alabama (193)

ROBERT K. NORRIS, *Department of Chemistry, Iowa State University, Ames, Iowa (423)

C. U. PITTMAN, JR., Department of Chemistry, University of Alabama, University, Alabama (193)

GLEN A. RUSSELL, Department of Chemistry, Iowa State University, Ames, Iowa (423)

D. W. SLOCUM, Department of Chemistry, Southern Illinois University, Carbondale, Illinois (337)

* Present address: Department of Chemistry, Imperial College of Science and Technology, London SW7 2AZ, England.

PREFACE

The study of reactive intermediates has been one of the most fruitful and exciting areas among the fields of chemical research that have blossomed as a result of instrumentation and related technological advances. Complementing discovery of knowledge is its dissemination. There has been no attempt to collectively transmit many important basic discoveries and facts about the intermediates and the methods used for their study since Professor J. E. Leffler's classic treatment appeared in 1956. It is the purpose of "Organic Reactive Intermediates" to present a modern treatment of all the major intermediates for those who have had a basic course in organic chemistry. The chapters were written with advanced undergraduates and graduate students as the intended audience since budding chemists at that level of training would most certainly need and appreciate a more thorough introduction to reactive intermediates than the scattered accounts presently available. This treatment should adequately bridge the gap between an elementary treatment and the advanced literature. Supplemented with appropriate problems, "Organic Reactive Intermediates" could serve as a textbook for the first graduate-level course. It could also serve as a useful book for industrial chemists either as a way of keeping abreast of new developments or as a handy reference work dealing with organic reaction mechanisms.

Each contributor is an expert in his field. Authors were asked to consider, after appropriate introductory comments, (1) methods of formation, (2) methods of investigation, (3) factors affecting the stability of the intermediate, (4) reactions of the intermediate, and (5) related species. To limit the book to reasonable size, the authors have been selective rather than comprehensive. This work contains examples from the literature through 1971 with a few references from 1972 added where possible.

There are many persons who deserve acknowledgment. Most of the authors were helpful in formulating the scope of the book. In addition, the following individuals presented me with helpful comments or words of encouragement: Drs. J. I. Brauman, D. J. Cram, R. W. Hoffmann, G. A. Olah, E. C. Steiner, M. Szwarc, and J. A. Zoltewicz. Acknowledgment is also due to the staff of Academic Press for their help at many stages of the development of this book and to Mrs Marie Williams for her able assistance in preparing much of the manuscript. Finally, my dear wife Nancy deserves acknowledgment for her patience and for proofing the entire manuscript.

SAMUEL P. MCMANUS

1

FREE RADICALS

Earl S. Huyser

I.	Introduction	1
	A. Historical Background and Scope	2
	B. Detection of Free Radicals	3
	C. Mechanisms of Free-Radical Reactions	7
II.	Formation of Free Radicals	10
	A. Thermolysis of Covalent Bonds	10
	B. Photochemical Processes	14
	C. Bimolecular Redox Reactions	16
III.	Reactions of Free Radicals	17
	A. Bimolecular Radical-Propagating Reactions	17
	B. Unimolecular Free-Radical-Propagating Reactions	21
	C. Termination Reactions	25
IV.	Reactivities of Free Radicals	26
	A. Resonance Factors	26
	B. Polar Effects	32
	C. Complexing	40
	D. Steric Effects	46
V.	Kinetic Aspects of Free-Radical Chain Reactions	50
	A. General Considerations	50
	B. Steady-State-Derived Rate Laws	53
	C. Determination of Rate Constants	56
	References	57

I. Introduction

Free radicals are chemical species having one or more unpaired electrons. Although some free radicals are stable enough to be isolated or maintained in solution at fairly high concentrations, most are chemically reactive species and encountered only as intermediates in chemical reactions. The purpose of this chapter is to examine the chemistry of free radicals as reaction intermediates and ascertain the unique aspects of such reactions.

A. HISTORICAL BACKGROUND AND SCOPE

The existence of organic free radicals was proposed in the earliest days of the development of organic chemistry. Reductions of simple alkyl halides with zinc or sodium, for example, were thought to yield free alkyl radicals as reaction products. Subsequent investigations, however, showed that stable

$$CH_3I + Na \longrightarrow CH_3\cdot + NaI \qquad (1)$$

dimeric or disproportionation products having the same empirical formulas as the radicals were the actual products formed in these reactions. Gomberg (1900a,b) reported that reactions of triarylmethyl halides with metals did indeed yield free radicals that were stable enough to be maintained in solution for extended periods of time. The intermediacy of free radicals in gas-phase

$$2(C_6H_5)_3CCl + Zn \longrightarrow ZnCl_2 + 2(C_6H_5)_3C\cdot \qquad (2)$$

reactions was brilliantly demonstrated by Paneth (Paneth and Hofeditz, 1929) in his classic lead alkyl pyrolysis experiment. The concept of free radicals as reaction intermediates in solution, however, did not receive careful attention by mechanistic organic chemists until the 1930's. The investigations of Kharasch (Kharasch et al., 1937) and his co-workers (Mayo, 1959) in the United States and Hey and Waters (1937) in England were particularly significant in bringing about the recognition of free radicals as reaction intermediates.

Free radicals are involved as reaction intermediates in a variety of reactions. Examples of reactions that involve free radicals are the light-induced halogenation of alkanes by the elemental halogens [Eq. (3)], the peroxide- or heat-induced formation of macromolecules from low molecular weight compounds

$$RH + X_2 \xrightarrow{h\nu} HX + RX \qquad (X = Cl \text{ or } Br) \qquad (3)$$

(e.g., styrene of methyl methacrylate) [Eq. (4)], and the autoxidation of alkyl aromatics such as cumene by molecular oxygen (Jores, 1954) in a reaction induced by azobisisobutyronitrile (AIBN) [Eq. (5)]. Other examples are as

$$n\text{-}CH_2{=}CHX \xrightarrow{\text{peroxide}} -(CH_2CHX)_n- \qquad (4)$$

$$C_6H_5CH(CH_3)_2 + O_2 \xrightarrow{\text{AIBN}} C_6H_5C(CH_3)_2OOH \qquad (5)$$

follows: peroxide-induced additions of polyhalomethanes to alkenes [Eq. (6)]

$$CX_4 + CH_2{=}CHR \longrightarrow X_3CCH_2CHXR \qquad (6)$$

(Walling and Huyser, 1963); rearrangements of cyclic acetals to esters [Eq. (7)] (Huyser and Garcia, 1962); fragmentation of tertiary alkyl hypohalites [Eq. (8)] (Chattaway and Backeberg, 1923); formation of pentaerithitol chloride in the reaction of diazomethane with carbon tetrachloride [Eq. (9)]

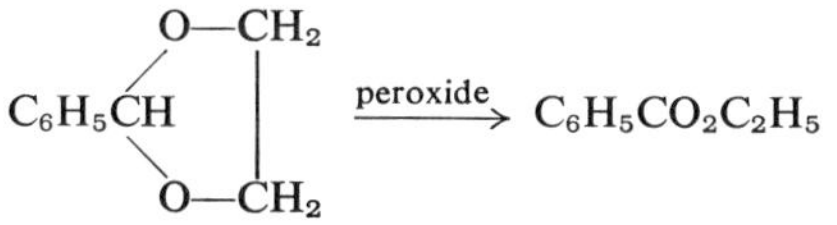

$$C_6H_5\overset{O-CH_2}{\underset{O-CH_2}{CH}} \xrightarrow{peroxide} C_6H_5CO_2C_2H_5 \quad (7)$$

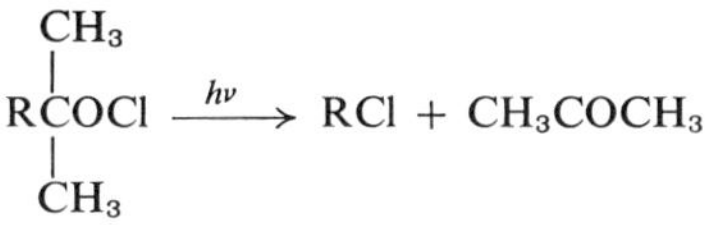

$$R\underset{CH_3}{\overset{CH_3}{\underset{|}{\overset{|}{C}}}}OCl \xrightarrow{h\nu} RCl + CH_3COCH_3 \quad (8)$$

(Urry and Eiszner, 1952); and oxidation of tetrachloroethylene to an acid

$$4\,CH_2N_2 + Cl_4C \longrightarrow C(CH_2Cl)_4 \quad (9)$$

chloride [Eq. (10)] (Dickinson and Leermakers, 1932).

$$Cl_2C{=}CCl_2 + \tfrac{1}{2}O_2 \longrightarrow Cl_3CCOCl \quad (10)$$

Although seemingly quite different, all of these reactions have many significant mechanistic similarities which arise from the intermediacy of free radicals. These similarities result not only from the nature of the behavior of free radicals as reaction intermediates, but also from the kinetic consequences of processes that involve reactive intermediates having the peculiar characteristics of free radicals.

B. DETECTION OF FREE RADICALS

Support for existence of transient species such as reaction intermediates often poses a formidable challenge. Indeed, much of the evidence for most reaction intermediates must come from an inference based on chemical observations. The detection of an intermediate by some physical means becomes a reality only if it can be made to exist long enough to allow for examination. In order to do so, the species may no longer qualify in the strictest sense as a reaction intermediate, since it either must be too stable to react readily or must be trapped in such a manner that its normal course of reaction is prohibited. In spite of these qualifications, some of the more exciting research in organic chemistry has involved the physical detection of reactive intermediates. The detection of free radicals has been a particularly productive area because of the unique properties rendered to such species owing to the unpaired electron(s).

1. *Magnetic Properties*

Free radicals have magnetic properties because of the spin of the unpaired electron(s). Magnetic susceptibility and electron spin resonance (ESR) are the principal means of detecting free radicals.

Magnetic susceptibility (Michaelis, 1949; Mulay, 1963) is determined by measuring the effect of placing a sample that contains free radicals in a strong magnetic field. Both solid and liquid samples can be examined by the Gouy method, which involves suspending the sample in a cylindrical container in the magnetic field and measuring the influence of this field by means of a sensitive balance (see Fig. 1). Liquids and, with some modifications, gases can be measured for magnetic susceptibility by the Quinke method in which the magnetic force is exerted on a sample placed in capillary tubes (see Fig. 2).

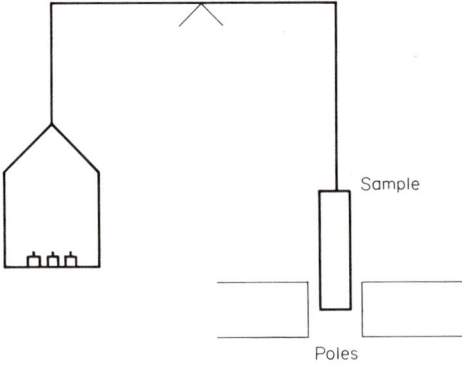

Fig. 1. Principle of the Gouy susceptibility method of detecting free radicals.

Although the sensitivity of magnetic susceptibility techniques in detecting free radicals is considerably lower than that of the ESR method described later in this section, organic reactions that involve free radicals as intermediates have been examined by these methods. The reduction of duroquinone with glucose and polymerization of 2,3-dimethyl-1,3-butadiene (Selwood, 1956) both involve free-radical concentrations that can be measured by these techniques. In the former case, a maximum radical concentration of 7×10^{-3} mole/liter was observed in the reduction of a duroquinone solution having an initial concentration of 1.4×10^{-2} mole/liter. The reaction most likely involves a comparatively stable semiquinone radical as an intermediate and does not qualify as a typical free-radical chain reaction involving a low, steady-state concentration of reactive free radicals. Magnetic susceptibility measurements give direct information only on the number of free radicals.

1. FREE RADICALS

Information concerning the structures of these intermediates cannot be obtained by this technique.

In ESR techniques (Kevan, 1969), the energy absorbed in exciting the unpaired electron to a higher energy level while in a magnetic field is measured. The energy requirements for the excitation of the unpaired electron and information concerning the structure of free radicals may be obtained. It is significant that the spin of the unpaired electron is coupled with the spins of various nuclei in the free radical, particularly with those of the protons. As a consequence, a high resolution ESR spectrum may show many of the same characteristics of a typical proton magnetic resonance spectrum. The number and intensities of the lines of an ESR spectrum of a free radical are determined by the number of nuclear spins coupled with the electron spin. [See Kevan (1969) and Carrington and McLachlan (1967) for examples of ESR spectra.]

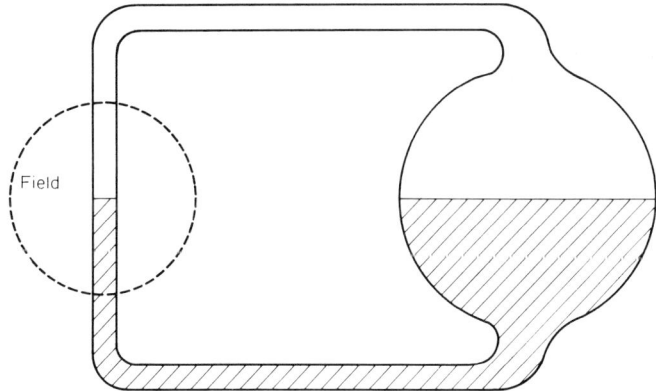

Fig. 2. Principle of the Quinke method for determining magnetic susceptibility.

Although significantly more sensitive than magnetic susceptibility techniques, ESR as a means of observing free radicals as reaction intermediates also has limitations. Radical concentrations as low as 10^{-8} mole/liter may be detected, but high resolution at lower concentrations of free radicals is not readily attained. The concentrations of free radicals in most chain reactions are often lower than the detection limits of the ESR equipment generally available. Rather special methods are required for generating free radicals in concentrations high enough to be investigated by ESR. Radiolysis and photolysis of compounds can be used to produce free radicals that may be trapped in frozen matrices or, in some cases, examined at the instant of their formation before they are destroyed by reaction. Chemical or electrochemical redox reactions that produce stable radical anions or cations have also been employed. In some instances, rapid-flow methods have been used with success.

2. Stable Free Radicals

Some free radicals are stable enough that solutions of them can be examined spectroscopically. Characteristic of most such free radicals is a high degree of resonance stabilization because of extensive delocalization of the unpaired electron throughout the species and/or steric properties that protect the radical site and, consequently, hinder reaction of the radicals with themselves. Other radicals that are less hindered at the radical site in many cases are able to react with the stable free radicals.

The electronic absorption characteristics of species having an extensive conjugative system as well as an unpaired electron are such that they not only absorb light in the visible region but generally have high extinction coefficients. Listed in Table I are some of the more familiar stable free radicals along with their principal canonical structures. These free radicals could not be classified as typical reaction intermediates, but they do serve an interesting role because

Table I
Stable Free Radicals

Name	Structure	Reference
Diphenylpicryhydrazyl	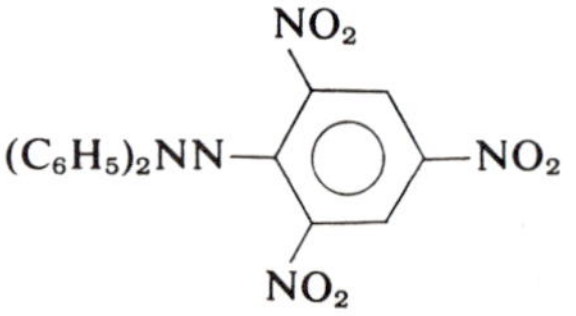 $(C_6H_5)_2NN$—⟨NO_2, NO_2, NO_2⟩	Muller et al. (1935); Baun and Miller (1951)
"Galvinoxyl"	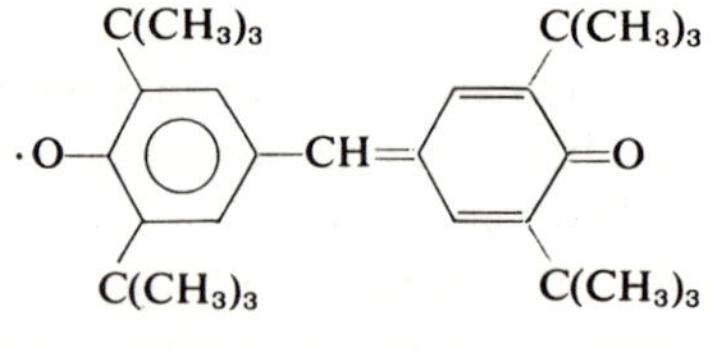 ·O—⟨$C(CH_3)_3$, $C(CH_3)_3$⟩—CH=⟨$C(CH_3)_3$, $C(CH_3)_3$⟩=O	Coppinger (1957)
2,4,6-Tri-*tert*-butylphenoxyl	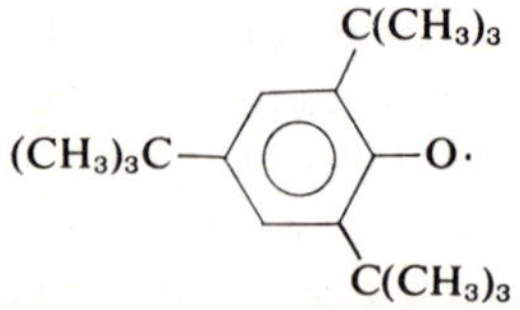 $(CH_3)_3C$—⟨$C(CH_3)_3$, $C(CH_3)_3$⟩—O·	Cook et al. (1956)
Di-*tert*-butylnitroxyl	$[(CH_3)_3C]_2NO$	Hoffmann and Henderson (1961, 1964)

of their optical properties and reactivity toward other radicals as a means of detecting the presence of other free radicals. The reaction with reactive free radicals results in the loss of the characteristic absorption spectra of the stable free radical and can be followed spectroscopically. Stable free radicals can serve, therefore, as a means of quantitatively measuring the rate of formation of other more reactive free radicals.

3. *Overhauser Effect*

A particularly facile means of detecting the intermediacy of free radicals using nuclear magnetic resonance (NMR) equipment is receiving increasing amounts of attention. Because of a weak coupling of electron and nuclear spins in a static magnetic field, the populations of nuclear spin energy levels are influenced by the direction of the electron spins. A change of the electron spin caused by absorption of energy in the radio-frequency range may cause an increased polarization of the nuclear spins. Deviations in the equilibrium populations of the nuclear spins can be observed either as an NMR absorption or emission (Hausser and Stehlik, 1968; Carrington and McLachlan, 1967).

C. Mechanisms of Free-Radical Reactions

1. *Chain Reactions*

One of the most characteristic aspects of most (but not all) free-radical reactions of interest to organic chemists is that the reactants are converted to products in a chain sequence of reactions involving free radicals as reactive intermediates. These free-radical chain reactions require a series of reactions of free radicals. These successive reactions ultimately yield not only the reaction products but also result in regeneration of the free radicals that participate in the chain sequence. For example, the bromination of the alkyl side chain of toluene with bromine proceeds by such a chain sequence of

$$Br\cdot \ + \ C_6H_5CH_3 \longrightarrow HBr \ + \ C_6H_5CH_2\cdot \quad (11)$$

$$C_6H_5CH_2\cdot \ + \ Br_2 \longrightarrow C_6H_5CH_2Br \ + \ Br\cdot \quad (12)$$

$$\overline{C_6H_5CH_3 \ + \ Br_2 \longrightarrow C_6H_5CH_2Br \ + \ HBr} \quad (13)$$

reactions. The bromine atom required for the first reaction of this two-step chain sequence is regenerated in the second step. The reactants toluene and bromine are converted to the products hydrogen bromide and benzyl bromide in the chain reaction. The algebraic sum of the reactions that comprise the chain sequence is the stoichiometry of the free-radical chain reaction. Some other free-radical chain reactions will help to illustrate this principle. Addition of carbon tetrachloride to an alkene, for example, proceeds by a two-step chain sequence, whereas a chain sequence of three reactions is

$$Cl_3C \cdot + CH_2 = CHR \longrightarrow Cl_3CCH_2\dot{C}HR \quad (14)$$

$$Cl_3CCH_2\dot{C}HR + Cl_4C \longrightarrow Cl_3CCH_2CHClR + Cl_3C \cdot \quad (15)$$

$$Cl_4C + CH_2 = CHR \longrightarrow Cl_3CCH_2CHClR \quad (16)$$

necessary to account for the reactions of trialkyl phosphites with mercaptans (Walling and Rabinowitz, 1959).

$$RS \cdot + P(OC_2H_5)_3 \longrightarrow R\dot{S}P(OC_2H_5)_3 \quad (17)$$

$$R\dot{S}P(OC_2H_5)_3 \longrightarrow R \cdot + SP(OC_2H_5)_3 \quad (18)$$

$$R \cdot + RSH \longrightarrow RH + RS \cdot \quad (19)$$

$$RSH + P(OC_2H_5)_3 \longrightarrow RH + (C_2H_5O)_3PS \quad (20)$$

The reactions that participate in the chain sequence must produce by necessity a free radical if the chain is to be propagated. These radical-propagating reactions are generally (but not always) unimolecular in free radicals. In principle, a single free radical should be capable of converting all of the available reactants to the desired products in the chain sequence. In practice, this desirable situation is not met because of the problem of introducing a single free radical in the reaction medium. Generally, initiation reactions are processes that result in the formation of two free radicals for each initiation event (e.g., homolysis or photolysis of a molecule). At this point, the prospects of a completely ideal situation in which the chain reaction proceeds to completion are already lost since, being very reactive species, free radicals prefer to react with themselves in bimolecular radical processes yielding nonradical products than to participate in chain-propagating reactions. Although the problem of a bimolecular radical reaction involving only two free radicals in volumes such as those encountered in most reactions is not a formidable one, the problem does become appreciable if the initiation process involves homolysis or photolysis of many molecules to produce free radicals. In such a case the facile bimolecular radical-destroying reactions may remove free radicals at rates too rapid to allow for appreciable amounts of reactions of the free radicals via the chain sequence. It is obvious that if the chain sequence is to be long and the stoichiometric reaction determined by the chain sequence is to occur, there must be a balance between the rates of the three processes, namely, the introduction of the free radicals into the medium (initiation), the rates of the reactions of the free radicals in the chain-propagating reactions (the chain sequence), and the bimolecular radical-destroying processes (termination). In this chapter we shall endeavor to determine some

1. FREE RADICALS

of the generalities concerning the initiation reactions that may be used to produce free radicals as well as the principles that determine the behavior of free radicals as reaction intermediates in both chain-propagating and bimolecular radical-destroying reactions.

2. Nonchain Reactions

In some cases the product of a free-radical reaction is formed in a bimolecular radical-destroying reaction rather than in the chain sequence. For example, the oxidative coupling of esters by reaction with an appropriate peroxide is such a reaction.

$$2\ CH_3CO_2C_2H_5 + Bz_2O_2 \longrightarrow C_2H_5O_2CCH_2CH_2CO_2C_2H_5 + 2\ BzOH \quad (21)$$
$$(Bz = C_6H_5CO)$$

via

$$Bz_2O_2 \longrightarrow 2\ BzO\cdot \quad (22)$$

$$BzO\cdot + CH_3CO_2C_2H_5 \longrightarrow BzOH + \cdot CH_2CO_2C_2H_5 \quad (23)$$

$$2\ \cdot CH_2CO_2C_2H_5 \longrightarrow C_2H_5O_2CCH_2CH_2CO_2C_2H_5 \quad (24)$$

Since the products of these reactions are not formed in a chain reaction, a stoichiometric amount of initiation (radical formation) is required for the amount of the product formed. In some cases the mode of initiation (e.g., certain photolysis reactions) may be such as to allow for efficient formation of a reaction product in a bimolecular radical reaction. The photochemical reductive dimerization of benzophenone in isopropyl alcohol (Pitts *et al.*, 1959) is one such process.

$$2\ (C_6H_5)_2C{=}O + (CH_3)_2CHOH \xrightarrow{h\nu} \underset{\underset{HO\ \ OH}{|\ \ \ |}}{(C_6H_5)_2C{-}C(C_6H_5)_2} + (CH_3)_2C{=}O \quad (25)$$

via

$$(C_6H_5)_2C{=}O \xrightarrow{h\nu} (C_6H_5)_2C{=}\overset{*}{O} \longrightarrow (C_6H_5)_2\dot{C}O\cdot \quad (26)$$

$$(C_6H_5)_2\dot{C}O\cdot + (CH_3)_2CHOH \longrightarrow (C_6H_5)_2\dot{C}OH + (CH_3)_2\dot{C}OH \quad (27)$$

$$(CH_3)_2\dot{C}OH + (C_6H_5)_2C{=}O \longrightarrow (CH_3)_2C{=}O + (C_6H_5)_2\dot{C}OH \quad (28)$$

$$2\ (C_6H_5)_2\dot{C}OH \longrightarrow \underset{\underset{HO\ \ OH}{|\ \ \ |}}{(C_6H_5)_2C{-}C(C_6H_5)_2} \quad (29)$$

If the chain-propagating reaction is not a product-forming reaction, as in the case of the addition of a free radical to an unsaturated linkage, product formation occurs in the bimolecular termination reaction. Vinyl polymerization is an excellent example of such a reaction. The formation of high molecular weight free radicals occurs in a succession of radical chain-propagating addition reactions, but the polymer molecules themselves are formed in the coupling or disproportionation (see Section III,C) reactions of these high molecular weight free radicals.

II. Formation of Free Radicals

Except for some of the unusually stable free radicals discussed earlier, most organic free radicals are too reactive to be isolated or maintained as radicals for any extended periods in solution. Reactions that involve free radicals as intermediates necessitate their formation at the reaction site by some free-radical-generating process. These processes can result in the initiation of a chain sequence and are, therefore, most often referred to as initiation reactions. The most common means of forming free radicals are homolytic cleavage reactions of covalent bonds by thermolysis or photolysis of appropriate compounds or by transfer of a single electron from one reactant to another in an oxidation–reduction reaction.

A. Thermolysis of Covalent Bonds

1. *Peroxides*

The σ bond between the two oxygens of a peroxide linkage is thermally labile. Decomposition of a peroxide yields two free radicals, each initially having an unpaired electron on oxygen. The temperature requirements for

$$RO-OR \longrightarrow 2\,RO\cdot \tag{30}$$

homolytic cleavage of the peroxide linkage depend on the nature of the R group as does the fate of the initially formed radical $RO\cdot$.

The most familiar of the dialkyl peroxides is *tert*-butyl peroxide. This compound is a liquid that boils without appreciable decomposition at 111°C. It decomposes in a unimolecular reaction to form two *tert*-butoxy radicals at slightly higher temperatures ($t_{1/2} = 34$ hr at 115°C and $t_{1/2} = 6.4$ hr at 130°C) (Raley et al., 1948). The activation energy for the decomposition in the gas

$$(CH_3)_3COOC(CH_3)_3 \longrightarrow 2\,(CH_3)_3CO\cdot \tag{31}$$

phase is 38.1 kcal per mole. The activation energy for the unimolecular decomposition in solution is similar to that in the gas phase for most solvents (Huyser and VanScoy, 1968).

The fate of the *tert*-butoxy radicals produced depends on their environment.

1. FREE RADICALS

In the gas phase, most of the radicals decompose to yield acetone and methyl radicals. The *tert*-butoxy radical is sufficiently reactive to remove hydrogen

$$(CH_3)_3CO\cdot \longrightarrow (CH_3)_2C{=}O + CH_3\cdot \tag{32}$$

atoms from many organic compounds and may do so rather than fragment. If fragmentation does occur, the resulting methyl radical is capable of reaction

$$(CH_3)_3CO\cdot + RH \longrightarrow (CH_3)_3COH + R\cdot \tag{33}$$

with the solvent to yield a solvent-derived radical. In either case, two free radicals are formed in the decomposition of one molecule of the peroxide.

$$CH_3\cdot + RH \longrightarrow CH_4 + R\cdot \tag{34}$$

The products of the thermal decomposition of acetyl peroxide, which decomposes at lower temperatures than *tert*-butyl peroxide ($t_{1/2} = 8.5$ hr at 70°C; $t_{1/2} = 1.1$ hr at 85°C), are methyl radicals and carbon dioxide (Kharasch *et al.*, 1945). The intermediate acetoxy radical apparently decomposes faster than it is able to escape the solvent cage in which it is formed. There is evidence, however, that the acetoxy radical may react in the solvent cage.

$$CH_3\overset{O}{\overset{\|}{C}}O{-}O\overset{O}{\overset{\|}{C}}CH_3 \longrightarrow \left(CH_3\overset{O}{\overset{\|}{C}}O\cdot + \cdot O\overset{O}{\overset{\|}{C}}CH_3 \right) \longrightarrow 2\,CH_3\cdot + CO_2 \tag{35}$$

The formation of methyl acetate and the scrambling of the labeled carbonyl oxygen in the recovered "unreacted" acetyl peroxide support the existence of the acetoxy radicals for a finite time and discount a concerted mechanism for the decomposition in which the C–C bonds rupture simultaneously with rupture of the O–O linkage (Taylor and Martin, 1966). For the most part, however, as an initiator, acetyl peroxide can be considered to yield carbon dioxide and two reactive methyl radicals.

$$\begin{array}{c} CH_3\overset{O^*}{\overset{\|}{C}}{\diagdown}\!\!\!\!\!{}\overset{O^*}{\overset{\|}{C}}CH_3 \\ O{-}O \end{array} \rightleftarrows \left(CH_3\overset{\delta*O}{\overset{\diagup}{C}}{}\overset{O^{\delta*}}{\overset{\diagdown}{C}}CH_3 \right) \begin{array}{c} \rightleftarrows \\ \\ \underset{\substack{\text{escape}\\\text{from}\\\text{cage}}}{\Big\downarrow} \end{array} \begin{array}{c} CH_3\overset{\delta*O{-}O^{\delta*}}{\overset{\diagup\quad\diagdown}{C}}{}\overset{}{C}CH_3 \\ \overset{\|}{\underset{\delta*O}{}}\quad\overset{\|}{\underset{O^{\delta*}}{}} \end{array} \tag{36}$$

$$\longrightarrow 2\,CH_3\cdot + 2\,CO_2$$

Benzoyl peroxide undergoes unimolecular decomposition at rates comparable to those of acetyl peroxide. The products of the homolytic fission are two benzyloxy radicals. In contrast to acetoxy radicals, the benzoyloxy radicals do not decompose readily and often react as such outside of the

$$C_6H_5\overset{O}{\overset{\|}{C}}O-O\overset{O}{\overset{\|}{C}}C_6H_5 \longrightarrow 2\, C_6H_5C\overset{O}{\underset{O\cdot}{\diagup\!\!\!\diagdown}} \qquad (37)$$

$$C_6H_5\overset{O}{\overset{\|}{C}}O\cdot + RH \longrightarrow C_6H_5CO_2H \qquad (38)$$

$$C_6H_5C\overset{\diagup O}{-}O\cdot \longrightarrow CO_2 + C_6H_5\cdot \qquad (39)$$
$$\underset{RH}{\longmapsto} C_6H_6 + R\cdot$$

solvent cage in which they are formed (Hammond and Soffer, 1950). If no suitable substrate is available, however, benzoyloxy radicals fragment, yielding phenyl radicals and carbon dioxide. Both benzoyloxy and phenyl radicals are reactive species, and the unimolecular decomposition of one molecule of benzoyl peroxide results in the formation of two reactive free radicals.

Peresters such as *tert*-butyl peracetate and *tert*-butyl perbenzoate, having attributes of both alkyl and acyl peroxides, decompose at rates intermediate to those of alkyl and acyl peroxides. The resulting radicals in both cases resemble those obtained from either the alkyl or acyl peroxides (Bartlett and Hiatt, 1958).

$$CH_3\overset{O}{\overset{\|}{C}}OOC(CH_3)_3 \longrightarrow CH_3\cdot + CO_2 + (CH_3)_3CO\cdot \qquad (40)$$

$$C_6H_5\overset{O}{\overset{\|}{C}}OOC(CH_3)_3 \longrightarrow C_6H_5CO_2\cdot + (CH_3)_3CO\cdot \qquad (41)$$
$$\longmapsto C_6H_5\cdot + CO_2$$

Hydrogen peroxide and the alkyl hydroperoxides require temperatures above ~150°C for cleavage of the O–O linkage to proceed at rates that would make them useful as initiators. They do have value in initiating chain reactions that take place readily only at higher temperatures. They are used more extensively for the generation of free radicals by various redox reactions.

One limitation of peroxides as a source of free radicals is their lability to reactions other than the unimolecular thermolysis. The most prevalent of these are interactions of the peroxides with other free radicals, reactions that

1. FREE RADICALS

yield a variety of products depending on the nature of the free radicals (Huyser, 1970a). These radical-induced decompositions of the peroxide yield a peroxide-derived radical, but being chain-propagating reactions they result in no net gain in free radicals (e.g., the reaction of an α-hydroxyalkyl radical with *tert*-butyl peroxide).

$$R_2\dot{C}OH + (CH_3)_3COOC(CH_3)_3 \longrightarrow R_2C=O + (CH_3)_3COH + (CH_3)_3CO\cdot \quad (42)$$

2. Azo Compounds

Compounds having the general formula R—N=N—R decompose thermally to yield nitrogen and two free radicals. The temperature requirements for the unimolecular decomposition of azo compounds depend largely on the resonance stabilities of the free radicals (R·) formed in the process. Thus,

$$RN=NR \longrightarrow 2R\cdot + N_2 \quad (43)$$

whereas the activation energy for the thermal decomposition of azomethane ($R = CH_3$) yielding methyl radicals is about 50 kcal/mole, thermal decomposition of azodiphenylmethane [$R = (C_6H_5)_2CH\cdot$], which yields the comparatively more stable diphenyl methyl radicals, has an activation energy of only 26.6 kcal/mole.

The azo compound most often used as an initiator is AIBN, which on decomposition yields two isobutyronitrile radicals. Although stabilized by resonance, these radicals are capable of reaction with many other species.

$$\underset{\underset{CN}{|}}{(CH_3)_2C}-N=N-\underset{\underset{CN}{|}}{C(CH_3)_2} \longrightarrow 2\,(CH_3)_2\dot{C}CN + N_2 \quad (44)$$

One limitation of AIBN as a source of free radicals is that the isobutyronitrile radicals, because of their resonance stabilization, are subject to recombination both in and out of the solvent cage in which they are formed, yielding either tetramethylsuccinonitrile or a keteneimine (Hammond *et al.*, 1960). The amount of recombination inside the cage depends on the reactivity of the

$$2\,(CH_3)_2\dot{C}CN \longrightarrow \underset{\underset{CN\ CN}{|\ \ |}}{(CH_3)_2C-C(CH_3)_2} \quad (45)$$

$$\searrow \underset{\underset{CN}{|}}{(CH_3)_2C=C=NC(CH_3)_2} \quad (46)$$

solvent toward reaction with the isobutyronitrile radicals. For example, in carbon tetrachloride, 90% of the radicals recombine, probably both in and out of the cage, to form tetramethylsuccinonitrile (Hammond *et al.*, 1955), whereas in liquid bromine, a species more reactive toward isobutyronitrile radicals, no tetramethylsuccinonitrile is formed (Trapp and Hammond, 1959).

B. Photochemical Processes

Absorption of electromagnetic radiation in the visible and ultraviolet regions by chemical species results in excitation of electrons to higher energy levels. The electronically excited species may undergo chemical reaction either by fragmentation or rearrangement to produce photochemically generated reaction products or dissipate its energy by different means. Emission of radiation (fluorescence) or transfer of its energy by molecular collisions producing thermal energy results in the return of the species to its original ground state. Routes are available, however, by which the excited species may use its absorbed energy to produce free radicals. The excited species may fragment by means of a homolytic cleavage of a covalent bond or it may "cross-over" to a different excited energy level (triplet state), one which amounts to the unpairing of the electrons that compose a π bond of the molecule in its ground state. The latter produces a diradical species that can participate in typical free-radical reactions or return to the ground state either by emission of radiation (phosphorescence) or molecular collisions producing thermal energy.

1. *Photolysis*

The halogens absorb energy in the visible region of the spectrum as evidenced by their color. The bond dissociation energies of the σ bonds of the halogens are low enough (see Table II) to allow the electronically excited

$$X_2 \xrightarrow{h\nu} X_2^* \longrightarrow 2\,X\cdot \tag{47}$$

molecule to fragment, thus yielding a pair of halogen atoms. Many free-radical reactions that involve either molecular chlorine or bromine are initiated by photolysis of the halogens since the process is efficient as a means of producing these atomic species.

Many organic compounds that have weak σ bonds have electromagnetic absorption characteristics that result in the formation of an excited state that is sufficiently energetic to result in homolytic cleavage of this bond. Both *tert*-butyl peroxide and AIBN decompose photolytically at temperatures well below those required for thermolysis. Another compound that undergoes facile photolysis is bromotrichloromethane. All these compounds have spectral absorptions above 300 nm and, therefore, illumination of reaction mixtures containing them in Pyrex vessels can result in the formation of an

$$(CH_3)_3COOC(CH_3)_3 \xrightarrow{h\nu} 2\,(CH_3)_3CO\cdot \qquad (48)$$

$$(CH_3)_2\underset{CN}{C}-N{=}N-\underset{CN}{C}(CH_3)_2 \xrightarrow{h\nu} 2\,(CH_3)_2\dot{C}CN + N_2 \qquad (49)$$

$$BrCCl_3 \xrightarrow{h\nu} Br\cdot + Cl_3C\cdot \qquad (50)$$

excited electronic state and subsequent formation of free radicals. Compounds that have absorptions below 300 nm must be illuminated by an ultraviolet source through quartz. The energies of the absorptions are greater at these higher frequencies, but radical production is often less efficient since the compounds do not have labile bonds that rupture readily to produce free radicals. In many cases, other photochemical transformations occur or the absorbed energy is dissipated either by fluorescence or molecular collisions.

2. Triplet Formation

Compounds with a carbonyl function have a weak ultraviolet absorption resulting from excitation of an electron from a nonbonding orbital of the oxygen to an antibonding π orbital ($n \to \pi^*$). In this excited state the species may lose some of its energy by passing to a lower energy excited state that involves the unpairing of the electrons that comprise the π bond of the carbonyl function. This process is referred to as an intersystem crossover.

$$\diagup\!\!\!\diagdown\!C{=}O \xrightarrow{h\nu} \left(\diagup\!\!\!\diagdown\!C{=}O\right)^* \longrightarrow \diagup\!\!\!\diagdown\!\dot{C}{-}O\cdot \qquad (51)$$

$$\text{Excited singlet} \qquad \text{Triplet}$$

The resulting diradical generally has a long enough lifetime before it decays to the ground state by phosphorescence (10^{-4} sec in contrast to about 10^{-9} sec for fluorescence) to participate in free-radical reactions. The free radicals formed in this manner structurally resemble those obtained by thermolysis of peroxides, but may differ in that they may be excited species and possibly more energetic than the corresponding alkoxy free radical in the ground state.

The efficiency of producing triplet free radicals depends on both the efficiency of the absorption that results in forming the $n \to \pi^*$ excited state and the intersystem crossover from this singlet excited state to the degenerate triplet state. The former, as measured by its extinction coefficient, is generally very inefficient compared to other ultraviolet absorptions (for benzophenone, $\epsilon_{n \to \pi^*} = 100$). The absorptions occur at longer wavelengths if the carbonyl function is conjugated with an aromatic group, and, in the case of benzophenone, can be effected by light at 331 nm. Illumination with ultraviolet light through quartz is required for similar excitations of carbonyl groups that are not extensively conjugated.

The efficiency of intersystem crossover to the triplet state from the singlet state is not readily measured. It might be expected, however, to depend to some extent on the amount of stabilization the triplet diradical attains by delocalization of the unpaired electron on the carbon of the carbonyl function. Again, aromatic groups can provide excellent stabilization of the triplet species and may, therefore, render the intersystem crossover process more efficiency than would be encountered in nonaromatic carbonyl-containing compounds.

C. Bimolecular Redox Reactions

Free radicals can be generated in oxidation–reduction reactions between appropriate species. For example, hydrogen peroxide and ferrous ion react to form a ferric ion and a hydroxy radical, both species having unpaired electrons (Uri, 1952). Similarly, cuprous ions react with acyl peroxides to

$$Fe(II) + H_2O_2 \longrightarrow Fe(III) + HO^{(-)} + HO\cdot \qquad (52)$$

yield-free radicals (Kochi, 1963). Classifying such reactions as these as radical forming may not be precisely correct since the metallic ions have

$$Cu(I) + R\overset{O}{\overset{\|}{C}}OO\overset{O}{\overset{\|}{C}}R \longrightarrow Cu(II) + RCO_2^- + R\cdot + CO_2 \qquad (53)$$

unpaired electrons and, therefore, are themselves free radicals, and the reactions are actually radical-propagating processes. On the other hand, the oxidation of *vic*-diols by ceric ions is a radical-producing reaction since the ceric ion is likely diamagnetic (Littler and Waters, 1960).

$$Ce(IV) + R_2C(OH)C(OH)R_2 \longrightarrow Ce(III) + R_2C{=}O + H^+ + R_2\dot{C}OH \qquad (54)$$

There are reactions in which nonradical species interact in bimolecular processes to produce free radicals. Acyl peroxides and amines react at temperatures considerably lower than those required for unimolecular decomposition of the peroxide-yielding free radicals. For example, benzoyl peroxide

$$Bz_2O_2 + C_6H_5N(CH_3)_2 \longrightarrow C_6H_5\overset{\cdot+}{N}(CH_3)_2 + BzO\cdot + BzO^- \qquad (55)$$

and dimethylaniline react at temperatures as low as 0°C to produce free radicals that cause the polymerization of styrene (Walling and Indictor, 1960; Horner and Sherf, 1951).

Bimolecular redox reactions yielding free radicals may be more prevalent than expected. In many instances the radicals formed in such processes initiate a chain sequence of reactions that account for the redox reaction that is observed in the stoichiometric reaction. Only if the radicals produced in the initiation step can be made to compete effectively with other substrates than the reagents involved in the initiation reaction can these reactions be of value as initiation processes for other reactions.

III. Reactions of Free Radicals

An appreciation of the mechanistic aspects of free-radical reactions is simplified by the fact that free radicals are limited in what they are able to do as reaction intermediates. Free radicals participate in two principal types of reactions. One type is unimolecular in free radicals and, in order to preserve a balance of matter, yields a free radical as a reaction product. These are the radical-propagating reactions, and they may involve an interaction of the free radical with another molecule and are, therefore, kinetically bimolecular or they may be unimolecular processes that involve only the free radical itself. These reactions are those that comprise the chain sequence of free-radical-propagating steps that are encountered in the conversion of reactants to products. The other types of reactions of free radicals are those that are bimolecular in free radicals and result in the removal of the radicals as reactive intermediates from the reaction medium. The reactions destroy the free radicals that participate in the chain-propagating reactions, thereby ending the chain sequence, and are generally referred to as termination reactions.

A. Bimolecular Radical-Propagating Reactions

1. *Displacement (Abstraction)*

The free-radical-displacement reaction is similar to the familiar S_N2 displacement reactions encountered in reactions of nucleophiles with appropriate substrates, and it is subject to many of the same limitations as S_N2 reactions in that steric, resonance, and polar factors play significant roles in determining the facility of the reactions. Although these factors will be taken up in more detail subsequently (see Section IV), it is pertinent to point out here that the steric requirements impose the most significant limitation of free-radical-displacement reactions. Most, but not all, free-radical displacements occur on univalent elements, namely, hydrogen and the halogens. The displacements take place on hydrogen by a variety of free radicals in reactions that are often referred to as hydrogen abstraction reactions. Obviously, the rate of any of

$$CH_3\cdot + HCH_2CO_2C_2H_5 \longrightarrow CH_4 + \cdot CH_2CO_2C_2H_5 \qquad (56)$$

$$Cl_3C\cdot + HCH_2C_6H_5 \longrightarrow Cl_3CH + \cdot CH_2C_6H_5 \qquad (57)$$

$$(CH_3)_3CO\cdot + HCR_2OH \longrightarrow (CH_3)_3COH + \cdot CR_2OH \qquad (58)$$

$$Cl\cdot + HCH_3 \longrightarrow HCl + CH_3\cdot \qquad (59)$$

these reactions depends on both the nature of the displacing radical and the nature of the free radical that is displaced from the hydrogen atom.

Displacement on a halogen by a free radical is similar to that of hydrogen, since approach to the univalent halogen by the attacking free radical is not subject to the same steric considerations that would be encountered with elements having higher coordination numbers. The rate of a propagating

$$CH_3\cdot + Cl-Cl \longrightarrow CH_3Cl + Cl\cdot \qquad (60)$$

$$C_6H_5CH_2\cdot + Cl-CCl_3 \longrightarrow C_6H_5CH_2Cl + \cdot CCl_3 \qquad (61)$$

$$CF_3CF_2CF_2\cdot + ICF_3 \longrightarrow CF_3CF_2CF_2I + \cdot CF_3 \qquad (62)$$

$$C_2H_5\cdot + ClOC(CH_3)_3 \longrightarrow C_2H_5Cl + \cdot OC(CH_3)_3 \qquad (63)$$

displacement reaction on a halogen is dependent on the halogen as well as on both the attacking radical and the radical that is displaced from the halogen.

Although considerably less common, free-radical-displacement reactions do occur on bivalent oxygen and sulfur. The nature of the attacking and displaced radicals are particularly significant in these reactions because polar and resonance factors must contribute significantly in order to outweigh the steric problems encountered in such reactions. The most common displacements on bivalent atoms are those that are observed in the reactions of peroxides and disulfides.

$$C_2H_5O\overset{\cdot}{C}HCH_3 + C_6H_5\overset{O}{\overset{\|}{C}}OO\overset{O}{\overset{\|}{C}}C_6H_5 \longrightarrow C_2H_5OCH\overset{O}{\overset{\|}{O}}CC_6H_5 + C_6H_5CO_2\cdot \qquad (64)$$
$$\underset{CH_3}{|}$$

$$\underset{C_6H_5}{\overset{\sim CH_2\overset{\cdot}{CH}\cdot}{|}} + n\text{-}C_4H_9SSC_4H_9\text{-}n \longrightarrow \underset{C_6H_5}{\overset{\sim CH_2CHSC_4H_9\text{-}n}{|}} + n\text{-}C_4H_9S\cdot \qquad (65)$$

Nucleophilic displacement reactions on carbon are common but free-radical displacements on carbon are rare. Those that have been reported involve the opening of a strained cyclopropane ring by the attacking free radical (see Section IV,D).

At first inspection, the limitation of facile displacements occurring only on univalent atoms may appear to be severe. However, the prevalence of hydrogen in organic compounds makes most of them candidates for displacement reactions. Not all hydrogen-containing molecules are subject to ready attack by free radicals, nor are the various hydrogens of a given molecule equally reactive toward displacement by all free radicals. Other factors rendered by the attacking free radical and the displaced free radicals must be appreciated in order to determine both the limitations and specificity of the reaction.

2. *Addition*

Free radicals add to unsaturated linkages, a reaction that results in the formation of an adduct radical. Although a variety of unsaturated linkages undergo addition by free radicals, the most familiar are the additions to alkenes. Generally, addition reactions in which the adding radical bonds to

$$Cl_3C\cdot + CH_2=CHR \longrightarrow Cl_3CCH_2\dot{C}HR \qquad (66)$$

$$Br\cdot + CH_2=CHR \longrightarrow BrCH_2\dot{C}HR \qquad (67)$$

$$(C_2H_5O)_2\dot{P}O + CH_2=CHR \longrightarrow (C_2H_5O)_2P(O)CH_2\dot{C}HR \qquad (68)$$

the terminal carbon of a terminal alkene are more facile than additions to nonterminal double bonds. The propensity for the bonding of a free radical to the less substituted of the two carbon atoms that comprise the π system can be ascribed, at least in part, to resonance and polar factors dictated both by the adding radical and the unsaturated compound (see Section IV). However, in view of the markedly lower reactivities of nonterminal double bonds compared to terminal double bonds toward addition, steric factors must also play a significant role in retarding the bonding of the adding free radical to the more substituted carbon of the unsaturated linkage.

Other unsaturations are also subject to addition by free radicals but are generally less reactive than alkenes. Additions of free radicals to acetylenes

$$Cl_3C\cdot + HC{\equiv}CR \longrightarrow Cl_3CCH={\dot{C}}R \qquad (69)$$

$$CH_3\cdot + HC{\equiv}CH \longrightarrow CH_3CH={\dot{C}}H \qquad (70)$$

yield vinyl radicals as the addition products. Aromatic rings react with free radicals to yield adduct radicals that no longer have aromatic character.

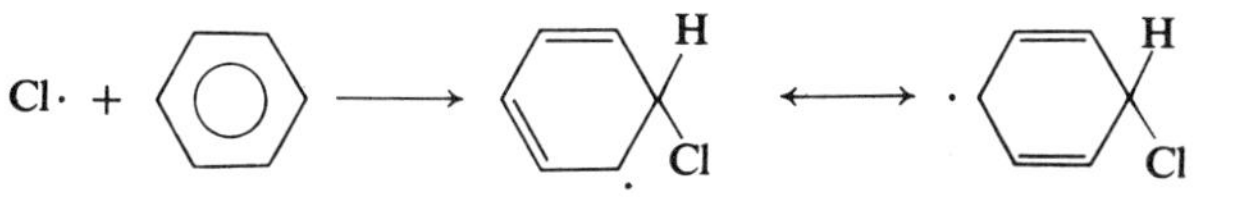

(71)

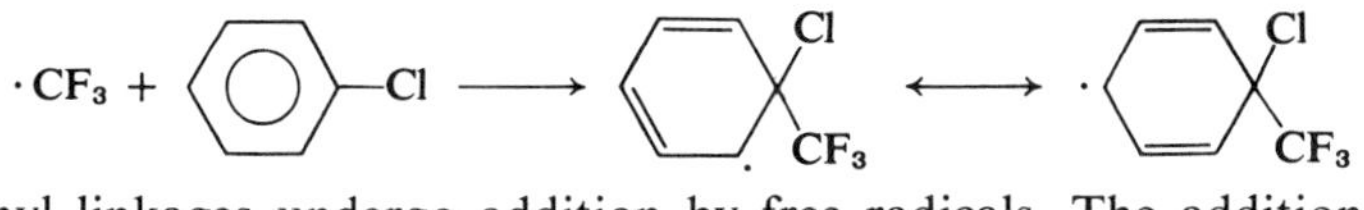

(72)

Carbonyl linkages undergo addition by free radicals. The addition to this heteroatomic linkage may occur in two directions, depending on the nature of both the adding radical and the carbonyl function. For example, alkyl radicals bond to the carbon atom of formaldehyde to yield alkoxy radicals as

$$\text{cyclopentyl}\cdot + H_2C=O \longrightarrow \text{cyclopentyl-}CH_2O\cdot \qquad (73)$$

reaction products (G. Fuller and Rust, 1958). On the other hand, benzoyl radicals add to benzaldehyde with bonding at the oxygen of the carbonyl function to form an α-benzoyloxybenzyl radical (Rust et al., 1948). Hydrogen

$$C_6H_5\overset{\cdot}{C}=O + C_6H_5\overset{H}{C}=O \longrightarrow C_6H_5\overset{\cdot}{C}HO\overset{O}{\overset{\|}{C}}C_6H_5 \qquad (74)$$

atoms are transferred from appropriate reducing free radicals (see redox reactions in Section III,A,3) and add both to carbonyl functions (Huyser and Neckers, 1963) and imines (Huyser et al., 1968). The hydrogen atom in each case bonds to the more electronegative atom of the two functionalities.

$$R_2\overset{\cdot}{C}OH + O=CR_2' \longrightarrow R_2C=O + HO\overset{\cdot}{C}R_2' \qquad (75)$$

$$R_2\overset{\cdot}{C}OH + HN=C(C_6H_5)_2 \longrightarrow R_2C=O + H_2N\overset{\cdot}{C}(C_6H_5)_2 \qquad (76)$$

Molecules which have no unsaturated linkages but do have atoms that are able to expand their valence shell may undergo addition of free radicals. Phosphorus, having d orbitals of sufficiently low energy, is such an element, and various reactions of phosphorus compounds indicate the intermediacy of an adduct radical resulting from addition of a free radical to the phosphorus. For example, thiyl radicals react with trialkyl phosphites to yield an adduct radical in which a S–P bond is formed (Walling and Rabinowitz, 1959).

$$RS\cdot + (C_2H_5O)_3P \longrightarrow (C_2H_5O)_3\overset{\cdot}{P}SR \qquad (77)$$

Additions of free radicals may be regarded as reversible processes. In some instances (e.g., additions of bromine atoms and thiyl radicals to alkenes) the reversibility of the addition reaction may significantly influence the course of the chain reaction. The subject of the reverse reaction (β elimination) will be discussed subsequently and its relationship to the addition reaction is indicated.

3. Reduction Reactions

Some free radicals are able to reduce a substrate with concurrent oxidation of the free radical to a nonfree-radical species. The reductions occur by the transfer of either a single electron or a hydrogen atom from the free radical. The reaction of an acyl-free radical with peroxydisulfate, for example, is a

1. FREE RADICALS

radical-propagating reaction in which the acyl radical is oxidized to the corresponding acylium ion and the peroxydisulfate ion is reduced to the sulfate ion and a sulfate radical anion (F. L. McMillan, 1965). The process is

$$R\dot{C}O + S_2O_8^{2-} \longrightarrow R\overset{+}{C}O + SO_4^{2-} + SO_4^{-}\cdot \qquad (78)$$

one involving electron transfer from the acyl radical to the persulfate ion. The transfer of hydrogen atoms to both peroxide linkages [Eq. (42)], carbonyl functions [Eq. (75)], and imines [Eq. (76)] by α-hydroxyalkyl radicals are reductions of these functionalities, although they might be regarded mechanistically as displacement or addition reactions also.

Reactions of metal ions that undergo one-electron oxidation or reduction changes can be regarded as free-radical-propagating processes provided the reacting ions are paramagnetic [see Eqs. (52) and (53)]. Reductions of the

$$Cu(II) + R\cdot \longrightarrow Cu(I) + R^{(+)} \qquad (79)$$

metal ions by free radicals such as the interaction of alkyl radicals with copper(II) (Kochi, 1962) are also radical-propagating reactions since the metal ion formed is paramagnetic.

B. Unimolecular Free-Radical-Propagating Reactions

1. *Fragmentation Reactions*

Some free radicals decompose yielding both a stable molecule and a different free radical. The most common fragmentations are the β elimination reactions that yield a free radical and an unsaturated compound. As pointed out previously (see Section III,A,2), the addition of a free radical to a π bond is a reversible reaction. It may be regarded as a system that may attain

$$CH_3CH_2\dot{C}H_2 \longrightarrow CH_3\cdot + CH_2{=}CH_2 \qquad (80)$$

$$BrCH_2\dot{C}(CH_3)_2 \longrightarrow Br\cdot + CH_2{=}C(CH_3)_2 \qquad (81)$$

$$C_2H_5C(CH_3)_2O\cdot \longrightarrow C_2H_5\cdot + (CH_3)_2C{=}O \qquad (82)$$

$$\underset{CH_2{-}CH\dot{C}HCH_3}{\overset{CH_2}{\diagup\diagdown}} \longrightarrow \cdot CH_2CH_2CH{=}CHCH_3 \qquad (83)$$

equilibrium. The position of the equilibrium depends to a great extent on the reaction temperature, favoring β elimination at higher temperatures. Since

$$X\overset{|}{\underset{|}{C}}{-}\dot{C}{\diagdown} \rightleftharpoons X\cdot + {\diagup}C{=}C{\diagdown} \qquad (84)$$

all addition reactions are reversible, β elimination becomes a significant factor only if the temperature is sufficiently high for the entropy of the two component system on the elimination side of the equilibrium to outweigh the enthalpy gained from the addition of the radical to the π bond.

The temperature at which the rate of addition of a free radical to an unsaturated system is equal to the rate of β elimination is referred to as the ceiling temperature ($\Delta F = \Delta H - T \Delta S = 0$ and $T_{ct} = \Delta H / \Delta S$). The ceiling temperature depends to a significant extent on the enthalpy of the addition reaction, since the entropy changes for all additions are generally similar (for two species with three degrees of translation freedom relative to one species, ΔS is generally about -30 eu). The enthalpies of addition reactions that result in the formation of C–C bonds (see Section IV,A) are in most instances sufficiently high (~ 20 kcal/mole) so that ceiling temperatures for such additions are in the area of 400°C. On the other hand, the enthalpies of addition of bromine atoms or thiyl radicals to alkene linkages are considerably lower (10 kcal/mole) and ceiling temperatures are also considerably lower (50°–80°C). Consequently, reactions that involve additions of either bromine atoms or thiyl radicals to alkene linkage show much evidence of reversibility of the addition reaction (Huyser, 1970b).

Similarly, the enthalpy of addition of an alkyl radical to a carbonyl linkage is considerably lower than that of an addition to an alkene. Both alkoxy radicals and α-alkoxyalkyl radicals fragment more readily than alkyl radicals in the temperature range in which most free-radical reactions are performed

$$R_3CO\cdot \longrightarrow R\cdot + R_2C{=}O \tag{85}$$

$$RO\dot{C}R_2' \longrightarrow R\cdot + O{=}CR_2' \tag{86}$$

(0°–200°C). Fragmentation of carboxy radicals may be regarded as the reverse of the less favorable addition of a free radical to carbon dioxide.

$$RCO_2\cdot \longrightarrow R\cdot + CO_2 \tag{87}$$

α Elimination reactions are also known. Acyl radicals fragment to yield an alkyl radical and carbon monoxide. The reverse process, namely, formation of

$$R\dot{C}O \longrightarrow R\cdot + CO \tag{88}$$

an acyl radical by addition of an appropriate free radical to carbon monoxide, is also known but requires a reactive free radical such as a vinyl radical in addition to high pressures (Saurer, 1957).

2. Rearrangement

Certain free radicals are capable of undergoing a 1,2-shift of a group from an atom in a β position relative to the radical site, thereby generating a different free radical. Rearrangements of this kind, although a relatively common process for cations, are comparatively rare in free-radical reactions. In the first place, the groups that are capable of migrating in such 1,2-shifts are limited by certain mechanistic requirements of the reaction that will be discussed subsequently. Second, the reactions are generally (but not always) exothermic and, consequently, occur only if the radical site is in a thermodynamically unfavorable situation. In spite of these limitations, there are examples of free-radical chain reactions in which one of the radical-propagating steps is a unimolecular isomerization of a free-radical that proceeds via a 1,2-shift.

Groups that have been observed to participate as the migrating moiety in 1,2-shifts in free radicals are the aryl, vinyl, and carbonyl groups and the halogen atoms, particularly chlorine. Some chain reactions in which 1,2-shifts have been observed as chain-propagating steps are shown below [Eqs. (89–92) (Skell *et al.*, 1961) and (93–96) (Wilt and Philip, 1959)]. For other examples, see Huyser (1970c) and Freidlina (1965).

$$Cl_2 + CH_3CHBrCH_3 \longrightarrow CH_3CHClCH_2Br + HCl \quad (89)$$

via

$$Cl\cdot + CH_3CHBrCH_3 \longrightarrow HCl + CH_3CHBr\dot{C}H_2 \quad (90)$$

$$CH_3CHBr\dot{C}H_2 \xrightarrow{1,2\text{-shift}} CH_3\dot{C}HCH_2Br \quad (91)$$

$$CH_3\dot{C}HCH_2Br + Cl_2 \longrightarrow CH_3CHClCH_2Br + Cl\cdot \quad (92)$$

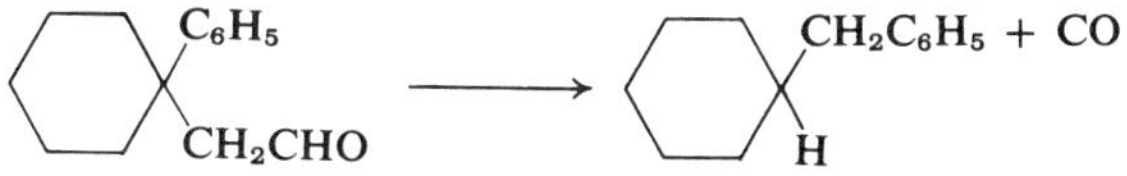

$$\text{Ph}(C_6H_5)(CH_2CHO)\text{C} \longrightarrow \text{Ph}(CH_2C_6H_5)(H)\text{C} + CO \quad (93)$$

via

$$\text{Ph}(C_6H_5)(CH_2\dot{C}O)\text{C} \longrightarrow \text{Ph}(C_6H_5)(\dot{C}H_2)\text{C} + CO \quad (94)$$

$$\text{Ph}(C_6H_5)(\dot{C}H_2)\text{C} \xrightarrow{1,2\text{-shift}} \text{Ph}\text{–}CH_2C_6H_5 \quad (95)$$

$$\text{Ph}\cdot\text{–}CH_2C_6H_5 + \text{Ph}(C_6H_5)(CH_2CHO)\text{C} \longrightarrow \text{Ph}(CH_2C_6H_5)(H)\text{C} + \text{Ph}(C_6H_5)(CH_2\dot{C}O)\text{C} \quad (96)$$

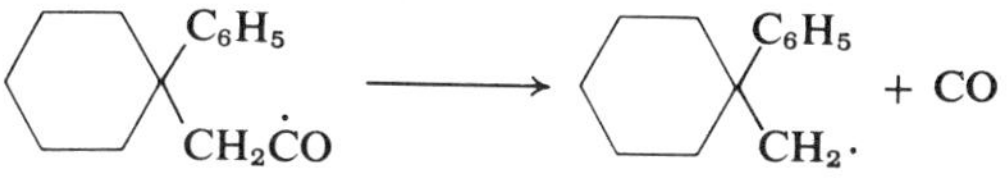

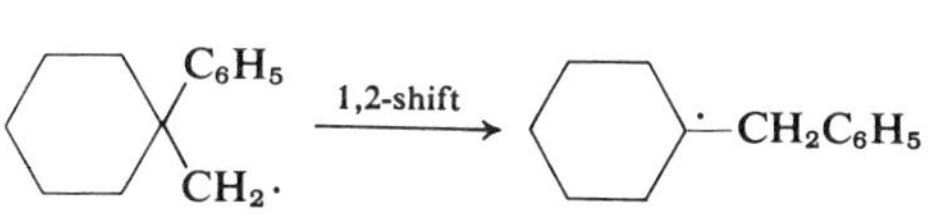

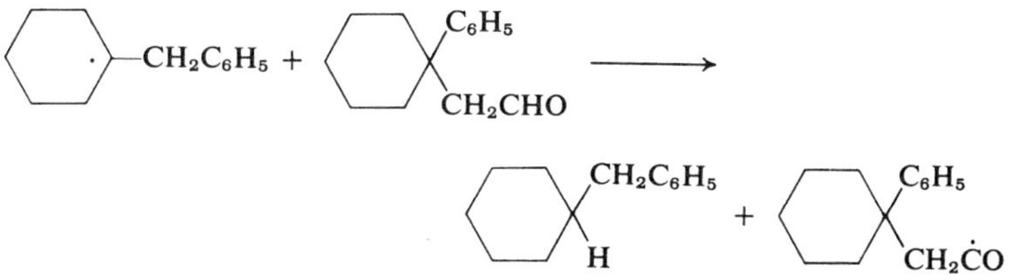

The lack of 1,2-shifts of hydrogen atoms and alkyl groups in free-radical rearrangements provides some insight into the mechanism of the observed migrations. The migrations may be either a one-step or a two-step process. In the one-step mechanism, bond-forming and bond-breaking processes occur in the transition state of the migration in a manner similar to the 1,2-shifts to electron-deficient centers. The two-step mechanism requires the

$$-\overset{S}{\underset{|}{C_1}}-\overset{\cdot}{\underset{|}{C_2}}- \longrightarrow \left[-\overset{S}{\underset{|}{C_1}}\cdots\overset{}{\underset{|}{C_2}}- \right] \longrightarrow -\overset{S}{\underset{|}{\overset{\cdot}{C_1}}}-\overset{}{\underset{|}{C_2}}- \qquad (97)$$

formation of an intermediate which on subsequent reaction yields the rearranged radical. This intermediate is an adduct radical resulting from an addition reaction of the parent radical site to the migrating group. The adduct

$$-\overset{S}{\underset{|}{C_1}}-\overset{\cdot}{\underset{|}{C_2}}- \rightleftharpoons -\overset{\overset{\cdot}{S}}{\underset{|}{C_1}}\diagup\overset{}{\underset{|}{C_2}}- \rightleftharpoons -\overset{S}{\underset{|}{\overset{\cdot}{C_1}}}-\overset{}{\underset{|}{C_2}}- \qquad (98)$$

radical intermediate, which has the unpaired electron localized on the migrating group, may undergo β elimination to yield either the rearranged radical or the parent radical. In order for an intermediate free radical of this nature to be formed, the migrating group should either have an unsaturated linkage, as in the case of aryl, vinyl, and carbonyl groups, or be an atom that has empty atomic orbitals of sufficiently low energy, as in the case of chlorine, to accommodate the electron and allow for the formation of a suitable intermediate that would lead to the rearranged radical. If the rearrangement occurs in a one-step process, transition states in which there is no incorporation of either the unsaturated group or the low energy d orbitals can readily be proposed. The available data suggest that the formation of an intermediate in a two-step reaction seems more likely than in the one-step mechanism.

The lack of any rearrangement of free radicals involving 1,2-shifts of either alkyl groups or hydrogen atoms is actually a valuable consideration in the study of free radicals as intermediates. The site of radical formation (e.g., in a displacement reaction) is not obscured by rearrangement, and subsequent reactions of the radical occur at its site of origination. If the free radical formed is thermodynamically less stable than one that could result from a 1,2-shift of either a hydrogen atom or an alkyl group, such migrations do not occur as they would in the corresponding carbonium ions (e.g., rearrangement of the *n*-propyl cation to the isopropyl cation). If a 1,2-shift of a group capable of migration in a free radical produces a more stable radical, the migration may occur readily provided that some bimolecular reaction of the unrearranged radical with an available substrate does not take place at a more rapid rate than the unimolecular rearrangement.

C. Termination Reactions

Two processes are known in which two free radicals react to produce stable, nonradical products. These bimolecular radical reactions are encountered in all reactions that involve free radicals as reaction intermediates. Since the chain-carrying free radicals that are part of the propagating reactions are generally involved in these radical-destroying processes, these bimolecular radical interactions are often referred to as "termination reactions" because the removal of free radicals from the medium (in contrast to initiation which introduces additional radicals) causes the termination of two sequences of chain-propagating reactions.

As might be expected, bimolecular reactions of reactive free radicals are very rapid. The high rates of termination reactions play an important role in the course of free-radical reactions, and the significance of these rapid bimolecular radical reactions is discussed in relation to the less rapid chain-propagating reactions (Section V).

1. Radical Coupling

Two free radicals may react to form a σ bond between the atoms that are the sites of the unpaired electrons. For example, two methyl radicals couple to form a molecule of ethane and two benzyl radicals may couple to form a

$$2\ CH_3 \cdot \longrightarrow CH_3CH_3 \tag{99}$$

$$2\ C_6H_5CH_2 \cdot \longrightarrow C_6H_5CH_2CH_2C_6H_5 \tag{100}$$

molecule of 1,2-diphenylethane. Cross coupling between two unlike radicals may occur as in the case of a trichloromethyl radical and a benzyl radical.

$$Cl_3C \cdot + C_6H_5CH_2 \cdot \longrightarrow C_6H_5CH_2CCl_3 \tag{101}$$

2. Disproportionation Reactions

Two free radicals may react in such a manner that one of them is oxidized with concurrent reduction of the other. Two ethyl radicals, for example, can react to yield a molecule each of ethane and ethylene. In some cases, two different radicals may be involved, as in the case of an α-hydroxyalkyl radical

$$2\ CH_3CH_2 \cdot \longrightarrow CH_3CH_3 + CH_2{=}CH_2 \tag{102}$$

and an alkoxy radical. Although in some instances disproportionation is not a likely route for termination because of the nature of the free radicals involved

$$R_2\dot{C}OH + R'O \cdot \longrightarrow R_2C{=}O + R'OH \tag{103}$$

(e.g., $CH_3 \cdot$, $C_6H_5\dot{C}H_2$), many free radicals are capable of both coupling and disproportionation processes. While coupling involves only σ bond formation, disproportionation involves both bond making and bond breaking. If a pair of free radicals has a choice, coupling generally may be expected if the free radicals are the same and are complex enough to allow for dissipation of the energy evolved in bond formation either by various vibrational modes of the resulting molecule or by collision with a third body. Disproportionation is more likely if the radicals involved are small, yield a pair of stable oxidation–reduction products, and the reaction occurs in the gas phase.

IV. Reactivities of Free Radicals

The factors that influence the reactivities of free radicals as reaction intermediates in both chain-propagating and termination reactions do not differ significantly from those encountered in the reactions of charged intermediates. These factors are steric, polar, resonance, and solvent effects. The importance of these effects is determined both by the free radicals themselves and the substrates with which they may react in chain-propagating reactions. An appreciation of the role these factors play makes possible an understanding of the facility with which some free-radical reactions occur. It also provides the ability to predict the feasibility of reactions that involve free radicals as reaction intermediates.

A. Resonance Factors

The effects of resonance stabilization both on the ease of forming a free radical and on its reactivity as a reaction intermediate are significant and, in some instances, may be the limiting factors in the feasibility of a free-radical reaction. Comparative resonance energies of free radicals can be estimated from bond dissociation energies, the energies required to break chemical bonds homolytically to yield two free radicals. Table II shows the bond dissociation energies (H_{dis}) of several C–H bonds encountered in some relatively simple organic compounds. Examination of the data in Table II shows that a free radical having a significant degree of resonance stabilization is formed with less expenditure of energy in a homolytic cleavage than one having little or no significant stabilization.

An examination of the radicals and their comparative resonance energies shows that the extent of this stabilization is related to the ability of the unpaired electron to be delocalized throughout the free radical. Thus, the alkyl radicals have resonance stabilizations that are related to the extent that hyperconjugative delocalization of the unpaired electron is possible. Nine structures of the type shown contribute to the hybrid structure of the *tert*

Table II
C–H Bond Dissociation Energies and Resonance Energies of Free Radicals

Bond	Dissociation energy	Radical formed	Resonance[a] energy
CH_3—H	104.0 ± 1	$CH_3\cdot$	0.0
CH_3CH_2—H	98.0 ± 1	$CH_3CH_2\cdot$	6.0 ± 2
$(CH_3)_2CH$—H	94.5 ± 1	$(CH_3)_2CH\cdot$	9.5 ± 2
$(CH_3)_3C$—H	91.0 ± 1	$(CH_3)_3C\cdot$	13.0 ± 2
$C_6H_5CH_2$—H	85 ± 1	$C_6H_5CH_2\cdot$	19 ± 2
$CH_2{=}CHCH_2$—H	85 ± 1	$CH_2{=}CHCH_2\cdot$	19 ± 2
CH_3COCH_2—H	92 ± 3	$CH_3COCH_2\cdot$	12 ± 4
$NCCH_2$—H	86 ± 3	$NCCH_2\cdot$	18 ± 4
$CH_2{=}CH$—H	104 ± 2	$CH_2{=}CH\cdot$	0.0
C_6H_5—H	104 ± 2	$C_6H_5\cdot$	0.0
Cl_3C—H	95.7 ± 1	$Cl_3C\cdot$	8.3 ± 2
F_3C—H	106 ± 1	$F_3C\cdot$	0.0

[a] Relative to $CH_3\cdot$ which is assigned a resonance energy of zero.
[b] Taken from tables compiled by Kerr (1966) and by Benson (1965).

butyl radical, whereas six and three are involved in delocalizing the unpaired electrons of isopropyl and ethyl radicals, respectively. Conjugation of a π bond with the site of bond cleavage lowers the energy requirement for bond dissociation because of significant delocalization of the unpaired electron with the π system. The more extensive the delocalization, as in the case of the

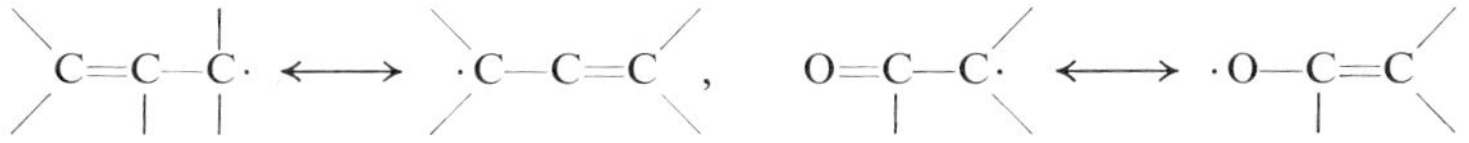

benzyl radical, or the more similar in energy the structures that contribute to the hybrid, the lower the observed bond dissociation energy. Radicals that do not have the capabilities of delocalizing the unpaired electron are formed less

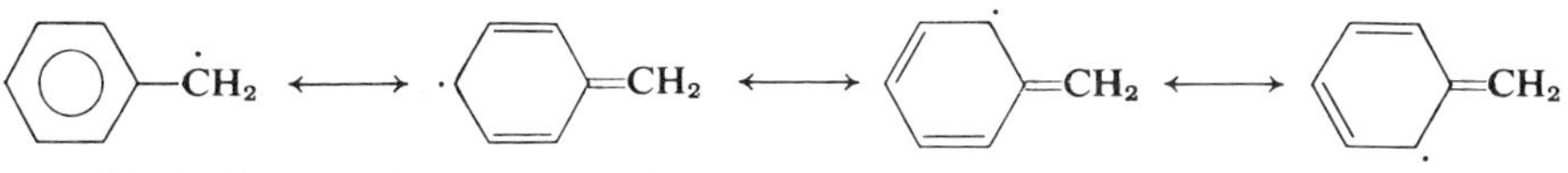

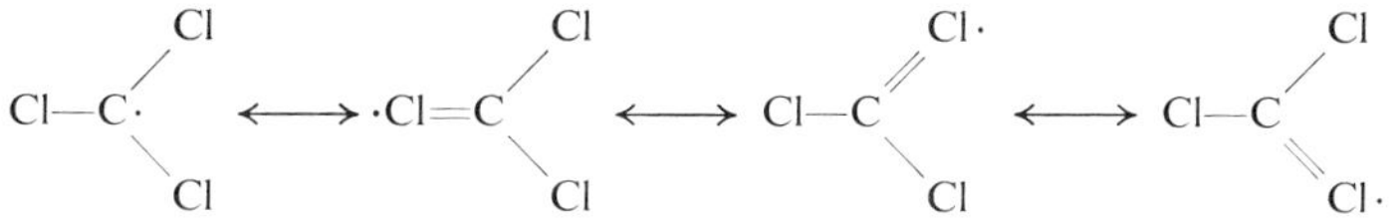

readily by homolytic cleavage of a C–H bond. This is evidenced by the bond dissociation energies of the processes that lead to methyl, vinyl, and phenyl radicals. Halogen atoms bonded to the free-radical site are capable of delocalizing the unpaired electron if empty d orbitals of low enough energy are available, which may account for the observed stability of the trichloromethyl radical. On the other hand, fluorine, having no readily attainable empty atomic orbitals, is not capable of delocalization of the unpaired electron as evidenced by the lack of any resonance stabilization of the trifluoromethyl radical.

The role (both its importance and limitations) of resonance factors in free-radical, chain-propagating reactions can be illustrated by an examination of the enthalpies, as calculated from bond dissociation energies, and the kinetics of various chain-propagating reactions. Table III lists several bond dissociation energies which will serve both for the discussion that follows as well as for polar effects in Section IV,B.

Table III
Bond Dissociation Energies[a]

Bond	ΔH (kcal/mole)	Bond	ΔH (kcal/mole)
$HOCH_2$—H	92 ± 2	CH_3—F	108 ± 3
$\dot{H}OCH(CH_3)$—H	90 ± 2	CH_3—Cl	84 ± 2
CH_3O—H	102 ± 2	CH_3—Br	70 ± 2
CH_3S—H	88 ± 2	Cl_3C—Cl	73 ± 2
CH_3—CH_3	$75 \pm ?$	CH_3—I	56.3 ± 1
$C_6H_5CH_2$—CH_3	70 ± 2	Cl_3C—Br	54 ± 2
F—F	38	H—F	135.8
Cl—Cl	58	H—Cl	103.0
Br—Br	46	H—Br	87.5
I—I	36.1	H—I	71.3
$\cdot CH_2CH_2$—H	40.0	C_2H_5—Br	69 ± 2

[a] Taken from tables compiled by Kerr (1966) and by Benson (1965).

The reaction of a methyl radical with isobutane to yield the *tert*-butyl radical and methane by displacement on the tertiary hydrogen is an exothermic reaction. Similarly, the reaction of a methyl radical with ethane producing

$$CH_3\cdot + (CH_3)_3CH \longrightarrow CH_4 + (CH_3)_3C\cdot \quad (104)$$

Make CH_3—H	$\Delta H =$	-104 kcal/mole
Break $(CH_3)_3C$—H	$\Delta H =$	$+91$ kcal/mole
	$\Delta H =$	-13 kcal/mole

methane and ethyl radicals is also exothermic but to a lesser extent. As a first approximation, the more exothermic formation of the *tert*-butyl radical would be expected to be faster than the less exothermic formation of the ethyl radical. Examination of the energy profiles for the two reactions illustrates that a certain degree of uncertainty exists, however, when attempts are made to relate in a direct manner the enthalpies of the exothermic reaction with the activation energies of the reactions (Fig. 3). It could be generalized that more

$$CH_3\cdot + CH_3CH_3 \longrightarrow CH_4 + CH_3CH_2\cdot \qquad (105)$$

Make CH_3—H	$\Delta H =$	-104 kcal/mole
Break C_2H_5—H	$\Delta H =$	$+98$ kcal/mole
	$\Delta H =$	-6 kcal/mole

exothermic reactions have lower activation energy requirements than those that are less exothermic, and all other factors being equal (steric, polar, and solvent interactions), this generalization has validity (*vide infra*). However, even a cursory examination of other reactions for which enthalpies can be calculated shows that the contribution of resonance factors to the reaction rate can be markedly outweighed by other considerations. For example, the reaction of a chlorine atom with ethane could proceed by either displacement at hydrogen or carbon [see Eqs. (106) and (107)]. Although the displacement on carbon is an overall more exothermic process, the activation energy requirement, because of steric factors, is prohibitively high and only the less exothermic displacement on hydrogen is observed.

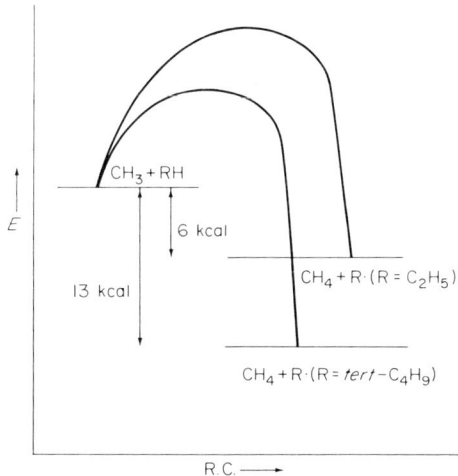

Fig. 3. Energy profile for reactions of methyl radicals with isobutane and ethane. R.C. is the reaction coordinate.

Two conditions do exist that allow for the meaningful use of reaction enthalpies in relating resonance factors to reaction rates. If the free-radical,

$$Cl\cdot + CH_3CH_3 \begin{array}{l} \nearrow HCl + CH_3CH_2\cdot \quad (\Delta H = -5 \text{ kcal/mole}) \quad (106) \\ \searrow CH_3Cl + CH_3\cdot \quad (\Delta H = \sim -9 \text{ kcal/mole}) \quad (107) \end{array}$$

chain-propagating reactions being compared are very similar in character, the reaction enthalpies are reflected in the activation energy requirements of the reactions. The order of reactivity of hydrogens in alkanes toward displacement by a free radical yielding alkyl radicals is generally tertiary > secondary > primary. Although polar effects may influence the reactivities of alkanes, as is the case with chlorine atom reactions (see Section IV, B, 1), the activation energies for the formation of alkyl radicals in propagating reactions often parallel their resonance stabilities (Fig. 4; see also Fig. 6).

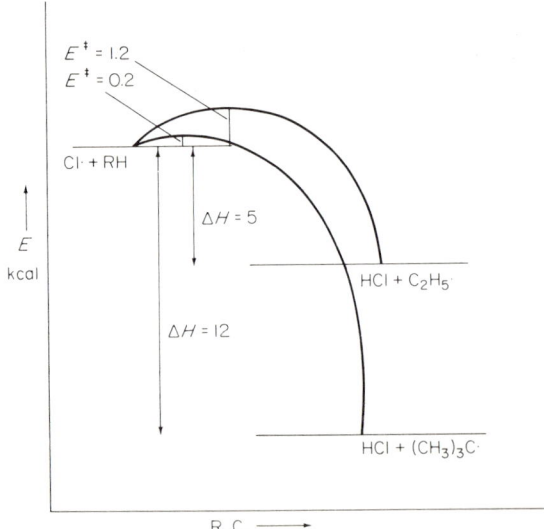

Fig. 4. Energy profile for reactions of alkyl hydrogens. R.C. is the reaction coordinate.

If the reaction is endothermic, the activation energy requirements for such processes must be at least the endothermicity of the reaction (barring tunneling). Other factors may influence the transition state of the reaction and may thereby cause the activation energy requirement for the reaction to be significantly higher than the minimum limitation established by the endothermicity of the reaction.

Information concerning the enthalpy of a chain-propagating reaction can also be useful in gaining some insight into the nature of the transition state of the reaction by use of the "Hammond postulate" (Hammond, 1955) (Fig. 5). This postulate is a useful generalization that predicts that exothermic reactions have low activation energy requirements and transition states that resemble the reactants more than the products. On the other hand, transition states that resemble the products more than the reactants are encountered in

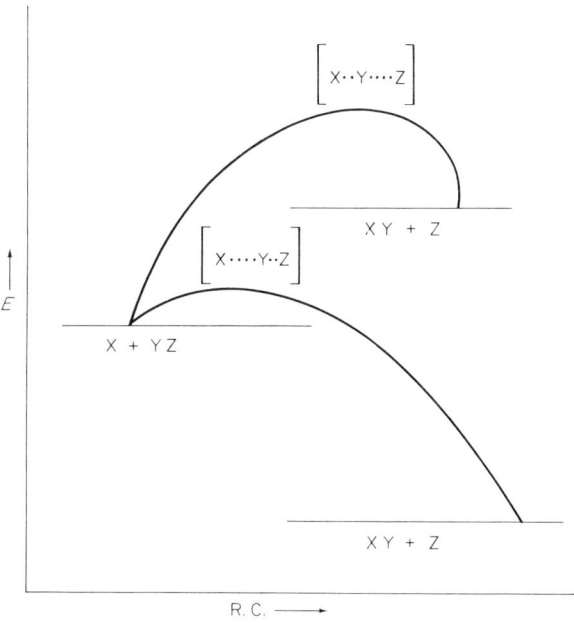

Fig. 5. Energy profiles illustrating the "Hammond postulate." R.C. is the reaction coordinate.

endothermic reactions having higher activation energies. Resonance stabilization of the product radical most likely does not play a significant role in determining the activation energy requirement of an exothermic reaction having a low activation energy. Only in less exothermic or endothermic reactions which have productlike transition states will the resonance factor of the radical produced in the reaction render a significant effect on the activation energy requirement and, hence, the reaction rate.

The stoichiometry of a free-radical chain reaction, as illustrated previously, is the algebraic sum of the radical-propagating reactions that comprise the chain sequence. The enthalpy of any chain reaction is a sum of the enthalpies of the radical-propagating reactions. Although it is possible that the enthalpy of a chain reaction involving two or more radical-propagating reactions may be appreciably exothermic, it does not necessarily follow that the reaction will proceed readily via the chain sequence. A single chain-propagating reaction having a sufficiently high activation energy requirement may be too slow to allow for the chain sequence to be maintained regardless of how exothermic the remaining radical-propagating reactions may be.

Estimations can be made of the enthalpies of additions of free radicals to unsaturated linkages. The difference in the bond dissociation energies of the C–H bond of ethane and that of a β C–H bond of an ethyl radical, namely, 58 kcal/mole, is the bond dissociation energy of the π bond of ethylene.

$$CH_3CH_2-H \longrightarrow CH_3CH_2\cdot + H\cdot, \quad \Delta H = +98 \text{ kcal/mole} \quad (108)$$

$$\cdot CH_2-CH_2-H \longrightarrow (\cdot CH_2CH_2\cdot + H\cdot) \longrightarrow CH_2=CH_2 + H\cdot, \quad (108a)$$
$$\Delta H = +40 \text{ kcal/mole}$$

$$\cdot CH_2-CH_2\cdot \longrightarrow CH_2=CH_2, \quad \Delta H = -58 \text{ kcal/mole} \quad (108b)$$

Breaking the π bond of an unsaturated linkage in a chain-propagating reaction is generally more facile than most displacement reactions. For example, addition of a bromine atom to ethylene in an exothermic process may be

$$Br\cdot + CH_2=CH_2 \longrightarrow BrCH_2CH_2\cdot \quad (109)$$

Make —C—Br	$\Delta H = -70$ kcal/mole	
Break C=C	$\Delta H = +58$ kcal/mole	
	$\Delta H = -12$ kcal/mole	

expected to proceed more readily than displacements on hydrogen by bromine. Similarly, additions of other radicals are generally exothermic, although the rates of addition are often influenced by steric factors in the reactions of nonterminal unsaturated linkages.

B. Polar Effects

Although most free radicals discussed in this chapter are uncharged neutral species, many display pronounced polar characteristics as reaction intermediates. Two kinds of polar effects are observed in free-radical reactions, the nature of the effect depending on the transition state of the particular reaction. Inductive polar effects are observed in the reactions of reactive electronegative free radicals in that they prefer to react at sites of high electron

1. FREE RADICALS

density in reactions that generally have "reactantlike" transition states. Somewhat more subtle, but far more common, are the donor–acceptor polar effects encountered in reactions that involve transition states having "productlike" character. An appreciation of these polar effects, particularly the latter, is essential to an understanding of many free-radical, chain-propagating reactions.

1. Inductive Effects

The reactions of reactive and electronegative free radicals are markedly influenced by differences in electron densities at various available reaction sites in the substrates with which they may react. The electron density factors are often more important than resonance considerations in determining the course of a free-radical-propagating reaction. For example, chlorine atoms react rapidly, thus displacing alkyl radicals from hydrogens of most alkanes (Fig. 6). In most cases (the exception being methane) the reactions are exothermic and, invoking the "Hammond postulate," the transition states of the reaction resemble the reactants more than the products. Resonance contributions from the product radical to the stabilization of the transition state complex are, therefore, small in such reactions. The degree of specificity observed in displacement reactions of chlorine atoms on various alkyl hydrogens,

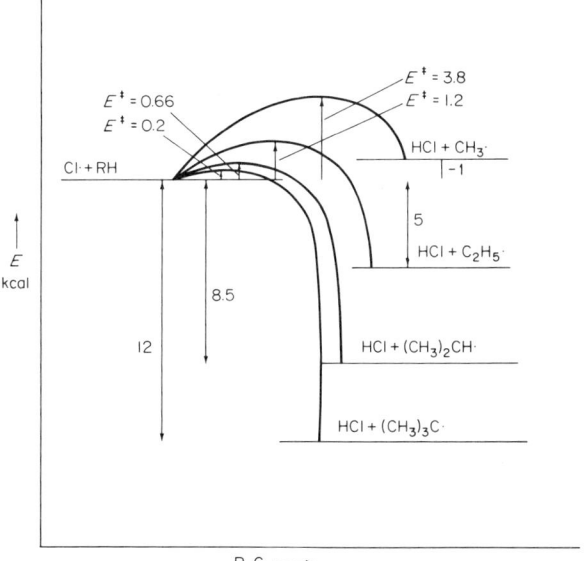

Fig. 6. Energy profiles for hydrogen abstraction by chlorine atoms. R.C. is the reaction coordinate.

namely, tertiary > secondary > primary, must be ascribed to some factor other than the relative degree of resonance stabilization of the radicals formed. The observed order of reactivity of various alkyl hydrogens can readily be explained in terms of the electron densities at the available reaction sites, namely, at the hydrogens that are subject to the displacement by the chlorine atom. The electron densities at various alkyl hydrogens vary depending on the number of electron-releasing alkyl groups bonded to the carbon bearing the hydrogen. For example, a primary alkyl hydrogen is bonded to a carbon having a single electron-releasing group, whereas a tertiary alkyl hydrogen is bonded to a more electron-rich carbon since it has three electron-releasing

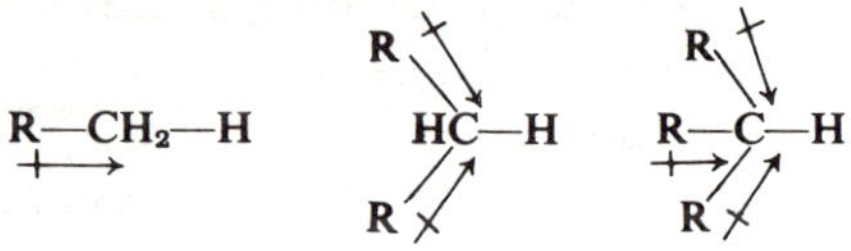

$$R-CH_2-H \qquad \begin{array}{c} R \\ \diagdown \\ HC-H \\ \diagup \\ R \end{array} \qquad \begin{array}{c} R \\ \diagdown \\ R-C-H \\ \diagup \\ R \end{array}$$

alkyl groups increasing the electron density at the reaction site. An electrophilic radical reacts more readily at sites of high electron density, and, consequently, tertiary alkyl hydrogens are attacked by chlorine atoms more readily than primary hydrogens (see Table IV).

Table IV
Relative Rates of Hydrogen Abstraction by Chlorine Atoms

Compound	Temperature (°C)	Reference
$\underset{\underset{CH_3}{\vert}}{CH_3CH}-\underset{\underset{CH_3}{\vert}}{\overset{4.2}{CH}CH_3}$ $\;\;1.0$	25	Russell (1958)
$\underset{\underset{CH_3}{\vert}}{CH_3-\overset{2.48}{CH}}-\overset{2.10}{CH_2}-\overset{1.0}{CH_2}-CH_2-CH_3$ $\;\;0.83\;\;\;\;2.16$	20	A. E. Fuller and Hickinbottom (1965)
$CH_3CH_2-H\;\;\;1.00$ $(CH_3)_2CH-H\;\;\;4.4$ $(CH_3)_3C-H\;\;\;6.7$	25	Poutsma (1969)

The inductive polar effect is also displayed in the reactions of chlorine atoms with substrates having electronegative substituents. Chlorination of n-butyronitrile, for example, yields mainly β-chloro-n-butyronitrile and some of the γ isomer, but none of the α-chloro isomer. Since the site of chlorination in the product is determined by the site of hydrogen atom abstraction in the free-radical chain sequence, the product distribution of this reaction illustrates the preference of the electronegative chlorine atom to react with hydrogens bonded on the β carbon rather than those of the α carbon. In terms of possible resonance contributions to the transition state of the reaction, the α hydrogens should be more labile to attack because of the resonance stabilization rendered by the cyano group. Since there is little product character in the transition state of the displacement reaction, resonance factors are relatively unimportant, and the electronegativity of the cyano groups apparently renders a more significant effect by withdrawing electrons from the α position and thereby decreasing its electron density relative to the other two possible sites of attack. The β hydrogens are possibly more reactive toward abstraction by the electronegative chlorine in this case [see Eq. (110) (Bruylants et al., 1952)] than the γ hydrogens, because the terminal methyl group is effective in releasing electrons to the β carbon.

$$Cl_2 + CH_3CH_2CH_2CN \xrightarrow{h\nu} \begin{cases} CH_3CH_2CHClCN\ (0\%) \\ + \\ CH_3CHClCH_2CN\ (69\%) \\ + \\ CH_2ClCH_2CH_2CN\ (31\%) \end{cases} \qquad (110)$$

Similar specificities in the site of chlorination are found in the reactions of other compounds having electronegative groups. The percentage of chlorination shown in each of the following compounds [Eqs. (111) (Vaughan and Rust, 1941), (112) (Henne and Hinkamp, 1945), and (113) (Bruylants et al., 1949)] indicates that the same phenomenon observed in the chlorination of n-butyronitrile is operative, namely, little or no displacement of the hydrogens of the carbon bonded to electron-withdrawing functionalities.

$$CH_3CH_2Cl + Cl_2 \xrightarrow[208°C]{h\nu} CH_2ClCH_2Cl\ (80\%) + CH_3CHCl_2\ (20\%) \qquad (111)$$

$$CH_3CH_2CH_2CF_3 + Cl_2 \xrightarrow{h\nu}$$
$$CH_2ClCH_2CH_2CF_3\ (55\%) + CH_3CHClCH_2CF_3\ (45\%) \qquad (112)$$

$$CH_3CH_2CH_2CH_2CO_2CH_3 + Cl_2 \xrightarrow{h\nu} \begin{cases} CH_3CHClCH_2CH_2CO_2CH_3\ (29\%) \\ + \\ CH_3CH_2CHClCH_2CO_2CH_3\ (71\%) \end{cases} \qquad (113)$$

Apparently, there is little product character in the hydrogen abstractions by *tert*-butoxy radicals, which are also electronegative and prefer to react at sites of high electron density. Hydrogen abstractions by this radical are observed in the light-induced chlorinations of various substrates with *tert*-butyl hypochlorite. Reactions of *tert*-butyl hypochlorites with *meta*- and

$$X\text{-}C_6H_4\text{-}CH_3 + (CH_3)_3CO\cdot \longrightarrow X\text{-}C_6H_4\text{-}CH_2\cdot + (CH_3)_3COH \quad (114)$$

$$X\text{-}C_6H_4\text{-}CH_2\cdot + (CH_3)_3COCl \longrightarrow X\text{-}C_6H_4\text{-}CH_2Cl + (CH_3)_3CO\cdot \quad (114a)$$

para-substituted toluenes indicate that benzylic hydrogens are abstracted by the *tert*-butyloxy radical more readily from methyl groups having higher electron densities. This is evidenced by the linear correlation observed by Walling and Jacknow (1960) when the log of the relative reactivity ratios is plotted against the σ values of the substituents ($\rho = -0.83$ at 40°C). In similar investigations of the chlorination of substituted toluenes with molecular chlorine, the electron density at the methyl group appears to be the determining factor, although much of the relative reactivity data correlates as well with the σ^+ parameters, which measure a cationic resonance factor (see Section IV,B,2), as with the σ values of the substituents (Russell and Williamson, 1964).

The inductive polar effect in free-radical reactions is not common since there are few electronegative (or, for that matter, nucleophilic) free radicals that take part in reactions with transition states that are reactantlike. Two electronegative radicals that might be expected to be influenced by inductive effects are the fluorine atom and the trifluoromethyl radical. Little evidence has been accumulated at this time to indicate that reactions of these radicals are influenced by inductive effects, since the radicals do not participate in simple chain reactions that lend themselves readily to detection of such effects.

2. Donor–Acceptor Polar Effects

Most free-radical, chain-propagating reactions have transition states that involve a considerable amount of bond making and breaking. As a consequence, characteristics of the product radical have significant influence on the activation energy requirements for the transition states of the reactions. Not only does the resonance stabilization of the product radical become evident, but a polar effect is also displayed in many free-radical-propagating reactions.

Almost all free radicals can be expected to show some polar characteristics depending on whether they might add an electron and become an anion or lose an electron and become a cation. Such oxidation and reduction reactions

1. FREE RADICALS

may actually be effected for many free radicals. Generally, neutral radicals do not undergo such changes in most of their chain-propagating reactions with

$$R^+ \underset{\text{oxidation}}{\xleftarrow{-e^{(-)}}} R \cdot \underset{\text{reduction}}{\xrightarrow{+e^{(-)}}} R^{(-)} \tag{115}$$

uncharged substrates. They can, however, display their propensity for oxidation or reduction because of the relative stability of the corresponding cation or anion in the transition states of reactions with appropriate substrates.

Free radicals that form a stabilized cation, because of delocalization of the positive charge, are referred to as donor radicals. Examples of donor radicals are alkyl, α-arylalkyl, allylic, α-hydroxyalkyl, α-alkoxyalkyl, acyl, and α-aminoalkyl since loss of an electron from such radicals yields comparatively stable carbonium ions.

Donor radicals

$$R_3C \cdot \longrightarrow R_3C^+ + e^-$$

$$ArCH_2 \cdot \longrightarrow ArCH_2^+ + e^-$$

$$R_2C{=}CHCH_2 \cdot \longrightarrow R_2C{=}CHCH_2^+ + e^-$$

$$R_2\dot{C}OH \longrightarrow R_2\overset{+}{C}OH + e^-$$

$$R_2\dot{C}OR \longrightarrow R_2\overset{+}{C}OR + e^-$$

$$R\dot{C}{=}O \longrightarrow R\overset{+}{C}{=}O + e^-$$

$$R_2\dot{C}NH_2 \longrightarrow R_2\overset{+}{C}NH_2 + e^-$$

Free radicals that form more stable anions by gain of an electron than cations by loss of an electron are called acceptor radicals. Halogen atoms and alkoxy, thiyl, trihalomethyl, α-carbonylalkyl, and α-cyanoalkyl radicals are examples of acceptor radicals. In these cases the corresponding anions are comparatively more stable than the cations. Both electronegativity and resonance factors that allow the negative charge to be delocalized render stable anionic character to these species.

A polar effect that lowers the activation energy requirement of a free-radical-propagating reaction is operative whenever the reaction of a donor or acceptor free radical produces as a reaction product an acceptor or donor free radical, respectively. The transition states of such reactions have a degree of charge separation resulting from the interaction of the moieties in the system that allow for both donating and accepting an electron. For example, the reaction of an acceptor radical such as the bromine atom with isobutane yields the *tert*-butyl radical which is a donor radical. The transition state of this reaction is a resonance hybrid of the canonical forms in which there is

Acceptor radicals

$$Cl\cdot + e^- \longrightarrow Cl^-$$
$$Br\cdot + e^- \longrightarrow Br^-$$
$$RO\cdot + e^- \longrightarrow RO^-$$
$$RS\cdot + e^- \longrightarrow RS^-$$
$$Cl_3C\cdot + e^- \longrightarrow Cl_3C^-$$
$$RCO\dot{C}H_2 + e^- \longrightarrow RCOCH_2^-$$
$$\underset{CO_2R}{R\dot{C}H} + e^- \longrightarrow \underset{CO_2R}{R\bar{C}H}$$
$$R_2\dot{C}-CN + e^- \longrightarrow R_2\bar{C}-CN$$

no charge separation and complete charge separation. In the former, only the resonance stabilization of the *tert*-butyl free radical contributes to the stabilization of the transition state. The activation energy requirement can be expected

$$Br\cdot + H-C(CH_3)_3 \longrightarrow [Br\cdots H\cdots C(CH_3)_3] \longleftrightarrow$$
$$[\overset{(-)}{Br}\cdots H\cdots \overset{(+)}{C(CH_3)_3}] \longrightarrow HBr + \cdot C(CH_3)_3 \qquad (116)$$

to decrease significantly if the charged structure is also contributing to the hybrid structure of the transition state of the reaction. The extent of the role that the charged structure plays in determining the activation energy requirement for the reaction depends on the donor and acceptor qualities of both reactant and product free radicals. Thus, there is a more significant polar contribution to the transition state in displacement reactions on hydrogen by acceptor radicals if *tert*-alkyl radicals are produced than if secondary or primary radicals are produced because of the greater stability of *tert*-alkyl relative to secondary and primary carbonium ions (pri:sec:tert = 1.00:160: 4713 at 100°C). The specificity noted in hydrogen abstraction by bromine

$$[Br\cdots H\cdots CH_2CH_3] \longleftrightarrow [\overset{(-)}{Br}\cdots H\cdots \overset{(+)}{C}H_2CH_3]$$
Little contribution

$$[Br\cdots H\cdots CH(CH_3)_2] \longleftrightarrow [\overset{(-)}{Br}\cdots H\cdots \overset{(+)}{C}H(CH_3)_2]$$
Significant contribution

$$[Br\cdots H\cdots C(CH_3)_3] \longleftrightarrow [\overset{(-)}{Br}\cdots H\cdots \overset{(+)}{C}(CH_3)_3]$$
Important contribution

atoms can, therefore, be attributed to a large extent to the relative resonance stabilities of the corresponding carbonium ions rather than solely to resonance stabilization of the free radicals formed. The formation of benzylic and allylic free radicals resulting from abstraction by bromine (as well as other acceptor free radicals) also mirrors the corresponding carbonium ion resonance stabilization. Indeed, the ease of benzylic hydrogen abstractions from *meta* and *para*-substituted toluenes by bromine atoms (Pearson and Martin, 1965) and trichloromethyl radicals (Huyser, 1960) is directly related to the benzylic carbonium ion stabilities of the corresponding free radicals. Linear correlations are observed when the log k/k_0 (k/k_0 being the relative reactivity ratio of the substituted with respect to the unsubstituted toluene toward benzylic hydrogen abstraction) is plotted against the σ^+ values of the substituents ($\rho = -1.38$ at 80°C for Br and $\rho = -1.40$ at 50°C for Cl$_3$C). The σ^+ values for the substituents serve as a measure of their ability to stabilize cationic character in the benzylic position (Brown and Okamoto, 1958).

Polar contributions to the transition states of other displacements as well as addition, β elimination, and, in some instances, rearrangement reactions are also important. The chain sequences of the following reactions [Eqs. (117)–(122)] illustrate these principles. It should be noted in these examples that donor radicals are produced from reactants that have donor properties and that acceptor radicals are produced from acceptor substrates in each case. It, therefore, follows that if one chain-propagating reaction of a two-step chain sequence has favorable polar characteristics, the other will also display favorable polar effects in its transition state.

$$RSH + CH_2=CHR' \longrightarrow RSCH_2CH_2R' \tag{117}$$

via

$$RS\cdot + CH_2=CHR' \longrightarrow [\overset{\delta-}{RS}\cdots CH_2\overset{\delta+}{\cdots}\dot{C}HR'] \longrightarrow RSCH_2\dot{C}HR' \tag{118}$$

$$RSCH_2\dot{C}HR' + RSH \longrightarrow \left[\begin{array}{c}\overset{\delta+}{RSCH_2CH}\cdots H\cdots \overset{\delta-}{SR}\\ |\\ R'\end{array}\right] \longrightarrow$$

$$RSCH_2CH_2R' + RS\cdot \tag{119}$$

$$C_2H_5C(CH_3)_2OCl \longrightarrow C_2H_5Cl + (CH_3)_2C=O \tag{120}$$

via

$$C_2H_5C(CH_3)_2O\cdot \longrightarrow \left[\begin{array}{c}CH_3\\ |\\ \overset{\delta+}{CH_3CH_2}\cdots \overset{\delta-}{C\cdots O}\\ |\\ CH_3\end{array}\right] \longrightarrow$$

$$C_2H_5\cdot + (CH_3)_2C=O \tag{121}$$

$$C_2H_5\cdot + C_2H_5C(CH_3)_2OCl \longrightarrow [C_2H_5C(CH_3)_2\overset{\delta-}{O}\cdots Cl\cdots \overset{\delta+}{CH_2CH_3}] \longrightarrow$$

$$C_2H_5Cl + C_2H_5C(CH_3)_2O\cdot \tag{122}$$

C. Complexing

The behavior of free radicals as reaction intermediates can be affected by their interactions with other species. Solvent effects are observed for some chain-carrying free radicals, and, in some cases, interactions of nonconjugated moieties within the free radical itself (bridging) affect its behavior.

1. *Solvent Effects*

Free radicals that are involved as intermediates in reactions that occur in solution almost certainly encounter some sort of interaction (aside from actual bond forming) with the molecules that comprise the liquid phase. The nature of these interactions is often different both in degree and kind from those observed for most charged species. The nature of the solvent effects in free-radical reactions is subtle and generally difficult to examine owing to both the limited extent of the interactions of uncharged species with the solvent and the fortuitous temperature range in which most free-radical chain reactions are investigated. The lack of any marked solvent effects in most free-radical reactions will become apparent after an examination of some of the data accumulated concerning the reactions of free radicals that do display solvent effects.

The site of chlorination of an alkane having different kinds of hydrogens available for displacement depends on factors that are involved in the displacement reactions by the chlorine atom on the hydrogens of the alkane. For example, chlorination of 2,3-dimethylbutane at 55°C in the absence of any solvent (except the alkane itself) yields a mixture of 1-chloro-2,3-dimethyl butane and 2-chloro-2,3-dimethylbutane, the composition of which indicates

$$\text{Cl}\cdot + \text{CH}_3\overset{\overset{\text{CH}_3}{|}}{\underset{\underset{\text{H}}{|}}{\text{C}}}-\overset{\overset{\text{CH}_3}{|}}{\underset{\underset{\text{H}}{|}}{\text{C}}}\text{CH}_3 \begin{array}{c} \overset{k_t}{\nearrow} \\ {\scriptstyle -\text{HCl}} \\ \\ \underset{-\text{HCl}}{\searrow} \\ {\scriptstyle k_p} \end{array} \begin{array}{l} \text{CH}_3\overset{\overset{\text{CH}_3}{|}}{\underset{\underset{\text{H}}{|}}{\text{C}}}-\overset{\overset{\text{CH}_3}{|}}{\underset{|}{\text{C}}}\text{CH}_3 \xrightarrow{\text{Cl}_2} \text{CH}_3\overset{\overset{\text{CH}_3}{|}}{\underset{\underset{\text{Cl}}{|}}{\text{C}}}-\overset{\overset{\text{CH}_3}{|}}{\underset{\underset{\text{H}}{|}}{\text{C}}}\text{CH}_3 \quad (123) \\ \\ \text{CH}_3\overset{\overset{\text{CH}_3}{|}}{\underset{\underset{\text{H}}{|}}{\text{C}}}-\overset{\overset{\text{CH}_3}{|}}{\underset{\underset{\text{H}}{|}}{\text{C}}}\text{CH}_2\cdot \xrightarrow{\text{Cl}_2} \text{CH}_3\overset{\overset{\text{CH}_3}{|}}{\underset{\underset{\text{H}}{|}}{\text{C}}}-\overset{\overset{\text{CH}_3}{|}}{\underset{\underset{\text{H}}{|}}{\text{C}}}-\text{CH}_2\text{Cl} \quad (124) \end{array}$$

the reactivity ratio $k_t/k_p = 3.7$ (statistically corrected). Russell (1958) has observed that this reactivity ratio is markedly altered if the chlorination reactions are performed in solvents other than the alkane (see Table V). In some solvents the reactivity ratio is larger, indicating that a greater degree of selectivity is displayed by the chlorine atom as a hydrogen abstractor. Such would be the case if the chlorine atom were complexed by the solvent mole-

Table V
Solvent Effects in Chlorination of 2,3-Dimethylbutane

Solvent	Molar concentration	k_t/k_p 25°C	k_t/k_p 55°C
2,3-Dimethylbutane	7.6	4.2	3.7
tert-Butyl alcohol	4.0	—	4.8
Dioxane	4.0	—	5.6
Carbon disulfide	12.0	225	—
Benzene	2.0	11	—
	4.0	20	—
	8.0	49	—
Nitrobenzene	4.0	9.0	—
Chlorobenzene	4.0	17.1	—
	6.0	27.5	—
Anisole	4.0	—	18.4

cules and, therefore, becomes less energetic. The less energetic, solvated chlorine atom displays a higher degree of selectivity as a hydrogen abstractor. The transition states of the reactions have more product character and resemble those of hydrogen abstraction by bromine atoms (see Fig. 7). As a consequence, both the donor characteristics and, to some extent, the radical resonance stabilization of the alkyl radicals formed play a more significant role in determining the reactivity of primary and tertiary alkyl hydrogens than the inductive factors.

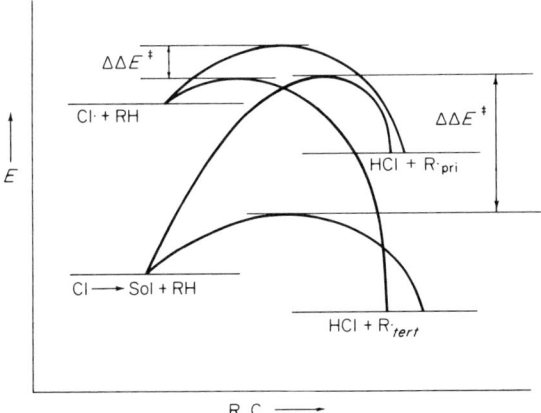

Fig. 7. Energy profiles for hydrogen abstraction by chlorine atoms and solvated chlorine atoms. R.C. is the reaction coordinate.

The nature of the interaction of the chlorine atom with the solvent varies. The electronegativity of the chlorine atom appears to be the primary reason for its ability to be solvated in most cases. The chlorine atom behaves as a Lewis acid and forms π complexes with various bases. The extent of the complexing depends on both the strength of the Lewis base and its concentration. Russell showed that substituted benzenes display solvent effects in the 2,3-dimethylbutane system that parallel the electron-releasing and -withdrawing abilities of the substituents as measured by their Hammett σ meta values. The nonbonding electrons of oxygen in both alcohols and ethers complex the chlorine atom as do the electrons of aromatic rings. The pronounced solvent effects observed in carbon disulfide possibly reflect a different kind of complexing, namely, the formation of a σ complex.

$$\text{Cl} \cdot + \text{R}\ddot{\text{O}}\text{R} \rightleftharpoons \text{Cl} \longrightarrow \text{O}\begin{smallmatrix}\text{R}\\\text{R}\end{smallmatrix} \tag{125}$$

$$\text{Cl} \cdot + \text{C}_6\text{H}_6 \rightleftharpoons [\text{C}_6\text{H}_6]\text{—Cl} \tag{126}$$

$$\text{Cl} \cdot + \text{S}=\text{C}=\text{S} \rightleftharpoons \text{ClS}\dot{\text{C}}=\text{S} \tag{127}$$

Chlorine atoms are monatomic and, therefore, may be complexed by a single molecule but still be able to participate in hydrogen atom abstraction reactions without encountering any severe steric problems. The transition states for the reactions most likely involve an approach to the hydrogen atom on which the displacement occurs from the nonsolvated side of the chlorine atom. If two solvent molecules are complexed with the chlorine atom, it will

$$\text{Solvent} \longleftarrow \text{Cl} + \text{HR} \longrightarrow [\text{solvent} \longleftarrow \text{Cl} \cdot \cdot \text{H} \cdot \cdot \text{R}] \longrightarrow$$
$$\text{solvent} + \text{HCl} + \text{R} \cdot \tag{128}$$

be "sandwiched" between them and sterically hindered from any direct interaction with an alkyl hydrogen. It is necessary that some degree of desolvation occur before the transition state can be reached.

Polyatomic free radicals can be expected to behave somewhat differently from the monosolvated chlorine atom. If the free radical is complexed at the radical site, desolvation is necessary (as in the case of the disolvated chlorine atom) to attain transition states of its reactions with other substrates. The energy required for this desolvation must be added to the activation energy requirement for the reaction. However, the desolvation of the free radical is accompanied by an increase in the entropy of the system. Quite possibly the reason that solvent effects are not generally observed in reactions of most polyatomic free radicals is that at the temperatures at which most free-radical reactions are performed the $T \Delta S$ factor counterbalances the increased enthalpy requirement for attainment of the transition state of the reactions. The above considerations have been observed in the reactions of *tert*-butoxy radicals (Walling and Wagner, 1964). The site of chlorination of 2,3-dimethylbutane with *tert*-butyl hypochlorite is determined by the hydrogen atom abstraction characteristics of the *tert*-butoxy free radical. Table VI lists the relative reactivity ratios (k_t/k_p) observed for chlorination of 2,3-dimethylbutane in different solvents at several temperatures. Although it appears that the solvents have little effect on k_t/k_p, an examination of the differences in the activation parameters of the reactions indicates that solvent effects are operative. The extent of desolvation depends on the nature of the transition state and most likely is more extensive in reactions of the complexed *tert*-butoxy radical with primary hydrogens since there is more productlike character in the transition state of this less exothermic reaction than reactions with the tertiary hydrogens. The higher activation energy for the reaction of the *tert*-butoxy radical at the primary hydrogens of 2,3-dimethylbutane relative to the tertiary hydrogens is reflected in the values of $E_{pri} - E_{tert}$. If the *tert*-butoxy radical is effectively solvated, as it apparently is by acetone and acetonitrile, the difference in the activation energies is greater because of the amount of energy required to desolvate the radical to enable the transition state for the primary hydrogen abstraction to be attained. Very significant, however, is the fact that accompanying the effect observed in $E_{pri} - E_{tert}$ is a concurrent change in log A_t/A_p (where A is the preexpotential factor which includes both the collision frequency and probability factor) of the Arrhenius rate expression $k = Ae^{-E/RT}$ and serves here as a measure of the entropy of activation. In each case where the effectiveness of the solvent is displayed as increasing the activation energy requirement for abstraction of a primary hydrogen, the value of log A_t/A_p decreases. The reason for the decrease is that a greater change in entropy is encountered in desolvation of an effectively solvated *tert*-butoxy radical in attaining the transition state for primary hydrogen abstraction (relative to tertiary hydrogen abstraction).

Table VI
Solvent Effects in Reactions of 2,3-Dimethylbutane with *tert*-Butyl Hypochlorite

Solvent	k_t/k_p					$E_p - E_t$	log (A_t/A_p)
	0°C	25°C	40°C	70°C	100°C		
None	89	70	55	—	—	1.99	0.35
Chlorobenzene	94	66	54	35	—	2.58	−0.08
Acetone	128	76	51	30	20	3.77	−0.92
Acetonitrile	—	47	33	17	10	4.57	−1.67

2. Bridging

Evidence has been accumulated to indicate that hydrogen abstraction can be effected by bromine atoms and divalent sulfur functions when bonded to the β carbon relative to the reaction site. For example, bromination of n-butyl bromide yields mainly 1,2-dibromobutane as the monobromination product (Thaler, 1963). The transition state for hydrogen abstraction of hydrogen from C-2 has been proposed to involve an anchimeric effect of the bromine bonded to C-1. The requirements for bridging of this type to play a

$$Br\cdot + C_2H_5CH_2CH_2Br \longrightarrow \left[Br\cdots H\cdots \underset{\underset{Br}{\diagdown\diagup}}{\overset{\overset{C_2H_5}{|}}{CH}}\!\!-\!\!CH_2 \right] \longrightarrow HBr + C_2H_5\dot{C}HCH_2Br \quad (129)$$

$$C_2H_5\dot{C}HCH_2Br + Br_2 \longrightarrow C_2H_5CHBrCH_2Br + Br\cdot \quad (130)$$

significant role in lowering the activation energy requirements for hydrogen atom abstraction are as follows: (1) The transition state must involve a considerable amount of C–H bond breaking so that sufficient radical character is developed, and (2) the bridging moiety apparently must have low energy, empty d orbitals. As a consequence of the first requirement, hydrogen abstraction must be performed by less energetic free radicals (e.g., Br· or RS·) rather than by very reactive radicals such as the chlorine atom, since little product character is developed in the transition states of the reactions of the latter. Bridging is observed, therefore, in bromination reactions using molecular bromine as the brominating agent but not in chlorinations with molecular chlorine. The bridging function may be a chlorine atom, a bromine atom, or a sulfide function (Huyser and Feng, 1970) since such moieties have energetically available d orbitals.

$$Br\cdot + H\!-\!\overset{|}{\underset{|}{C}}\!-\!\overset{|}{\underset{|}{C}}\!-\!Cl \longrightarrow \left[Br\cdots H\cdots \underset{\underset{Cl}{\diagdown\diagup}}{C}\!-\!C\!- \right] \longrightarrow HBr + \cdot C\!-\!C\!-\!Cl \quad (131)$$

$$RS\cdot + H\overset{|}{\underset{|}{C}}\!-\!\overset{|}{\underset{|}{C}}SR \longrightarrow \left[RS\cdots H\cdots \underset{\underset{S}{\diagdown\diagup}}{C}\!-\!C \right] \longrightarrow RSH + \cdot C\!-\!C\!-\!SR \quad (132)$$

The bridging concept has been questioned by Tanner *et al.* (1969). An alternate explanation, particularly in the bromination reactions of alkyl bromides, has been suggested to account for the apparently high reactivity of hydrogens on the β carbon relative to a bromine atom. Dehydrohalogenation of the alkyl bromide catalyzed by hydrogen bromide generated as a reaction product yields an alkene that on stereospecific ionic addition of molecular bromine would produce the same product that would be predicted by the bridging concept. Compelling evidence supporting the bridged radical is

$$CH_3CH_2CH_2CH_2Br \xrightarrow{HBr} CH_3CH_2CH=CH_2 + HBr \qquad (133)$$
$$\xrightarrow{Br_2} CH_3CH_2CHBrCH_2Br$$

found, however, in the bromination reactions of (+)-1-bromo-2-methylbutane and (+)-1-chloro-2-methylbutane (Skell et al., 1963). The observed formation of optically active products in these reactions can best be explained in terms of the formation and existence of an asymmetric intermediate.

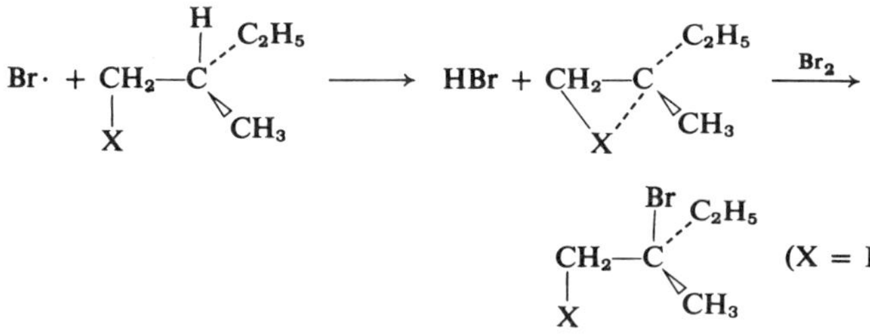

$$Br\cdot + CH_2\text{—}\overset{H}{\underset{X}{C}}\overset{C_2H_5}{\diagdown CH_3} \longrightarrow HBr + CH_2\text{—}\overset{}{\underset{X}{C}}\overset{C_2H_5}{\diagdown CH_3} \xrightarrow{Br_2}$$

$$CH_2\text{—}\overset{Br}{\underset{X}{C}}\overset{C_2H_5}{\diagdown CH_3} \qquad (X = Br \text{ or } Cl) \qquad (134)$$

Bromination of *cis*-4-*tert*-butylcyclohexyl bromide yields a single dibromination product that is formed in a reaction 15 times faster than brominations of the trans isomer which yields a mixture of dibromination products (Skell and Readio, 1964). In the case of the cis compound, the bromine atom in its most stable conformer is in an axial position, and a transition state involving the bridged species is readily attainable. The bridged radical apparently is formed as an intermediate and reacts stereospecifically with molecular bromine to yield *axial*-1,2-dibromo-4-*tert*-butylcyclohexane as the only reaction product. Bridging is not available as a low energy process

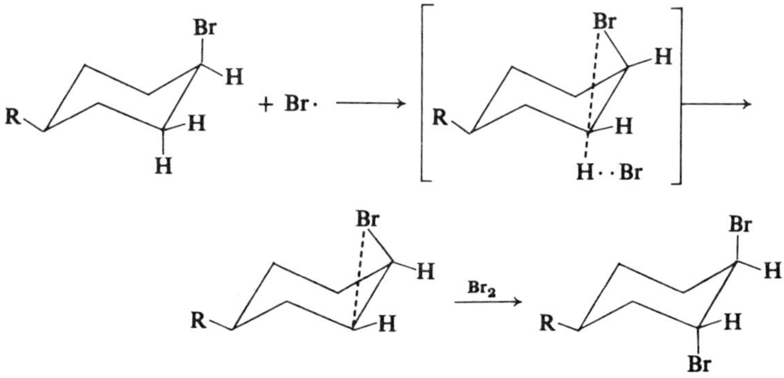

(135)

in the hydrogen abstraction from the trans isomer unless the less stable conformer, having the *tert*-butyl group in an axial position, is involved, or the reaction must involve the considerably energetic twist-boat conformation [Eq. (136)]. As a consequence, hydrogen atom abstraction from the trans isomer does not occur exclusively from C-2, and a mixture of dibromination reaction products is produced in the bromination of this isomer.

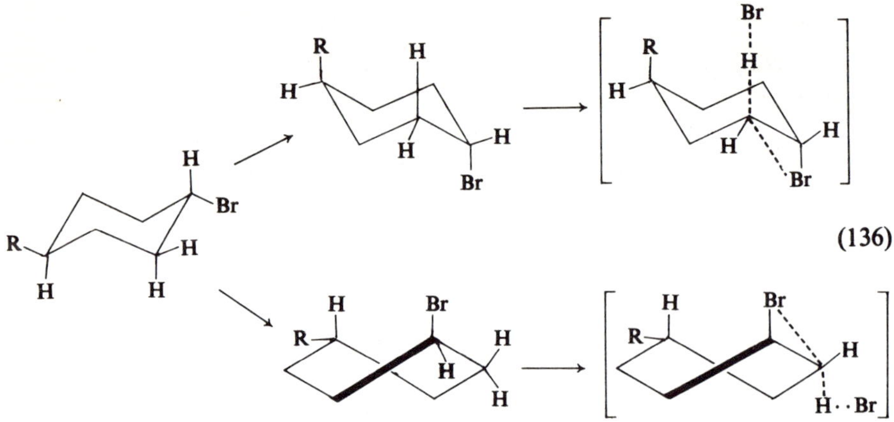

(136)

D. STERIC EFFECTS

Certain aspects of the effects of steric inhibition of free-radical, chain-propagating reactions have been discussed previously (see Section III,A). It was pointed out, and is reemphasized at this point, that chain-propagating reactions of free radicals that require their interaction with a substrate, namely, displacement and addition reactions, are markedly influenced by steric factors. Most observed displacement reactions occur on univalent hydrogen or halogen atoms and only reluctantly on bivalent oxygen and sulfur atoms. This is evidence of the steric limitations of these reactions. Displacement reactions on tetravalent carbon by nucleophilic reagents are comparatively common, but similar reactions on carbon by free radicals are rare (see Ingold and Roberts, 1971). Cyclopropane derivatives are labile to ring opening by chlorine atoms in reactions [Eqs. (137) (Applequist *et al.*, 1960) and (138) (Poutsma, 1965)] that most likely do involve displacement on carbon. These reactions reflect not only the reactivity of the chlorine atom but

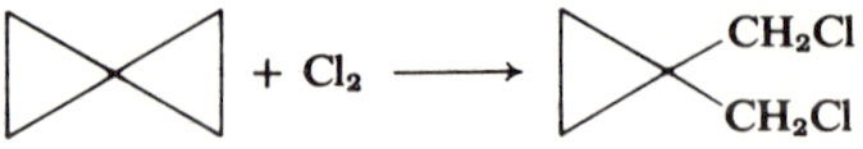

(137)

(138)

also the somewhat different steric environment of the carbon atoms of the strained cyclopropane ring system compared to the carbon atoms encountered in less strained compounds. In some compounds the steric situation about a given carbon atom may be such that it is largely exposed to attack, and ready displacements may occur. The bridgehead carbons of the hydrocarbon tricyclo[3.2.1.0]octane are exposed to attack by free radicals As a consequence, both bromotrichloromethane and oxygen react readily with this hydrocarbon to yield products that can best be explained as resulting from a chain sequence that involves a displacement on carbon as a chain-propagating reaction (Wiberg, 1972).

$$\text{structure} + BrCCl_3 \longrightarrow \text{structure with } CCl_3 \text{ and } Br \tag{139}$$

$$n\,\text{structure} + nO_2 \longrightarrow \text{structure with } OO\sim \text{ and } \sim OO \tag{140}$$

Free radicals add to nonterminal unsaturated linkages less readily than to a terminal unsaturation where the free radical is able to bond to a methylene carbon. Since the approach of a free radical to a π bond is most likely perpendicular to the plane described by the atoms that comprise the unsaturated linkage, it is not immediately obvious how the substitution of a hydrogen on one of the sp^2-hybridized carbons with some other group (halogen, alkyl, aryl, alkoxy, carboalkoxy, etc.) inhibits the rate of the addition reaction. The observed reluctance for addition of a free radical to a nonterminal unsaturated carbon may reflect the steric inhibition that groups larger than hydrogen exert on the formation of a σ bond between the attacking free radical and a carbon atom. The significance of the role of steric inhibition of the σ bond between the adding radical and a substituted carbon of an unsaturated linkage becomes evident in comparing the relative rates of formation of adduct radicals that have similar structures and, therefore, most likely involve similar polar and resonance factors in their formation. The relative rates of additions of free radicals to chlorine-substituted ethylenes illustrated the reluctance of addition to occur at nonterminal double bonds (Table VII).

In some cases the free radical itself may have a structure that sterically inhibits its reaction in chain-propagating reactions and, in some instances, bimolecular radical-destroying reactions. Trichloromethyl radicals have been

Table VII
Reactivities of Haloalkenes toward Radical Addition at 60°C[a]

Haloalkene	Adding radical		
	~CH$_2$ĊHCN	~CH$_2$ĊHOAc	~CH$_2$ĊHC$_6$H$_5$
CH$_2$=CHCl	1.0	1.0	1.0
CH$_2$=CCl$_2$	3.6	10	9.2
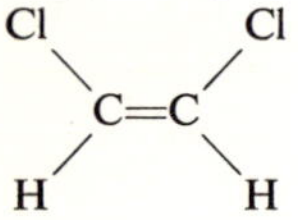 cis-ClHC=CHCl	—	0.05	0.08
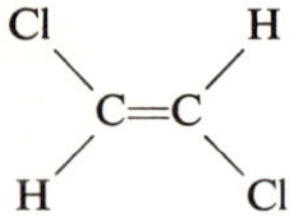 trans-ClHC=CHCl	—	0.3	0.46
ClHC=CCl$_2$	0.05	0.45	1.0
Cl$_2$C=CCl$_2$	0.007	0.04	0.09

[a] From Alfrey and Greenberg (1948) and Doak (1948).

observed to add to the unsaturated linkages of both α,o,o,p-tetramethylstyrene (Huyser and Kim, 1968) and 2,3-di-*tert*-butyl-1,3-butadiene (Huyser et al., 1968). The adduct radicals are sterically hindered, however, from reacting with a reactive reagent such as bromotrichloromethane.

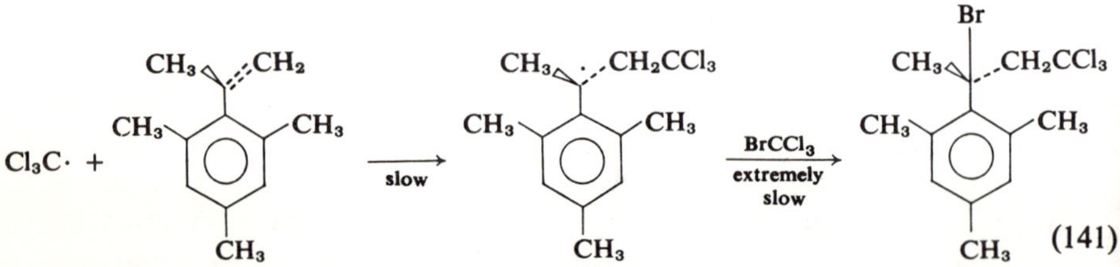

$$\text{Cl}_3\text{C}\cdot + \text{(mesitylene with =CH}_2\text{)} \xrightarrow{\text{slow}} \text{(adduct with CH}_2\text{CCl}_3\text{)} \xrightarrow[\text{slow}]{\text{BrCCl}_3\text{ extremely}} \text{(Br-adduct)} \quad (141)$$

Steric factors may prevent a substrate from attaining the necessary conformations that allow for maximum delocalization of the unpaired electron,

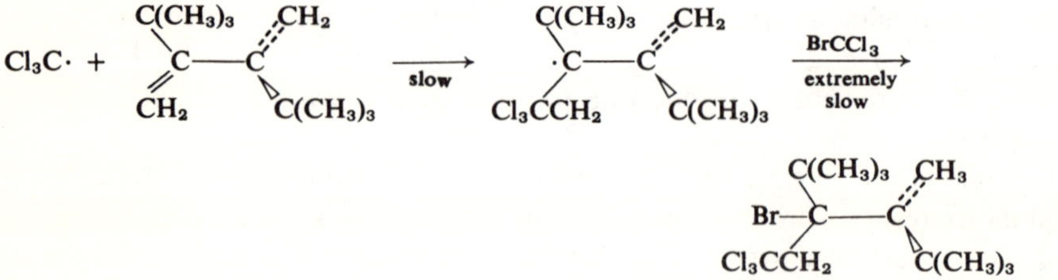

$$\text{Cl}_3\text{C}\cdot + \underset{\text{CH}_2}{\overset{\text{C(CH}_3)_3}{\text{C}}}=\underset{\text{C(CH}_3)_3}{\overset{\text{CH}_2}{\text{C}}} \xrightarrow{\text{slow}} \cdot\underset{\text{Cl}_3\text{CCH}_2}{\overset{\text{C(CH}_3)_3}{\text{C}}}-\underset{\text{C(CH}_3)_3}{\overset{\text{CH}_2}{\text{C}}} \xrightarrow[\text{slow}]{\text{BrCCl}_3\text{ extremely}} \text{Br}-\underset{\text{Cl}_3\text{CCH}_2}{\overset{\text{C(CH}_3)_3}{\text{C}}}-\underset{\text{C(CH}_3)_3}{\overset{\text{CH}_3}{\text{C}}} \quad (142)$$

thereby lowering the reactivity of the substrate in chain-propagating reactions. For example, the rates of addition of trichloromethyl radicals to α,o,o,p-tetramethylstyrene and 2,3-di-*tert*-butyl-1,3-butadiene are decidedly slower than the rates of addition to 1,3-butadiene and α-methylstyrene, respectively. The latter compound yields radicals that are not prohibited from attaining conformations that allow for delocalization of the unpaired electron into the unsaturated linkages, whereas the orbitals of the radical sites in the former are sterically prohibited from being coplanar with the π electron systems in the other parts of the radical.

A marked difference in the reactivities of diethyl fumarate and diethyl maleate toward addition of free radicals is observed because of steric inhibition of resonance in the latter compound (Lewis and Mayo, 1948). Although the ultimate adduct radical is the same from both compounds, the transition state for the addition reaction is attained more readily if the carbonyl function is coplanar with the double bond, as in the case of the fumarate ester. The maleate ester reacts slower because the two carbonyls are sterically prohibited from becoming completely coplanar with the unsaturated linkage, and, hence, resonance contributions to the transition state of the addition reaction are less important.

$$R\cdot + \underset{C_2H_5OC}{\overset{H}{\underset{\parallel}{C}}}\!\!=\!\!\underset{H}{\overset{COC_2H_5}{C}} \longrightarrow \left[-R\text{--}\underset{C_2H_5OC}{\overset{H}{C}}\!\!=\!\!\underset{O}{\overset{COC_2H_5}{C}} \right]$$

$$\underset{\underset{CO_2C_2H_5}{|}}{RCHCHCO_2C_2H_5} \quad (143)$$

$$R\cdot + \underset{H}{\overset{C_2H_5OC}{\underset{\parallel}{C}}}\!\!=\!\!\underset{H}{\overset{COC_2H_5}{C}} \longrightarrow \left[\underset{H}{\overset{C_2H_5OC}{R\text{--}C}}\!\!=\!\!\underset{H}{\overset{O\ \ O}{\underset{\parallel\ \ \parallel}{C\text{--}O}}}\ C_2H_5 \right]$$

"Galvinoxyl" is a stable free radical that can be maintained in solution because the radicals are sterically prohibited from interacting with each other at the oxygens which are protected by the flanking *tert*-butyl groups. Other smaller radicals can, however, approach the radical site. Loss of the characteristic spectral absorption of the "galvinoxyl" radical when it does react with another radical has been proved valuable in that the stable radical can be used to count other free radicals.

$$\cdot O{-}\underset{(CH_3)_3C}{\overset{(CH_3)_3C}{\langle}}{-}CH{=}\underset{C(CH_3)_3}{\overset{C(CH_3)_3}{\langle}}{=}O \;+\; R\cdot \;\longrightarrow\; RO{-}\underset{(CH_3)_3C}{\overset{(CH_3)_3C}{\langle}}{-}CH{=}\underset{C(CH_3)_3}{\overset{C(CH_3)_3}{\langle}}{=}O$$

(144)

V. Kinetic Aspects of Free-Radical Chain Reactions

A. General Considerations

The rate laws of free-radical chain reactions have unique features that can be ascribed to a combination of the kinetic aspects of radical-forming processes (initiation), free-radical, chain-propagating reactions (the chain sequence), and radical-destroying reactions (termination). The agreement observed between the derived and observed rate laws for free-radical chain reactions supports many of the conclusions that have been reached concerning the behavior of free radicals as reaction intermediates.

Listed in Table VIII are reaction rate constants that have been reported for a variety of initiation processes, radical-propagating reactions, and termination reactions.

Initiation reactions that involve a unimolecular decomposition are generally much slower than either chain-propagating or termination reactions. In a typical chain reaction (slow radical formation, rapid chain-propagating reactions, and extremely rapid radical-destroying reactions) certain interesting and useful concepts concerning the concentrations of free radicals can be reached. After a suitable induction period the concentration of free radicals in such a system is both small and essentially constant. This concentration of the free radicals is referred to as their steady-state concentration and is typical of the steady-state concentration of any reactive intermediates that are formed in a slow process (initiation) and consumed in a rapid reaction (termination).

If it is assumed that the concentrations of free radicals in a free-radical chain reaction attain a steady state, useful conclusions can be made about the rates of the various steps in the overall reaction. Consider as an example the following hypothetical, (and somewhat ideal) chemically initiated reaction [Eq. (145)] which may proceed by the mechanism shown in Eqs. (146)–(150). In a good free-radical chain reaction, most of the reactants are converted to the products in the chain sequence. As a consequence, few of the reactants are converted to products such as R–B that result from interactions of initiator fragments with a reactant [Eq. (147)] to produce chain-carrying free

Table VIII
Rate Constants for Free-Radical Reactions

Reaction	Temp. (°C)	k	Reference
$(CH_3)_3COOC(CH_3)_3 \rightarrow 2\,(CH_3)_3CO\cdot$	125	1.1×10^{-5} sec^{-1}	Raley et al. (1948)
$(CH_3)_3CO\cdot + C_6H_5CH_3 \rightarrow (CH_3)_3COH + C_6H_5CH_2\cdot$	30	7.2×10^4 liters mole^{-1} sec^{-1}	Walling and Kurkov (1967)
$(CH_3)_3CO\cdot + C_6H_{12} \rightarrow (CH_3)_3COH + C_6H_{11}\cdot$	24	2.6×10^4 liters mole^{-1} sec^{-1}	Howard and Ingold (1966)
$Cl_3C\cdot + (CH_3)_3COCl \rightarrow ClCCl_3 + (CH_3)_3CO\cdot$	24	1.2×10^3 liters mole^{-1} sec^{-1}	Howard and Ingold (1966)
$\sim CH_2\dot{C}H\underset{C_6H_5}{\vphantom{C}} + CH_2{=}CHC_6H_5 \rightarrow \sim CH_2CH{-}CH_2\dot{C}H$ with C_6H_5, C_6H_5	30	5.5×10^1 liters mole^{-1} sec^{-1}	Matheson et al. (1951)
$(CH_3)_3CO\cdot \rightarrow CH_3COCH_3 + CH_3\cdot$	51	1×10^4 sec^{-1}	G. R. McMillan (1960)
$2\,Cl_3C\cdot \rightarrow Cl_3CCCl_3$	24	5×10^7 liters mole^{-1} sec^{-1}	Howard and Ingold (1966)
$2\sim CH_2\dot{C}H\underset{C_6H_5}{\vphantom{C}} \rightarrow \sim CH_2CH{-}CHCH_2\sim$ with C_6H_5, C_6H_5	30	2.5×10^7 liters mole^{-1} sec^{-1}	Matheson et al. (1951)
$2\,(CH_3)_3CO\cdot \rightarrow (CH_3)_3COOC(CH_3)_3$	24	7×10^7 liters mole^{-1} sec^{-1}	Howard and Ingold (1966)

radicals or termination products such as A–X that result from the radical-destroying reaction (150). Indeed, the amount of termination product produced must be equivalent to the amount of initiator used. Not only is this

$$AB + XY \xrightarrow{\text{Initiator}} AY + BX \tag{145}$$

$$\text{In} \xrightarrow{k_i} 2\,R\cdot \tag{146}$$

$$R\cdot + AB \longrightarrow RB + A\cdot \tag{147}$$

$$A\cdot + XY \xrightarrow{k_a} AY + X\cdot \tag{148}$$

$$X\cdot + AB \xrightarrow{k_x} BX + A\cdot \tag{149}$$

$$A\cdot + X\cdot \xrightarrow{k_t} AX \tag{150}$$

stoichiometric relationship between the initiator consumed and termination product formed required, but the steady-state assumption concerning free-radical concentrations necessitates that the rate of initiation (R_i) must be equal to the rate of termination (R_t). This is true since these are the only processes in which there are changes in radical concentrations.

$$R_i = k_i[\text{In}] = k_t[A\cdot][X\cdot] = R_t \tag{151}$$

Each chain-carrying free radical has a steady-state concentration. In order for this condition to be maintained, each step in the chain sequence must proceed at the same rate. Thus, either of these reactions can be described as the rate of propagation [Eq. (152)].

$$k_a[A\cdot][XY] = k_x[X\cdot][AB] = R_p \quad (R_p = \text{rate of propagation}) \tag{152}$$

In order for a free-radical reaction to approach the characteristics of an ideal reaction (complete conversion of reactants to desired products), the reaction must have a long kinetic chain length (γ). The latter can be described either as the number of times the chain repeats itself before termination occurs or as the ratio of the rate of propagation with respect to the rate of termination. Equation (153) shows that the kinetic chain length is directly related to

$$\gamma = \frac{R_p}{R_t} = \frac{k_a[A\cdot][XY]}{k_t[A\cdot][X\cdot]} = \frac{k_a[XY]}{k_t[X\cdot]}$$

or (153)

$$\gamma = \frac{k_x[AB]}{k_t[A\cdot]}$$

the concentration of the reactants but inversely proportional to the concentration of the chain-carrying free radicals. In view of the large difference in the values of the rate constants for chain-propagating reactions with respect to termination ($k_a/k_t = \sim 10^{-4}$), it follows that the concentration of chain-carrying free radicals must be small if a long kinetic chain length is to be expected. For example in a reaction having the above ratio of rate constants and concentrations of [XY] and [AB] equal to 10 M, the steady-state concentration of A must be 10^{-4} for a kinetic chain length of 10, 10^{-5} for a chain length of 100, and 10^{-6} for a chain length of 1000. The amounts of the reactants converted to termination products plus the initiator fragment derived product are 10, 1, and 0.1%, respectively, for these chain lengths.

Long kinetic chain lengths are obtained if the initiation rate is slow. Slow rates of initiation can be achieved either by using small concentrations of the initiator or by employing reaction temperatures that do not result in rapid radical formation or, in the case of light-induced reaction, by decreasing the intensity of the illumination.

B. Steady-State-Derived Rate Laws

Since reactants are converted to products in the chain sequence of the reaction, it follows that the rate law for the reaction, if the steady state has been reached, can be described as a rate of any chain-propagating reaction. Thus, for the hypothetical reaction (145) the rate law for the reaction is either (154) or (155).

$$\text{Rate} = k_a[\text{A}\cdot][\text{XY}] \qquad (154)$$

$$\text{Rate} = k_x[\text{X}\cdot][\text{AB}] \qquad (155)$$

In the form of Eqs. (154) or (155), the rate law has little intrinsic value since simple and accurate methods are not available for determining the steady-state concentrations of the chain-carrying free radicals. However, assuming that the steady state has been reached [Eq. (151)], Eq. (156) is obtained:

$$[\text{A}\cdot] = \frac{k_i[\text{In}]}{k_t[\text{X}\cdot]} \qquad (156)$$

From (154) and (155), one also obtains Eqs. (157)–(159).

$$k_a[\text{A}\cdot][\text{XY}] = k_x[\text{X}\cdot][\text{AB}] \qquad (157)$$

and

$$[\text{X}\cdot] = \frac{k_a[\text{A}\cdot][\text{XY}]}{k_x[\text{AB}]} \qquad (158)$$

or

$$[\text{A}\cdot] = \frac{k_x[\text{X}\cdot][\text{AB}]}{k_a[\text{XY}]} \qquad (159)$$

Substituting the values of either [A·] or [X·] from (158) and (159) into (156), values for the steady-state concentrations of the chain-carrying free radicals are obtained in terms of nonradical species. These values can be

$$[A\cdot] = \left(\frac{k_i k_x [\text{In}][AB]}{k_t k_a [XY]}\right)^{1/2} \quad (160)$$

$$[X\cdot] = \left(\frac{k_i k_a [\text{In}][XY]}{k_t k_x [AB]}\right)^{1/2} \quad (161)$$

substituted in either (154) or (155) to give the rate law for the reaction. An inspection of this derived rate law for an "ideal" free-radical chain reaction reveals some characteristics that can be ascribed to all derived rate laws for

$$\text{Rate} = \left(\frac{k_i k_a k_x}{k_t} [\text{In}][AB][XY]\right)^{1/2} \quad (162)$$

free-radical chain reactions based on steady-state assumptions. First, the rate is proportional to the square root of the rate of initiation ($R_i = k_i[\text{In}]$) This phenomenon is observed regardless of the mode of initiation. Thus, reactions that are chemically initiated by a unimolecular decomposition reaction of an appropriate initiator have rate laws that include a square-root term of the initiator concentration, and rates of light-induced reactions are proportional to the square root of the intensity of the illumination. The rate laws also include in the denominator a square-root term of the reaction rate constant of the termination reaction.

The nature of concentration terms of the reactants and rate constants of the chain-propagating reactions that appear in steady-state-derived rate laws depend on the particular mode of termination that is operative for the reaction. In the "ideal" case illustrated, termination involves both chain-carrying radicals, which must be present at the steady state in comparable concentrations. As a consequence, the derived rate law has a concentration term for each reactant and square-root terms for each radical-propagating reaction. If a chain reaction is terminated by a dimerization of either of the two chain-carrying free radicals, the rate law assumes a somewhat different form. For example, termination may occur by reaction (163), the dimerization of A· radicals. In order for termination to occur by this route, the

$$2\,A\cdot \xrightarrow{k_{t(a)}} A_2 \quad (163)$$

steady-state concentration of A· must be comparatively larger than that of Y·. Two factors can cause such an imbalance of the steady-state concentrations of the chain-carrying radicals. One of these is a markedly low reactivity of chain-carrying radicals with the substrate with which it must react ($k_a < k_x$ in the case illustrated). The steady-state concentration of that radical will be necessarily greater than that of the more reactive chain-carrying radical in order that the rates of the chain-propagating reactions are the same. Also, if the reaction rate constants for the chain-propagating reactions are similar but the concentration of one reactant (e.g., XY) is much smaller than the other, the steady-state concentration of the chain-carrying free radical that reacts with that substrate must be greater in order to maintain equal reaction rates for both chain-propagating reactions. In either case, termination will occur by reaction (163), and the rate law for the reaction is (164). Only the rate constant and substrate concentration for the rate-limiting step [Eq. (148)] appear

$$\text{Rate} = \left(\frac{k_i[\text{In}]}{2k_{t(a)}}\right)^{1/2} k_a[\text{XY}] \quad (164)$$

in this rate law. If termination were to occur in a bimolecular reaction of X· [Eq. (165)], the rate law becomes Eq. (166). In this case, reaction (149) is the rate-limiting step, and the rate law involves the substrate concentration and reaction rate constant for that particular chain-propagating reaction.

$$2\,\text{X}\cdot \xrightarrow{k_{t(x)}} \text{X}_2 \quad (165)$$

and

$$\text{Rate} = \left(\frac{k_i[\text{In}]}{2k_{t(x)}}\right)^{1/2} k_x[\text{AB}] \quad (166)$$

The reaction rate laws must assume a different form if the chain-carrying radical of a unimolecular radical-propagating reaction (rearrangement or elimination) is involved in the termination reaction. Consider the following hypothetical reaction in which the chain-carrying free radical A· undergoes rearrangement to C·. If termination occurs only by reactions of C· and Y·,

$$\text{AB} + \text{XY} \xrightarrow{\text{In}} \text{CX} + \text{YB} \quad (167)$$

$$\text{In} \xrightarrow{k_i} 2\,\text{R}\cdot \quad (168)$$

$$\text{R}\cdot + \text{AB} \longrightarrow \text{RB} + \text{A}\cdot \quad (169)$$

$$\text{A}\cdot \xrightarrow{k_a} \text{C}\cdot \quad (170)$$

$$\text{C}\cdot + \text{XY} \xrightarrow{k_c} \text{CX} + \text{Y}\cdot \quad (171)$$

$$\text{Y}\cdot + \text{AB} \xrightarrow{k_y} \text{YB} + \text{A}\cdot \quad (172)$$

$$\text{Y}\cdot + \text{C}\cdot \xrightarrow{k_t} \text{YC} \quad (173)$$

the rate laws for the reactions will have the same form as those outlined above, since these radicals participate in bimolecular chain-propagating reactions. On the other hand, if the termination process involves A·, the steady-state-derived rate laws have a different form. For example, the rate law for the reaction, if terminated by the cross termination reaction (174), is given in (175).

$$A\cdot + Y\cdot \xrightarrow{k_{ay}} AY \qquad (174)$$

$$\text{Rate} = \left(\frac{k_a k_y k_i [\text{In}][\text{AB}]}{k_{ay}}\right)^{1/2} \qquad (175)$$

The reaction is overall first order in reactants, including the initiator, rather than overall three-halves order. If termination reaction (176) is operative, the rate law for the reaction is (177).

$$2\,A\cdot \xrightarrow{k_{t(a)}} A_2 \qquad (176)$$

$$R = k_a \left(\frac{k_i[\text{In}]}{2k_{t(a)}}\right)^{1/2} \qquad (177)$$

Note that the concentration of neither reactant appears in the rate law since the reactants are not involved in the rate-limiting step of the chain sequence.

The observed laws of free-radical chain reactions often deviate from ideality. In the first place, the efficiency of the initiation reaction in starting chains may be considerably less than unity because of other reactions of the initiator (e.g., solvent cage recombination of initiator-derived radicals, ionic reactions, or radical-induced decompositions of the initiator). The lack of complete efficiency of the initiator in starting chains, however, does not markedly alter the observed rate laws from the derived rate laws since an efficiency factor (generally designated as f and ranging from 0 to 1) can be incorporated in the rate of initiation.

$$R_i = f k_i[\text{In}] \qquad (178)$$

$$\text{Rate} = \left(\frac{k_a k_y (f k_i)[\text{In}][\text{AB}][\text{XY}]}{k_t}\right)^{1/2} \qquad (179)$$

A somewhat more important deviation from ideality is the fact that more than one of the possible termination reactions may be operative. If this is the case, the observed rate law will have the proportional character of the derived rate laws for those termination reactions.

C. Determination of Rate Constants

Although valuable information can be obtained concerning the relative values of rate constants for chain-propagating reactions from the observed rate laws of chain reactions, steady-state kinetics do not allow for the determination of the absolute values for the rate constants of either chain-propagating or terminating reactions. A value of the reaction rate constant for the initiation process can be determined by some independent study of its rate of decomposition. With this information, along with the observed reaction rate and the concentrations of the reactants, a ratio of the rate constants for the propagating reaction(s) over the square root of the rate constant for the termination reaction can be calculated. In order to determine the absolute values for either of these constants, other kinetic information involving these rate constants must be found.

Most often, the average lifetime (T_s) of the free-radical chain sequence is determined. This quantity generally is not obtained as readily as the steady state reaction rate. The rate of the reaction must be determined under nonsteady-state conditions. One of the most frequently used methods involves examining the reaction rate of a photoinitiated reaction which is allowed to be illuminated only intermittently (rotating-sector method). The observed rate is then related to the time intervals of illumination and nonillumination. Other methods involve determination of the rate of either the attainment of the steady-state concentration or its rate of decay. The reader is encouraged to examine other sources for more explicit details concerning both the experimental and theoretical aspects of these methods of determining absolute rate constants for the reactions of free radicals (Walling, 1957; Ferrington, 1959).

References

Alfrey, T., Jr., and Greenberg, S. (1948). *J. Polym. Sci.* **3**, 297.
Applequist, D. E., Fanta, G. F., and Hendrickson, B. W. (1960). *J. Amer. Chem. Soc.* **82**, 2368.
Bartlett, P. D., and Hiatt, R. R. (1958). *J. Amer. Chem. Soc.* **80**, 1398.
Baun, C. E. H., and Miller, S. F. (1951). *Trans. Faraday. Soc.* **47**, 1216.
Benson, S. W. (1965). *J. Chem. Educ.* **42**, 502.
Brown, H. C., and Okamoto, Y. (1958). *J. Amer. Chem. Soc.* **79**, 1913.
Bruylants, A., Tits, M., and Danby, R. (1949). *Bull. Soc. Chim. Belg.* **58**, 210.
Bruylants, A., Tits, M., Dieu, C., and Gaulhier, R. (1952). *Bull. Soc. Chim. Belg.* **61**, 266.
Carrington, A., and McLachlan, A. D. (1967). "Introduction to Magnetic Resonance," Chapter 6. Harper, New York.
Chattaway, F. D., and Backeberg, O. G. (1923). *J. Chem. Soc., London* **123**, 2999.
Cook, C. D., Kuhn, D. A., and Fianu, P. (1956). *J. Amer. Chem. Soc.* **78**, 2002.
Coppinger, G. (1957). *J. Amer. Chem. Soc.* **79**, 501.
Dickinson, R. A., and Leermakers, P. A. (1932). *J. Amer. Chem. Soc.* **54**, 3852.
Doak, K. W. (1948). *J. Amer. Chem. Soc.* **70**, 1525.
Ferrington, T. E. (1959). *J. Chem. Educ.* **36**, 179.
Freidlina, R. Kh. (1965). *Advan. Free-Radical Chem.* **1**, 211.
Fuller, A. E., and Hickinbottom, W. J. (1965). *J. Chem. Soc., London* p. 3228.
Fuller, G., and Rust, F. F. (1958). *J. Amer. Chem. Soc.* **80**, 6148.
Gomberg, M. (1900a). *J. Amer. Chem. Soc.* **22**, 757.
Gomberg, M. (1900b). *Chem. Ber.* **33**, 3150.
Hammond, G. S. (1955). *J. Amer. Chem. Soc.* **77**, 334.

Hammond, G. S., and Soffer, L. M. (1950). *J. Amer. Chem. Soc.* **72**, 4711.
Hammond, G. S., Sen, J. N., and Boozer, C. E. (1955). *J. Amer. Chem. Soc.* **77**, 3244.
Hammond, G. S., Wu, S. W., Trapp, O. D., Wardenten, J., and Keys, R. T. (1960). *J. Amer. Chem. Soc.* **82**, 5394.
Hausser, K. H., and Stehlik, D. (1968). *Advan. Magn. Resonance* **3**, 79.
Henne, A. L., and Hinkamp, F. B. (1945). *J. Amer. Chem. Soc.* **67**, 1194.
Hey, D. H., and Waters, W. A. (1937). *Chem. Rev.* **21**, 169.
Hoffmann, A. V., and Henderson, A. T. (1961). *J. Amer. Chem. Soc.* **83**, 4671.
Hoffmann, A. V., and Henderson, A. T. (1964). *J. Amer. Chem. Soc.* **86**, 639.
Horner, L., and Sherf, K. (1951). *Justus Liebigs Ann. Chem.* **537**, 35.
Howard, J. A., and Ingold, K. U. (1966). *J. Amer. Chem. Soc.* **88**, 4725.
Huyser, E. S. (1960). *J. Amer. Chem. Soc.* **82**, 394.
Huyser, E. S. (1970a). "Free Radical Chain Reactions," Chapter 10. Wiley, New York.
Huyser, E. S. (1970b). "Free Radical Chain Reactions," pp. 364–365. Wiley, New York.
Huyser, E. S. (1970c). "Free Radical Chain Reactions," Chapter 9. Wiley, New York.
Huyser, E. S., and Bredeweg, C. J. (1964). *J. Amer. Chem. Soc.* **86**, 2401.
Huyser, E. S., and Feng, R. (1970). *J. Org. Chem.* **36**, 731.
Huyser, E. S., and Garcia, Z. (1962). *J. Org. Chem.* **27**, 2716.
Huyser, E. S., and Kim, L. (1967). *J. Org. Chem.* **32**, 618.
Huyser, E. S., and Neckers, D. C. (1963). *J. Amer. Chem. Soc.* **85**, 3641.
Huyser, E. S., and VanScoy, R. M. (1968). *J. Org. Chem.* **33**, 3524.
Huyser, E. S., Wang, R. H. S., and Short, W. T. (1968). *J. Org. Chem.* **33**, 4323.
Ingold, K. U., and Roberts, B. P. (1971). "Free Radical Substitution Reactions." Wiley, New York.
Jores, G. G. (1954). U.S. Pat. 2,681,936.
Kerr, J. A. (1966). *Chem. Rev.* **66**, 465.
Kevan, L. (1969). *In* "Methods in Free-Radical Chemistry" (E. S. Huyser, ed.), Vol. I, Chapter 1, p. 1. Dekker, New York.
Kharasch, M. S., Englemann, H., and Mayo, F. R. (1937). *J. Org. Chem.* **2**, 288.
Kharasch, M. S., Jensen, E. V., and Urry, W. H. (1945). *J. Org. Chem.* **10**, 386.
Kochi, J. K. (1962). *Tetrahedron* **18**, 483.
Kochi, J. (1963). *J. Amer. Chem. Soc.* **85**, 1958.
Lewis, F. M., and Matheson, M. S. (1949). *J. Amer. Chem. Soc.* **71**, 747.
Lewis, F. M., and Mayo, F. R. (1948). *J. Amer. Chem. Soc.* **70**, 1533.
Littler, J. S., and Waters, W. A. (1960). *J. Chem. Soc., London* p. 2767
McMillan, F. L. (1965). *Diss. Abstr.* **27**, 1819-B.
McMillan, G. R. (1960). *J. Amer. Chem. Soc.* **82**, 2422.
Matheson, M. S., Aver, E. E., Bevilacqua, E. B., and Hart, E. J. (1951). *J. Amer. Chem. Soc.* **73**, 1700.
Mayo, F. R. (1959). *In* "Vistas in Free-Radical Chemistry" (W. A. Waters, ed.), pp. 139–142. Pergamon, Oxford.
Michaelis, L. (1949). *Tech. Org. Chem.* **1**, Part II, 1885.
Mulay, L. N. (1963). "Magnetic Susceptibility." Wiley (Interscience), New York.
Müller, E., and Ley, K. (1954). *Chem. Ber.* **87**, 922.
Müller, E., Muller-Rodloff, I., and Bung, W. (1935). *Justus Liebigs Ann. Chem.* **520**, 235.
Nozaki, K., and Bartlett, P. D. (1946). *J. Amer. Chem. Soc.* **68**, 1686.
Paneth, F., and Hofeditz, W. (1929). *Chem. Ber.* **62**, 1335.
Pearson, R. E., and Martin, J. C. (1965). *J. Amer. Chem. Soc.* **85**, 3142.
Pitts, J. N., Jr., Letsinger, R., Taylor, R., Patterson, S., Rectenwald, G., and Martin, R. (1959). *J. Amer. Chem. Soc.* **81**, 1068.

Poutsma, M. (1965). *J. Amer. Chem. Soc.* **87**, 4293.
Poutsma, M. L. (1969). *In* "Methods in Free-Radical Chemistry" (E. S. Huyser, ed.), Chapter 3, p. 79. Dekker, New York.
Raley, J. H., Rust, F. F., and Vaughan, W. E. (1948). *J. Amer. Chem. Soc.* **70**, 88 and 1338.
Russell, G. A. (1958). *J. Amer. Chem. Soc.* **80**, 4987.
Russell, G. A., and Williamson, R. C., Jr. (1964). *J. Amer. Chem. Soc.* **86**, 2357.
Rust, F. F., Seubold, F. H., and Vaughn, W. E. (1948). *J. Amer. Chem. Soc.* **70**, 3258.
Saurer, J. C. (1957). *J. Amer. Chem. Soc.* **79**, 5314.
Selwood, P. W. (1956). "Magneto-Chemistry," 2nd ed. Wiley (Interscience), New York.
Skell, P. S., and Readio, P. D. (1964). *J. Amer. Chem. Soc.* **86**, 3334.
Skell, P. S., Allen, R. G., and Gilmour, N. D. (1961). *J. Amer. Chem. Soc.* **83**, 504.
Skell, P. S., Tuleen, D. L., and Readio, P. D. (1963). *J. Amer. Chem. Soc.* **85**, 2849.
Tanner, D. D., Mosher, M. W., and Bunce, N. J. (1969). *J. Amer. Chem. Soc.* **91**, 7389.
Taylor, J. W., and Martin, J. C. (1966). *J. Amer. Chem. Soc.* **88**, 3650.
Thaler, W. A. (1963). *J. Amer. Chem. Soc.* **85**, 2607.
Trapp, O. D., and Hammond, G. S. (1959). *J. Amer. Chem. Soc.* **81**, 4876.
Uri, N. (1952). *Chem. Rev.* **50**, 375.
Urry, W. H., and Eiszner, J. R. (1952). *J. Amer. Chem. Soc.* **75**, 5822.
Vaughan, W. E., and Rust, F. F. (1941). *J. Org. Chem.* **6**, 479.
Walling, C. (1957). "Free Radicals in Solution," p. 85. Wiley, New York.
Walling, C., and Huyser, E. S. (1963). *Org. React.* **13**, 122–131.
Walling, C., and Indictor, N. (1960). *J. Amer. Chem. Soc.* **80**, 5814.
Walling, C., and Jacknow, B. B. (1960). *J. Amer. Chem. Soc.* **82**, 6108.
Walling, C., and Kurkov, V. P. (1967). *J. Amer. Chem. Soc.* **89**, 4895.
Walling, C., and Rabinowitz, R. (1959). *J. Amer. Chem. Soc.* **81**, 1243.
Walling, C., and Wagner, P. (1964). *J. Amer. Chem. Soc.* **86**, 3368.
Wiberg, K. B. (1972). To be published.
Wiberg, K. B., and Burgmaier, G. J. (1972). *J. Amer. Chem. Soc.* **94**, 7396.
Wilt, J. W., and Philip, H. (1959). *J. Org. Chem.* **24**, 1335.

2

CARBENES AND RELATED SPECIES

D. Bethell

I.	Introduction	61
	A. Definitions and Nomenclature	61
	B. History and Development of the Subject	62
	C. Sources of Carbenes and Carbenoids	64
II.	Electronic States and Structures of Carbenes	65
	A. Theoretical Aspects	65
	B. Spectroscopic Observation of Carbenes	72
III.	The Formation of Transient Carbenes and Carbenoids in Solution	77
	A. Types of Evidence	77
	B. Principal Sources of Carbenes and Carbenoids in Solution	79
IV.	Mechanism and Reactivity in Carbene and Carbenoid Reactions	89
	A. Introduction	89
	B. Coordination to Nucleophiles Having Nonbonded Electron Pairs	91
	C. Insertion	92
	D. Addition to Multiple Bonds	101
	E. Rearrangements	113
V.	Related Species	116
	A. Introduction	116
	B. Atomic Carbon	116
	C. Carbynes	119
VI.	Conclusion	119
	References	120

I. Introduction

A. DEFINITIONS AND NOMENCLATURE

Carbenes are electrically neutral organic species which contain, at least in one of their valence-bond representations, a divalent carbon atom associated with only six valence electrons. Four of these valence electrons are involved in the two covalent bonds, leaving two nonbonding electrons which may be paired with spins opposed in one of the two available orbitals of the central carbon atom (singlet state) or may occupy separate orbitals, usually with parallel spins (triplet state).

In cases where their multiplicity and geometry are unimportant, carbenes will be represented by the symbol RR'C:. When multiplicity is significant, singlet states will be represented as RR'C↑↓ and triplet states as RR'C↑↑.

The simplest carbene is H_2C: (methylene), and other carbenes will be named as its derivatives. Thus, $(C_6H_5)_2C$: will be referred to as diphenylmethylene rather than diphenylcarbene. The word carbene will be retained only as a generic term. The naming of carbenes in which the divalent carbon atom is situated in a ring presents some problems. The suffix -ylidene can be appended, as it can in acyclic systems, or the prefix carbena-, used in the manner of oxa- and aza- in heterocyclic systems, can be employed.

B. History and Development of the Subject

The possibility that stable organic compounds might be prepared in which carbon has a formal covalency of 2, as in carbon monoxide, rather than the usual 4, as in carbon dioxide, has been considered for about 150 years. Early attempts to prepare methylene from compounds of the type CH_3X (methanol or methyl chloride) by elimination of HX were undertaken in the belief that the desired product would be a stable, isolable compound. Although carbenes have now been prepared in stable form, albeit dispersed in an inert solid matrix at low temperatures, early investigators recognized quite soon that in the liquid and gas phases carbenes would have very short lifetimes. Indeed, as early as 1862 Geuther suggested that dichloromethylene might be an intermediate in the formation of carbon monoxide from chloroform and alkali, a conclusion borne out by detailed mechanistic studies almost a century later. Following Geuther's suggestion, the idea that carbenes might be transient intermediates was advanced, most notably by Nef (1897), for a variety of organic reactions, not always on the basis of sound evidence. Indeed, as late as 1927 carbenes were still being seriously considered as intermediates in substitution reactions of diphenylmethyl chloride (Ward, 1927).

Carbene chemistry was placed on firmer ground by investigations on the photochemical decomposition of diazomethane by Hantzsch and Lehmann (1901) and especially by Staudinger (Staudinger and Kupfer, 1912) in the early decades of this century. Similarities between carbenes and trivalent carbon free radicals were subsequently recognized. But it was the clear demonstration that halogenomethylenes could be readily produced in solution by the action of bases on haloforms (Hine, 1950) and that such carbenes could be made to react with olefins (Doering and Hoffmann, 1954) which triggered the remarkable increase of interest in divalent carbon compounds in recent years.

The period since 1950 has seen very rapid development and refinement of our understanding of carbenes. Their discrete if fleeting existence has now been established beyond doubt by a large body of experimental evidence, which in some cases includes direct spectroscopic observation. Most of the evidence, however, refers to systems in which carbenes have such a short lifetime that spectroscopic detection is not possible. Less direct evidence has to be sought, and the simplest comes from examination of the products of reactions thought to involve carbenes. Reactions most commonly associated with carbenes are shown in Fig. 1: insertion into σ bonds between carbon and

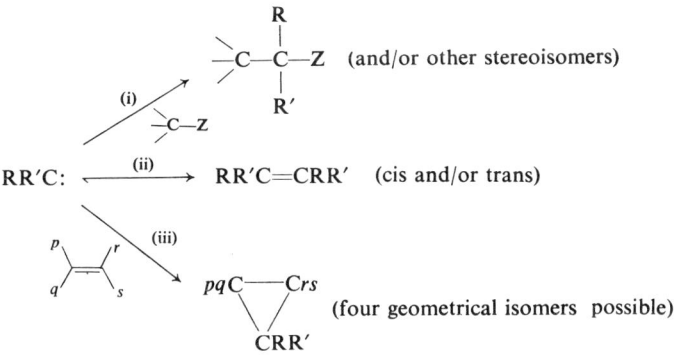

Fig. 1. Typical reactions of carbenes.

other elements [path (i)], reaction with carbene precursors or, exceptionally, another carbene to form dimeric olefins [path (ii)], and addition across a C–C or other multiple bond [path (iii)]. Identification of the products of such reactions has been used from the earliest times as evidence for intermediate carbene formation.

More recently, it has been recognized that intermediates other than free carbenes are capable of yielding products of this type. Such intermediates are often grouped together and referred to as *carbenoids*. In some instances it has proved possible to specify more precisely the nature of these species. For example, an α-halogenoalkyl lithium can show carbenoid reactivity, and a number of stable transition metal–carbene complexes have been characterized. In other cases the detailed constitution and structure of the carbenoid intermediate are less clear, and the noncommittal term carbenoid (symbolized in this chapter as RR'Coid) seems most appropriate, at least as a temporary expedient. The term will be taken, however, to imply a structurally more specific interaction of the RR'C moiety with other molecules or groups than is implied by the term "solvation."

C. Sources of Carbenes and Carbenoids

Most carbenes are endothermic species and their formation from stable organic molecules requires the input of substantial amounts of energy. This energy may be supplied by heat or irradiation. Alternatively, treatment with an external reagent may convert the reactant directly or by way of a metastable intermediate into a carbene. The common routes from three general reactant structures are shown in Fig. 2. All the routes may be classified as α eliminations in that two bonds to the central carbon atom are broken. The bond cleavage may be synchronous or consecutive; in the latter case the intermediate (e.g., RR'C-n) is often a carbanion or its organometallic equivalent which may behave as a carbenoid. As expected, the reactions of Fig. 1 find their reverse in Fig. 2.

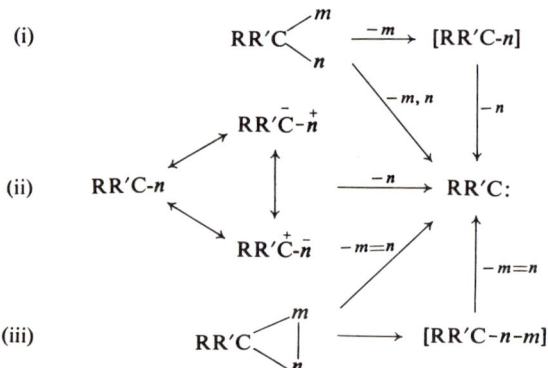

Fig. 2. General routes to carbenes. Note that in the routes from reactants (i) and (iii) the intermediates (which for simplicity have been written without charges) could be regarded as belonging to type (ii).

Addition, insertion, and rearrangement processes can yield carbenes as well. Thus, carbon atoms can by insertion into σ bonds or addition to π systems yield carbenes directly (see Section V). Carbene formation by rearrangement of other carbenes is relatively rare, but instances of rearrangement of cyclic ketones to alkoxymethylenes have been described (see Table I).

Compounds which have been reported to yield carbenes or carbenoids are grouped in Table I according to the categories distinguished in Fig. 2. The reaction conditions are indicated, together with the nature of the intermediate and the by-products. In most cases where metal-containing reagents are

2. CARBENES AND RELATED SPECIES

required, the evidence points to the formation of a carbenoid rather than a free carbene. This matter is discussed in detail for the most important carbene sources, diazoalkane decomposition and α eliminations of organic halides induced by bases and organometallic reagents, in Section III.

It should be noted that carbene formation is not a common reaction even of the types of compounds mentioned in Table I. Special structural and environmental requirements often need to be satisfied if the production of carbenes is to compete with other modes of reaction. Carbene formation should be favored if the carbene is particularly stable (e.g., $F_2C:$), if a stable by-product is produced with the carbene (e.g., N_2, CO, aromatics), or if strain in the reactant is relieved in the carbene-producing fragmentation (e.g., di-α-naphthylmethylene from tetra-α-naphthylethylene). Reaction conditions favoring modes of decomposition other than to carbenes may need to be avoided. For example, the presence of proton donors often leads to carbonium ion formation from diazoalkanes.

II. Electronic States and Structures of Carbenes

A. THEORETICAL ASPECTS

1. *Methylene*

Methylene, the simplest carbene, will serve to illustrate the theoretical principles governing the structures of carbenes.

The three-atom system H–C–H possesses six valence electrons of which four are to be associated with the two C–H σ bonds. The two electrons remaining are nonbonding and are located essentially on the divalent carbon atom which has two low energy orbitals available to accommodate them. If these orbitals are degenerate, then in the ground state of the carbene the two electrons are assigned one to each orbital with parallel spins as prescribed by Hund's rule. The carbene thus has a total spin quantum number (S) of 1, a multiplicity ($= 2S + 1$) of 3, and is said to be in a triplet electronic state. If the available orbitals are nondegenerate, then the nonbonding electrons can be accommodated in the orbital of lower energy provided that their spins are opposed: S is then zero, the multiplicity is 1, and the electronic state is a singlet. However, for nondegenerate orbitals which are fairly close in energy, the ground state of the carbene may still be a triplet, since electrostatic interactions between electrons are reduced if the electrons occupy different orbitals.

The spatial arrangement of the three nuclei in methylene, the energies of the nonbonding orbitals, and, hence, the electronic state of the carbene are intimately related. In methylene the nuclei may be arranged in a linear fashion

Table I
Sources of Carbenes and Carbenoids

Reactant type	Other reagent	Phase	Product	By-product	Reference
Category (i)					
Alkyl halide, RR'CHX	LiR"	*l*	RR'C: or RR'Coid	R"H + LiX	Köbrich (1967)
	Other base, B	*l*	RR'C: or RR'Coid	BH$^+$	Kirmse (1965)
gem-Dihalide, RR'CX$_2$	LiR"	*l*	RR'C: or RR'Coid	R"X + LiX	Köbrich (1967)
	CrSO$_4$	*l*	RR'Coid	—	Castro and Kray (1966)
	Zn/Cu (X = I)	*l*	RR'Coid	ZnI$_2$	Simmons et al. (1964)
	^{3}H atoms (R,R' = H)	*g*	X^{3}HC:	HX	Tang and Rowland (1968)
gem-Diazide, RR'C(N$_3$)$_2$	$h\nu$	*s*	RR'C:	N$_2$	Barash et al. (1967)
Trihalomethyl metal, R"MCX$_3$ (M = Hg, Sn, Si)	Δ	*l*	X$_2$C:	R"MX	Seyferth et al. (1965a,b, 1967b); Clark and Willis (1960)
Trihaloacetic acid derivative					
X$_3$C·CO$_2^-$Na$^+$	Δ	*l*	X$_2$C:	NaX, CO$_2$	
X$_3$C·CO$_2$R	NaOR'	*l*	X$_2$C:	ROR', NaX, CO$_2$	
Category (ii)					
Diazoalkane, RR'CNN^a	Δ, $h\nu$	*g, l, s*	RR'C:	N$_2$	Huisgen (1955); Zollinger (1961)
	Metallic (e.g., Cu, Zn, Ni, Ir) derivatives	*l*	RR'Coid	N$_2$	Müller et al. (1966); Cowell and Ledwith (1970)
Ketene, RR'C=CO	$h\nu$	*g, l*	RR'C:	CO	Ho and Noyes (1967)
Olefin, RR'C=CRR'	Δ	*l*	RR'C:	—	Franzen and Joschek (1960)
	$h\nu$ (R, R' = F)	*g*	F$_2$C:	—	Heicklen et al. (1965)
Ylid, RR'C$^-$—$\overset{+}{\text{ZR}}_n{''}$	Δ	*l*	RR'C:	ZR$_n{''}$	Hruby and Johnson (1962)
(e.g., ZR$_n{''}$ = SMe$_2$)	$h\nu$	*l*	RR'C:	ZR$_n{''}$	Trost (1966)

Compound	Conditions	Carbene	Phase	Product	Reference
Ketone, $RR'C=O$	C atoms (^1S or ^1D)	$RR'C:$	s	CO	Skell and Plonka (1970a)
	$(CF_3)_2C:$	$RR'C:$	l	$(CF_3)_2CO$	Mahler (1968)
	$Zn]/BF_3$	$RR'C:$	l	Zn^{2+}, F^-, OBF_2	Elphimoff-Felkin and Sarda (1969)
	$h\nu$ (Cyclic ketone)	$R\ddot{C}OR'$	l	—	Yates and Kilmurry (1966)
Thiocarbonyl compound, $RR'C=S$	$(RO)_3P$	$RR'C:$		$(RO)_3PS$	Corey et al. (1965)
	$h\nu$	$RR'C:$		S	Schmidt and Kabitzke (1964)
Category (iii)					
Diazirine, $RR'CN_2$[a]	$\Delta, h\nu$	$RR'C:$	g, l	N_2	Schmitz (1967)
Oxirane, 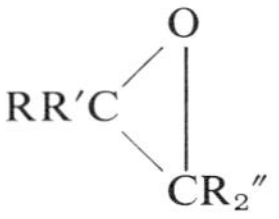 $RR'C\!\!-\!\!\overset{O}{\underset{CR_2''}{\diagdown}}\!\!-\!\!CR_2''$	$h\nu$	$RR'C:$	l, s	$R_2''C=O$	Becker et al. (1970)
	LiR'' ($R' = H$)	$R\ddot{C}\!\!-\!\!CR_2''O^-Li^+$	l	$R'''H$	Crandall and Lin (1967)
Cyclopropane, $RR'C\!\!-\!\!\overset{CR_2''}{\underset{CR_2''}{\diagdown}}$ 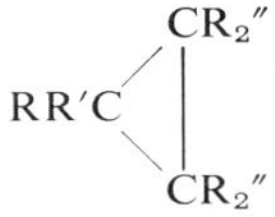	$h\nu$	$RR'C:$	l	$R_2''=CR_2''$	Richardson et al. (1965); M. Jones et al. (1966); Swenton and Krubsack (1969)
Bridged aromatic 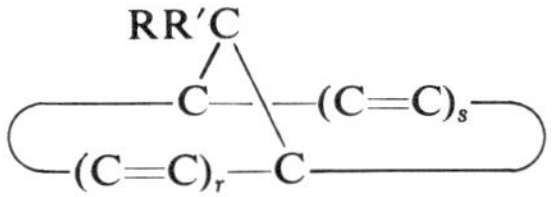 $RR'C\!\!<\!\!\overset{(C=C)_s}{\underset{(C=C)_r}{}}$	Δ	$RR'C:$	l	Aromatic $((C=C)_{r+s+1})$	Lemal et al. (1966)
Dioxolan, 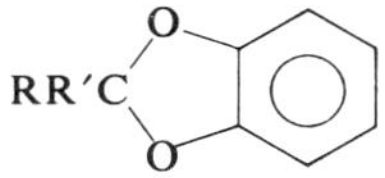 $RR'C\!\!<\!\!\overset{O}{\underset{O}{\diagdown}}\!\!\bigcirc$	$h\nu$	$RR'C:$	l	Quinone, $\bigcirc\!\!=\!\!O / \bigcirc\!\!=\!\!O$	Nair et al. (1968)
Aziridine, 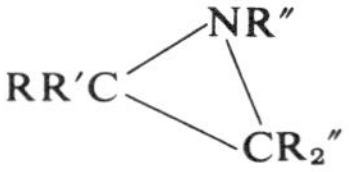 $RR'C\!\!<\!\!\overset{NR''}{\underset{CR_2''}{\diagdown}}$	$h\nu$	$RR'C:$	l	$R_2''C=NR''$	Nozaki et al. (1968a)

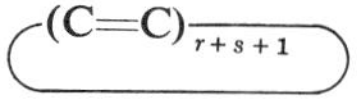

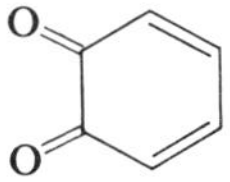

[a] Throughout this chapter, diazoalkanes are represented $RR'CNN$ to distinguish them from diazirines (cyclodiazoalkanes), $RR'CN_2$.

($D_{\infty h}$ symmetry) or nonlinear fashion (C_{2v} symmetry). Binding is achieved with suitable combinations of hydrogen 1s and carbon 2s and 2p atomic orbitals, the symmetry of which is indicated by symbols (a_1, b_1, etc., in C_{2v} and σ_g, π_u, etc., in $D_{\infty h}$). These combinations, their energies, and the way they are correlated in the two types of nuclear arrangements are shown schematically in Fig. 3 (Walsh, 1953; Closs, 1968).

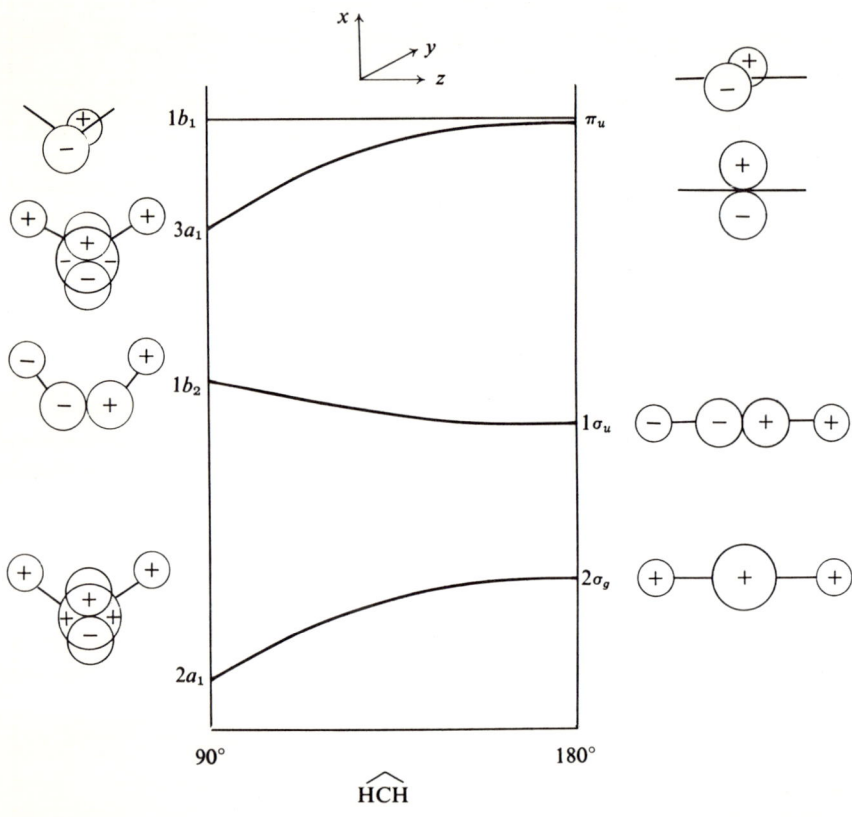

Orbitals in C_{2v} Orbitals in $D_{\infty h}$

Fig. 3. Molecular orbitals of methylene and their energy as a function of bond angle.

$D_{\infty h}$ Symmetry requires that the C–H bonds of methylene consist of overlapping hydrogen 1s and carbon 2s and $2p_z$ atomic orbitals ($2\sigma_g$ and $1\sigma_u$). The orbitals available to the two nonbonding electrons are then a pair of degenerate orbitals (π_u), essentially carbon $2p_x$ and $2p_y$ atomic orbitals. Linear methylene thus should be a triplet (designation $^3\Sigma_g^-$). Reducing HĈH from

180°, as indicated in Fig. 3, has a negligible effect on the out-of-plane π_{uy} orbital which now becomes $1b_1$ in C_{2v}. The p-type orbital in the plane of bending (π_{ux}), however, acquires an increasing s character because of the mixing in of the carbon $2s$ atomic orbital, thereby reducing its energy. In the limit the in-plane orbital could be closely similar to the carbon $2s$ atomic orbital with C–H bonds constructed from carbon $2p_x$ and $2p_z$ and hydrogen $1s$ orbitals. $\widehat{HCH}$ would then be 90° and the ground state of methylene would probably be a singlet since the energy separation of the in-plane and out-of-plane nonbonding orbitals would be maximized. The electronic arrangements and term symbols of the most important states of methylene are given in Table II.

Table II
Electronic States of Methylene

			Occupied orbitals		
State	Geometry	Term symbol	Carbon $1s$	C–H bonds	Non-bonding
Lowest triplet	Bent (C_{2v})	$^3B_1$	$(1a_1)^2$	$(2a_1)^2(1b_2)^2$	$(3a_1)(1b_1)$
	Linear ($D_{\infty h}$)	$^3\Sigma_g^-$	$(1\sigma_g)^2$	$(2\sigma_g)^2(1\sigma_u)^2$	$(\pi_{ux})(\pi_{uy})$
Lowest singlet	Bent (C_{2v})	$^1A_1$	$(1a_1)^2$	$(2a_1)^2(1b_2)^2$	$(3a_1)^2$
First excited singlet	Bent (C_{2v})	$^1B_1$	$(1a_1)^2$	$(2a_1)^2(1b_2)^2$	$(3a_1)(1b_1)$

On the basis of such considerations, a large number of quantum-mechanical calculations of widely varying sophistication have been carried through. Results from some of the best are summarized in Table III. (For reviews, see Gaspar and Hammond, 1964; Harrison and Allen 1969.) The latest calculations point to the $^3B_1$ state of methylene being the ground state with a bond angle in the range 130°–155°. However, the effect on the energy of increasing $\widehat{HCH}$ to 180° is calculated to be only 4–7 kcal mole^{-1}; therefore quoted bond angles should be regarded with caution. The energy separation between the ground state and the lowest singlet state is probably in the range 40–60 kcal mole^{-1}, although a lower limit of 33.4 kcal mole^{-1} has been suggested. This value is close to the 29.1 kcal mole^{-1} obtained spectroscopically for the energy separation of the ^3P and ^1D states of atomic carbon, which is equated with the energy for pairing electrons in orbitals on carbon. However, recent kinetic and thermodynamic interpretations of experimental data suggest that triplet methylene lies less than 10 kcal mole^{-1} below the singlet (Halberstadt and McNesby, 1967; Carr et al., 1970).

Table III
Calculated Geometries and Relative Energies of the Lowest Electronic States of Methylene[a]

	Method			
	CI[b]	Ab initio VB[c]	EHMO[d]	SCF CI[e]
r_{C-H} (Å)	1.11	1.17	1.06[f]	1.10[f]	1.096
$^3B_1$ HĈH	129°	—	138°	155°	135°
Δ[g]	0	—	0	13.8[h]	0
$^1A_1$ HĈH	—	90°	108°	115°	—
Δ[g]	—	44	58	0[h]	—
$^1B_1$ HĈH	132°	—	148°	—	—
Δ[g]	109	—	93	—	—

[a] Abbreviations: CI, configuration interaction; VB, valence bond; EHMO, extended Hückel molecular orbital; SCF, self-consistent field.
[b] From Foster and Boys (1960).
[c] From Harrison and Allen (1969).
[d] From Hoffman et al. (1968).
[e] From Bender and Schaefer (1970).
[f] Assumed values.
[g] Energy (kcal mole^{-1}) above ground state.
[h] EHMO method ignores electron–electron interaction. Since Δ for $^3B_1$ is less than the electron pairing energy (~ 30 kcal mole^{-1}), $^3B_1$ is the ground state.

2. Substituted Methylenes

In addition to bending, the degeneracy of the two π_u orbitals of linear methylene can be lifted by replacing hydrogen atoms with substituent groups capable of electronic interaction with only one of the degenerate orbitals. The necessary specificity is achieved if the substituent has vacant or filled orbitals of a symmetry suitable for overlap with π_{uy}. The interaction is shown diagrammatically in Fig. 4 for low-lying vacant orbitals [Fig. 4(a)] and high-lying filled orbitals [Fig. 4(b)]. In order to be effective in rendering the ground state of the carbene a singlet, the interaction must produce an energy separation between the highest occupied and lowest vacant orbitals greater than the electron-pairing energy (about 30 kcal mole^{-1}). Limited EHMO calculations indicate that this is rarely achievable when the substituent has vacant orbitals because the energy of the in-plane orbital (π_{ux}) is lowered by bending, thus offsetting the lowering of the energy of the out-of-plane orbital (Gleiter and Hoffmann, 1968; Hoffmann et al., 1968). However, for substituents with filled orbitals, bending increases the energy gap since the in-

2. CARBENES AND RELATED SPECIES

plane orbital is now the highest occupied level. Thus, for example, difluoro- and dichloromethylene are predicted to be ground-state singlet carbenes, in line with experimental evidence (Section II,B). On the other hand, arylmethylenes appear to have triplet ground states.

The same lines of reasoning lead to the conclusion that the interaction of the nonbonding orbitals on the divalent carbon with vacant orbitals on the substituents should enhance the electrophilic nature of carbenes, whereas substituents with filled orbitals should lead to more nucleophilic behavior. However, less specific polar and steric effects of substituents are superimposed, and these may be dominant. Thus, the dihalomethylenes show electrophilic reactivity (Section IV) despite the stabilizing influence of the filled orbitals of the halogens on the divalent center. Steric effects could interfere with the conjugation and, hence, affect the electronic state and pattern of reactivity of carbenes.

Interaction of substituents orbitals with orbitals on the divalent carbon atom has long been envisaged by organic chemists. The observed modes of formation of conjugated carbenes, such as that in Eq. (1) (Olofson *et al.*, 1968), as well as the chemical reactivity of carbenes such as **1** and **2** can be conveniently suggested by the zwitterionic structures. Similarly, the behavior of cyclic-conjugated carbenes like cycloheptatrienylidene (**3**) lends support to

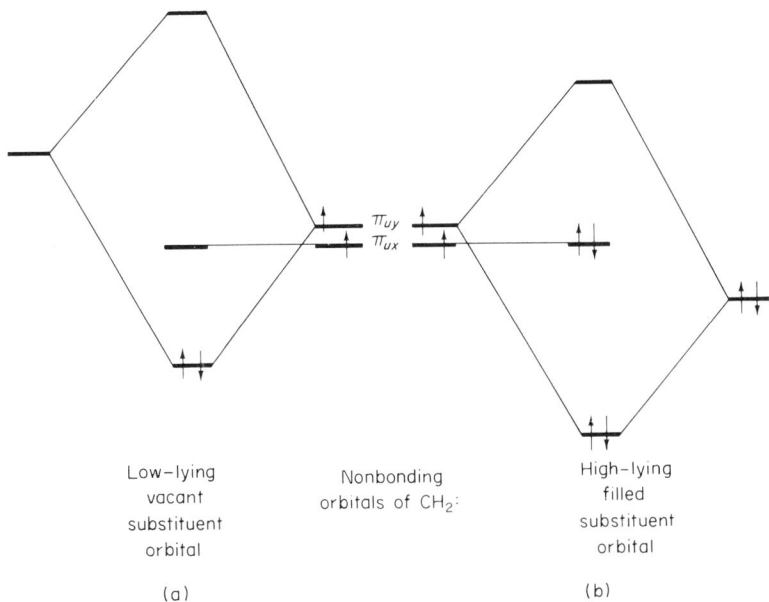

Fig. 4. Interaction diagram for orbitals capable of conjugation with the π_{uy} orbital of linear methylene.

the view that interaction to produce aromatic conjugation is important (W. M. Jones and Ennis, 1969; M. Jones et al., 1969).

$$(CH_3S)_2CH^+BF_4^- \xrightarrow{NR_3} \begin{matrix} CH_3S \\ \diagdown \\ CH_3S \end{matrix} C: \longleftrightarrow \begin{matrix} CH_3S^+ \\ \diagdown \\ CH_3S \end{matrix} C^- \longrightarrow (CH_3S)_2C=C(SCH_3) \quad (1)$$
(1)

$$(CH_3)_2C=C=C: \longleftrightarrow (CH_3)_2\overset{+}{C}-C\equiv\overset{-}{C}$$
(2)

(3)

B. Spectroscopic Observation of Carbenes

1. *Electronic and Infrared Spectroscopy*

The most detailed structural information has come from the analysis of electronic spectra, both in emission and absorption, of the simplest carbenes, methylene and the mono- and dihalomethylenes. Because of the reactivity of carbenes, spectra are recorded immediately after flash photolysis in the gas phase or in some cases by isolating the carbene in a solid matrix at low temperature. Great care is essential to ensure that the spectral data refer to the carbene rather than a precursor. Generation of the same carbene from more than one source can be useful, and the use of isotopes has proved invaluable in making assignments.

Herzberg's flash photolysis studies on diazomethane are particularly instructive, since they reveal important information about conditions under which singlet and triplet states of methylene are formed as well as structural parameters (Herzberg and Shoosmith, 1959; Herzberg, 1961a,b; Herzberg and Johns, 1966). Using gaseous mixtures of nitrogen with CH_2NN, $CHDNN$, and CD_2NN, absorption by the photolytically generated methylene was observed at 1414.5, 1415.5, and 1415.8 Å, the isotopic shifts confirming that the absorbing species contains only two hydrogen atoms. Analysis of the rotational fine structure indicated a linear or nearly linear triplet state for methylene. An alternative analysis suggesting a bond angle of 140° was thought less probable. When the proportion of nitrogen in the mixture with diazomethane was reduced, the photolytically generated species showed more prominent absorption at 5500–9500 Å. Rotational analysis indicated that this arose from the $^1A_1-^1B_1$ transition of methylene and yielded the molecular dimensions. The results suggested that the flash photolysis produces singlet methylene, which by collision with the inert nitrogen molecules undergoes intersystem crossing to the more stable triplet state.

Table IV contains structural information derived from electronic and

2. CARBENES AND RELATED SPECIES

Table IV

Structural Parameters from Electronic and Infrared Spectra of Carbenes (RR'C:)

R	R'	Carbene source	Phase	Spectral data	Electronic state (and transition)	r_{C-R} (Å)	$r_{C-R'}$ (Å)	$\widehat{RCR'}$ [a]	Reference
H	H	$CH_2NN/h\nu$	g	1414.5 Å	$^3\Sigma_g^- (\to ^3\Sigma_u^-)$	1.03	1.03	180°(155)	Herzberg (1961a); Herzberg and Johns (1966)
				5500–9500 Å	$^1A_1(\to ^1B_1)$	1.11	1.11	102.4(155)	
					$^1B_1(\leftarrow ^1A_1)$	1.05	1.05	140	
H	F	$CHFBr_2/h\nu$	g	4000–9000 Å	$^1A_1(\to ^1A_1'')$	1.12	1.31	102(103)	Merer and Travis (1966b)
					$^1A_1''(\leftarrow ^1A_1)$	1.12	1.30	127	
H	Cl	$CHClBr_2/h\nu$	g	4000–9000 Å	$^1A_1(\to ^1A_1'')$	1.12	1.69	103(110)	Merer and Travis (1966a)
					$^1A_1''(\leftarrow ^1A_1)$			~134	
F	F	$CF_2N_2/h\nu$	Ar matrix	668, 1102, 1222 cm^{-1} 2300–2670 Å	$^1A_1$	1.32	1.32	100–8(98)	Milligan et al. (1964)
		$CF_2=CFCl$/discharge	g	2340–2660 Å	$^1A_1$	1.30	1.30	104.9[b]	Laird, et al. (1950) Mathews (1966)
Cl	Cl	$Li + CCl_4$ $C + Cl_2$	Ar matrix	719.5, 745.7 cm^{-1}	$^1A_1$	—	—	~100(112)	Andrews (1968); Milligan and Jacox (1967)
CN	CN	$(NC)_2CNN/h\nu$	Matrix	392, 1158, 1756 cm^{-1}	?	—	—	180°(180°)	Smith and Leroi (1969)

[a] Values in parentheses are bond angles predicted by the EHMO method assuming bond lengths close to those deduced from spectra (Hoffmann et al., 1968).

[b] Similar structure deduced from the microwave spectrum of F_2C: (Powell and Lide, 1966).

vibration–rotation spectra of a variety of simple carbenes. Noteworthy features of the data are the good agreement between experimental and theoretical (EHMO) values of $\widehat{RCR'}$ and the absence of any marked influence of the phase on the spectrum and structure of difluoromethylene. The similarity of the C–Cl stretching frequencies in dichloromethylene and dichloromethane have led to the suggestion that, unlike the situation in difluoromethylene (Simons, 1965), there is little π interaction between the halogen and the divalent carbon in Cl_2C: (Andrews, 1968).

Electronic absorption spectra of triplet diarylmethylenes have also been observed (e.g., Gibbons and Trozzolo, 1966; Moritani *et al.*, 1968a). Detailed structural information has not been derived, but polarized optical absorption studies on triplet diphenylethylene oriented in single crystals of 1,1-diphenylethylene confirm the expectation that it is the electron in the out-of-plane ($1b_1$) orbital which is excited (Hutchison, 1967).

2. *Electron Paramagnetic Resonance (EPR) Spectroscopy*

The other major source of structural information, particularly $\widehat{RCR'}$, is EPR spectroscopy on carbenes, generated usually by photolysis of diazoalkanes is solid matrices at low temperatures, typically 77°K. This spectroscopic technique is restricted to carbenes in triplet states (or dicarbenes in quintet states) and, because of the low temperature employed, the EPR spectra observed are taken to be associated with the ground state of the divalent carbon species.

For a carbene in a triplet state, the total electron-spin quantum number S can take the values 1, 0, and -1. In general, the two spins interact. Thus the corresponding states of the carbene are nondegenerate even in the absence of a magnetic field, and the energy separation of the states is further increased in a magnetic field. Transitions between the states can then be induced by radiation of the appropriate frequency, the selection rule being $\Delta S = \pm 1$. The fine structure resulting from such transitions is dependent on the orientation of the carbene with respect to the direction of the radiation and can be observed for carbenes in dilute solid solution in single crystals of a convenient host. However, a weaker absorption at lower magnetic fields corresponding to $\Delta S = \pm 2$ (i.e., simultaneous flipping of both spins) as well as $\Delta S = \pm 1$ transitions can be observed for carbenes randomly oriented in glasses, and most of the structural information comes from EPR spectra of this type.

The splitting of the energy levels of triplet carbenes in the absence of a magnetic field, characterized by the so-called zero-field-splitting parameters D and E, is closely related to the geometry of the carbenes. Values of D and

2. CARBENES AND RELATED SPECIES

E are obtained by fitting the observed EPR spectrum to the Hamiltonian,

$$\mathcal{H} = g\beta HS + DS_z^2 + E(S_x^2 - S_y^2)$$

where S is the spin operator having components S_x, S_y, and S_z, β is the Bohr magneton, and g is the Landé splitting factor. D is a measure of the magnetic dipole interaction along the z axis (Fig. 3), whereas E measures the difference in interactions along the x and y axes, which are the directions of the two nonbonding orbitals on the divalent carbon atom. The value of D is roughly proportional to R^{-3}, where R is the average separation of the unpaired spins and thus gives an indication of the electron delocalization. For a given value of $\widehat{RCR'}$, D and E can be taken to be proportional to the electron-spin densities in the out-of-plane and in-plane nonbonding orbitals, respectively. The ratio E/D is often equated with the fractional s character of the in-plane orbital. Thus, $E/D = 0$ corresponds to a linear carbene, and $E/D = 0.33$ would indicate sp^2 hybridization of the divalent carbon atom and a bond angle of 120°.

The reliability of structural information obtained from zero-field-splitting parameters is not above suspicion, however. The validity of the assumptions underlying the relation drawn between E and D and molecular geometry can be criticized, and E and D values do vary with the nature of the matrix, though the effect is fairly small (Trozzolo et al., 1964). Table V gives zero-field splitting parameters and derived bond angles for a selection of carbenes.

More precise information about the geometry and electron distribution in triplet carbenes comes from examination of the hyperfine structure in the EPR spectra resulting from magnetic interaction of the unpaired electrons with nuclei. This more difficult procedure is made possible by the introduction of ^{13}C into the carbene, for example, as the divalent carbon in $(C_6H_5)_2C$: (Brandon et al., 1965). Proton hyperfine structure is normally obscured by the broadness of the lines in the EPR spectrum, but the splittings can be obtained by the ENDOR (electron nuclear double resonance) technique (Hutchison, 1967; Hutchison and Kohler, 1969).

The data in Table V show that although some triplet carbenes are linear, many are bent. Bond angles generally agree well with the predictions of EHMO theory. For some cyclic carbenes such as fluorenylidene (9-carbenafluorene), the bond angle derived from D and E is greater than can be accommodated in the ring without distortion of the normal carbon skeleton. In these cases, it has been suggested that the bonds to the divalent carbon are bent (Wasserman et al., 1964a,b; Moritani et al., 1967). Where $\widehat{RCR'}$ is less than 180° and the line of the R–C: bond is not an axis of the symmetry of R, the possibility of geometrical isomerism arises for carbenes, provided that rotation about the bond is slow on the EPR time scale. Such is the case for

Table V

Zero-Field Splitting Parameters from EPR Spectra of Triplet Carbenes, RR'C:, with Derived Bond Angles

R	R'	Matrix	D (cm^{-1})	E (cm^{-1})	$\widehat{RCR'}$ Derived	$\widehat{RCR'}$ EHMO calculation[a]	Reference
H	H	Xe	0.688	0.0035	172°(136)[b]	155°	Wasserman et al. (1970)
H	C≡CH	PCTFE[c]	0.628	0.000	180	180	Bernheim et al. (1965)
H	C≡N	PCTFE	0.863	0.000	180	180	Bernheim et al. (1965)
C≡N	C≡N	"Fluorolube"	1.002	<0.002	180	180	Wasserman et al. (1965a)
		C_6F_6	1.002	0.0033	165–70	—	Wasserman et al. (1965a)
H	CF_3	—	0.712	0.021	160	155	Wasserman et al. (1965b)
CF_3	CF_3	$CFCl_3$	0.744	0.0437	140	165	Wasserman et al. (1965b)
H	C_6H_5	—	0.515	0.0251	155	143[d]	Trozzolo et al. (1962)
C_6H_5	C_6H_5	$(C_6H_5)_2CO$	0.405	0.194	150	165[e]	Trozzolo et al. (1964)
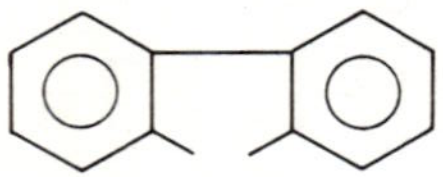 ⌬–⌬ (fluorenylidene)		C_6F_6	0.408	0.0283	135	(109)	Wasserman et al. (1964a)
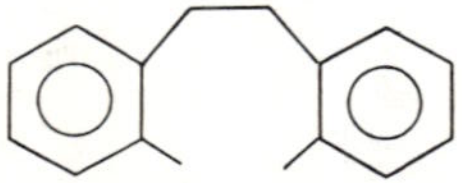 dibenzocycloheptenylidene		—	0.393	0.0170	150	(129)	Moritani et al. (1967)

[a] Values in parentheses are geometrically derived for a planar five- or seven-membered ring (Hoffmann et al., 1968).
[b] Derivation assumes slightly hindered rotation of CH_2: about its z axis in the matrix.
[c] Polychlorotrifluoroethylene.
[d] Calculation assumes bending in the plane of the phenyl ring.
[e] Calculation assumes phenyl rings rotated 45° about C_6H_5–C: bond. ENDOR experiments give angle of rotation as 34°, $\widehat{RCR'}$ as 148° (Hutchison and Kohler, 1969).

2. CARBENES AND RELATED SPECIES

α- and β-naphthylmethylenes, both of which have EPR spectra indicating the presence of two triplet species, designated syn and anti forms [(**4**) and (**5**)]. Conjugation of the aromatic substituent with the out-of-plane orbital on the divalent carbon atom leads to restricted rotation and, hence, to different interactions between the two unpaired electrons in the syn and anti isomers. The spectra are assigned on the basis of the quantum-mechanical calculation of spin-density distributions and resultant values of D and E (Trozzolo et al., 1965).

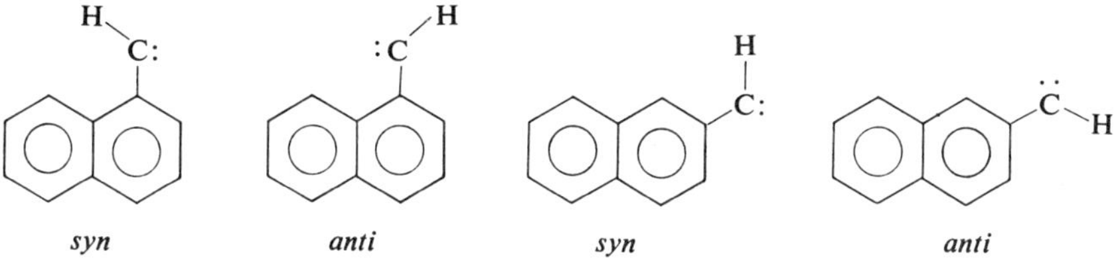

syn   anti   syn   anti
(**4**)   (**5**)

3. Mass Spectrometry

Mass spectrometry provides an entirely different method of direct observation of carbenes, which has not been exploited extensively. Attention thus far has been concentrated on the ionization potentials of carbenes generated by pyrolysis, for example, Cl_2C: from $Cl_3C \cdot SiCl_3$. A combination of ionization potentials with heats of formation of precursors permits the calculation of the heat of formation of the carbene. Values of the gas-phase heats of formation (kcal mole^{-1} at 25°C) are CH_2:, 86 (Kerr, 1966); Cl_2C:, 57 (J. S. Shapiro and Lossing, 1968); and F_2C:, -39 (Fisher et al., 1965; Zmbov et al., 1968). The exothermic nature of difluoromethylene stems presumably from strong π interaction between the halogen atoms and the p-type orbital on the divalent carbon. Infrared evidence indicates that this is much less important in dichloromethylene.

III. The Formation of Transient Carbenes and Carbenoids in Solution

A. TYPES OF EVIDENCE

Direct observation of carbenes is much more difficult in solution than in the gas or solid phases owing to the very short lifetimes and low standing concentrations produced when there are frequent collisions between carbenes and potential reaction partners. Flash photolysis, the most appropriate technique, has been used infrequently in the solution chemistry of carbenes. The transient formation of **6** from the corresponding diazoalkane has been demonstrated, however, by comparison of the solution spectrum (λ_{max} = 340 nm) a few microseconds after photolysis with that of the carbene generated in a solid matrix and known from EPR spectroscopy to be the triplet (Moritani et al., 1968a).

(6)

Usually carbene formation is inferred from the structures of the reaction products. This approach can be justified in part since when solid matrices containing carbenes, as shown spectroscopically, are allowed to liquefy the carbenes disappear and insertion, addition, and other products are formed. In most cases, however, spectroscopic evidence of carbene formation is lacking, and the structures of reaction products do not distinguish carbenes from carbenoids. Additional types of evidence are necessary in order to make this further distinction. The basis of such evidence may be indicated by reference to Eq. (2).

$$RR'Cmn \xrightarrow{l} \begin{bmatrix} \text{intermediate carbene} \\ \text{or carbenoid} \end{bmatrix} \begin{matrix} \xrightarrow{x} \text{product, } P_x \\ \xrightarrow{y} \text{product, } P_y \end{matrix} \quad (2)$$

i. The nature of the products P_x, P_y, etc., and their relative proportions should be independent of the identity of m and n and of the reagent (l) if the intermediate is a carbene. Conversely, if the intermediate is a carbenoid, then l, m, and n may profoundly influence the products.

ii. The rate of disappearance of the reactant RR'Cmn should be independent of the nature and concentration of the trapping reagents x, y, etc., if the intermediate is a carbene but may not be necessarily so if the intermediate is a carbenoid. The premise here is that carbenes are high energy intermediates, whose formation is the rate-limiting step of the conversion of RR'Cmn to P_x, P_y, etc. Carbenoids are regarded as less energetic and their reactions with x, y, etc., could be rate determining. In the limit, a carbenoid might be isolable even in a reaction in solution.

Thus, evidence of the reaction steps in which carbenes and carbenoids are formed is used in conjunction with evidence of steps in which they are consumed in order to define the nature of the intermediate.

The application of this approach (which is common to the study of other classes of transient intermediates) to three of the principal sources of carbenes and carbenoids in solution follows. As a broad generalization, "free" carbenes are formed in thermal or photochemical reactions not involving

2. CARBENES AND RELATED SPECIES

metallic or organometallic reagents, and carbenoids are intermediates in reactions induced by bases, especially metal alkyls. There are exceptions to both parts of this generalization.

B. PRINCIPAL SOURCES OF CARBENES AND CARBENOIDS IN SOLUTION

1. *Diazoalkanes and Related Compounds*

Because they are readily prepared and can be induced fairly easily to lose a molecule of nitrogen with heat or irradiation, diazoalkanes (RR'CNN) shown in their canonical forms in **7** are widely used as sources of carbenes and carbenoids. In some instances the diazoalkane itself is too thermally labile to be isolated and characterized. In such cases it is often prepared from a stable precursor, such as an alkali–metal salt of the toluene-*p*-sulfonylhydrazone (tosylhydrazone) of the corresponding carbonyl compound in the appropriate solvent and decomposed *in situ* [Bamford–Stevens reaction; Eq. (3)].

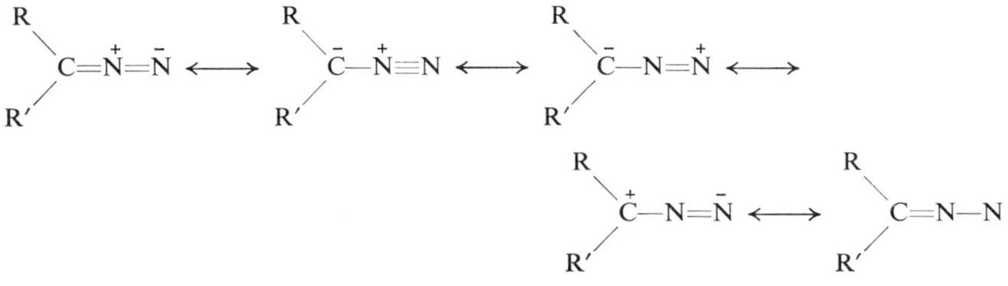

$$R_2C=N^+=N^- \leftrightarrow R_2C^- -N^+\equiv N \leftrightarrow R_2C^- -N=N^+ \leftrightarrow R_2C^+ -N=N^- \leftrightarrow R_2C=N-N$$

(7)

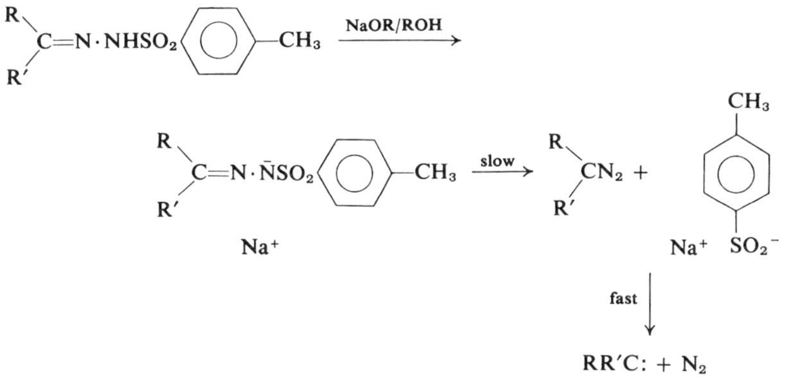

$$R_2C=N\cdot NHSO_2\text{-}C_6H_4\text{-}CH_3 \xrightarrow{NaOR/ROH}$$

$$R_2C=N\cdot \bar{N}SO_2\text{-}C_6H_4\text{-}CH_3 \; Na^+ \xrightarrow{slow} R_2CN_2 + CH_3\text{-}C_6H_4\text{-}SO_2^- \; Na^+ \quad (3)$$

$$\xrightarrow{fast} RR'C: + N_2$$

In interpreting reactions in which diazoalkanes are used, it must be remembered that these compound can react with loss of nitrogen by ways not involving the production of carbenes (Huisgen, 1955; Zollinger, 1961; Cowell and Ledwith, 1970). Proton donors, in particular, can convert diazoalkanes to carbonium ions [Eq. (4)] (More O'Ferrall, 1967), and in some reaction conditions it is not easy to decide which of the final reaction products have arisen from carbenes and which from carbonium ions. This is particularly so for simple diazoalkanes (such as 1-diazobutane) in protic solvents like alcohols where the position of the diazoalkane–alkyldiazonium ion equilibrium can be important in determining the course of reaction (Kirmse and Rinkler, 1967; Friedman et al., 1969). Thus, the proportions of products in Bamford–Stevens reactions [Eq. (3)] can show a dependence on the base concentration used.

$$RR'CNN + HX \rightleftarrows RR'CHN_2^+ X^- \xrightarrow{-N_2} RR'CH^+ X^- \qquad (4)$$

These problems can be conveniently illustrated by the reaction of camphor tosylhydrazone (**8**) with sodium methoxide in diglyme. The products are tricyclene (**9**), which might be regarded as the intramolecular insertion product of the carbene derived from diazocamphor, and camphene (**10**), which might be taken to be derived from the corresponding carbonium ion since its formation is favored in protic solvents of high protonating ability. However, the ratio of camphene to tricyclene decreases with increasing base concentration, and, when deuterium oxide is present, the tricyclene produced contains

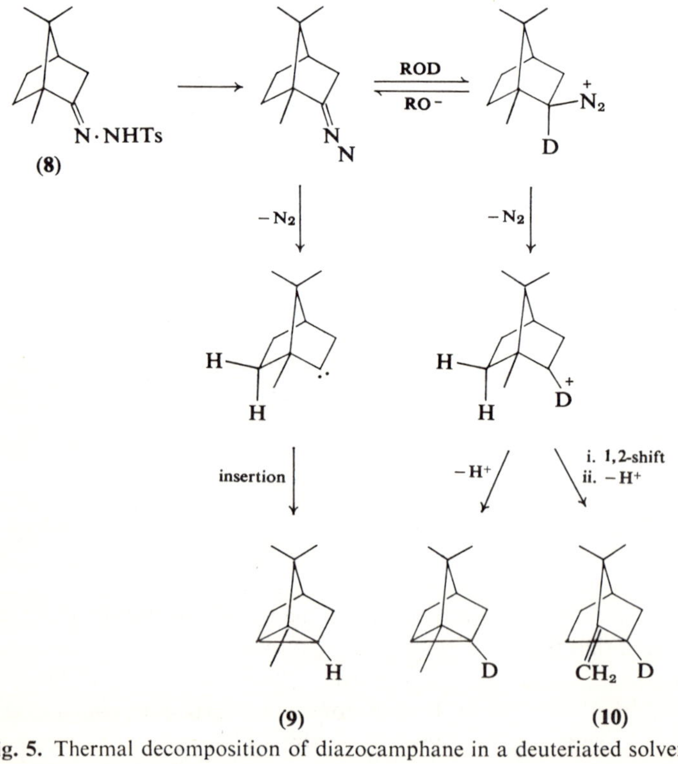

Fig. 5. Thermal decomposition of diazocamphane in a deuteriated solvent.

2. CARBENES AND RELATED SPECIES

deuterium in an amount which also decreases as the sodium methoxide concentration goes up (R. H. Shapiro et al., 1967). Figure 5 shows the simplest scheme which will account for these results, with tricyclene as the product of both carbene and carbonium ion routes and camphene arising only from the bicyclic cation.

The thermal decomposition of diazoalkanes in aprotic solvents is less ambiguous. Carbene formation is the principal reaction path as evidenced by the kinetic form of the decomposition of, for example, phenyl- and diphenyldiazomethane in acetonitrile, which indicates a primary unimolecular process (Bethell et al., 1965; Bethell and Whittaker, 1966). Care is still essential in interpreting the reaction products since modes of decomposition (which may be heterogeneous in some cases) not involving carbenes may still lead to products typical of carbene formation (see, for example, Bethell et al., 1969). When cyclopropanes are produced from olefins and diazoalkanes, it is always necessary to check that the olefin does not enhance the rate of consumption of the diazo compound. A kinetic dependence on the olefin concentration is indicative of the intermediate formation of a pyrazoline by 1,3-dipolar cycloaddition (Cowell and Ledwith, 1970) rather than a carbene as shown in Fig. 6.

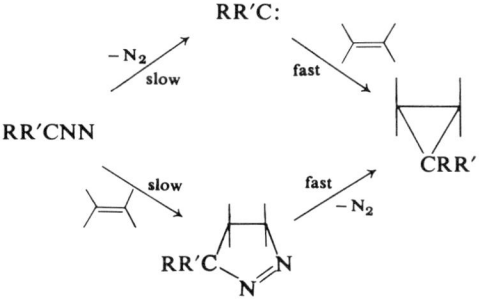

Fig. 6. Routes to cyclopropanes from diazoalkanes and olefins.

The spectroscopic studies outlined in Section II demonstrate that carbenes can be readily produced by photolysis of diazoalkanes. In contrast to thermal decomposition, in which spin conservation determines that carbenes produced from the ground state of a diazoalkane be initially in the singlet state, photolysis of diazo compounds can yield directly either triplet or singlet carbenes, depending on whether or not a triplet photosensitizer [e.g., Hg (^3P) atoms, benzophenone] is used. The first excited state of, for example, diazomethane in n-hexane is bonding, however (Bradley et al., 1964; Hoffmann, 1966), and might be capable of reacting with another molecule directly rather than lose nitrogen unimolecularly to give the carbene. Diazomethane is known to act as a photosensitizer in certain reactions (Bradley and Ledwith, 1967).

Diazoalkanes undergo decomposition in the presence of a variety of metallic derivatives (Cowell and Ledwith, 1970; Lappert and Poland, 1970). Halides of metals such as mercury, zinc, and cadmium, and Lewis acids, such as boron compounds, lead to the formation of α-haloalkyl metal halides [Eq. (5)] which can be isolated in some cases (Seyferth, 1955) and can show carbenoid reactivity (Section III, B, 3). If nitrogen loss from the dipolar intermediate in Eq. (5) is not concerted with the 1,2-shift of the group X, then $RR'C^+M^-X_n$, a carbene–metal halide complex, could be an intermediate.

$$RR'CNN \xrightarrow{MX_n} [RR'C(N_2^+)\bar{M}X_n] \xrightarrow{-N_2} RR'C(X)MX_{n-1} \quad (5)$$

Complexes of the type $RR'C^+MX_n^-$ have long been regarded as intermediates in the decomposition of diazoalkanes induced by metallic copper and copper derivatives (Yates, 1952; Müller et al., 1966). Cyclopropanes can be formed when the decomposition is carried out in the presence of olefins, and the sequence of olefin reactivities suggests an electrophilic intermediate, but one showing somewhat greater selectivity than the corresponding "free" carbene generated by photolysis. Thus, the carbenoid involved shows little tendency to insert into C–H bonds, and its stereoselectivity in cyclopropane formation from olefins (Section IV,D) differs from that of the corresponding carbene. Strong support for the suggestion that the intermediate is a copper-containing carbene complex comes from the observation (Nozaki et al., 1968b) that the chiral copper complex (11) induces optical activity in the mixture of cyclopropanes produced from styrene and ethyl diazoacetate [Eq. (6)]. The size and electronic nature of ligands attached to copper can also modify the steric course of such carbenoid additions (Moser, 1969).

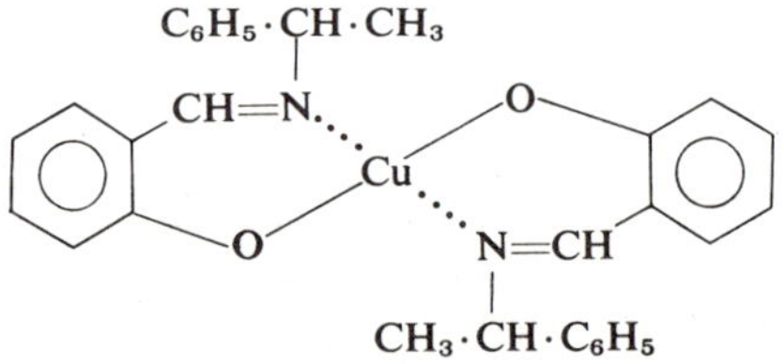

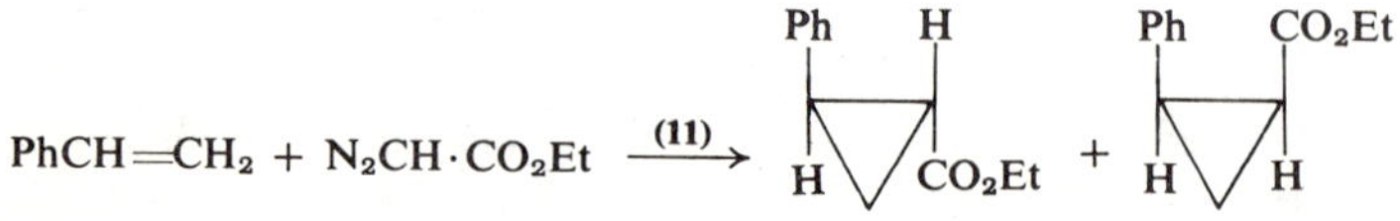

However, since the reactive intermediates in copper-catalyzed decompositions of diazoalkanes have not been isolated thus far, the precise nature of the carbene–metal bond is inadequately characterized.

Complexes of other transition metals [e.g., **12** (Mango and Dvoretzky, 1966), **13** (Werner and Richards, 1968), and **14** (Moritani et al., 1969)] promote decomposition of diazoalkanes, and relatively stable intermediates,

2. CARBENES AND RELATED SPECIES

in which the corresponding carbene may act as an additional ligand to the metal (e.g., as in **15**), have been isolated. These intermediates can behave as carbenoids.

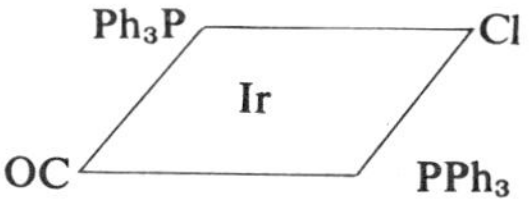

(12) Ph_3P, Ir, OC, Cl, PPh_3

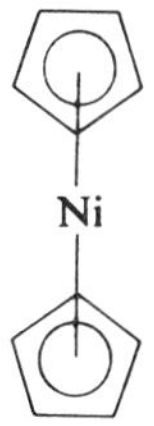

(13) Ni with two cyclopentadienyl rings

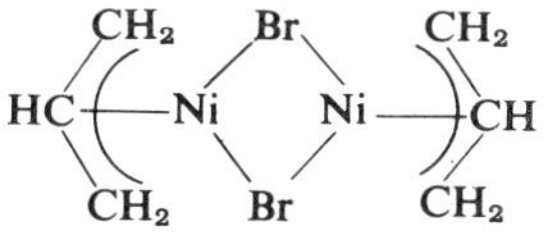

(14) $HC(CH_2)(CH_2)$–Ni–Ni–$CH(CH_2)(CH_2)$ with bridging Br, Br

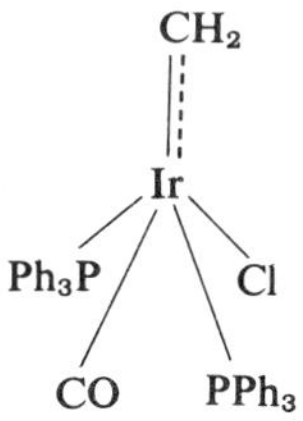

(15) CH_2=Ir with Ph_3P, CO, Cl, PPh_3

Diazirines (Schmitz, 1967), the cyclic isomers of diazoalkanes, can also yield carbenes, but they have not been exploited so widely. Though less accessible than diazoalkanes, they have the advantage of being unaffected by proton donors which would convert their linear isomers into carbonium ions. However, there is some evidence that certain diazirines can be isomerized thermally to the diazoalkane (Overberger and Anselme, 1963; Amrich and Bell, 1964) so that ambiguities in the interpretation of diazoalkane reactions may not always be removed by using diazirines as the carbene source.

Ketenes which are isoelectronic with diazoalkanes also yield carbenes when irradiated in the gas phase. Polymerization is a competing process, however, especially on thermal excitation in the liquid phase.

2. Base-Induced α Elimination

This type of reaction is represented in a general way in Fig. 7. The basic reagent MB, examples of which could be potassium *tert*-butoxide or lithium butyl, generates a carbanion (or its organometallic equivalent) from the substrate, typically an organic halide or gem dihalide. Loss of M^+X^- then gives the carbene. Route (ii) is favored for organometallic reagents with X as bromine or iodine.

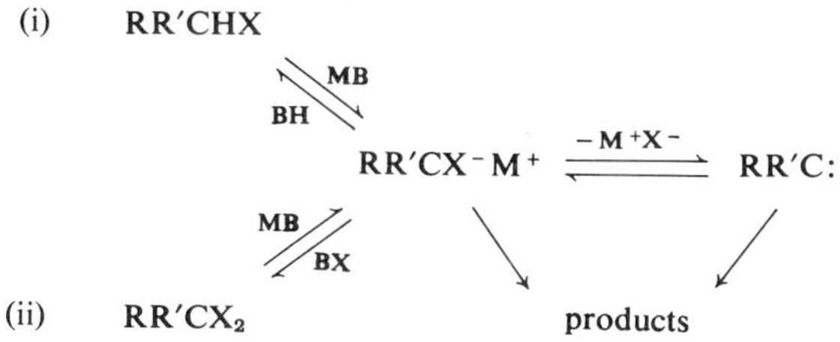

(i) $RR'CHX$ $\xrightarrow{MB, -BH}$ $RR'CX^-M^+$ $\xrightleftharpoons{-M^+X^-}$ $RR'C:$

(ii) $RR'CX_2$ $\xrightarrow{MB, -BX}$ → products

[X = halogen
MB = basic reagent, e.g., $K^+Bu^tO^-$, LiC_4H_9]

Fig. 7. α Elimination from organic halides.

Carbene formation by this type of mechanism was first conclusively demonstrated in the hydrolysis of chloroform to carbon monoxide and formate ion. The mass of evidence (Hine, 1964) can be summarized as follows.

a. Hydrolysis of chloroform in the presence of powerful bases is remarkably rapid compared with hydrolysis of the other chloromethanes, CH_3Cl, CH_2Cl_2, and CCl_4. Using weakly basic nucleophiles, the normal sequence of reactivity, $CH_3Cl > CH_2Cl_2 > CHCl_3 > CCl_4$, is found.

b. The hydrogen in chloroform exchanges with the hydrogen in the solvent much more rapidly than hydrolysis occurs.

c. The rate of hydrolysis of chloroform, $V = k[CHCl_3][OH^-]$, is reduced by added halide ions. Noncommon ions permit isolation of a mixed haloform.

These observations are consistent with a rate-limiting loss of chloride ion from a reversibly formed trichloromethyl anion. This anion, when generated from starting materials other than chloroform (e.g., $Cl_3C \cdot CO_2Me$ + sodium methoxide), is always capable of producing dichloromethylene. Other haloforms behave in a manner similar to chloroform, although difluoromethylene is formed from $CHBrF_2$ by a *synchronous* loss of H and Br.

By treating haloforms with strong bases in the presence of olefins, dihalocyclopropanes can be readily prepared. The obvious inference is that dihalomethylenes are produced and that these add across the double bond, the observed cis stereospecificity further suggesting that the carbenes are in the singlet state (Section IV,D). Caution is necessary, however, in interpreting these and similar observations. Treatment of CCl_3Br with methyl lithium in ethereal solution at $-115°C$ gives $Li^+CCl_3^-$, which is stable at $-100°C$, but which reacts with cyclohexene at the same temperature to form the corresponding dichlorocyclopropane (7,7-dichloronorcarane) (Miller and Whalen, 1964). Obviously, the intermediate salt may act as a carbenoid, reacting with the olefin to give the product expected from the carbene (see, however, Skell and Cholod, 1969a). More extensive studies (Hoeg *et al.*, 1965) indicate that α-haloalkyl lithium reagents can be produced in tetrahydrofuran from a variety of alkyl halides. These reagents are stable at low temperatures, but will react with carbon dioxide or reactive alkyl halides (e.g., CH_3I). On raising the temperature, irreversible exothermic decomposition giving polymeric material begins at about $-65°C$. Both above and below this temperature, olefins can be converted into cyclopropanes, and at $-78°C$ the sequence of reactivity of the olefins parallels their nucleophilicity. Thus, carbenoid addition is probably involved, and the carbenoids are evidently electrophiles. The alternative interpretation would be that free carbenes are generated from the α-haloalkyl lithium and add to the olefin [Eq. (7)]. The olefin reactivity sequence would then require that k_b [olefin] be less than or comparable to k_{-a}. Such rate-determining consumption of the free carbene is, as we have seen, generally thought unlikely, although Cl_2C: may be a special case (Skell and Cholod, 1969a).

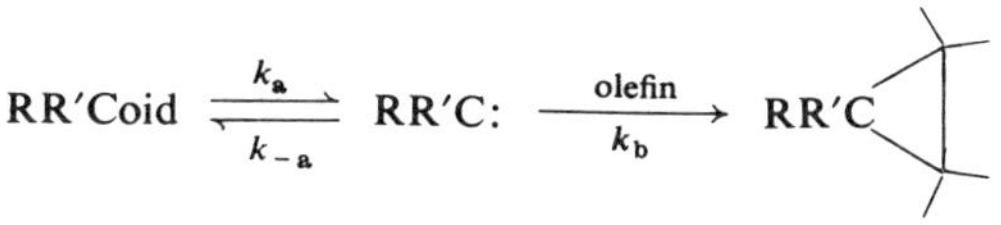

2. CARBENES AND RELATED SPECIES

$$\text{RR'Coid} \underset{k_{-a}}{\overset{k_a}{\rightleftarrows}} \text{RR'C:} \xrightarrow[k_b]{\text{olefin}} \text{RR'C}\hspace{-2pt}\triangleleft \tag{7}$$

Further evidence that the intermediates in alkyl lithium-induced reactions of organic halides are carbenoids can be summarized as follows.

1. Cyclopropanes formed by the reaction of organic halides and alkyl lithium with olefins often have a different stereochemistry from those produced by photochemical or thermal decomposition of the analogous diazoalkane in the presence of the same olefins. Examples are given in Eqs. (8) (Closs and Closs, 1962) and (9) (Closs and Moss, 1964; Goh et al., 1969).

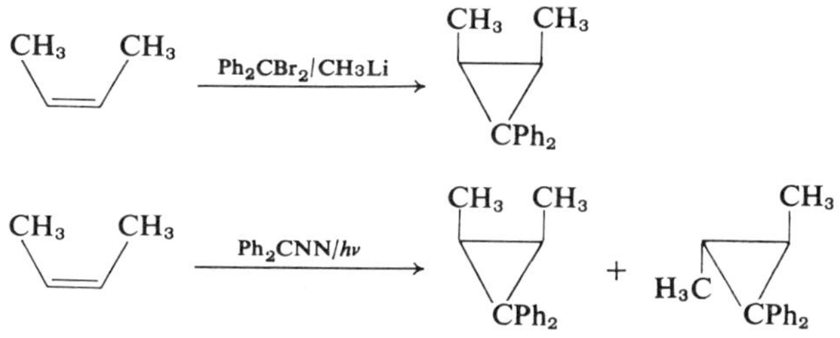

(8)

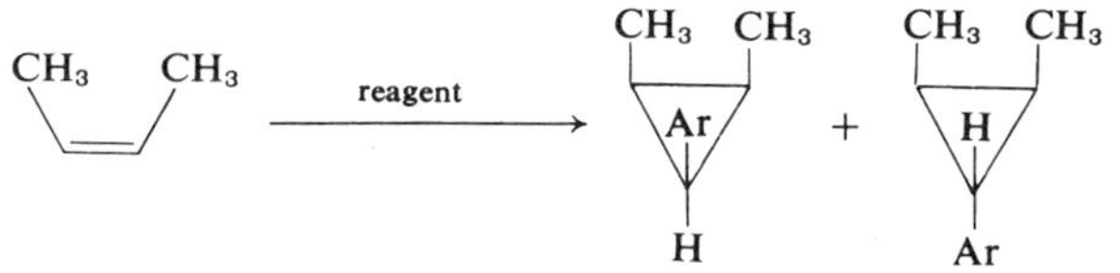

(9)

Reagent	syn/anti
ArCHBr$_2$/CH$_3$Li	2.4 Ar = C$_6$H$_5$; 2.3 Ar = p-CH$_3$·C$_6$H$_4$
ArCHNN/$h\nu$	1.1 Ar = C$_6$H$_5$; 1.7 Ar = p-CH$_3$·C$_6$H$_4$
ArCHNN/LiBr	7.5 Ar = p-CH$_3$·C$_6$H$_4$

2. The magnitude of the intramolecular kinetic hydrogen isotope effect (k_H/k_D) observed in reaction (10) increases from 1.7 to 2.4 as the halogen X is changed from Cl to Br to I (Goldstein and Dolbier, 1965). There should be no such variation if the intermediate were a free carbene.

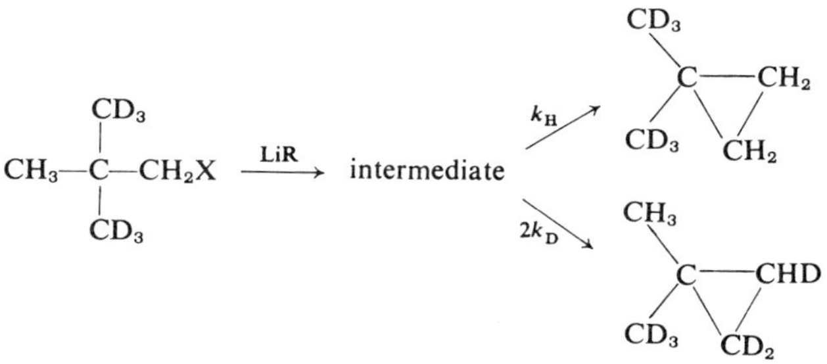

(10)

The formation of free carbenes by base-induced α elimination is thus rather rare. α-Haloalkyl anions or the related organometallic reagents are usually responsible for the carbene-like reactions. In α eliminations leading to a dimeric olefin [Eq. (11)], intermediate formation of a carbene should be indicated by a first-order dependence of the rate of reaction or the concentration of the reactant halide or -onium compound. Second-order rate laws are usually found (Bethell, 1963; Bethell *et al.* 1967; Hanna and Wideman, 1968). This is consistent with the rate-limiting displacement of halide ion from a second molecule of the reactant by the rapidly formed α-haloalkyl anions, the olefin being formed by subsequent β elimination from the intermediate dimeric halide so formed. First-order kinetics have been observed with compounds of only two structural types, namely, 9-halofluorenes with powerfully electron-withdrawing substituents in the 2-position (**16**) and 4-nitrobenzyl halides and -onium salts (**17**). It turns out that for **16** in *tert*-butyl alcohol solutions of sodium *tert*-butoxide the first-order kinetic law arises from rate-limiting dissociation of an intermediate 9-halofluorenyl sodium ion pair, the free carbanion reacting with unchanged **16** to give the dimeric halide much more rapidly than the ion pair (Bethell *et al.*, 1967). The mechanisms of conversion of **17** into 4,4′-dinitrostilbene is still not completely clear. 4-Nitrophenylmethylene may be an intermediate as the kinetic form suggests (Hanna *et al.*, 1961; Swain and Thornton, 1961), but it has never been trapped using olefins (Rothberg and Thornton, 1964). Pathways involving radical anions could also account for the observations, including that of photocatalysis, but EPR spectroscopy has failed to detect the appropriate radicals (Russell and Danen, 1968).

$$RR'CHX \xrightarrow[-BH]{M^+B^-} RR'CX^-M^+ \xrightarrow{RR'CHX} RR'C(X)CHRR' \xrightarrow{-HX} RR'C=CRR' \qquad (11)$$

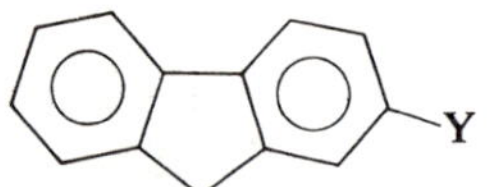

$X = Br; Y = Br, CN, NO_2$ $X = Cl, SMe_2^+$

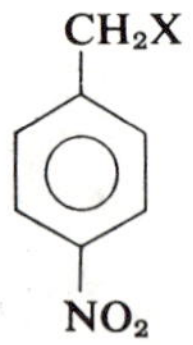

(**16**)   (**17**)

3. Organometallic Reagents

The transient carbenoids formed in solution from diazoalkanes and metal derivatives and by base-induced α elimination from organic halides are usually thought to have in common a metal atom bonded to a carbon atom in an α-haloalkyl moiety or a carbene ligand. This section is concerned with more stable organometallic compounds which show some carbenoid behavior, but ones whose structures are more clearly defined than those of the fugitive intermediates mentioned so far. The fundamental mechanistic question here is whether the organometallic compound reacts directly with another molecule in a carbenoid manner ("methylene transfer") or free carbenes are formed as intermediates which might then be trapped by suitable reagents. Both types of behavior are observed. In general, the evidence for methylene transfer consists of observations of a kinetic dependence on the concentration of the methylene acceptor coupled sometimes with stereospecific cis addition to olefins.

Studies of the well-known Simmons–Smith (1959) reaction in which olefins are converted to cyclopropanes by treatment with methylene iodide and zinc–copper couple [Eq. (12)] exemplify these approaches. The addition is exclusively cis and the reaction rate is proportional to the olefin concentration. The evidence indicates that methylene transfer occurs, and the transition state is formulated as **18** (Simmons *et al.*, 1964; Blanchard and Simmons, 1964). The simplest constitution of the reactive organometallic species is ICH_2ZnI, although this may be involved in a Schlenk-type equilibrium with $Zn(CH_2I)_2 \cdot ZnI_2$. gem-Diiodides other than methylene iodide are ineffective under Simmons–Smith conditions, but ethylidene iodide with diethyl zinc can be used for stereospecific ethylidene transfer (Nishimura *et al.*, 1969). Diazoalkanes with zinc iodide gives α-iodoalkyl zinc iodides identical with those produced from the corresponding gem diiodides and zinc–copper couple (Wittig and Wingler, 1962; Bethell and Brown, 1967; see also Goh *et al.*, 1969).

$$CH_2I_2 \xrightarrow[CH_3 \ CH_3]{Zn/Cu} \underset{(18)}{\begin{array}{c} I-Zn-----I \\ \diagdown \ \diagup \\ CH_2 \\ \diagup \ \diagdown \\ CH_3 \ \ CH_3 \end{array}} \longrightarrow \begin{array}{c} CH_2 \\ \triangle \\ CH_3 \ CH_3 \end{array} \quad (12)$$

Mercury forms well-characterized α-haloalkyl compounds, $R''HgCRR'X$, some of which are capable of converting olefins into the corresponding RR'-substituted cyclopropane on heating in an inert solvent such as benzene. The rate of decomposition of $PhHgCCl_2Br$ is independent of the concentration of olefin at the start of the reaction but decreases in the later stages more markedly for the less reactive olefins than for the more nucleophilic ones. Coupled with the observation that PhHgBr, one of the reaction products, retards the decomposition, this is consistent with intermediate formation of dichloromethylene (Eq. 13) for which the rate law is Eq. (14) (Seyferth *et al.*,

$$\text{PhHgCCl}_2\text{Br} \underset{k_{-1}}{\overset{k_1,\text{ slow}}{\rightleftarrows}} \text{PhHgBr} + \text{Cl}_2\text{C:} \overset{\text{olefin}}{\underset{k_2}{\longrightarrow}} \underset{\text{CCl}_2}{\triangle} \quad (13)$$

$$\frac{-d[\text{PhHgCCl}_2\text{Br}]}{dt} = \frac{k_1[\text{PhHgCCl}_2\text{Br}]}{(1 + k_{-1}[\text{PhHgBr}]/k_2[\text{olefin}])} \quad (14)$$

1967b). Since reagents known to trap trihalomethyl anions are ineffective in this reaction, it seems that dichloromethylene is ejected directly from PhHgCCl$_2$Br (Seyferth et al., 1968). However, the decomposition of PhHgCCl$_3$, which is slow in acetone or glyme, is accelerated by addition of the stoichiometric amount of sodium iodide, although the pattern of olefin reactivity in dichlorocyclopropane formation is not changed. The iodide ion here displaces the trichloromethyl anion which then affords dichloromethylene (Seyferth et al., 1967a, 1969a). Difluoromethylene can be made similarly by the action of sodium iodide on the organotin compound (CH$_3$)$_3$SnCF$_3$ (Seyferth et al., 1965a; Seyferth and Armbrecht, 1969), whereas organosilicon halides such as Cl$_3$SiCCl$_3$ decompose unimolecularly to give silicon tetrachloride and the carbene directly (Bevan et al., 1961).

Except for trihalomethyl derivatives of metals, organometallic compounds are not very fruitful sources of free carbenes. Cyclopropane formation from olefins, using (BrCH$_2$)$_2$Hg, for example (Seyferth et al., 1969b), is usually a carbenoid reaction (methylene transfer) by way of a transition state analogous to **18**. This is not the only alternative to the carbene mechanism; however, the 1,2-addition 1,3-elimination route in Eq. (15) has been proposed for α-haloalkyl aluminum compounds to account for the occurrence of **19** among the products after hydrolysis (Hoberg, 1962). However, this mechanism is difficult to reconcile with stereospecific cyclopropane formation.

$$\underset{X}{\overset{R}{\underset{R'}{>}}}\text{C—Al}\overset{C=C}{\longrightarrow}\underset{\text{Al} \quad \text{CRR'X}}{>\text{C—C}<}\longrightarrow\underset{R'}{\overset{R}{\triangle}}+\underset{\text{CRR'X}}{>\text{CH—C}<} \quad (15)$$

$$(19)$$

A series of stable transition metal complexes in which a carbene acts as a ligand has been fully characterized (Ryang, 1970). The general preparative procedure is in Eq. (16). In these complexes the carbene behaves in a fashion analogous to carbon monoxide, providing two electrons for the metal–carbon bond and having available a vacant orbital on the central carbon atom for back donation from the metal. The strong interaction between the central carbon atom of the carbene ligand and the metal results in chemical behavior which is often quite different from that expected of a carbene. Thus, the alkoxy group in **20** readily undergoes nucleophilic displacement on treatment with, for example, amines, hydroxylamine, hydrazine, isocyanides, and thiophenol, the product being a different carbene complex. Thermal decomposition of some transition metal–carbene complexes gives products which might have arisen from carbenes. The zero-valent chromium complex (**21**) yields the dimeric olefins (**22**) in this manner, but no cyclopropanes result when the reaction is carried out in tetramethylethylene, a good trap for carbenes. Stereospecific cyclopropane formation occurs with methyl *trans*-crotonate, but there is no evidence for the formation of free carbenes (Fischer and Dotz, 1970).

$$M(CO)_n \xrightarrow{RLi} \left[(CO)_{n-1}M-C \begin{smallmatrix} R \\ \\ OLi \end{smallmatrix} \right] \xrightarrow[\substack{2.\ diazo \\ alkane\ or \\ R_3'O^+BF_4^-}]{1.\ H^+} (CO)_{n-1}M-C \begin{smallmatrix} R \\ \\ OR' \end{smallmatrix}$$

(20)

M = W, Cr, Re (16)

$(CO)_5CrC(OPh)CH_3$

(21)

$\begin{smallmatrix} CH_3 \\ \\ PhO \end{smallmatrix} C=C \begin{smallmatrix} CH_3 \\ \\ OPh \end{smallmatrix} + \begin{smallmatrix} CH_3 \\ \\ PhO \end{smallmatrix} C=C \begin{smallmatrix} OPh \\ \\ CH_3 \end{smallmatrix}$

(22)

IV. Mechanism and Reactivity in Carbene and Carbenoid Reactions

A. Introduction

The very high reactivity of carbenes toward practically all classes of organic compounds, including paraffins, makes direct kinetic investigation difficult. Flash photolysis of the diazoalkane in solution has permitted hydrogen abstraction by the triplet carbene (**6**) to be followed directly (the reaction is complete in a few microseconds) but the technique has not been widely applied. Instead, indirect methods of study, particularly competition methods, have to be employed, and, as usually practiced, these entail assumptions, for example, about the nature of the intermediate and the kinetic order of its reactions with the compounds under study. Because of the exothermic nature of reactions of carbenes (cyclopropane formation from methylene and ethylene has $\Delta H = -86$ kcal mole^{-1}), products are formed with excess energy ("hot"). At low pressures in the gas phase where collisional deactivation is inefficient, this excess energy can lead to the formation of rearranged products, the nature and proportions of which are not directly related to the reactivity of the carbene. In solution, however, the products generally reflect the competitive situation in the primary reactions of carbenes.

Any discussion of mechanism and reactivity in reactions of carbenes is complicated, at least potentially, by the existence of low-lying singlet and triplet states of divalent carbon intermediates. The problem can be posed in the form of two questions. (i) Can carbenes react with other molecules in both their singlet and triplet states irrespective of which is the ground state? (ii) If both states of a given carbene can react, are these two types of reactions experimentally distinguishable? The consensus of opinion is that the answer to both questions is a qualified yes. However, it must be added that although carbene reactions can be interpreted in terms of the multiplicity of the electronic states of the carbene, many of the features of the reactions can be interpreted equally well in terms of varying degrees of excitation of the carbene, without explicit consideration of spin. These ambiguities have been discussed at length (Gaspar and Hammond, 1964; DeMore and Benson, 1964; Closs, 1968). In this chapter we shall cautiously relate reactivity to multiplicity because this approach permits useful analogies to be drawn between the behavior of carbenes and other reactive carbon species such as carbonium ions and free radicals.

The electronic configuration of carbenes in their lowest singlet state bears a formal similarity to both carbonium ions and carbanions. The electronic sextet endows most carbenes with a predominantly electrophilic nature, although this can be modified by substituents (Section II,A,2). However, the analogy between singlet carbenes and carbonium ions is likely to be imperfect because of the presence of the nonbonding electron pair on the divalent carbon atom. In Pearson's terminology, this renders the carbene "softer" than the related carbonium ion (Pearson and Songstad, 1967). Solvation, too, could modify the reactivity of singlet carbenes, enhancing or decreasing their electrophilic character, provided that the lifetime of the carbene is sufficiently long for solvent molecules to become appropriately oriented. Some experimental support for solvation of carbenes is available (Russell and Hendry, 1963; Rando, 1970).

Whereas singlet carbenes are expected to react with electron *pairs*, triplet carbenes are usually regarded as having chemical properties dominated by their unpaired (though not completely uncoupled) electrons. Under suitable circumstances, carbenes can show radical-like behavior (one-electron processes), and this is plausibly ascribed to the triplet state. Less probable is that radical-like behavior might also be shown by the first excited singlet state of carbenes (e.g., the $^1B_1$ state of H_2C:) since the nonbonding electrons occupy different orbitals.

These electronic analogies lead us to expect chemically distinguishable behavior of carbenes in different electronic states. In a real situation, however, it must be remembered that interconversion of electronic states may take place at rates comparable with the reactions in which the carbenes are consumed. The final product mixture will be rendered still more complex, because a carbene will generally be capable of reacting with a given organic compound at a number of different points, each of which leads to a different product. In the following discussion we shall, for convenience, treat the different types of reactions, such as insertion and addition, separately.

B. Coordination to Nucleophiles Having Nonbonded Electron Pairs

The simplest way by which a *singlet* carbene can complete the electronic octet around its central carbon atom is by coordination to a nucleophile (N) possessing nonbonded electron pairs. This yields a carbanion if N is negatively charged or an ylid when N is electrically neutral [Eq. (17)]. Such a process may be one step in other reactions of carbenes such as insertions, but here we shall concentrate our attention on the simple coordination.

$$RR'C\uparrow\downarrow + N \longrightarrow RR'\overset{-}{C}-\overset{+}{N} \tag{17}$$

The retarding effect of halide and other anions on the rate of hydrolysis of chloroform indicates the sequence of reactivity of the nucleophiles toward dichloromethylene to be $I^- > Br^- > Cl^- > NO_3^- > H_2O$, paralleling the reactivity of the nucleophiles in displacement of bromide ion from methyl bromide (Hine and Dowell, 1954). Mixed haloforms formed by protonation of the product carbanion can in some cases be isolated from such reactions. As expected on the basis of their "softness," carbenes readily react, particularly with polarizable nucleophiles. Dichloromethylene reacts, for example, with the carbanion derived from methylmalonic ester (Krapcho, 1962) and also with thiophenoxide ion, giving $(PhS)_3CH$ after protonation (Hine, 1950). In contrast, phenoxide and alkoxide ions do not lead to simple coordination at oxygen. Dichloromethylene attacks phenoxide ion at the ortho and para positions of the aromatic ring to give after hydrolysis, the corresponding *o*- and *p*-hydroxybenzaldehydes (Reimer–Tiemann reaction) as shown in Eq. (18) (Hine and van der Veen, 1959). Alkoxides, having no very polarizable position, can show entirely different behavior. Thus, dibromomethylene with alkoxide ions suffers displacement of bromide ion to give an alkoxybromomethylene, which produces a carbonium ion on ionic fragmentation [Eq. (19)] (Skell and Starer, 1959).

$$RO^- + Br_2C: \longrightarrow RO\ddot{C}Br \longrightarrow R^+ + CO + Br^- \quad (19)$$

Triphenylphosphine and organic sulfides are polarizable nucleophiles and can be used to trap carbenes as the corresponding phosphonium or sulfonium ylid (Wittig and Schlosser, 1962; Ando et al., 1969a). Oxonium and ammonium ylids similarly formed tend to rearrange, giving insertion products (Section IV,C). Ylids themselves can react with carbenes to give olefins [e.g., Eq. (20)] presumably by decomposition of the initial dipolar adduct (Okano et al., 1964). Diazoalkanes are structurally related to nitrogen ylids (Section III,B,1), and they, too, can react with carbenes [Eq. (21)], giving olefins by attack on carbon followed by nitrogen loss and azines by simple coordination at the terminal nitrogen (Reimlinger, 1964).

$$\text{(fluorenyl)}=\overset{+}{P}Ph_3 \xrightarrow{Cl_2C:} \left[\text{(fluorenyl)}\overset{\overset{+}{P}Ph_3}{\underset{\bar{C}Cl_2}{\diagup}}\right] \longrightarrow \text{(fluorenyl)}=CCl_2 + PPh_3 \quad (20)$$

$$\left(\overset{-}{\diagdown}C-\overset{+}{N}\equiv N \longleftrightarrow \overset{\diagdown}{\diagup}C=\overset{+}{N}=\bar{N}\right) \xrightarrow{RR'C:} \overset{\diagdown}{\diagup}C=CRR' + \overset{\diagdown}{\diagup}C=N-N=CRR' \quad (21)$$

C. Insertion

1. Mechanisms of Insertion

It is a generally held belief that the insertion of carbenes into covalent bonds [Eq. (22)] is the most characteristic reaction of free divalent carbon species. Insertion into inter alia C–H, C–C, C–X, N–H, O–H, S–H, M–C, M–X, and M–M bonds have been observed. Nevertheless, insertion is not observed in all reactions involving intermediate formation of carbenes. For example, dihalomethylenes produced from haloforms with a base react so readily with nucleophiles and in other ways that C–H bond insertion is not usually observed. Diphenylmethylene only inserts into C–H bonds of low dissociation energy, such as those in benzylic systems. Insertion is not usually observed with carbenoids.

$$RR'C: + X-Y \longrightarrow X \cdot CRR' \cdot Y \quad (22)$$

Insertion can be envisaged as occurring by one or more of three possible general mechanisms. These are as follows: (i) concerted or direct insertion by way of a triangular transition state [Eq. (23a)]; (ii) stepwise abstraction–recombination involving a radical or ion pair intermediate [Eq. (23b)]; and

(iii) coordination to form an ylid, followed by rearrangement [Eq. (23c)]. Singlet carbenes should be capable of reacting by any of these mechanisms, although the abstraction–recombination mechanism (ii) would involve an ionic rather than radical pair intermediate, whereas mechanism (iii) would be restricted to the reaction of the carbene with molecules containing lone-pair electrons. Triplet carbenes should favor mechanism (ii) with intermediate formation of a radical pair, the collapse of which would give the insertion product.

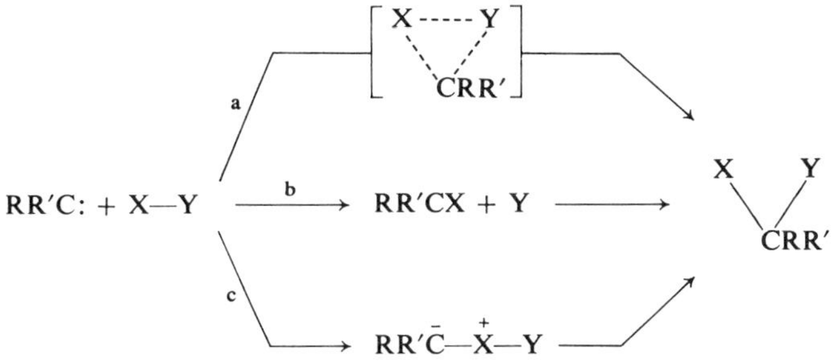

$$RR'C: + X-Y \xrightarrow{a,b,c} \begin{cases} [X\cdots Y / \dot{C}RR'] \\ RR'CX + Y \\ RR'\bar{C}-\overset{+}{X}-Y \end{cases} \longrightarrow \overset{X\diagdown\diagup Y}{\underset{CRR'}{}} \quad (23)$$

Experimental evidence concerning the mechanism of insertion comes mainly from the following areas: stereochemistry of insertion, reactivity of bridgehead C–H bonds in rigid bicyclic aliphatic systems, and detection of intermediates, particularly those formed by abstraction processes, by the occurrence of allylic rearrangements, or, in the case of radicals, by spectroscopy.

Under conditions favoring reaction through the singlet state, methylene undergoes C–H insertion in **23** with retention of configuration (Franzen, 1962; but see Franzen and Edens, 1969), and the bridgehead C–H bond in **24** is not unusually unreactive (Doering, 1964a). Both of these results indicate direct insertion. On the other hand, singlet ethoxycarbonylmethylene, which inserts into **25** with retention, is unusually unreactive toward the bridgehead C–H bond of nortricyclene (**26**) (Sauers and Kiesel, 1967). This may indicate charge separation in the transition state for direct insertion in this instance (**27**), the rigid carbon skeleton preventing attainment of the planarity necessary for optimum stabilization of the carbonium center (Doering and Knox, 1961).

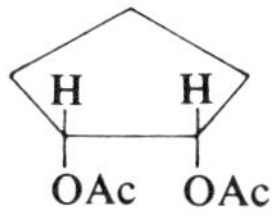

OCH_3
$CH_3 \cdot CH \cdot CO \cdot OCH_3$

(23) 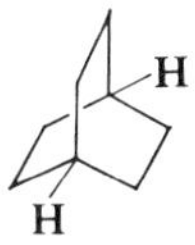   (24) 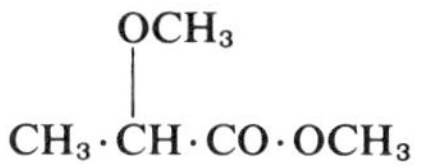   (25)    (26)

$$\text{(a)} \quad \overset{\text{\tiny{||||}}}{C}\text{-----}H \quad\longleftrightarrow\quad \overset{\text{\tiny{||||}}}{C^+} \quad \underset{^-CH \cdot CO \cdot OC_2H_5}{H} \quad \text{(b)}$$

$$\text{CH} \cdot \text{CO} \cdot \text{OC}_2\text{H}_5$$

(27)

Dichloromethylene inserts into C–Hg and Si–H bonds with retention (Landgrebe and Thurman, 1968; Sommer *et al.*, 1968) as expected for direct insertion by a ground-state singlet carbene. However, insertion into the benzylic C–H bond of 2-butylbenzene takes place with racemization (Franzen and Edens, 1969), indicating an abstraction–recombination mechanism. This could plausibly involve hydride transfer to the divalent carbon, though charge separation in the transition state is probably small (Seyferth *et al.*, 1968, 1970a,b).

Reaction of photochemically generated methylene with 2-methylpropane-1-^{14}C gives the allylic insertion product, 2-methylbut-1-ene. In the gas phase, 8% of the product had the radioactive label rearranged to the 3-position in the product, whereas in solution only 2% rearrangement was observed (Doering and Prinzbach, 1959). This can be interpreted as evidence that 16% of the gas-phase reaction and 4% of the reaction in solution occur by the abstraction–recombination mechanism, the rest of the reaction taking place by direct insertion [Eq. (24)]. However, it is worth noting that, for example, abstraction followed by recombination within the solvent cage could lead to an insertion product with an unrearranged label.

$$\text{CH}_2\text{:} + \text{CH}_3-\underset{\underset{\text{CH}_3}{|}}{C}=^{14}\text{CH}_2 \longrightarrow \left[H\text{-----}\underset{\underset{\text{CH}_2}{\diagdown\diagup}}{\text{CH}_2}-\underset{\underset{\text{CH}_3}{|}}{C}=^{14}\text{CH}_2 \right] \longrightarrow \text{CH}_3 \cdot \text{CH}_2-\underset{\underset{\text{CH}_3}{|}}{C}=^{14}\text{CH}_2$$

$$\downarrow$$

$$\text{CH}_3 \cdot\ +\ \cdot\text{CH}_2-\underset{\underset{\text{CH}_3}{|}}{C}=^{14}\text{CH}_2 \quad (24)$$

$$\updownarrow$$

$$\text{CH}_2=\underset{\underset{\text{CH}_3}{|}}{C}-^{14}\text{CH}_2\cdot \longrightarrow \text{CH}_2=\underset{\underset{\text{CH}_3}{|}}{C}-^{14}\text{CH}_2\cdot\text{CH}_3$$

More clear-cut evidence exists for the abstraction–recombination mechanism of insertion by triplet carbenes. In such reactions an intermediate (triplet) radical pair is formed. This, after spin inversion, can collapse to give the insertion product. Alternatively, the component radicals, if they are particularly stable, can diffuse apart and yield characteristic dimerization products. For example, diphenylmethylene reacts, presumably in its triplet ground state, with toluene to give not only the insertion product 1,1,2-triphenylethane but also the radical dimers *sym*-tetraphenylethane and bibenzyl [Eq. (25)] (Bethell *et al.*, 1965; Nozaki *et al.*, 1966). Products containing unpaired electrons have been detected by EPR spectroscopy in similar reactions (Singer and Lewis, 1968).

$$Ph_2C\uparrow\uparrow + PhCH_3 \longrightarrow \begin{bmatrix} Ph_2CH\uparrow & \longrightarrow & Ph_2CH\cdot CHPh_2 \\ + & \searrow & Ph_2CH\cdot CH_2Ph \\ PhCH_2\uparrow & \longrightarrow & PhCH_2\cdot CH_2Ph \end{bmatrix} \quad (25)$$

The observation of *chemically induced dynamic nuclear polarization* (CIDNP) provides strong evidence for the formation of radical pairs in carbene reactions and is potentially applicable to a wide variety of systems (Closs, 1971). Qualitatively, the phenomenon is thought to arise as follows. The triplet radical pair produced when a triplet carbene abstracts an atom from another molecule may dissociate into free radicals or combine to form the insertion product. For combination to occur, the radical pair must cross into the singlet state (by mixing of T_o and S states) and, because of the interaction of the odd electrons of the radicals with nearby magnetically active nuclei, the nuclear spin states influence the probability of crossing. Thus nuclear spin selection occurs and, as a result, products are formed with a nonequilibrium distribution of nuclear spins. The reestablishment of the equilibrium distribution can take place over a period of several minutes and manifests itself as emission or enhanced absorption signals in the NMR spectrum of the reaction products during and immediately after the reaction. Figure 8 shows the ^{1}H NMR spectrum of the methine protons of $Ph_2CH\cdot CHPh\cdot CO_2CH_3$ produced by thermal decomposition of diphenyldiazomethane in solvent methyl phenylacetate (Closs and Closs, 1969). The combination of enhanced absorption and emission signals within a multiplet is a characteristic feature of systems exhibiting CIDNP. Nuclear spin polarization can also occur with singlet radical pairs and, since the pattern of absorptions and emissions is related to the spin state of the radical pair formed initially, CIDNP studies are a valuable indicator of the precursors of products in carbene reactions (Closs and Trifunac, 1969, 1970; see also Kaptein *et al.*, 1970).

2. Structure and Reactivity

Methylene generated in alkane solution inserts almost randomly into the available C–H bonds, the proportions of products being determined

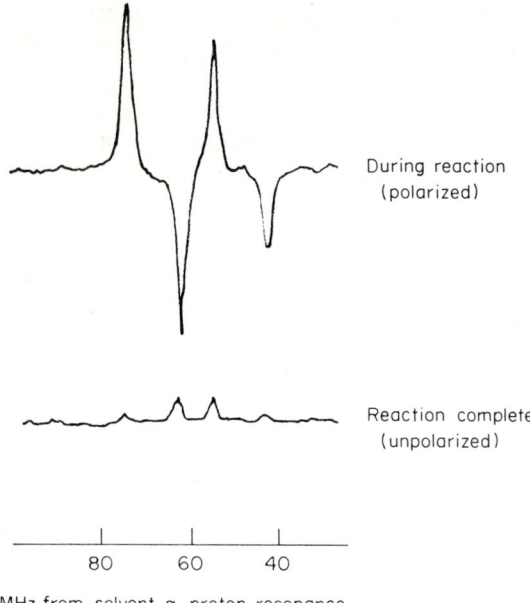

Fig. 8. ^{1}H NMR spectrum (60 MHz) of the methine protons of Ph$_2$CH·CHPh; CO$_2$Me showing CIDNP during thermolysis (140°) of diphenyldiazomethane in methyl phenylacetate.

largely by the relative numbers of primary, secondary, and tertiary hydrogens (Doering *et al.*, 1956). In the gas phase, methylene shows somewhat higher selectivity (Richardson *et al.*, 1960, 1961). The reaction in solution is plausibly ascribed to the singlet state of methylene which reacts in the first few collisions with solvent molecules before intersystem crossing can occur. In the gas phase, collisional deactivation of the singlet competes more effectively with chemical reaction, the resultant triplet methylene being more selective than its precursor. Analysis of gas-phase reactions along these lines suggests, however, that the discrimination of singlet methylene can be increased by decreasing its excess energy (Ho and Noyes, 1967).

Values of the relative reactivities of primary, secondary, and tertiary C–H bonds in some simple alkanes toward a selection of carbenes are assembled in Table VI. The substituents attached to the divalent carbon atom of these carbenes all have filled orbitals capable of overlapping with the vacant *p*-type orbital of the singlet state. All substituents are then likely to stabilize the singlet state and also make the divalent center less reactive than that in methylene itself, thus enhancing selectivity. However, the ability of the substituents to delocalize negative charge would favor reaction through charge-separated transition states like **27**. Because of the carbonium character of the carbon atom which is attacked, the reactivity order for C–H bonds predicted on electronic grounds is tertiary > secondary > primary as observed. Steric influences, though not explicitly tested, are expected to affect reactivity in the reverse direction. The relative reactivities of C–H bonds toward alkylmethylenes, which should promote polarization of the transition state in the opposite sense to that in **27b**, are not available because of the ease with which these carbenes undergo intramolecular insertion and rearrangement [e.g., Eq. (26) (Mansoor and Stevens, 1966)]. In intramolecular insertion, cyclic products are produced, and additional steric and conformational influences are superimposed on the usual pattern of reactivity (Gutsche *et al.*, 1962).

$$(CH_3)_3C\cdot\ddot{C}\cdot CH_3 \longrightarrow \underset{H_3C}{\overset{CH_3\ H}{\triangle}}_{CH_3} + (CH_3)_3C\cdot CH=CH_2 + (CH_3)_2C=C(CH_3)_2$$

(26)

Carbene insertion is a highly exothermic process, and in accordance with Hammond's postulate the transition state should resemble the reactants rather than the reaction products. This explains the small values observed for kinetic hydrogen isotope effects in C–H insertion (Table VII) and may also explain the fairly low sensitivity of relative rates of insertion to the structure of the molecule attacked.

Intermolecular insertion of carbenes into bonds other than those between carbon and hydrogen is often difficult to observe, and the evidence seems to suggest that the direct mechanism does not operate. For example, the insertion of methylene into C–Cl bonds may involve abstraction of chlorine atoms after a preliminary coordination of the divalent carbon through lone-pair electrons on the halogen [as in the ylid mechanism, Eq. (23c)] (Bamford *et al.*, 1968; Bamford and Casson, 1969). Consistently, methyl insertion into the C–Cl bond of 2-butyl chloride occurs with racemization (Doering, 1964b). Photochemically generated methylene inserts into the C–Cl bonds of carbon tetrachloride with a quantum efficiency of about 300, suggesting a radical-chain mechanism initiated by chlorine abstraction by the carbene (Urry and Eiszner, 1952). CIDNP studies suggest that singlet methylene is responsible (Roth, 1971).

When singlet carbenes insert into molecules which contain atoms having nonbonded electron pairs, intermediate ylid formation invariably occurs. Thus, diphenylmethylene inserts into the O–H bond of water, giving diphenylmethanol, at almost the same rate as it inserts into the O–D bond of deuterium oxide. However, a large intramolecular kinetic isotope effect is observed when diphenylmethylene reacts with tritiated water molecules (HTO), much more protium than tritium becoming attached to the divalent carbon in the

Table VI
Relative Reactivities of C–H Bonds in Carbene Insertion Reactions

Carbene source	Conditions and phase	Alkane	Relative reactivity			Reference
			1°	2°	3°	
CH_2NN	$h\nu, l$	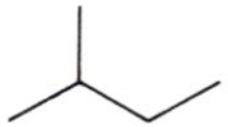 isobutane	(1)	1.22	1.51	Herzog and Carr (1967)
		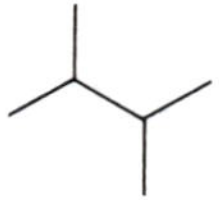 2,3-dimethylbutane	(1)	—	1.2	Doering et al. (1956)
	$h\nu, g$	 n-pentane	(1)	1.20	—	Frey (1958)
		 isobutane	(1)	—	1.48	Frey (1958)
CH_2CO	$h\nu$ (3800 Å), g	 propane	(1)	3.75^a	—	Ho and Noyes (1967)
	$h\nu$ (2800 Å), g	 propane	(1)	1.15^a	—	Ho and Noyes (1967)
	$h\nu, g$	 propane	(1)	13.6^b	—	Ho and Noyes (1967)

2. CARBENES AND RELATED SPECIES

Carbene	Conditions	Product	Ratio		Reference
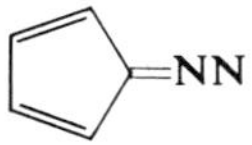 =NN (cyclopentadienylidene)	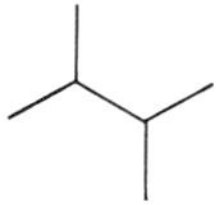 hv, l	(isopropyl branched) (1)	—	7.3	Moss (1966)
PhCHNN	hv, l	 (1)	6.3	—	Gutsche et al. (1962)
MeO·CO·CHNN	hv, l	 (1)	2.3	—	Doering and Knox (1961)
		 (1)	—	3.1	Doering and Knox (1961)
(EtO·CO)₂CNN	hv, l	 (1)	8.4	—	Doering and Knox (1961)
		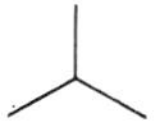 (1)	—	21.0	Doering and Knox (1961)
(NC)₂CNN	Δ(80°), l	 (1)	4.6	—	Ciganek (1966)
		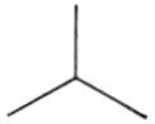 (1)	—	12.0	Ciganek (1966)
ClCHNN	hv (−50°), l	 (1)	20	—	Closs and Coyle (1965)
BrCHNN	hv (−50°), l	 (1)	25	—	Closs and Coyle (1965)

[a] Singlet methylene.
[b] Triplet methylene.

Table VII

Kinetic Hydrogen Isotope Effects in C–H Insertion Reactions of Carbenes and Carbenoids

Bond	Carbene source	Conditions	k_H/k_D	Reference
R_2CH_2	CH_2NN	$h\nu, g$	1.23	Chesick and Willcott (1963)
–CH_3[a]	$(CH_3)_3C\cdot CHX_2$; X = Cl	$LiC_4H_9, -44°, l$	1.71	Goldstein and Dolbier (1965)
	X = Br	$LiC_4H_9, -44°, l$	2.06	Goldstein and Dolbier (1965)
	X = I	$LiC_4H_9, -44°, l$	2.43	Goldstein and Dolbier (1965)
R_3CH[b]	$(CH_3)_2CH\cdot CHNN$	$h\nu, 25°, l$	1.18	Kirmse et al. (1968)
$PhCHR_2$	$PhHgCCl_3$	$100°, l$	1.8	Franzen and Edens (1969)
>C=CR·CH₃	CH_2CO	$h\nu, g$	1.96	Simon and Rabinovitch (1964)
>C=CH·R	CH_2CO	$h\nu, g$	1.55	Simon and Rabinovitch (1964)
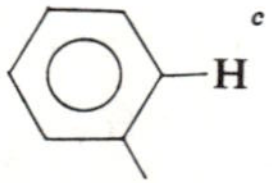 o-tolyl–H[c]	$o\text{-}PhC_6H_4CHNN$	$h\nu, l$	1.12	Denney and Klemchuk (1958)

[a] Intramolecular insertion; product 1,1-dimethylcyclopropane.
[b] Intramolecular rearrangement; product isobutene.
[c] Intramolecular insertion; product fluorene.

final product. These results indicate that the carbene attacks the oxygen atom first, giving the ylid $Ph_2\overset{-}{C}\overset{+}{O}H_2$ which rearranges to the final product in a subsequent rapid step (Bethell et al., 1969). A similar mechanism probably applies to the insertion of diarylmethylenes into the O–H bond of alcohols (Bethell et al., 1971), even though the acidity of alcohols seems to parallel their reactivity (Kirmse, 1963; Bethell and Howard, 1969).

Unstable ylids when formed do not always give rise to insertion products. Olefin elimination can compete with rearrangement in reactions of carbenes with amines and ethers [Eq. (27)] (Kirmse and Arold, 1968). Ylid formation is a plausible interpretation of the insertion of methylene or ethoxycarbonylmethylene into allylic halides [Eq. (28)] which occurs in competition with addition to the double bond (Kirmse and Kapps, 1965, 1968; Ando et al., 1969a). Significantly, carbenes generated photochemically using triplet sensitizers give predominantly addition, supporting the view that only singlet carbenes form ylids (Ando et al., 1969a,b).

$$RCH_2 \cdot CH_2OR' + :CHR'' \longrightarrow$$

$$\begin{bmatrix} RCH_2 \cdot CH_2\overset{+}{O}R' \\ | \\ {}^-CHR'' \end{bmatrix} \begin{array}{l} \longrightarrow RCH_2 \cdot CH_2 \cdot CHR''OR' \\ \longrightarrow R''CH_2OR' + RCH=CH_2 \end{array} \quad (27)$$

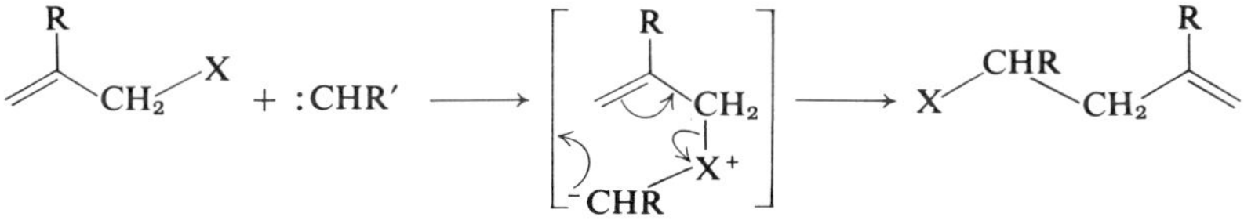

(28)

D. Addition to Multiple Bonds

1. Addition to Olefins

The most common addition of carbenes to multiple bonds is cyclopropane formation from olefins. The reaction is shown in its most general form in Eq. (29). In principle, four products are possible, namely, a pair of cyclopropanes formed by cis addition to the double bond and another pair

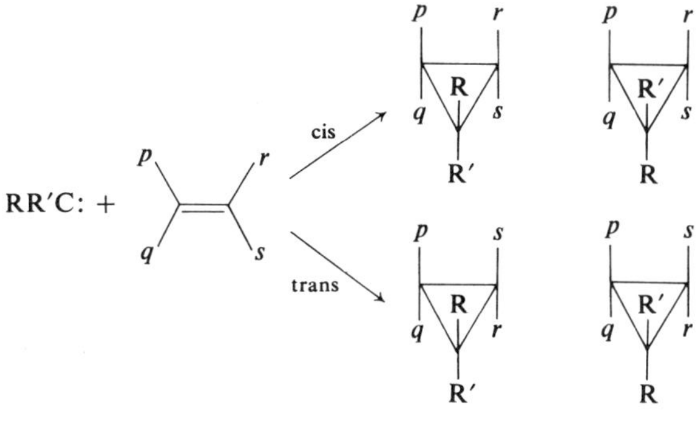

(29)

resulting from trans cycloaddition. Each pair consists of a syn and anti isomer, depending on the relative dispositions of the substituents on the reactants in the final product.

A satisfactory theory of carbene addition to olefins must explain the general experimental features summarized as follows (Closs, 1968; Moss, 1970).

i. Addition may produce both cis and trans addition products or it may be stereospecifically cis; stereospecific trans addition is never observed.

ii. Carbenoids and carbenes whose ground state is singlet add stereospecifically cis.

iii. Nonstereospecific addition is always associated with carbenes whose ground state is triplet, but some ground-state triplets give stereospecifically cis addition.

iv. Addition usually gives rise to the thermodynamically less stable one of a pair of syn–anti related cyclopropanes.

v. The most nucleophilic olefins are, in general, the most reactive toward carbenes.

2. Theoretical Considerations

An early, widely adopted rationalization of the stereochemical course of the cycloaddition of carbenes to olefins is due to Skell (Skell and Garner, 1956; Skell and Woodworth, 1956). According to this, cycloaddition of a singlet carbene is expected to be concerted since the two new σ bonds can be formed without an intermediate change in multiplicity and, hence, are stereospecifically cis [Eq. (30)]. On the other hand, the addition of a triplet carbene to a singlet olefin giving singlet cyclopropane should not be concerted. Instead, a stepwise mechanism is envisaged in which addition of the carbene at one end of the double bond gives a triplet trimethylene which can collapse to cyclopropane only after inversion of the spin of one of the electrons [Eq. (31)]. If rotation about the C–C bonds of the diradical intermediate is more rapid than spin inversion and collapse, the overall cycloaddition should be nonstereospecific.

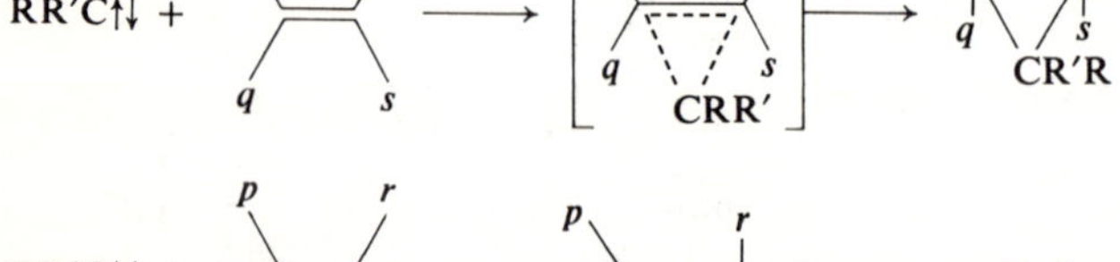

$$RR'C\uparrow\downarrow + \quad \text{olefin} \longrightarrow [\text{intermediate}] \longrightarrow \text{cyclopropane} \quad (30)$$

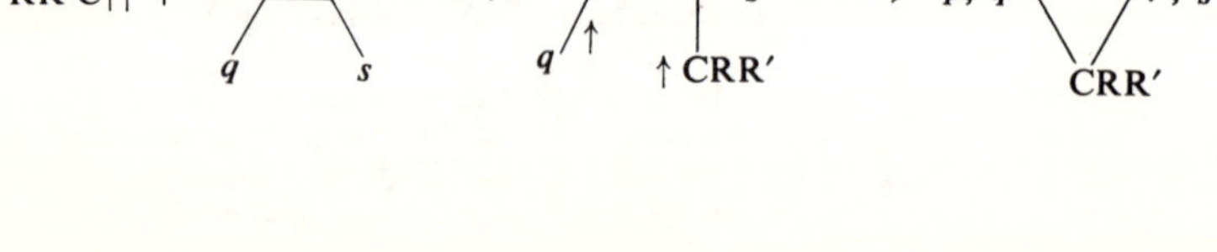

$$RR'C\uparrow\uparrow + \quad \text{olefin} \longrightarrow \text{diradical} \longrightarrow \text{cyclopropane} \quad (31)$$

An alternative interpretation not involving explicit consideration of the multiplicity of the carbene uses the principle of conservation of orbital symmetry (Woodward and Hoffmann, 1970). It can be illustrated by considering the addition of methylene to ethylene (Hoffmann, 1968; Closs, 1968).

We assume that the nonlinear carbene approaches one lobe of the olefinic π orbital symmetrically so that the nuclear arrangement in the transition state (**28**) is as close as possible to that in the product. A plane of symmetry, indicated in **28**, is maintained throughout the cycloaddition, and the reactant and combinations of product orbitals are classified symmetric or antisymmetric with respect to this. These orbitals and their correlation are indicated in Fig. 9 from which the state correlation diagram (Fig. 10) is readily constructed.

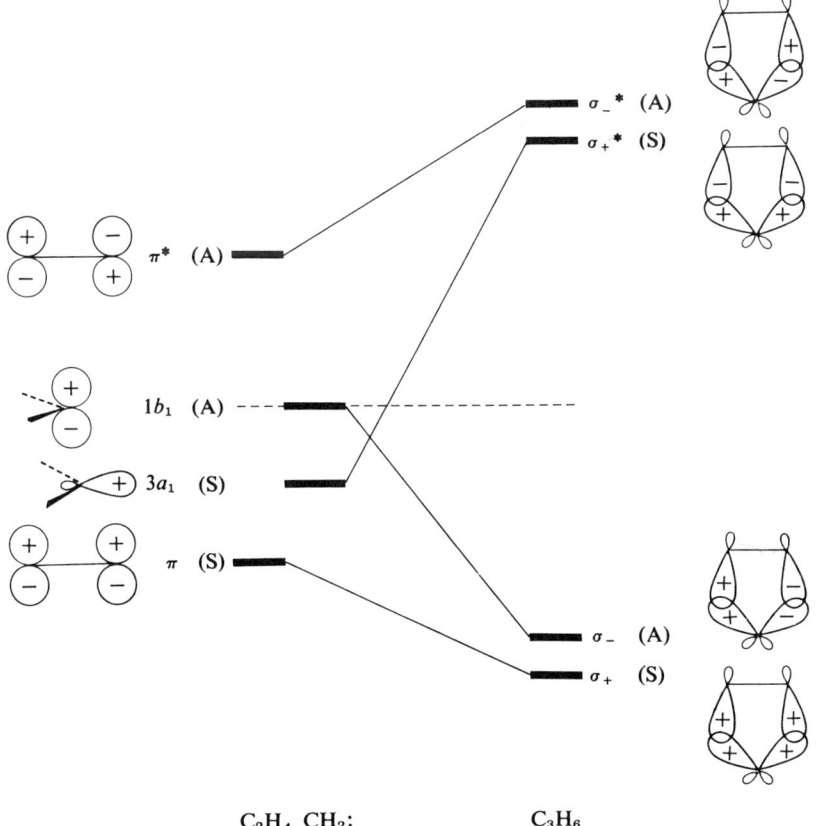

Fig. 9. Orbital correlation diagram for the addition of methylene to ethylene through transition state **28**.

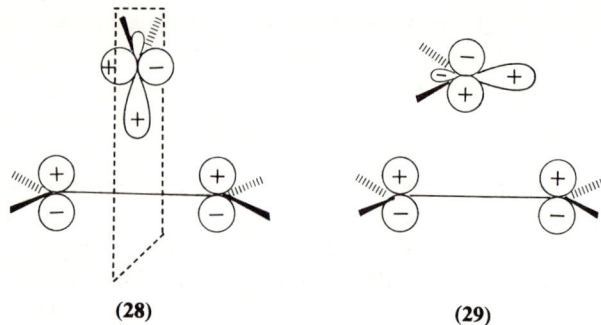

(28)   (29)

From Fig. 10 it can be seen that the ground state of ethylene and the $^1A_1$ state of methylene correlate with the ground state of cyclopropane only by the operation of the noncrossing rule for states of the same overall symmetry. The energy requirements of such a reaction are prohibitive, and the symmetrical cycloaddition of singlet methylene to ethylene is disallowed. On the other hand, the diagram indicates that ethylene and ground state $^3B_1$ methylene

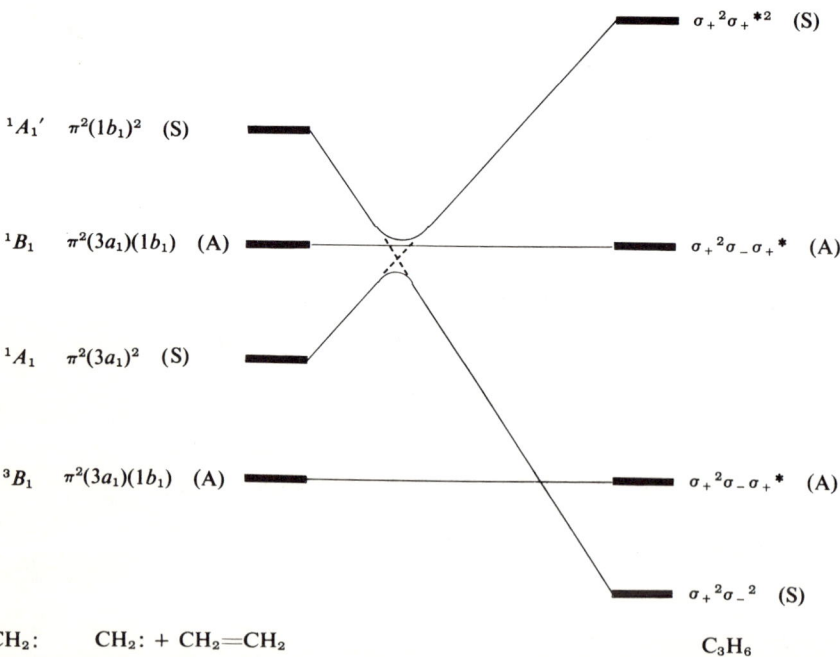

Fig. 10. Schematic state correlation diagram for the addition of methylene to ethylene. Bracketed letters indicate the total symmetry (S) or antisymmetry (A) of each state.

correlate with the first excited triplet state of cyclopropane, which calculation suggests is in fact the ground state of the trimethylene diradical (Hoffmann, 1968). Similarly, ethylene and the first excited singlet ($^1B_1$) methylene correlate with the first excited singlet state of cyclopropane, again probably singlet trimethylene. Both symmetrical cycloadditions are thus formally allowed, but because of the predicted facility of bond rotation in trimethylene (Hoffmann, 1968), they should lead to a nonstereospecific cyclopropane product.

The concerted cycloaddition of $^1A_1$ methylene to ethylene is allowed if the reactants approach in a less symmetrical way than that shown in **28**. Possible modes of interaction are exemplified by **29** in which the only element of symmetry is a mirror plane represented by the plane of the paper. All the orbitals in both reactants and products are symmetrical with respect to this, and the process is allowed. Stereospecific cis addition occurs, but it can be seen that some adjustment of the position of the hydrogens initially on the divalent carbon must occur after the transition state is reached. Note that the primary interaction in **29** is between the vacant p-type orbital of methylene and the filled π orbital of ethylene, indicating an electrophilic attack of the carbene on the double bond as deduced from structural effects on the reactivity of olefins. The disallowed cycloaddition of singlet methylene through transition state **28** would imply nucleophilic attack. Similar conclusions can be reached by analogous treatments for linear methylene (Anastassiou, 1968).

3. Structure and Reactivity

a. Stereospecificity. When diazomethane is photolyzed in the gas phase in the presence of cis-2-butene at low total pressures, the reaction products are the isomeric pentenes resulting from insertion and the addition products, cis- and trans-1,2-dimethylcyclopropanes. When allowance is made for the geometrical and structural isomerization of excited product cyclopropanes, it can be shown that methylene under these conditions adds almost stereospecifically cis (Frey, 1960). If the total pressure in the system is increased by the addition of an inert gas such as argon until it is greater than that required to deactivate completely all the excited reaction products formed initially, then the stereospecificity of addition drops dramatically. The parallelism between these observations and the spectroscopic observation which Herzberg made on methylene produced under similar conditions provides circumstantial evidence that the stereospecific cis addition at low pressures involves the singlet ($^1A_1$) state and the nonstereospecific addition involves the triplet ($^3B_1$) state of the carbene. Moreover, high pressure experiments in which a small partial pressure of oxygen was included led to a marked increase in the proportion of cis-1,2-dimethylcyclopropane produced. Oxygen, a ground-state triplet molecule, traps preferentially the triplet carbenes, thus increasing the proportion of addition product arising from the singlet carbene. The amount of cis addition product can be decreased by using triplet photosensitizers such as mercury vapor which promote the generation of triplet methylene.

Carbenes, whose ground states are believed (from theoretical calculations, spectroscopy, or chemical considerations) to be singlet, all add stereospecifically cis to olefins. Examples are the mono- and dihalomethylenes generated by a variety of methods. Some carbenes whose ground states are known to be triplet, such as phenylmethylene, diphenylmethylene, and fluorenylidene (9-carbenafluorene), all produced from the corresponding diazoalkane, give nonstereospecific addition but with widely varying ratios of cis and trans products (Table VIII). Methylene itself and the carbene **30**, which also has a triplet ground state, show limiting behavior in solution, adding stereospecifically.

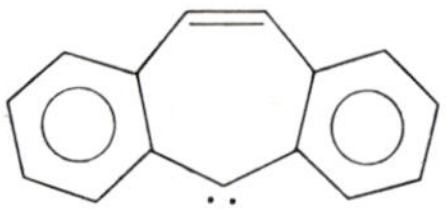

(30)

Table VIII
Stereochemistry of Addition of Carbenes with Triplet Ground States to cis- and trans-but-2-enes in Solution[a]

Carbene	Generation[b]	Percent trans addition		Reference[f]
		cis-but-2-ene	trans-but-2-ene	
CH_2:	$h\nu$	0(34)[c]	0(Trace)[c]	(i)
PhCH:	$h\nu$	3	3	(ii)
Ph_2C:	$h\nu$	13	Trace	(iii)
Fl:[d,e]	$h\nu$	33	—	(iv)
$MeO \cdot CO \cdot CH$:	$h\nu$	0	0	(v)
$(MeO \cdot CO)_2C$:	$h\nu$	8(90)[c]	10(14)[c]	(vi)
$(NC)_2C$:[e]	80°	70	30	(vii)
PhCOCH:	$h\nu$	50(55)[c]	27(26)[c]	(viii)
	$CuSO_4$	28	0	(viii)

[a] Taken largely from a more extensive compilation of Kirmse (1969).
[b] From the corresponding diazoalkane.
[c] Photolysis in the presence of benzophenone.
[d] Fluorenylidene.
[e] Product stereochemistry dependent on olefin concentration.
[f] References: (i) Skell and Woodworth, 1956, 1959; Doering and La Flamme, 1956; Kopecky et al., 1962; (ii) Gutsche et al., 1962; (iii) Closs and Closs, 1962; (iv) M. Jones and Rettig, 1965; (v) Doering and Mole, 1960; (vi) M. Jones et al., 1967a,b; (vii) Ciganek, 1966; (viii) Cowan et al., 1964.

Insight into this pattern of behavior is provided by the reaction scheme in Fig. 11. In the absence of triplet photosensitizers, photochemical or thermal decomposition of a diazoalkane should give rise initially to the singlet carbene. This can either add immediately to the olefin in a stereospecific manner or undergo spin inversion ("intersystem crossing"), giving the triplet carbene. The triplet can in turn react nonstereospecifically with the olefin or with other species such as oxygen which are capable of trapping it. Clearly, if intersystem crossing is irreversible, then a reduction in the concentration of olefin should favor triplet formation and lead to a reduction in the fraction of addition product which is cis. Such is the case with fluorenylidene (M. Jones and Rettig, 1965), bismethoxycarbonylmethylene (M. Jones *et al.*, 1967a), and dicyanomethylene (Ciganek, 1966). No change in stereoselectivity with olefin concentration is observed with diphenylmethylene, and this suggests that singlet–triplet interconversion may be reversible (Closs, 1968), a conclusion reached independently from experiments not involving addition to olefins (Bethell *et al.*, 1970). Stereoselectivity can be increased by additives which compete with the olefin for the triplet carbene and decreased by the use of triplet photosensitizers (e.g., benzophenone) to generate the carbene (Kopecky *et al.*, 1962; M. Jones *et al.*, 1967a,b).

Although it accounts plausibly for the experimental observations, the scheme in Fig. 11 is probably an oversimplification because of its neglect of possible stereoselection in the ring closure of the diradical intermediate produced by the addition of the triplet carbene to the olefin. The possibility of intersystem crossing of photoexcited diazoalkane molecules may provide a further complication.

Questions of multiplicity do not arise, of course, in discussing the addition of carbenoids. The experimental observation is of cis stereospecificity as implied by formulations of the transition state such as **18**.

b. Olefin Reactivity. The relative reactivities of a wide selection of olefins toward dichloro- and dibromomethylene are collected in Table IX. The sequence of olefin reactivities is qualitatively similar to that found in addition

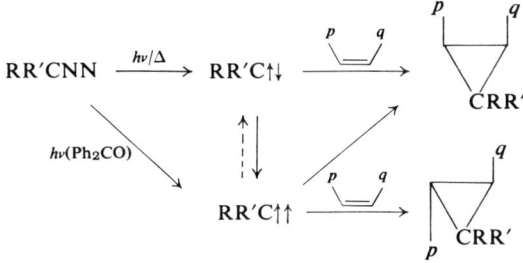

Fig. 11. The addition to olefins of carbenes derived from diazoalkanes.

Table IX
Relative Reactivity of Olefins toward Dihalomethylenes

Olefin	k_{rel} $Cl_2C:$ $(0°)^a$	k_{rel} $Br_2C:$ $(-10°)^b$	Olefin	k_{rel} $Cl_2C:$ $(0°)^a$	k_{rel} $Br_2C:$ $(-10°)^b$
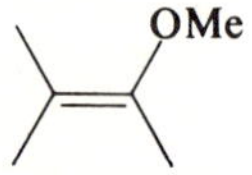 (OMe, trisubstituted)	10	—	 (cis-2-butene)	0.32	—
 (tetramethylethylene)	7.6	3.5	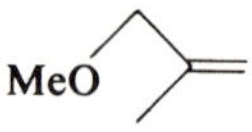 MeO—	0.28	—
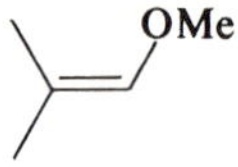 (OMe)	7.8	—	 (propene)	—	0.49
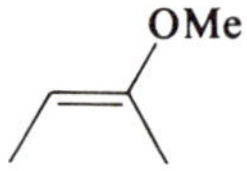 (OMe)	5.6	—	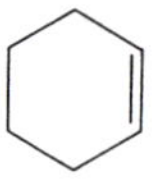 (cyclohexene)	0.19	0.40
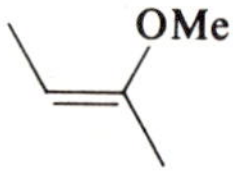 (OMe)	4.0	—	$PhCH=CH_2$	—	0.40
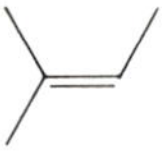	3.0	3.2	 (trans-2-butene)	0.038	—
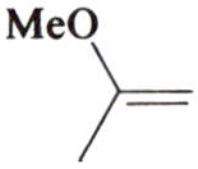 MeO—	2.0	—	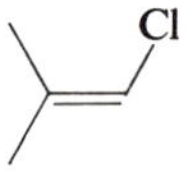 (Cl-substituted)	0.038	—
 (isobutene)	(1.0)	(1.0)	 (propene)	0.029	—
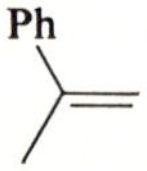 Ph—	—	0.79	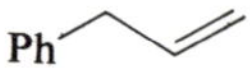 Ph—	—	0.05

[a] n-C_4H_9Li + $CHCl_3$, THF solution (Skell and Cholod, 1969b).
[b] $tert$-C_4H_9OK + $CHBr_3$ (Skell and Garner, 1956).

of other electrophiles, for example, bromine, although the spread of reactivities is somewhat smaller. This electrophilic behavior is general for carbenes as can be seen from the results for a wide selection of carbenes, reacting as singlets, and carbenoids assembled in Table X.

In general, then, the reactivity of a C–C double bond with a particular carbene increases with an increase in the number of alkyl substituents and

Table X
Structural Effects on the Relative Reactivity of Olefins toward Carbenes and Carbenoids in Solution[a]

RR'C:/oid	Source	Me₂C=CMe₂	Me₂C=CHMe	Me₂C=CH₂	trans-MeCH=CHMe	1-alkene	Reference
CH_2:	$CH_2NN/h\nu$	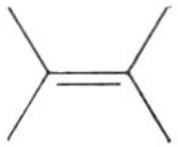 1.10	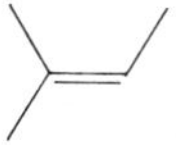 1.08	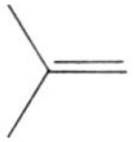 (1.0)	 0.70	 0.71	 0.83	Moss (1969)
FCH:	$^3H. + CH_2F_2$	2.10	1.48	(1.0)	1.08	1.40	0.73	Tang and Rowland (1967)
ClCH:	$ClCHNN (-30°)$	1.20	1.18	(1.0)	0.99	1.09	0.74	Closs and Coyle (1965)
	$Bu^nLi + CH_2Cl_2 (-35°)$	2.81	1.78	(1.0)	0.91	0.45	0.23	Closs and Coyle (1965)
BrCH:	$BrCHNN (-30°)$	1.18	—	(1.0)	1.02	1.10	0.75	Closs and Coyle (1965)
F_2C:	$F_2CN_2/h\nu (36°, g)$	13.1	3.53	(1.0)	0.065	0.074	0.0105	Mitsch and Rodgers (1969)
ClFC:	$Bu^tOK + (CCl_2F)_2CO (-12°)$	31.0	6.5	(1.0)	0.14	0.097	0.0087	Moss and Gerstl (1967a,b)
Cl_2C:	$Bu^nLi + CHCl_3 (0°)$	7.6	3.0	(1.0)	0.32	0.15^b	0.029	Skell and Cholod (1969b)
	$Bu^nLi + CHCl_3 (-78°)$	5.2	3.0	(1.0)	0.26	—	0.010	Skell and Cholod (1969b)
Br_2C:	$Bu^tOK + CHBr_3 (-10°)$	3.5	3.2	(1.0)	—	—	—	Skell and Garner (1956)
PhCH:	$PhCHNN/h\nu (-10°)$	—	3.6	(1.0)	2.0	1.1	1.1	Closs and Moss (1964)
	$MeLi + PhCHBr_2 (-10°)$	—	2.0	(1.0)	1.8	0.59	0.96	Closs and Moss (1964)
PhFC:	$Bu^tOK + PhCHBrF (25°)$	2.7	1.2	(1.0)	0.12	0.10	—	Moss and Przybyla (1969)
PhClC:	$PhClCN_2/h\nu (25°)$	5.1	3.2	(1.0)	0.37	0.20	—	Moss et al. (1969)
	$Bu^tOK + PhCHCl_2 (25°)$	2.6	1.6	(1.0)	0.31	0.11	—	Moss et al. (1969)
PhBrC:	$PhBrCN_2/h\nu (25°)$	4.4	2.5	(1.0)	0.53	0.26	—	Moss (1967)
	$Bu^tOK + PhCHBr_2 (25°)$	1.6	1.3	(1.0)	0.29	0.15	—	Moss and Gerstl (1966)
CH_3ClC:	$CH_3ClCN_2/h\nu (25°)$	3.9	2.4	(1.0)	0.74	0.52	—	Moss and Mamantov (1970)
EtOCOCH:	$EtOCOCHNN/CuSO_4$	1.8	1.8	—	—	—	$(1.0)^c$	Skell and Etter (1958)

[a] Taken in part from a compilation by Bethell (1969).
[b] $Bu^tOK + CHCl_3 (\sim 0°)$.
[c] 1-Hexene.

other groups capable of electron donation into the π bond. Electron withdrawal by halogen, for example, and conjugation with attached π systems, as in butadiene and styrene, tend to reduce the reactivity. Superimposed on electronic influences are steric effects. This is shown by the sequence of reactivities toward dichloromethylene: $EtCH:CH_2$, 1.0; $Pr^iCH:CH_2$, 0.43; $Bu^tCH:CH_2$, 0.029 (Moss and Mamantov, 1968). Furthermore, in the reaction of diphenylmethylene with olefins, allylic insertion competes with cycloaddition most successfully in the case of tetramethylethylene (2,3-dimethylbut-2-ene) and least successfully in the case of propene, suggesting important steric effects on the formation of cyclopropane (M. Jones *et al.*, 1970). Clearly, the balance between steric and electronic effects will vary from one carbene to another, depending on its structure and probably also on its spin state, though little is known about this.

Relative reactivities of olefins seem to be rather insensitive to temperature. The sequence of olefin reactivities, toward addition of dichloromethylene at least, is largely entropy controlled (Skell and Cholod, 1969b). The addition of dichloromethylene to meta- and para-substituted styrenes, however, involves constant steric influences, and polar effects can be studied in isolation. The reactivities of these olefins relative to styrene itself fit the modified Hammett equation $\log k_{rel} = \rho\sigma^+$ with ρ values of -0.62 (styrenes, 80°C) (Seyferth *et al.*, 1968) and -0.38 (α-methylstyrenes, 0°C) (Sadler, 1969). For comparison, the zinc carbenoid from diethyl zinc and methylene iodide gives a ρ value of -1.61 at 79°C (Nishimura *et al.*, 1971). The ρ values confirm the electrophilic nature of dichloromethylene but indicate little charge separation in the transition state. The fact that σ^+ values give better correlations than σ values further suggests some accumulation of positive charge at the benzylic position, and this might imply that the bonds between the divalent carbon atom and the two olefinic carbons are formed to different extents in the transition state (**29**). Steric interactions between substituents in the olefin and carbene could also affect the symmetry of the bonding in **29** in other systems.

c. Carbene Structure. Whereas the qualitative sequence of olefin reactivities is almost invariant with changing carbene structure, the data in Table X indicate quantitative differences. Thus, difluoromethylene is the most selective of the carbenes, being close to molecular bromine in its ability to discriminate between olefins. Methylene, on the other hand, shows very low selectivity. These variations can be put on a numerical basis by means of a linear free energy correlation. Using $\log k_{rel}$ for dichloromethylene addition at 0°C as a standard measure of the reactivity of olefins (r), then for addition of other singlet carbenes it is found that $\log k_{rel} = sr$, where s, a measure of selectivity, is greater or less than unity according to whether the carbene is more or less discriminating than dichloromethylene. Representative s values are as follows: $F_2C:$, 1.9; $ClFC:$, 1.5; $Cl_2C:$, (1.0); $PhClC:$, 0.83; $Br_2C:$, 0.82; $CH_3ClC:$, 0.50; and $CH_2:$, ~0.1 (Moss and Mamantov, 1970; Skell and Cholod, 1969b). The parameter s seems to be clearly related to the ability of substituents to conjugate with the vacant p-type orbital on the divalent carbon atom. Surprisingly, carbenoids are only slightly more selective than the corresponding carbene; for example, s for the Simmons–Smith reagent is only about 0.25.

d. Syn/Anti Product Ratios. Values of the ratio of syn to anti addition product proportions for a variety of unsymmetrically substituted carbenes reacting with 2-methylbut-2-ene, cis-but-2-ene, and but-1-ene are given in Table XI. Considerable variation in syn/anti stereoselectivity is found, and this cannot be a consequence only of the smaller substituent on the divalent carbon atom being placed syn to the greater number of alkyl substituents in the product cyclopropane. For example, chloromethylene ($s = 0.1$) shows no stereoselection, whereas chlorofluoromethylene ($s = 1.5$), which has similar steric requirements, gives rise to the highest syn/anti ratios in Table XI. The more reactantlike transition state in additions of highly reactive carbenes should minimize steric and polar influences in the case of chloromethylene. However, for the more selective chlorofluoromethylene the larger chloro substituent is placed syn to the greater number of alkyl groups. This has been attributed to the greater polarizability of chlorine than fluorine, permitting a greater degree of charge dissipation in the transition state by mutual polarization of the olefinic alkyl groups and the syn substituent on the carbene. The stereoselectivities of the phenylhalomethylenes provide a consistent sequence.

In carbenoid additions, the polar and steric effects of the associated moieties must also be considered. Thus, in the decomposition of ethyl diazoacetate in the presence of cyclohexene induced by copper (I) derivatives of the type $(RO)_3PCuCl$, the syn/anti product ratio increases with increasing size and electron withdrawal of the group R (Moser, 1969). Interaction of the methylene acceptor with the metal atom of the carbenoid can also modify the stereoselectivity (e.g., Poulter *et al.*, 1969).

Where α-haloalkyl lithium is the carbenoid, a knowledge of the degree of aggregation of the organometallic intermediate under the reaction conditions is desirable before discussing the factors controlling stereoselectivity. By analogy with simple alkyl lithiums in ethereal solvents, this is likely to be at least 2, a number for which there is some experimental support (Closs and Lin, quoted by Moss, 1969; Friedman *et al.*, 1970). Not surprisingly, the syn/anti ratio shows a considerable solvent dependence and in some cases a pronounced increase with decreasing olefin concentration (Schlosser and Heinz, 1970).

Table XI
Syn/Anti Product Ratios from Additions of Carbenes and Carbenoids to Olefins in Solution[a]

RR′C:/oid	Source	Syn/Anti Product Ratio[b,c] >=/	_=/	/_/
ĊlCH:	ClCHNN	1.0	1.0	1.0
	Bu^{n}Li + CH$_2$Cl$_2$	1.6	5.5	3.4
BrCH:	BrCHNN	1.0	1.0	1.0
ClFC:	Bu^{t}OK + (CCl$_2$F)$_2$CO	2.35	3.08	—
CH$_3$ClC:	CH$_3$ClCN$_2$/hν	1.45	2.84	—
PhCH:	PhCHNN/hν	1.1	1.1	1.0
	MeLi + PhCHBr$_2$	1.3	2.4	2.1
FPhC:	Bu^{t}OK + PhCHBrF	0.76	1.23	—
ClPhC:	PhClCN$_2$/hν	1.28	1.97	—
	Bu^{t}OK + PhCHCl$_2$	1.5	3.0	1.7
BrPhC:	PhBrCN$_2$/hν	1.31	1.55	—
	Bu^{t}OK + PhCHBr$_2$	1.28	1.35	—

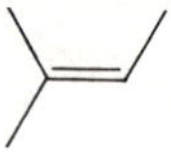

[a] Taken largely from a compilation by Bethell (1969).
[b] The syn isomer is that in which the substituent written first in column 1 lies on the same side of the plane of the cyclopropane ring as the larger number of alkyl substituents.
[c] References are the same as for the corresponding reactions in Table X.

4. Other Cycloadditions of Carbenes

Carbenes and carbenoids will undergo cycloaddition reactions with multiple bonds other than C–C double bonds, but in general these reactions have not received the depth of study accorded to cyclopropane formation.

Additions to *acetylenes* give cyclopropenes in fairly low yield. As in other electrophilic additions, acetylenes are somewhat less reactive than olefins. However, with certain conjugated enynes, dichloromethylene favors attack on the triple bond rather than on the double bond [Eq. (32)] (Dehmlow, 1968).

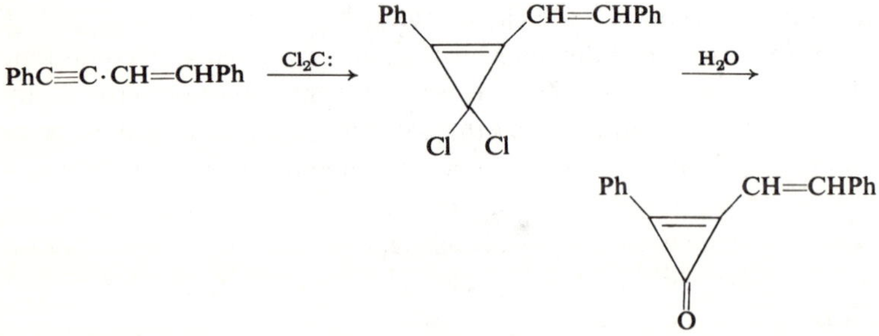

$$PhC\equiv C\cdot CH=CHPh \xrightarrow{Cl_2C:} \underset{Cl\ \ Cl}{\underset{}{\triangle}}{\overset{Ph\ \ \ \ CH=CHPh}{}} \xrightarrow{H_2O} \underset{O}{\underset{}{\triangle}}{\overset{Ph\ \ \ \ CH=CHPh}{}}$$

(32)

Addition of carbenes to *aromatic compounds* will occur under suitable circumstances. The usual requirement is a fairly reactive carbene such as methylene, chloromethylene, or ethoxycarbonylmethylene, but less reactive carbenes will add to aromatic compounds which are activated, for example, by methoxyl groups or hetero atoms or by having positions with high double-bond character as in phenanthrene. The effect of substituents in the benzene nucleus on the ease of addition of carbenes (for example, ethoxycarbonylmethylene) or carbenoids is relatively small (Hammett's ρ for $EtO \cdot CO \cdot CH$: is -0.38) but indicative of electrophilic attack (Müller *et al.*, 1963; Baldwin and Smith, 1967; Nishimura *et al.*, 1970). The initial adduct often undergoes valence isomerization, giving a cycloheptatrienyl product [Eq. (33)]. The intermediate norcaradienes have been isolated in the addition of dicyanomethylene to benzene, *p*-xylene, and naphthalene (Ciganek, 1965, 1966).

$$\text{benzene} \xrightarrow{\text{BrCH:}} \text{norcaradiene} \longrightarrow \text{cycloheptatriene} \rightleftharpoons \text{tropylium} + Br^- \quad (33)$$

Multiple bonds between carbon and atoms other than carbon can add carbenes to give heterocyclic products. Because of the polarization of such bonds brought about by the electronegative hetero atom, adduct formation can be brought about by carbene precursors such as α-halocarbanions rather than by the carbene itself. Where carbenes are involved, the details of the reaction mechanisms are not known with certainty. The presence of non-bonded electron pairs on the hetero atom makes intermediate ylid formation a possibility. Equations (34) (Cook and Fields, 1962) and (35) give two examples of carbene addition to C–N multiple bonds. Note that benzoylmethylene adds in a 1,3-manner to give the five-membered isoxazole ring (Huisgen *et al.*, 1964).

$$p\text{-}ClC_6H_4N=CHPh \xrightarrow{Cl_2C:} p\text{-}ClC_6H_4N\underset{CCl_2}{\overset{CHPh}{<}} \quad (34)$$

$$PhC\equiv N + PhCO \cdot CH: (\longleftrightarrow Ph\overset{+}{C}=\overset{|}{\underset{O^-}{C}}H) \longrightarrow \text{isoxazole} \quad (35)$$

E. REARRANGEMENTS

The analogy between rearrangements of carbenes [e.g., Eq. (36)] and carbonium ions [Eq. (37)] has long been recognized. In both cases, migration of a suitably oriented group attached to the β-carbon atom with retention of configuration is thought to occur by interaction with the vacant *p* orbital at the reactive center. Carbene rearrangements should, therefore, be reactions of the singlet state. Indeed, in cases where rearrangement competes with other reactions of the carbene, the use of triplet photosensitizers in generating the divalent carbon species tends to reduce the proportion of rearranged product (Moritani *et al.*, 1968b). While a 1,2-shift in a carbonium ion to produce a second, more stable cation is common, rearrangement of a carbene to form another carbene is relatively rare. The known examples are all arylmethylenes which, under pyrolytic conditions, undergo reversible ring expansion to form an intermediate seven-membered cyclic carbene [Eq. (38)] (Crow and Wentrup, 1968; Joines *et al.*, 1969; Myers *et al.*, 1970; Baron *et al.*, 1970). More often, as shown in Eq. (36), a 1,2-shift in a carbene gives directly a stable molecule, commonly an olefin.

$$\underset{R'}{\overset{R}{>}}C - \ddot{C}R' \longrightarrow \underset{R'}{\overset{R}{>}}C = C \underset{R'}{\overset{R}{<}} \qquad (36)$$

$$\underset{}{\overset{R}{>}}C - \overset{+}{C}R'R'' \longrightarrow \overset{+}{>}C - CRR'R'' \qquad (37)$$

$$\text{Ar-}\ddot{C}R \rightleftharpoons \text{(cycloheptatrienylidene)} \rightleftharpoons \text{Ar-}\ddot{Z}: \qquad Z = CH, N \quad (38)$$

The most common reaction formulated as involving a carbene rearrangement is the Wolff rearrangement of diazoketones to ketenes [Eq. (39a)] (Franzen, 1958). The reaction can be brought about thermally, photochemically, or "catalytically" by the action of silver compounds. Products corresponding to attack by an unrearranged ketocarbene on nucleophiles can be produced [Eq. (39b)] without affecting the overall reaction rate (Jugelt and Schmidt, 1969). Other evidence suggests that rearrangement may be concerted with nitrogen loss in some cases (Kaplan and Meloy, 1966). In the photochemical rearrangement of azibenzil to diphenylketene, the oxirene intermediate (**31**) is formed (Frater and Strausz, 1970), and this suggests, though it does not demand, the formation and interconversion of free carbenes [Eq. (40)].

$$RCO \cdot C(NN)R' \longrightarrow RCO \cdot \ddot{C}R' \begin{array}{c} \overset{(a)}{\nearrow} O = C = CRR' \; (\xrightarrow{MeOH} RR'CH \cdot CO_2Me) \\ \underset{MeOH}{\searrow} \\ RCO \cdot CH(OMe)R' \end{array} \qquad (39)$$

$$\text{PhCO·C(NN)Ph} \longrightarrow \text{PhCO·}\ddot{\text{C}}\text{Ph} \longrightarrow \text{PhC}\overset{\overset{\displaystyle O}{\triangle}}{=}\text{CPh} \longrightarrow \text{Ph}\ddot{\text{C}}\text{·COPh} \quad (40)$$
$$(31)$$

The migrating group in Eq. (39a) can be alkyl, aryl, or alkoxy (Strausz et al., 1968a). The rearrangement of phenyl groups is facilitated by electron-repelling substituents in a qualitatively similar way to carbonium ion rearrangements.

A similar substituent effect on the migratory aptitude of phenyl groups is found in the conversion of 1,1-diaryl-2-haloethylenes into diarylacetylenes brought about by the action of lithium alkyls or other basic reagents [Fritsch–Buttenberg–Wiechell rearrangement; Eq. (41)] (Köbrich, 1967). An intermediate α-halovinyl lithium derivative (carbenoid) can be trapped by reaction with carbon dioxide or olefins, and this, by the loss of lithium halide, could produce a carbene capable of rearrangement. However, since the aryl group trans to the departing halide is the one which migrates, it is more likely that rearrangement is synchronous with loss of halide.

$$\underset{\text{Ar}'}{\overset{\text{Ar}}{>}}\text{C}=\text{C}\underset{\text{X}}{\overset{\text{H}}{<}} \xrightarrow[-\text{RH}]{\text{LiR}} \left[\underset{\text{Ar}'}{\overset{\text{Ar}}{>}}\text{C}=\text{C}\underset{\text{X}}{\overset{\text{Li}}{<}} \right] \xrightarrow{-\text{LiX}} \text{Ar}'\text{C}\equiv\text{CAr} \quad (41)$$

Rearrangement of free carbenes is more likely in thermal and photochemical decomposition of diazoalkanes. The sequence of migratory aptitudes is generally H > aryl > alkyl. Hydrogen migration takes place in preference to Wolff rearrangement in diazoketones of the type $\text{RCO·C(NN)CHR}_2'$. For 1,2-aryl migration in carbenes of the types $p\text{-XC}_6\text{H}_4\text{·CPh}_2\text{·CH:}$ and $p\text{-XC}_6\text{H}_4\text{CMe}_2\ddot{\text{C}}\text{Ph}$, the influence of the para substituents X relative to H on the migratory aptitude is correlated by the equation $\log k_{\text{rel}} = \rho\sigma^+$ as in carbonium ion rearrangements (Sargeant and Shechter, 1964; Landgrebe and Kirk, 1967). The reaction constant has a small negative value, implying little charge separation in the transition state for rearrangement, which is expected to be reactantlike since it is exothermic. The low value of kinetic hydrogen isotope effects in H migration in carbenes (e.g., the fifth entry in Table VII) is also consistent with a reactantlike transition state (Kirmse et al., 1968). The conformation of the reactant carbene in its singlet state seems to control the proportion of cis- and trans-olefin produced by rearrangement (Kirmse and Buschhoff, 1967; Yamamoto and Moritani, 1970). Thus, in rearrangements of carbenes of the type $\text{R}\ddot{\text{C}}\text{CH}_2\text{R}'$, the thermodynamically more stable trans-olefin is formed when R and R′ are alkyl or aryl, but the cis-olefin is produced when $\text{R} = \text{R}' = \text{CO}_2\text{Et}$.

V. Related Species

A. Introduction

A number of reactive intermediates are closely related to carbenes in that they possess a reactive carbon atom or hetero atom (such as nitrogen) having nonbonded electrons and one or more low-lying vacant orbitals. Nitrenes (Chapter 3) are perhaps the most widely studied of these analogs of carbenes. Here we shall consider only those species having reactive carbon centers. Kirmse (1969) has given a more comprehensive review.

B. Atomic Carbon

Carbon forms a number of discrete atomic species, C_1, C_2, C_3, ..., those containing an odd number of atoms being predicted to be rather more stable than those with an even number.

Monatomic carbon can be produced in a highly energetic ("hot") form by nuclear recoil in transformations such as $^{14}N(n, p)^{14}C$ or $^{12}C(n, 2n)^{11}C$. These energetic atoms are highly reactive, inserting into C–H bonds in nearby molecules to produce carbenes and abstracting hydrogen atoms to give methyne (HC:) and methylene. The study of such reactions is experimentally difficult, and little is known about the electronic states of the carbon atoms involved (Wolf, 1964; Nicholas et al., 1965).

All the atomic carbon species, C_1–C_5, are more conveniently prepared by resistive heating of carbon or by striking an arc between graphite electrodes in a high vacuum. The composition of carbon vapor varies with conditions, but at 4100°K, C_1–C_5 species have been identified in the ratio 1:2.8:4.5:0.35:0.5 (Drowart et al., 1959). These reactive entities can be condensed on a cold surface (77°K) together with other substances and allowed to react. Subsequent separation and identification of the products then allows the behavior of the individual components of carbon vapor to be examined separately. C_1 and C_3 species are stable for long periods on a neopentane-covered surface at 77°K, and by delaying the condensation of other reactants the time dependence of the reactivity of these species can be investigated (Skell et al., 1965).

Table XII shows the three electronic states of monatomic carbon, which are known from spectroscopic studies, and their electronic arrangement. Monatomic carbon inserts into C–H bonds in a variety of organic molecules to give initially a carbene, the fate of which appears to depend on its multiplicity. Ground-state (^3P) atoms convert C–H into C–CH$_3$ in low yield ($\leq 5\%$). The reaction is inhibited by butadiene, which is capable of reacting with free radicals, but unaffected by methanol, a trap for singlet carbenes

2. CARBENES AND RELATED SPECIES

Table XII
Structure and Properties of Monatomic Carbon

Electronic state	Occupied orbitals			Excitation[a] (eV)	Half-life (calc.)[b] (sec)
	$2s$	$2p_x$	$2p_y$		
^3P	↑↓	↑	↑	0	—
^1D	↑↓	↑↓	—	1.3	2000 ($\to {}^3P$)
^1S	↑↓	↑	↓	2.7	2 ($\to {}^1D$)

[a] Herzberg (1944).
[b] Yilmaz (1955).

(Skell et al., 1970). A triplet carbene is thought to be the first-formed intermediate [Eq. (42)]. Time-delay studies suggest that the ^1S state of carbon will insert into C–H bonds in aliphatic hydrocarbons (Skell and Engel, 1966b). Here, however, the major pathway for subsequent reaction is intramolecular insertion or rearrangement of the intermediate carbene which is plausibly singlet [Eq. (43)].

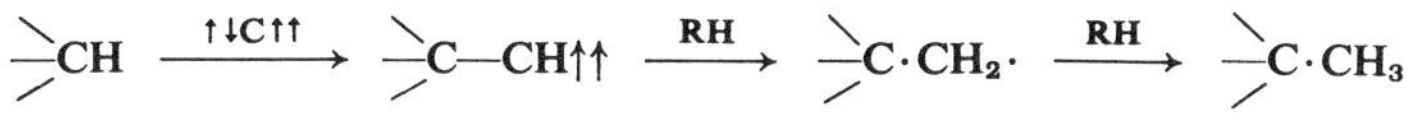

$$\ce{>CH} \xrightarrow{\uparrow\downarrow C\uparrow\uparrow} \ce{>C-CH\uparrow\uparrow} \xrightarrow{RH} \ce{>C\cdot CH_2\cdot} \xrightarrow{RH} \ce{>C\cdot CH_3} \quad (42)$$

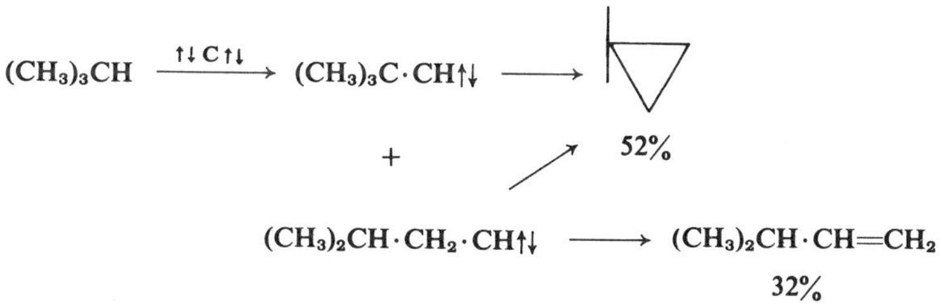

$$(CH_3)_3CH \xrightarrow{\uparrow\downarrow C\uparrow\downarrow} (CH_3)_3C\cdot CH\uparrow\downarrow \longrightarrow \triangle \quad 52\%$$

$$+ \quad (CH_3)_2CH\cdot CH_2\cdot CH\uparrow\downarrow \longrightarrow (CH_3)_2CH\cdot CH=CH_2 \quad 32\% \quad (43)$$

^3P Carbon atoms do not appear to be reactive toward the olefinic π bond (Skell et al., 1971) as it was thought at one time (Skell and Engel, 1966a). Metastable singlet carbon atoms, however, give rise to allenes by rearrangement of the intermediate singlet cyclopropylidene [Eq. (44)]. Accompanying the allene are conjugated dienes resulting from allylic insertion and hydrogen migration. The geometrical configuration of the π bond is predominantly retained.

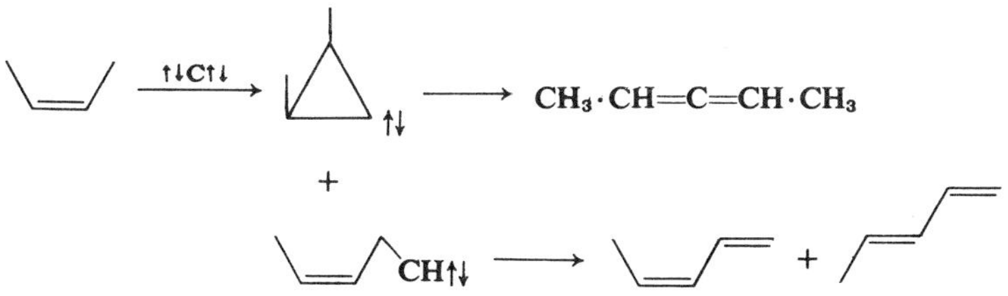

$$\ce{\diagdown\!=\!/} \xrightarrow{\uparrow\downarrow C\uparrow\downarrow} \triangle_{\uparrow\downarrow} \longrightarrow CH_3\cdot CH=C=CH\cdot CH_3$$

$$+ \quad \ce{\diagdown\!=\!/}^{CH\uparrow\downarrow} \longrightarrow \ce{\diagdown\!=\!/\!=} + \ce{/\!=\!/} \quad (44)$$

Singlet carbon atoms are also responsible for deoxygenation of epoxides to olefins and ketones to carbenes (Skell and Plonka, 1970a). Benzene can be converted to a mixture of toluene and cycloheptatriene by monatomic carbon (Sprung et al., 1965).

Spectroscopic and theoretical studies on C_3 reviewed by Skell and co-workers (1965) indicate that its ground state is a symmetrical linear singlet. Triatomic carbon does not insert into C–H bonds, and time-delay studies show that in its ground state it adds stereospecifically cis to two molecules of olefin to give a bisethanoallene [Eq. (45)] by way of the singlet carbene **32**. The relative reactivities of olefins toward C_3 and $C_1(^1S)$ are given in Table XIII. Co-condensation of C_3 and cis-but-2-ene gives small amounts of **33**,

Table XIII
Relative Reactivities of Atomic Carbon and
Related Species with Olefins at 77°K

	Relative reactivity	
Olefin	$C_1\,(^1S)^a$	$C_3^{\,b}$
 ⌐⌐	1.06	1.7
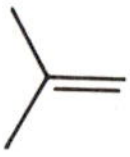 ⌐⌐	1.92	3.8
 ⌐⌐	—	(1.0)
 ⌐⌐	(1.00)	0.62
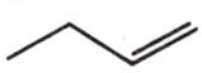 ⌐⌐	5.21	2.0
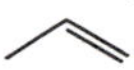 ⌐⌐	1.44	1.6

[a] Skell and Engel (1967).
[b] Skell et al. (1965).

especially at high arc voltages. This suggests the existence of a short-lived metastable triplet state of C_3. Co-condensed with alcohols, C_3 gives acetals of propargyl aldehyde, $HC{\equiv}C\cdot CH(OR)_2$, together with a mixture of propene, allene, and propyne from hydrogenation of excited C_3 (Skell and Harris,

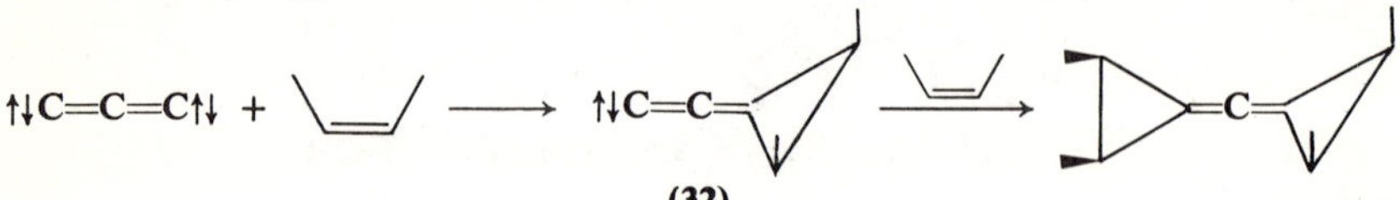

(32)

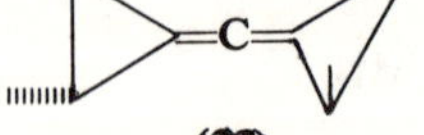

(33)

(45)

C. CARBYNES

Carbynes are electrically neutral, monovalent carbon species of the general formula $RC\!:\!$. As such, they lie between carbenes, on the one hand, and monatomic carbon, on the other. The parent species methyne, $HC\!:\!$, is known spectroscopically (e.g., Braun et al., 1967). Having three nonbonding electrons and three available carbon orbitals, carbynes can exist in doublet, $RC\uparrow(\uparrow\downarrow)$, and quartet, $RC\uparrow\uparrow\uparrow$, states analogous to singlet and triplet states of carbenes. Radical properties as well as carbenelike behavior are expected.

Ethoxycarbonylmethyne, $EtO\cdot CO\cdot C\!:\!$, can be generated by photolysis ($\lambda < 290$ nm) of the diazomercurial, $(EtO\cdot CO\cdot CNN)_2Hg$. Its EPR spectrum at 77°K indicates a doublet ground state. With cyclohexene at normal temperatures, addition and insertion take place with stable products being formed by subsequent hydrogen abstraction from the solvent [Eq. (46)]. Additions to cis- and trans-but-2-enes are stereospecifically cis as expected of a reaction through the ground state (DoMinh et al., 1967; Strausz et al., 1968b).

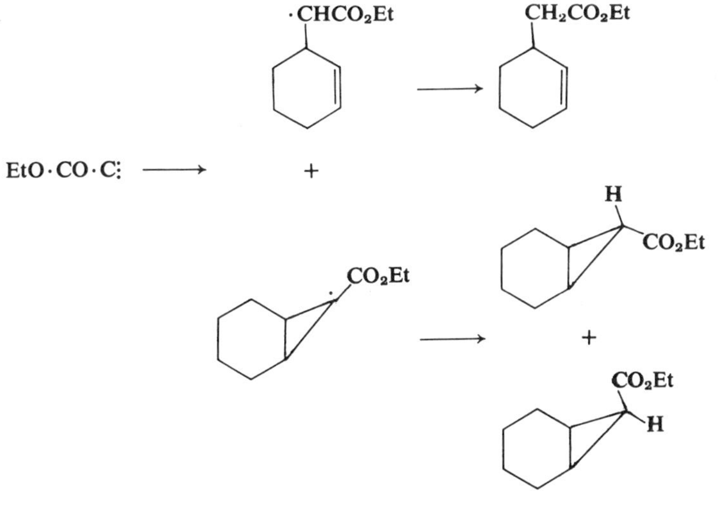

1969). Diatomic and tetratomic carbon are also partially hydrogenated when co-condensed with organic compounds as disparate as hexane, methanol, acetaldehyde, and tert-butyl chloride (Harris and Skell, 1968). There is evidence for the involvement of both singlet and triplet states of C_2 in such hydrogenations (Skell and Plonka, 1970b).

VI. Conclusion

The study of carbenes and carbenoids is an area of intense activity. Much remains to be investigated, especially the quantitative relations between structure and reactivity, but the broad theoretical foundation on which subsequent development of the subject can be built has now been laid. In particular, the chemistry of triplet molecules has been placed on a broader basis than purely photochemical studies would permit.

As synthetic procedures, reactions involving carbenes are not usually sufficiently specific. The high reactivity of free carbenes can lead to complex product mixtures. Intramolecular reactions of carbenes, because of their special steric and conformational constraints, can, however, be useful sometimes. However, it is in the use of the much greater selectivity of carbenoids that carbene chemistry finds its most important synthetical application. Further developments seem likely here, especially in the area of transition metal complexes of carbenes.

References

Amrich, M. J., and Bell, J. A. (1964). *J. Amer. Chem. Soc.* **86**, 292.
Anastassiou, A. G. (1968). *Chem. Commun.* p. 992.
Ando, W., Kondo, S., and Migita, T. (1969a). *J. Amer. Chem. Soc.* **91**, 6516.
Ando, W., Nakayama, K., Ichibori, K., and Migita, T. (1969b). *J. Amer. Chem. Soc.* **91**, 5164.
Andrews, L. (1968). *J. Chem. Phys.* **48**, 979.
Baldwin, J. E., and Smith, R. A. (1967). *J. Amer. Chem. Soc.* **89**, 1886.
Bamford, C. H., and Casson, J. E (1969). *Proc. Roy. Soc., Ser. A* **312**, 141 and 163.
Bamford, C. H., Casson, J. E., and Hughes, A. N. (1968). *Proc. Roy. Soc., Ser. A* **306**, 135.
Barash, L., Wasserman, E., and Yager, W. A. (1967). *J. Amer. Chem. Soc.* **89**, 3931.
Baron, W. J., Jones, M., and Gaspar, P. P. (1970). *J. Amer. Chem. Soc.* **92**, 4739.
Becker, R. S., Bost, R. O., Kolc, J., Bertoniere, N. R., Smith, R. L., and Griffin, G. W. (1970). *J. Amer. Chem Soc* **92**, 1302.
Bender, C. F., and Schaefer, H. F. (1970). *J. Amer. Chem. Soc.* **92**, 4984.
Bernheim, R. A., Kempf, R. J., Gramas, J. V., and Skell, P. S. (1965). *J. Chem. Phys.* **43**, 196.
Bethell, D. (1963). *J. Chem. Soc., London* p. 666.
Bethell, D. (1969). *Advan. Phys. Org. Chem.* **7**, 153.
Bethell, D., and Brown, K. C. (1967). *Chem. Commun.* p. 1266.
Bethell, D., and Howard, R. D. (1969). *J. Chem. Soc., B* p. 745.
Bethell, D., and Whittaker, D. (1966). *J. Chem. Soc., B* p. 778.
Bethell, D., Whittaker, D., and Callister, J. D. (1965). *J. Chem. Soc., London* p. 2466.
Bethell, D., Cockerill, A. F., and Frankham, D. B. (1967). *J. Chem Soc., B* p. 1287.
Bethell, D., Newall, A. R., Stevens, G., and Whittaker, D. (1969). *J. Chem. Soc., B* p. 749.
Bethell, D., Stevens, G., and Tickle, P. (1970). *Chem. Commun.* p. 792.
Bethell, D., Newall, A. R., and Whittaker, D. (1971). *J. Chem. Soc., B* p. 23.
Bevan, W. I., Haszeldine, R. N., and Young, J. C. (1961). *Chem. Ind. (London)* p. 789.
Bird, C. W., and Wong, D. Y. (1969). *Chem. Commun.* p. 932.
Blanchard, E. P., and Simmons, H. E. (1964). *J. Amer. Chem. Soc.* **86**, 1337.
Boldt, P., Schulz, L., and Etzemüller, J. (1967). *Chem. Ber.* **100**, 1281.

2. CARBENES AND RELATED SPECIES

Bradley, J. N., Cowell, G. W., and Ledwith, A. (1964). *J. Chem. Soc., London* p. 353.
Bradley, J. N., and Ledwith, A. (1967). *J. Chem. Soc., B* p. 96.
Brandon, R. W., Closs, G. L., Davoust, C. E., Hutchison, C. A., Kohler, B. E., and Silbey, R. (1965). *J. Chem. Phys.* **43**, 2006.
Braun, W., McNesby, J. R., and Bass, A. M. (1967). *J. Chem. Phys.* **46**, 2071.
Carr, R. W., Eder, T. W., and Topor, M. G. (1970). *J. Amer. Chem. Soc.* **89**, 3417.
Castro, C. E., and Kray, W. C. (1966). *J. Amer. Chem. Soc.* **88**, 4447.
Chesick, J. P., and Willcott, M. R. (1963). *J. Phys. Chem.* **67**, 2580.
Ciganek, E. (1965). *J. Amer. Chem.* **87**, 652.
Ciganek, E. (1966). *J. Amer. Chem. Soc.* **88**, 1979.
Clark, H. C., and Willis, C. J. (1960). *J. Amer. Chem. Soc.* **82**, 1888.
Closs, G. L. (1968). *Top. Stereochem.* **3**, 193.
Closs, G. L. (1971). XXIIIrd I.U.P.A.C. Congress, Special Lectures, Vol. 4, p. 19. Butterworths, London.
Closs, G. L., and Closs, L. E. (1962). *Angew. Chem.* **74**, 431; *Angew Chem., Int. Ed Engl.* **1**, 334 (1962).
Closs, G. L., and Closs, L. E. (1969). *J. Amer. Chem. Soc.* **91**, 4549.
Closs, G. L., and Coyle, J. J. (1965). *J. Amer. Chem. Soc.* **87**, 4270.
Closs, G. L., and Moss, R. A. (1964). *J. Amer. Chem. Soc.* **86**, 4042.
Closs, G. L., and Trifunac, A. D. (1969). *J. Amer. Chem. Soc.* **91**, 4554.
Closs, G. L., and Trifunac, A. D. (1970). *J. Amer. Chem. Soc.* **92**, 2186 and 7227.
Cook, A. G., and Fields, E. K. (1962). *J. Org. Chem.* **27**, 3686.
Corey, E. J., Carey, F. A., and Winter, R. A. E. (1965). *J. Amer. Chem. Soc.* **87**, 935.
Cowan, D. O., Couch, M. M., Kopecky, K. R., and Hammond, G. S. (1964). *J. Org. Chem.* **29**, 1922.
Cowell, G. W., and Ledwith, A. (1970). *Quart. Rev. Chem. Soc.* **24**, 119.
Crandall, J. K., and Lin, L.-H. C. (1967). *J. Amer. Chem. Soc.* **89**, 4526.
Crow, W. D., and Wentrup, C. (1968). *Tetrahedron Lett.* p. 6149.
Dehmlow, E. V. (1968). *Chem. Ber.* **101**, 410 and 427.
DeMore, W. B., and Benson, S. W. (1964). *Advan. Photochem.* **2**, 219.
Denney, D. B., and Klemchuk, P. P. (1958). *J. Amer. Chem. Soc.* **80**, 3289.
Doering, W. von E. (1964a). Quoted by Kirmse (1964, p. 24).
Doering, W. von. E. (1964b). Quoted by Kirmse (1964, p. 39).
Doering, W. von E., and Hoffmann, A. K. (1954). *J. Amer. Chem. Soc.* **76**, 6162.
Doering, W. von E., and Knox, L. H. (1961). *J. Amer. Chem. Soc.* **83**, 1989.
Doering, W. von E., and LaFlamme, P. (1956). *J. Amer. Chem. Soc.* **78**, 5447.
Doering, W. von E., and Mole, T. (1960). *Tetrahedron* **10**, 65.
Doering, W. von E., and Prinzbach, H. (1959). *Tetrahedron* **6**, 24.
Doering, W. von E., Buttery, R. G., Laughlin, R. G., and Chaudhuri, N. (1956). *J. Amer. Chem. Soc.* **78**, 3224.
DoMinh, T., Gunning, H. E., and Strausz, O. P. (1967). *J. Amer. Chem. Soc.* **89**, 6785.
Drowart, J., Burns, R. P., DeMaria, G., and Inghram, M. G. (1959). *J. Chem. Phys.* **31** 1131.
Elphimoff-Felkin, I., and Sarda, P. (1969). *Chem. Commun.* p. 1065.
Fischer, E. O., and Dotz, K. H. (1970). *Chem. Ber.* **103**, 1273.
Fisher, I. P., Homer, J. B., and Lossing, F. P. (1965). *J. Amer. Chem. Soc.* **87**, 957.
Foster, J. M., and Boys, S F. (1960). *Rev. Mod. Phys.* **32**, 305.
Franzen, V. (1958). "Reaktionsmechanismen," p. 53. Huthig Verlag, Heidelberg.
Franzen, V. (1962). *Abstr. Pap., 141st Meet., Amer. Chem. Soc.* p. 230.
Franzen, V., and Edens, R. (1969). *Justus Liebigs Ann. Chem.* **729**, 33.

Franzen, V., and Joschek, H. I. (1960). *Justus Liebigs Ann. Chem.* **633**, 7.
Frater, G., and Strausz, O. P. (1970). *J. Amer. Chem. Soc.* **92**, 6654.
Frey, H. M. (1958). *J. Amer. Chem. Soc.* **80**, 5005.
Frey, H. M. (1960). *J. Amer. Chem. Soc.* **82**, 5947.
Friedman, L., Jurewicz, A. T., and Bayless, J. H. (1969). *J. Amer. Chem. Soc.* **91**, 1795.
Friedman, L., Honour, R. J., and Berger, J. G. (1970). *J. Amer. Chem. Soc.* **92**, 4640.
Funakubo, E., Moritani, I., Nagai, T., Nishida, S., and Murahashi, S.-I. (1963). *Tetrahedron Lett.* p. 1069.
Gaspar, P. P., and Hammond, G. S. (1964). *In* W. Kirmse's, "Carbene Chemistry," 1st ed., Chapter 12, p. 235. Academic Press, New York.
Geuther, A. (1862). *Ann. Chem. Pharm.* **123**, 121.
Gibbons, W. A., and Trozzolo, A. M. (1966). *J. Amer. Chem. Soc.* **88**, 172.
Gleiter, R., and Hoffmann, R. (1968). *J. Amer. Chem. Soc.* **90**, 5457.
Goh, S. H., Closs, L. E., and Closs, G. L. (1969). *J. Org. Chem.* **34**, 25.
Goldstein, M. J., and Dolbier, W. J. (1965). *J. Amer. Chem. Soc.* **87**, 2293.
Gutsche, C. D., Bachman, G. L., and Coffey, R. S. (1962). *Tetrahedron* **18**, 617.
Halberstadt, M. L., and McNesby, J. R. (1967). *J. Amer. Chem. Soc.* **89**, 3417.
Hanna, S. B., and Wideman, L. G. (1968). *Chem. Ind. (London)* p. 486.
Hanna, S. B., Iskander, Y., and Riad, Y. (1961). *J. Chem. Soc., London* p. 217.
Hantzsch, A., and Lehmann, M. (1901). *Chem. Ber.* **34**, 2522.
Harris, R. F., and Skell, P. S. (1968). *J. Amer. Chem. Soc.* **90**, 4172.
Harrison, J. F., and Allen, L. C. (1969). *J. Amer. Chem. Soc.* **91**, 807.
Heicklen, J., Knight, V., and Green, S. A. (1965). *J. Chem. Phys.* **42**, 221.
Herzberg, G. (1944). "Atomic Spectra and Atomic Structure," p. 142. Dover, New York.
Herzberg, G. (1961a). *Proc. Roy. Soc., Ser. A* **262**, 291.
Herzberg, G. (1961b). *Can. J. Phys.* **39**, 1511.
Herzberg, G., and Johns, J. W. C. (1966). *Proc. Roy. Soc., Ser. A* **295**, 107.
Herzberg, G., and Shoosmith, J. (1959). *Nature (London)* **183**, 1801.
Herzog, B. M., and Carr, R. W. (1967). *J. Phys. Chem.* **71**, 2688.
Hine, J. (1950). *J. Amer. Chem. Soc.* **72**, 2438.
Hine, J. (1964). "Divalent Carbon." Ronald Press, New York.
Hine, J., and Dowell, A. M. (1954). *J. Amer. Chem. Soc.* **76**, 2688.
Hine, J., and van der Veen, J. M. (1959). *J. Amer. Chem. Soc.* **81**, 6446.
Ho, S.-Y., and Noyes, W. A. (1967). *J. Amer. Chem. Soc.* **89**, 5091.
Hoberg, H. (1962). *Justus Liebigs Ann. Chem.* **656**, 1 and 15.
Hoeg, D. F., Lusk, D. I., and Crumbliss, A. L. (1965). *J. Amer. Chem. Soc.* **87**, 4147.
Hoffmann, R. (1966). *Tetrahedron* **22**, 539.
Hoffmann, R. (1968). *J. Amer. Chem. Soc.* **90**, 1475.
Hoffmann, R., Zeiss, G. D., and Van Dine, G. W. (1968). *J. Amer. Chem. Soc.* **90**, 1485.
Hruby, V. J., and Johnson, A. W. (1962). *J. Amer. Chem. Soc.* **84**, 3586.
Huisgen, R. (1955). *Angew. Chem.* **67**, 439.
Huisgen, R., Binsch, G., and Ghosez, L. (1964). *Chem. Ber.* **97**, 2628.
Hutchison, C. A. (1967). *J. Phys. Chem.* **71**, 203.
Hutchison, C. A., and Kohler, B. E. (1969). *J. Chem. Phys.* **51**, 3327.
Joines, R. C., Turner, A. B., and Jones, W. M. (1969). *J. Amer. Chem. Soc.* **91**, 7754.
Jones, M., and Rettig, K. R. (1965). *J. Amer. Chem. Soc.* **87**, 4013 and 4015.
Jones, M., Sachs, W. H., Kulczycki, A., and Waller, F. J. (1966). *J. Amer. Chem. Soc.* **88**, 3167.
Jones, M., Kulczycki, A., and Hummel, K. F. (1967a). *Tetrahedron Lett.* p. 183.
Jones, M., Ando, W., and Kulczycki, A. (1967b). *Tetrahedron Lett.* p. 1391.

Jones, M., Harrison, A. M., and Rettig, K. R. (1969). *J. Amer. Chem. Soc.* **91**, 7462.
Jones, M., Baron, W. J., and Shen, Y. H. (1970). *J. Amer. Chem. Soc.* **92**, 4745.
Jones, W. M., and Ennis, C. L. (1969). *J. Amer. Chem. Soc.* **91**, 6391.
Jugelt, W., and Schmidt, D. (1969). *Tetrahedron* **25**, 969.
Jurewicz, A. T., and Friedman, L. (1967). *J. Amer. Chem. Soc.* **89**, 149.
Kaplan, F., and Meloy, G. K. (1966). *J. Amer. Chem. Soc.* **88**, 950.
Kaptein, R., den Hollander, J. A., Antheunis, D., and Oosterhoff, L. J. (1970). *Chem. Commun.* p. 1687.
Kerr, J. A. (1966). *Chem. Rev.* **66**, 465.
Kirmse, W. (1963). *Justus Liebigs Ann. Chem.* **666**, 9.
Kirmse, W. (1964). "Carbene Chemistry," 1st ed. Academic Press, New York.
Kirmse, W. (1965). *Angew. Chem.* **77**, 1; *Angew. Chem. Int. Ed. Engl.* **4**, 1 (1965).
Kirmse, W. (1969). "Carbene, Carbenoide, und Carbenanaloge," Chem. Taschenbüche No. 7. Verlag Chemie, Weinheim.
Kirmse, W., and Arold, H. (1968). *Chem. Ber.* **101**, 1008.
Kirmse, W., and Buschhoff, M. (1967). *Chem. Ber.* **100**, 1491.
Kirmse, W., and Kapps, M. (1965). *Angew. Chem.* **77**, 679; *Angew. Chem. Int. Ed Engl.* **4**, 691 (1965).
Kirmse, W., and Kapps, M. (1968). *Chem. Ber.* **101**, 994.
Kirmse, W., and Rinkler, H. A. (1967). *Justus Liebigs Ann. Chem.* **707**, 57.
Kirmse, W., von Stolz, H.-D., and Arold, H. (1968). *Justus Liebigs Ann. Chem.* **711**, 22.
Köbrich, G. (1967). *Angew. Chem.* **79**, 15; *Angew. Chem. Int. Ed. Engl.* **6**, 41 (1967).
Kopecky, K. R., Hammond, G. S., and Leermakers, P. A. (1962). *J. Amer. Chem. Soc.* **84**, 1015.
Krapcho, A. P. (1962). *J. Org. Chem.* **27**, 2375.
Laird, R. K., Andrews, E. B., and Barrow, R. F. (1950). *Trans. Faraday Soc.* **46**, 803.
Landgrebe, J. A., and Kirk, A. G. (1967). *J. Org. Chem.* **32**, 3499.
Landgrebe, J. A., and Thurman, D. E. (1968). *J. Amer. Chem. Soc.* **90**, 6256.
Lappert, M. F., and Poland, J. S. (1970). *Advan. Organometal. Chem.* **9**, 397.
Lemal, D. M., Gosselink, E. P., and McGregor, S. D. (1966). *J. Amer. Chem. Soc.* **88**, 582.
Mahler, W. (1968). *J. Amer. Chem. Soc.* **90**, 523.
Mango, F. D., and Dvoretzky, I. (1966). *J. Amer. Chem. Soc.* **88**, 1654.
Mansoor, A. M., and Stevens, I. D. R. (1966). *Tetrahedron Lett.* p. 1733.
Marchand, A. P., and Brockway, N. M. (1970). *J. Amer. Chem. Soc.* **92**, 5801.
Mathews, C. W. (1966). *J. Chem. Phys.* **45**, 1068.
Merer, A. J., and Travis, D. N. (1966a). *Can. J. Phys.* **44**, 525.
Merer, A. J., and Travis, D. N. (1966b). *Can. J. Phys.* **44**, 1541.
Miller, W. T., and Whalen, D. M. (1964). *J. Amer. Chem. Soc.* **86**, 2089.
Milligan, D. E., and Jacox, M. E. (1967). *J. Chem. Phys.* **47**, 703.
Milligan, D. E., Mann, D. E., Jacox, M. E., and Mitsch, R. A. (1964). *J. Chem. Phys.* **41**, 1199.
Mitsch, R. A., and Rodgers, A. S. (1969). *Int. J. Chem. Kinet.* **1**, 439.
More O'Ferrall, R. A. (1967). *Advan. Phys. Org. Chem.* **5**, 331.
Moritani, I., Murahashi, S.-I., Nishino, M., Yamamoto, Y., Itoh, K., and Mataga, N. (1967). *J. Amer. Chem. Soc.* **89**, 1259.
Moritani, I., Murahashi, S.-I., Ashitaka, H., Kimura, K., and Tsubomura, H. (1968a). *J. Amer. Chem. Soc.* **90**, 5918.
Moritani, I., Yamamoto, Y., and Murahashi, S.-I.(1968b). *Tetrahedron Lett.* pp. 5755 and 5697.

Moritani, I., Yamamoto, Y., and Konishi, H. (1969). *Chem. Commun.* p. 1457.
Moser, W. R. (1969). *J. Amer. Chem. Soc.* **91**, 1135 and 1141.
Moss, R. A. (1966). *J. Org. Chem.* **31**, 3296.
Moss, R. A. (1967). *Tetrahedron Lett.* p. 4905.
Moss, R. A. (1969). *Chem. Eng. News* **47**, No. 25, 60; No. 27, 50.
Moss, R. A. (1970). *In* "Selective Organic Transformations" (B.S. Thyagarajan, ed.), Vol. I, p. 35. Wiley (Interscience), New York.
Moss, R. A., and Gerstl, R. (1966). *Tetrahedron* **22**, 2637.
Moss, R. A., and Gerstl, R. (1967a). *J. Org. Chem.* **32**, 2268.
Moss, R. A., and Gerstl, R. (1967b). *Tetrahedron* **23**, 2549.
Moss, R. A., and Mamantov, A. (1968). *Tetrahedron Lett.* p. 3425.
Moss, R. A., and Mamantov, A. (1970). *J. Amer. Chem. Soc.* **92**, 6951.
Moss, R. A., and Przybyla, J. R. (1969). *Tetrahedron* **25**, 647.
Moss, R. A., Whittle, J. R., and Freidenreich, P. (1969). *J. Org. Chem.* **34**, 2220.
Müller, E., Fricke, H., and Kessler, H. (1963). *Tetrahedron Lett.* 1501.
Müller, E., Kessler, H., and Zech, B. (1966). *Fortschr. Chem. Forsch.* **7**, 128.
Myers, J. A., Joines, R. C., and Jones, W. M. (1970). *J. Amer. Chem. Soc.* **92**, 4740.
Nair, R. M. G., Meyer, E., and Griffin, G. W. (1968). *Angew. Chem., Int. Ed. Engl.* **7**, 462.
Nef, J. U. (1897). *Justus Liebigs Ann. Chem.* **298**, 202.
Nicholas, J. E., Mackay, C., and Wolfgang, R. (1965). *J. Amer. Chem. Soc.* **87**, 3008.
Nishimura, J., Kawabata, N., and Furukawa, J. (1969). *Tetrahedron* **25**, 2647.
Nishimura, J., Furukawa, J., Kawabata, N., and Fujita, T. (1970). *Tetrahedron* **26**, 2229.
Nishimura, J., Furukawa, J., Kawabata, N., and Kitayama, M. (1971). *Tetrahedron* **27**, 1799.
Nozaki, H., Nakano, M., and Kondo, K. (1966). *Tetrahedron* **22**, 477.
Nozaki, H., Fujita, S., and Noyori, R. (1968a). *Tetrahedron* **24**, 2193.
Nozaki, H., Takaya, H., Moriuti, S., and Noyori, R. (1968b). *Tetrahedron* **24**, 3655.
Okano, M., Ito, Y., and Oda, R. (1964). *Bull. Inst. Chem. Res., Kyoto Univ.* **42**, 217.
Olofson, R. A., Walinsky, S. W., Marino, J. P., and Jernow, J. L. (1968). *J. Amer. Chem. Soc.* **90**, 6554.
Overberger, C. G., and Anselme, J.-P. (1963). *Tetrahedron Lett.* p. 1405.
Pearson, R. G., and Songstad, J. (1967). *J. Amer. Chem. Soc.* **89**, 1827.
Poulter, C. D., Friedrich, E. C., and Winstein, S. (1969). *J. Amer. Chem. Soc.* **91**, 6892.
Powell, F. X., and Lide, D. R. (1966). *J. Chem. Phys.* **45**, 1067.
Rando, R. R. (1970). *J. Amer. Chem. Soc.* **92**, 6706.
Reimlinger, H. (1964). *Chem. Ber.* **97**, 339 and 3503.
Richardson, D. B., Simmons, M. C., and Dvoretzky, I. (1960). *J. Amer. Chem. Soc.* **82**, 5001.
Richardson, D. B., Simmons, M. C., and Dvoretzky, I. (1961). *J. Amer. Chem. Soc.* **83**, 1934.
Richardson, D. B., Durett, L. R., Martin, J. M., Putnam, W. E., Slaymaker, S. C., and Dvoretzky, I. (1965). *J. Amer. Chem. Soc.* **87**, 2763.
Roth, H. D. (1971). *J. Amer. Chem. Soc.* **93**, 1527 and 4935.
Rothberg, I., and Thornton, E. R. (1964). *J. Amer. Chem. Soc.* **86**, 3296, 3302.
Russell, G. A., and Danen, W. C. (1968). *J. Amer. Chem. Soc.* **90**, 347.
Russell, G. A., and Hendry, D. G. (1963). *J. Org. Chem.* **28**, 1933.
Ryang, M. (1970). *Organometal. Chem. Rev., Sect. A* **5**, 67.
Sadler, I. H. (1969). *J. Chem. Soc., B* p. 1024.
Sargeant, P. B., and Shechter, H. (1964). *Tetrahedron Lett.* p. 3957.

2. CARBENES AND RELATED SPECIES

Sauers, R. R., and Kiesel, R. J. (1967). *J. Amer. Chem. Soc.* **89**, 4695.
Schlosser, M., and Heinz, G. (1970). *Chem. Ber.* **103**, 3543.
Schmidt, U., and Kabitzke, K. H. (1964). *Angew. Chem., Int. Ed. Engl.* **3**, 641.
Schmitz, E. (1967). "Dreiringe mit Zwei Heteroatomen," pp. 114–167. Springer-Verlag, Berlin and New York.
Seyferth, D. (1955). *Chem. Rev.* **55**, 1155.
Seyferth, D., and Armbrecht, F. M. (1969). *J. Amer. Chem. Soc.* **91**, 2616.
Seyferth, D., Mui, J. Y.-P., Gordon, M. E., and Burlitch, J. M. (1965a). *J. Amer. Chem. Soc.* **87**, 681.
Seyferth, D., Burlitch, J. M., Minasz, R. J., Mui, J. Y.-P., Simmons, H. D., Treiber, A. J.-H., and Dowd, S. R. (1965b). *J. Amer. Chem. Soc.* **87**, 4259.
Seyferth, D., Gordon, M. E., Mui, J. Y.-P., and Burlitch, J. M. (1967a). *J. Amer. Chem. Soc.* **89**, 959.
Seyferth, D., Mui, J. Y.-P., and Burlitch, J. M. (1967b). *J. Amer. Chem. Soc.* **89**, 4953.
Seyferth, D., Mui, J. Y.-P., and Damrauer, R. (1968). *J. Amer. Chem. Soc.* **90**, 6182.
Seyferth, D., Hopper, S. P., and Darragh, K. V. (1969a). *J. Amer. Chem. Soc.* **91**, 6536.
Seyferth, D., Turkel, R. M., Eisert, M. A., and Todd, L. J. (1969b). *J. Amer. Chem. Soc.* **91**, 5027.
Seyferth, D., Burlitch, J. M., Yamamoto, K., Washburne, S. S., and Attridge, C. J. (1970a). *J. Org. Chem.* **35**, 1989.
Seyferth, D., Mai, V. A., and Gordon, M. E. (1970b). *J. Org. Chem.* **35**, 1993.
Shapiro, J. S., and Lossing, F. P. (1968). *J. Phys. Chem.* **72**, 1552.
Shapiro, R. H., Duncan, J. H., and Clopton, J. C. (1967). *J. Amer. Chem. Soc.* **89**, 471 and 1442.
Simmons, H. E., and Smith, R. D. (1959). *J. Amer. Chem. Soc.* **81**, 4256.
Simmons, H. E., Blanchard, E. P., and Smith, R. D. (1964). *J. Amer. Chem. Soc.* **86**, 1347.
Simon, J. W., and Rabinovitch, B. S. (1964). *J. Phys. Chem.* **68**, 1322.
Simons, J. P. (1965). *J. Chem. Soc., London* p. 5406.
Singer, L. S., and Lewis, I. C. (1968). *J. Amer. Chem. Soc.* **90**, 4212.
Skell, P. S., and Cholod, M. S. (1969a). *J. Amer. Chem. Soc.* **91**, 6035.
Skell, P. S., and Cholod, M. S. (1969b). *J. Amer. Chem. Soc.* **91**, 7131.
Skell, P. S., and Engel, R. R. (1966a). *J. Amer. Chem. Soc.* **88**, 3749.
Skell, P. S., and Engel, R. R. (1966b). *J. Amer. Chem. Soc.* **88**, 4883.
Skell, P. S., and Engel, R. R. (1967). *J. Amer. Chem. Soc.* **89**, 2912.
Skell, P. S., and Etter, R. M. (1958). *Chem. Ind. (London)*, p. 624.
Skell, P. S., and Garner, A. Y. (1956). *J. Amer. Chem. Soc.* **78**, 5430.
Skell, P. S., and Harris, R. F. (1969). *J. Amer. Chem. Soc.* **91**, 699.
Skell, P. S., and Plonka, J. H. (1970a). *J. Amer. Chem. Soc.* **92**, 836.
Skell, P. S., and Plonka, J. H. (1970b). *J. Amer. Chem. Soc.* **92**, 5620.
Skell, P. S., and Starer, I. (1959). *J. Amer. Chem. Soc.* **81**, 4117.
Skell, P. S., and Woodworth, R. C. (1956). *J. Amer. Chem. Soc.* **78**, 4496.
Skell, P. S., and Woodworth, R. C. (1959). *J. Amer. Chem. Soc.* **81**, 3383.
Skell, P. S., Wescott, L. D., Golstein, J. P., and Engel, R. R. (1965). *J. Amer. Chem. Soc.* **87**, 2829.
Skell, P. S., Plonka, J. H., and Wood, L. S. (1970). *Chem. Commun.* p. 710.
Skell, P. S., Villaume, J. E., Plonka, J. H., and Fagone, F. A. (1971). *J. Amer. Chem. Soc.* **93**, 2699.
Smith, W. H., and Leroi, G. E. (1969). *Spectrochim. Acta, Part A* **24**, 1917.
Sommer, L. H., Ulland, L. A., and Ritter, A. (1968). *J. Amer. Chem. Soc.* **90**, 4486.

Sprung, J. L., Winstein, S., and Libby, W. F. (1965). *J. Amer. Chem. Soc.* **87**, 1812.
Staudinger, H., and Kupfer, O. (1912). *Chem. Ber.* **45**, 501.
Strausz, O. P., DoMinh, T., and Gunning, H. E. (1968a). *J. Amer. Chem. Soc.* **90**, 1660.
Strausz, O. P., DoMinh, T., and Font, J. (1968b). *J. Amer. Chem. Soc.* **90**, 1930.
Swain, C. G., and Thornton, E. R. (1961). *J. Amer. Chem. Soc.* **83**, 4033.
Swenson, J. S., and Renaud, D. J. (1965). *J. Amer. Chem. Soc.* **87**, 1394.
Swenton, J. S., and Krubsack, A. J. (1969). *J. Amer. Chem. Soc.* **91**, 786.
Tang, Y.-N., and Rowland, F. S. (1967). *J. Amer. Chem. Soc.* **89**, 6420.
Tang, Y.-N., and Rowland, F. S. (1968). *J. Amer. Chem. Soc.* **90**, 574.
ter Borg, A. P., Razenberg, E., and Kloosterziel, H. (1966). *Rec. Trav. Chim. Pays-Bas* **85**, 774.
Trost, B. M. (1966). *J. Amer. Chem. Soc.* **88**, 1587.
Trozzolo, A. M., Murray, R. W., and Wasserman, E. (1962). *J. Amer. Chem. Soc.* **84**, 4990.
Trozzolo, A. M., Wasserman, E., and Yager, W. A. (1964). *J. Chim. Phys.* **61**, 1663.
Trozzolo, A. M., Wasserman, E., and Yager, W. A. (1965). *J. Amer. Chem. Soc.* **87**, 129.
Urry, W. H., and Eiszner, J. R. (1952). *J. Amer. Chem. Soc.* **74**, 5822.
Walsh, A. D. (1953). *J. Chem. Soc., London* p. 2660.
Ward, A. M. (1927). *J. Chem. Soc., London* p. 2285.
Wasserman, E., Barash, L., Trozzolo, A. M., Murray, R. W., and Yager, W. A. (1964a). *J. Amer. Chem. Soc.* **86**, 2304.
Wasserman, E., Trozzolo, A. M., Yager, W. A., and Murray, R. W. (1964b). *J. Chem. Phys.* **40**, 2408.
Wasserman, E., Barash, L., and Yager, W. A. (1965a). *J. Amer. Chem. Soc.* **87**, 2075.
Wasserman, E., Barash, L., and Yager, W. A. (1965b). *J. Amer. Chem. Soc.* **87**, 4974.
Wasserman, E., Kuck, V. J., Hutton, R. S., and Yager, W. A. (1970). *J. Amer. Chem. Soc.* **92**, 7491.
Werner, H., and Richards, J. H. (1968). *J. Amer. Chem. Soc.* **90**, 4976.
Weyerstahl, P., Klamann, D., Finger, C., Nerdel, F., and Buddrus, J. (1967). *Chem. Ber.* **100**, 1858.
Willis, C., and Bayes, K. D. (1966). *J. Amer. Chem. Soc.* **88**, 3203.
Wittig, G., and Schlosser, M. (1962). *Tetrahedron* **18**, 1023.
Wittig, G., and Wingler, F. (1962). *Justus Liebigs Ann. Chem.* **656**, 18.
Wolf, A. P. (1964). *Advan. Phys. Org. Chem.* **2**, 201.
Woodward, R. B., and Hoffmann, R. (1970). "The Conservation of Orbital Symmetry." Verlag Chemie, Weinheim.
Yamamoto, Y., and Moritani, I. (1970). *Tetrahedron* **26**, 1235.
Yates, P. (1952). *J. Amer. Chem. Soc.* **74**, 5376.
Yates, P., and Kilmurry, L. (1966). *J. Amer. Chem. Soc.* **88**, 1563.
Yilmaz, H. (1955). *Phys. Rev.* **100**, 1148.
Zmbov, K. F., Uy, O. M., and Margrave, J. L. (1968). *J. Amer. Chem. Soc.* **90**, 5090.
Zollinger, H. (1961). "Azo and Diazo Chemistry." Wiley (Interscience), New York.

3

NITRENES

R. A. Abramovitch

I. Introduction: Nomenclature and Historical Background	. .	127
A. Nomenclature	. . . .	128
B. Historical Background.	. . . .	128
II. Intermediacy of Nitrenes in Reactions	. . . . .	129
III. Electronic Structure and Spectra	. . . . . .	131
A. Theoretical Calculations	. . . . . .	131
B. Infrared and Ultraviolet Spectra	. . . . .	132
C. Electron-Spin Resonance	. . . . . .	132
IV. Methods of Generation of Nitrenes	. . . . .	135
A. From Azides	. . . .	135
B. Deoxygenation	. . . .	152
C. α Eliminations	. . . .	153
D. Oxidative Routes	. . . .	155
E. Other Sources of Nitrenes	. . . . . .	156
V. Reactions	. . . .	157
A. Hydrogen Abstraction .	. . . . . .	157
B. Insertion into Aliphatic C–H Bonds	. . . .	160
C. Rearrangement	. . . .	163
D. Addition to Olefins	. . . .	167
E. Dimerization	. . . .	169
F. Aromatic Substitution .	. . . . . .	171
G. Trapping by Nucleophiles	. . . . . .	179
VI. Nitrenium Ions	. . . .	181
VII. Applications	. . . .	184
References	. . . .	188

I. Introduction: Nomenclature and Historical Background

Nitrenes (R—N) are electron-deficient monovalent nitrogen species, the parent of which is NH. The nitrogen atom has a sextet of electrons in its outer shell, and the species may exist either in the triplet diradical state (I) or the electrophilic singlet state (II).

$$R-\overset{\cdot\uparrow}{\underset{\cdot\cdot}{N}}\cdot\uparrow \qquad R-\underset{\cdot\cdot}{N}{:}\uparrow\downarrow$$
$$\text{(I)} \qquad\qquad \text{(II)}$$

A. Nomenclature

Many names have been proposed for these intermediates, including imene (Horner and Christmann, 1963a; Lüttringhaus et al., 1959); imine radicals (Rice and Luckenbach, 1960; Heacock and Edmison, 1960); azenes (Bunyan and Cadogan, 1963; Smith and Hall, 1962; Smolinsky, 1961a); azylenes (Smith et al., 1963); phenylstickstoff (for PhN) (Appl and Huisgen, 1959); azacarbenes (Knunyants and Bykhovskaya, 1960); and imin (Kirmse, 1959). The name nitrene was proposed by analogy with the isoelectronic carbene intermediates but was objected to because the term had previously been used to designate azomethine ylides (Staudinger and Miescher, 1919; Hassall and Lippman, 1953), and the ending "ene" is reserved for olefins and aromatic hydrocarbons (IUPAC nomenclature rules). On the basis of this last objection, the editors of *Chemical Reviews* and *Chemical Abstracts Index* insisted on the name "Imidogen" for NH and an "imido intermediate" or "an imidogen" for R—N (Abramovitch and Davis, 1964). The term nitrene has, however, found almost universal acceptance (including its use in the biweekly indexes of *Chemical Abstracts*!) and is now in common usage. Its use will also be adopted here. The protonated or alkylated nitrene, R—N$^+$—R', is called a nitrenium ion.

B. Historical Background

The parent compound NH is of great interest to the astrophysicists since it has been observed as a component of comet heads and tails and of the sun, and its spectral bands have been observed in the night sky spectrum. It has been suggested that some of the colors on Jupiter may be due to condensed reactive species such as $(NH)_n$ (from $N + MH \rightarrow NH + M$).

The proposal that nitrenes may be formed as reactive intermediates was first made by Tiemann (1891) in connection with the mechanism of the Lossen rearrangement. Stieglitz (1896) postulated their formation to account for the mechanisms of the Hofmann, Curtius, Lossen, and Beckmann rearrangements as well as for a number of rearrangements involving compounds of the type $(C_6H_5)_3CNHX$ (Morgan, 1916; Stagner, 1916; Vosburgh, 1916). Curtius (1913) compared the intermediates formed in the decomposition of sulfonyl azides in aromatic hydrocarbons with those from ethyl diazoacetate. The rejection of the intermediacy of a monovalent nitrogen species in the Beckmann rearrangement cast a simultaneous shadow on its postulated participation in other reactions. The subject lay relatively dormant until interest in the preparation and properties of nitrenes was stimulated by the renaissance in carbene chemistry. Since the first reviews on the subject (Kirmse, 1959; Horner and Christmann, 1963a; Abramovitch and Davis, 1964) it has blossomed enormously, and a number of reviews (Lwowski, 1967; Abramo-

3. NITRENES

vitch, 1970; Abramovitch and Sutherland, 1970; Belloli, 1971) and two books on these intermediates (Gilchrist and Rees, 1969; Lwowski, 1970) have come to light. Indeed, there has been a tendency to postulate the intermediacy of nitrenes in many reactions without much evidence for their formation. By analogy with carbene chemistry, one might better refer to a number of reactions as involving nitrenoïd species in which the nitrogen is not free but is bound in some way to a metal, for instance, or a leaving group from which it is not completely detached when reaction with the substrate occurs.

Since the behavior of the nitrene depends markedly on the nature of the group attached to nitrogen, it has been customary to discuss and subdivide the chemistry of these species in terms of the various types of substituents R in R—N. This practice will not be followed here, instead the various types of intermediates will be compared and contrasted.

II. Intermediacy of Nitrenes in Reactions

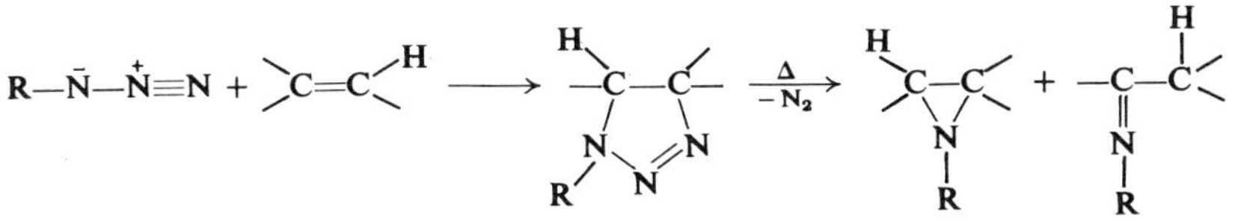

Since nitrenes are not isolable under normal reaction conditions, the question often arises as to whether or not they are intermediates in a given reaction. For instance, though it is known that photolysis of an azide in a frozen matrix at low temperatures leads to loss of nitrogen and formation of the nitrene which drops down to the triplet ground state, this does not mean that nitrenes are necessarily involved in the photodecomposition of the same azide at room temperature (see alkyl azides, for instance) or that nitrenes are formed as discrete intermediates in the thermal decomposition of an azide in the presence of a suitable substrate. The final product could be formed in a two-step process involving the substrate and azide to give an intermediate product which subsequently loses nitrogen on heating. Alternatively, the substrate may participate in the rate-determining elimination of nitrogen to give product, and, thus, a nitrene is not formed as a discrete intermediate (which would require the unimolecular loss of nitrogen followed by attack of the nitrene on the substrate). A well-known example of these alternate pathways is the thermal reaction of aryl azides with suitable olefins. With strained olefinic double bonds, for instance, Δ^2-triazolines are formed under mild conditions which on heating lose nitrogen to give aziridines or imines, the products that would have also been expected from the reaction of the nitrenes with the olefins (L'abbé, 1969). In other instances, though no triazoline may actually form, the olefinic double bond participates in the rate-determining elimination of nitrogen. For example, thermolysis of aryl azides in styrene or indene indicated participation of the double bond in the nitrogen elimination, since the rates were faster than in nitrobenzene or decalin and energies of activation were lower by 15–20 kcal mole^{-1} (Walker and Waters, 1962).

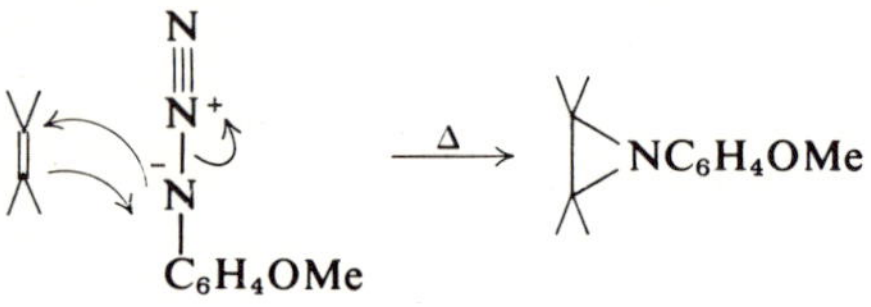

It is obvious, therefore, that in such cases kinetic studies have to be carried out to determine whether or not the substrate participates in the rate-determining step. If it does, then a free nitrene is clearly not an intermediate (for a more detailed discussion of this point, see Lwowski, 1970).

Evidence that a nitrene is an intermediate in a reaction has been drawn from a variety of observations. The most common (but mechanistically not very satisfactory perhaps) is based on the known, or anticipated, chemical properties of the species and the nature of the product formed. These, in turn, depend on such factors as the nature of R in R—N, the spin state, the phase in which the reaction is carried out, the temperature, and the presence of metals or metal compounds. Thus, the intervention of a nitrene intermediate is rendered highly probable by generating the species by two or more independent routes and by showing that the same products or product mixtures result. For example, the formation of a nitrene intermediate has been established kinetically (Smith and Hall, 1962) and spectroscopically (Reiser *et al.*, 1966a) in the thermolysis and photolysis of 2-azidobiphenyl (**1**) to give carbazole (**2**). Sauer and Engels (1969) studied the cyclization of **3**

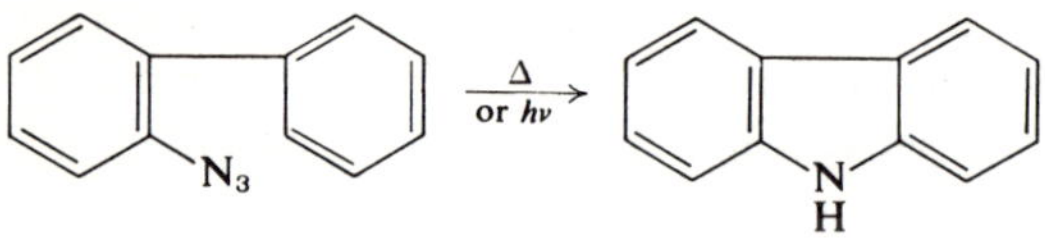

(**1**)   (**2**)

(X = NO_2, NO, N_3, and N═C═O) which can give two isomeric carbazoles, **4** and **5**. They thus found that the ratio of **4**:**5** was the same (52–56:48–44) whether the reactive species was produced from **3** (X = NO_2) and triethyl phosphite in various solvents at various temperatures, from **3** (X = NO) and triethyl phosphite, from **3** (X = N_3) by thermolysis or photolysis, or from **3** (X = N═C═O) on photolysis, implicating the same intermediate in all

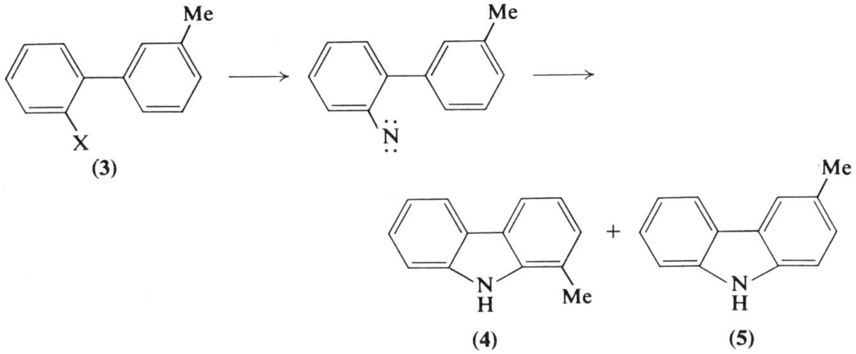

3. NITRENES

cases, the free singlet nitrene. On the other hand, it had been suggested (Abramovitch and Davis, 1964, 1968) that when nitrobenzene derivatives are heated with a transition metal oxalate, a nitrene intermediate bound to the surface of the metal derivative (a nitrenoïd as opposed to a free nitrene) was probably generated (this could be a nitrene–metal carbonyl complex). In agreement with this, Sauer and Engels (1969) found that when **3** (X = NO$_2$) was heated with ferrous oxalate the ratio of **4:5** was 35–30:65–70 and depended on the temperature at which the reaction was carried out.

Another example of this general approach relates to carbethoxynitrene (N—CO$_2$Et) which can be generated by photolysis (Lwowski and Mattingly, 1962) or thermolysis (Cotter and Beach, 1964; Sloan et al., 1964) of ethyl azidoformate (N$_3$CO$_2$Et) or by α elimination of p-nitrobenzenesulfonate from the anion of N-(p-nitrobenzenesulfonyloxy)urethane (p-NO$_2$C$_6$H$_4$-SO$_2$ONHCO$_2$Et) (Lwowski et al., 1963). The species produced in every case shows almost the same selectivity for C–H insertion (Lwowski, 1970).

III. Electronic Structure and Spectra

A. Theoretical Calculations

The electronic structure of the parent nitrene NH has been calculated (for a review, see Berry, 1970). The ground state is a $^3\Sigma^-$ triplet, and the lowest excited state is the $^1\Delta$ state, 1.76 eV above the ground state. The configuration of both states is $1\sigma^2 2\sigma^2 3\sigma^2 1\pi^2$. In the ground state, one electron occupies each of the two degenerate π orbitals, whereas in the singlet $^1\Delta$ state, two electrons are present in the same π orbital. Only a qualitative treatment has been accorded aryl nitrenes (Reiser et al., 1966a). It was concluded that the splitting of the $3b_2$ and $1b_2$ orbitals derived from the $2p\pi$ orbitals of nitrogen must be smaller than the exchange interaction between these orbitals in order that the ground state (or a very low-lying state) of Ph—N be a triplet, as it is experimentally observed to be. Extensive calculations have been carried out on cyanonitrene (NCN) (Anastassiou et al., 1970). It was concluded that the linear structure (III) was preferred over the bent (IV). While extended

$$\cdot N{=}C{=}N\cdot \qquad \underset{\cdot}{N}\diagdown\overset{C}{\diagup}\underset{\cdot}{N}$$

(III)   IV

Hückel calculations proved incapable of predicting the C–N bond length of minimum energy (1.23 Å experimentally), it does predict correctly the ground-state configuration as a triplet $(2\sigma_1 2\sigma_2 2\sigma_3 4\pi_1 2\sigma_4 2\pi_2)$.

B. Infrared and Ultraviolet Spectra

The UV and IR spectra of NH and ND have been measured and analyzed both for the free species and the intermediates trapped in a solid matrix at low temperatures (reviewed by Abramovitch and Davis, 1964; Berry, 1970) as have the IR spectra of NCl, NBr, and NF (Comeford and Mann, 1965; Diesen, 1964). The UV spectra of a number of aryl nitrenes have been described (Reiser et al., 1966a), the triplet nitrene giving a spectrum resembling those of aromatic radicals. Thus, phenylnitrene in an organic glass at 77°K exhibits two bands at 314 and 402 nm which correspond to the bands of the benzyl radical at 318 and 463 nm. Flash photolysis of 1-azidoanthracene in ethanol at room temperature gave a short-lived intermediate whose UV spectrum (342 nm) is the same as that of the nitrene produced in the matrix photolysis of the same azide at 77°K. The lifetime of the absorption at 342 nm indicated that the half-life of this intermediate is between 3 and 10 msec under these conditions (Reiser et al., 1966b).

The work on the spectra of NCN has been reviewed (Anastassiou et al., 1970). Flash photolysis of N_3CN produces a high concentration of the nitrene as indicated by the characteristic absorption in the 3290 Å region owing to this species. Photolysis of the azide in a solid matrix at 14° or 20°K gave further information. In addition to the 3290 Å band, a progression of bands appears between 3000 and 2400 Å. In the IR, the mixture from filtered photolysis (to avoid rearrangement of the nitrene) displayed a medium band at 423 cm^{-1} as well as intense absorption at 1475 cm^{-1}, both attributed to NCN. From such studies, stretching force constants and a number of thermodynamic properties of the nitrene could be calculated.

C. Electron-Spin Resonance

Spin conservation dictates that the nitrene intermediate produced by a thermal process has to be formed initially in the singlet state. The photodecomposition of an azide must also occur with overall conservation of spin

$$(R-\bar{\ddot{N}}-\overset{+}{\ddot{N}}_2) \xrightarrow{\Delta} {}^1(R-\ddot{\ddot{N}}) + {}^1(N_2)$$

in the absence of external perturbation so that, in principle, four possibilities can be envisioned [Eqs. (1)–(4)]. Equations (2) and (4) are unlikely since the

$$R-N_3 \xrightarrow{h\nu} {}^1(R-N_3)^* \xrightarrow{ISC} {}^3(R-N_3)^* \begin{cases} \to {}^1(R-\ddot{\ddot{N}}) + {}^1(N_2) & (1) \\ \to {}^3(R-\dot{\ddot{N}}\cdot) + {}^3(N_2) & (2) \\ \to {}^3(R-\dot{\ddot{N}}\cdot) + {}^1(N_2) & (3) \\ \to {}^1(R-\ddot{\ddot{N}}) + {}^3(N_2) & (4) \end{cases}$$

light employed in the photolyses does not have the necessary energy to generate molecular nitrogen in an excited triplet state. On the other hand, equilibrium is theoretically possible between the singlet and triplet nitrene species formed according to Eqs. (1) and (3), respectively, and, given time, the species initially formed could drop to the ground state if it is not generated in that state and if it does not react before undergoing spin inversion.

$$R-\ddot{\ddot{N}} \rightleftharpoons R-\dot{\ddot{N}}\cdot$$
$$\text{(6a)} \qquad \text{(6b)}$$

The theoretical calculations mentioned above predict a triplet ground state **6b** for nitrenes, and this has been confirmed in many cases by ESR measurements, whereby the triplet diradical species can be observed.

The ESR spectra of nitrenes generated at 77°K in a fluorolube matrix by UV irradiation of the corresponding azide were detected and assigned to the triplet ground state, or to a state just above the ground state, for the following species: C_6H_5N, $o\text{-}CF_3C_6H_4N$, $C_6H_5SO_2N$, $p\text{-}MeC_6H_4SO_2N$, and NCN. No resonance was observed for $C_6H_5CH=CHN$, EtOCON, and PhOCON probably because they react further too quickly to permit a sufficient stationary concentration to be detected (Smolinsky et al., 1962; Wasserman et al., 1965). The triplet ground state of carbethoxynitrene can be observed by photolysis of ethyl azidoformate in a rigid matrix at liquid helium temperatures (Wasserman, quoted in Lwowski, 1970).

Phenyl- and o-trifluoromethylphenylnitrene both exhibited $\Delta m = 2$ and $\Delta m = 1$ transitions. The zero-field parameters D and E for phenylnitrene (whose magnitude give a measure of interaction between the unpaired spins—the smaller the values of D and E, the smaller the interaction between the unpaired electrons, i.e., one of them is delocalized away from the nitrogen atom) were $D = 0.99 \text{ cm}^{-1}$ and $E < 0.002 \text{ cm}^{-1}$, suggesting significant delocalization of one unpaired electron into the aromatic system.

Attempts to observe triplet alkyl nitrenes by direct irradiation of alkyl azides (e.g., n-PrN$_3$, 2-OctN$_3$, and $C_6H_{11}N_3$) at 77°K were unsuccessful. At a lower temperature (4°K), however, weak signals were observed by the photolysis of alkyl azides in hexafluorobenzene or perfluorodimethylbutane (Wasserman et al., 1964; Barash et al., 1967). As will be discussed later, this would suggest that at 77°K a concerted photodecomposition is occurring in which no nitrene is formed (migration of alkyl group and expulsion of nitrogen are synchronous), but at 4°K there may be insufficient molecular bending motion to permit migration. The singlet azide could then undergo intersystem crossing to the triplet azide and then the triplet nitrene would result. On the other hand, triplet-sensitized (benzophenone) photolysis of mono- and diazides did give the triplet nitrenes at 77°K which were stable for days at this temperature (Barash et al., 1967). The D values are appreciably larger for the alkyl nitrenes than for Ar—N, since delocalization of one of the unpaired electrons is no longer as important: n-PrN, $D = 1.607 \text{ cm}^{-1}$, $E = 0.0034 \text{ cm}^{-1}$; tert-BuN, $D = 1.625$, $E < 0.002$. Since E is a measure of the difference in the unpaired electron distribution along the x and y axes, the small values are compatible with the approximately cylindrical distribution about the C—N bond expected for such nitrenes. The fact that $E \neq 0$ for the symmetrical tert-Bu—N is probably due to distortions resulting from the matrix. The quintet state of dinitrene (7) has been observed (Wasserman et al., 1967), whereas the triplet spectrum of 8 has been recorded (Trozzolo et al., 1963).

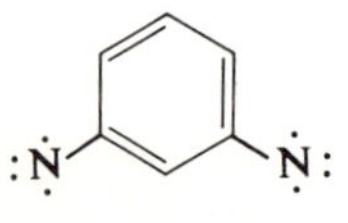

(7)    (8)

Electron-spin-resonance studies of the photolysis of the gem-dialkyl azide (9) at 77°K showed that its photosensitized decomposition gave a triplet nitrene (10) (Barash et al., 1967); prolonged irradiation gave the carbene (11). The direct photolysis of 9 in benzene at ambient temperature (Moriarity and

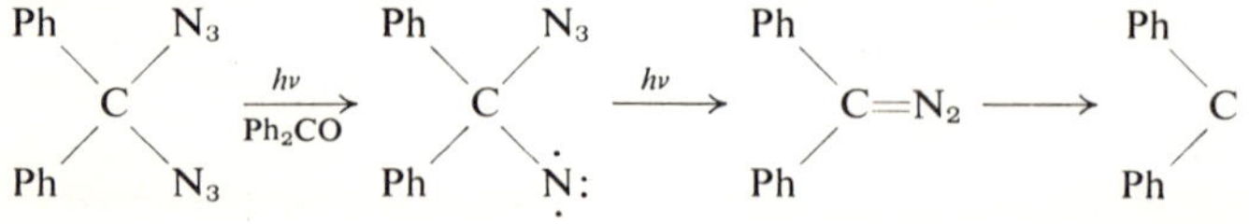

(9)   (10)   (11)

3. NITRENES

Kliegman, 1967) gave **12**, **13**, and **14** but no product derived from diphenylcarbene (**11**). This illustrates the danger of comparing photosensitized with direct photolyses, on the one hand, and low temperature solid matrix studies with ambient temperature solution studies, on the other.

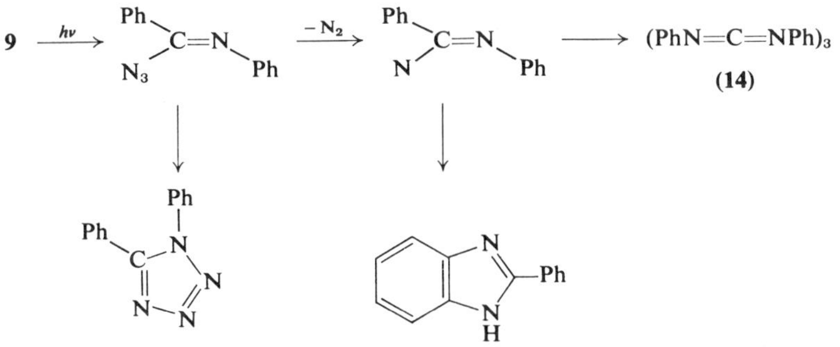

$$\mathbf{9} \xrightarrow{h\nu} \underset{N_3}{\overset{Ph}{>}}C=N-Ph \xrightarrow{-N_2} \underset{N}{\overset{Ph}{>}}C=N-Ph \longrightarrow (PhN=C=NPh)_3 \quad (\mathbf{14})$$

(**12**) — Ph–C(=N–N=N–N)–Ph with N–Ph

(**13**) — benzimidazole with Ph substituent

Electron-spin-resonance measurements have also been reported on the photoproducts of aryl- and alkylsulfonyl azides (for a review, see Abramovitch, 1970). The magnitudes of the zero-field parameters indicate that the unpaired electrons are essentially localized on nitrogen, e.g., for $MeSO_2N$, $D = 1.569 \text{ cm}^{-1}$, and $E = 0 \text{ cm}^{-1}$.

Although the ESR evidence indicates a triplet ground state for these nitrenes, it tells us nothing about the electronic state of the species at the moment at which it is undergoing reaction. Evidence bearing on the latter comes mainly from a consideration of the nature and relative rates of the reactions undergone by nitrenes and by the nature (and isomer ratios) of the products so formed. Much of the effort in nitrene chemistry research (as in carbene chemistry) has gone into trying to determine whether the nitrene reacts as the singlet (**6a**) or triplet (**6b**) or as a mixture of both.

IV. Methods of Generation of Nitrenes

A. FROM AZIDES

1. *Thermolysis*

Azides are the most common source of nitrene intermediates. If there is no group adjacent to the azide function which can participate in the nitrogen elimination or if the group undergoing attack does not assist in the decomposition of the azide, then thermolysis of the azide usually leads to singlet nitrene.

$$RN_3 \xrightarrow{\Delta} R\ddot{N} + N_2$$

That this is the case with aryl azides in solvents that do not interact with the starting azide has been shown by studies of the kinetics of the decomposition of suitable molecules. For example, the rates of unimolecular decomposition of phenyl azides in both nitrobenzene and tetralin solution are the same, indicating that the rate-determining step does not involve the solvent and that phenylnitrene was formed (Russell, 1955). Other examples include the thermolysis of hydrazoic acid at 1000°C and a pressure of 0.05–0.2 mm to give the parent NH (Rice and Freamo, 1951; Rice and Grelecki, 1957; Rice and Ingalls, 1959),

$$HN_3 \xrightarrow[0.05-0.2\,mm]{1000°C} [NH] + N_2$$

the generation of fluoronitrene on evaporation of FN_3 (Haller, 1942),

$$FN_3 \longrightarrow N_2 + FN$$

the thermolysis of sulfonyl azides at 120°C,

$$RSO_2N_3 \xrightarrow{120°C} RSO_2\ddot{N} + N_2$$

the decomposition of alkyl azidoformates,

$$RO\overset{O}{\overset{\|}{C}}N_3 \xrightarrow{\Delta} RO\overset{O}{\overset{\|}{C}}-\ddot{N} + N_2$$

and the particularly facile decomposition of cyanogen azide at 40–50°C,

$$N_3CN \xrightarrow{40-50°C} \ddot{N}-C\equiv N \longleftrightarrow N\equiv C-\ddot{N}$$

the latter undoubtedly being due to the symmetry and resonance stabilization of the nitrene formed.

If a functional group in the azide-bearing residue or the substrate can participate in the nitrogen-elimination process, then the picture can change drastically. For example, in the case of aryl azides, if a nucleophilic substituent is present ortho to the azide function, nitrogen elimination may either be concerted with or follow reaction of the azido group with the nucleophilic center so that a discrete nitrene intermediate is never formed. The thermal decompositions of *o*-nitrophenyl azide (**14**) to give benzofurazan (**15**) (Patai and Gotshal, 1966), of *o*-azidobenzophenone (**16**) to give 3-phenylanthranil (**17**) (Smith *et al.*, 1953), and the first step of the formation of dibenzo-1,3a,4,6a-tetrazapentalene (**19**) from 2,2'-diazidoazobenzene (**18**) (Carboni and Castle, 1962) probably do not involve nitrene intermediates. For example, a lower temperature is needed for the decomposition of **14** than of phenyl azide, and $\Delta H^{\ddagger}$ for the process is appreciably lower (65–80°C) than in unassisted decompositions (140°–170°C) (Birkhimer *et al.*, 1960; Fagley *et al.*, 1956). This, together with the negative entropy of activation (Patai and Gotshal, 1966), suggests a concerted mechanism involving rupture of the N–N$_2$ bond of the azide with concomitant cyclization. A kinetic study of the thermal decomposition of *o*-azidobenzophenones of the type *o*-N$_3$C$_6$H$_4$-

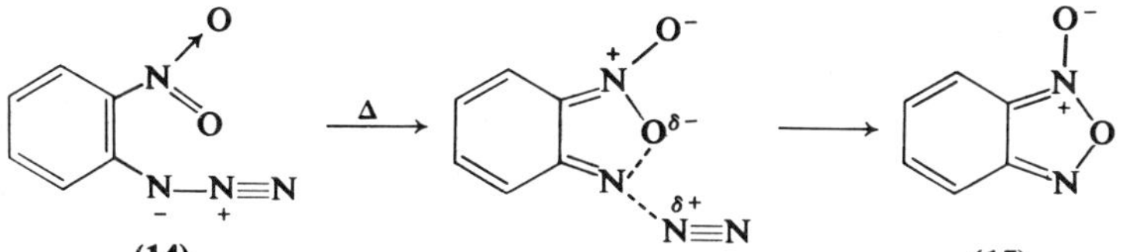

COC$_6$H$_4$R-*p* to give 3-arylanthranils showed that the rate of nitrogen loss was accelerated by electron-withdrawing and retarded by electron-donating groups R. The enthalpies and entropies of activation varied over a wide range: $\Delta H^\ddagger = 20.5$ to 27.0 kcal mole^{-1} and $\Delta S^\ddagger = -5.9$ to -21.4 cal deg^{-1} mole^{-1} (Hall and Behr, 1969, 1972). The data are inconsistent with either a nitrene or a concerted nitrogen-elimination cyclization process. A mechanism involving an intramolecular 1,3-dipolar addition **16 → 17** appears more

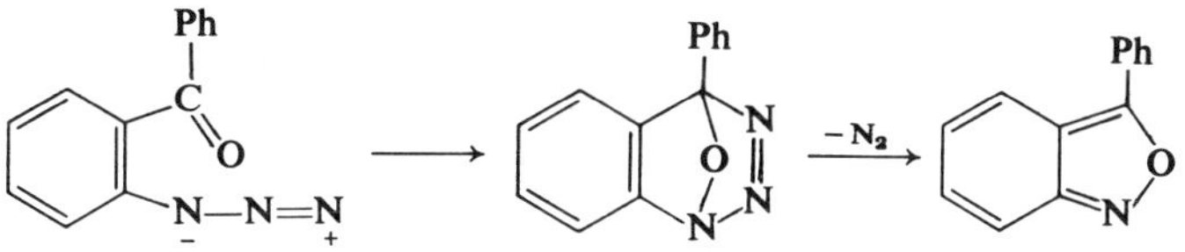

likely. The double cyclization of 2,2′-diazidoazobenzene (**18**) proceeds in two distinct steps: The first mole of nitrogen is eliminated at 58°C in what is undoubtedly a concerted process, the second comes off at about 170°C and probably involves the formation of a nitrene intermediate (Abramovitch and Davis, 1964; Carboni and Castle, 1962; Carboni *et al.*, 1967). This last step is similar to the formation of pyrido[1,2-*b*]indazole reported earlier (Abramovitch and Adams, 1961). Other examples of concerted azide decompositions have been summarized (L'abbé, 1969).

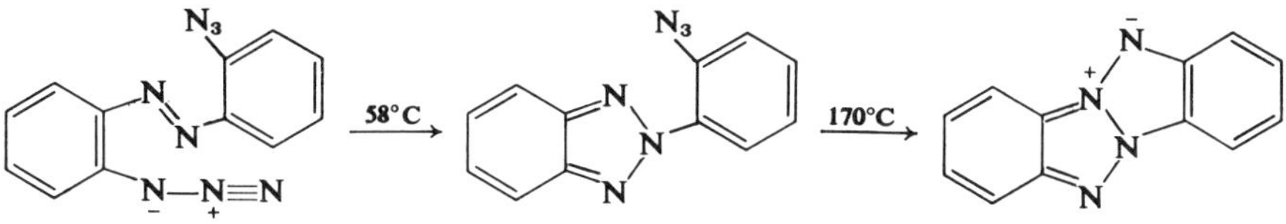

As discussed in Section II, nitrenes, are, again, not involved in the thermal decomposition of azides if the latter undergo 1,3-dipolar addition to the substrate or if the substrate (e.g., an olefin) participates in the rate-determining step.

Little definitive work has been carried out on whether or not nitrenes are formed in the thermal decomposition of alkyl azide (for a summary, see Abramovitch and Kyba, 1971a). Cyclohexyl azide decomposes at an appreciable rate in inert solvents only at elevated temperatures (Walker and Waters, 1962). The pyrolysis of α-azidocarbonyl compounds (20) at 200°C to give imines (22) may proceed via a nitrene intermediate (21) (Abramovitch and Davis, 1964). In all cases of thermal decompositions of alkyl azides the main products are those of molecular rearrangement.

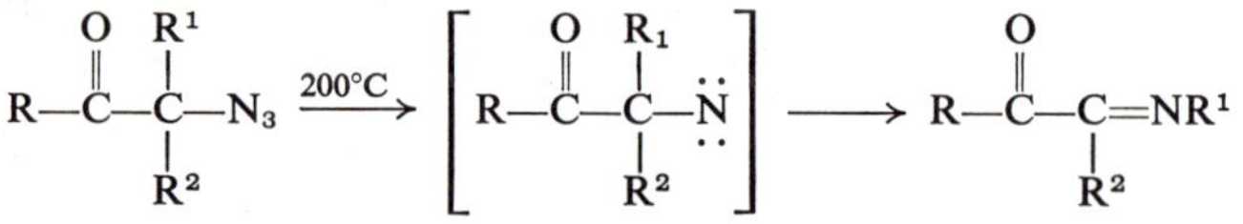

$$R-\underset{\underset{R^2}{|}}{\overset{\overset{O}{\|}}{C}}-\underset{\underset{R^2}{|}}{\overset{R^1}{C}}-N_3 \xrightarrow{200°C} \left[R-\underset{\underset{R^2}{|}}{\overset{\overset{O}{\|}}{C}}-\underset{\underset{R^2}{|}}{\overset{R_1}{C}}-\overset{..}{\underset{..}{N}} \right] \longrightarrow R-\underset{\underset{R^2}{|}}{\overset{\overset{O}{\|}}{C}}-C=NR^1$$

(20)　　　　　　　(21)　　　　　　　(22)

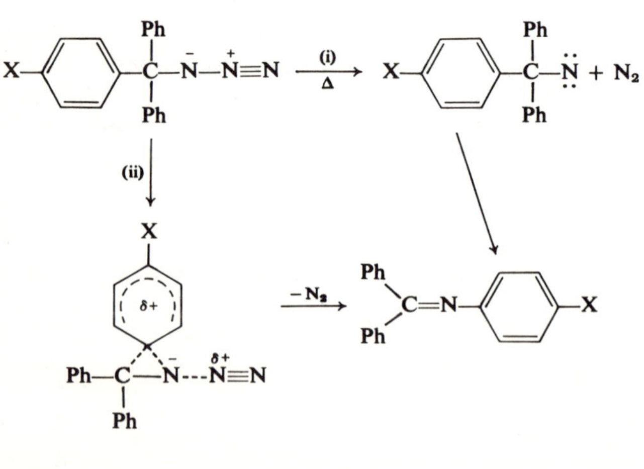

The thermal decomposition of triarylmethyl azide at 170°–190°C to give benzophenone anils has been studied carefully (Saunders and Ware, 1958; Saunders and Caress, 1964; Lewis and Saunders, 1967). A strong variation of the enthalpy and entropy of activation with a para substituent X in diphenyl (4-X-phenyl)methyl azide in the thermolysis was in the direction expected for rearrangement to an electron-deficient nitrogen intermediate, but it was felt to be too large to explain in terms of just an inductive effect [path (i)] and more in agreement with a concerted process [path (ii)]. The thermal migratory aptitudes varied from Ar/Ph = 6.7 for X = p-NMe$_2$ to Ar/Ph = 0.20 for X = p-NO$_2$. These are appreciably larger than in the photoinduced decompositions for which a nitrene mechanism was proposed. It was argued (Abramovitch and Davis, 1964), however, that these rate differences are indeed small compared to the corresponding migratory aptitudes in assisted migrations involving carbonium ion intermediates, and that if one considers only the effect of substituents on $\Delta H^{\ddagger}$, the inductive order is indeed followed.

From the thermolysis of 1,1-diphenylethyl and 2-phenyl-2-propyl azide, Saunders and co-workers (Saunders and Caress, 1964; Lewis and Saunders, 1968) obtained the following migratory aptitudes: Ph/Me = 2.36 or 4.05, depending on which azide was used. In the photolyses the migratory aptitude was Ph/Me = 0.96 (a value of Ph/Me = 0.75 can be estimated by extrapolation of Saunders' data to low conversions), which was taken as evidence for the intervention of a highly reactive and unselective triplet nitrene. A careful and quantitative study of the rearrangement products and migratory aptitudes (Abramovitch and Kyba, 1969; Abramovitch and Kyba, 1971a) has yielded the ratios shown in the following tabulation.

Migratory Aptitudes in the Thermolysis of Tertiary Alkyl Azides[a]

Biphenyl-2-yl/Me	Biphenyl-2-yl/Ph	Ph/Me
1.9	1.1	1.9

[a] Abramovitch and Kyba, 1971a.

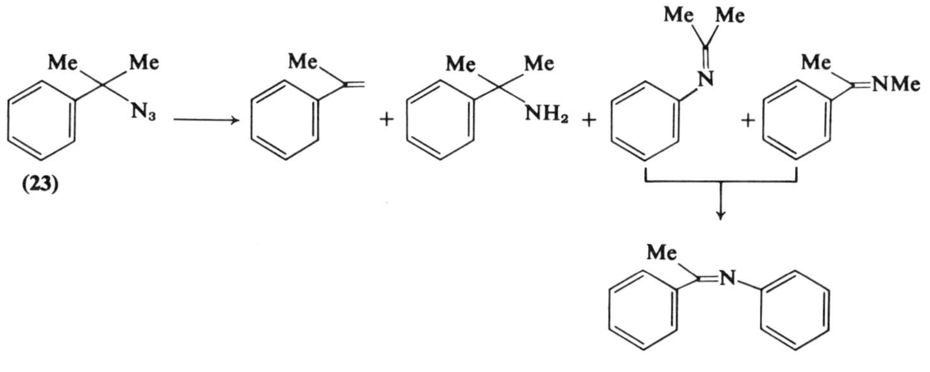

It was concluded (Abramovitch and Kyba, 1971a) that these ratios were more consistent with the intervention of a free nitrene than with a concerted migration elimination process. In a number of thermolyses, e.g., 2-phenyl-2-propyl azide (23) and 2-azido-2-methyl-4-phenylbutane, small yields of the corresponding primary amine were isolated, suggesting once again the intermediacy of a free nitrene. The formation of biphenyl-2-yldiphenylmethyl radicals competes to a small extent with nitrene formation in the thermolysis of biphenyl-2-yldiphenylmethyl azide (Abramovitch and Kyba, 1969).

The gas-phase pyrolysis of some alkyl azides at 350°–410°C or the thermolysis in hydrocarbon solvents at about 400°C also gave some of the corresponding primary amines together with rearrangement products. The results were interpreted in terms of a free nitrene intermediate (Pritzkow and Timm, 1966). It is clear that much more work is required on the mechanism of the thermal decomposition of alkyl azides.

Acylnitrenes (Lwowski, 1970) have been proposed as intermediates in the Curtius rearrangement of carbonyl azides to isocyanates.

$$RCON_3 \xrightarrow{\Delta \text{ or } h\nu} RN=C=O + N_2$$

Indeed, such intermediates have been produced by photolysis of the azides and trapped by insertion into suitably oriented C–H groups or by addition to olefinic double bonds (in the latter case the reactions are carried out at temperatures below those at which the azides could themselves interact thermally with the double bonds) [Eqs. (5) (Apsimon and Edwards, 1962)

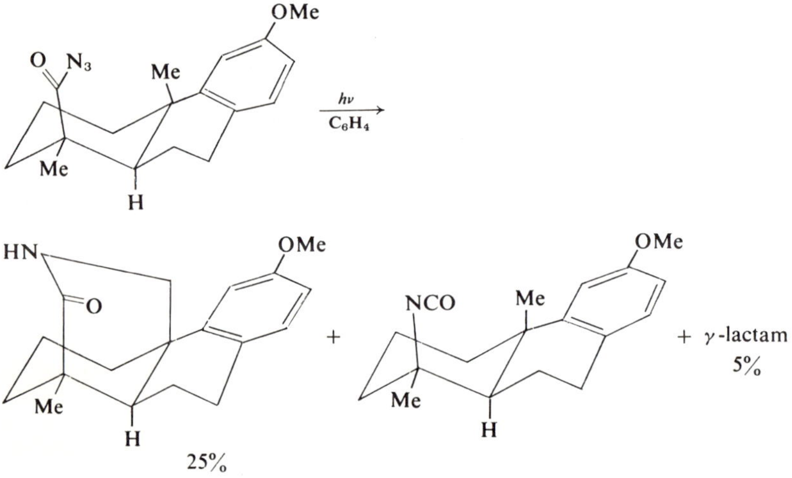

and (6) (Linke et al., 1967)]. In contrast, no acylnitrenes have been trapped in the thermolysis of the corresponding azides. Also, whereas the volume of activation of the Curtius reaction ($\Delta V^* = 2\text{--}5 \text{ cm}^3/\text{mole}$) corresponds to a

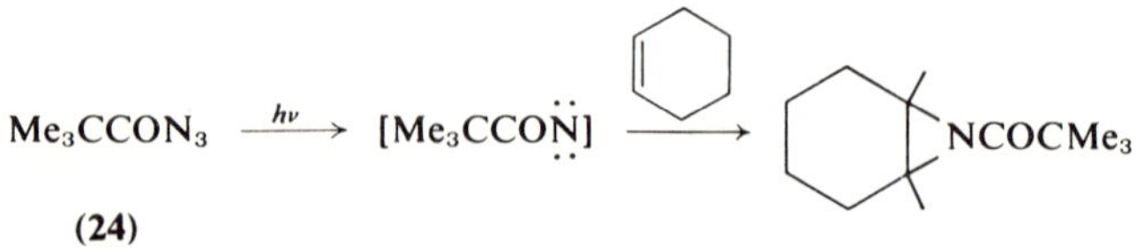

3. NITRENES

large measure of N–N bond rupture in the transition state (Brower, 1963), large kinetic isotope effects have been observed in the thermal Curtius rearrangement of suitably labeled azides (—^{13}C—CO—^{15}N—^{15}N≡N) (Fry and Wright, 1966, 1968), so that the bonds of the migrating groups must be broken in the rate-determining step, and free nitrenes cannot be intermediates in the thermolyses studied thus far.

Thermolysis of azidoformates in various solvents has been studied. Despite wide differences in the chemical nature of the solvents used, the thermolysis rates of the azides change little (for a summary, see Lwowski, 1970). Thus, the carbalkoxy nitrene is implicated as the intermediate formed in the rate-determining step.

$$R-O-\overset{\overset{O}{\|}}{C}N_3 \xrightarrow[\substack{\text{slow} \\ -N_2}]{\Delta} R-O-\overset{\overset{O}{\|}}{C}-\overset{..}{\underset{..}{N}} \xrightarrow[\text{solvent}]{\text{fast}} \text{products}$$

The uncatalyzed thermal decomposition of sulfonyl azides in the absence of olefins, metal compounds, or nucleophilies is believed to give nitrene intermediates and nitrogen.

$$RSO_2N_3 \xrightarrow{\Delta} RSO_2N + N_2$$

The evidence for this came mainly from trapping experiments, the temperature required to effect decomposition, and the nature of the products formed. Thus, the rate of nitrogen evolution from toluene-p-sulfonyl azide at 130°–135°C is independent of the nature of the solvent (Horner and Christmann, 1963b), and the decomposition of benzenesulfonyl azide follows first-order kinetics at 127°C in chlorobenzene, nitrobenzene, or p-xylene (Leffler and Tsuno, 1963) with $k_1 = 1.48 \times 10^{-3}$ min^{-1}. The decomposition of alkyl- and arylsulfonyl azides in a variety of solvents has been studied recently and found to be first order with a negligible substituent effect ($\rho = -0.1$) (Breslow, 1970).

A radical side reaction accompanies the normal decomposition of some alkylsulfonyl azides, leading to the evolution of some SO_2 and the formation of alkyl radicals and to the departure from first-order kinetics. This can be suppressed in the presence of radical inhibitors, e.g., hydroquinone. A radical chain decomposition has been proposed [Eq. (7), (Breslow et al., 1969)]. Most of the SO_2 evolved in the thermal decomposition of arylsulfonyl

$$\begin{aligned} R^1\cdot + RSO_2N_3 &\longrightarrow RSO_2\cdot + R^1N_3 \\ RSO_2\cdot &\longrightarrow R\cdot + SO_2 \\ R\cdot + R^2H &\longrightarrow RH + R^2\cdot \end{aligned} \quad (7)$$

azides undoubtedly arises from the breakdown of the sulfonylanilines formed through a Curtius-type rearrangement (see later in this chapter) (Abramovitch and Holcomb, 1969). Some radical-induced decomposition undoubtedly also takes place in a few instances, e.g., the formation of dodecyl azides from the thermolysis of mesitylene-2-sulfonyl azide in *n*-dodecane (Abramovitch and Holcomb, 1969) and in the decomposition of diphenyl sulfone-2-sulfonyl azide (25) in dodecane at 150°C when diphenyl sulfone (26) and diphenyl sulfone-2-sulfonamide (27) were obtained. When the thermolysis was carried out in Freon E-4 at 150°C, the products were diphenylene sulfone (28) and 27 (Abramovitch *et al.*, 1969a) (28 is the product of a Pschorr-type cyclization). Ferrocenylsulfonyl azide also gives some ferrocene on thermolysis in cyclohexane under pressure (Abramovitch *et al.*, 1971a).

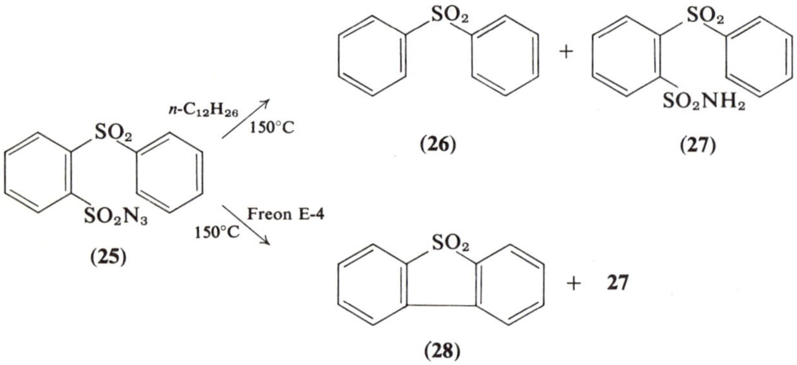

Definitive evidence that cyanonitrene is formed in the thermolysis of N_3CN was obtained by using ^{15}N-labeled azide as shown in reaction (8) (Anastassiou and Simmons, 1967). Since it could be shown that scrambling

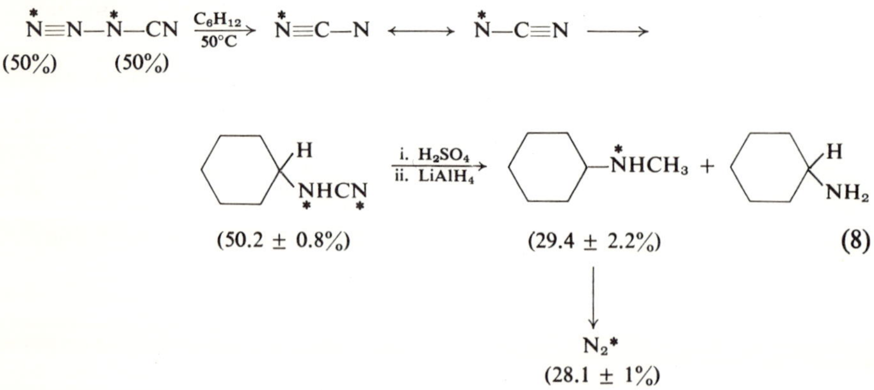

$$\overset{*}{N}\equiv N-\overset{*}{N}-CN \xrightarrow[50°C]{C_6H_{12}} \overset{*}{N}\equiv C-N \longleftrightarrow \overset{*}{N}-C\equiv N \longrightarrow$$

(50%)   (50%)

$$\underset{(50.2 \pm 0.8\%)}{\text{C}_6\text{H}_{11}-\overset{*}{\text{N}}\text{HCN}} \xrightarrow[\text{ii. LiAlH}_4]{\text{i. H}_2\text{SO}_4} \underset{(29.4 \pm 2.2\%)}{\text{C}_6\text{H}_{11}-\overset{*}{\text{N}}\text{HCH}_3} + \underset{\text{NH}_2}{\text{C}_6\text{H}_{11}-\text{H}} \quad (8)$$

$$\downarrow$$

$$N_2^*$$
$$(28.1 \pm 1\%)$$

of the label did not occur prior to reaction, this proved that the free nitrene was formed in which the isotopic label is nearly equally distributed among its two nitrogen atoms.

2. *Photolysis*

NH and the halogenated nitrenes can be obtained by photolysis of the appropriate azides (for reviews, see Abramovitch and Davis, 1964; Berry, 1970).

Photochemical decomposition of aryl azides to aryl nitrenes is firmly established (Reiser and Marley, 1968). For example, irradiation of 2-azidobiphenyl (1) in a solid matrix at 77°K produces the nitrene (recognized by its UV spectrum) with a quantum yield of 0.43 ± 0.05. At room temperature in cyclohexane solution, carbazole (2) is formed directly (97%) with a quantum yield of 0.44 ± 0.02. The close agreement between these values indicates that nitrene formation is the primary step of the reaction, even in solution and at room temperature. Photolysis of phenyl azide in the presence of secondary amines gives the 3H-azepine derivative (Doering and Odum, 1966), and similar products are obtained when nitrosobenzene is treated with triphenylphosphine and secondary amines. A common intermediate, the aryl nitrene, is reasonably assumed (Odum and Brenner, 1966). Most of the reaction products obtained on photolysis of aryl azides are best rationalized in terms of a nitrene being formed initially [e.g., the photolysis of 2,2′diazidobiphenyl (Boyer and Mikol, 1969)] (for a review, see Smith, 1970).

The situation in regard to the photolysis of alkyl azides has only been clarified recently and has been reviewed (Abramovitch and Kyba, 1971a). Currie and Darwent (1963) investigated the vapor-phase photolysis of methyl azide at low conversions and over various ranges of pressure, temperature, intensity, and wavelength. Their results were consistent with a quantum yield of unity for the formation of the nitrene in the initial step. The photolysis of a number of alkyl azides and α-azido acids has similarly been explained on the basis of a nitrene being formed, albeit an activated one (Moriarity and Rahman, 1965). On the other hand, Koch (1967) reported that the quantum yield of nitrogen evolution increases with the concentration of HN_3 and MeN_3, respectively, in the photolysis of these compounds, reaching a maximum value of 2. Monomolecular decay of MeN_3^* to a free nitrene was said to be an insignificant process, the formation of reaction products being better explained by a primary attack of the excited MeN_3 upon MeN_3, solvents, or intermediate products. In the photolysis of ethyl azide, however, the quantum yield of N_2 was about unity and independent of EtN_3 concentration so that a nitrene was thought to be formed in this case (see, however, the discussion by Abramovitch and Kyba, 1971a).

The sensitized and unsensitized photodecomposition of tertiary alkyl azides has been studied much more extensively (Lewis and Saunders, 1968; Abramovitch and Kyba, 1969, 1971b). The lack of marked selectivity in migration aptitudes in triarylmethyl and dialkylarylmethyl azides and the formation of triphenylmethylamine in photolysis of triphenylmethyl azide in the presence of efficient hydrogen donors tended, at first glance, to support the existence of a discrete nitrene intermediate (Tables I and II).

Table I

Migratory Aptitudes[a] in the Direct and Triphenylene Sensitized[b] Photolysis of p-$XC_6H_4(C_6H_5)_2CN_3$

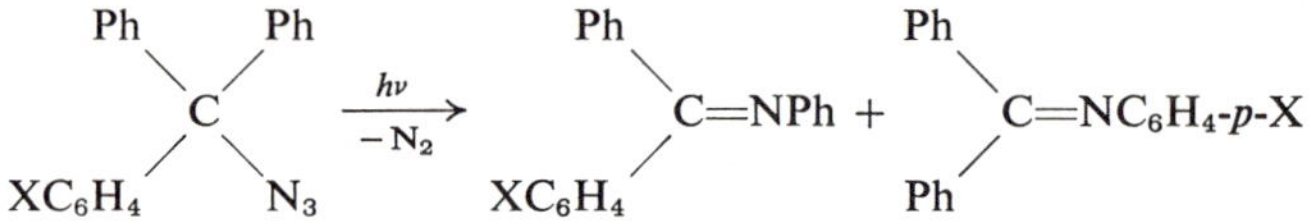

$$\begin{array}{c} Ph \\ \diagdown \\ XC_6H_4 \end{array} \!\! C \!\! \begin{array}{c} Ph \\ \diagup \\ N_3 \end{array} \xrightarrow[-N_2]{h\nu} \begin{array}{c} Ph \\ \diagdown \\ XC_6H_4 \end{array} \!\! C\!\!=\!\!NPh \; + \; \begin{array}{c} Ph \\ \diagdown \\ Ph \end{array} \!\! C\!\!=\!\!NC_6H_4\text{-}p\text{-}X$$

X in p-XC_6H_4	Direct photolysis	Sensitized photolysis
NO_2	1.03	1.07
Cl	1.26[c]	0.97
CH_3	—	0.89
OCH_3	1.16	1.11
NMe_2	—	1.08

[a] p-XC_6H_4/Ph, corrected for statistical preference.
[b] The triphenylene used is a singlet sensitizer.
[c] Average of two values from different extents of conversion.

Table II

Migratory Aptitudes in the Photolysis of *tert*-Alkyl Azides

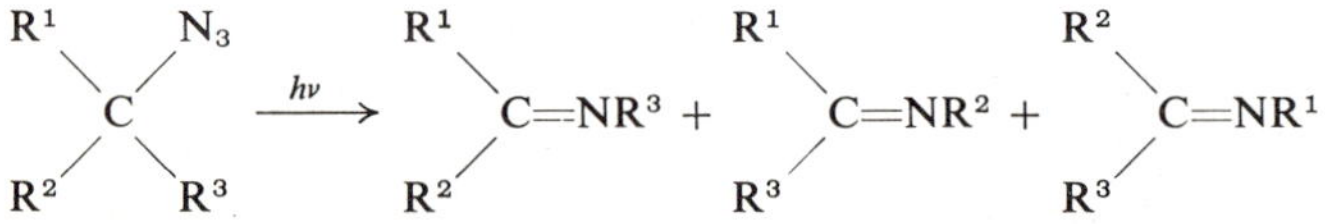

$$\begin{array}{c} R^1 \\ \diagdown \\ R^2 \end{array} \!\! C \!\! \begin{array}{c} N_3 \\ \diagup \\ R^3 \end{array} \xrightarrow{h\nu} \begin{array}{c} R^1 \\ \diagdown \\ R^2 \end{array} \!\! C\!\!=\!\!NR^3 \; + \; \begin{array}{c} R^1 \\ \diagdown \\ R^3 \end{array} \!\! C\!\!=\!\!NR^2 \; + \; \begin{array}{c} R^2 \\ \diagdown \\ R^3 \end{array} \!\! C\!\!=\!\!NR^1$$

Ph/Me	Ar/Me[a]	Ar/Ph[a]	$PhCH_2CH_2$/Me
0.75	0.69	0.44	0.89

[a] Ar = 2-biphenylyl.

In contrast to the results summarized in Table I, those in Table II show clearly that the reactive intermediate is not free to choose which of the groups it would, for electronic reasons, prefer to attack. Thus, had a discrete nitrene (singlet or triplet) intermediate been formed, it would have been expected that phenyl would have migrated faster than methyl (as it does in the thermolyses) and that the more nucleophilic 2-biphenylyl would have migrated faster than phenyl, which was not the case. Thus, it appeared that the ground-state conformation of the tertiary alkyl azide played an important role in determining the migration ratio, and a theory has been proposed to explain all of these results (including those in Table I) which does not allow for the formation of a free nitrene in these photolyses but takes into account the preferred ground-state conformations of the azides and the geometry of the orbitals in the photoexcited state of the azido group (Abramovitch and Kyba, 1971b).

The electronic transition (287 nm) normally involved in the photoexcitation of alkyl azides is $\pi_y \to \pi_x^*$ (29 → 30) (Closten and Gray, 1963). This would leave the p_y orbital on the α nitrogen atom electron deficient. If this is so, then a concerted migration–elimination would not involve backside

(trans) attack (Moriarity and Reardon, 1970) (which would require migration to a filled sp_x^2 orbital), but instead the migrating and departing groups would be orthogonal to each other so that the bonding orbital of the migrating group would overlap with the electron-deficient p_y orbital. We can illustrate this by means of an example of a tertiary alkyl azide LM$_2$C—N$_3$ (31) in which L and M are large and medium groups, respectively. 31 is a Newman projection looking down the C–N' bond. The choice of 31 as the preferred conformation is based on the assumption that the L–N$_2$ nonbonded interaction is greater than that between M and N$_2$, and that these, in turn, are greater than the corresponding interactions between those substituents and the N' lone pair (sp_x^2 in 29). If one further assumes the Franck–Condon principle to hold in these photolyses, then the preferred ground-state conformation of the alkyl azide would determine which group would be suitably oriented to migrate. Thus, in the photolysis of 2-phenyl-2-propyl azide (Ph = L, Me = M) it would be expected that the methyl groups migrate more readily than phenyl since conformation 31 should be more populated than conformation 33. This is what is actually observed in photolysis (Table II). This would also explain why in the photolysis of several para-substituted triarylmethyl azides both possible imines are obtained in statistical amounts (Table I), since the para substituent would not affect the ground-state conformation of the azide, whereas an *o*-phenyl group has an appreciable influence on the migration aptitude (Table II).

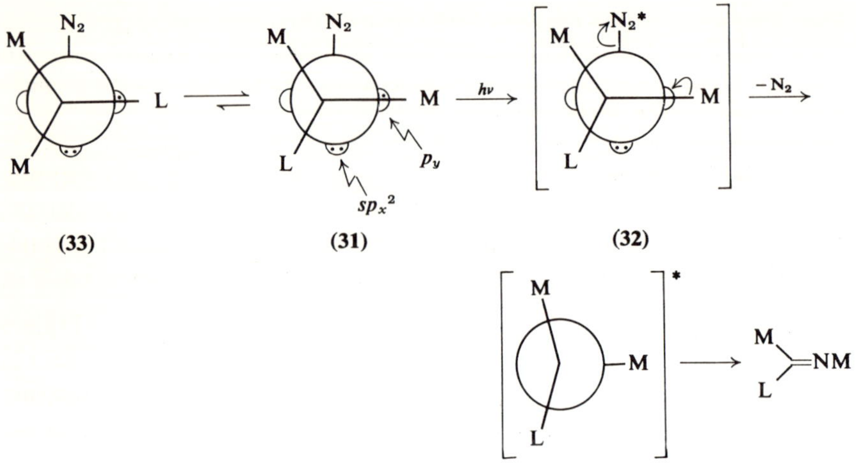

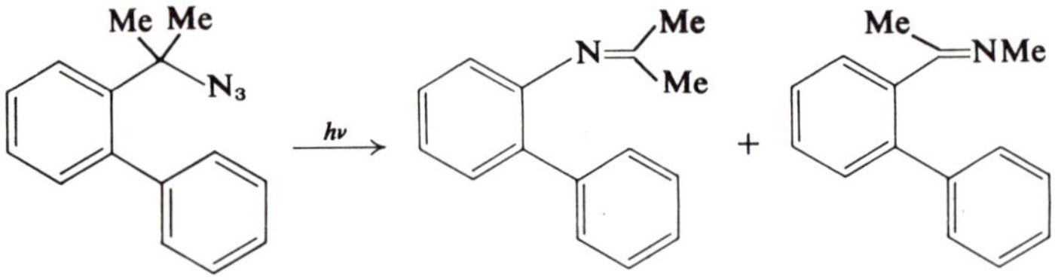

The primary importance of the ground-state conformation has been established conclusively by studying the influence of temperature on the migration aptitudes in the direct photolysis of **34** (Table III) and α-methylbenzyl azide (**37**) (Table IV). The former gives two products **35** and **36** at low conversions, whereas the latter gives the imines **38**, **39**, and **40**. Thus, both

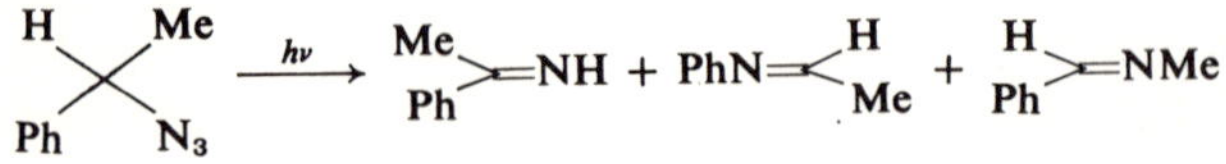

sets of reactions approach a statistical distribution of migratory aptitudes as the temperature is raised, which is expected if the higher energy conformations are more populated or, conversely, if the more stable conformation predominates at low temperatures. This clearly eliminates the highly

Table III
Photolysis of 34 at Various Temperatures

Temperature (°C)	Percentage migration		
	Me	2-Biphenylyl	2-Biphenylyl/methyl
−117	64.6	35.4	0.55
22	59.1	40.9	0.69
101[a]	56.9	43.1	0.76

[a] Azide does not undergo thermolysis at this temperature.

Table IV
Photolysis of 37 at Various Temperatures

Temperature (°C)	Percentage migration		
	H (38)	Ph (39)	Me (40)
−117	25.0	25.1	49.9
−77	26.2	26.9	47.0
22	29.0	29.2	41.8
101	31.1	31.4	37.7

reactive free nitrene intermediate as a contender, since the latter should be cylindrically symmetrical (see Section III,C) and have no conformational preference. If the relevant photoexcited state of the azide is indeed the $\pi_y\pi_x^*$, then the conformations of 37 which would explain the results would be 37a, 37b, and 37c, with the order of stability 37a > 37b ≃ 37c.

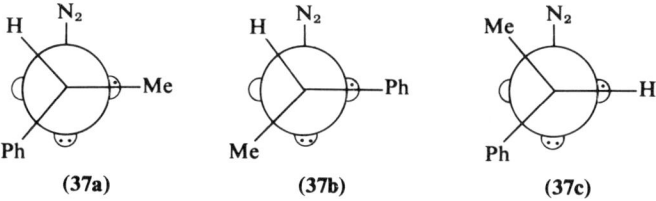

The fact that no ESR spectrum can be observed when alkyl azides are photolyzed in a frozen matrix at 77°K but weak signals can be seen at 4°K is in agreement with the concept that a free nitrene is not formed in the photolyses at room temperature. Only when the C–C bending motions are sufficiently frozen (4°K) can N–N bond cleavage occur with a small degree of migration of the group attached to the tertiary carbon. On the other hand, the very small selectivity observed in the migration ratios indicates that N–N bond cleavage *has* proceeded to a considerable extent, and group migration to N only to a very small extent, in the transition state of the photolyses.

In contrast to these smooth decompositions, the photolysis of sulfonyl azides in nonprotic nonpolar solvents usually gives rise to tar formation with little or no evidence for the intervention of nitrenes. Low temperature photolysis in a frozen matrix (Smolinsky *et al.*, 1962) or as single crystals (Moriarity *et al.*, 1966) results in the production of a triplet nitrene. On the other hand, photolysis of sulfonyl azides in benzene, cyclohexene, pyridine, and thiophene generally gives rise to insoluble polymerlike products. Photolysis of methanesulfonyl azide in benzene gave a yellow amorphous material which did not melt below 290°C and formed a gum with boiling ethanol. If deposition of a polymer on the sides of the quartz flask is prevented, however, a very small amount of nitrene is formed on photolysis either at 28° or 80°C, and this reacts with the benzene (Abramovitch and Uma, 1967).

Photolysis of sulfonyl azides in dimethyl sulfoxide (DMSO) using 2537 Å radiation gives *N*-sulfonylsulfoximines (**41**) in 15–50% yield (Horner and Christmann, 1963b). Better yields (48–55%) of trapped product (**42**) were obtained with dimethyl sulfide. Though the reaction with DMSO was formulated as proceeding via a nitrene intermediate, the possibility exists, which has not been disposed of, that prior complexing of the sulfonyl azide occurs with the solvent acting as an electrophile followed by nitrogen elimination. This cannot apply to the dimethyl sulfide reaction, however, which, thus, probably does go through a sulfonyl nitrene.

$$RSO_2N_3 + Me_2SO \xrightarrow{h\nu} RSO_2N{=}\overset{\overset{O}{\uparrow}}{S}Me_2 + N_2$$
(**41**)

$$RSO_2N_3 + Me_2S \xrightarrow{h\nu} RSO_2N{=}SMe_2 + N_2$$
(**42**)

$$RSO_2\overset{-}{N}{-}\overset{+}{N}{\equiv}N + Me_2\overset{+}{S}{-}\overset{-}{O} \longrightarrow \left[\begin{array}{c} RSO_2{-}N{-}\overset{+}{N}{\equiv}N \\ \mid \\ O{-}SMe_2 \end{array} \right]$$

or

$$\left[\begin{array}{c} RSO_2{\diagdown}{\diagup}N \\ N N \\ \mid \mid \\ Me_2S O \end{array} \right] \xrightarrow{h\nu} RSO_2N{=}\overset{\overset{O}{\uparrow}}{S}Me_2 + N_2$$

Photolysis of sulfonyl azides in alcohols occurs readily to give products of insertion into alcohols, hydrogen abstraction, and Curtius-type rearrangement (Horner and Christmann, 1963b; Lwowski and Scheiffele, 1965; Reagan and Nickon, 1968). Hydrogen bonding by the solvent appears to be important. Irradiation of benzenesulfonyl azide in methanol may not involve a free nitrene intermediate. A Curtius-type rearrangement of the hydrogen-bonded azide is possible (Lwowski and Scheiffele, 1965). On the other hand, the reaction of $MeSO_2N_3$ is isopropanol with 2537 Å light or using 3660 Å

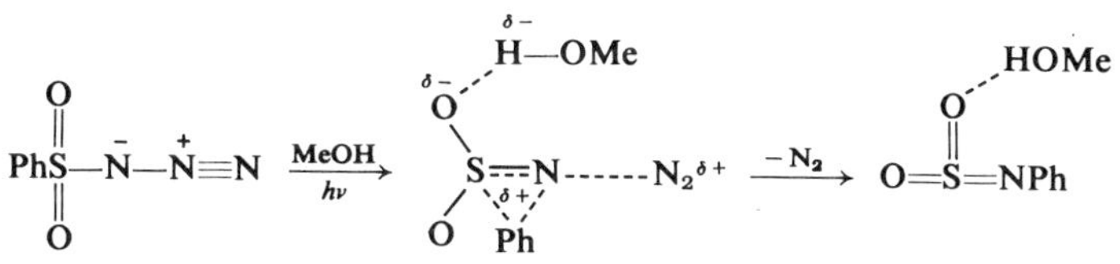

$$PhS(O)_2-\overset{-}{N}-\overset{+}{N}\equiv N \xrightarrow[h\nu]{MeOH} \underset{Ph}{\overset{H-OMe}{\underset{O}{S}}}{=}N \text{------} N_2^{\delta+} \xrightarrow{-N_2} O{=}\overset{HOMe}{\underset{O}{S}}{=}NPh$$

radiation and a benzophenone sensitizer proceeds differently and leads to a quantitative yield of methanesulfonamide and acetone. A radical chain mechanism consisting of two propagation sequences was proposed (Reagan and Nickon, 1968).

In contrast to the problems encountered on photolysis of alkyl- and arylsulfonyl azides, ferrocenylsulfonyl azide (**43**) is smoothly decomposed by 3500 Å light in cyclohexane or benzene to give ferrocene (**44**), ferrocenylsulfonamide (**45**), and the bridged [2]ferrocenophanethiazine-1,1-dioxide (**46**) (Abramovitch et al., 1969b). The yield of **46** varied with the nature of the solvent, being 67% in benzene, 13.3% in cyclohexane, and 0% in dimethyl sulfoxide or DMSO benzene (Abramovitch et al., 1971a). None was formed on thermolysis of **43**; instead the main products were **44**, the unsubstituted

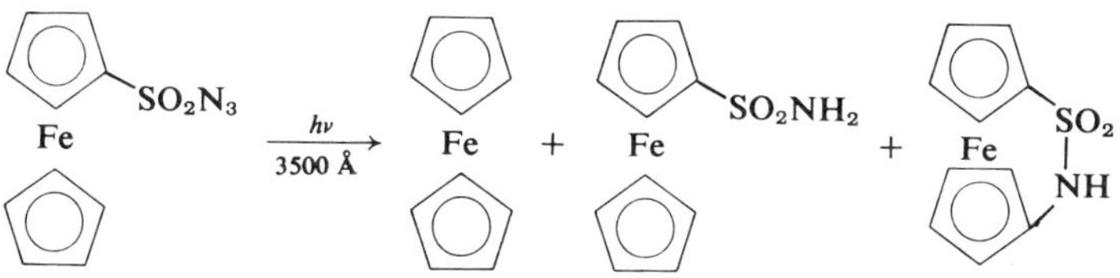

amide **45**, and the substituted amide. It is conceivable that the thermolysis involves a metal–nitrene complex, whereas the photolysis involves the free nitrene. Since ferrocene itself is an efficient triplet quencher as well as a

$$43 \xrightarrow[\Delta]{RH} 44 + 45 + FcSO_2NHR$$

RH = C_6H_{12}	20%	48.4%	24%
RH = C_6H_6	17%	76.8%	6.5%
RH = C_6H_{10}	2%	85.4%	8%

sensitizer, it is difficult to decide what the spin state of the ferrocenylsulfonylnitrene is at the moment of reaction. The cyclization appears to be a singlet reaction since the yield of bridged product (**46**) in benzene solution is essentially unaffected by oxygen or hydroquinone.

Carbalkoxynitrenes are readily produced by the photodecomposition of azidoformates using low pressure mercury lamps and fused silica or Vycor 7912 vessels (for a review, see Lwowski, 1970). The electronic state of the carbalkoxynitrenes so formed has been elucidated by the elegant studies of Lwowski and his school mainly on the basis of the Skell postulated for the stereospecifity of the addition of singlet and triplet carbenes to olefins (singlet adds stereospecifically, triplet does not) as well as by dilution experiments (at lower concentrations of substrate the singlet nitrene has a greater chance to drop to the triplet ground state before reacting with the substrate). These studies will be discussed in Section V,D. They did show that about 30% of the photolytically formed nitrene is in the triplet state.

As indicated earlier (Section IV,A,I), carbonyl nitrenes can be produced and intercepted by photolysis of carbonyl azides but not by thermolysis.

3. Metal-Catalyzed Decompositions

The decomposition of azides catalyzed by Lewis acids such as $AlCl_3$ and $SbCl_5$ has been the object of recent study. Phenyl azide forms a hydrocarbon-soluble complex at $-70°C$ with Et_3Al, Et_2AlCl, or $EtAlCl_2$ (Hoergerlee and Butler, 1964) which decomposes slowly on warming to a phenylnitrene–aluminium complex $Ph\overset{+}{N}\overset{-}{A}lEt_3$. This rearranges to a variety of amidoalkylaluminium products. No free nitrene is apparently formed. Again, no nitrene is formed in the $AlBr_3$-catalyzed decomposition of PhN_3 in toluene at $0°C$ in the presence of traces of HBr (L'abbé, 1969). Alkyl azides are similarly decomposed by Lewis acid via azide–metal complexes (Kreher and Jäger, 1965; Goubeau et al., 1964).

Diiron nonacarbonyl causes the decomposition of methyl azide to give mainly **47** and **48** (Dekker and Knox, 1967).

$$MeN_3 + Fe_2(CO)_9 \longrightarrow \underset{(47)}{\begin{array}{c} Me\diagdown \underset{\|}{N}\diagup\overset{O}{\underset{\|}{C}}\diagdown N\diagup Me \\ | \diagup\!\!\!\diagdown | \\ (CO)_3Fe\text{------}Fe(CO)_3 \end{array}} + \underset{(48)}{\begin{array}{c} N\text{---}N \\ MeN\diagup\diagdown\diagup NMe \\ Fe \end{array}}$$

Aryl, alkoxylcarbonyl, and sulfonyl azides undergo decomposition under very mild conditions with $Fe_2(CO)_9$ to give iron complexes, and less readily with $Fe(CO)_5$ to give the same complexes [Eqs. (9) (Campbell and Rees, 1969a) and (10) (Abramovitch and Knaus, 1971)]. Sulfonyl azides, e.g.,

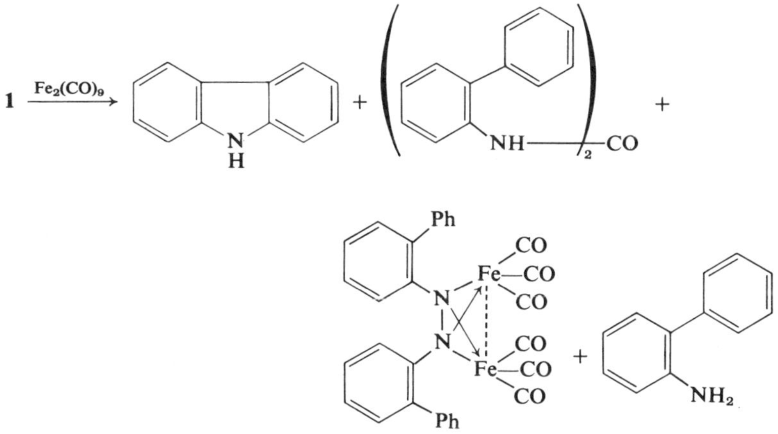

3. NITRENES

$$1 \xrightarrow{Fe_2(CO)_9} \text{carbazole} + (\text{biphenyl-NH})_2\text{-CO} + \text{Fe complex} + \text{2-aminobiphenyl} \quad (9)$$

$$\underset{\text{Me}}{\overset{\text{Me}}{\text{MeC}}}-\overset{\text{O}}{\overset{\|}{\text{OCN}_3}} + Fe(CO)_5 \xrightarrow[80°C]{C_6H_6} \text{complex} \xrightarrow{Al_2O_3}$$

$$(tert\text{-BuO}\overset{O}{\overset{\|}{C}})_2\text{NH} + tert\text{-BuO}\overset{O}{\overset{\|}{C}}\text{NH}_2 + (tert\text{-BuO}\overset{O}{\overset{\|}{C}}\text{NH})_2\text{CO} + \text{Me-oxazolidinone} \quad (10)$$

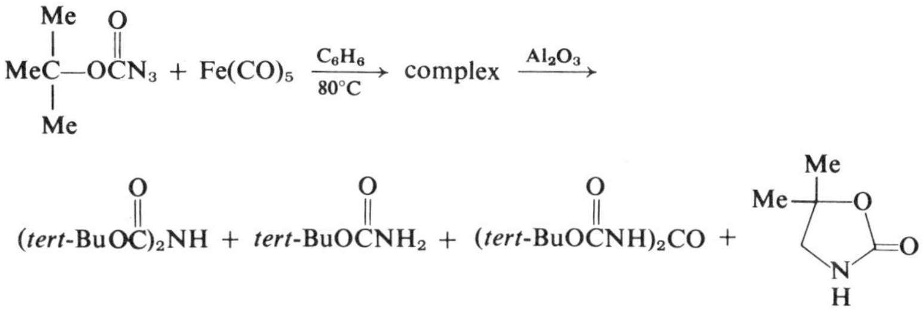

$MeSO_2N_3$, gave novel iron complexes devoid of CO groups bound to iron. It is not known whether these are dimers (**49**) or polymers (**50**) [Eq. (11)

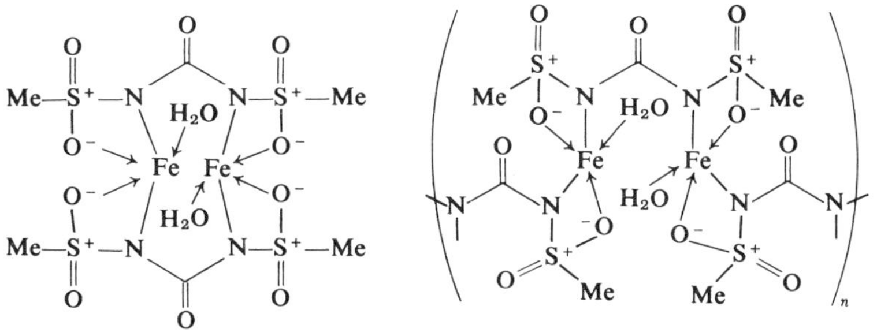

$$\text{Structures (49) and (50)} \quad (11)$$

(Abramovitch and Knaus, 1971)]. Other structures are also possible. In neither of these decompositions nor in the copper metal-catalyzed decompositions of sulfonyl azides is there any evidence for the intermediacy of free nitrenes. In the latter reaction, formation of a copper–nitrene complex has been postulated (Kwart and Kahn, 1967).

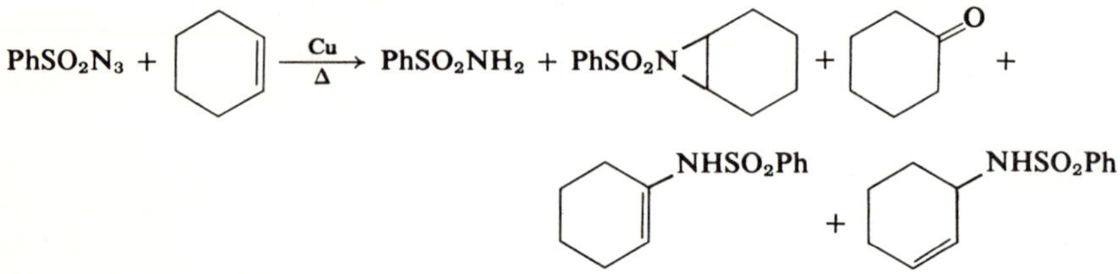

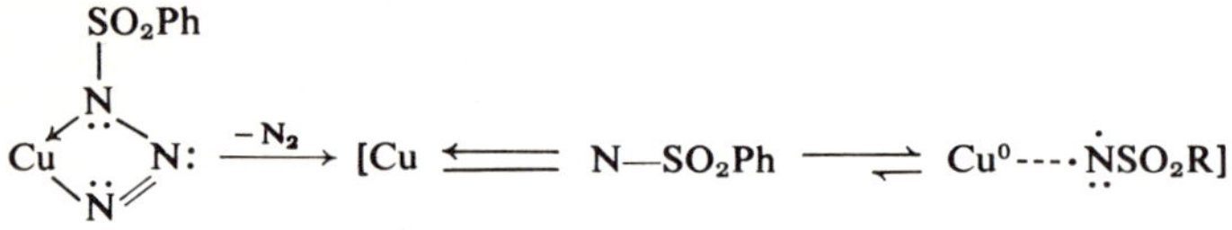

$$PhSO_2N_3 + \text{cyclohexene} \xrightarrow{Cu, \Delta} PhSO_2NH_2 + PhSO_2N\text{-cyclohexene} + \text{cyclohexanone} +$$

plus aziridine with NHSO$_2$Ph and allylic NHSO$_2$Ph products.

$$Cu\overset{SO_2Ph}{\underset{N}{\overset{N}{\diagdown}}}N: \xrightarrow{-N_2} [Cu = N-SO_2Ph \rightleftharpoons Cu^0 \cdots \overset{..}{N}SO_2R]$$

B. Deoxygenation

Nitro and nitroso compounds can serve as sources of nitrenes and nitrenoïd intermediates (Boyer, 1970). The synthesis of carbazole (2) from 2-nitrobiphenyl and ferrous oxalate or triethyl phosphite is a good example.

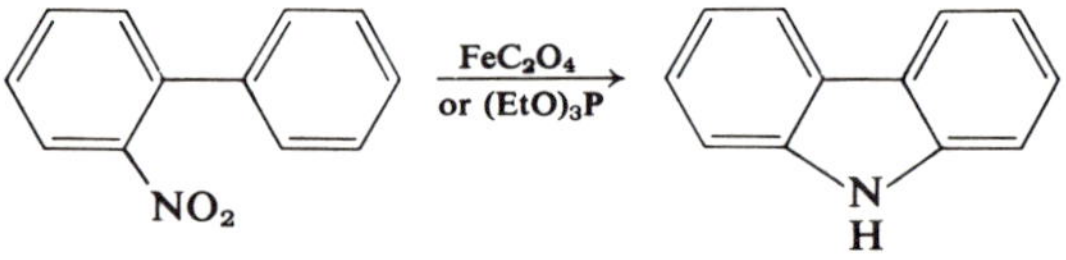

$$\text{2-nitrobiphenyl} \xrightarrow[\text{or (EtO)}_3P]{FeC_2O_4} \text{carbazole}$$

As has been discussed earlier (Section II), the ferrous oxalate deoxygenations do not involve free nitrenes; instead, the intermediate is bound to the surface of the catalyst. In view of the isolation of aryl nitrene–iron carbonyl complexes, it seems likely that the iron (II) oxalate reactions involve similar intermediates.

$$ArNO_2 \xrightarrow[FeC_2O_4]{\Delta} ArN\text{---}Fe(II) \text{ complex} \longrightarrow \text{products}$$

Deoxygenation with triethyl phosphite and triphenyl phosphine is cleaner and better and, in all likelihood, does involve aryl nitrene intermediates (Odum and Brenner, 1966; Cadogan, 1968; Sundberg, 1966a,b; Cadogan and Cooper, 1969). Deoxygenation of nitrosobenzene derivatives with triethyl phosphite, though less useful owing to the difficulty in obtaining starting

3. NITRENES

nitroso compounds, takes place under very mild conditions ($-75°$–$20°C$). Aprotic solvents such as benzene can be used.

$$ArNO_2 + R_3P \longrightarrow Ar\overset{O^-}{\underset{|}{N}}-O-\overset{+}{P}R_3 \longrightarrow Ar-N=O + R_3P \to O$$

$$ArN=O + R_3P \longrightarrow Ar-\overset{-}{N}-O-\overset{+}{P}R_3 \longrightarrow Ar-\ddot{N} + R_3P \to O$$

Evidence for the formation of a nitrene in these deoxygenations comes mainly from the similarity of the products formed with those from the corresponding azides.

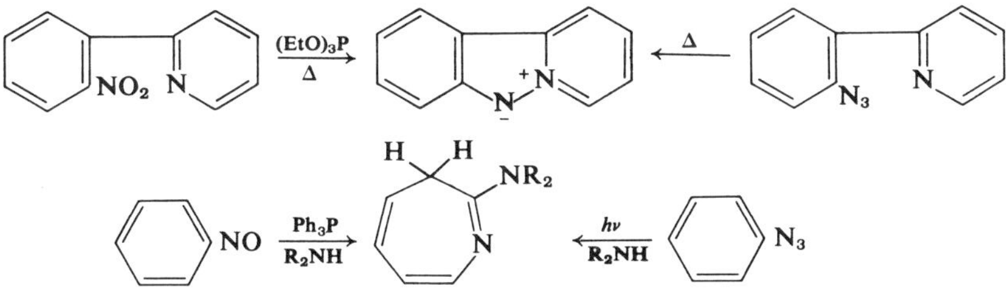

One can conceive, however, of cases in which a free nitrene is not formed in the phosphite-mediated deoxygenations and that a concerted attack by the nucleophilic substrate and elimination of phosphate could occur.

$$Ar-\overset{-}{N}-O-\overset{+}{P}R_3 \longrightarrow Ar-N=Y + R_3PO$$
$$Y^-$$

C. α Eliminations

By analogy with the usually base-induced α eliminations yielding carbenes, one can envisage a similar process leading to nitrenes.

$$R-\underset{\underset{Y}{|}}{N}-H + :B \longrightarrow R-\ddot{N} + BH^+ + Y^-$$

Indeed, some examples of this type of reaction have been documented. Thus, treating *N*-(*p*-nitrobenzenesulfonyloxy)carbamates (**51**) with base gives the *N*-anion, which eliminates the *p*-nitrobenzenesulfonate anion to form singlet carbalkoxynitrenes (Lwowski and Maricich, 1965).

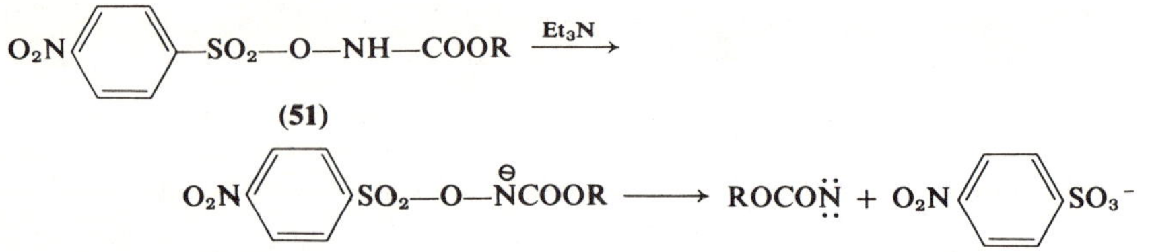

$$O_2N-C_6H_4-SO_2-O-NH-COOR \xrightarrow{Et_3N}$$
(51)

$$O_2N-C_6H_4-SO_2-O-\overset{\ominus}{\underset{..}{N}}COOR \longrightarrow RO\overset{..}{C}O\overset{..}{N} + O_2N-C_6H_4-SO_3^-$$

The selectivity of the nitrene so generated toward insertion in C–H bonds, e.g., 2-methylbutane, is similar to that of the intermediate produced by the thermolysis of azidoformates. Also, the rate of decomposition of the N-anion in dichloromethane solution in the presence of cyclohexene is first order in the anion and independent of the cyclohexene concentration, confirming the nitrene mechanism (Maricich, 1965).

Similarly, N-(p-nitrobenzenesulfonyloxy)benzenesulfonamide (52) gives nitrene-type products on treatment with base in methanol or ethanol (Lwowski and Scheiffele, 1965). Since, however, no benzenesulfonanilides (53) were

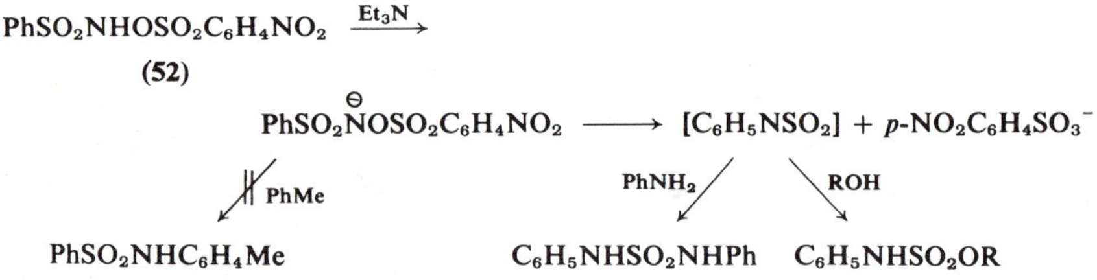

$$PhSO_2NHOSO_2C_6H_4NO_2 \xrightarrow{Et_3N}$$
(52)

$$PhSO_2\overset{\ominus}{N}OSO_2C_6H_4NO_2 \longrightarrow [C_6H_5NSO_2] + p\text{-}NO_2C_6H_4SO_3^-$$

PhMe ↙   PhNH$_2$ ↙   ↘ ROH

PhSO$_2$NHC$_6$H$_4$Me   C$_6$H$_5$NHSO$_2$NHPh   C$_6$H$_5$NHSO$_2$OR
(53)

obtained when the decomposition was carried out in toluene–methylene chloride (as are usually formed in the thermal decomposition of sulfonyl azides in toluene at 120°C), it was felt that the rearrangement did not involve a free nitrene but rather a hydrogen-bonded or protonated species. It has recently been shown, however, that the intermediate from the base-initiated decomposition of N-arylsulfonyloxysulfonamides can be trapped with dimethyl sulfide or DMSO (Okahara and Swern, 1969). These latter reactions could also involve a concerted process between the trapping agent and the anion. It should be pointed out that N-sulfonylazepines (54) and not the anilides would be expected from the reaction of aromatic compounds with sulfonyl nitrenes at room temperature (see Section V,F).

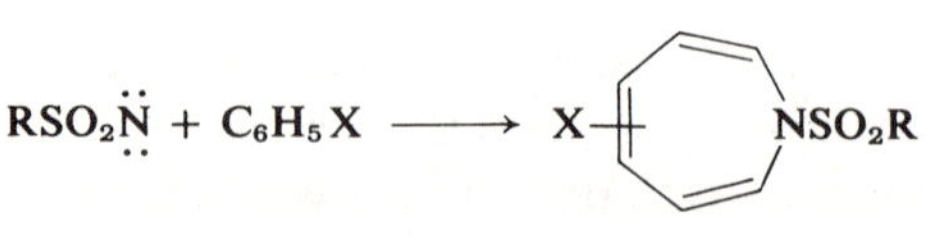

$$RSO_2\overset{..}{N} + C_6H_5X \longrightarrow X-\text{azepine}-NSO_2R$$
(54)

3. NITRENES

The Hofmann and Lossen rearrangements have often been pictured as proceeding via a free nitrene mechanism.

$$\underset{O}{\overset{R}{\underset{\|}{C}}}\text{—NHBr} \xrightarrow{\text{OH}^-} \underset{O}{\overset{R}{\underset{\|}{C}}}\text{—}\overset{\ominus}{N}\text{—Br} \xrightarrow[\text{stepwise}]{\text{Br}^-} \underset{O}{\overset{R}{\underset{\|}{C}}}\text{—}\overset{..}{\underset{..}{N}} \longrightarrow R\text{—}N\text{=}C\text{=}O$$

$$\xrightarrow{\text{concerted}}$$

$$\underset{O}{\overset{R}{\underset{\|}{C}}}\text{—NHOCOR}' \xrightarrow{\text{OH}^-} \underset{O}{\overset{R}{\underset{\|}{C}}}\text{—}\overset{\ominus}{N}\text{—OCOR}' \xrightarrow{-R'CO_2^-} \underset{O}{\overset{R}{\underset{\|}{C}}}\text{—}\overset{..}{\underset{..}{N}} \longrightarrow R\text{—}N\text{=}C\text{=}O$$

The concerted mechanisms are now generally favored since no nitrenes have been trapped from these reactions, and, as in the Curtius rearrangement, the observation of substantial isotope effects (Fry and Wright, 1966, 1968) shows that the group R participates in the rate-determining step. On the other hand, substituents in the group R have little effect upon the rates of rearrangement; e.g., *m*-nitro-*N*-bromobenzamide and *p*-nitro-*N*-bromobenzamide rearrange at the same rate, and the heats of activation for the rearrangement of *N*-bromobenzamide and *m*-chloro-*N*-bromobenzamide are the same within the limits of experimental error. It has been suggested that the small (indeed negligible, occasionally) substituent effects can be attributed to a rather lopsided transition state, the energy of which is largely determined by the formation (in the case of N_2) or solvation (Br^- in the Hofmann rearrangement) of the leaving group (Lwowski, 1970). Although this may indeed be the case, we believe that one can visualize a whole continuum of mechanisms ranging from those cases where substituents facilitate participation by the migrating group in the rate-determining step to those in which participation is unlikely because of the highly electron-attracting nature of the substituents. In these latter cases, it might become possible to trap a nitrene in competition with the intramolecular rearrangement, particularly if a nucleophilic solvent is used.

D. Oxidative Routes

In principle, it should be possible to oxidize a primary amino group to the nitrene level.

$$R\text{—}NH_2 \xrightarrow{[O]} R\overset{..}{\underset{..}{N}} + H_2O$$

Such oxidations have been reported to be particularly successful with hydrazine derivatives to give aminonitrenes (for a review, see Lemal, 1970). Mercuric oxide and, more recently, lead tetraacetate have been used extensively. If the oxidation of hydrazine with $Pb(OAc)_4$ is carried out in the

$$R_2N-NH_2 \xrightarrow{HgO} R_2N-\ddot{N} \longleftrightarrow R_2\overset{+}{N}=\overset{-}{N}$$

presence of suitable olefins, aziridines may be formed in high yield (Atkinson and Rees, 1967; Anderson et al., 1969). Although such reactions may well

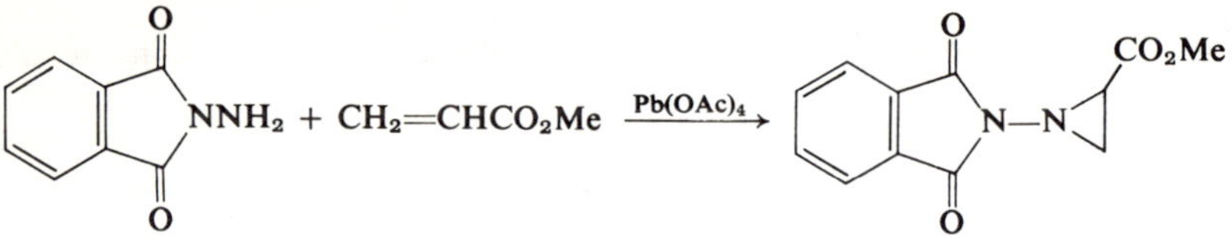

[reaction: phthalimide-NNH₂ + CH₂=CHCO₂Me —Pb(OAc)₄→ phthalimide-N—N aziridine with CO₂Me]

involve the formation of nitrenes, a concerted process between the substrate and a metal salt of amine anion could explain the observed products just as well, and studies of the kinetic involvement of the substrate in the rate-determining step will have to be carried out before the mechanism of these oxidations can be settled.

E. OTHER SOURCES OF NITRENES

Photolysis of starting materials other than azides has occasionally been a source of nitrene intermediates. Thus, irradiation of oxaziridines leads to nitrenes which can be intercepted (Meyer and Griffin, 1967; Splitter and Calvin, 1968).

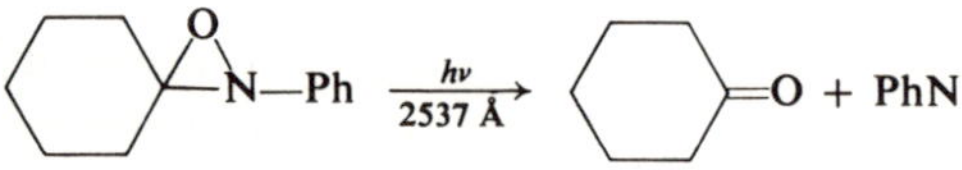

[reaction: oxaziridine —hν, 2537 Å→ cyclohexanone + PhN]

Phosphinimines are also said to give nitrenes on photolysis (Zimmer and Jayawant, 1966).

$$Ph_3P=NCMe_3 \xrightarrow{h\nu} \begin{array}{l} Ph_3P + Me_3C-N \\ Ph_3P=N\cdot + Me_3C\cdot \end{array}$$

Some isocyanates undergo photolysis to the desired nitrenes and CO [Eqs. (12) (Boyer et al., 1967), (13) (Swenton, 1967; Sauer, 1970), and (14) (Bamford and Bamford, 1941)].

3. NITRENES

$$PhCH=CHN=C=O \xrightarrow{h\nu} CO + PhCH=CH-N \quad (12)$$

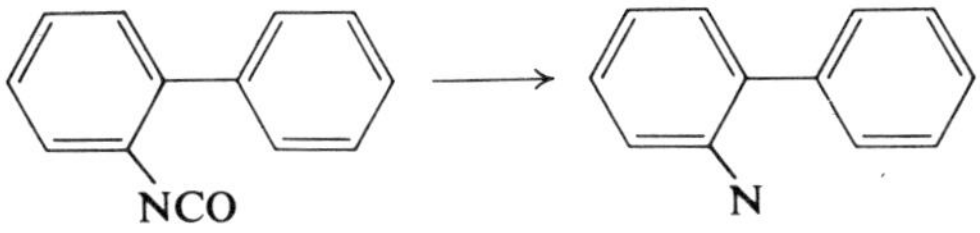

$$\text{(2-biphenylyl-NCO)} \longrightarrow \text{(2-biphenylyl-N)} \quad (13)$$

$$MeNCO \xrightarrow{h\nu} \begin{array}{l} Me\cdot + \cdot NCO \\ MeN + CO \end{array} \quad (14)$$

Irradiation of 3-aryl-1,4,2-dioxazolidin-5-one (**55**) is believed to give aroyl nitrenes (Sauer and Mayer, 1968).

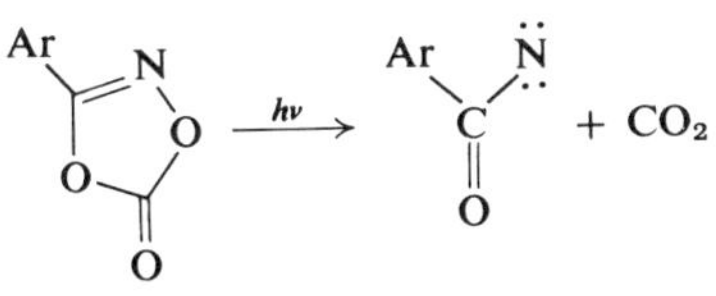

$$\underset{(55)}{\text{Ar-C(=N-O-C(=O)-O)}} \xrightarrow{h\nu} \text{Ar-C(=O)-N:} + CO_2$$

Other possible photochemical and thermal sources of nitrenes have been discussed in the various reviews on nitrene chemistry that have appeared, and these should be consulted for more information on the subject.

V. Reactions

The reactions undergone by nitrenes fall under seven general headings: hydrogen abstraction, insertion into aliphatic C–H bonds, rearrangement, additions to olefins, dimerization, aromatic substitution, and trapping by nucleophiles (other than olefins). Not all nitrenes undergo all of these modes of stabilization, the reactions observed depending on the nature of the group on nitrogen to a major extent.

A. Hydrogen Abstraction

This is a common side reaction (in some cases it becomes the dominant if not exclusive reaction as, for example, in the photolysis of $MeSO_2N_3$ in isopropanol), particularly with aryl and sulfonyl nitrenes, and has often been used as a diagnosis for the participation of nitrenes in a reaction, e.g., in the thermolysis of 2-phenyl-2-propyl azide. It is perhaps important to emphasize again that hydrogen abstraction may become the dominant reaction for sulfonyl nitrene–metal complexes in aromatic solvents. In aliphatic solvents, hydrogen abstraction probably involves the triplet nitrene.

$$R\ddot{\dot{N}}\cdot + R^1H \longrightarrow R\ddot{\dot{N}}H + R^1\cdot$$

$$R\ddot{\dot{N}}H \xrightarrow{[H]} RNH_2$$

Examples include the thermolysis of phenyl azide in hydrocarbon solvents (Hall *et al.*, 1968),

$$PhN_3 \xrightarrow[RH]{160°C} PhNH_2 + \text{other products}$$

the photolysis of *N-tert*-butyltriphenylphosphinimine in cyclohexene (Zimmer and Jayawant, 1966),

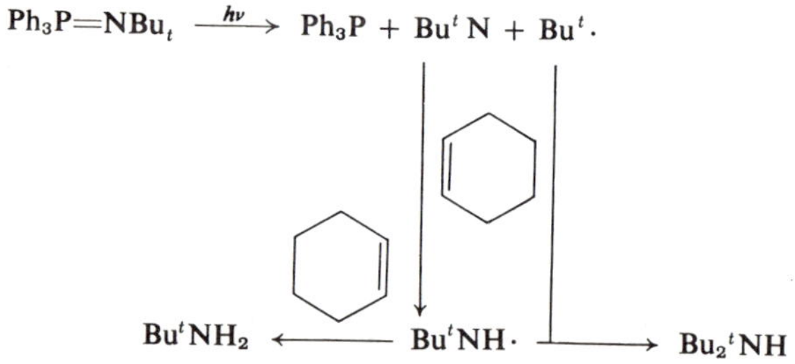

Ph₃P=NBu_t
hν
Ph₃P + Bu^t N + Bu^t·
Bu^tNH₂
Bu^tNH·
Bu₂^tNH

and the thermolysis of 2-azido-2′,4′,6′-trimethylbiphenyl (**56**) (Smolinsky, 1960) or the reaction of 2-nitro-2′,4′,6′-trimethylbiphenyl (**57**) with iron (II) oxalate (Abramovitch *et al.*, 1961).

Me
Me
N₃ Me
Δ
Me
Me + other products
NH₂ Me
Me
Δ
FeC₂O₄
Me
NO₂ Me

(**56**)

(**57**)

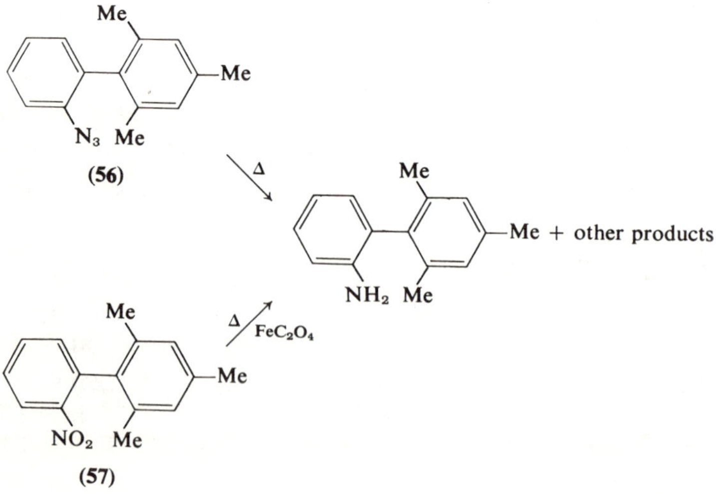

In all the above cases, numerous other products are also formed.

Alkylsulfonyl nitrenes undergo hydrogen abstraction to varying extents when generated in the presence of aliphatic hydrocarbons. Thus, 2-propanesulfonyl azide gave a maximum of 3% of 2-propanesulfonamide on thermolysis in cyclohexane (Breslow et al., 1969), whereas α-toluenesulfonyl azide gave a 26.5% yield of α-toluenesulfonamide by abstraction from n-dodecane (Abramovitch et al., 1969a). Arylsulfonyl nitrenes also abstract hydrogen from aliphatic hydrocarbons: Toluene-p-sulfonyl azide gave toluene-p-sulfonamide in cyclohexane (Breslow et al., 1969) and diphenyl sulfide 2-sulfonyl azide gave a 19% yield of the primary amide in dodecane (Abramovitch et al., 1969a). On the other hand, thermolysis of ferrocenylsulfonyl azide (43) in aliphatic solvents led to a predominant formation of the amide (45) (Abramovitch et al., 1971a). Lower yields of 45 were obtained on photolysis of 43 in cyclohexane and cyclohexene. This suggests, again, that a metal nitrene complex is an intermediate in the thermolysis of 43. Benzenesulfonamide was the main product (37%) in the copper-catalyzed decomposition of $PhSO_2N_3$ in cyclohexane, and the yield was not decreased but increased slightly to 49% in the presence of hydroquinone (Kwart and Kahn, 1967).

Sulfonyl nitrenes abstract hydrogen from aromatic solvents readily. This is not thought to be a triplet sulfonyl nitrene reaction since none of the expected radical products formed by abstraction of a single hydrogen atom from the aromatic nucleus were ever observed. It was concluded, therefore, that abstraction of both hydrogens was taking place in a concerted, or almost concerted, process, and it was suggested that the aziridine intermediate formed initially (see Section V,F) may collapse to the sulfonamide and benzyne, the latter going to tars (Abramovitch et al., 1965). No evidence was presented or found for the intervention of benzynes in these reactions.

$$\left[\begin{array}{c} H \\ \\ H \end{array} \bigcirc NSO_2Me \right] \longrightarrow \left[\bigcirc \| \right] + MeSO_2NH_2$$

2-Pyridyl- and 4,6-dimethyl-2-pyrimidylnitrene–copper complexes abstract hydrogen from cyclohexane and cyclohexene at 140°C to give the corresponding 2-amino derivatives in good yield (von Fraunberg and Huisgen, 1969). Uncatalyzed thermolysis also led to hydrogen abstraction products.

Sulfonyl nitrene–metal complexes also abstract hydrogen from aromatic solvents. For example, whereas the decomposition of $MeSO_2N_3$ in degassed benzotrifluoride under nitrogen at 120°C gave $MeSO_2NHC_6H_4CF_3$ (20.4%) and $MeSO_2NH_2$ (21.9%), the addition of $Fe_3(CO)_{12}$ led to a drop in the anilide yield to 0.75% and an increase in the $MeSO_2NH_2$ yield to 61.5% (Abramovitch et al., 1969a).

Triplet carbonyl azides also abstract hydrogen, for example from cyclohexane, to give cyclohexene, and it appears that both hydrogens at the adjacent carbons are removed simultaneously (Breslow et al., 1967).

B. Insertion into Aliphatic C–H Bonds

This reaction is usually thought to be typical of singlet nitrenes and to take place with retention of configuration (Lwowski, 1967; Breslow et al., 1969), though hydrogen abstraction by triplet nitrene followed by radical coupling would also lead to products of apparent insertion. In the latter case, coupling would have to be faster than rotation about a single bond to explain retention

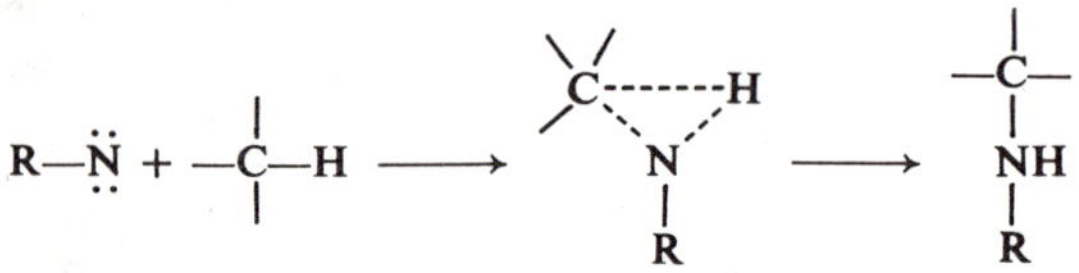

of stereochemistry.

Examples of intramolecular insertions by aryl nitrenes and nitrenoïds are known. Some of these are outlined below [Eqs. (15) (Smolinsky and Feuer, 1966), (16) (Sundberg, 1966b), (17) (Abramovitch et al., 1969c), and (18) (Meth-Cohn et al., 1963)].

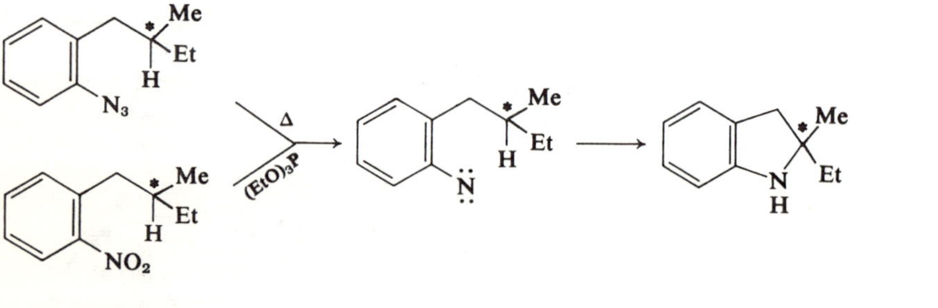

(15)

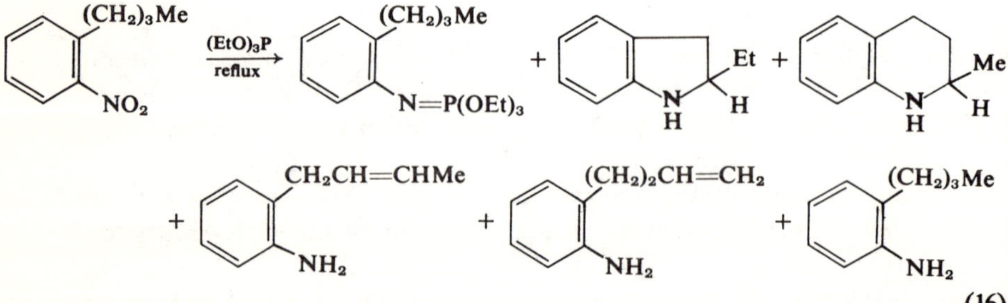

(16)

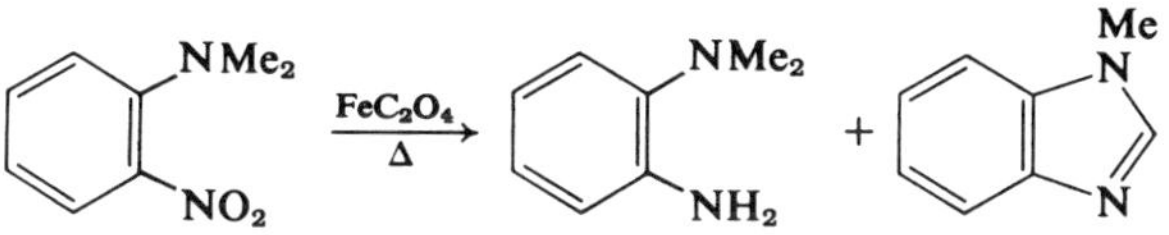

[Structure diagram: 2-NMe₂-nitrobenzene → (FeC₂O₄, Δ) → 2-aminodimethylaniline + 1-methylbenzimidazole] (17)

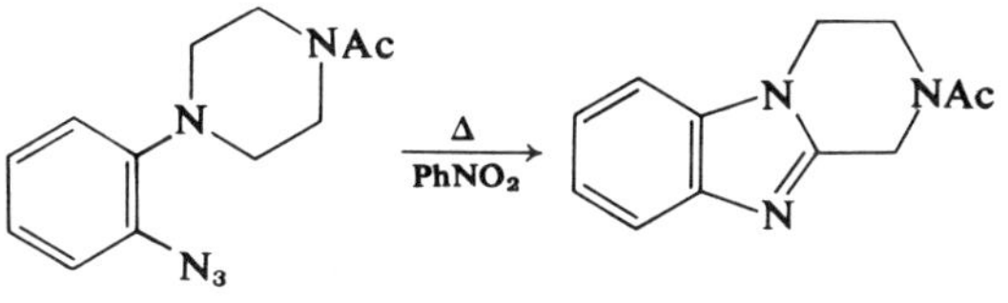

[Structure diagram: N-(2-azidophenyl)-N'-acetylpiperazine → (Δ, PhNO₂) → fused benzimidazole-NAc product] (18)

Some evidence has been presented, however, that the intermolecular insertion of phenyl nitrene into the C–H bonds of aliphatic hydrocarbons (a rather inefficient process) is actually due to the triplet (Hall *et al.*, 1968).

Contrary to the implication in a recent review (Belloli, 1971), no clear-cut examples of intramolecular C–H insertion by alkyl nitrenes have been reported. Numerous attempts to reproduce the earlier reported examples (Barton and Morgan, 1962) of such intramolecular insertions were unsuccessful (Barton and Starratt, 1965; Smolinsky and Feuer, 1964), the main product in all cases being the imine arising by a 1,2-hydrogen shift. Moriarity and Rahman (1965) reported that intramolecular photocyclization of alkyl azides to form pyrrolidine derivatives was a highly irreproducible process, this pathway accounting for only a very minor amount of product or none at all. 1,2-Hydrogen migration was the main reaction pathway, but a small amount of primary amine was formed by hydrogen abstraction, probably from solvent, and 1–2% of secondary amines were formed by insertion into the solvent cyclohexane.

In contrast, alkyl- and arylsulfonyl nitrenes insert readily into aliphatic C–H bonds [Eqs. (19) (Breslow *et al.*, 1969; Abramovitch *et al.*, 1971a) and (20) (Abramovitch *et al.*, 1969a)]. Intramolecular C–H insertion has also been reported [Eq. (21) (Abramovitch *et al.*, 1969a)].

The determination of the relative insertion selectivity into primary,

$$RSO_2N_3 + C_6H_{12} \xrightarrow{\Delta} RSO_2NHC_6H_{11}$$ (19)

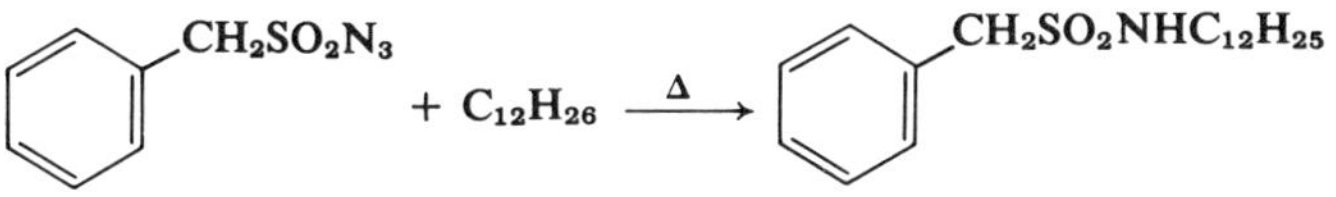

[Structure: PhCH₂SO₂N₃ + C₁₂H₂₆ → (Δ) → PhCH₂SO₂NHC₁₂H₂₅] (20)

[Structure: 2-(phenoxy)benzenesulfonyl azide + C₁₂H₂₆ → (Δ) → 2-(phenoxy)benzenesulfonamide-C₁₂H₂₅ + other products]

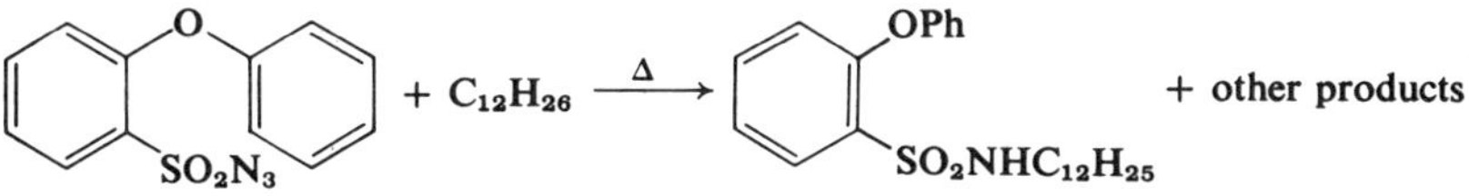

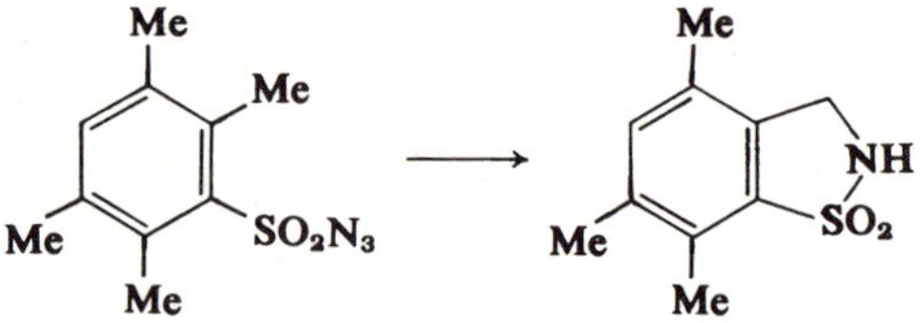

[Structure (21): 2,3,4,6-tetramethylbenzenesulfonyl azide → cyclized benzosultam]

$$\text{(21)}$$

secondary, and tertiary C–H bonds is complicated by the fact that the SO_2 liberated in the reaction decomposes some of the substituted sulfonamides formed. Insertion of toluene-*p*-sulfonyl nitrene into 2-methylbutane gave a mixture of products which could not be completely resolved but from which the ratio of (primary):(secondary + tertiary) = 1.53 could be established (Breslow et al., 1969). This is lower than the corresponding ratio for carbethoxynitrene (see below), indicating the lowered selectivity of the sulfonyl nitrene relative to carbethoxynitrene since the latter has the possibility of resonance stabilization (**58**). It has not been established yet whether only singlets or both singlets and triplets (as found in some aromatic substitution reactions) are involved in these insertion selectivity studies.

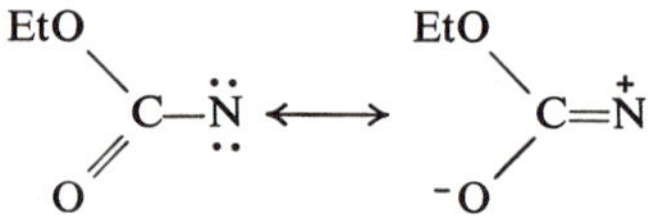

$$\text{EtO} \diagdown \atop O \diagup \!\!\! = \!\!\! C-\ddot{\underset{..}{N}} \longleftrightarrow \text{EtO} \diagdown \atop ^{-}O \diagup C = \overset{+}{N}$$

(**58**)

As just indicated above, carbalkoxynitrenes insert readily into aliphatic C–H bonds, both inter- and intramolecularly. For example, photolysis of $EtOCON_3$ in 2-methylbutane gave a mixture of insertion products in which the ratio of insertion into 3° to 2° to 1° C–H bonds (after statistical correction)

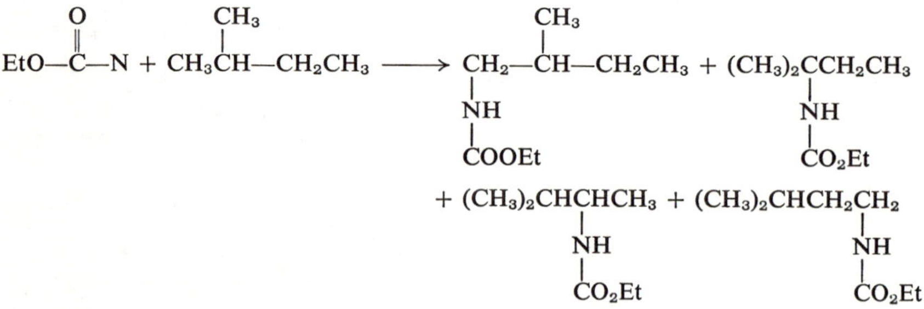

$$\text{EtO}-\underset{\text{O}}{\overset{\|}{C}}-N + CH_3\underset{CH_3}{\overset{|}{C}}H-CH_2CH_3 \longrightarrow CH_2-\underset{CH_3}{\overset{|}{C}}H-CH_2CH_3 + (CH_3)_2CCH_2CH_3$$

with NH–COOEt and NH–CO$_2$Et substituents

$$+ (CH_3)_2CHCHCH_3 + (CH_3)_2CHCH_2CH_2$$

with NH–CO$_2$Et substituents

was found to be 34:9:1 (Lwowski and Maricich, 1965). As indicated above, this selectivity is greater than that of *p*-$MeC_6H_4SO_2N$. The intermolecular insertion of carbethoxynitrene into the tertiary C–H bond of optically active 3-methylhexane gave the urethane with 100% retention of configuration (Lwowski and Simson, 1967). The intramolecular cyclization of $R(+)$-2-

methylbutyl azidoformate (**59**) gave 4-ethyl-4-methyloxazolidin-2-one (**60**) with total retention of configuration (Smolinsky and Feuer, 1964; Yamada et al., 1965). From these and other experiments it was concluded that it is

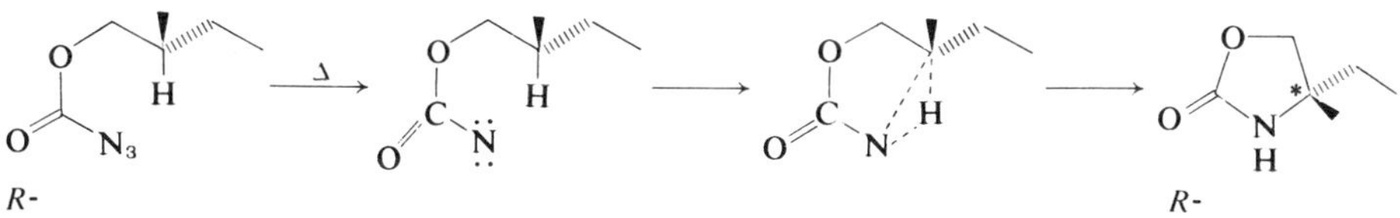

the singlet carbalkoxynitrene which inserts into inactivated C–H bonds. Had hydrogen abstraction followed by radical coupling occurred, one would have expected loss of stereospecificity unless coupling took place faster than did inversion about the trigonal carbon radical (i.e., within a solvent cage before the radicals had time to diffuse apart). This conclusion is supported by experiments with cyclohexene at concentrations varying from 0.2 to 100 mole%. Both singlet and triplet nitrene add to the double bond (*vide infra*), but only the singlet inserts into saturated C–H bonds. As the concentration of cyclohexene decreases, the amount of addition relative to insertion increases, since the nitrene now has more time to drop to the triplet state which adds to the olefin but does not insert (Lwowski and Woerner, 1965).

C. Rearrangement

As more and more examples are being uncovered, it is clear that this is an important mode of stabilization of nitrenes. Singlet aryl nitrenes probably undergo the most interesting intramolecular bond reorganizations to give a number of other reactive intermediates which can be trapped at various stages, depending on the reaction conditions, to give stable compounds. These may be summarized as follows for phenylnitrene itself:

(**61**)   (**62**)   (**63**)   (**64**)   (**65**)

If the aryl azide is decomposed in the presence of a good nucleophile, e.g., a secondary amine, **62** is trapped as the 2-substituted amino-3*H*-azepine (**66**) (Doering and Odum, 1966; Ogata et al., 1968; Appl and Huisgen, 1959), and similar results are obtained in the thermal or photochemical deoxygenation of nitro (Sundberg et al., 1969; Cadogan and Todd, 1967) and nitroso (Odum and Brenner, 1966) compounds with triethyl phosphite. In the photochemical

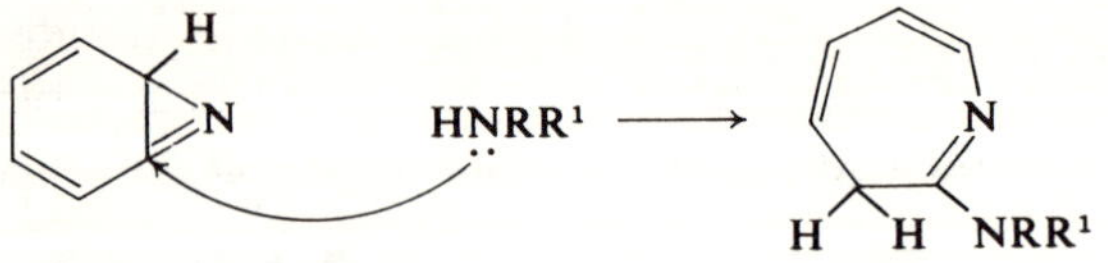

deoxygenation of nitro compounds with triethyl phosphite the reaction can go further, and pyridine derivatives (**67**) are also formed (Sundberg et al., 1969). Alternatively, a solvent-stabilized intermediate of the type **65** could be trapped directly by the nitrene or its precursor (Wentrup, 1969). Such intermediates as **63** have been postulated to explain the reversible isomerization of 2-pyridylcarbenes (**68**) and phenylnitrenes (**69**) at elevated temperatures ($\approx 900°C/0.92$ mm) (Crow and Wentrup, 1968). Indeed, the gas-phase pyrolysis of 2,6-dimethylphenyl azide gives an 8% yield of 6-methyl-2-vinylpyridine (**70**) together with 2,6-xylidine (30%), benzonitrile (18%), indole (15%), and a methylindole. Interestingly, it has also been shown by (Wentrup, 1969) ^{15}N labeling in cases where no ortho methyl groups are present

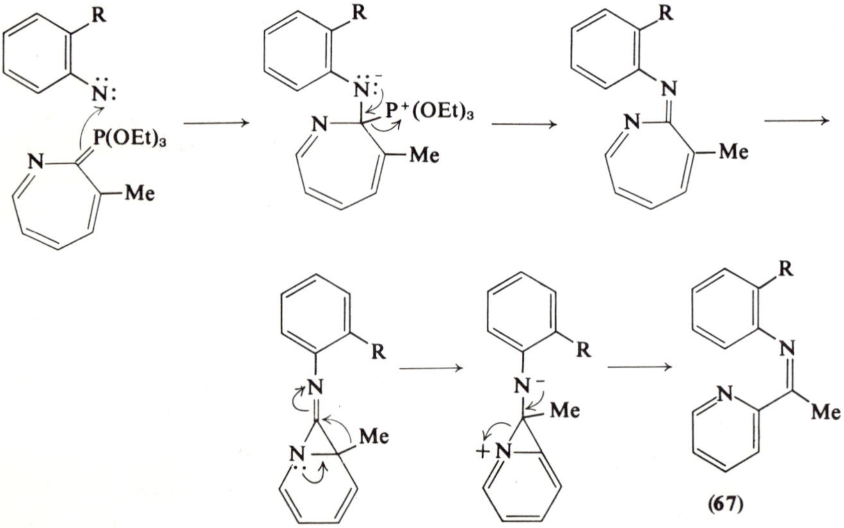

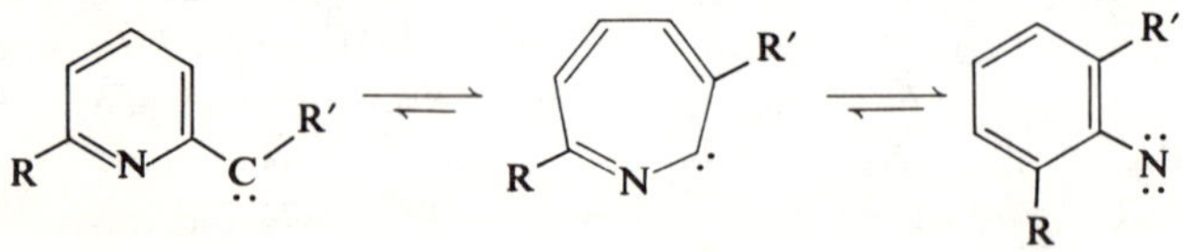

that the nitrene nitrogen was retained in the pyridines formed. Labeling of the ortho positions in **69** with deuterium demonstrated intermolecular scrambling of ortho hydrogens in the formation of pyridines, and a mechanism

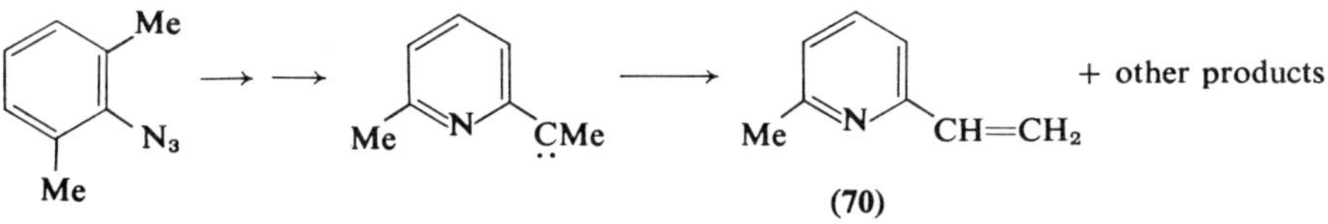

for this was proposed (Wentrup, 1969). Ring-opening of aryl nitrenes to give rearranged or ring-contracted products can occur. The thermolysis of *o*-diazidobenzene gives *cis,cis*-mucononitrile (Hall, 1965).

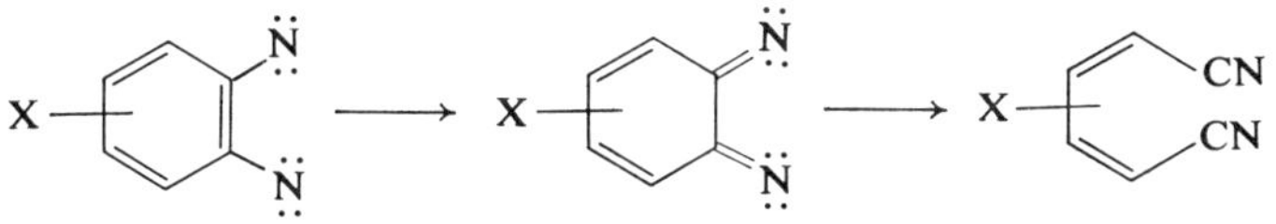

The ring expansion of the triazole (**71**) to the tetrazine (**72**) may proceed as follows (Takimoto and Denault, 1966):

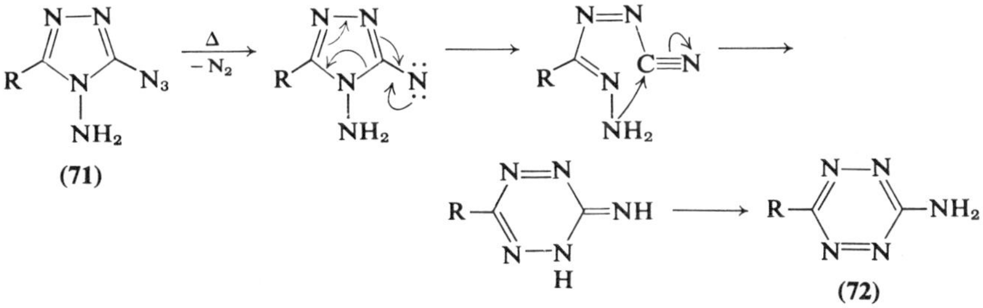

Ring contraction of aromatic and heterocyclic nitrenes occurs readily at high pressure in the gas phase to give the corresponding nitrile. ^{15}N labeling in

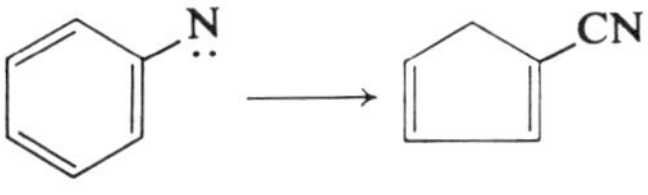

2-pyridylnitrene demonstrates nitrogen scrambling in the gas phase via an intermediate such as **73** in which the two nitrogens become equivalent (Crow and Wentrup, 1969).

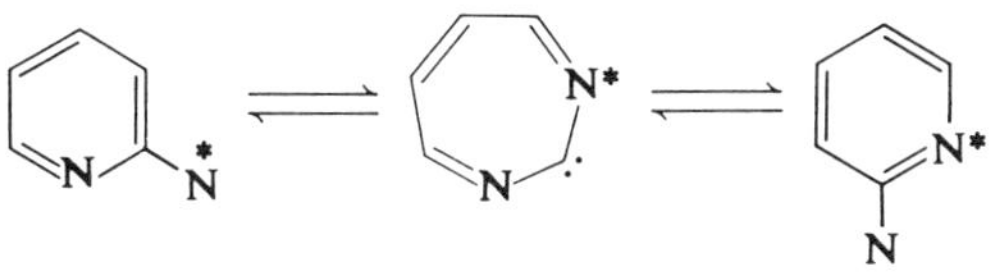

The rearrangements of alkyl nitrenes are much more straightforward, and 1,2-hydrogen, -alkyl, or -aryl shifts occur. This is an

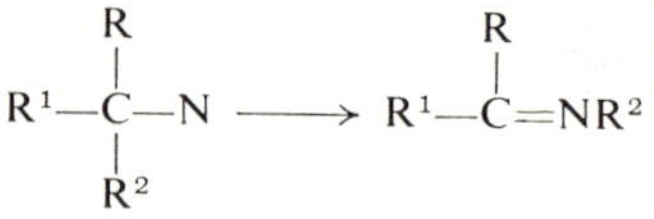

$$R^1-\underset{R^2}{\overset{R}{C}}-N \longrightarrow R^1-\overset{R}{C}=NR^2$$

extremely ready process and is the main (if not exclusive) mode of stabilization of these reactive intermediates. Migratory aptitudes have already been discussed in Sections IV,A,1 and IV,A,2.

Whereas carbonyl azides (and the related species in the Hofmann and Lossen rearrangements) undergo 1,2-shifts readily via a nitrene intermediate or in a concerted process (see discussion above), carbalkoxy and sulfonyl azides were regarded by Curtius as being "rigid" (*starre*), i.e., they did not undergo rearrangement on thermolysis.

Sulfonyl azides have recently been shown not to be as "rigid" as they were previously thought to be. Photolysis of benzenesulfonyl azide in alcohols and amines probably involves the protonated species, and no free nitrene is formed (Lwowski and Schieffele, 1965). The vapor-phase pyrolysis of benzenesulfonyl azide at 625°C gave a 17.5% yield of azobenzene (Reichle, 1964), and a trace of the latter was also formed when the azide was heated in boiling cyclohexanone (Balabanov *et al.*, 1966).

When mesitylene-2-sulfonyl azide (74) was heated to 150°C in dodecane, the products included those of hydrogen abstraction (75), intermolecular C–H insertion (76), intramolecular C–H insertion (77), radical transfer (78), and Curtius-type rearrangement [(79) and (80)]. Durene-3-sulfonyl azide behaves similarly (Abramovitch and Holcomb, 1969).

Cyanonitrene generated photolytically from the azide in an argon matrix

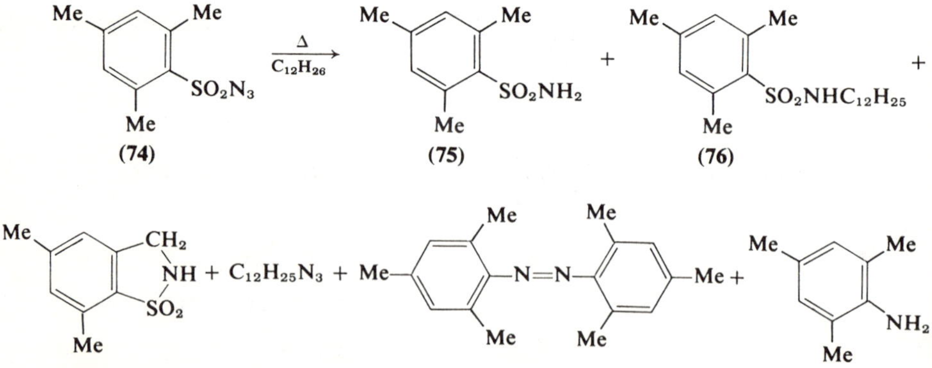

3. NITRENES

also isomerizes and gives diazomethylene (**81**) (Anastassiou et al., 1970). It is proposed that this rearrangement involves the formation of atomic carbon

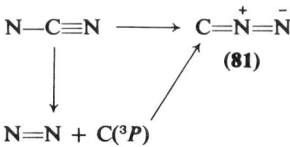

(^3P) and nitrogen which recombine to give **81** (recombination can also occur between the carbon atom and the nitrogen molecule originally lost from the azide).

D. Addition to Olefins

Although this is a very common process in carbene chemistry, it has only been shown to occur readily with carbethoxy- and cyanonitrenes. Additions of aryl, alkyl, and sulfonyl nitrenes to such linkages are relatively unknown. The reason is that the nitrene precursor, usually the azide, undergoes 1,3-dipolar addition to give a triazoline, which then loses nitrogen and yields the same aziridine as would have been obtained by the direct addition of the nitrene to the olefin. Alternatively, as discussed previously, the olefinic double bond participates in the nitrogen elimination, and a free nitrene is never involved. Should methods be developed to generate sulfonyl nitrenes at temperatures below those at which the azide adds to the olefin, it should perhaps become possible to observe 2 + 1 additions as with NCN and EtOCON.

In the addition of carbethyoxynitrene to olefins, both the singlet and the triplet species add, the former stereospecifically and the latter nonstereospecifically, as expected on the basis of Skell's postulate concerning the addition of carbenes (see Chapter 2). This was studied by Lwowski and McConoghy (see Lwowski, 1970) who formulated reaction (22) (in which cis- and trans-4-methyl-2-pentene are used as examples of the olefins).

Singlet nitrene adds in one step to give the aziridine in which the original geometry of the olefin is retained. With the triplet nitrene, both **82** and **83** are formed in relative amounts which depend on the relative rates of spin inversion in the intermediate diradicals and rotation about the C-2 to C-3 bond. The faster the latter, the less stereospecificity will be observed.

With 1,3-dienes, only 1,2-addition is observed, but the vinylaziridines formed can be thermally rearranged to the apparent 1,4-addition products (Mishra et al., 1968).

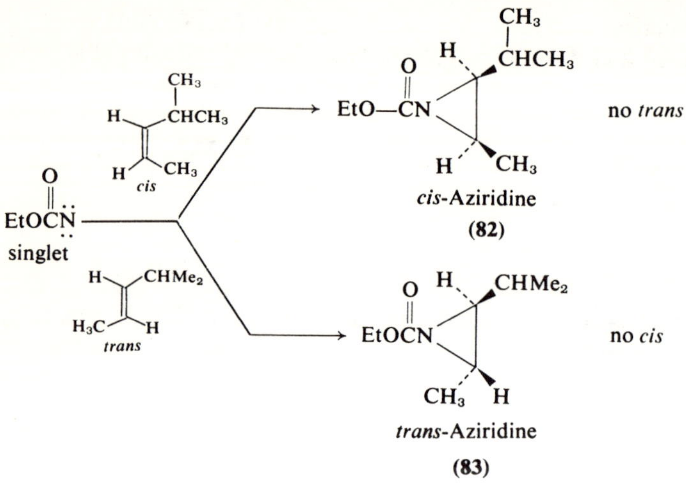

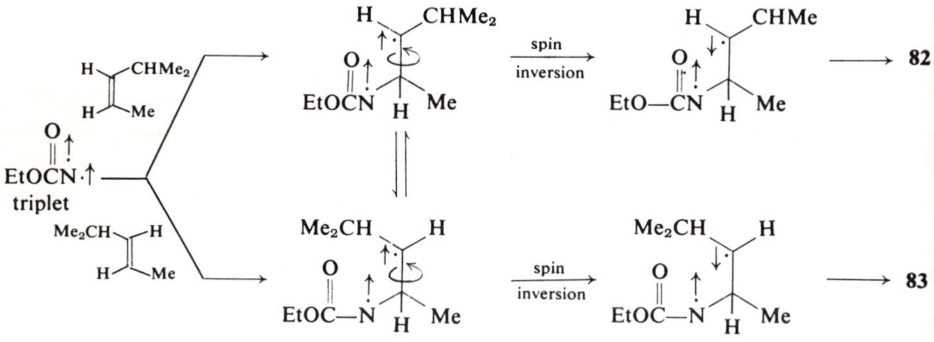

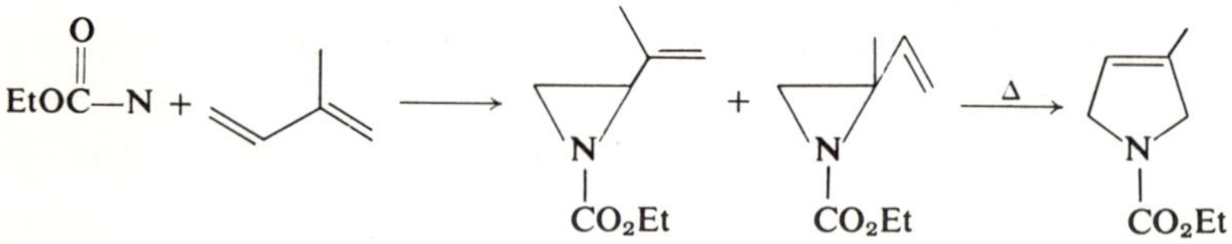

Cyanogen azide reacts with double bonds very readily, probably in a concerted process not involving a nitrene. It reacts sluggishly with cyclo-octatetraene, however, and at 80°C in ethyl acetate (a temperature higher than that normally necessary to decompose this azide), addition products **84, 85,** and **86** are formed, presumably via the nitrene (Anastassiou, 1965, 1968).

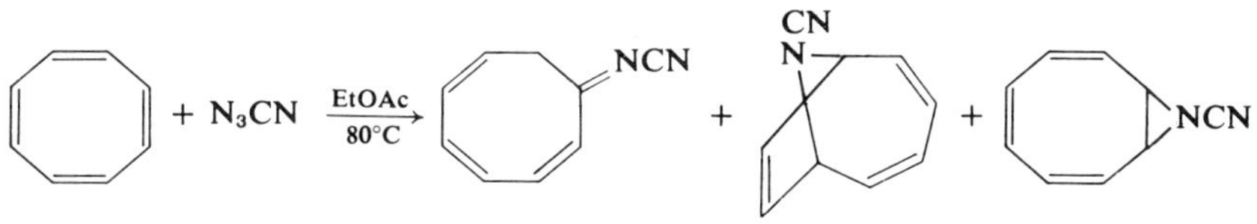

3. NITRENES

[Scheme showing cyclooctatetraene + N₃CN → EtOAc, 80°C → products (84), (85), (86)]

(84)   (85)   (86)

Nitrene–metal complexes appear to add to olefinic double bonds. For instance, the 4,6-dimethyl-2-pyrimidylnitrene–copper complex adds stereospecifically *cis* to *trans*-stilbene, but no addition to the double bond of cyclohexene was reported (von Fraumberg and Huisgen, 1969). The copper-catalyzed decomposition of benzenesulfonyl azide in cyclohexene did give the aziridine (Kwart and Kahn, 1967). Photolysis of ferrocenylsulfonyl azide in cyclohexene gave the corresponding aziridine derivative (9%) but thermolysis did not (Abramovitch et al., 1971a).

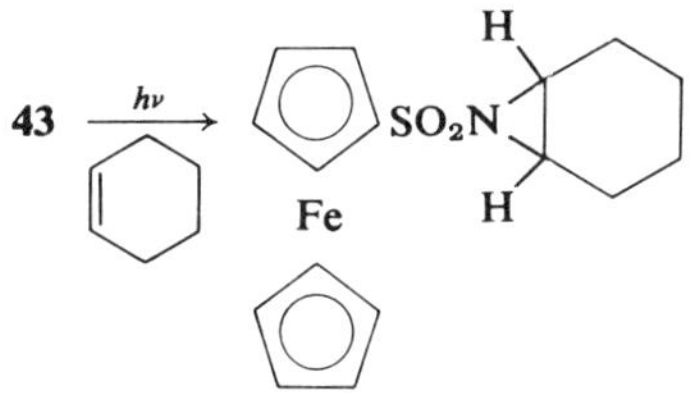

[Scheme: 43 →hv→ ferrocenyl-SO₂NH-cyclohexyl]

Photolysis or thermolysis of ferrocenyl azide (87) in cyclohexene gives a small amount of the aziridine derivative (88) together with other products. This photolysis is probably the first authenticated example of the addition of an aryl nitrene (metal complex?) to an olefinic double bond (Abramovitch et al., 1971b).

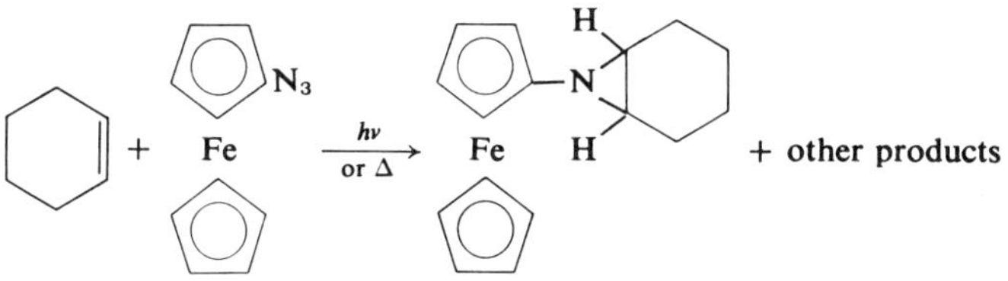

[Scheme: cyclohexene + ferrocenyl-N₃ (87) →hv or Δ→ ferrocenyl-NH-cyclohexyl (88) + other products]

(87)   (88)

E. Dimerization

This is a common reaction of aryl nitrenes leading to azo compounds, and is often diagnostic of the formation of an aryl nitrene. It is most probably a reaction of the triplet species (if it is a dimerization and not an attack by the nitrene on the nitrene precursor*) since this would not require a spin inversion

$$2 \text{ Ar}\ddot{\text{N}}\cdot \longrightarrow \text{Ar}-\text{N}=\text{N}-\text{Ar}$$

or uncoupling. There is no definitive evidence that this is a dimerization except that vapor-phase pyrolyses of azides gave very good yields of the azo compounds (Smolinsky, 1961b). Also, sensitized (acetophenone) irradiation of 2-azidobiphenyl gives mainly the azo compound (**89**) (43%) together with less than 8% carbazole (**2**), whereas direct photolysis of the azide in a concentrated ether solution using a Pyrex filter and in the presence of 2 M

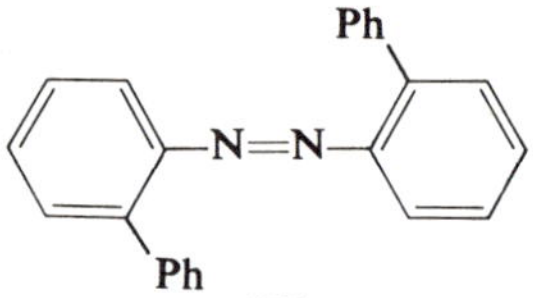

<img: structure of compound 89, an azo compound with two ortho-phenyl substituted phenyl groups connected by N=N>

(**89**)

piperylene gave an 89% yield of carbazole and only a 4% yield of **89**. Support for the above suggestion (Swenton, 1968) that cyclization involves the singlet and azo compound formation the triplet comes from the observation that **89** is formed in the photolysis of 2-azidobiphenyl in the presence of aromatic hydrocarbons which transfer singlet energy; the cyclization product is obtained under these conditions (Swenton *et al.*, 1969).

This type of reaction does not appear to have been observed with alkyl or sulfonyl nitrenes, though azomethane may be an intermediate in the formation of some $(\text{MeN})_n$ in the photolysis of methyl azide in the vapor phase (Currie and Darwent, 1963).

Tetrazenes are often the major products in the oxidation of 1,1-disubstituted hydrazines.

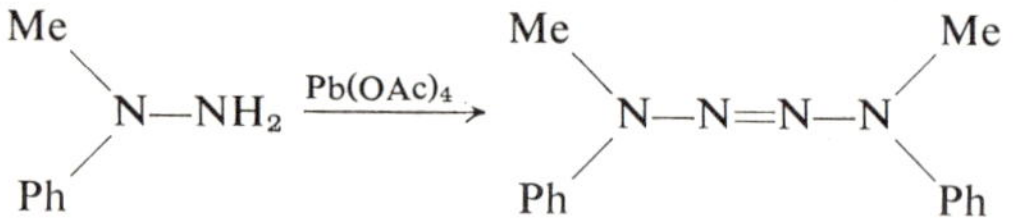

$$\underset{\text{Ph}}{\overset{\text{Me}}{\diagdown}}\text{N}-\text{NH}_2 \xrightarrow{\text{Pb(OAc)}_4} \underset{\text{Ph}}{\overset{\text{Me}}{\diagdown}}\text{N}-\text{N}=\text{N}-\text{N}\underset{\text{Ph}}{\overset{\text{Me}}{\diagup}}$$

Whereas dimerization of the aminonitrene (R_2N-N) can be visualized, formation of a tetrazane followed by oxidation is more likely (Lemal, 1970).

$$R_2N-\ddot{N} + H_2N-NR_2 \longrightarrow R_2N-NH-NH-NR_2 \xrightarrow{[O]} R_2N-N=N-NR_2$$

---

* When phenylnitrene is generated from nitrosobenzene and triethyl phosphite in benzene at 5°C in the presence of 1 equivalent of PhN_3, the azide is recovered almost quantitatively. Some azoxybenzene and much tar are formed (Abramovitch and Challand, 1971).

F. Aromatic Substitution

Intramolecular aromatic substitutions by aryl nitrenes are well known and have been reviewed (Abramovitch and Davis, 1964; Cadogan, 1968; Abramovitch, 1967). It seems very likely (see later) that it is the singlet species that is the effective intermediate.

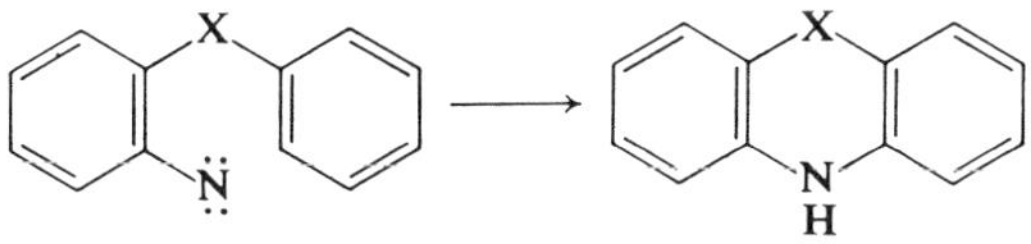

A comparison of intramolecular aromatic substitution and aliphatic C–H insertion can be made. 2-Azido-2′-methylbiphenyl (**90**) gives the carbazole (**91**) exclusively in 91% yield (Coffin and Robbins, 1965), and only when aromatic substitution becomes much more difficult, as in the case of 2,4,6-trimethyl-2′-azidobiphenyl, does C–H insertion become dominant (Smolinsky, 1960).

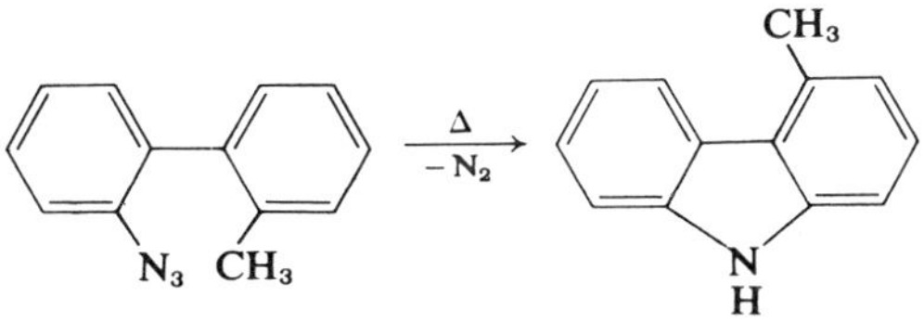

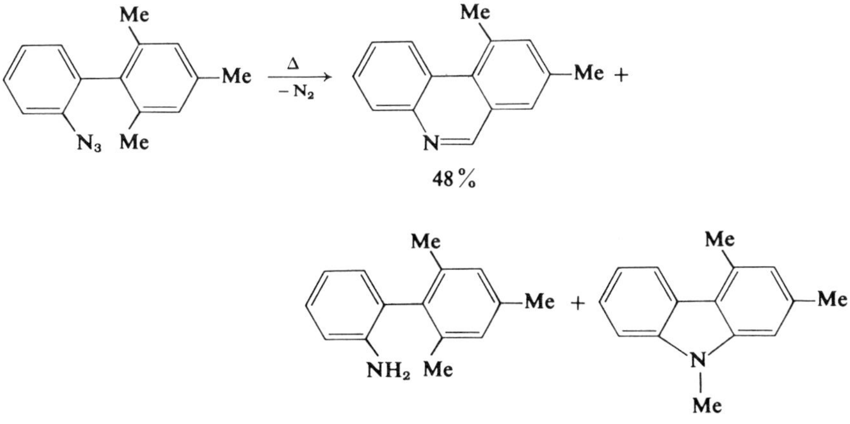

In some instances, molecular rearrangements accompany intramolecular aromatic substitution. This is the case in the cyclizations leading to phenothiazines (Cadogan, 1968; Messer and Farge, 1968).

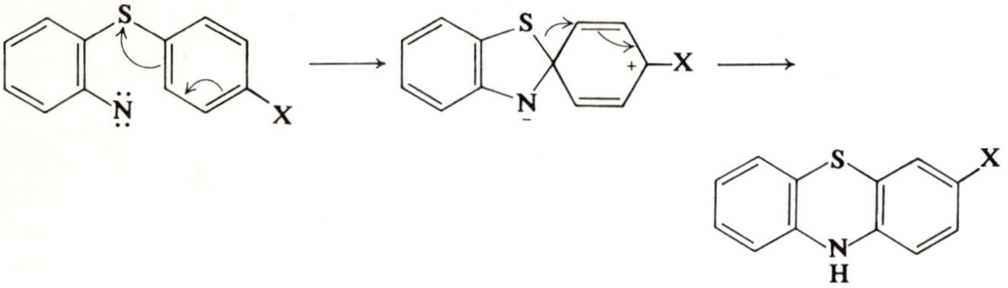

Until recently, intermolecular aromatic substitutions by aryl nitrenes were unknown. Even when phenylnitrene was generated thermally from the azide in a very large excess of benzene, no diphenylamine was formed (Abramovitch and Knaus, 1967). Since alkoxycarbonylnitrenes, cyanonitrenes, and sulfonyl nitrenes all undergo intermolecular aromatic "substitution" (see below), it was suggested that a possible explanation for the absence of a similar reaction with singlet phenylnitrene could be that the latter was not electrophilic enough [(92) ↔ (93) (X = H)]. Thermolysis of aryl azides

(92)   (93)

bearing an electron-withdrawing group (X = CN, NO_2, CF_3, which would destabilize **93**) in aromatic solvents bearing electron-donating groups, e.g., N,N-dimethylaniline and sym-trimethoxybenzene, did give rise to products of aromatic substitution (Abramovitch and Scriven, 1970). For example, p-cyanophenyl azide and N,N-dimethylaniline gave a mixture of o-(**94**) (25.1%) and p-(**95**) (3.4%) substitution products together with the hydrogen abstraction product (**96**) (20.3%). Aromatic substitution products were also obtained from p-cyanophenylnitrene and 1,3,5-trimethoxybenzene (19.2%) and mesitylene (13.2%), but not with the following solvents: benzene, anisole, and p-dimethyoxybenzene. p-Nitrophenylnitrene gave diphenylamine derivatives with N,N-dimethylaniline (13.5%) and 1,3,5-trimethoxybenzene (18.9%); p-trifluoromethylphenylnitrene and N,N-dimethylaniline gave 4-dimethylamino-4-trifluoromethyldiphenylamine (13.4%).

The alternative possibility that these reactions were electrophilic attacks of the undecomposed azides upon the nucleophilic aromatic substrates was

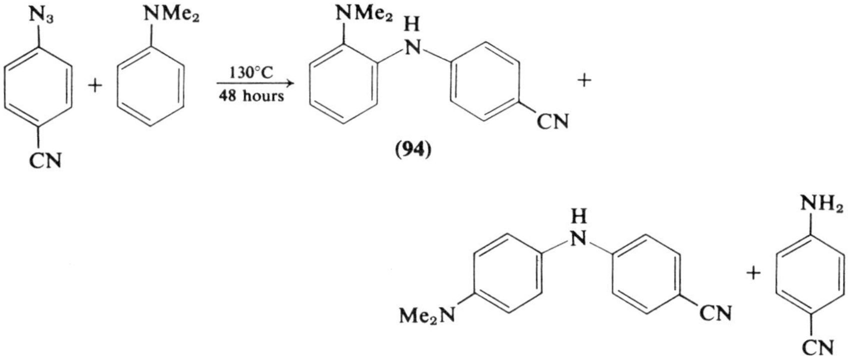

3. NITRENES

[Scheme showing p-cyanophenyl azide + N,N-dimethylaniline reacting at 130°C, 48 hours to give products (94), (95), and (96)]

disposed of by kinetic runs on the decomposition of *p*-cyanophenyl azide in chlorobenzene at 136°C, when it was shown that 1 to 3 molar equivalents of *N,N*-dimethylaniline had no influence on the rate of decomposition.

Huisgen and von Fraunberg (1969) have also reported a number of aromatic substitutions by 2-pyridyl- and 4,5-dimethyl-2-pyrimidylnitrene into reactive aromatic nuclei, e.g., naphthalene, anthracene, and anisole.

Aryl nitrenes, bearing the appropriate electron-attracting substituent and generated from the corresponding nitroso compound and triethyl phosphite, attack reactive aromatic nuclei similarly. The isomer ratio from, say, *p*-cyanophenylnitrene produced in this manner and dimethylaniline differs slightly from that observed using the azide as the precursor (Abramovitch and Scriven, 1970).

If the substituents in the aryl nitrene are sufficiently electron attracting, then intermolecular substitution into benzene itself can be achieved. Thus, when pentafluorophenylnitrene is generated in benzene, some pentafluorodiphenylamine is obtained together with the azoxy compound (Abramovitch *et al.*, 1972).

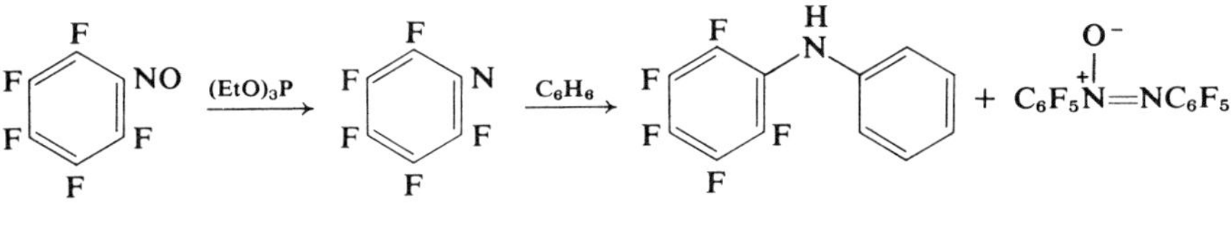

[Scheme: pentafluoronitrosobenzene + (EtO)₃P → pentafluorophenyl nitrene, then + C₆H₆ → pentafluorodiphenylamine + $C_6F_5\overset{+}{N}(O^-)=NC_6F_5$]

There is only one authenticated example of an aromatic substitution by an alkyl nitrene. 1,2-Shifts are usually very much faster than inter- or intramolecular attack by the nitrene on an aromatic nucleus. The nitrene produced by the thermolysis of 1-biphenyl-2-yl-1-methylethyl azide (97) does undergo intramolecular substitution in addition to the expected molecular rearrangements (Abramovitch and Kyba, 1969).

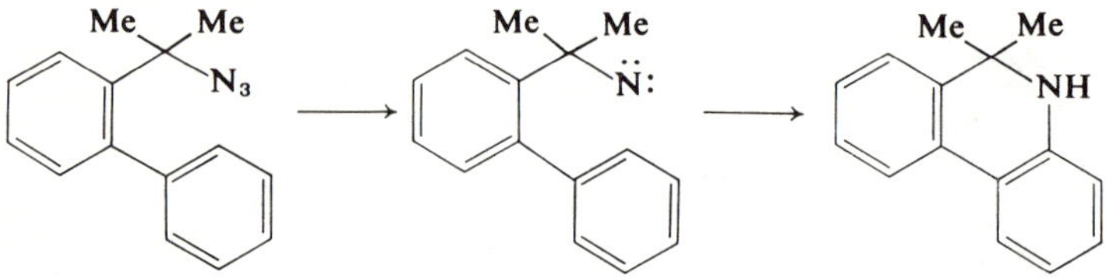

Carbethoxynitrene adds to benzene to give the benzaziridine, which ring-expands to *N*-carbethoxyazepine (**98**). Some *N*-carbethoxyaniline is also formed. Cyanonitrene behaves similarly. It has been shown that it is the

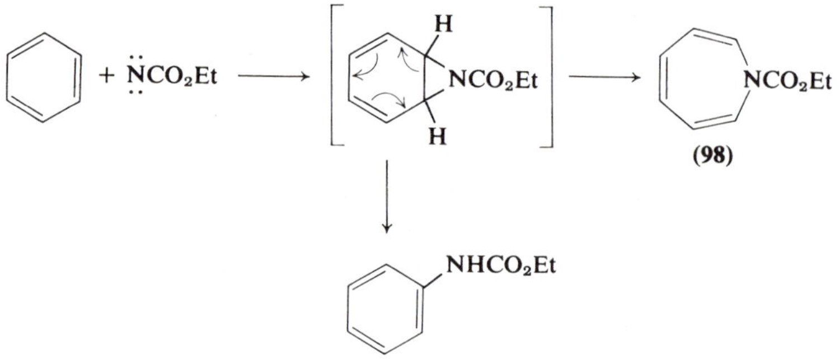

singlet nitrene that adds to the benzene "double bond" (Lwowski, 1970). With more reactive aromatic residues, e.g., anthracene, phenanthrene, and pyrene, as with mesitylene, azepines are not obtained; instead, the *N*-arylurethanes are isolated. The latter can be formed either from the benzaziridine via a dipolar intermediate or directly through a σ complex, perhaps (but not necessarily; see the discussion of substitution by aryl nitrenes below) involving some triplet nitrene (Beckwith and Redmond, 1968).

Hafner and Kaiser (1964) have studied the addition of carbethoxynitrene to pyrrole and thiophene. Their results are summarized in reaction (23).

Aromatic substitution is the best known of the reactions of sulfonyl nitrenes, and many examples were studied by Curtius and his co-workers. The topic has been reviewed in some detail recently (Abramovitch and Sutherland, 1970).

Thermal decomposition (120°C) of a sulfonyl azide in an aromatic solvent gives the singlet sulfonyl nitrene (**99**) which adds to the aromatic nucleus, if the latter is sufficiently reactive, to give an aziridine intermediate (**100**) in the

3. NITRENES

(23)

rate- but not product-determining step. At 120°C, this bicyclic intermediate opens to give a dipolar product (**101**) in a relatively fast, thermodynamically controlled process, the direction of ring opening being controlled by the electronic effect (particularly +M) of the substituent. Thus, whereas total rate ratios $_H^X K$ are a correct measure of the relative reactivities of substituted and unsubstituted benzenes, "partial rate factors" (determined from the isomer ratios and total rate ratio) are not meaningful as a measure of the relative reactivities of the various positions in PhX and C_6H_6 (Abramovitch et al., 1965).

Unlike the reaction of carbethoxynitrene with benzene which gives N-carbethoxyazepine, no N-methanesulfonylazepine (**102**) could be detected in the thermolysis of $MeSO_2N_3$ in benzene or substituted benzenes at 120°C. This was shown (Abramovitch and Uma, 1968) to be due to the fact that **102**

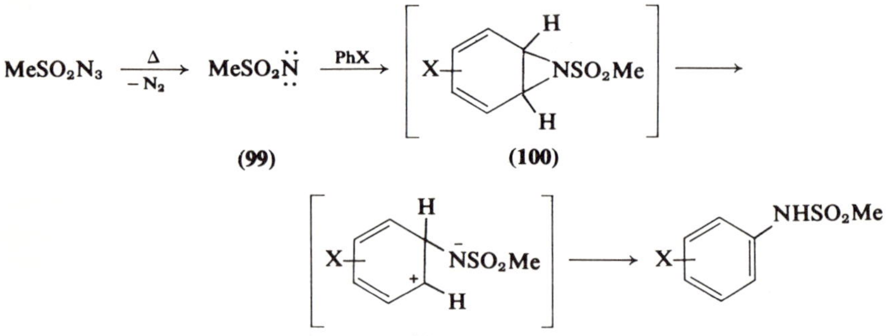

was the product of kinetic control of the rearrangement of **103**, whereas **104** was the product of thermodynamic control [Eq. (24)]. At 120°C, the equilibrium **104** ⇌ **103** ⇌ **102** lies almost exclusively on the side of **104**, but the activation energy for **103** → **102** is lower than for **103** → **104**. Thus, if a good

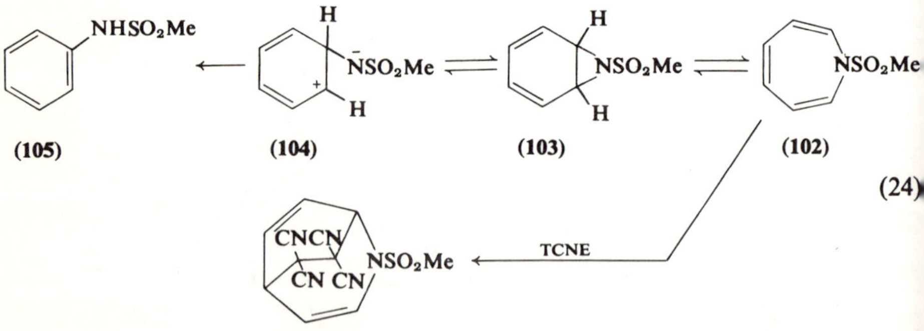

(24)

3. NITRENES

dienophile [tetracyanoethylene(TCNE)] is present in the reaction mixture, the traces of **102** formed first (kinetic control) can be trapped to give the 4 + 2 adduct (**106**) at the expense of **104**. Although $MeSO_2N_3$ did not thermolyze appreciably below 120°C, when its solution in benzene was heated at 80°C for a long time, a trace (0.5%) of **93** could be detected by thin layer chromatography together with mostly undecomposed azide, but no N-mesylaniline (**105**) (Abramovitch and Uma, 1968). Photolysis of $MeSO_2N_3$ in benzene at room temperature or 80°C gave traces of **102** together with much tar and unchanged azide, confirming that azepine formation is the kinetically controlled process.

The sulfonyl nitrene, generated initially in the singlet state, does not react as such with nitrobenzene (which is highly deactivated toward electrophilic addition) and drops down to the triplet ground state before it attacks the aromatic nucleus in a manner typical of a highly electrophilic free radical. If the aromatic nucleus is not as deactivated toward electrophilic attack as is nitrobenzene ($PhCO_2Me$, $PhCN$, $PhCF_3$), then the singlet → triplet conversion is less complete, and the pattern of substitution observed is that expected of a mixture of these two reactive intermediates. If a transition metal complex is present, the triplet is either trapped or sidetracked to an intermediate that mainly abstracts hydrogen, and the yield of substitution product drops markedly. The pattern of substitution, however, now becomes typical of what might have been expected of an attack by a singlet nitrene [**100** → **101**], and the meta isomer predominates (Abramovitch et al., 1969b).

The above results obtained with carbalkoxynitrenes and sulfonyl nitrenes raise the question of the mechanism of aromatic "substitution" by aryl nitrenes. Is the rate-determining step the direct formation of a σ complex or is a benzaziridine first formed which then ring-opens to give the observed substitution product?

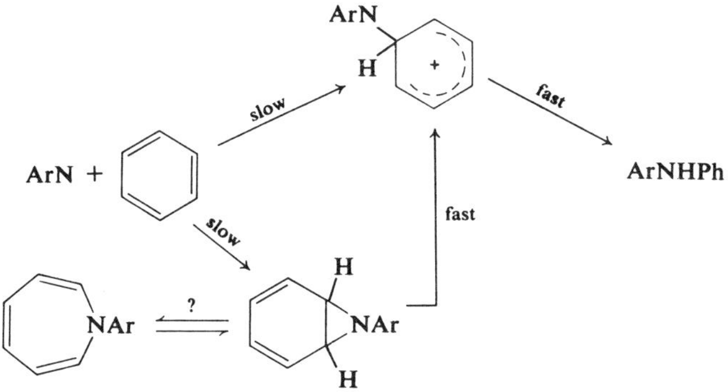

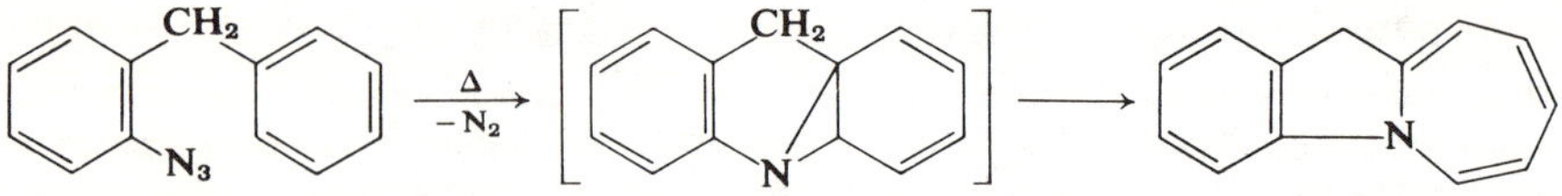

(108) → [structure] → **(109)**

Evidence that a benzaziridine is an intermediate in some cases of intramolecular aromatic substitution has been forthcoming. Thus, when 2-azidodiphenylmethane (108) was pyrolyzed in trichlorobenzene, azepino[2,1-a]-11H-indole (109) was obtained in 66% yield (Krbecheck and Takimoto, 1968). Other examples are known. In the intermolecular case, the answer is still not clear-cut, but a number of N-arylazepines have either been isolated or trapped by the reaction of pentafluoronitrosobenzene with a phosphite in the presence of an aromatic substrate (e.g., benzene, toluene, and anisole) at 0°C or lower (Abramovitch et al., 1972).

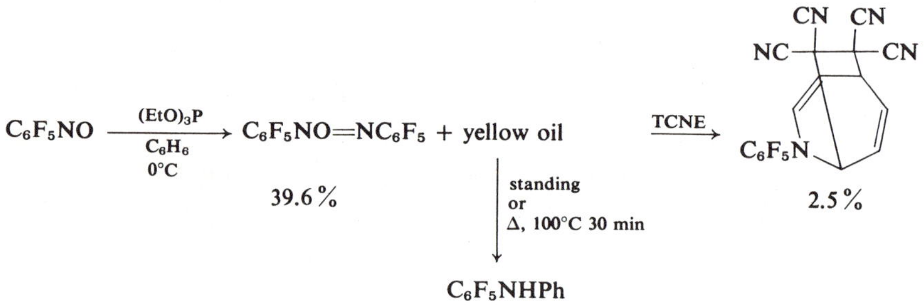

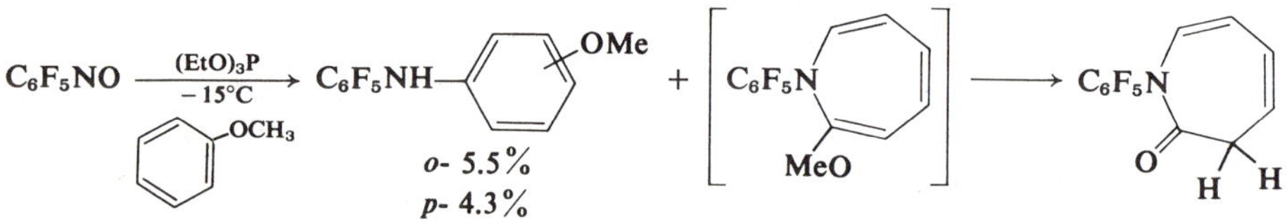

$C_6F_5NO \xrightarrow[C_6H_6]{(EtO)_3P} C_6F_5NO=NC_6F_5$ + yellow oil $\xrightarrow{TCNE}$ C_6F_5N—[adduct]

39.6%   standing or Δ, 100°C 30 min   2.5%

C_6F_5NHPh

$C_6F_5NO \xrightarrow[-15°C]{(EtO)_3P}$ C_6F_5NH—[anisole-OMe/OCH₃] + [C_6F_5N-cycloheptatriene-MeO] → C_6F_5N—[azepinone]

o- 5.5%
p- 4.3%

From these and other results, it has been tentatively concluded that a spectrum of transition states is possible, ranging from the fully formed σ complex to the aziridine intermediate, through partially formed aziridines (110) in which one bond is more developed than the other, and that the intermediate which is formed depends, among other things, on the nature of the substituents in both the nitrene residue and the aromatic substrate and on the temperature, solvent, and relative thermodynamic stabilities of the azepines and the σ complexes. These factors remain to be sorted out.

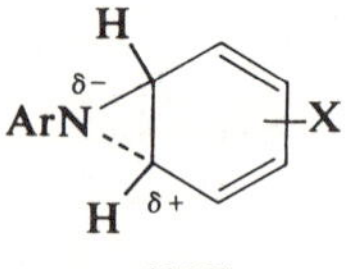

(110)

3. NITRENES

G. TRAPPING BY NUCLEOPHILES

A number of examples of intermolecular nucleophilic trapping of nitrenes are known. Aryl nitrenes may be intercepted by high pressures (> 136 atm) of carbon monoxide to give isocyanates on a preparative scale (Bennett and

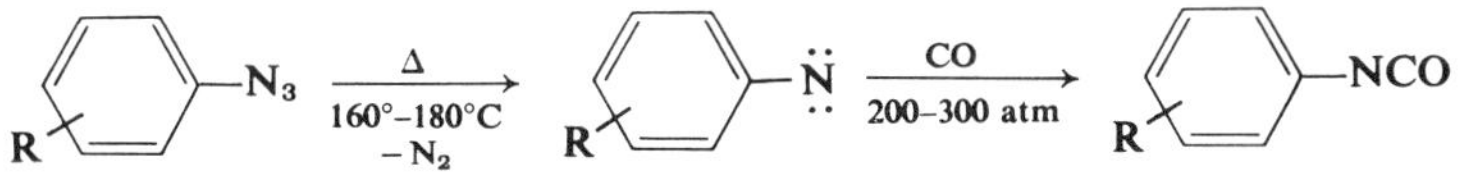

$$R\text{-}C_6H_4\text{-}N_3 \xrightarrow[-N_2]{\Delta \; 160°-180°C} R\text{-}C_6H_4\text{-}\ddot{N}\text{:} \xrightarrow[200-300 \text{ atm}]{CO} R\text{-}C_6H_4\text{-}NCO$$

Hardy, 1968). At lower CO pressures, increasing amounts of azobenzenes were obtained. More recently, singlet *p*-cyanophenylnitrene was trapped with dimethylamine to give 1,1-dimethyl-2-(4-cyanophenyl)hydrazine (**111**) (70%) together with some *p*-cyanoaniline (**112**) (5%) (Odum and Aaronson, 1969).

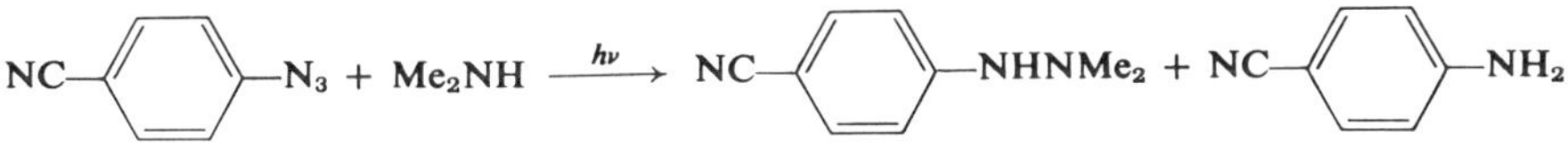

$$NC\text{-}C_6H_4\text{-}N_3 + Me_2NH \xrightarrow{h\nu} NC\text{-}C_6H_4\text{-}NHNMe_2 + NC\text{-}C_6H_4\text{-}NH_2$$
$$\qquad\qquad\qquad\qquad\qquad\qquad\qquad (\mathbf{111}) \qquad\qquad (\mathbf{112})$$

In the presence of a triplet sensitizer, xanthen-9-one which absorbs essentially all the light, inversion in the product yields was observed: **111** (6%), **112** (70%).

Intramolecular trapping of an aryl nitrene by a nucleophilic center is well known. Some examples are given in Eqs. (25) (Abramovitch and Adams, 1961), (26) (Lynch and Hung, 1965), and (27) (Cadogan *et al.*, 1965).

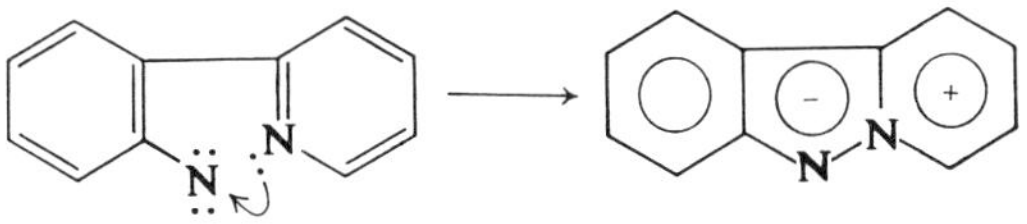

$$\text{(carbazole nitrene)} \longrightarrow \text{(cyclized product)} \qquad (25)$$

$$\text{(o-pyrazolyl aniline nitrene)} \longrightarrow \text{(cyclized product)} \qquad (26)$$

$$\text{(o-nitro benzylidene ArN)} \xrightarrow{(EtO)_3P} \text{(2-aryl-2H-indazole)} \qquad (27)$$

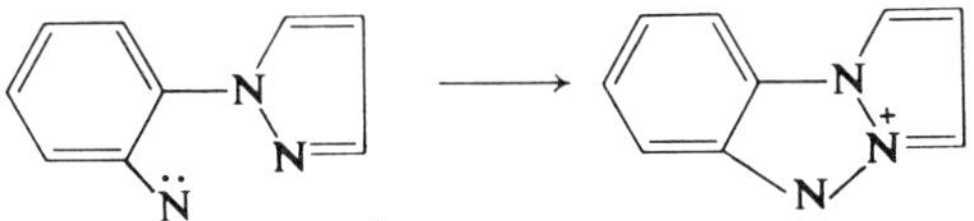

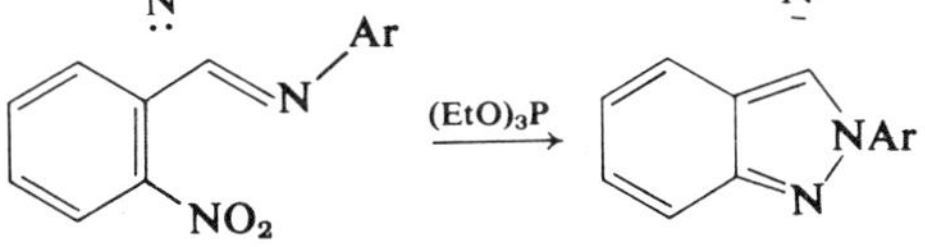

Sulfonyl nitrenes have been trapped with a variety of nucleophiles, e.g., dimethyl sulfide. An example of an intramolecular capture is **113** → **114**

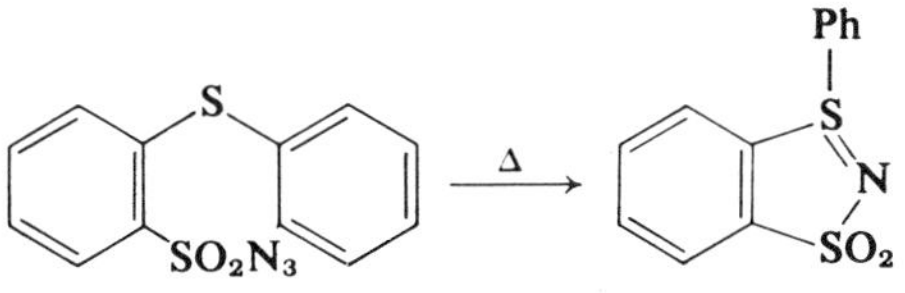

(**113**) → (**114**)

(Abramovitch et al., 1969a). In most cases, this could either be a direct acid–base type reaction of nitrene and electron donor or, in some instances, a reaction of the nitrene precursor with the nucleophile or 1,3-dipolarophile

$$RSO_2\ddot{N} + :SMe_2 \longrightarrow RSO_2N=SMe_2$$

followed by loss of nitrogen. Thus, the reaction of benzenesulfonyl azide with pyridine to give **115** (Curtius and Risson, 1930; Buchanan and Levine, 1950) could either be a concerted process or involve a sulfonyl nitrene. That some

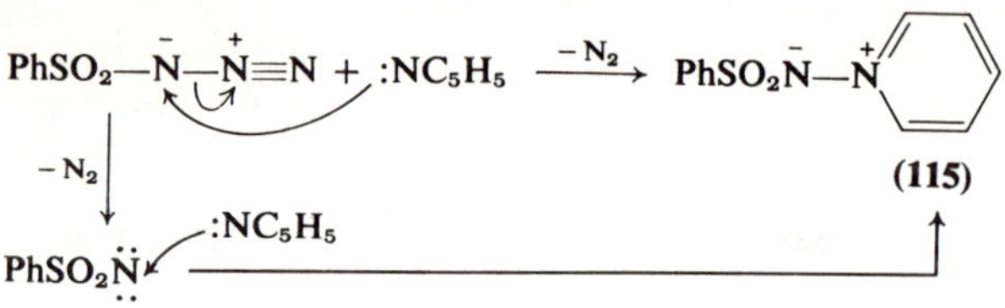

free nitrene can be involved in these reactions is demonstrated by the fact that some 3-benzenesulfonamido-2,6-lutidine (**116**) is formed in the reaction of $PhSO_2N_3$ with 2,6-lutidine, and similar products are obtained with 2-picoline and 2,4,6-collidine (Abramovitch and Takaya, 1972).

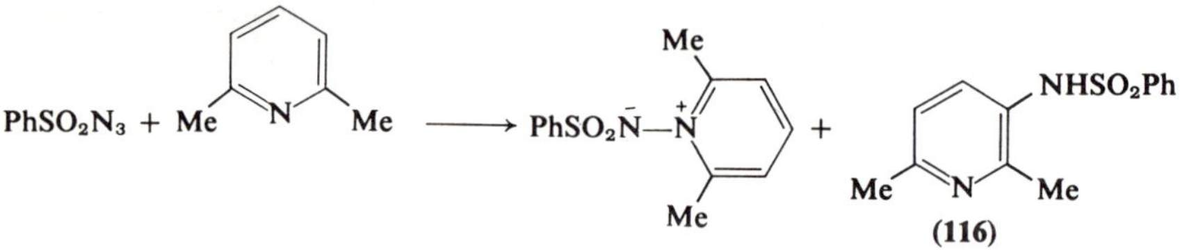

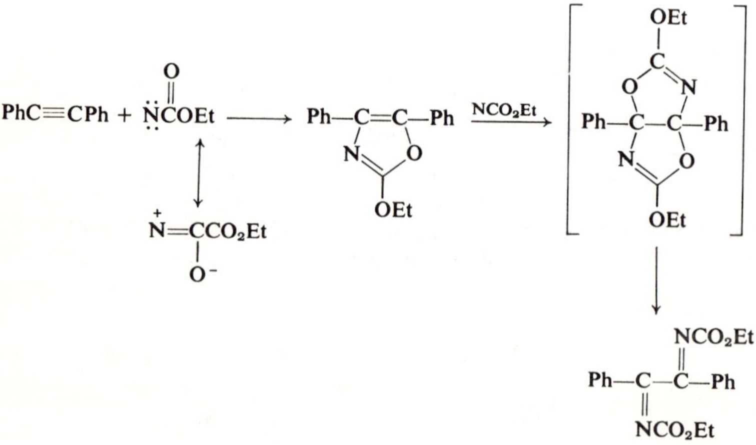

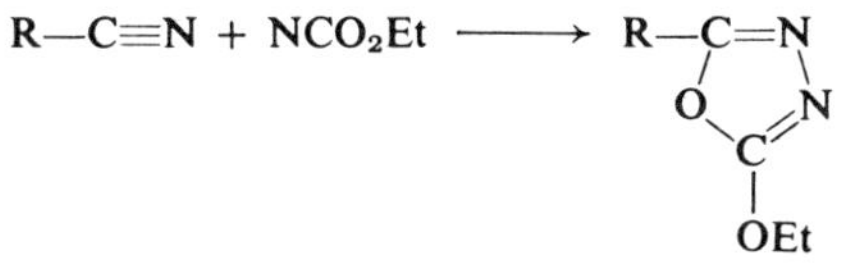

3. NITRENES

$$R-C\equiv N + NCO_2Et \longrightarrow \underset{OEt}{\underset{|}{R-C=N}} \atop {O\diagdown \!\!\!\!\diagup N \atop C}$$

$$EtOCON_3 + C_6H_5NH_2 \xrightarrow{\Delta} EtOCONHNHPh$$

$$PhCON_3 + C_6H_5NHMe \xrightarrow{h\nu} \underset{\underset{Me}{|}}{PhCONHNPh} + PhNHCONPh$$
(15%)

Carbethoxynitrene has been trapped by a wide variety of electron-rich centers including alkynes, nitriies, isonitriles, sulfoxides, amines, and CO (Lwowski, 1970). With alkynes and nitriles, 1,3-dipolar addition can occur.

NH itself has also been trapped. For example, photolysis of HN_3 in a matrix of solid N_2 in the presence of molecular O_2 gives *cis*- and *trans*-HONO (Baldeschwieler and Pimentel, 1960). More recently, aryl nitrenes have been trapped with molecular oxygen (Abramovitch *et al.*, 1971b), e.g.,

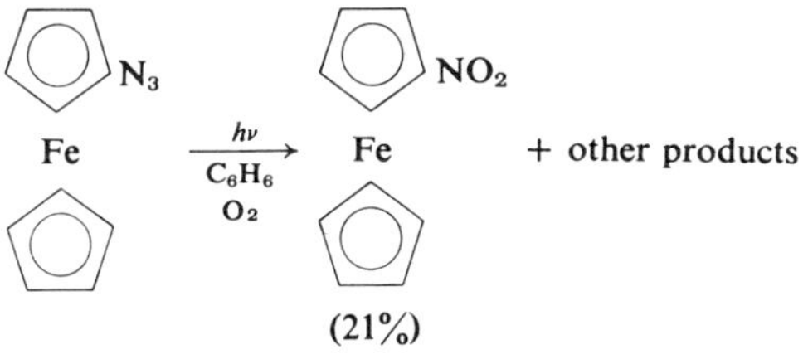

$$\underset{Fe}{\underset{\bigcirc}{\overset{\bigcirc\!\!-\!\!N_3}{}}} \xrightarrow[\substack{C_6H_6 \\ O_2}]{h\nu} \underset{Fe}{\underset{\bigcirc}{\overset{\bigcirc\!\!-\!\!NO_2}{}}} + \text{other products}$$

(21%)

VI. Nitrenium Ions

Carbenes and nitrenes differ qualitatively in one major aspect: Protonation or alkylation of a carbene would lead to a carbonium ion with no unshared electrons on the central carbon, whereas a protonated or alkylated nitrene would still have a pair of unshared electrons which could be spin paired or unpaired. Nitrenium ions can, therefore, in principle, exist either as singlets or triplets.

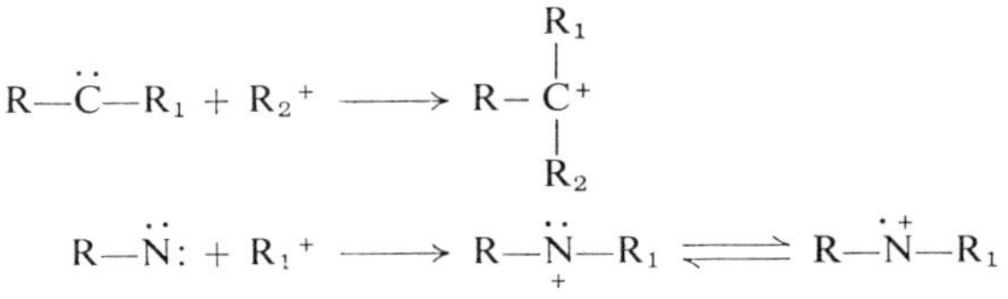

$$R-\overset{..}{C}-R_1 + R_2{}^+ \longrightarrow R-\underset{R_2}{\overset{R_1}{\underset{|}{\overset{|}{C^+}}}}$$

$$R-\overset{..}{N}: + R_1{}^+ \longrightarrow \underset{+}{R-\overset{..}{N}-R_1} \rightleftharpoons R-\overset{.\,+}{N}-R_1$$

Since the subject was originally surveyed in 1964 (Abramovitch and Davis, 1964), two excellent reviews have appeared (Gassman, 1970, 1971; Lansbury, 1970), and these should be consulted for a detailed discussion of the chemistry of this species.

Nitrenium ions have been generated mainly by one of two routes: by an S_N1 cleavage of the N–X bond in R_2NX:

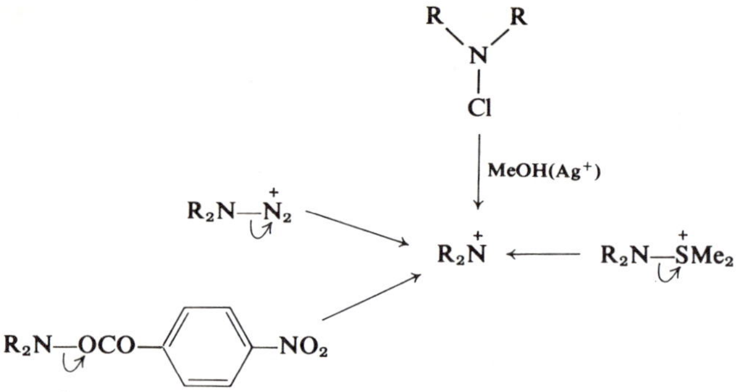

or, less often, from an appropriate oxime which is of suitable configuration so that acid treatment gives rise to the nitrenium ion rather than to a Beckmann rearrangement:

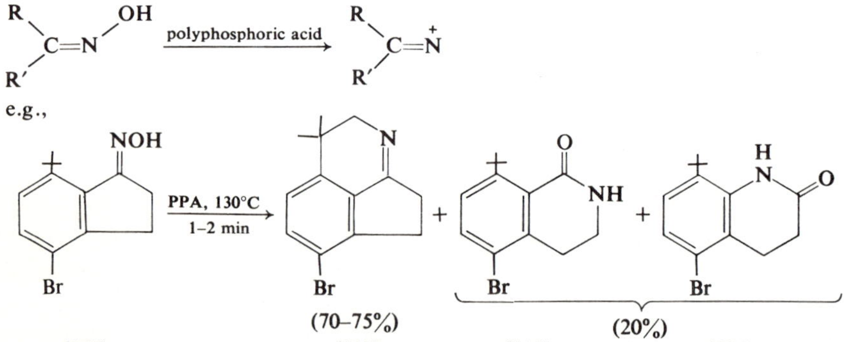

119 and **120** are the normal Beckmann products, whereas **118** is formed by the insertion of the nitrenium ion into one of the *tert*-butyl methyl groups (Lansbury *et al.*, 1964).

The most systematically investigated reactions involve the solvolysis of *N*-chloroamines studied by Gassman and his co-workers. The examples given in Eqs. (28)–(30) illustrate some of the reactions undergone by the intermediate nitrenium ions.

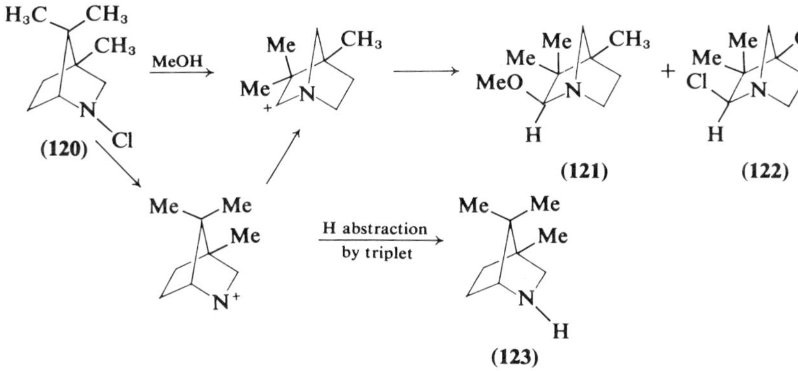

3. NITRENES

(28)

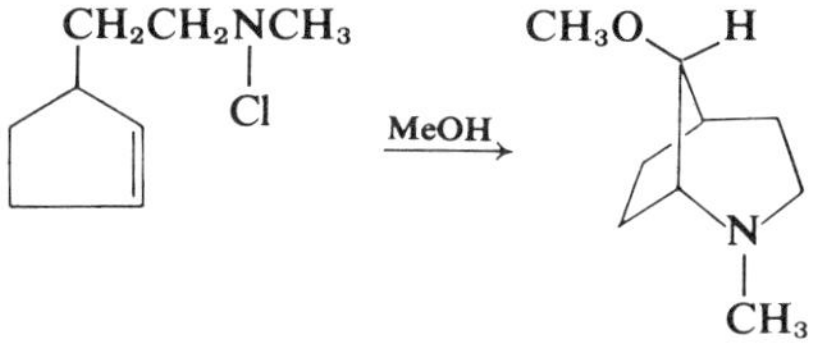

(29)

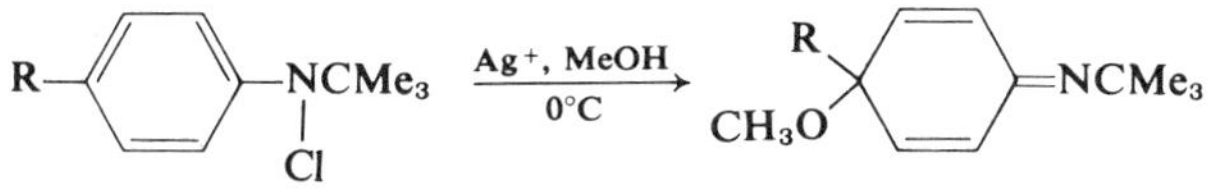

(30)

Though most of these reactions could be formulated as involving a nitrenium ion, they could also be visualized as going through a concerted elimination–migration or –addition step in which the bond-breaking step could have proceeded to a greater or lesser extent according to the reaction conditions. The most convincing evidence for the intermediacy of a nitrenium ion in such reactions comes from (1) the study of the relative rates of solvolysis of substituted N-chloroaziridines and (2) the observation of heavy atom catalysis of spin inversion from the initially formed singlet to the triplet ground state.

1. N-Chloroaziridines have been shown to solvolyze in a stereospecific fashion with cleavage of the C–C bond of the three-membered ring in a concerted disrotatory fashion [Eq. (31)].

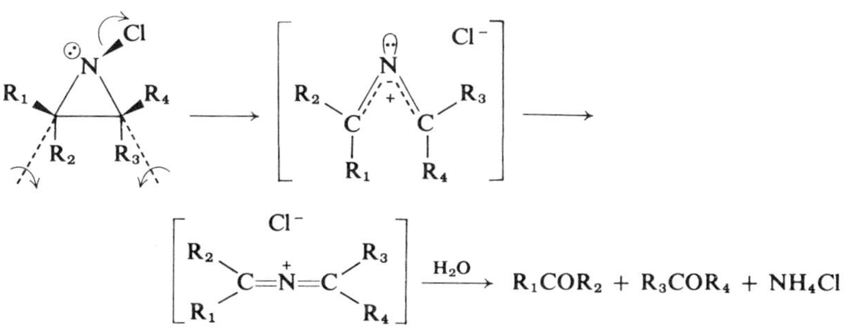

$$\left[R_2\underset{R_1}{\overset{+}{C}}=N=\underset{R_4}{\overset{}{C}}R_3 \right] \xrightarrow{H_2O} R_1COR_2 + R_3COR_4 + NH_4Cl \quad (31)$$

		k_{rel}
$R_1 = R_2 = R_3 = R_4 = H$		1
$R_1 = R_2 = R_3 = H;$	$R_4 = Me$	15
$R_1 = R_2 = R_4 = H;$	$R_3 = Me$	210
$R_1 = R_3 = H;$	$R_2 = R_4 = Me$	1490
$R_1 = R_2 = H;$	$R_3 = R_4 = Me$	1860
$R_1 = R_4 = H;$	$R_2 = R_3 = Me$	155,000

The results can be explained on the basis of increasing steric strain in the transition state or relief of steric compression in the ground state as well as the effect of the methyl groups in stabilizing the developing positive charge on carbon in the disrotatory C–C cleavage of the polar species (Gassman, 1970).

2. In the solvolysis of **120** in methanol–cyclohexane solution, the ratio of products formed is **121:122:123** = 10:56:1. If the cyclohexane is replaced by bromoform, the relative amounts formed as **121:122:123** = ≪1: ∼1:45 (Gassman and Cryberg, 1969). This has been interpreted as follows: **121** and **122** are products of the alkyl group migration to a singlet nitrenium ion (**124**), whereas **123** is the result of hydrogen abstraction by the triplet nitrenium ion

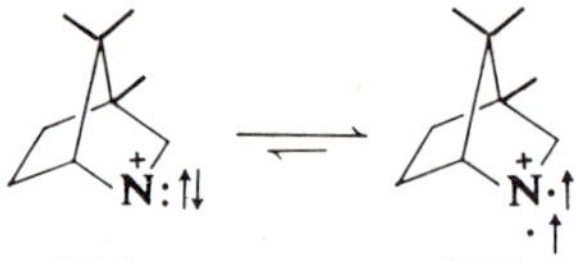

(124)   **(125)**

(**125**). Bromoform, acting as a heavy atom solvent, can catalyze the singlet → triplet conversion **124** → **125** with the result that a 500-fold change in the product ratio is observed, with the singlet products predominating in the absence of bromoform and the triplet product being formed almost exclusively in its presence. This is not due to an increase in the polarity of the solvent, since the ratio in pure methanol is almost the same as in methanol–cyclohexane. The interpretation of the effect as being due to a heavy atom catalysis by the halogenated solvent requires the discrete existence of nitrenium ions in this solvolysis.

VII. Applications

The synthetic possibilities of nitrene intermediates have been adumbrated throughout this chapter. Of particular value has been the applications to the synthesis of heterocyclic compounds and a few examples are given below:

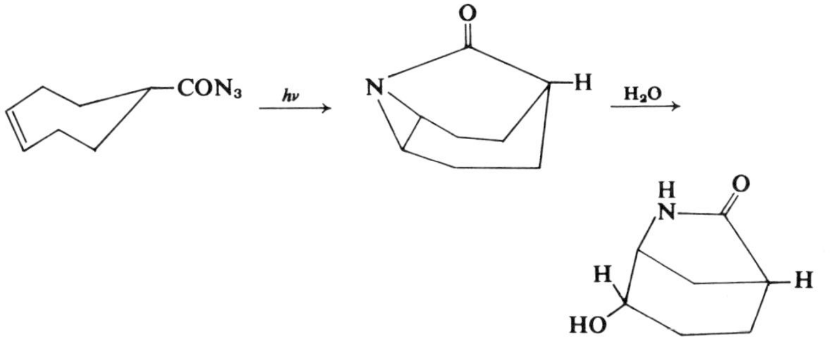

3. NITRENES

(Edwards, 1970)

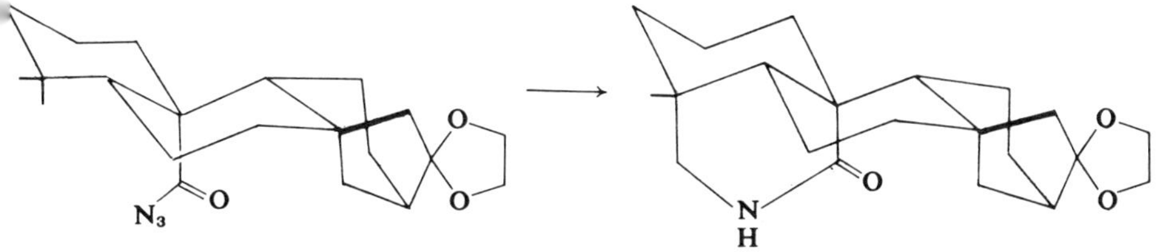

(Masamune, 1964)

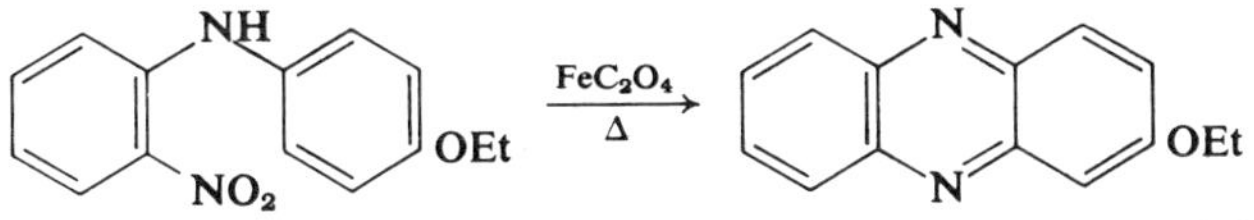

(Vivian et al., 1951)

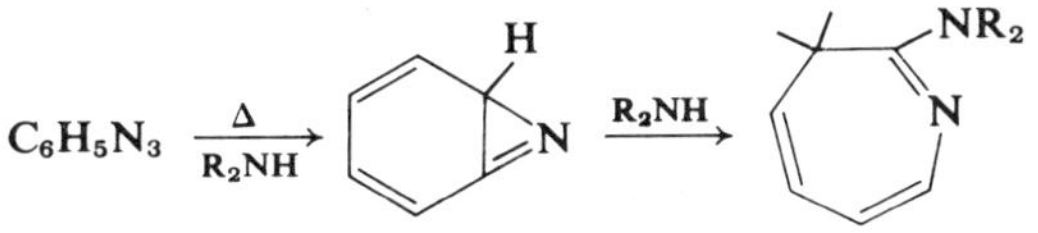

(Doering and Odum, 1966)

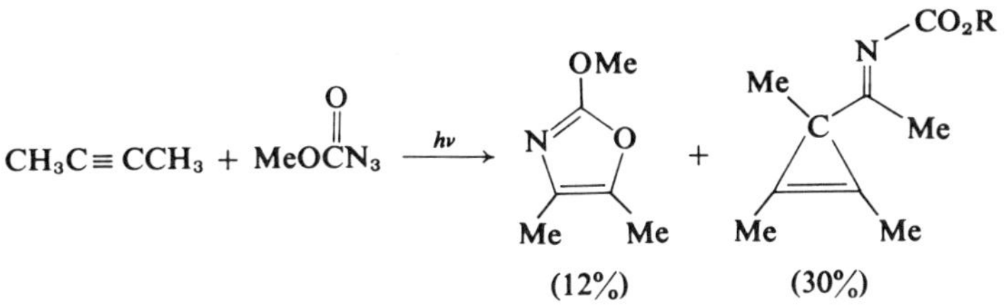

(Meinwald and Aue, 1966)

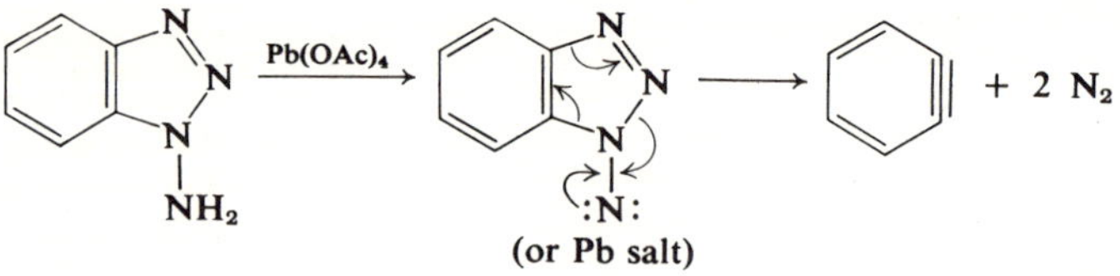

(Campbell and Rees, 1969b)

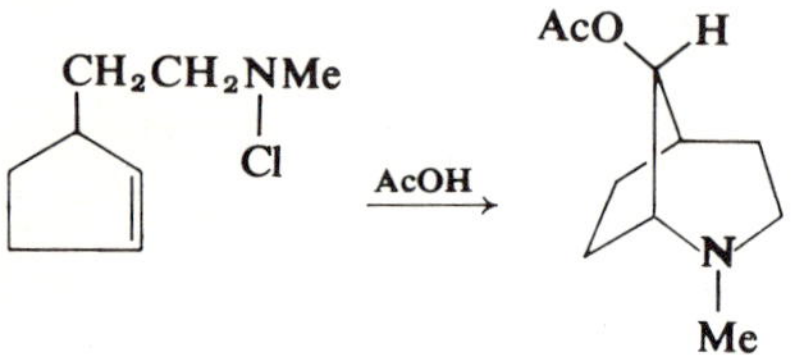

(Gassman, 1970)

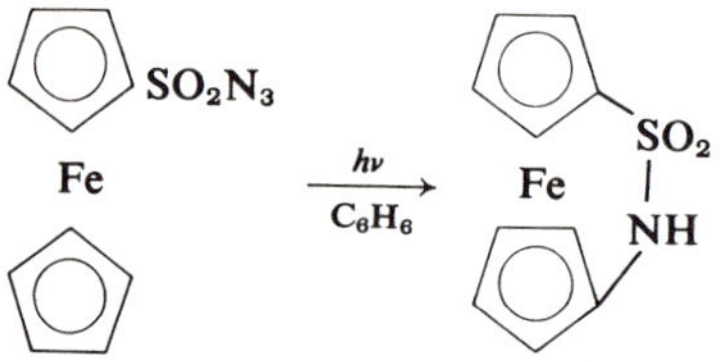

(Abramovitch et al., 1969b)

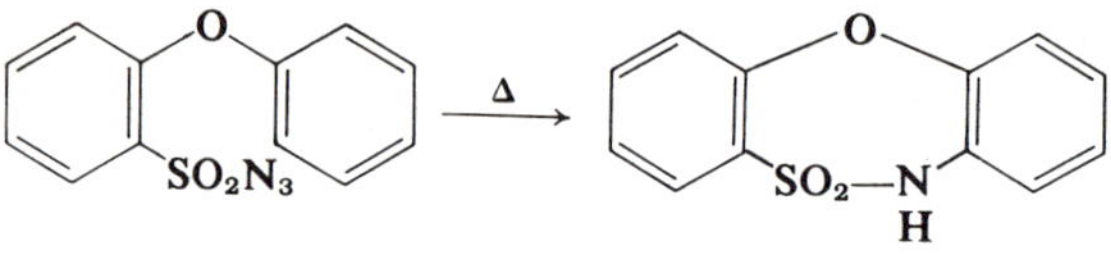

(Abramovitch et al., 1969a)

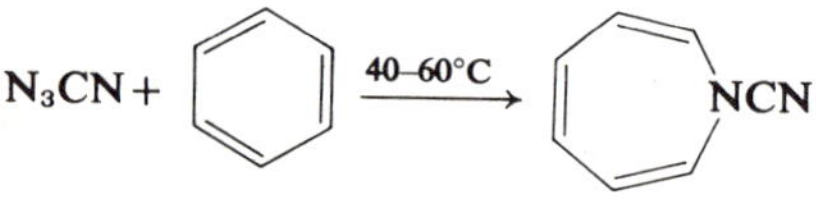

(Marsh and Simmons, 1965)

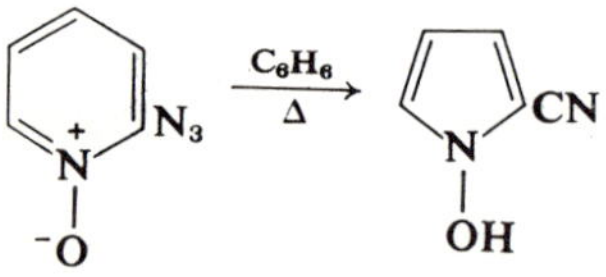

(Abramovitch and Cue, 1973)

Disulfonyl azides have been used as cross-linking and -foaming agents for cellulose acetate and a number of polymeric materials and as blowing agents (for a review, see Breslow, 1970). A number of patents have been granted on applications which make use of the ability of sulfonyl nitrenes to insert into aliphatic C–H bonds. Thus, mono- and poly(sulfonyl) azides have been used for the modification or cross-linking of hydrocarbon polymers such as polyvinyl ethers and chlorides and polypropylene and polyisobutylene.

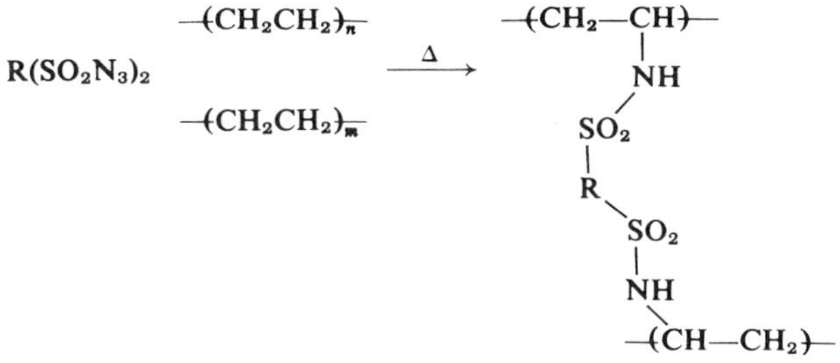

3. NITRENES

$$R(SO_2N_3)_2 \quad \begin{matrix} -(CH_2CH_2)_n- \\ -(CH_2CH_2)_m- \end{matrix} \xrightarrow{\Delta} \begin{matrix} -(CH_2-CH)- \\ | \\ NH \\ | \\ SO_2 \\ | \\ R \\ \quad \diagdown SO_2 \\ \quad\quad | \\ \quad\quad NH \\ \quad\quad \diagdown \\ \quad\quad -(CH-CH_2)- \end{matrix}$$

A particularly novel application of nitrene chemistry has been its use in identifying the site of interaction of a γ-globulin with the antibody binding site (Porter et al., 1970). 4-Fluoro-3-nitrophenyl azide was bound covalently to bovine γ-globulin by displacement of the fluorine atom by a primary amino group of lysine residues. This labeled γ-globulin was now injected into rabbits when the antibody protein bonded with it. Thus labeled, the antibody was isolated. Treatment of this with tritiated N-(4-azido-2-nitrophenyl)-L-lysine caused the latter to fit into the 4-azido-2-nitrophenyl specific binding slots of the antibody protein. Exposure to light produced the nitrene which inserted into C–H groups at the binding site. Polypeptide chain separation followed by degradation leads to labeled peptides identifying the binding site in the antibody protein.

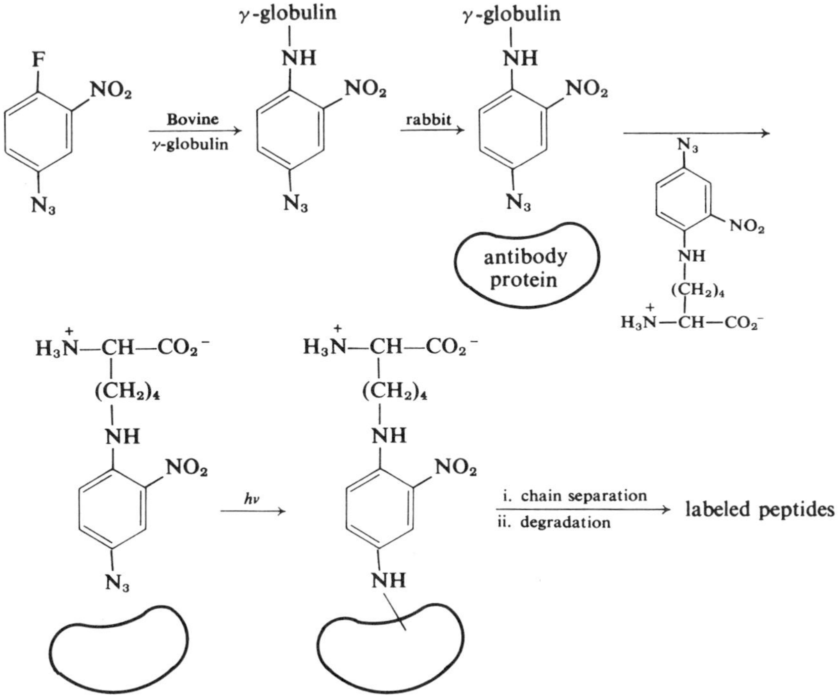

References

Abramovitch, R. A. (1967). *Advan. Free-Radical Chem.* **2**, 128.
Abramovitch, R. A. (1970). *Chem. Soc., Spec. Publ.* **24**, 323.
Abramovitch, R. A., and Adams, K. A. H. (1961). *Can. J. Chem.* **39**, 2576.
Abramovitch, R. A., and Challand, S. R. (1971). Unpublished results.
Abramovitch, R. A., and Cue, B. W. (1973). *J. Org. Chem.* **38**, 173.
Abramovitch, R. A., and Davis, B. A. (1964). *Chem. Rev.* **64**, 149.
Abramovitch, R. A., and Davis, B. A. (1968). *J. Chem. Soc.*, C p. 119.
Abramovitch, R. A., and Holcomb, W. D. (1969). *Chem. Commun.* p. 1298.
Abramovitch, R. A., and Knaus, E. E. (1967). Unpublished results.
Abramovitch, R. A., and Knaus, G. N. (1971). *Abstr. Pap., IUPAC Congr. 1971* Paper OC-8-306, p. 128.
Abramovitch, R. A., and Kyba, E. P. (1969). *Chem. Commun.* p. 265.
Abramovitch, R. A., and Kyba, E. P. (1971a). In "The Chemistry of the Azido Group" (S. Patai, ed.), Chapter 5, p. 221. Wiley, New York.
Abramovitch, R. A., and Kyba, E. P. (1971b). *J. Amer. Chem. Soc.* **93**, 1537.
Abramovitch, R. A., and Scriven, E. F. V. (1970). *Chem. Commun.* p. 787.
Abramovitch, R. A., and Sutherland, R. G. (1970). *Fortsch Chem. Forsch.* **16**, 1.
Abramovitch, R. A., and Takaya, T. (1972). *J. Org. Chem.* **37**, 2022.
Abramovitch, R. A., and Uma, V. (1967). Unpublished results.
Abramovitch, R. A., and Uma, V. (1968). *Chem. Commun.* p. 797.
Abramovitch, R. A., Ahmad, Y., and Newman, D. (1961). *Tetrahedron Lett.* p. 752.
Abramovitch, R. A., Roy, J., and Uma, V. (1965). *Can. J. Chem.* **43**, 3407.
Abramovitch, R. A., Azogu, C. I., and McMaster, I. T. (1969a). *J. Amer. Chem. Soc.* **91**, 1219.
Abramovitch, R. A., Azogu, C. I., and Sutherland, R. G. (1969b). *Chem. Commun.* p. 1439.
Abramovitch, R. A., Brown, R. A., and Davis, B. A. (1969c). *J. Chem. Soc.*, C p. 1146.
Abramovitch, R. A., Knaus, G. N., and Uma, V. (1969d). *J. Amer. Chem. Soc.* **91**, 7532.
Abramovitch, R. A., Azogu, C. I., and Sutherland, R. G. (1971a). *Tetrahedron Lett.* p. 1637.
Abramovitch, R. A., Azogu, C. I., and Sutherland, R. G. (1971b). *Chem. Commun.* p. 134.
Abramovitch, R. A., Challand, S. R. and Scriven, E. F. V. (1972). *J. Amer. Chem. Soc.* **95**, 1374.
Anastassiou, A. G. (1965). *J. Amer. Chem. Soc.* **87**, 5512.
Anastassiou, A. G. (1968). *J. Amer. Chem. Soc.* **90**, 1527.
Anastassiou, A. G., and Simmons, H. E. (1967). *J. Amer. Chem. Soc.* **89**, 3177.
Anastassiou, A. G., Shepelavy, J. N., Simmons, H. E., and Marsh, F. D. (1970). *In* "Nitrenes" (W. Lwowski, ed.), pp. 305–344. Wiley (Interscience), New York.
Anderson, D. J., Gilchrist, T. L., Horwell, D. C., and Rees, C. W. (1969). *Chem. Commun.* p. 146.
Appl, M., and Huisgen, R. (1959). *Chem. Ber.* **92**, 2961.
Apsimon, J. W., and Edwards, O. E. (1962). *Can. J. Chem.* **40**, 896.
Atkinson, R. S., and Rees, C. W. (1967). *Chem. Commun.* p. 1230.
Balabanov, G. P., Dergunov, Y. I., and Gal'perin, V. A. (1966). *J. Org. Chem. USSR* **2**, 1797.
Baldeschwieler, J. D., and Pimentel, G. C. (1960). *J. Chem. Phys.* **33**, 1008.
Bamford, D. A., and Bamford, C. H. (1941). *J. Chem. Soc., London* **30**, 30.

Barash, L., Wasserman, E., and Yager, W. A. (1967). *J. Amer. Chem. Soc.* **89**, 3931.
Barton, D. H. R., and Morgan, L. R., Jr. (1962). *J. Chem. Soc., London* p. 622.
Barton, D. H. R., and Starratt, A. N. (1965). *J. Chem. Soc. London* p. 2444.
Beckwith, A. L. J., and Redmond, J. W. (1968). *J. Amer. Chem. Soc.* **90**, 1351.
Belloli, R. (1971). *J. Chem. Educ.* **48**, 423.
Bennett, R. P., and Hardy, W. B. (1968). *J. Amer. Chem. Soc.* **90**, 3295.
Berry, R. S. (1970). In "Nitrenes" (W. Lwowski, ed.), pp. 13–45. Wiley (Interscience), New York.
Birkhimer, E. A., Norup, B., and Bak, T. A. (1960). *Acta Chem. Scand.* **14**, 1894.
Boyer, J. H. (1970). In "Nitrenes" (W. Lwowski, ed.), Chapter 5, pp. 163–184. Wiley (Interscience), New York.
Boyer, J. H., and Mikol, G. J. (1969). *Chem. Commun.* p. 734.
Boyer, J. H., Krueger, W. E., and Mikol, G. J. (1967). *J. Amer. Chem. Soc.* **89**, 5504.
Breslow, D. S. (1970). In "Nitrenes" (W. Lwowski, ed.), Chapter 8, pp. 245–303. Wiley (Interscience), New York.
Breslow, D. S., Prosser, T. J., Marcantonio, A., F. and Genge, C. A. (1967). *J. Amer. Chem. Soc.* **89**, 2384.
Breslow, D. S., Sloan, M. F., Newburg, N. R., and Renfrow, N. B. (1969). *J. Amer. Chem. Soc.* **91**, 2273.
Brower, K. R. (1963). *J. Amer. Chem. Soc.* **85**, 1401.
Buchanan, G. L., and Levine, R. M. (1950). *J. Chem. Soc., London* p. 2248.
Bunyan, P. J., and Cadogan, J. I. G. (1963). *J. Chem. Soc., London* p. 42.
Cadogan, J. I. G. (1968). *Quart. Rev., Chem. Soc.* **22**, 222.
Cadogan, J. I. G. (1971). Unpublished results.
Cadogan, J. I. G., and Cooper, A. (1969). *J. Chem. Soc., B* p. 883.
Cadogan, J. I. G., and Todd, M. J. (1967). *Chem. Commun.* p. 178.
Cadogan, J. I. G., Cameron-Wood, M., Mackie, R. K., and Searle, R. J. G. (1965) *J. Chem. Soc., London* p. 4831.
Campbell, C. D., and Rees, C. W. (1969a). *Chem. Commun.* p. 537.
Campbell, C. D., and Rees, C. W. (1969b). *J. Chem. Soc., C* pp. 748 and 752.
Carboni, R. A., and Castle, J. E. (1962). *J. Amer. Chem. Soc.* **84**, 2453.
Carboni, R. A., Kauer, J. C., Castle, J. E., and Simmons, H. E. (1967). *J. Amer. Chem. Soc.* **89**, 2618.
Closson, W. D., and Gray, H. B. (1963). *J. Amer. Chem. Soc.* **85**, 290.
Coffin, B., and Robbins, R. F. (1965). *J. Chem. Soc., London* p. 1252.
Comeford, J. J., and Mann, D. E. (1965). *Spectrochim. Acta* **21**, 197.
Cotter, R. J., and Beach, W. F. (1964). *J. Org. Chem.* **29**, 751.
Crow, W. D., and Wentrup, C. (1968). *Tetrahedron Lett.* pp. 5569 and 6149.
Crow, W. D., and Wentrup, C. (1969). *Chem. Commun.* p. 1387.
Currie, C. L., and Darwent, B. de B. (1963). *Can. J. Chem.* **41**, 1552.
Curtius, T. (1913). *Z. Angew. Chem.* **26**, III, 134.
Curtius, T., and Risson, J. (1930). *J. Prakt. Chem.* [2] **125**, 311.
Dekker, M., and Knox, G. R. (1967). *Chem. Commun.* p. 1243.
Diesen, R. W. (1964). *J. Chem. Phys.* **10**, 3256.
Doering, W. von E., and Odum, R. A. (1966). *Tetrahedron* **22**, 81.
Edwards, O. E. (1970). In "Nitrenes" (W. Lwowski, ed.), p. 233. Wiley (Interscience), New York.
Fagley, T. F., Sutter, J. R., and Oglukian, R. L. (1956). *J. Amer. Chem. Soc.* **78**, 5567.
Fry, A., and Wright, J. C. (1966). J. C. Wright, Dissertation, University of Arkansas, Little Rock; Fry, A., and Wright, J. C. (1968), *Chem. Eng. News* p. 28.
Gassman, P. G. (1970). *Accounts Chem. Res.* **3**, 26.

Gassman, P. G. (1971). *Abstr., 22nd Nat. Org. Symp., Amer. Chem. Soc., 1971* pp. 84–91.
Gassman, P. G., and Cryberg, R. L. (1969). *J. Amer. Chem. Soc.* **91**, 5176.
Gilchrist, T. L., and Rees, C. W. (1969). "Carbenes, Nitrenes, and Arynes." Nelson, London.
Goubeau, J., Allenstein, E., and Şchmidt, A. (1964). *Chem. Ber.* **97**, 884.
Hafner, K., and Kaiser, W. (1964). *Tetrahedron Lett.* p. 2185.
Hall, J. H. (1965). *J. Amer. Chem. Soc.* **87**, 1147.
Hall, J. H., and Behr, F. E. (1969). *Abstr. Int. Congr. Heterocycl. Chem.*, *2nd, 1969* Paper B-1, p. 60; Hall, J. H., and Behr, F. E. (1972). *J. Amer. Chem. Soc.* **94**, 4952.
Hall, J. H., Hill, J. W., and Fargher, J. M. (1968). *J. Amer. Chem. Soc.* **90**, 5313.
Haller, J. F. (1942). Thesis, Cornell University, New York.
Hassall, C. H., and Lippman, A. E. (1953). *J. Chem. Soc., London* p. 1059.
Heacock, J. F., and Edmison, M. T. (1960). *J. Amer. Chem. Soc.* **82**, 3460.
Hoergerlee, K., and Butler, P. E. (1964). *Chem. Ind. (London)* p. 933.
Horner, L., and Christmann, A. (1963a). *Angew. Chem.* **75**, 707.
Horner, L., and Christmann, A. (1963b). *Chem. Ber.* **96**, 388.
Huisgen, R., and von Fraunberg, K. (1969), *Tetrahedron Lett.* p. 2595.
Kirmse, W. (1959). *Angew. Chem.* **71**, 450.
Knunyants, I. L., and Bykhovskaya, E. G. (1960). *Proc. Acad. Sci. USSR* **132**, 513.
Koch, E. (1967). *Tetrahedron* **23**, 1747.
Krbechek, L., and Takimoto, H. (1968). *J. Org. Chem.* **33**, 4286.
Kreher, R., and Jäger, G. (1965). *Angew. Chem., Int. Ed. Engl.* **4**, 706.
Kwart, H., and Kahn, A. A. (1967). *J. Amer. Chem. Soc.* **89**, 1951.
L'abbé, G. (1969). *Chem. Rev.* **69**, 345.
Lansbury, P. T. (1970). *In* "Nitrenes" (W. Lwowski, ed.), Chapter 11, pp. 405–419. Wiley (Interscience), New York.
Lansbury, P. T., Colson, J. G., and Mancuso, N. R. (1964). *J. Amer. Chem. Soc.* **86**, 5225.
Leffler, J. E., and Tsuno, Y. (1963). *J. Org. Chem.* **28**, 190 and 902.
Lemal, D. M. (1970). *In* "Nitrenes" (W. Lwowski, ed.), Chapter 10, pp. 345–404. Wiley (Interscience), New York.
Lewis, F. D., and Saunders, W. H., Jr. (1967). *J. Amer. Chem. Soc.* **89**, 645.
Lewis, F. D., and Saunders, W. H., Jr. (1968). *J. Amer. Chem. Soc.* **90**, 7031.
Linke, S., Tisue, G. T., and Lwowski, W. (1967). *J. Amer. Chem. Soc.* **89**, 6308.
Lüttringhaus, A., Jander, J., and Schneider, R. (1959). *Chem. Ber.* **92**, 1756.
Lwowski, W. (1967). *Angew. Chem. Int. Ed. Engl.* **6**, 897.
Lwowski, W., ed. (1970). "Nitrenes." Wiley (Interscience), New York.
Lwowski, W., and Maricich, T. J. (1965). *J. Amer. Chem. Soc.* **87**, 3630.
Lwowski, W., and Mattingly, T. W., Jr. (1962). *Tetrahedron Lett.* p. 277.
Lwowski, W., and Scheiffele, E. (1965). *J. Amer. Chem. Soc.* **87**, 4359.
Lwowski, W., and Simson, J. (1967). *Abstr. Pap., 153rd Meet., Amer. Chem. Soc.* Abstract 0163.
Lwowski, W., and Woerner, F. P. (1965). **87**, 5491.
Lwowski, W., Marichich, T. J., and Mattingly, T. W., Jr. (1963). *J. Amer. Chem. Soc.* **85**, 1200.
Lynch, B. M., and Hung, Y.-Y. (1965). *J. Heterocycl. Chem.* **2**, 218.
Maricich, T. J. (1965). Thesis, Yale University, New Haven, Connecticut.
Marsh, F. D., and Simmons, H. E. (1965). *J. Amer. Chem. Soc.* **87**, 3529.
Masamune, S. (1964). *J. Amer. Chem. Soc.* **86**, 288.

Meinwald, J., and Aue, D. H. (1966). *J. Amer. Chem. Soc.* **88**, 2849.
Messer, M., and Farge, D. (1968). *Bull. Soc. Chim. Fr.* [5] p. 2832.
Meth-Cohn, O., Smalley, R. K., and Suschitzky, H. (1963). *J. Chem. Soc., London* p. 1666.
Meyer, E., and Griffin, G. W. (1967). *Angew. Chem.* **79**, 648.
Mishra, A., Rice, S. N., and Lwowski, W. (1968). *J. Org. Chem.* **33**, 481.
Morgan, A. F. (1916). *J. Amer. Chem. Soc.* **38**, 2095.
Moriarity, R. M., and Kliegman, J. M. (1967). *J. Amer. Chem. Soc.* **89**, 5959.
Moriarity, R. M., and Rahman, M. (1965). *Tetrahedron* **21**, 2877.
Moriarity, R. M., and Reardon, R. C. (1970). *Tetrahedron* **26**, 1379.
Moriarity, R. M., Rahman, M., and King, G. J. (1966). *J. Amer. Chem. Soc.* **88**, 842.
Odum, R. A., and Aaronson, A. M. (1969). *J. Amer. Chem. Soc.* **91**, 5680.
Odum, R. A., and Brenner, M. (1966). *J. Amer. Chem. Soc.* **88**, 2074.
Ogata, M., Kanō, H., and Matsumoto, H. (1968). *Chem. Commun.* p. 398.
Okahara, M., and Swern, D. (1969). *Tetrahedron Lett.* p. 3301.
Patai, S., and Gotshal, Y. (1966). *J. Chem. Soc., B* p. 489.
Porter, R. R., Knowles, J. R., and Fleet, G. W. J. (1970). *Chem. Eng. News* July 27, p. 28.
Pritzkow, W., and Timm, D. (1966). *J. Prakt. Chem.* [4] **32**, 178.
Reagan, M. T., and Nickon, A. (1968). *J. Amer. Chem. Soc.* **90**, 4096.
Reichle, W. T. (1964). *Inorg. Chem.* **3**, 402.
Reiser, A., and Marley, R. (1968). *Trans. Faraday Soc.* **64**, 1806.
Reiser, A., Bowes, G., and Horne, R. J. (1966a). *Trans. Faraday Soc.* **62**, 3162.
Reiser, A., Terry, G. C., and Willets, F. W. (1966b). *Nature (London)* **211**, 410.
Rice, F. O., and Freamo, M. (1951). *J. Amer. Chem. Soc.* **73**, 5523.
Rice, F. O., and Grelecki, C. J. (1957). *J. Amer. Chem. Soc.* **79**, 1880.
Rice, F. O., and Ingalls, R. B. (1959). *J. Amer. Chem. Soc.* **81**, 1856.
Rice, F. O., and Luckenbach, T. A. (1960). *J. Amer. Chem. Soc.* **82**, 2681.
Russell, K. E. (1955). *J. Amer. Chem. Soc.* **87**, 3487.
Sauer, J. (1970). Private communication.
Sauer, J., and Engels, J. (1969). *Tetrahedron Lett.* p. 5175.
Sauer, J., and Mayer, K. K. (1968). *Tetrahedron Lett.* p. 319.
Saunders, W. H., and Caress, E. A. (1964). *J. Amer. Chem. Soc.* **86**, 861.
Saunders, W. H., and Ware, J. C. (1958). *J. Amer. Chem. Soc.* **80**, 3328.
Sloan, M. F., Prosser, T. J., Newburg, N. R., and Breslow, D. S. (1964). *Tetrahedron Lett.* p. 2945.
Smith, P. A. S. (1970). *In* "Nitrenes" (W. Lwowski, ed.), Chapter 4, pp. 99–162. Wiley (Interscience), New York.
Smith, P. A. S., and Hall, J. H. (1962). *J. Amer. Chem. Soc.* **84**, 480.
Smith, P. A. S., Brown, B. B., Putney, R. K., and Reinisch, R. F. (1953). *J. Amer. Chem. Soc.* **75**, 6335.
Smith, P. A. S., Krbechek, L. O., and Resemann, W. (1963). *Abstr. Pap., 144th Nat. Meet., Amer. Chem. Soc.* p. 35M.
Smolinsky, G. (1960). *J. Amer. Chem. Soc.* **82**, 4717.
Smolinsky, G. (1961a). *J. Amer. Chem. Soc.* **83**, 2489.
Smolinsky, G. (1961b). *J. Org. Chem.* **26**, 4108.
Smolinsky, G., and Feuer, B. I. (1964). *J. Amer. Chem. Soc.* **86**, 3085.
Smolinsky, G., and Feuer, B. I. (1966). *J. Org. Chem.* **31**, 3882.
Smolinsky, G., Wasserman, E., and Yager, W. A. (1962). *J. Amer. Chem. Soc.* **84**, 3220.
Splitter, J. S., and Calvin, M. (1968). *Tetrahedron Lett.* p. 1445.
Stagner, B. A. (1916). *J. Amer. Chem. Soc.* **38**, 2069.

Staudinger, H., and Miescher, K. (1919). *Helv. Chim. Acta* **2**, 554.
Stieglitz, J. (1896). *Amer. Chem. J.* **18**, 751.
Sundberg, R. J. (1966a). *Tetrahedron Lett.* p. 477.
Sundberg, R. J. (1966b). *J. Amer. Chem. Soc.* **88**, 3781.
Sundberg, R. J., Das, B. P., and Smith, R. H., Jr. (1969). *J. Amer. Chem. Soc.* **91**, 658.
Swenton, J. S. (1967). *Tetrahedron Lett.* p. 2855.
Swenton, J. S. (1968). *Tetrahedron Lett.* p. 3421.
Swenton, J. S., Ikeler, T. I., and Williams, B. H. (1969). *Chem. Commun.* p. 1263.
Takimoto, H. H., and Denault, G. C. (1966). *Tetrahedron Lett.* p. 5369.
Tiemann, F. (1891). *Chem. Ber.* **24**, 4162.
Trozzolo, A. M., Murray, R. W., Smolinsky, G., Yager, W. A., and Wasserman, E. (1963). *J. Amer. Chem. Soc.* **85**, 2526.
Vivian, D. L., Greenberg, G. Y., and Hartwell, J. L. (1951). *J. Org. Chem.* **16**, 1.
von Fraunberg, K., and Huisgen, R. (1969). *Tetrahedron Lett.* p. 2599.
Vosburgh, I. (1916). *J. Amer. Chem. Soc.* **38**, 2081.
Walker, P., and Waters, W. A. (1962). *J. Chem. Soc., London* p. 1632.
Wasserman, E., Smolinsky, G., and Yager, W. A. (1964). *J. Amer. Chem. Soc.* **86**, 3166.
Wasserman, E., Barash, L., and Yager, W. A. (1965). *J. Amer. Chem. Soc.* **87**, 2075.
Wasserman, E., Murray, R. W., Yager, W. A., Trozzolo, A. M., and Smolinsky, G. (1967). *J. Amer. Chem. Soc.* **89**, 5076.
Wentrup, C. (1969). *Chem. Commun.* p. 1386.
Yamada, S. I., Terashima, S., and Achiwa, K. (1965). *Chem. Pharm. Bull.* **13**, 751.
Zimmer, H., and Jayawant, M. (1966). *Tetrahedron Lett.* p. 5061.

4

CARBONIUM IONS

S. P. McManus and C. U. Pittman, Jr.

I. General Aspects and Historical Background 194
   A. Definition of Terms and Bonding 194
   B. Other Onium Ions and Related Species 198
   C. Analogy between Boranes and Carbonium Ions . . . 199
   D. Nomenclature 200
II. Methods of Investigating Carbonium Ions 202
   A. Gas Phase versus Solution Phase 202
   B. Concentration, Lifetime Effects, and Stereochemistry . . 203
   C. Spectroscopic Methods 209
   D. Electrolytic Methods: Conductivity, EMF Measurements, and Polarography 229
   E. Kinetic Methods 233
   F. Chemical Methods (Product Analysis) 241
III. Methods of Formation 248
   A. Electron Removal from an Electrically Neutral Molecule or Radical 248
   B. Heterolytic Fission of a Larger Molecule 250
   C. Addition of a Cation to a Neutral Species 256
   D. Interaction of Neutral Molecules 258
IV. Factors Affecting the Stability of Carbonium Ions . . . 258
   A. Medium Effects; Kinetic and Thermodynamic Considerations . 259
   B. Inductive and Resonance Effects 266
   C. Steric Effects 273
V. Reactions of Carbonium Ions 287
   A. Elimination of a β Proton 289
   B. Elimination of an α Proton 290
   C. Rearrangement to Form Another Cation 290
   D. Addition to an Alkene (Alkylation) 295
   E. Intermolecular Hydride Transfer 295
   F. Fragmentation 297
   G. Friedel–Crafts Alkylation of an Aromatic Ring . . . 298
   H. Combination with a Nucleophile 299
   I. Photochemical Reactions 300
VI. Bridged Carbonium Ions 302
   A. Evidence for Bridged Ions 302

B. Electron-Sufficient Bridged Ions 307
　　C. Electron-Deficient Bridged Ions 314
　　D. Summary 320
　VII. Related Species 321
　　A. Nitrenium and Oxenium Ions 322
　　B. Silicenium Ions 323
　　References 324

I. General Aspects and Historical Background

A. DEFINITION OF TERMS AND BONDING

1. General Aspects

Of the intermediates discussed in this treatment, carbonium ions are among the most important and most studied. A carbonium ion is an organic cation containing a positive charged carbon which has a sextet of electrons surrounding the charged carbon. The parent member is the methyl cation (**1**) with three hydrogens attached to carbon. The charged carbon of the acetyl cation (**2**) has only two atoms directly bonded to the charged center but maintains the sextet of electrons around the formally charged carbon.

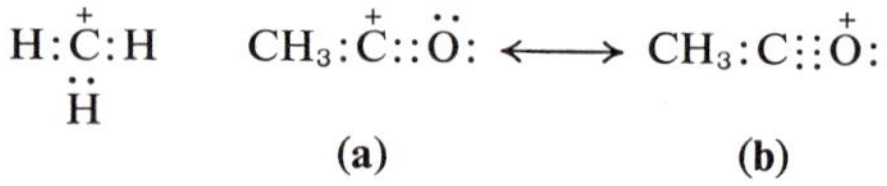

$$H:\overset{+}{C}:H \qquad CH_3:\overset{+}{C}::\overset{..}{O}: \longleftrightarrow CH_3:C::\overset{+}{O}:$$
$$\overset{..}{H}$$
$$\qquad\qquad\qquad (a) \qquad\qquad\qquad (b)$$
$$(1) \qquad\qquad\qquad (2)$$

Although the charge is formally written on a particular carbon atom, this does not require that the charge be localized at that position. Resonance hybrid structures of the benzyl cation [(**3a**)–(**3c**)] and the propenyl (allylic)

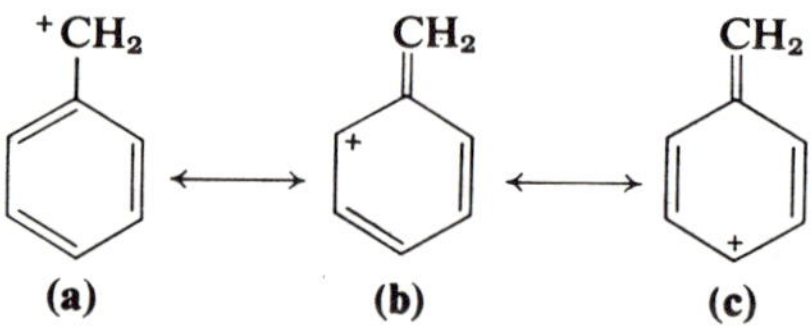

$$\text{(a)} \qquad \text{(b)} \qquad \text{(c)}$$
$$(3)$$

cation [(**4a**)–(**4c**)] are invoked to illustrate the dispersal of charge in these ions. However, even these structures do not satisfactorily represent the

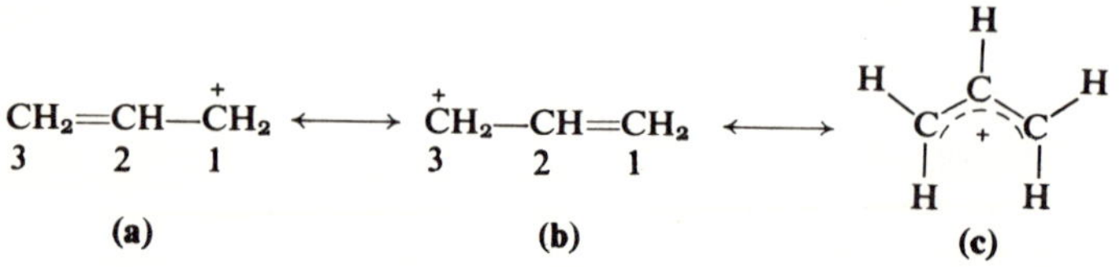

$$CH_2{=}CH{-}\overset{+}{C}H_2 \longleftrightarrow \overset{+}{C}H_2{-}CH{=}CH_2$$
$$\quad 3 \quad\; 2 \quad\; 1 \qquad\qquad 3 \quad\; 2 \quad\; 1$$
$$\text{(a)} \qquad\qquad\qquad \text{(b)} \qquad\qquad\qquad \text{(c)}$$
$$(4)$$

distribution of charge since some positive charge density is present at C-2 of the propenyl cation and at the meta carbons of the benzyl cation. Structure **4c** is somewhat better, but it does not indicate the greater concentration of charge at C-1 and C-3. Some charge density is also found at the hydrogen atoms. In the ethyl cation (**5**) some dispersal of charge may be represented by

$$\underset{\underset{H}{H}}{\overset{H}{\underset{|}{}}}C\overset{+}{-}\underset{H}{\overset{H}{\underset{|}{}}}C\underset{H}{\overset{H}{}} \longleftrightarrow \underset{H}{\overset{H^+}{}}C=\underset{H}{\overset{H}{}}C\underset{H}{\overset{H}{}}$$

(a)   (b)

(5)

invoking hybrid structures **5a** and **5b**. Hyperconjugation (see Dewar, 1962) is the term used to describe this method of charge dispersal onto the hydrogens. Hyperconjugation, as illustrated in the ethyl cation, does not imply that a double bond has been formed between the carbons or that a C–H bond has been broken. Instead, hybrid structure **5b** attempts to illustrate that the electron density in all three C_2–H bonds is somewhat decreased, and the bonds are slightly lengthened (weakened) as a consequence of this loss of electron density.

There is a marked tendency of the ligands attached to the charged carbon to arrange themselves in a planar conformation. For example, the *tert*-butyl cation (**6**) is sp^2 hybridized at the central carbon. The concentration of the bonding electrons into sp^2 orbitals in this manner maximizes nuclear electron

$$CH_3-\overset{+}{C}\underset{CH_3}{\overset{CH_3}{\diagdown}}$$

(6)

attractions and causes the closer electrons to be drawn to the central carbon's nucleus, thereby promoting stronger bonding. The van der Waals radius of the central carbon is reduced as a consequence of the excess charge, and the p_z orbital is vacant.

In many carbonium ions, atoms other than carbons or hydrogens may be bonded to the charged carbon, and classification of such ions as carbonium ions may be ambiguous. For instance, ions **7–9** have significant positive

$$CH_3-\ddot{\underset{..}{O}}-\overset{+}{C}H_2 \longleftrightarrow CH_3-\overset{+}{O}=CH_2$$

(a)   (7)   (b)

$$\underset{CH_3}{\overset{CH_3}{\diagdown}}\ddot{N}-\overset{+}{C}H_2 \longleftrightarrow \underset{CH_3}{\overset{CH_3}{\diagdown}}\overset{+}{N}=CH_2$$

(a)   (8)   (b)

charge density on both the carbon and the adjacent heteroatom. The hybrid structures **7a**, **8a**, and **9a** emphasize the carbonium ion nature, whereas **7b**, **8b**, and **9b** represent, respectively, the oxonium, ammonium, and sulfonium ion nature of these species.

$$\underset{(a)}{CH_3\ddot{\underset{..}{S}}-\overset{+}{C}H_2} \underset{(9)}{\longleftrightarrow} \underset{(b)}{CH_3\overset{+}{\underset{..}{S}}=CH_2}$$

2. Historical Background

Shortly after Gomberg described his discovery of the first stable free radical, Norris (1901) and Kehrmann and Wentzel (1901) independently observed that colorless derivatives of triphenylmethane, such as triphenylcarbinol and triphenylmethyl chloride, form deep yellow solutions in sulfuric acid. Several similar observations were made by these groups and by Gomberg (1902) and Walden (1902). The saltlike character of these conducting solutions in sulfuric acid and some organic solvents was noted by Gomberg (1902) and Baeyer and Villiger (1902). Baeyer, in studying these colored ("halochromic") solutions, suggested that a correlation existed between the formation of the salt and the appearance of color. Baeyer called these salts *carbonium salts*. Studies by Hantzsch (1907, 1921), which indicate that the cationic species formed from the different triphenylmethyl derivatives were identical, placed the structure of triarylmethyl cations in a position of little doubt.

From the beginning the study of organic cations has been a topic that has attracted the interest of many great organic chemists. No one has dominated this field of study although a few individuals have occupied a dominant role during key periods in the 70-year history. Unlike most areas of pursuit in organic chemistry, carbonium ion chemistry has continued to produce lively debate and controversy. The history of the field is best treated in a chapter by Nenitzescu (1968) in the four-volume set entitled "Carbonium Ions," edited by Olah and Schleyer (1968–1972). A more condensed but complete treatment of carbonium ions has been written by Bethell and Gold (1967).

On the strength of kinetic studies, Meerwein and Van Emster (1922) suggested the intermediacy of carbonium ions in the conversion of camphene hydrochloride into isobornyl chloride, a reaction discovered by Wagner (1899). Their kinetic results indicated that the fastest rearrangements were in SO_2 and nitromethane solvents which had previously been shown to be good media for the detection of carbonium ions by conductivity methods. That the reactions were slow in poor solvents, such as benzene, corroborated the theory on solvent effects. The second key observation of Meerwein and Van Emster was that metal halides of the type that catalyzed ionization of triphenylmethyl chloride accelerate the rearrangement of camphene hydrochloride. Thus, the authors concluded that "... the rearrangement takes place only after a preceding ionization. The conversion of camphene hydrochloride to isobornyl chloride actually does not consist of a migration of a chloride atom but of a rearrangement of the cation."

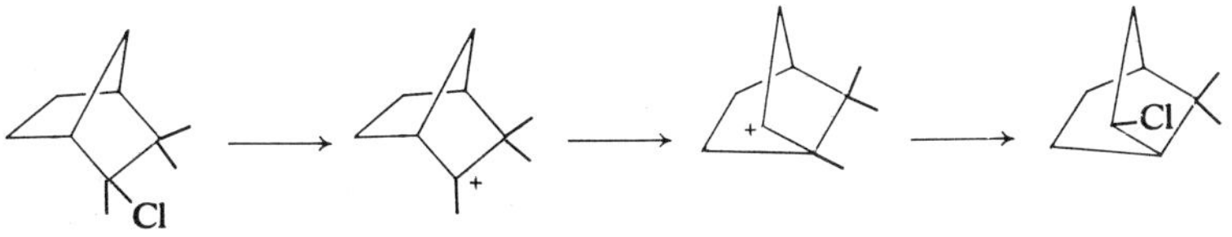

$$\text{(structures shown)} \quad (1)$$

Meerwein and his group (1927) later generalized the rearrangement to include several other terpene systems. Whitmore and his school extensively studied rearrangements in acyclic systems in an attempt to provide a unifying theory for several seemingly unrelated reactions (Whitmore, 1932). He considered that the Hofmann, Lossen, Curtius, Beckmann, and Demjanov rearrangements involve, like the Wagner–Meerwein and pinacol rearrangements, the initial formation of an electron-deficient carbon with a "sextet of electrons" by expulsion of a stable anion or molecule. It was evident in his

$$:\!\ddot{\text{A}}\!:\!\ddot{\text{B}}\!:\!\ddot{\text{X}}\!: \longrightarrow :\!\ddot{\text{A}}\!:\!\ddot{\text{B}} + :\!\ddot{\text{X}}\!: \quad (2)$$

early papers that Whitmore used such terms as "sextet of electrons" and "positive organic fragment" as opposed to "carbonium ion" in order to appease the referees. An excellent review by Whitmore of his work and aliphatic carbonium ions in general appeared in 1948. Of the greatest significance from Whitmore's studies was the recognition of the effects of alkyl substituents on the stability of aliphatic cations.

Before Whitmore's ideas or results were published, Ingold and Rothstein (1928) proposed a duality of mechanism (Ingold, 1927) for the reaction of benzyl chloride with hydroxide ion. The terminology S_N1 (substitution nucleophilic, unimolecular) and S_N2 (substitution, nucleophilic, bimolecular) were introduced (Gleave et al., 1935) to describe these mechanisms. Ward (1927) had independently observed that the rate of hydrolysis of α-phenylethyl chloride in 80% ethanol and the rate of ethanolysis of benzhydryl chloride are not affected by the addition of base. He invoked a mechanism involving the ionization of the C–Cl bond in the rate-determining step to account for these facts. Between 1933 and 1935, Hughes, Ingold, and their co-workers, published detailed accounts of their work on nucleophilic substitution at carbon. A rather unexpected result, but one that was complimentary to the work of Whitmore's group, was that alkaline hydrolysis of the simple alkyl bromides occurs in accordance with different mechanisms. There is a moderate rate decrease in going from methyl to ethyl, and the reaction follows second-order kinetics over a range of alkali concentrations. There is a large rate increase in going from isopropyl to *tert*-butyl, with the latter reaction being first order and independent of the concentration of alkali (Gleave *et al.*, 1935). Since an aliphatic alkyl halide was not expected to ionize, the latter result was of some interest.

The stereochemical course of nucleophilic substitution reactions was also of interest to Ingold. That a stereochemically inactive (racemic) product resulted from the S_N1 process and that inversion of configuration (rather than retention or racemization) resulted from the S_N2 mechanism were proved conclusively (see Ingold, 1969).

The foregoing discussions of some of the major early developments in carbonium ion chemistry serve to introduce some of the topics which will be treated in depth subsequently.

B. OTHER ONIUM IONS AND RELATED SPECIES

Unlike the special cases of **7–9**, many organic cations have no carbonium ion character. Oxonium ions, for example, may contain a positive charged trivalent oxygen with an electron octet. The trimethyloxonium ion (**10**), an important methylating agent, is one example.

$$CH_3 \overset{\overset{..}{\overset{+}{O}}}{\underset{CH_3}{|}} CH_3 \rightleftharpoons CH_3 \overset{CH_3}{\underset{\overset{..}{\overset{+}{O}}}{|}} CH_3$$

(**10**)   pyramidal geometry

In saturated oxonium ions, such as **10**, the positive charge is localized largely on the oxygen, whereas in the related cations formulated with a C–O double bond it is delocalized. Ion **11** has both carbonium and oxonium ion character. As the oxonium ion center becomes more unsaturated, oxonium ion character decreases. The acetyl cation (**2**) can be represented as an oxinium

$$\underset{R_1}{\overset{R_2}{\diagdown}}C\overset{+}{=}\overset{..}{O}-R_3 \longleftrightarrow \underset{R_1}{\overset{R_2}{\diagdown}}\overset{+}{C}-\overset{..}{\overset{..}{O}}-R_3$$

(**11**)

Oxonium ion   Alkoxycarbonium ion

ion (**2b**) (Jennen, 1966) or an oxocarbonium ion (**2a**). The contribution of oxinium ion character (i.e., **2b**) in the acetyl cation (**2**) is less than that of oxonium ion character in **11** (Boer, 1966). Cations with a positive charged oxygen having a monovalent sextet ($R-\overset{..}{O}^+$) are called oxenium ions (Jennen, 1966), and these should be carefully distinguished from oxonium ions. Perst (1971) has presented a detailed discussion of the classification, nomenclature, and chemistry of all known classes of oxonium ions, and Pittman *et al.* (1972b) have exhaustively reviewed the chemistry of 1,3-dioxolan-2-ylium (i.e., **12**) and related cations.

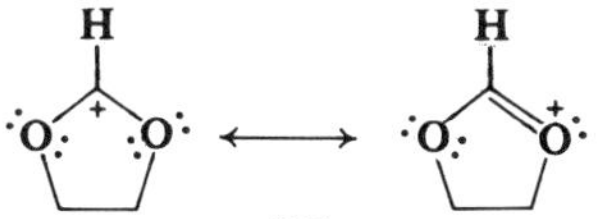

(**12**)

Sulfonium ions contain a positive, trivalent sulfur with an electron octet and as such are analogous to oxonium ions. Similarly, sulfenium ions are positive and monovalent with a sextet of electrons. They are analogs of oxenium ions. Onium compounds of oxygen with hydrogen as a ligand are

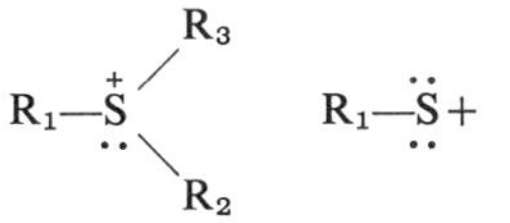

$$R_1 - \overset{+}{S} \begin{matrix} R_3 \\ \\ R_2 \end{matrix} \qquad R_1 - \overset{..}{\underset{..}{S}} +$$

Sulfonium   Sulfenium

thermodynamically more stable than their sulfonium ion counterparts, but the stability of trialkyloxonium ions is much lower than that of trialkylsulfonium ions.

C. Analogy between Boranes and Carbonium Ions

Carbonium ions are formally isoelectronic with boranes, both having a sextet of electrons surrounding the central atom. This sextet causes both carbonium ions and boranes to behave as Lewis acids, although carbonium ions by virtue of their excess positive charge are stronger (harder) Lewis acids. Trimethylboron (**13**) and boron trihalides are known to have the boron atom at the center of a planar sp^2-hybridized system (Levý and Brockway, 1937). Since the electronic ground state of the boron atom is $1s^2$, $2s^2$, $2p^1$, some energy input is required to form the sp^2 hybrid. Like carbon, this energy is

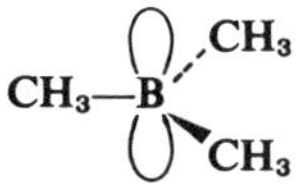

(**13**)

imparted by the formation of extra bonds to the boron which more than compensates for this promotional energy.

The B–C bond distance in **13** is greater than the C–C distances in the *tert*-butyl cation (**6**). In part, this is because of the shrinkage of the carbon atomic radius owing to the excess positive charge. The charged carbon of a carbonium ion is far more electronegative than the boron atom. Thus, the magnitude of bond dipoles are expected to differ in these systems. For example, the trifluoromethyl cation (**14**) should exhibit much smaller C–F bond dipoles than the B–F dipoles in trifluoroborane (**15**). This increased electronegativity should result in a greater contribution of back-donation

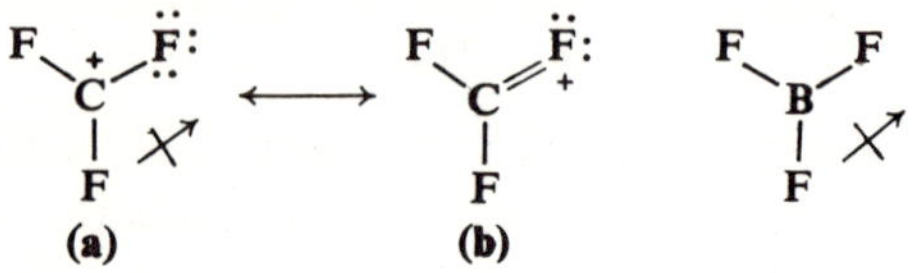

$$\begin{array}{ccc} \text{(a)} & \text{(b)} & \\ \text{(14)} & & \text{(15)} \end{array}$$

(i.e., hybrid **14b**) in carbonium ions than in boranes. This phenomenon has been supported by theoretical calculations (Kispert *et al.*, 1971) and ^{19}F NMR observations (Olah and Comisarow, 1969).

D. Nomenclature

The nomenclature of carbonium ions is a topic which currently must certainly depress and confuse young students of chemistry. The term "carbonium" was first coined by Baeyer (1905) who had other cations (onium ions such as ammonium, oxonium, and sulfonium) in mind. However, as we have seen, the parallel between trivalent carbonium ions with an electron sextet and the other onium ions with an electron octet is inconsistent. Other onium ions increase their covalency by 1 on formation from their precursors, whereas carbonium ions decrease their covalency by 1 from the normal value of 4 for its precursor to 3 for the ion. Gomberg (1902) immediately pointed out this discrepancy and proposed, without success, that these ions be called "carbyl salts." Since oxenium and sulfenium ions contain an electron sextet and are isoelectronic with carbonium ions, the term "carbenium" is a more systematic name. Furthermore, the "carbenium" nomenclature relates these cations directly to carbenes. For example, the protonation of carbene (methylene), $:CH_2$, gives CH_3^+, the carbenium ion. On this basis the *tert*-butyl cation (**6**) would be the trimethylcarbenium ion. This system has been advocated several times by such workers as Dilthey and Dinklage (1929), Arndt and Lorenz (1930), and Jennen (1966) but without success. The voluminous literature of this field almost entirely employs the term "carbonium ion," and it appears here to stay since the major current works on this topic bear the partial or full title "Carbonium Ions" (Olah and Schleyer, 1968–1972; Bethell and Gold, 1967).

4. CARBONIUM IONS

It should be pointed out that Olah (1971, 1972) has now begun to use the "carbenium ion" nomenclature. He proposes the use of the term "carbocations" to cover all cations of carbon. Trivalent carbocations (i.e., R_3C^+) are termed carbenium ions, and penta- (or tetra-) coordinated carbocations are called carbonium ions. Furthermore, this system would be extended to silicocations. Thus, according to Olah and Mo (1971), trivalent and pentavalent silicocations would be called silicenium and siliconium ions, respec-

System	Ph₃C⁺	CH_5^+	$^+SiF_3$	SiH_5^+
Carbonium	Triphenylcarbonium		Trifluorosiliconium	
Carbenium	Triphenylcarbenium	Carbonium	Trifluorosilicenium	Siliconium

tively. Whether or not this most recent and serious advocacy of the carbenium ion terminology will be accepted remains to be seen. Although we urge carbonium ion chemists to follow Olah's lead, we will use the term "carbonium ion" here because it is the term which is now predominant in the literature. [To complicate matters, the IUPAC 1965 Rules (*Pure Appl. Chem.* **11**, 1 [Rule C83]) also allow names ending in ylium and enium. Thus, the *tert*-butyl cation could also be called the 2-methylpropylium or 2-methylpropenium ion.]

Carbonium ion nomenclature may best be illustrated by the consideration of several examples. There are two general methods which can be employed. For example, the ion derived from *tert*-butyl alcohol (trimethylcarbinol) may be called either the *tert*-butyl cation or the trimethylcarbonium ion (see structure **6**). Naming **6** the *tert*-butylcarbonium ion only introduces ambiguities leading to needless confusion. The ion $CH_3CH_2^+$ may be called the methyl carbonium ion or the ethyl cation. Likewise, C_6H_5—CH_2^+ is the phenyl carbonium ion or the benzyl cation. In general, the cation nomenclature is superior because the name preceding the cation immediately places a picture of the structural system in the reader's mind. Using the term carbonium ion has a parallel in the carbinol method of naming alcohols which is no longer accepted usage. However, especially in the triaryl carbonium ion series, this method is still very often used, and it therefore must be considered. The following tabulated examples are instructive:

Structure	Cation	Carbonium ion
$CH_3-\overset{+}{C}H-CH_3$	2-Propyl	Dimethyl
$CH_2=CH-\overset{+}{C}H_2$	Propenyl Allyl	Vinyl
$CH_3-CH=\underset{\underset{CH_3}{\|}}{C}-\overset{+}{C}H-CH_3$	3-Methyl-2-pentenyl	Methyl 2-(but-2-enyl)
 (2,5-dimethylheptadienyl cation structure)	2,5-Dimethylheptadienyl	
$Ph_2-\overset{+}{C}H$	Benzhydryl Diphenylmethyl	Diphenyl
Ph_3-C^+	Triphenylmethyl	Triphenyl
$CH_3O\overset{+}{C}H_2$	Methoxymethyl	Methoxy

In this chapter we shall use the cation nomenclature almost invariably because it does not hide the generic relationship between the carbonium ion and its precursor. Common names will only be employed where systematic names are too unwieldy.

II. Methods of Investigating Carbonium Ions

A. Gas Phase versus Solution Phase

To determine if a carbonium ion is an intermediate in a given reaction frequently is a difficult problem. The lifetime of the carbonium ion may be very short, and its concentration at any instant might be very low. In the past, this required the use of indirect methods such as kinetics or product analysis to implicate the intervention of carbonium ions. In the gas phase, mass spectrometers are able to separate positive charged ions according to their charge/mass (e/m) ratio, and carbonium ion reactions in the gas phase are now well known. In addition, ion cyclotron resonance techniques are being increasingly employed to such gas-phase reactions. The advantage of studying ions in the gas phase is the absence of solvent effects. Thus, orders of carbonium ion stability determined in the gas phase give a better picture of the intrinsic properties of each ion without complications owing to solvation. For example, the stabilizing or destabilizing effects of substituents measured in the gas phase can be thought of as the standard state in which these interactions may be viewed. In solution, the carbonium ion center might be highly solvated, thereby reducing the electron demand this center would make on the substituents in the molecule. In effect, the substituent effects will not be as pronounced in the liquid or solid phases as they are in the gas phase. Although mass spectrometric techniques offer this great advantage, only the e/m ratio is elucidated and not the specific structure of the fragment being investigated. Furthermore, the majority of the reactions which interest the organic chemist take place in solution where the interaction of the cation with its counter ion and with solvent is of paramount interest. Finally, ions produced in a mass spectrometer are the result of the decomposition of radical ions formed from electron bombardment of molecules at 70 eV. Thus, about 1000 kcal/mole of excitation energy are theoretically possible in the ions formed. The excited radical ion produced initially decomposes to an ion and a radical. However, the processes observed may represent only a small fraction of the total processes occurring since neutral and negative species remain undetected. The use of mass spectrometry in elucidating carbonium ion decompositions in the gas phase is a huge topic which has been reviewed (Field and Franklin, 1957; McLafferty, 1963).

B. Concentration, Lifetime Effects, and Stereochemistry

In solution, lower energies are needed to generate carbonium ions because of strong solvation both of the transition state and the final ionic products. However, a carbonium ion in solution, unless it is particularly stable, usually has a very short lifetime. In general, the ion's lifetime depends on two major factors: (1) the ion's intrinsic stability and (2) the nature of the solvent. The more basic and nucleophilic the solvent, the shorter the ion's lifetime; the less nucleophilic and more acidic the solvent, the longer the ion's lifetime becomes. For convenience, we will define the two limiting processes. In the first, a "free ion" is formed, and in the second, no carbonium ion intermediate exists. These are the familiar limiting S_N1 and S_N2 processes. All degrees of intermediate behavior exist. Considering processes where only the S_N1 mode is operative, a tremendous range of carbonium ion lifetimes is possible. For

$$S_N1 \qquad RX \longrightarrow R^+ + X^- \xrightarrow{Y^-} RY \qquad (3)$$

$$S_N2 \qquad RX + Y^- \longrightarrow [^{\delta-}Y\cdots R\cdots X^{\delta-}]^\ddagger \longrightarrow RY + X^- \qquad (4)$$

example, the deoxidation reaction of alcohols in basic solution with haloform generates carbonium ions with lifetimes that are very short, $\sim 10^{-10}$ second (Skell and Maxwell, 1962; Keating and Skell, 1970).

$$HCX_3 + OH^- \longrightarrow CX_3^- + H_2O \qquad (5)$$

$$CX_3^- \longrightarrow :CX_2 + X^- \qquad (6)$$

$$RO^- + :CX_2 \longrightarrow RO\ddot{C}X + X^- \qquad (7)$$
$$\phantom{RO^- + :CX_2 \longrightarrow RO\ddot{C}X}\longrightarrow R^+ + CO + X^-$$

At the other extreme, on dissolving either *tert*-butyl fluoride or *tert*-butyl alcohol in FSO_3H—SbF_5—SO_2 (or SOClF) solutions the *tert*-butyl cation is formed quantitatively, and 1 M solutions can be kept at room temperature for weeks (see Olah and Olah, 1970).

Most early studies of carbonium ions involved a study of the products produced during solvolysis reactions at an asymmetric carbon. Ingold formulated the S_N1 rule which stated: "Mechanism S_N1 proceeding through a carbonium ion, involves racemization, together in general, with an excess of inversion, unless a configuration-holding group is present when configuration is predominantly retained" (see Ingold, 1969). In this view the transformation of an sp^3-hybridized carbon to an sp^2-hybridized carbonium ion generates an intermediate with a plane of symmetry which on reaction with a nucleophile gives a racemic product. However, these reactions are frequently coupled with a significant amount of net inversion. In fact, the percent inversion usually increases as the carbonium ion formed becomes less and less stable. For example, inversion increases in the series secondary arylalkyl < *tert*-alkyl < *s*-alkyl as seen in Table I. Predominant inversion can be explained in terms of lifetime effects. As the leaving group departs it acquires a

Table I
Stereochemistry of S_N1 Reactions[a]

Substrate	Substituting group	Medium	Percent net inversion
2-Octyl bromide	H_2O	60% aq. EtOH	66
2-Octyl bromide	EtOH	60% aq. EtOH	74
3,7-Dimethyl-3-octyl chloride	H_2O	80% aq. Acetone	21
α-Phenylethyl chloride	H_2O	H_2O	17
α-Phenylethyl chloride	H_2O	80% aq. Me_2CO	2

[a] Data from Ingold (1969).

solvent shell which is increasingly bonded to it. During this period of time the front face of the carbonium ion is shielded from solvent attack. Solvent molecules bound to the leaving group cannot attack the carbonium ion. If the ion is relatively unstable, rapid solvent attack will occur. This must occur from the backside unless the leaving group and its solvent shell have moved far away from the front face. If the ion is quite stable, its lifetime in a given solvent will be greater. Now the leaving group and its solvent shell have time to move farther away. This leaves a symmetrical solvent shell around the carbonium ion, and reaction with the solvent nucleophiles may occur equally as well from either side leading to racemic product. Hammett (1940) introduced the term ion pair to describe the initial S_N1 reaction product. This ion pair can be pictured as subsequently separating further to form a solvent-separated ion pair and finally a set of completely separated ions. This process, represented in Eq. (9), has been supported in extensive studies of solvolysis reactions by Winstein *et al.* (1956, 1965).

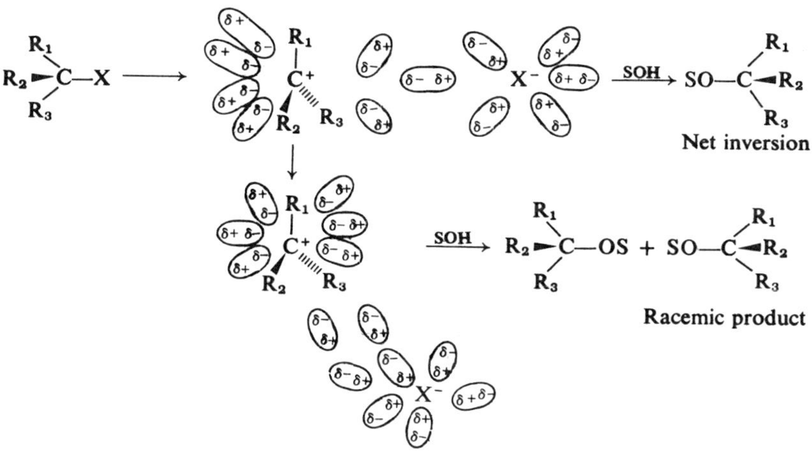

$$RX \rightleftharpoons \underset{\text{tight ion pair}}{R^+X^-} \rightleftharpoons \underset{\text{solvent-separated ion pair}}{R^+\|X^-} \rightleftharpoons \underset{\text{separated ions}}{R^+ + X^-} \quad (9)$$

In any S_N1 reaction, great difficulties exist in elucidating the exact degree of association of the carbonium ion with the counterion or solvent and the interrelation of this solvation with rearrangement, racemization, ion-pair recombination, and finally with the observed rate law for the given reaction. Consider a simplified kinetic description of the S_N1 process. In Eq. (10) we

$$RX \underset{k_{-1}}{\overset{k_1}{\rightleftharpoons}} R^+ + X^- \xrightarrow[k_2]{Y^-} RY \quad (10)$$
$$ \downarrow_{k_3}^{Z} P$$

see that the carbonium ion R^+ can recombine with X^- (a process usually called ion-pair return or internal return) or it can react with the solvent media in two ways. It can combine to give RY or it can undergo reaction by a parallel route to give P. An example of this type of sequence would be the ionization of *tert*-butyl chloride in an aqueous solvent system. The *tert*-butyl cation could lose a proton to give an olefin (i.e., P) or combine with water to give *tert*-butyl alcohol (i.e., RY). If R^+ is envisioned as a very short-lived species, we can assume the concentration of R^+ is always small in comparison with the concentration of other species present. Then applying the steady-state hypothesis we can say the rate of formation of R^+ equals its rate of consumption. In order to further simplify the kinetic analysis, assume the reaction of R^+ with Y^- predominates and no P is formed. In this case the rate of disappearance of RX is given by Eq. (11).

$$\text{Rate} = -d[\text{RX}]/dt = k_1[\text{RX}] - k_{-1}[\text{R}^+][\text{X}^-] \tag{11}$$

The lifetime of R^+ is very short under most circumstances, and we cannot measure the concentration of R^+. However, if the concentration of R^+ does not change with time (i.e., steady-state assumption) during the course of the reaction, one can set $d[R^+]/dt = 0$. The equation for the rate of formation of R^+ is

$$d[\text{R}^+]/dt = k_1[\text{RX}] - k_{-1}[\text{R}^+][\text{X}^-] - k_2[\text{R}^+][\text{Y}^-] \tag{12}$$

Since the concentration of R^+ is constant,

$$0 = k_1[\text{RX}] - k_{-1}[\text{R}^+][\text{X}^-] - k_2[\text{R}^+][\text{Y}^-] \tag{13}$$

or

$$[\text{R}^+](k_{-1}[\text{X}^-] + k_2[\text{Y}^-]) = k_1[\text{RX}]$$

therefore

$$[\text{R}^+] = \frac{k_1[\text{RX}]}{k_{-1}[\text{X}^-] + k_2[\text{Y}^-]} \tag{14}$$

Substituting $[R^+]$ into the original rate expression gives

$$\frac{-d[\text{RX}]}{dt} = k_1[\text{RX}] - \left(\frac{k_{-1}k_1[\text{RX}][\text{X}^-]}{k_{-1}[\text{X}^-] + k_2[\text{Y}^-]}\right) \tag{15}$$

Clearing this fraction, we obtain:

$$\frac{-d[\text{RX}]}{dt} = \frac{k_1 k_{-1}[\text{RX}][\text{X}^-] + k_1 k_2[\text{RX}][\text{Y}^-] - k_{-1} k_1[\text{RX}][\text{X}^-]}{k_{-1}[\text{X}^-] + k_2[\text{Y}^-]} \tag{16}$$

or

$$\text{Rate} = \frac{k_1 k_2[\text{RX}][\text{Y}^-]}{k_{-1}[\text{X}^-] + k_2[\text{Y}^-]} \tag{17}$$

In S_N1 reactions where $k_{-1}[\text{X}^-]$ is negligible compared to $k_2[\text{Y}^-]$, Eq. (17) reduces to Eq. (18). This can be the case where the leaving group has a weak

$$\text{Rate} = k_1[\text{RX}] \tag{18}$$

affinity for R^+ and the $[X^-]$ is low. However, many reactions have been observed where the rate decreases as the reaction proceeds. This is due to the increase in the $[X^-]$ which causes $k_{-1}[\text{X}^-]$ to contribute significantly in Eq. (17). The solvolysis of benzhydryl halides in aqueous acetone exhibits this decrease in rate (Benfey et al., 1952), but tert-butyl halides do not, and obeys Eq. (18) throughout the entire reaction (Bateman et al., 1940).

The *tert*-butyl cation is far more reactive (a much harder acid) and, therefore, is less selective. The more stable benzhydryl cation does not react with water as rapidly. Thus, owing to its longer lifetime, it survives long enough to permit collisions with the more nucleophilic halide ion, resulting in internal return, i.e.,

$$R^+ \xrightarrow{X^-} RX$$

The rate of an S_N1 reaction can be retarded by adding an excess of X^- to the reaction solution. This retardation of the rate is called the "mass-law effect" or the "common ion effect." In general, the more stable the carbonium ion, the more pronounced the mass-law effect becomes, because the $k_{-1}[X^-]$ term [Eq. (17)] is larger, relative to $k_2[Y^-]$, in stable ions than it is in unstable ions.

The term $k_2[Y^-]$ becomes increasingly large with respect to $k_{-1}[X^-]$ as Y^- becomes increasingly nucleophilic with respect to X^-. Thus, if X^- equaled Cl^- and Y^- were I^-, the rate law would be Eq. (18) at much lower concentrations of I^- than it would be if more weakly nucleophilic F^- were the nucleophile Y^-. In weakly nucleophilic solvents such as CF_3CO_2H, the term $k_2[Y^-]$ is smaller than for the same solvolysis reactions conducted in more nucleophilic solvents such as acetic acid or water.

The phenomenon of ion-pair return (internal return) has been studied in several nonrearranging systems including benzhydryl (Goering and Levy, 1962, 1964), 2-phenyl-2-butyl (Goering and Chang, 1965), *trans*-α-methyl-δ-phenylallyl (Goering and Linsay, 1969), and α-*p*-anisylethyl and α-phenylethyl *p*-nitrobenzoates (Goering *et al.*, 1970). The ion-pair return associated with the solvolyses [Eq. (19)] of these substrates results in randomization of the carbonyl oxygen atoms [Eq. (20)] and sometimes in partial racemization of optically active substrates [Eq. (21)]. For example, solvolysis of α-*p*-anisylethyl and α-phenylethyl *p*-nitrobenzoates in aqueous acetone involves alkyl oxygen cleavage [Eq. (19)], but this is accompanied by ion-pair return that

$$R-O-\overset{\overset{^{18}O}{\|}}{C}-Ar \xrightarrow[H_2O/acetone]{k_t} R-OH + HOOC-Ar \quad (19)$$

or R' some ^{18}O some ^{18}O

solvolysis with alkyl–oxygen cleavage and ^{18}O randomization

$$R-O-\overset{\overset{O^{18}}{\|}}{C}-Ar \xrightarrow{k_{eq}} R-^{18}O-\overset{\overset{O}{\|}}{C}-Ar + R-O-\overset{\overset{^{18}O}{\|}}{C}-Ar \quad (20)$$

(*d*) (*d*) (*d*)

randomization of ^{18}O via ion-pair return without racemization

$$R'-O-\overset{^{18}O}{\underset{\|}{C}}-Ar \longrightarrow R'-^{18}O-\overset{O}{\underset{\|}{C}}-Ar + R'-O-\overset{^{18}O}{\underset{\|}{C}}-Ar \quad (21)$$
(d) \qquad\qquad (dl) \qquad\qquad\qquad (dl)

randomization of ^{18}O via ion-pair return with racemization

$$R = \underset{d \,=\, \text{optically active}}{\text{Ph}-\overset{CH_3}{\underset{|}{CH}}-} \qquad R' = CH_3O-\underset{dl \,=\, \text{optically inactive}}{\text{Ar}-\overset{CH_3}{\underset{|}{CH}}-}$$

results in randomization of the carboxyl oxygen atoms in the unsolvolyzed esters [Eq. (20) and (21)] as well as incorporation of an ^{18}O label in both the alcohol and products. At 60°C the anisylethyl ester solvolyzes more than 30,000 times faster than the phenylethyl ester. Despite this huge difference, the internal return to solvolysis (i.e., k_{eq}/k_t) ratios were remarkably similar. On the other hand, the stereochemistry of ion-pair return was strikingly different for these two cases. In the anisylethyl ester, ion-pair return results in substantial racemization of the unsolvolyzed ester [Eq. (21)], but ion-pair return in the phenylethyl ester does not result in detectable racemization. Furthermore, that fraction of ion-pair return in the anisylethyl system which results in racemization is not eliminated by the addition of excess, highly nucleophilic azide ion. These results can be neatly correlated by the Winstein mechanism for solvolysis [Eq. (22)]. According to this mechanism the solvent-separated ion-pair intermediate is intercepted by azide ion, thus eliminating external ion-pair return to the starting p-nitrobenzoate. Ion-pair return from the tight ion pair can result in carboxyl oxygen randomization, but not

$$R-p\text{-NB} \rightleftharpoons R^+ \, p\text{-NB}^- \rightleftharpoons R^+ \| \, p\text{-NB}^- \longrightarrow \text{solvolysis products}$$
$$\underset{N_3^-}{\Big\downarrow} \quad RN_3 \qquad (22)$$

racemization when R = phenyl. In the p-anisyl case the solvent-separated ion pair can, in the absence of azide ion, undergo ion-pair return (via the tight ion pair) which results in racemization as well as carboxyl oxygen randomization. Racemization may also occur in the tight ion pair for the p-anisyl case. In this view, the forces of attraction in the tight ion pair are not great enough to preserve optical configuration. Thus, even though azide ions capture R^+ effectively from the solvent-separated ion pair, a large proportion of the ion-pair return which results in oxygen randomization is also causing racemization in the p-anisylethyl system. This must be taking place in the tight ion pair. It also appears that solvent capture of R^+ in the tight ion pair is important. There are other solvolysis reactions, such as hydrolysis, where that fraction of internal return which causes racemization comes only from return via the solvent-separated ion pair. In these cases, addition of azide prevents racemization of the unsolvolyzed starting material, but it does not prevent ^{18}O randomization which takes place during return from the tight ion pair.

C. Spectroscopic Methods

Several excellent reviews of the application of spectroscopic techniques to the direct investigation of carbonium ions have appeared, and the reader is encouraged to read these to supplement the introduction presented here. Olah and Pittman (1966) generally reviewed the literature through 1965. In addition, UV spectra (Olah et al., 1968a), vibrational spectra (Evans, 1968), NMR (Fraenkel and Farnum, 1968), cryoscopic, and conductimetric measurements (Gillespie and Robinson, 1968; Lichtin, 1968) have been reviewed.

1. Solution: NMR, Raman, UV, and IR

a. Nuclear Magnetic Resonance. The most successful spectroscopic technique for studying the structure of carbonium ions in solution has been NMR spectroscopy (Jackman and Sternhell, 1969). This technique has provided detailed structural information in recent years. Olah and coworkers, for example, have combined ^{1}H, ^{13}C, ^{2}H, and ^{19}F NMR studies on a huge variety of ions. We will illustrate the power of these methods with some alkyl cations. More detail can be found in the recent review by Olah and Olah (1970) and in the specific reviews cited earlier in this section.

Isopropyl fluoride in SbF$_5$, SbF$_5$—SO$_2$, or other such media ionizes to form the 2-propyl cation.

$$\text{CH}_3\text{CH}-\text{CH}_3 \xrightarrow{\text{SbF}_5} \text{CH}_3-\overset{+}{\text{CH}}-\text{CH}_3 + \text{SbF}_6^- \qquad (23)$$
$$\phantom{\text{CH}_3\text{CH}}|\phantom{-\text{CH}_3}$$
$$\phantom{\text{CH}_3\text{CH}}\text{F}$$

The following evidence in the proton magnetic resonance (PMR) spectrum is indicative of carbonium ion formation.

1. The large downfield shift of the lone proton at C-2 (δ13.5 ppm from TMS), relative to its position in the starting fluoride (δ4.64 ppm), indicates a large withdrawal of electron density from the vicinity of that proton.

2. The coupling between the F and the C-2 proton, which was present in the starting fluoride ($J_{\text{H-F}} = 48$), disappears.

3. The substantial downfield shift of the methyl protons from δ1.23 ppm in the fluoride to δ5.06 ppm in the cation indicates a large electron withdrawal from the vicinity of the methyl protons. This can be viewed as a hyperconjugative interaction.

4. The methyl groups in the cation which are split only by the proton at C-2 appear as a doublet and the C-2 proton appears as a septuplet. These

assignments are further confirmed by deuterium magnetic resonance measurements made at 9.2 MHz. Perdeuterated isopropyl fluoride was examined in SbF$_5$ media with ^{2}H resonance, and it was possible to reproduce the ^{1}H chemical shifts after correcting for the ratio of the magnetogyric ratios. In CD$_3{}^+$CDCD$_3$ the CD$_3$ groups appeared at δ4.90 and the –C$^+$D– at δ13.48 ppm in good agreement with the ^{1}H data.

The ^{19}F resonance spectra further confirmed these assignments. In the ^{1}H spectra the disappearance of the J_{F-H} suggested a complete dissociation of the C–F bond in the SbF$_5$ solutions, although fast fluorine exchange in a highly polar, donor–acceptor complex could result in the absence of H–F coupling. The ^{19}F spectra also suggested the absence of covalent C–F bonds and the presence of SbF$_6{}^-$ forms, perhaps highly coordinated with more SbF$_5$ (Olah et al., 1964).

Another instructive example is the *tert*-butyl cation (Olah et al., 1964). The PMR spectrum of this cation in SbF$_5$ solution exhibits a single peak at δ4.35 ppm because of the methyl protons. In *tert*-butyl fluoride, the cation's precursor, the methyl groups appear at δ1.30 ppm as a doublet (split by fluorine) with J_{H-F} = 20 Hz. As was the case with the 2-propyl cation, this downfield shift ($\Delta\delta'H$) of 3.05 ppm indicates that a very powerful electron-withdrawing effect is operating to decrease the electron density about the protons. This, along with the disappearance of the C–F resonance in the

$$\text{CH}_3\text{—}\underset{\underset{\text{CH}_3}{|}}{\overset{\overset{\text{CH}_3}{|}}{\text{C}}}\text{—F} \xrightarrow{\text{SbF}_5} \text{CH}_3\text{—}\text{C}^+\overset{\text{CH}_3}{\underset{\text{CH}_3}{\diagdown}}\mspace{-10mu}\diagup + \text{SbF}_6{}^- \quad (24)$$

^{19}F spectrum, supports the formation of the *tert*-butyl cation. The PMR spectra of several ions, representative of a number of classes, are illustrated in Table II with all shifts relative to TMS. These are not absolutely comparable since internal or secondary internal standards and different solvents were used, however, they are instructive. These spectra allow some general conclusions to be made about the carbonium ion structure using NMR. The effect of charge on chemical shift falls off rapidly with distance in aliphatic cations. For example, in the 2-methyl-2-butyl cation (**16**) the CH$_3$ group (C-4), insulated by a single methylene, appears at δ2.27 ppm verses δ4.93 ppm for the methylene group. In the 2-cyclopropyl-2-propyl cation (**19**) the two methyl groups are nonequivalent and separated by 0.54 ppm. Hydrogens lying in the face of cyclopropane rings experience upfield shifts of from 0.3 to 0.5 ppm. This led Pittman and Olah (1965) to conclude that rotation about the cyclopropyl to carbinyl carbon bond has an activation barrier of about 14 kcal/mole. Furthermore, they showed that the plane of the cyclopropane ring was parallel to the axis of the vacant p orbital of the carbinyl carbon.

Table II
NMR Data for Some Stable Carbonium Ions[a]

Cation structure	Cation no.	Proton(s)	Chemical shift, δ[b]
CH_3 (a), $\overset{+}{C}$—CH_2CH_3 (b,c), CH_3 (a)	16	a b c	4.50, t 4.93, m 2.27, t
Phenyl-$\overset{+}{C}(CH_3)_2$ with ring H$_c$, H$_b$, H$_d$	17	a b c d	3.60 8.30 7.95 8.55
Ph$_2\overset{+}{C}$H (diphenyl cation with H$_a$, H$_b$, H$_c$)	18	a b c	7.69 7.87 8.24
Cyclopropyl cation with CH_3 (a), CH_3 (b), H$_c$, H$_d$, H$_e$ (side view)	19	a b c d e	2.50, s 3.14, s 3.95–3.70, m 3.44–3.85, m
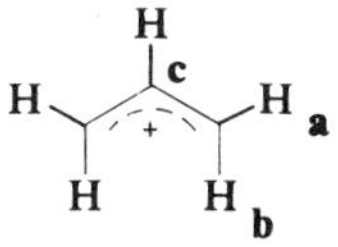 Cyclopentadienyl cation (H$_a$, H$_b$, H$_c$)	4	a and b c	8.97 broad doublet (not well resolved) 9.64
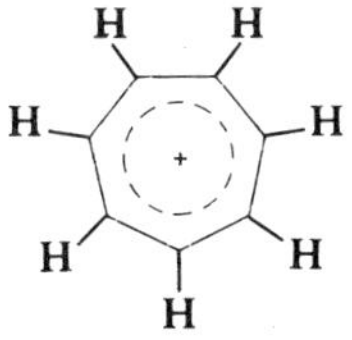 Tropylium cation	20	a	9.28, s
Benzenium ion (H$_a$, H$_b$, H$_c$, H$_d$)	21	a b c d	9.20 8.40 9.42 5.84

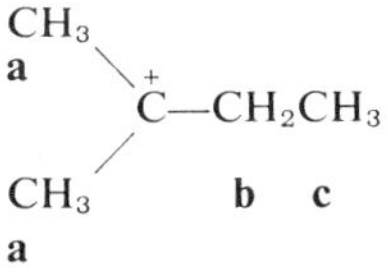

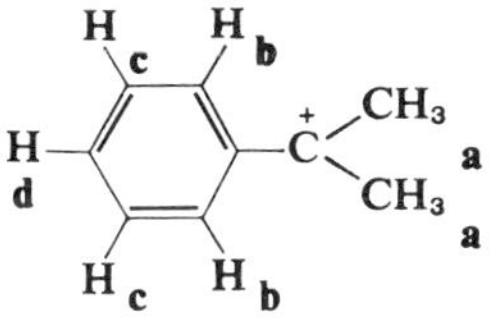

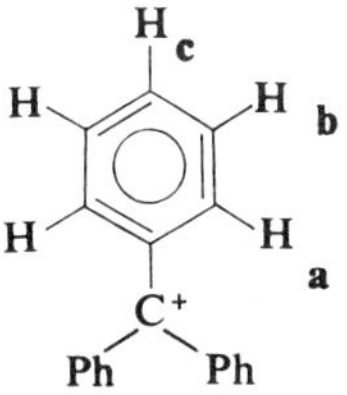

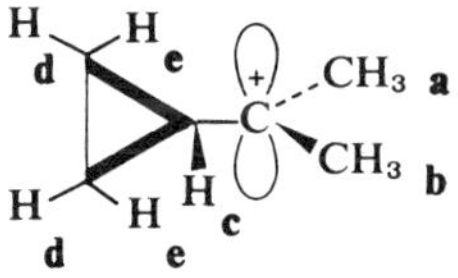

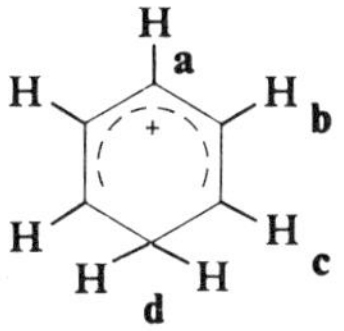

[a] See Olah and Schleyer (1968, 1970, Vols. I and II).
[b] In various solvents with all peaks downfield relative to TMS; s = singlet; t = triplet; m = multiplet.

This places one methyl group cis and one trans to the ring. This accounts for the nonequivalence of the methyl groups. The dicyclopropyl- and tricyclopropylmethyl cations fit this pattern (Deno et al., 1962d; Pittman and Olah, 1965).

The 2-phenyl-2-propyl cation's (17) chemical shifts exhibit the pattern which is expected if more charge is delocalized into the ortho and para positions of the ring in accord with the predictions of writing simple valence-bond resonance structures. On the other hand, in the triphenylmethyl cation (18), the meta protons are further downfield than the ortho protons (para protons are most deshielded). However, the phenyl rings in 18 are in the propellar conformation with each ring tilted out of the plane. Each ring also undergoes a rotation which is synchronized with each of the other rings (a fact which was finally definitively shown using temperature-dependent decoupled ^{19}F NMR kinetic studies (Schuster et al., 1968). Thus, complicated anisotropy effects of the neighboring rings complicate the chemical shifts.

^{13}C NMR provides a very powerful tool for examining carbonium ion structure. Although ^{13}C appears in only 1.1% natural abundance, the use of ^{13}C enrichment, white noise decoupling techniques, and the INDOR method have permitted wide application of ^{13}C NMR to the study of carbonium ion structure. Inherently, there are great advantages of ^{13}C over ^{1}H NMR in studying carbonium ions. First, the charge is on the carbon skeleton primarily, allowing a more direct look at the charged skeleton. Second, when carbonium ions are generated there is usually a change in hybridization at the carbon in question (i.e., sp^3 to sp^2 during the formation of the tert-butyl cation from tert-butyl fluoride). Simply changing hybridization results in large chemical shift changes in ^{13}C NMR. Another advantage of ^{13}C NMR is that the scale of chemical shifts is much larger for ^{13}C than it is for ^{1}H, making ^{13}C resonance a far more sensitive probe of the effect of charge. Finally, the ^{13}C–^{1}H coupling constants provide a valuable index to the state of hybridization of that carbon. (This principle has long been exploited in bridgehead bicyclic and other strained alicyclic compounds to help determine the structure of these species.)

Table III
Illustration of the Usefulness of ^{13}C Magnetic Resonance

Precursor (sp^3)	Precursor chemical shift, δ^{13}C (ppm)[a]	Cation (sp^2)	Cation chemical shift, δ^{13}C (ppm)	$\Delta\delta^{13}$C (ppm)	$\Delta\delta^{1}$H (ppm)
$(CH_3)_3\ ^{13}$C—Cl	+126.1	$(CH_3)_3\ ^{13}$C$^+$	−146.9	−273	−3.05
$Ph_3\ ^{13}$C—OH	+111.5	$Ph_3\ ^{13}$C$^+$	−18.1	−126.6	∼−1

[a] All ^{13}C shifts listed are relative to ^{13}CS$_2$.

4. CARBONIUM IONS

The use of ^{13}C magnetic resonance for studying carbonium ions may best be illustrated with two simple examples shown in Table III. In the *tert*-butyl cation the ^{13}C resonance of the charged carbon appears at 146.9 ppm downfield from $^{13}CS_2$ (the reference standard used in all the following examples). In the triphenylmethyl cation the charged carbon's shift is -18.1. This indicates a far greater charge density on the central carbon of the *tert*-butyl cation. Furthermore, the change in chemical shift ($\Delta\delta^{13}C$) going from the precursor to the cation is much larger for the *tert*-butyl cation (-273 ppm) than for the triphenylmethyl cation (-126.6 ppm). The huge range of the observed $\Delta\delta^{13}C$ is emphasized by comparing these values to those of the hydrogens in these ions. The $\Delta\delta^{13}C$ are about 100 times greater than $\Delta\delta^1H$.

The large downfield shifts ($\Delta\delta^{13}C$) exhibited by the carbinyl carbon, going from the precursor to the cation, is due only in part to the charge generated on the carbon. Part of this downfield shift is due to rehybridization of this carbon to sp^2. To assess qualitatively how much of the observed shift is due to charge effects and how much is due to rehybridization, it has been instructive to compare the chemical shifts of the cation to the ^{13}C chemical shifts of a series of sp^2-hybridized compounds (Olah and Pittman, 1966). The ^{13}C shifts of a large variety of sp^2-hybridized compounds have been studied by Lauterbur (1957, 1962) and appear to be independent of the nature of the molecules. Thus (see Table IV), a base value of about $+65$ ppm can be used for the chemical shift of an sp^2-hybridized carbon. By comparing the observed shift in the cation with this value, a new value reflecting the influence of the charge is obtained. This treatment neglects solvent and anisotropy effects. For all the cations listed in Table IV the effect of the positive charge is large.

Table IV
^{13}C NMR Chemical Shifts of Typical sp^2 Systems and Some Carbonium Ions

Species	$\delta^{13}C$	$\Delta\delta^{13}C$ from 65
$(CH_3)_3{}^{13}C^+$	-146.9	-211.9
$H-{}^{13}C^+(CH_3)_2$	-125.0	-190.0
$\triangleright\!\!-{}^{13}C^+(CH_3)_2$	-86.8	-151.8
$Ph-{}^{13}C^+(CH_3)_2$	-61.1	-126.1
$HO-{}^{13}C^+(CH_3)_2$	-55.7	-120.7
$Ph_2-{}^{13}C^+-H$	-5.6	-70.6
$Ph_3{}^{13}C^+$	-18.1	-83.1
$CH_3-{}^{13}C^+=O$	$+44.3$	-21.7
Cyclohexene (C-1)	$+67$	—
Acrylic acid (C-2)	$+64$	—
Benzene	$+65$	—

For example, of the total $\Delta\delta^{13}C$ of 273 ppm observed going from *tert*-butyl chloride to the *tert*-butyl cation about 212 ppm can be attributed to charge and 61 ppm to the sp^3–sp^2 change. For the triphenylmethyl cation, both contributions are smaller.

Table IV also shows that the cyclopropyl group ($\delta^{13}C$ at -86.8 ppm) stabilizes an electron-deficient carbon more than a hydrogen (2-propyl cation) but less than a hydroxy group (protonated acetone). The carbinyl carbon appears 31.1 ppm to the low field of protonated acetone and 38.3 ppm to the high field of the 2-propyl cation. The same comparison when made for phenyl ($\delta^{13}C = -61.1$ for the 2-phenyl-2-propyl cation) indicates the phenyl is more effective than cyclopropyl in stabilizing the cation. A somewhat surprising result is that the ^{13}C shift for the triphenylmethyl cation is at lower field (-18.1 ppm) than for the diphenylmethyl cation (-5.6 ppm). This is explained by realizing that the phenyl rings cannot come close to being coplanar in the former. Thus, conjugation is not as efficient per ring as in the diphenylmethyl cation. This result agrees with the conclusion (Deno and Schriesheim, 1955) that the resonance energies of the diphenylmethyl and triphenylmethyl cations are similar with the large pK_a difference of 6.7 being a manifestation of the release of steric (back) strain during ionization of the triphenyl precursor.

A comparison of the $C(OH)_3{}^+$, $CH_3C(OH)_2{}^+$, $(CH_3)_2COH^+$, and $(CH_3)_3C^+$ ions (and their uncharged precursors) established that a linear correlation of the ($\sigma^{13}C$) chemical shifts with the calculated π electron densities existed with the dependence being 360 ppm per electron (Olah and White, 1968).

The magnitude of the ^{13}C–H coupling constant is a valuable indication of the hybridization at carbon in carbonium ions as well as in neutral compounds. Numerous investigations with neutral compounds (Stothers, 1965; Grant, 1964) have shown that a general linearity exists between the value of $J_{^{13}C-H}$ and the fractional s character of the ^{13}C hybrid atomic orbital in the C–H bond. This supports the conclusion that the Fermi contact term is almost the sole contributor to the coupling as suggested from valence-bond theory (Karplus and Grant, 1959).

The Muller–Pritchard (1959, 1962) relationship, $\%s = 0.20 J_{^{13}C-H}$, correlates a large amount of the reported data. The $J_{^{13}C-H}$ values for the "charged carbon" for a number of carbonium ions in strong acid solutions are listed in Table V with some of their corresponding hydrocarbons. These studies establish that the linear relationship between $J_{^{13}C-H}$ and $\%s$ character is also valid for carbonium ions, and this provides another very useful probe of carbonium ion structure. It is true that the hyperfine contact term is also sensitive to bond polarization and C–H bond distances (Grant and Litchman,

1965; Muller, 1962). In a carbonium ion the bond (I) has a different polarity than its precursor, i.e., (II).

$$\begin{array}{cc} \diagdown \\ {}^+\!C\!-\!H & =\!C \diagup \\ \diagup & \diagdown H \\ (I) & (II) \end{array}$$

However, these effects do not appear to invalidate the %s correlation in the examples reported to date.

Table V
^{1}H—^{13}C Spin-Spin Coupling Constants of Carbonium Ions and Some Parent Hydrocarbons[a]

Species	$J_{^{13}C-H}$ (Hz)	Calculated s character (%)	State of hybridization
$(CH_3)_2\,^{13}CH_2$	128	25.6	sp^3
$CH_3-^{13}\overset{+}{C}H-CH_3$ SbF_5Cl^-	168	33.6	sp^2
$(Ph)_2\,^{13}CH_2$	126	25.2	sp^3
$Ph-^{13}\overset{+}{C}H-Ph$	164	32.8	sp^2
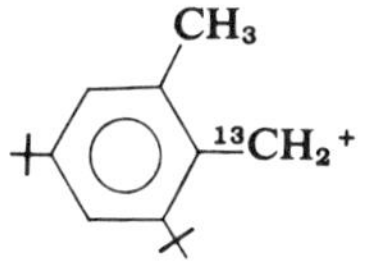 $\overset{+}{\underset{X}{\bigodot}}\!\!-\!\!^{13}CH_2^+$ with CH_3	169	33.8	sp^2

[a] From Olah and Comisarow, 1966; Olah and White, 1969.

^{19}F Magnetic resonance has been useful in studying cations with fluorine bound to the charged carbon and adjacent to this carbon. Olah and Comisarow (1969) reported data for a series of alkyl and aryl fluorocarbonium ions. Like ^{13}C, ^{19}F has a much larger range of chemical shifts than ^{1}H. The fluorine resonance moves progressively to lower field as the delocalizing ability of the remaining ligands on the central atoms decreases. This was interpreted as evidence for back π donation of electron density from F to

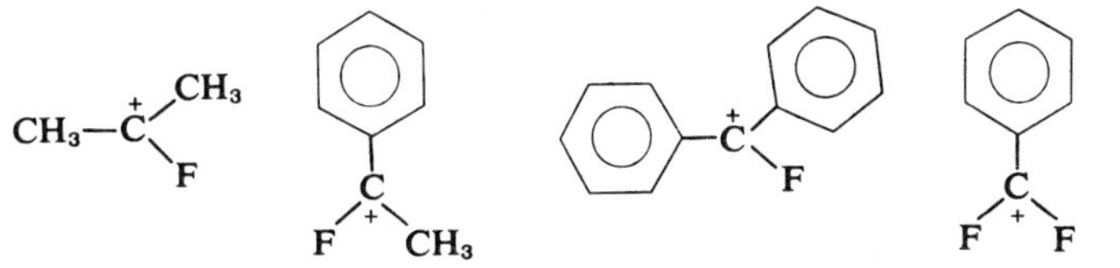

	$CH_3-\overset{+}{C}(CH_3)F$	$Ph-\overset{+}{C}(F)(CH_3)$	$Ph_2\overset{+}{C}F$	$Ph\overset{+}{C}F_2$
$\delta^{19}F$ (relative to CCl_3F)	−181.5	−51.5	−11.26	−11.99
$\Delta\delta^{19}F$ (relative to chloride precursor)	−266.8	−140.2	−100.87	−75.6

C^+, the strength of which increased as the electron demand at carbon increased. Thus, F stabilizes carbonium ions as depicted below. Pittman *et al.* (1972a) have provided a strong theoretical justification of this view. Detailed

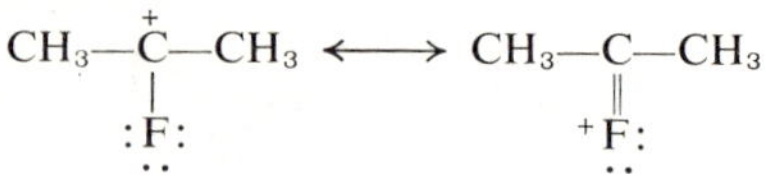

$$CH_3-\overset{+}{\underset{:\ddot{F}:}{C}}-CH_3 \longleftrightarrow CH_3-\underset{^+\ddot{F}:}{\overset{\|}{C}}-CH_3$$

INDO calculations on a series of fluorocarbonium ions showed that as the electron demand by carbon increased, the amount of F to C π donation markedly increased. However, this was accompanied by an increasingly stronger C to F polarization of the σ bond. The net result was that the net charge density at fluorine did not change substantially.

b. Dynamic Nuclear Magnetic Resonance. Magnetic resonance has shown considerable usefulness in examining dynamic processes. For example, in $FSO_3H-SbF_5-SO_2$, solution ion **22** has a single resonance at $\delta 2.90$ in the ^1H spectrum (Olah and Lukas, 1967). This could be due to either (1) a rapidly equilibrating structure (**22a** and **22b**) or (2) a rapidly equilibrating protonated cyclopropane structure (**22c**) equilibrating through classic intermediates. The ^1H NMR spectrum excludes a static classic or bridged ion (either static structure of types **22a, 22b,** or **22c**). Attempts to lower the temperature enough to "freeze out" the equilibration process (where it would

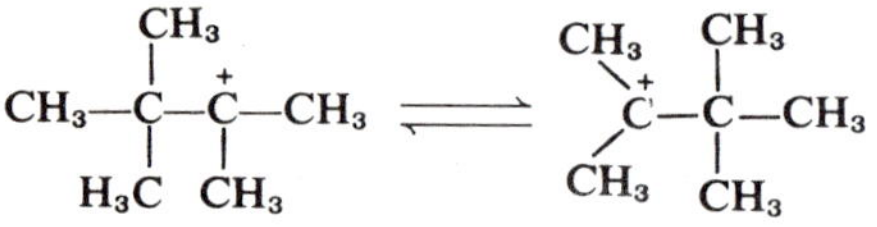

$$\underset{\text{(a)}}{CH_3-\underset{\underset{H_3C}{|}}{\overset{\overset{CH_3}{|}}{C}}-\overset{+}{\underset{\underset{CH_3}{|}}{C}}-CH_3} \rightleftharpoons \underset{\text{(b)}}{\underset{CH_3}{\overset{CH_3}{\diagdown}}\overset{+}{C}-\underset{CH_3}{\overset{CH_3}{|}}-CH_3} \tag{25}$$

(22)

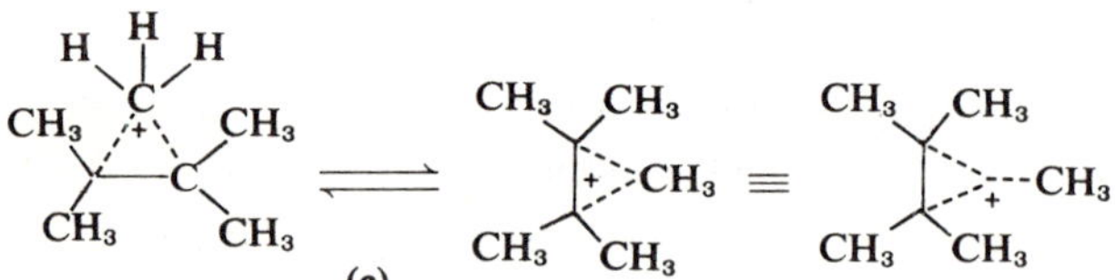

$$\text{(c)} \qquad \rightleftharpoons \qquad \equiv \qquad \tag{26}$$

Indicates three-center two-electron bond (see Olah, 1972)

occur slowly on the NMR time scale) were unsuccessful because of the very rapid equilibration rate. At $-150°$ to $-160°C$ the process was still rapid. Viscosity broadening prevented studies at lower temperatures. These studies indicate that the activation barrier for the equilibration is less than 5 kcal/ mole. Since it was impossible to prevent the rapid equilibration, both the ^1H and ^{13}C chemical shifts were compared to established structures. In the well-known tetramethylethylene halonium ions (23) the methyl protons appear in the region δ0.95 to δ0.75 ppm from TMS). This is about 2 ppm to

$$\underset{CH_3}{\overset{CH_3}{C}} \overset{\overset{X}{+}}{\underset{}{\text{—}}} \underset{CH_3}{\overset{CH_3}{C}} \quad X = I, Br, Cl$$

(23)

higher field than the proton resonance observed for ion 22. Thus, if 22 were the bridged–protonated tetramethylcyclopropane, the methyl groups would be expected at the higher fields observed in 23 because of charge delocalization on the ring and an expected anisotropic effect. Thus, the classic equilibrating structure 22a ⇌ 22b is favored. The ^{13}C spectrum of this ion provided further support for this assignment (Olah and White, 1969). The ion was generated from 2,3,3-trimethyl-2-butanol with 56% ^{13}C enrichment at C-2. The value of δ^{13}C for this atom was $\delta -11.5$ ppm, which is consistent only with the rapidly equilibrating structure 22a ⇌ 22b. This is made especially clear when one compares the ^{13}C spectrum of 22 with that of the nonclassic norbornyl cation (24) which is now well established as a bridged structure from ^{1}H, ^{13}C NMR, laser Raman, and IR studies (see Section VI,B,3). In 24 the rapid degenerate rearrangement shown in Eq. (27) is fast on the NMR time scale at 70°C.

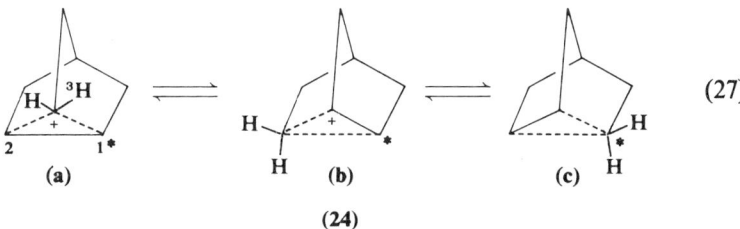

(24)

Here the ^{13}C resonance of the singly labeled three equivalent cyclopropyl carbons (C-1) was +101.8 ppm. At −150°C the "frozen-out" ^{13}C INDOR spectrum δ^{13}C of the bridging carbon (i.e., C-3 in 24a) is 173 ppm, and that of the two cyclopropane-type carbons (i.e., C-1 and C-2 in 38a) is 70 ppm (Olah and White, 1969). This further demonstrates that high field ^{13}C resonances are expected in bridged-type ions. Thus, it is clear that the δ^{13}C of −11.5 found in the dimethyl-*tert*-butyl carbonium ion is strong support for the equilibration shown for 24.

Another example is the 2-propyl cation which when generated in SO$_2$ClF—SbF$_5$ from 2-chloropropane with a 50% ^{13}C enrichment of the 2-carbon showed the ^{13}C was scrambled with a half-life of 1 hr at −60°C (Saunders and Hagen, 1968). This confirmed the interpretation of the temperature dependence of the proton spectrum of this ion in the −20 to −40 range studied by Saunders and suggested the intervention of a protonated cyclopropane mechanism [Eq. (28)].

$$\begin{array}{c} CH_3 \\ \diagdown \\ {}^*\overset{+}{CH} \\ \diagup \\ CH_3 \end{array} \rightleftharpoons \begin{array}{c} CH_2 \text{—} \overset{*}{CH_2} \\ \diagdown \overset{+}{\diagup} H \\ CH_2 \end{array} \rightleftharpoons \begin{array}{c} {}^+CH \text{—} \overset{*}{CH_3} \\ \diagdown \\ CH_3 \end{array} \qquad (28)$$

c. IR and Raman. In comparison to NMR, IR and Raman spectroscopy have not been extensively employed for studying carbonium ions. Since the time scale of vibrational processes is so short (10^{-12} sec), both IR and Raman offer a spectroscopic technique which should be able to distinguish directly between equilibrating classic and nonclassic ions where these questions arise. However, these studies usually must be carried out in solution where solvent bands also complicate the spectra. When complete Raman spectra with depolarization data and complete IR spectra are available, it is often possible to deduce the molecular symmetry, especially where it is of high order. This has been done for the cycloheptatrienyl (tropylium) cation.

Doering and Knox (1954) first recognized that the simple IR spectra of tropylium bromide was consistent with the presumed high symmetry of the ion. The most complete vibrational study of any carbonium ion to date was then provided by Fateley *et al.* (1957) who combined IR and Raman methods to show that the tropylium cation resembled benzene in many respects. They deduced that the ring was planar, aromatic, and had D_{7h} symmetry. Force constants were derived and found similar to benzene. For example, the C–C stretching force constants were 8.29 mdyn/Å for the tropylium ion and 8.41 for benzene. This indicated slightly less double bond character in the tropylium ion which subsequently agreed with x-ray crystallographic (Kitaigorodskii *et al.*, 1960) bond lengths of 1.47 ± 0.03 Å (verses 1.397 for benzene). Mahler *et al.* (1964) have studied metal carbonyl cations complexed to the tropylium system (20). The group VI B metals (Cr, Mo, W) bond symmetrically to each of the seven carbons via the six π electrons. However, differences in the C–H stretching region for the analogous iron compound (three bands verses a single intense 731 cm^{-1} in the group VI B complexes) suggested that structure 25 with a cis double bond would be more appropriate. Olah *et al.* (1964) reported the first IR studies of alkyl carbonium ions (see Evans, 1968).

(20)

(25)

4. CARBONIUM IONS

Solvents and concentration problems limited the study to above 800 cm^{-1} in the infrared. The observed bands were compatible with a planar C_{3h} structure for the *tert*-butyl cation (6), although C_{3v} symmetry was also

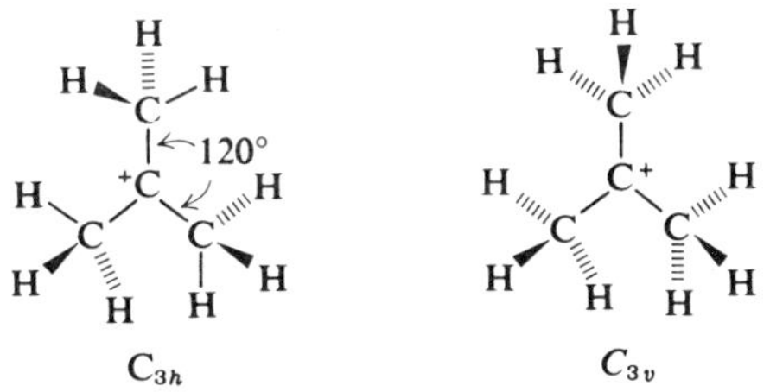

(6)

possible. Calculations using Urey–Bradley force fields required the C–C stretching force constant to increase from 2.2 (in paraffins) to 3.0 mdyn/Å to obtain agreement with the observed spectra. Similarly, the C–H force constants were reduced. The strong 1290 cm^{-1} band was assigned to the antisymmetric stretching mode of the carbon skeleton.

Recently, more detailed IR and laser Raman spectra for the *tert*-butyl (6), *tert*-amyl, 2,3-dimethyl-2-butyl, and 2,3,3-trimethyl-2-butyl (26) cations have been published (Olah *et al.*, 1971). The Raman data permitted unambiguous assignment of the bands in 6 as 2947, 2850, 1450, 1295, 667, 347, and 306 cm^{-1}. These are compared in Table VI with isoelectronic trimethylboron (13) and the perdeutero analogs. In the Raman and IR spectra of 6, the ν_3, ν_4, ν_5, ν_8, ν_{11}, ν_{15}, ν_{16}, ν_{20}, ν_{21}, and ν_{22} modes were not detected owing to overlap with bands of the acid media or the rather low cation concentrations used. Correlation of the IR and Raman frequencies of the four fundamental skeletal modes of 13 and 6 is entirely consistent with a planar *tert*-butyl cation skeleton. Careful examination led to the assignment of the C_{3v} molecular symmetry, where one hydrogen from each methyl group is above the $^+CC_3$ plane lying in the σ_v planes. This is the geometry in which hyperconjugation would be maximized.

The Raman spectra of 26 resembles that of the *tert*-butyl cation. It exhibits an intense 683 cm^{-1} band characteristic of a neopentyl-like CC_4 fundamental skeletal vibration. This is strong evidence in favor of structure 26a as opposed to the static bridged structure 26b. The Raman spectra of 26b would be

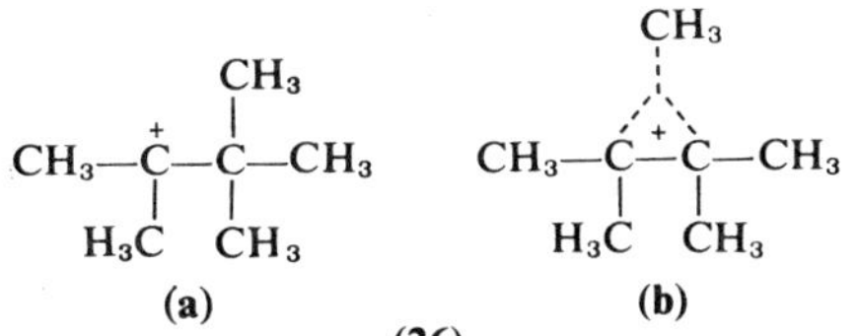

(26)

Table VI

Raman and Infrared Frequencies of the *tert*-Butyl Cation (6) and Trimethylboron (13)

Species	$\nu_1, \nu_{12}, \nu_7, \nu_{19}$	ν_2, ν_{13}	ν_{21}	ν_{14}	ν_{15}	$\nu_{17}{}^a$	ν_5	ν_{16}	$\nu_6{}^b$	ν_9	$\nu_{10}{}^c$	$\nu_{18}{}^d$
$(CH_3)_3C^+$	2947	2850	—	1450	—	1295	—	—	667	—	347	306
$(CH_3)_3B$	2975	2875	1060	1440	1300	1150	906	886	675	973	336	320
$(CD_3)_3C^+$	2187	2090	—	1075	—	980	—	—	720	—	347	300
$(CD_3)_3B$	2230	2185	—	1033	1018	1205	—	—	620	870	289	276

[a] XC str. asymmetric.
[b] XC_3 str. symmetric.
[c] XC_3 symmetric out-of-plane bending.
[d] C–X–C def. in-plane.

4. CARBONIUM IONS

expected to exhibit a "cyclopropane"-type C–H stretching vibration at ~ 3100 cm^{-1} which is exhibited by protonated nortricyclene (Olah et al., 1970b).

In general, carbonium ions all possess a very strong absorption band in the 1250–1550 cm^{-1} range with the exact location dependent on the type of structure studied. This band is very strong because the C–C bonds are highly polarized in cations and undergo a relatively large change in dipole moment during vibration. The tricyclopropyl carbonium ion has three strong bands at 837, 1279, and 1445 cm^{-1} (Deno et al., 1962b). The 1279 cm^{-1} band is probably best described as the antisymmetric stretching mode of the central carbon skeleton (Evans, 1968). A review of the early IR spectra of alkyl cations by Olah and Pittman (1966) has appeared.

Infrared studies have been used to clearly differentiate oxocarbonium ions (27) from donor–acceptor complexes (28) in the products produced by treating acyl halides with Lewis acids in both liquid and solid states (Cooke et al., 1954; Oulevey and Susz, 1965; Cook, 1959; Olah et al., 1962, 1963; Susz and Cassimatis, 1961; Susz and Wuhrmann, 1957). The donor–acceptor complexes exhibit a carbonyl stretching frequency at ~ 1550 cm^{-1} which is

$$R-\overset{+}{C}=O \quad MX_{n+1}^{-} \qquad R-C\overset{\delta^{+}\!\!\nearrow O}{\underset{\searrow X}{}} \longrightarrow \overset{\delta^{-}}{MX_n}$$

(27)   **(28)**

at significantly lower energy than the corresponding frequency in the starting acyl halides (1850–1700 cm^{-1}). Oxocarbonium ions (27), however, exhibit the carbonyl absorption at much higher frequencies (2200–2300 cm^{-1}), and they are accompanied by the absorptions characteristic of the symmetrical anion MX_n^{-} derived from the Lewis acid (i.e., SbF_6^{-}, BF_4^{-}, $AlCl_4^{-}$, etc.). The change in hybridization from sp^2 to sp at the carbonyl carbon, when coupled with some increase in back π bonding from oxygen, contributes to the high frequency in the oxocarbonium ions.

The structure of the norbornyl cation either as a rapidly equilibrating set of secondary classic ions [(29a) $\rightleftharpoons$ (29b)] (Brown, 1967) or as a nonclassical, σ delocalized ion (30) (Winstein and Trifan, 1949; see review by Sargent, 1966) has sparked one of the greatest debates in organic chemistry since

(29)

(a)   **(b)**
(29)   **(30)**

Kekulé proposed his structure for benzene. The structure was not settled by ^{1}H NMR because rapid equilibration of the 6-, 1-, and 2-proton occurred on the NMR time scale even at $-120°C$ in super-acid media. Vibrational spectroscopy does not have this drawback. Olah and co-workers (1970b) have now combined ^{1}H and ^{13}C NMR studies on the "frozen-out" norbornyl cation at $-154°C$ in SbF_5—SO_2ClF, SO_2F_2 with Raman studies at $-70°$ and $-80°C$ to demonstrate that the nonclassic structure containing a pentacoordinated bridging carbon atom involved in three-center bond formation is the correct structure (in this media). This structure is the corner-protonated cyclopropane type (**31**). In this study it was demonstrated that the vibrational spectra of the norbornyl cation were similar to nortricyclene (**32**) and its 1-bromo and 1-methyl derivative but quite different than that of norbornane

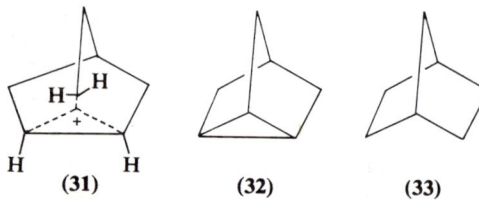

(**31**)   (**32**)   (**33**)

(**33**). The spectral data are summarized in Table VII. Nortricyclene and 1-bromonortricyclene have only one very strong polarized Raman line at 950 cm^{-1}. In contrast, norbornane and its derivatives consistently display

Table VII
Main Raman Spectral Lines (cm^{-1}) of the Norbornyl Cation, Nortricyclene, and Norbornane[a]

Species	C–H Region (3100–2700)	C–C Region (1000–250)
Norbornyl cation	3110 3030 3010 2947 2860	972 (vs, p) 796 (w) 390
Nortricyclene	3087 3006 2975 2955 2875	959 (vs, p) 732 (m) 300
Norbornane	2964 2936 2873	920 (s, p) 871 (s, p) 753 (m)

[a] vs = very strong; s = strong; m = medium; w = weak; p = polarized.

two intense bands at ~920 and ~875 as well as a moderately intense band at ~750 cm^{-1}. In the C–H stretching region, norbornane and its derivatives have only three lines, all at frequencies below 3000 cm^{-1}, whereas nortricyclene and its 3-bromo and 3-methyl derivatives each display a minimum of five lines in this region. The line at 3080 cm^{-1} is characteristic of cyclopropane rings. Furthermore, the Raman spectra of the norbornyl cation is identical to the spectra of protonated nortricyclene. The vibrational spectral data reflect the close skeletal similarity of the norbornyl cation with nortricyclene and argue strongly for structure **31** without the difficulty of possible equilibration during the time scale of the measurement.

d. Ultraviolet Spectroscopy. The use of UV spectroscopy to study carbonium ions has been continuous, intense, and full of "red herrings" during the past 40 years. To do justice to this topic is far beyond the scope of this book. However, a comprehensive review by Olah *et al.* (1968a) is available to the interested reader. As a tool to identify carbonium ions, UV spectroscopy offers the advantage of being able to detect very tiny concentrations of ions. Carbonium ion concentrations of 10^{-6} to 10^{-4} M are readily measured. Solution phase UV spectra, however, do not exhibit vibrational fine structure, and it is difficult to obtain structural information on the carbonium ions being observed. Thus, a major use of UV spectroscopy now is to monitor second-order carbonium ion reactions, where the kinetic processes occur too rapidly to follow at the concentrations necessary for NMR spectroscopy. Ultraviolet spectroscopy also plays a leading role in the measurement of the position of equilibrium in carbonium ion processes. Finally, information on carbonium ion excited states is provided since the energy difference between the ground and excited state is obtained in the UV measurement.

After a long history of misinterpretation and speculation (see Olah *et al.*, 1968a for a complete discussion) it was conclusively demonstrated by Olah *et al.* (1966b) that alkyl carbonium ions do not absorb in the UV above 210 nm. This was accomplished by generating solutions of ions which could be positively identified by NMR and then measuring the UV spectra of these same solutions at both high and low concentrations. It had long been argued that the strong hyperconjugative interactions in alkyl cations might give rise to a long-wavelength "hyperconjugative transition." However, the simple view that alkyl cations concentrate their electrons in σ bonds and that $\sigma \to \sigma^*$ transitions are very high energy transitions (thus, not observed in the accessible UV region) adequately explains the data obtained.

When a cyclopropyl group is attached to the cation center, intense bands in the 270–290 nm region are detected (Pittman and Olah, 1965). Successive substitution by cyclopropyl groups does not alter the band maxima greatly, but the extinction coefficient increases in a manner similar to the effect observed with mono-, di-, and triphenylcarbonium ions. Table VIII contains UV data on a variety of organic cations.

A series of cyclopentenyl, cyclohexenyl, and acyclic allylic cations have been reported by Deno and co-workers (see Deno *et al.*, 1965b and papers referenced therein). In addition, Katz and Gold (1964) obtained the UV spectra of the 1,2,3,4-tetramethyl- and pentamethylcyclobutenyl cations. The energies at which allylic cations absorb decrease in the series cyclobutenyl > cyclopentenyl > acyclic allylic > cyclohexenyl. This can be illustrated by

Table VIII
The Ultraviolet Spectra of Representative Carbonium Ions

No.	Cation	λ_{max} (nm)	ϵ
6	tert-Butyl	<210	—
31	2-Norbornyl	<210	—
2	$CH_3C^+=O$	<215	—
19	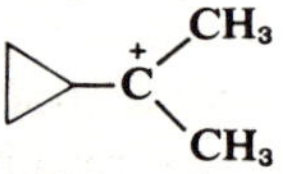 cyclopropyl-$C^+(CH_3)_2$	289	10,800
18	Ph_3C^+	331	39,600
		404	39,600
34	Dipropylcyclopropenyl	185	—
20	Tropylium	275	4,350
		247	1,150
35	 (cyclobutenyl cation, dimethyl)	305	10,500
36	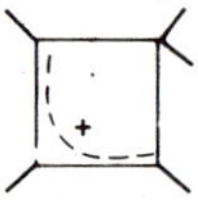 (cyclobutenyl cation)	245	2,880
		242	3,100
37	 (cyclopentenyl cation)	275	11,600
38	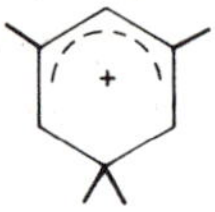 (cyclohexenyl cation)	315	9,100
39	 dienyl $n = 1$	397	10,000
40	39 ($n = 2$) Trienyl	473	109,000
41	39 ($n = 3$) Tetraenyl	550	154,000
42	39 ($n = 4$) Pentaenyl	625	169,000
43	39 ($n = 5$) Hexaenyl	702	79,000

examining the increase in wavelength in the series of cations 35–38. This relationship holds even with cyclopropyl and phenyl substituents which increase the length of the conjugated system.

The particularly high energy absorption for cyclobutenyl cations can be explained in terms of significant 1,3-π overlap. The closer the 1 and 3 carbon atoms are, the greater the overlap. The first π level (π_1) between C-1 and C-3 is bonding, whereas π_2 is antibonding. Thus, as C-1 and C-3 get closer, the π_1 level decreases in energy (more strongly bonding), whereas that of π_2 increases (Mulliken, 1958; Walsh, 1953). Thus, the energy gap between them increases, and λ_{max} will decrease.

The polyenyl cation series 39–43 is extremely interesting (Deno and Pittman, 1964b; Deno et al., 1965b; Sorensen, 1965a). Simple Hückel molecular orbital (HMO) theory predicts a 1.4β energy gap between the highest filled and lowest unfilled molecular orbitals in allylic ions. Extending the conjugation to dienyl cations, Hückel theory predicts a lowering of the energy of both the highest energy filled (π) and lowest energy unfilled (π^*) molecular orbitals, but a greater lowering of the π^* orbital is predicted. Thus, the predicted $\pi \rightarrow \pi^*$ transition energy is now only 1.0β units. Thus, the increase in λ_{max} from ~ 275–315 nm in allylic cations to ~ 400 nm in dienylic cations **39** is exactly what is expected. The UV spectra for ions **39–43** in 80% H_2SO_4 fit the empirical relationship $\lambda_{max} = (330.5 + 65.6n)$ nm. This shows the progressive decrease in energy for the $\pi \rightarrow \pi^*$ transition as the length of the conjugated system increases.

The long-wavelength absorption of benzene is observed at 198 nm, whereas that for ethylene is found at 193 nm. Both have $\pi \rightarrow \pi^*$ transition energies of 2β units predicted by HMO theory. Using benzene's observed 198 nm bond to evaluate β, one predicts that allylic cations (1.4β for the $\pi \rightarrow \pi^*$ transition energy) should absorb at 293 nm. The agreement with experiment (see **35**, Table VIII) is remarkable. The HMO $\pi \rightarrow \pi^*$ transition (1β) for dienyl cations is predicted to occur at 410 nm, which is in excellent agreement with the observed value of 397 nm for ion **39**. The HMO prediction for trienyl cations **40** is 513 nm verses the observed values of 473 nm. The trivinylmethyl cation absorbs at 402 nm, which is in superb agreement with the transition energy of 1β predicted for this cross-conjugated system (Sorensen, 1965b). Hafner and Pelster (1961) prepared the series of polyenyl cations

$$\text{Ph}-\overset{+}{\text{CH}}-(\text{CH}=\text{CH})_n-\text{Ph}$$

as stable salts. Like the aliphatic series, the addition of each double bond causes a 67 nm increase in wavelength [i.e., $\lambda_{max}(\text{nm}) = 420 + 67n$]. Thus, the predictions of simple HMO theory have been followed in the allylic $\rightarrow$ polyenyl cation series.

Can HMO be used to provide a rationale for the observed UV spectra of aromatic systems such as the cyclopropenyl and cycloheptatrienyl cations? The lowest energy transition possible in the cyclopropenyl cation is 3β by HMO. Thus, the λ_{max} would be expected below 170 nm. In accord with this prediction, the dipropylcyclopropenyl cation (**34**) exhibits only end absorption at 185 nm (Breslow *et al.*, 1962). Comparing the UV spectra of triphenylcyclopropene ($\lambda_{max} = 300$ nm; $\epsilon = 30,000$) with that of the triphenylcyclopropenyl cation ($\lambda_{max} = 307$ nm; $\epsilon = 48,000$) suggests that there is little charge delocalization into the aryl rings (Breslow and Yuan, 1958). This is expected if one invokes a loss in aromatic stabilization of the cyclopropenyl cation on donation of electron density from the phenyl rings to the cyclopropenyl ring. Figure 1 shows that the donation of an electron would require placing the third electron in an antibonding orbital, thus resulting in a loss of stability. Relative to alkyl groups, phenyl groups are inductively electron

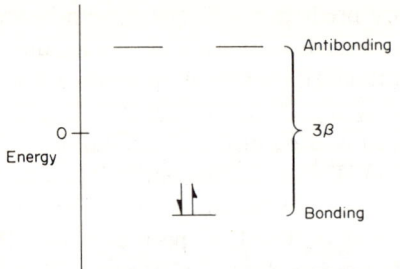

Fig. 1. HMO diagram of the cyclopropenyl cation.

withdrawing. This agrees with the observation that the tri-n-propylcyclopropenyl cation (pK_{R^+} = 7.2) is several orders of magnitude more stable than the triphenylcyclopropenyl cation (pK_{R^+} = 3.1). This emphasizes that the phenyl rings cannot use all their $+R$ (resonance) effect to stabilize the cation.

HMO theory predicts the $\pi \rightarrow \pi^*$ transition of energy of the cycloheptatrienyl cation (**20**) to be 1.7β. This energy is slightly larger than that predicted for allylic cations (1.4β; λ_{max} = 293 nm) and suggests a transition should be found at about 250 nm. This agrees reasonably well with the observed 275 and 247 nm transitions (for a discussion, see Doering and Knox, 1954, and references therein).

The triphenylcarbonium ion (**18**) absorbs at higher energy than its diphenyl analog (**44**) (λ_{max} = 440 nm; ϵ = 44,500). This has been shown to be due to the propellar geometry of **18**, whereas it is possible for a closer approach to planarity to be achieved in **44** (Deno et al., 1955, 1959a; Gold et al., 1952). Substitution of groups which increase the length of the conjugated system (such as dimethylamino groups) causes a decrease in the transition energy with a corresponding increase in λ_{max}.

Ultraviolet has been used to establish the position of equilibrium for solutions of triarylmethyl halides in organic solvents with added Lewis acids such as $HgCl_2$. Both ionization (K_1) and separation of ion pairs to free ions (K_2) have been obtained (Bayles et al., 1955).

$$RCl + HgCl_2 \rightleftharpoons R^+ HgCl_3^- \quad K_1 \quad (30)$$

$$R^+ HgCl_3^- \rightleftharpoons R^+ + HgCl_3^- \quad K_2 \quad (31)$$

Other studies on aryl-substituted carbonium ions have been made (Olah et al., 1966b; Waack and Doran, 1963; Velthorst and Hoijtink, 1965).

2. Crystalline Salts: X-Ray, IR, and Photoelectron Spectroscopy

The preparation of triaryl carbonium ion salts was accomplished early, and reviews of carbonium ion salts are available (Pfeiffer, 1927; Olah and Meyer, 1963). Most carbonium ion salts which have been prepared are those of very stable ions. However, salts of a wide variety of more reactive oxocarbonium ions have now been isolated and studied (see Olah, 1963).

The preparation of crystalline carbonium ion salts permitted several x-ray crystal structures to be performed which provide detailed structural information. One of the most reactive cations to be studied by x-ray crystallography is the acetyl cation (**2**) (Boer, 1966) as its hexafluoroantimonate salt. The linear geometry was proved, and the C–Me bond distance (1.378 Å) was unusually short. This was attributed to localization of positive charge at the carbonyl carbon. This distance can be compared to the C–C distances of 1.458 and 1.460 Å in the isoelectronic molecules $CH_3-C\equiv N$ and $CH_3-C\equiv CH$, respectively. The C–O distance of 1.116 Å is close to the $C\equiv O$

$$CH_3-\overset{+}{C}=O \qquad SbF_6 \qquad CH_3-C\equiv \overset{+}{O}$$
$$\underbrace{}_{1.378}\underbrace{}_{1.116}$$

(a) \qquad\qquad (b)

(2)

distance in carbon monoxide (1.128 Å) and somewhat shorter than the $C=O$ distances of about 1.20 Å in aldehydes. The contribution of hybrid structure (**2b**) was not thought to be substantial, and the C—O distance was that expected in view of the shrinkage of the carbon's van der Waals radius.

In the triphenylmethyl salts, x-ray determinations (Gomes de Mesquita et al., 1965) confirmed the propellar geometry. In the tri(p-aminophenyl)-

$$\underset{Ph}{\overset{Ph}{\diagdown}}\overset{+}{C}-Ph \qquad 1.44\ \text{Å average}$$

methyl perchlorate (Eriks and Koh, 1963) the three rings were twisted out of plane by 29, 34, and 34 deg. Furthermore, the bond lengths confirmed that the quinonoid representation best fit the observed values. The geometry about the

central carbon was planar. A detailed IR study of the splittings observed in the individual rings of a series of triphenylmethyl salts provided clear evidence for D_3 molecular symmetry which requires the central carbon to be planar (Sharp and Sheppard, 1957). The crystal structure (Bryan, 1964) of the product obtained from the reaction of tin (IV) chloride and dibromotetraphenylcyclobutadiene proved that this was the 3-chloro-1,2,3,4-tetraphenylcyclobutenyl pentachlorostannate (**45**) and not the tetraphenylcyclobutadienyl dication previously suggested (Freedman and Frantz, 1962). Ring **45b** was inclined 57 deg to the planes of rings **45a** and **45c**. The charge is delocalized primarily into rings **45a** and **45c** as would be expected from the allylic nature of the cation. The cycloheptatrienyl cation is planar and appears to participate in a charge-transfer interaction with the iodide ion in the solid state (Kitaigorodskii *et al.*, 1960). It is well known that the color of this cation is very sensitive to its anion.

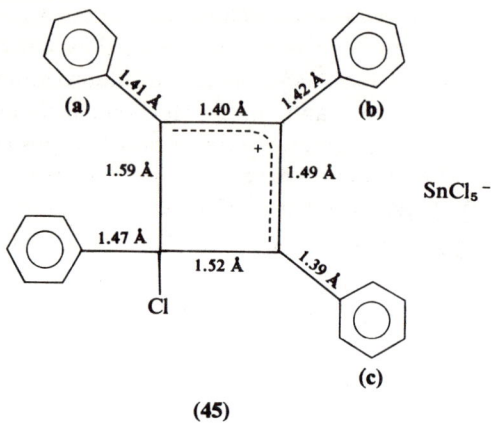

(45)

An extremely interesting crystal structure of the *sym*-triphenylcyclopropenyl perchlorate has appeared (Sundaralingam and Jensen, 1966). In this ion the phenyl rings are twisted in a propellarlike arrangement making angles of 7.6, 12.1, and 21.2 deg with respect to the plane of the cyclopropenyl ring. The cyclopropenyl ring and the three phenyl carbon atoms attached to it lie in one plane. The C–C bond lengths (average) in the three-membered ring were 1.373 Å, appreciably shorter than the benzene C–C bond. The average exocyclic bond length was 1.44 Å, which is less than normal for $C_{(sp^2)}$–$C_{(sp^2)}$ single bonds.

An analysis of the IR spectra of trichlorocyclopropenyl tetrachloroaluminate and tribromocyclopropenyl tetrabromaluminate (West *et al.*, 1966; Tobey and West, 1964) led to the conclusion that the cation had trigonal planar D_{3h} symmetry. Urey–Bradley force field calculations resulted in assigning the ring C–C force constant a value of 6.3 mdyn/Å, which is surprisingly higher than the corresponding value for benzene (5.6 mdyn/Å). This agrees with the x-ray structure of the triphenylcyclopropenyl cation which showed the ring C–C bond distance (1.373 Å) was shorter than that of benzene.

The IR spectra of several oxocarbonium ion salts have been measured (Susz and Wuhrmann, 1957; Cook, 1963; Olah *et al.*, 1962). The spectra of carbonium ions in solution were summarized in the previous section.

Olah *et al.* (1970a) have published the first report of the use of photoelectron spectroscopy of carbonium ions. In this technique the sample is bombarded with soft x rays, and the energy of the ejected electrons are resolved and measured. The difference in energy between the x ray and the excess energy of the ejected electrons represents the binding energy of those particular electrons. This technique has been reviewed by Hollander and Jolly (1970), Betteridge and Baker (1970), and Hercules (1970). Since x rays of 1486 eV were used, enough energy is available to eject both valence and inner-shell electrons. Using this method, the carbon 1s electron-binding energies of the *tert*-butyl, triphenylmethyl, and cycloheptadienyl cations were measured. These binding energies are listed in Table IX along with some representative neutral compounds. The binding energy of the 1s electron increases sharply for the central carbon of the *tert*-butyl cation. Decreased screening results in an increased nuclear attraction in this case. In the triphenylmethyl and

Table IX

Carbon 1s Electron-Binding Energies of Some Representative Organic Cations and Neutral Compounds

Compound	Carbon 1s electron-binding energy (eV)
$(CH_3)_3C-Cl$	284.1
Graphite	284.0
$(CH_3)_3C^+ \, SbF_6^-$	288.6 Central carbon
	285.2 Methyl carbon
$Ph_3C^+ \, SbF_6^-$	284.7 Central carbon
$C_7H_7^+ \, SbF_6^-$	284.7

tropylium ions the charge is highly delocalized, and no significant increase in the binding energy is observed.

The combined use of carbon and nitrogen 1s binding energy measurements may prove to be a powerful new tool in assessing charge densities (see Cox *et al.*, 1972).

D. Electrolytic Methods: Conductivity, EMF Measurements, and Polarography

1. *Conductivity*

A solution which contains dissolved salts has a higher conductivity than the pure solvent. Conductivity studies of triphenylmethyl chloride and bromide in liquid SO_2 by Walden (1902) and Gomberg (1902) provided some of the first evidence ever compiled for the existence of carbonium ions. Compared to spectroscopic techniques, conductivity has been used less in carbonium ion studies, probably because it does not provide direct information on the geometry and charge distribution about the ions. Although conductivity measurements register only free ions, the concentrations of associated ions (i.e., tight ion pairs) can be determined by difference. Consider Eq. (32) where ion pairs and free ions are present.

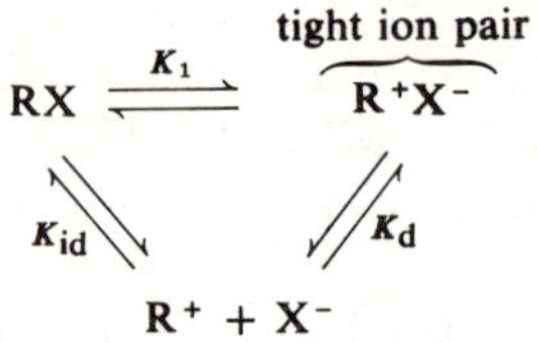

$$RX \underset{}{\overset{K_1}{\rightleftarrows}} \overbrace{R^+X^-}^{\text{tight ion pair}}$$

$$K_{id} \searrow \qquad \swarrow K_d \qquad (32)$$

$$R^+ + X^-$$

From this scheme we see that

$$K_{eq} = \frac{(a_{R^+})(a_{X^-})}{a_{RX} + a_{\widehat{R^+X^-}}} = \frac{K_1 K_d}{1 + K_1} \qquad (33)$$

It is K_{eq} which may be obtained from conductivity measurements; K_i and K_d are not directly obtained using this technique. Chemically, K_i values are very useful, and in many cases these are obtained using values of K_d calculated from Bjerrum's equation and K_{eq} from conductivity measurements (Harned and Owen, 1958).

The choice of a solvent for conductivity studies on carbonium ions is of central importance. The solvent must not react with the ions, and it must promote ionization by virtue of its solvating ability or high dielectric constant. If precise thermodynamic data are to be obtained, the solvent must have a conductivity which is much less than the solute conductivity throughout the concentration range studied. In order to observe the effect of structural changes on carbonium ion stability, ionization must be *incomplete* for the series of ions to be studied so that equilibrium constants may be obtained. If one only desires to prove that ionization has taken place, complete ionization is permitted. In this situation, strongly conducting solvents such as H_2SO_4 may be used.

Table X lists the specific conductances and dielectric constants for some typical solvents. All of these solvents, except sulfuric acid, have low enough specific conductances to allow their use in the measurement of low concen-

Table X
Example Solvents Used in Conductivity Studies of Carbonium Ions

Solvent	Specific conductance (mho cm^{-1})	Dielectric constant
Nitrobenzene	10^{-10}	34.8
Acetonitrile	4×10^{-7}	36.7
Antimony trichloride	5×10^{-6}	33.2
Sulfur dioxide	10^{-7} or less	15.4
Sulfuric acid	10^{-2}	101

4. CARBONIUM IONS

trations of free ions. One must consider the objectives of the measurement and the chemistry of the system prior to selection of a solvent.

A few representative values of K_{eq}, K_{id}, and K_d for triarylchloromethanes are listed in Table XI from a larger study in which a ρ value of -4.41 was obtained on correlating the data by the Hammett equation (Lichtin and Vignale,

Table XI
Ionization and Dissociation Equilibria of Triphenylmethyl Chloride Derivatives in Liquid SO_2

Substituents on each ring			$K_{eq}(\times 10^5)$	$K_d(\times 10^3)$	$K_i(\times 10^2)$
H	H	H	4.1	2.88	1.46
H	H	ortho Cl	1.06	3.41	0.31
H	H	ortho CH_3	63	2.98	27
H	H	meta Cl	0.15	3.25	4.7×10^{-2}
H	H	meta CH_3	9.5	2.92	3.4
H	H	para Cl	1.26	3.36	0.38
para Cl	para Cl	para Cl	0.12	3.71	0.03
H	H	para $C-C_3H_7$	170	3.70	85
H	H	para CH_3	64	3.36	23

1957). By studying K_{eq} values as a function of temperature, thermodynamic quantities (ΔG^0, ΔH^0, and ΔS^0) have been obtained and are summarized in a recent review by Lichtin (1968). Even primary carbon and secondary deuterium isotope effects in triphenylchloromethane ionization have now been obtained with high precision conductivity measurements (K_{12_C}/K_{13_C}) = 0.9833, see Kresge et al., 1965). Conductivity studies of SO_2 solutions of $CH_3COCl \cdot SbCl_5$ from $-70°$ to $0°C$ exhibit similarity with $KSbCl_6$ which indicated the dissociation of these complexes to CH_3CO^+ and $SbCl_6^-$ was complete (Seel, 1943).

2. EMF Measurements and Polarography

The objective of EMF measurements is to devise a galvanic cell in which the half-cell potential of the reaction

$$R^+ + e^- \rightleftharpoons \tfrac{1}{2}R_2$$

may be measured. Conant et al. (1925; Conant and Chow, 1933) made pioneering efforts to study quantitatively reaction (34).

$$Ph_3C^+ + e^- \longrightarrow Ph_3C \cdot \qquad (34)$$

Jensen and Taft (1964) reported a galvanic cell with a liquid junction through an asbestos fiber of the following type:

$$Pt|R_2, R^+BF_4^-, CH_3CN\|AgNO_3, CH_3CN|Ag \qquad (35)$$
$$\text{(solvent)} \qquad \text{(solvent)}$$

This promising source of quantitative data on carbonium ion equilibria allowed the standard free energy change ΔG^0 and the equilibrium constant K_{eq} of Eq. (36) to be measured for the substituted triphenylmethyl cation

$$R^+ + \tfrac{1}{2}(Ph_3C-CPh_3) \longrightarrow Ph_3{}^+C + \tfrac{1}{2}(R-R) \tag{36}$$

series given in Table XII. The huge range of stabilities measured attests to the potential of this method.

Polarography and related methods such as cyclic voltametry could be used to probe both carbonium ion stability and kinetic processes. Several groups have recorded polarographic observations of the reduction of carbonium ions. The reduction potential of a series of carbonium ions, compared to the same standard electrode, should reflect their stability with the greater reduction potential reflecting greater cation stability. Equation (37) represents a typical process. For example, in 10.2 M H_2SO_4 the first reduction potential

Table XII

Values of ΔG^0 and K_{eq} for the Reaction (36) of Substituted Triphenylmethyl Cations Determined from EMF Measurements

p-Substituents in phenyl rings of R^+			K_{eq}	ΔG^0
H	H	H	(0.0)	(1)
F	H	H	0.98	0.1
CH_3	H	H	3.6×10^{-2}	2.0
CH_3O	H	H	2.0×10^{-4}	5.1
CH_3O	CH_3O	H	3.7×10^{-8}	8.8
CH_3O	CH_3O	CH_3O	1.7×10^{-9}	12.0
$(CH_3)_2N$	H	H	6.5×10^{-12}	15.2
$(CH_3)_2N$	$(CH_3)_2N$	$(CH_3)_2N$	3.1×10^{-18}	23.9

(half-wave potential, $E_{1.2}$) of the triphenylmethyl cation was 0.58 V, whereas the tris-p-anisylmethyl cation was 1.09 V (verses $Hg|Hg_2SO_4$, 17 M H_2SO_4) (Feldman and Flythe, 1969). However, the second reduction potential merges

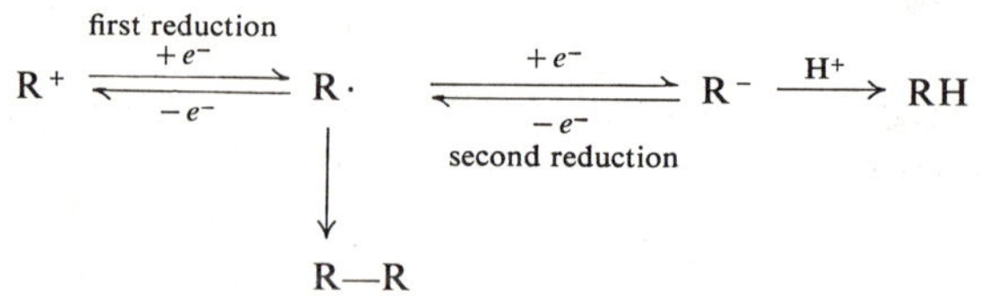

$$R^+ \underset{-e^-}{\overset{+e^-}{\rightleftarrows}} R \cdot \underset{-e^-}{\overset{+e^-}{\rightleftarrows}} R^- \xrightarrow{H^+} RH \tag{37}$$

(first reduction / second reduction, with R· → R—R branch)

into the first at higher acidity, and it appears as if reaction at the mercury electrode surface is occurring in both triaryl carbonium and 1,3-dioxolan-2-ylium ions in H_2SO_4 (Carney and Pittman, 1971). Often the reduction process

4. CARBONIUM IONS

appears to be complex, and interpretation of half-wave potentials may be complicated. Two waves are observed in the reduction of triphenylmethanol in methanesulfonic acid in the presence of water, which have been attributed to both one- and two-electron processes (Wawzonek et al., 1956); but in the absence of water only the two-electron process is observed.

The energy difference between the cation and the radical for the cyclopropenyl cation is 3β according to simple MO theory. For the tropylium ion (**20**) this energy difference is 1.7β. Thus, it is not surprising that the values of $E_{1/2}$ of substituted cyclopropenyl cations (Breslow et al., 1961) are larger than the tropylium ion, which in turn has a greater $E_{1/2}$ than the triphenylmethyl cation. The tropylium ion has a greater $E_{1/2}$ than its less stable homotropylium (**47**) analog (Feldman and Flythe, 1971). It is obviously much more difficult to reduce cyclopropenyl cations than the triphenylmethyl cation because the loss of resonance energy resulting when cyclopropenyl cations are converted to radicals corresponds to about 24 kcal/mole. Cyclic voltametry has been

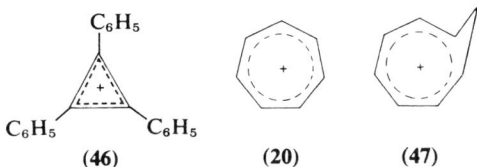

(**46**)　　(**20**)　　(**47**)

applied to **46**. When reduced at 10 cps, the resulting radical dimerized irreversibly, but at 150 cps the anodic and cathodic peak currents were equal, indicating the radical can be reoxidized quantitatively before dimerization.

Breslow and Chu (1970) and Volz and Lotsch (1969) have now used the half-wave reduction potentials of a series of para-substituted triphenylmethyl cations to determine relative pK_a values. Precise stability measurements of many unusual carbonium ions by polarographic techniques should appear in the near future.

E. KINETIC METHODS

The kinetic equation for carbonium ion formation by a general S_N1 process was developed in Section II,B. The generation and reaction of a carbonium ion may be represented on reaction coordinate diagrams. For

$$RX \underset{k_{-1}}{\overset{k_1}{\rightleftarrows}} R^+ + X^- \xrightarrow{\underset{k_2}{Y^-}} RY \qquad (38)$$
$$\phantom{RX \underset{k_{-1}}{\overset{k_1}{\rightleftarrows}} R^+ + X^- } \xrightarrow{k_3} P$$

the general process, see Eq. (38). Figure 2 illustrates a relationship between starting materials, intermediates, and products. The generation of the carbonium ion R^+ is a very endothermic process, and the unstable carbonium ion can revert back to RX, it can react with Y^- to give RY, or it can react to give P. In Fig. 2 the free energy of activation to form RY from R^+ ($\Delta G_3^{\ddagger}$) is less than that ($\Delta G_4^{\ddagger}$) to form P. Thus, RY will predominate over P (provided a suitable Y^- concentration is present). The ratio of RY to P will decrease as the overall energy available to the system increases (Boltzman Distribution Factor) or as $\Delta G_4^{\ddagger} - \Delta G_3^{\ddagger}$ decreases. In Fig. 2 products RY and P are stable under the reaction conditions and do not revert back to R^+. Thus, the product distribution is controlled by the relative values of $\Delta G_3^{\ddagger}$ and $\Delta G_4^{\ddagger}$ (and the concentrations of reagents reacting with R^+ in these processes).

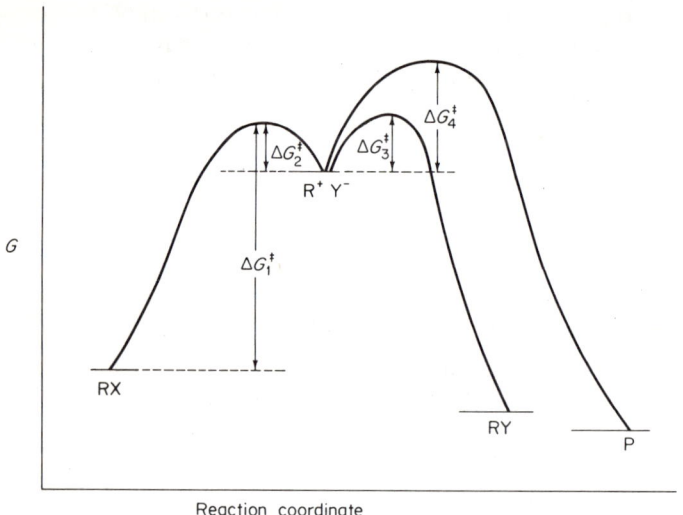

Fig. 2. Reaction path of an S_N1 process.

The difference between kinetic and thermodynamic control may be illustrated by Fig. 3. Here, a carbonium ion R^+ (whose formation is not represented) may react by path **a** to generate RZ or RY or path **b** to give R'Z. The free energies of activation by path **a** are much lower; thus, these processes are kinetically favored, and RY and RZ are formed much faster than R'Z which must surmount the large ($\Delta G_2^{\ddagger}$) activation barrier. However, products RY and RZ may be relatively easily converted back to R^+ by surmounting barriers of $\Delta G_3^{\ddagger}$ and $\Delta G_4^{\ddagger}$, respectively. Whenever R^+ does react to form R'Z, this species is stable because it must surmount the very large $\Delta G_5^{\ddagger}$ in order to revert back to R^+. Thus, although R'Z is formed more slowly, it is far more stable and is termed the "thermodynamic" product. When enough energy is available, R^+ eventually is completely converted to R'Z because RY and RZ rapidly revert back to R^+ when formed [Eq. (39)]. In some cases, one of the kinetic products might be stable enough to be formed under the

$$RY \underset{k_{-1}}{\overset{k_1}{\rightleftharpoons}} R^+ \xrightarrow{k_3} R'Z \quad (39)$$

$$R^+ \underset{k_{-2}}{\overset{k_2}{\rightleftharpoons}} RZ$$

reaction conditions without reverting to R^+ (and eventually to $R'Z$). Thus, isolation of RZ is possible, and this is an isolation of "kinetic product." The

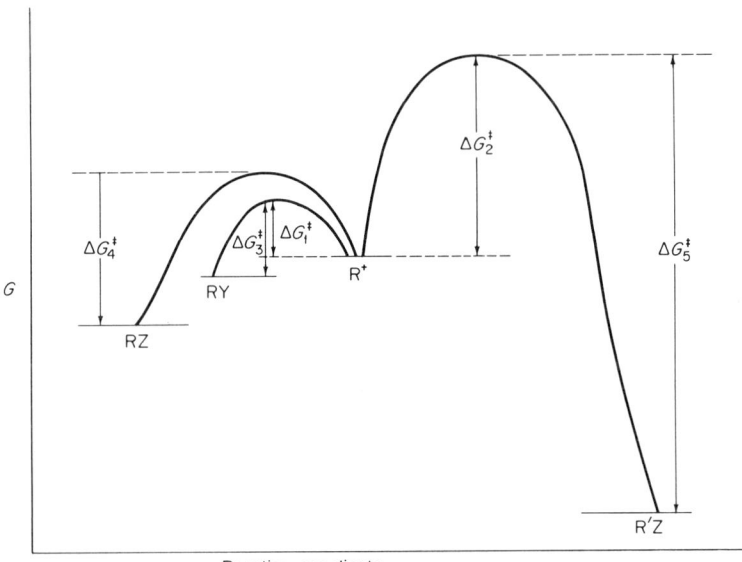

Fig. 3. Thermodynamic and kinetic control in the reaction of a carbonium ion.

reaction of the 2-methyl-1,3-dioxolan-2-ylium cation with ethoxide and iodide [Eq. (40)] illustrates a case where the kinetic product is formed in one case and the thermodynamic product in another (Meerwein et al., 1960b; for reviews, see Pittman et al., 1972b; Hünig, 1964).

Rate studies have been the most important single means for developing a scale of carbonium ion stabilities in solution. The technique has been to

$$(40)$$

solvolyze a series of related compounds with the same leaving group and compare the rates. The rates give a measure of the energy difference between the starting material and the transition state. However, solvolysis reactions are endothermic, and the Hammond postulate can now be invoked. For highly endothermic reactions, it states that the transition state is reached at a point along the reaction coordinate near the product (the carbonium ion in an S_N1 reaction). For a series of related reactions, the difference in energy between the reactants and the transition state then approximately reflects the difference in energy between the reactants and the carbonium ion. This is illustrated in Fig. 4 where the solvolyses of *tert*-butyl chloride and 2-phenyl-2-propyl

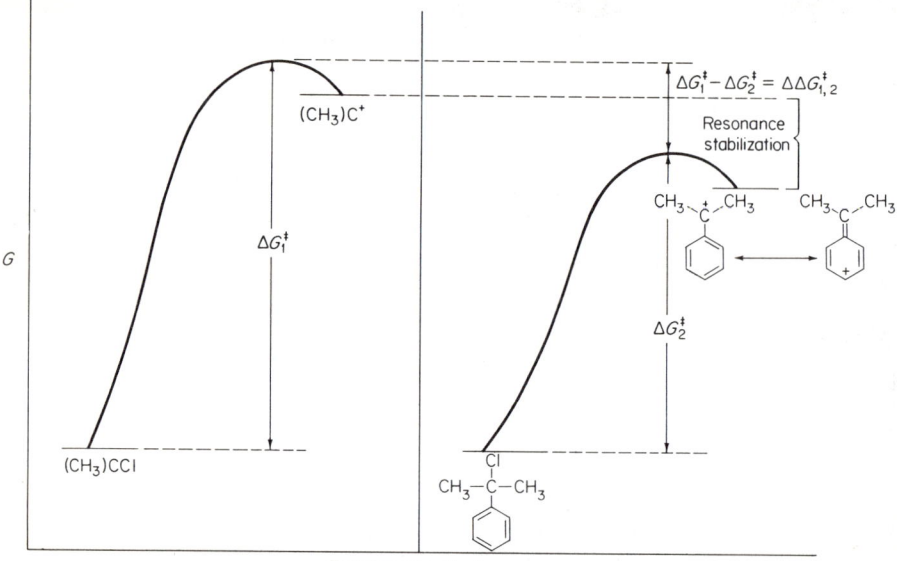

Fig. 4. Comparison of the solvolysis of *tert*-butyl and 2-phenyl-2-propyl chlorides.

chloride are compared. The solvolysis of 2-phenyl-2-propyl chloride is 600 times as fast as that of *tert*-butyl chloride. This is attributed largely to resonance delocalization of charge by the ring, which stabilizes the resulting 2-phenyl-2-propyl cation relative to the *tert*-butyl cation. Thus, the transition state leading to the 2-phenyl-2-propyl cation is lower in energy because effects which stabilize the product ion also act to stabilize the transition state (since the transition state greatly resembles the resultant ion). Since the transition state leading to the 2-phenyl-2-propyl cation is stabilized, $\Delta G_2^{\ddagger}$ is less than $\Delta G_1^{\ddagger}$, and this accounts for the rate difference of 600. It is very important to point out that this comparison is simplified since the reactants were assigned the same energy.

4. CARBONIUM IONS

It is quite possible that the *reactants* may have very different energies owing to factors such as steric strain, whereas the carbonium ions derived from them may have very similar energies. This situation is illustrated in Fig. 5. Tri-*tert*-butylmethyl-*p*-nitrobenzoate solvolyzes markedly faster than

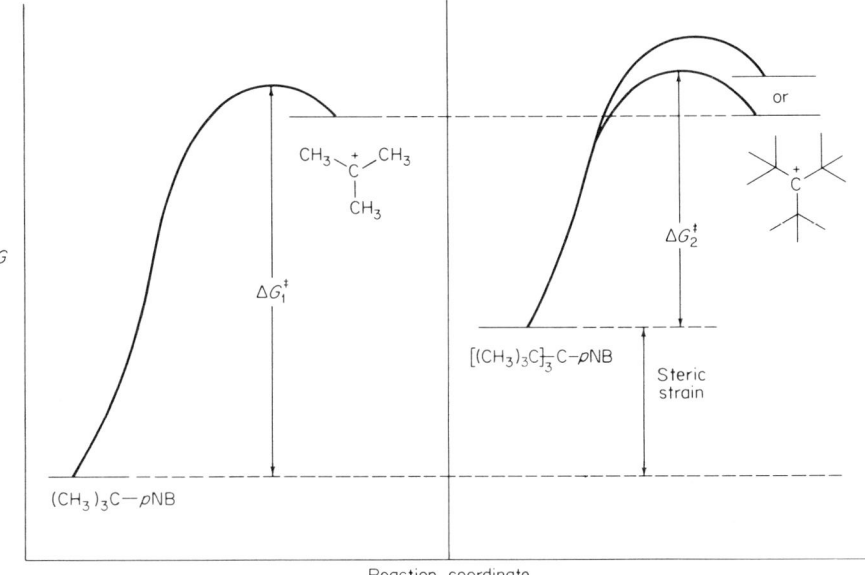

Fig. 5. The effect of releasing steric strain in going to the transition state.

tert-butyl-*p*-nitrobenzoate under identical conditions (Bartlett and Stiles, 1955; Bartlett and Swain, 1955). Since the carbonium ions resulting in either of these processes are tertiary alkyl cations, their stability would not be expected to differ very much. If the tri-*tert*-butyl carbonium ion was so sterically hindered that it could not be effectively solvated, it might be even less stable than the *tert*-butyl cation. The much faster rate of the tri-*tert*-butylmethyl-*p*-nitrobenzoate solvolysis must be due to its much higher energy relative to *tert*-butyl-*p*-nitrobenzoate. Thus, it starts out nearer the transition state that it must reach in the solvolysis process. This increase in energy is due to the severe steric compression which results from three bulky *tert*-butyl groups in addition to the *p*-nitrobenzoate groups which are attached to a single tetrahedral carbon. This steric strain is largely released on going to the sp^2-hybridized carbonium ion where the *p*-nitrobenzoate group is removed and the *tert*-butyl groups can spread out to 120 deg of one another.

A general order of carbonium ion stabilities is listed in Table XIII as derived from their relative solvolysis rates. These rates are approximate since

Table XIII
Effects of Delocalization on Solvolysis Rates

Compound	Relative k
$CH_3CH_2CH_2Cl$	0.04[a]
$CH_3—CH(Cl)CH_3$	0.1
$HC\equiv C—CH_2Cl$	0.01
$CH_2=CH—CH_2Cl$	1.0
trans-$CH_3CH=CHCH_2Cl$	2×10^3
trans-$C_6H_5—CH=CHCH_2Cl$	2×10^5
$(CH_3)_2C=CHCH_2Cl$	3×10^6
$CH_2=CH—C(CH_3)_2Cl$	1×10^7
$(CH_3)_3CCl$	4×10^4
$C_6H_5CH(CH_3)_2Cl$	8×10^4
$C_6H_5C(CH_3)_2Cl$	2×10^7
$(C_6H_5)_2CHCl$	4×10^7
$(C_6H_5)_3CCl$	2×10^{10}
$(c\text{-}C_3H_5)_3CCl$	4×10^{11}

[a] Undoubtedly not a limiting S_N1 process.

experimental work was not, in all cases, done under identical conditions, and the values listed are estimates based on comparisons given in a major review of this topic (Streitwieser, 1962). This order brings up several interesting questions. The rate of solvolysis of benzhydryl chloride is much slower than that of triphenylmethyl chloride (4×10^7 versus 2×10^{10}), yet the amount of delocalization into the two rings in the diphenylcarbonium ion should be comparable to the delocalization in the triphenylcarbonium ion where out-of-plane twisting into the propellar geometry is well established. Thus, the question arises: How much of this rate difference is due to steric compression in the starting chlorides? Although the major rate enhancement in the solvolysis of tricyclopropyl derivatives is due to delocalization, how important a role does release of steric strain play?

The energy diagram for the solvolysis of 2,2,2-triphenylethyl chloride (or tosylate) versus that of neopentyl chloride (or tosylate) is very interesting.

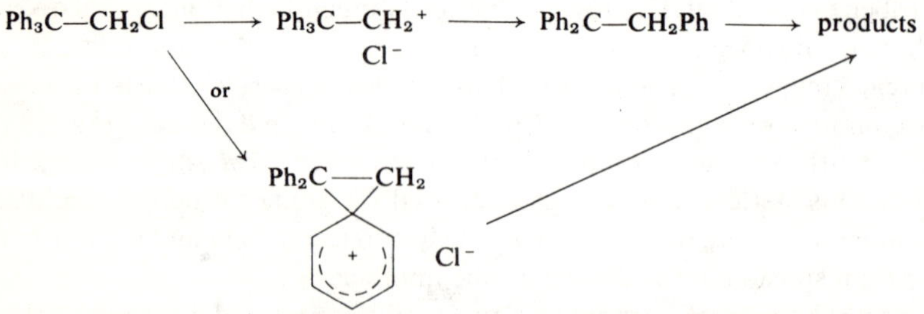

$$Ph_3C—CH_2Cl \longrightarrow Ph_3C—CH_2^+ \longrightarrow Ph_2C—CH_2Ph \longrightarrow \text{products} \quad (41)$$
$$Cl^-$$

or

Ph_2C—CH_2 bridged phenonium ion, Cl^-

$$(CH_3)_3C—CH_2Cl \longrightarrow (CH_3)_3C—CH_2^+ \longrightarrow (CH_3)_2\overset{+}{C}—CH_2CH_3 \longrightarrow \text{product} \quad (42)$$

4. CARBONIUM IONS

2,2,2-Triphenylethyl chloride solvolyzes at a rate about 10^4 times faster than the neopentyl chloride (Winstein et al., 1952). The rate difference was interpreted as strong evidence for phenyl participation in the solvolysis. However, this view has been questioned, and the possibility that steric strain is relieved on going to the transition state has been suggested as an alternative explanation (Brown, 1962). Figure 6 represents the various possibilities. If 2,2,2-triphenylethyl chloride is sterically compressed and a significant release of this strain is obtained on going to the 2,2,2-triphenyl-1-ethyl cation, then path 1 would account for the rate data. Path 2 would predict no rate difference between the systems. Path 3 predicts the large rate difference is due both to strain present in the 2,2,2-triphenylethyl reactant (relative to the neopentyl reactant) and stabilization of the transition state due to phenyl participation (anchimeric assistance) at the site of the developing carbonium ion (i.e., C-1). Path 4 attributes all the rate enhancement found in the 2,2,2-triphenylethyl system to phenyl participation. A clear analysis of these possible effects is often difficult.

A powerful test of whether or not positive charge is being generated at a carbon in the transition state of a reaction is the use of substituent effects. The most frequently employed technique is to perform the reaction in question with a series of substituents ranging from electron withdrawing, i.e.

$$-NO_2, \quad -CN, \quad -CF_3$$

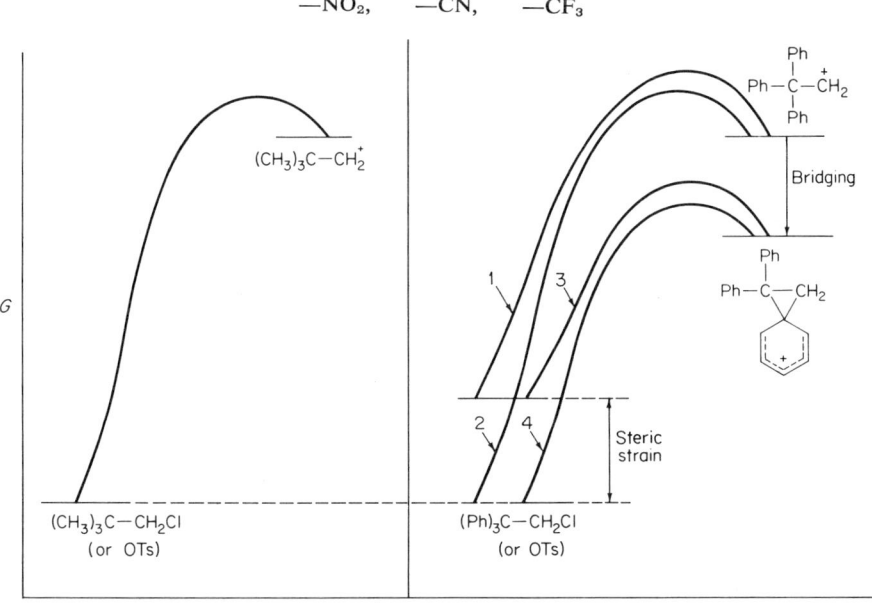

Fig. 6. Steric strain versus anchimeric assistance in 2,2,2-triphenylethyl chloride or tosylate solvolyses.

to electron donating, i.e.

$$CH_3, \quad -N\begin{smallmatrix}CH_3\\ \\CH_3\end{smallmatrix}$$

and to correlate the rate measurements by the Hammett equation (see Gould, 1959; Hammett, 1940; Jaffe, 1953; Taft, 1956) $\log (k/k_0) = \rho\sigma$. The σ values are positive for electron-withdrawing groups and are negative for electron-donating groups. Plotting the $\log k$ versus σ for the series of substrate compounds studied gives a straight line when $\Delta\Delta G^{\ddagger}$ values are proportional to the $\Delta\Delta G^0$ values (the requirement necessary for a linear free energy relationship to exist). The slope of this line is ρ. If ρ is negative, this means electron-donating groups facilitate the reaction and electron-withdrawing groups retard it. This is strong evidence that positive charge is being generated on the carbon atom at the reaction center. The position of equilibrium may also be used in place of rate constants. The Hammett equation becomes $\log (K/K_0) = \sigma\rho$. An obvious example of this method is the ionization of triarylmethyl chlorides in SO_2 at 0°C (see Fig. 7). The value of ρ was found to be -3.97 for this series of reactions, a sure indication that a substantial

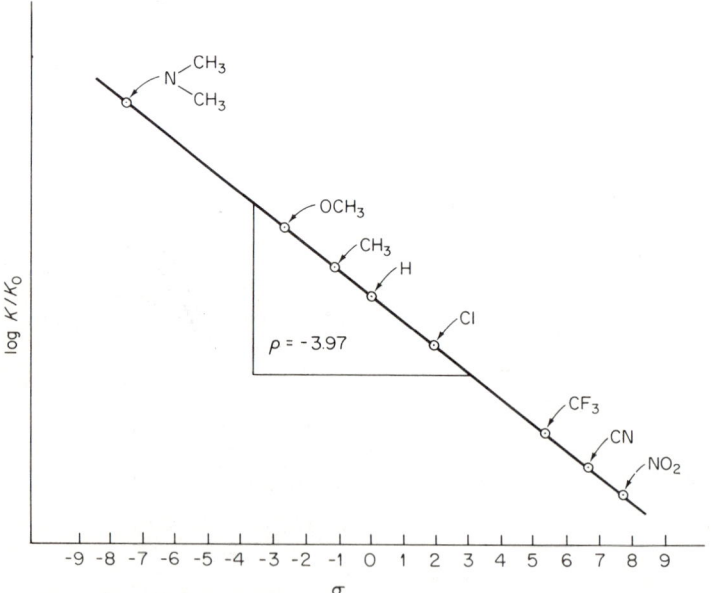

Fig. 7. Hammett plot of the equilibrium constants for the ionization of mono para-substituted triarylmethyl chlorides in SO_2 at 0°C.

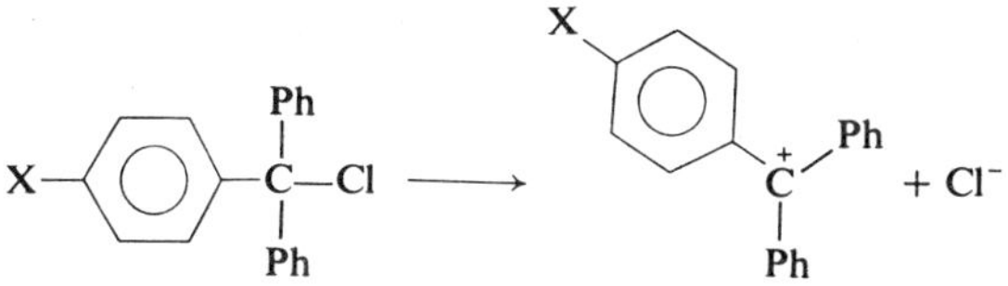

4. CARBONIUM IONS

$$X-\underset{Ph}{\underset{|}{C}}-Cl \longrightarrow \underset{Ph}{\underset{|}{\overset{+}{C}}} + Cl^- \quad (43)$$

buildup of positive charge had occurred in the transition state. The use of σ^+ constants in place of σ constants is preferred for reactions in which the electron demand on the substituent is so strong that the resonance contribution of the substituent is greater than usual (Brown and Okamoto, 1958). The σ^+ constants were defined utilizing the solvolysis of 2-phenyl-2-propyl chlorides in acetone–water (see Section II,B).

Another technique, the use of isotope effects, is very powerful in probing carbonium ion reactions. In the solvolysis of medium ring tosylates, Prelog and co-workers discovered 1–5 hydride shifts were taking place. For example, in the solvolysis of 5,6-d_4-cyclodecyltosylate [Eq. (44)], 1,2,2,7-tetradeuterocyclodecanol was obtained. However, hydrogen (deuterium) bridging in the solvolysis transition state was ruled out because no isotope effect was observed (i.e., $k_H/k_D = 1$) (Prelog, 1957; see, also, Prelog and Traynham, 1963).

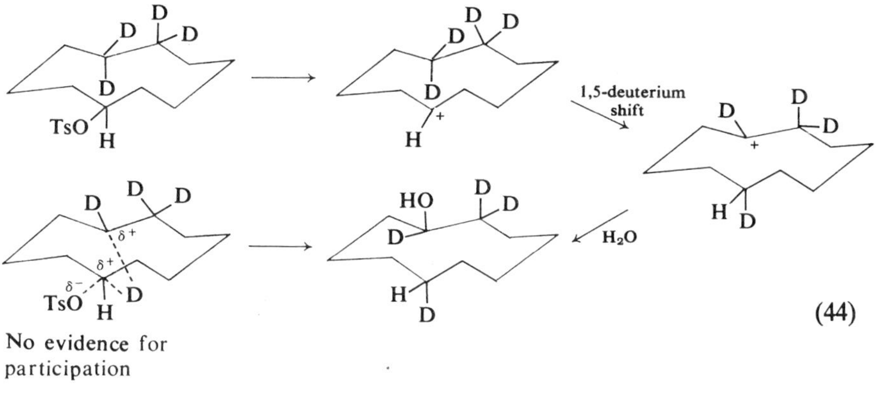

(44)

No evidence for participation

F. CHEMICAL METHODS (PRODUCT ANALYSIS)

By examining the reaction conditions and then analyzing the reaction products, it is possible to infer a mechanism through which the reaction proceeds. Just because the mechanism chosen fits the products, this does not assure that the proposed mechanism is true. Historically, most reaction mechanisms invoking carbonium ions were based on an analysis of the products. Several examples are instructive.

The addition of chlorine to ethylene in aqueous solution produces ethylene chlorohydrin and β,β'-dichlorodiethyl ether. This can be explained by invoking the intermediate cations illustrated in Eq. (45). The first intermediate

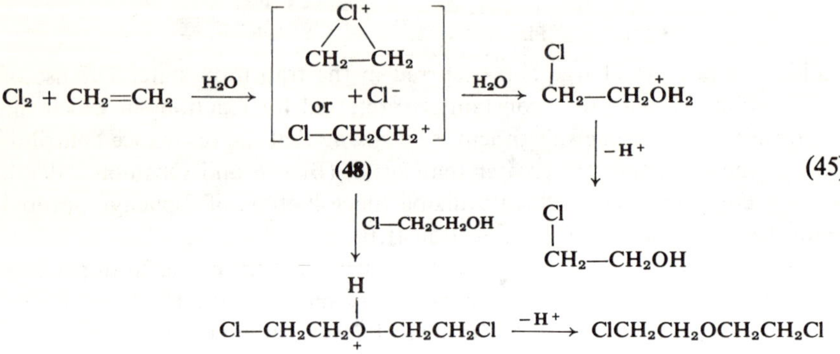

$$Cl_2 + CH_2=CH_2 \xrightarrow{H_2O} \begin{bmatrix} \overset{Cl^+}{\overset{|}{CH_2-CH_2}} \\ \text{or} \quad +Cl^- \\ Cl-CH_2CH_2^+ \end{bmatrix} \xrightarrow{H_2O} \underset{CH_2-CH_2\overset{+}{O}H_2}{\overset{Cl}{|}}$$

$$(48) \qquad \qquad \qquad \downarrow -H^+$$

$$\downarrow Cl-CH_2CH_2OH \qquad \underset{CH_2-CH_2OH}{\overset{Cl}{|}}$$

$$\underset{Cl-CH_2CH_2\overset{+}{O}-CH_2CH_2Cl}{\overset{H}{|}} \xrightarrow{-H^+} ClCH_2CH_2OCH_2CH_2Cl \qquad (45)$$

cation (**48**) can be attacked by either of two nucleophiles, water or ethylene chlorohydrin, and this accounts for the observed products. The Markownikoff addition of HBr or the acid-catalyzed addition of water to unsymmetrical olefins may also be rationalized as proceeding through carbonium ion intermediates. Both the bromine and hydroxy groups end up on the most substituted carbon of the original double bond. Since tertiary carbonium ions are more stable than secondary ions, this suggests that initial protonation takes place to give the more stable tertiary cation **16**. This cation is then

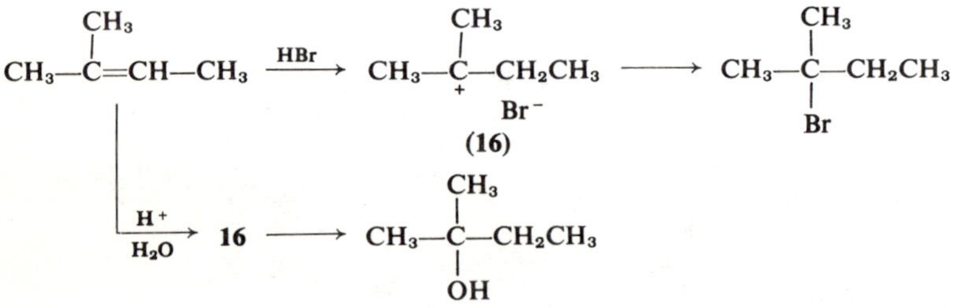

$$CH_3-\underset{|}{\overset{CH_3}{C}}=CH-CH_3 \xrightarrow{HBr} CH_3-\underset{Br^-}{\overset{CH_3}{\underset{|}{\overset{+}{C}}}}-CH_2CH_3 \longrightarrow CH_3-\underset{Br}{\overset{CH_3}{\underset{|}{C}}}-CH_2CH_3$$

$$\downarrow \overset{H^+}{H_2O} \quad \mathbf{16} \longrightarrow CH_3-\underset{OH}{\overset{CH_3}{\underset{|}{C}}}-CH_2CH_3 \qquad (46)$$

captured either by bromide or water. The acid-catalyzed dimerization of isobutene gives 2,4,4-trimethyl-2-pentene (**49**) and 2,4,4-trimethyl-1-pentene (**50**), which are readily explained by invoking the carbonium ion intermediates **6** and **51** shown in Eq. (47) (Whitmore, 1932, 1934, 1948; McCubbin and Adkins, 1930; Pepper, 1964).

Analysis of neutral products obtained after reacting phenyl-substituted allylic alcohols in H_2SO_4 requires [Eq. (48)] mechanisms involving phenyl-

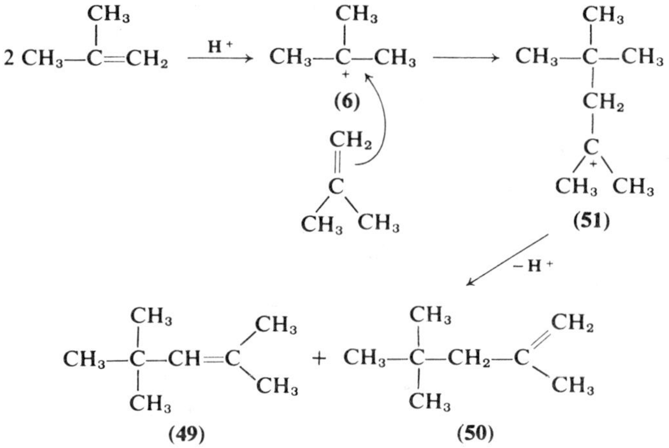

$$2\ CH_3-\underset{CH_3}{\underset{|}{C}}=CH_2 \xrightarrow{H^+} CH_3-\underset{CH_3}{\underset{|}{\overset{+}{C}}}-CH_3 \longrightarrow CH_3-\underset{\underset{\underset{\underset{CH_3\ CH_3}{\diagup^+\diagdown}}{C}}{\underset{|}{CH_2}}}{\underset{|}{C}}-CH_3$$

(6), with CH$_2$ branch leading to C(CH$_3$)$_2$ (double bond), (51) (47)

$$\downarrow -H^+$$

$$CH_3-\underset{CH_3}{\underset{|}{\overset{CH_3}{\overset{|}{C}}}}-CH=C\underset{CH_3}{\overset{CH_3}{\diagdown}} + CH_3-\underset{CH_3}{\underset{|}{\overset{CH_3}{\overset{|}{C}}}}-CH_2-C\underset{CH_3}{\overset{CH_2}{\diagup\!\!\!=}}$$

(49) (50)

substituted acyclic, allylic, and indanyl cations (Deno *et al.*, 1965b; Pittman and Miller, 1972). Two particularly interesting examples are alcohols **52a** and

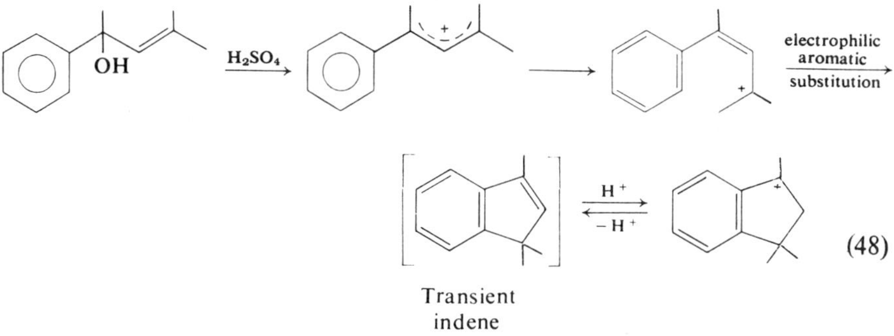

[Reaction scheme: PhCH(OH)CH=C(CH$_3$)$_2$ $\xrightarrow{H_2SO_4}$ cation $\longrightarrow$ cation $\xrightarrow{\text{electrophilic aromatic substitution}}$]

[Transient indene $\underset{-H^+}{\overset{H^+}{\rightleftarrows}}$ indanyl cation]   (48)

Transient
indene

52b which rearrange to give indanyl cations **53** and **54** in H$_2$SO$_4$ (Pittman and Miller, 1972). The mechanisms invoked in these two rearrangements account for the formation of ions **53** and **54** which have the trans arrangement of the substituents at the C-2 and C-3 positions. Furthermore, it is now recognized that allylic cations cyclize from the terminal end on which the charge is least stabilized.

Several reactions of 3-methyl-2-butyl derivatives can be rationalized as taking place through carbonium ions. These include solvolysis of the tosylate in acetic acid (Winstein and Takahashi, 1958), deamination of the amine in acetic acid (Silver, 1963), deoxidation of the alcohol in water, and anodic oxidation of the carboxylic acid in aqueous KOH (Keating and Skell, 1970). The initial product of each reaction is the 3-methyl-2-butyl cation [Eq. (50)], and the final product distributions are listed in Table XIV. From these products it is possible to write mechanisms which account for every product. The appearance of 3-methyl-1-butene and 3-methyl-2-butene results from the elimination of a proton from the 3-methyl-2-butyl cation. A 1,2-hydride shift followed by proton elimination gives 2-methyl-1-butene. Capture of the 3-methyl-2-butyl cation by acetate (acetic acid) gives 3-methyl-2-acetoxy-

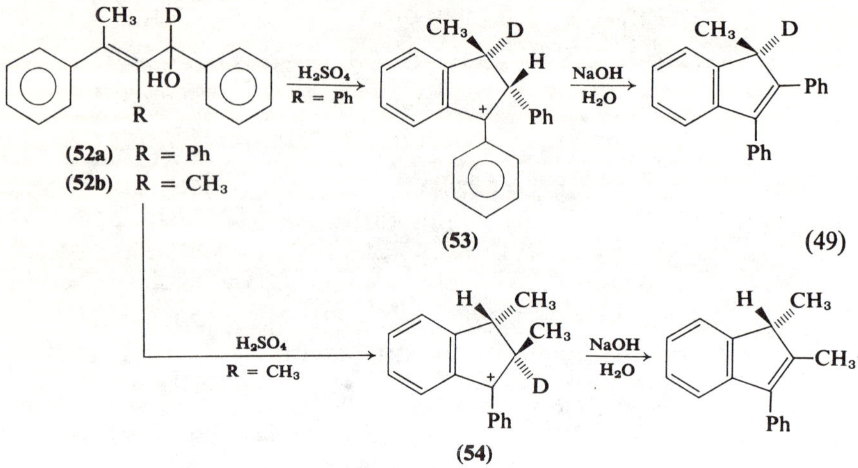

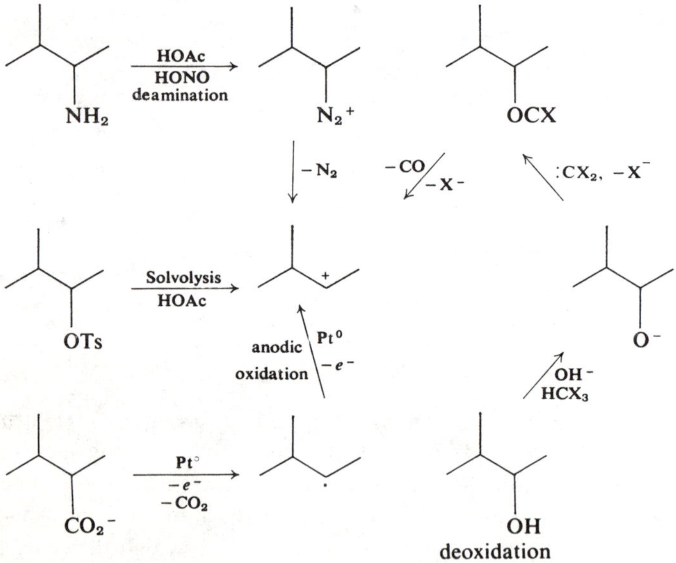

Table XIV
3-Methyl-2-butyl System

Products	Solvolysis 75°C HOAc	Deamination 55°C HOAc	Deoxidation H_2O	Anodic oxidation 1 M KOH
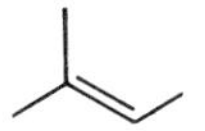 (2-methyl-2-butene)	54	12.4	24	16
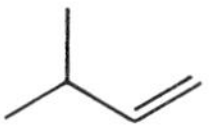 (3-methyl-1-butene)	1.3	14.5	48	43
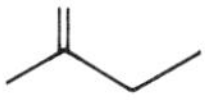 (2-methyl-1-butene)	14.7	7.9	23	20[a]
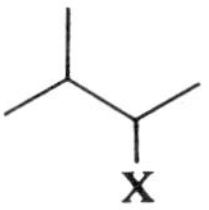 (3-methyl-2-butyl-X)	3	33	—	—
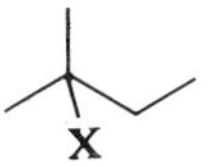 (2-methyl-2-butyl-X)	27	10.2	—	—
 (1,1-dimethylcyclopropane)	—	11.3	3	17
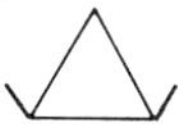 (1,2-dimethylcyclopropane)	—	5.7	2	4

[a] Ratio of *cis*- to *trans*-1,2-dimethylcyclopropane is about 1:2.

butane; but if a 1,2-hydride shift occurs before acetate capture, 2-methyl-2-acetoxybutane is formed. The formation of *cis*- and *trans*-dimethylcyclopropanes results from cyclization via protonated cyclopropane intermediates.

Although all these reactions proceed through the same intermediate 3-methyl-2-butyl cation, the product distribution is markedly different for each reaction. The solvolysis reactions produce a much larger amount of the more substituted olefin. Furthermore, solvolysis does not result in any cyclopropyl products. It is quite clear that the method of generating the cation and the solvent play a major role in determining the product distribution. Skell has termed ions produced by solvolysis "encumbered" cations, whereas those produced by deamination, deoxidation, and anodic oxidation were called "unencumbered" ions ("encumbered" and "solvated" may be assumed to be synonyms). Careful product analyses on a large number of systems have indicated (1) unencumbered ions rearrange to a greater extent and do not

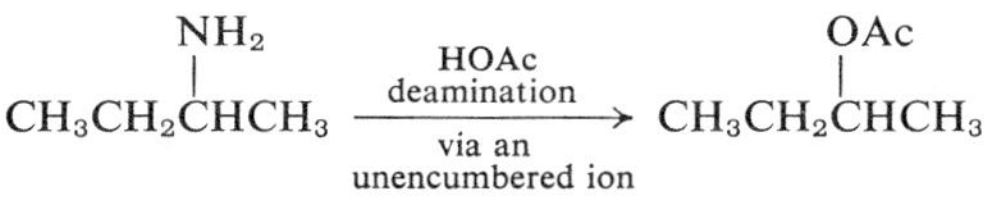

$$CH_3CH_2\overset{NH_2}{\underset{|}{C}}HCH_3 \xrightarrow[\text{via an unencumbered ion}]{\text{HOAc deamination}} CH_3CH_2\overset{OAc}{\underset{|}{C}}HCH_3 \quad \begin{array}{l} 64\% \text{ inversion} \\ 36\% \text{ retention} \end{array}$$

(51)

$$CH_3-(CH_2)_5-\underset{OBs}{CH}-CH_3 \xrightarrow[\text{via an encumbered ion}]{HOAc} CH_3(CH_2)_5-\underset{OAc}{CH}CH_3 \quad 100\% \text{ inversion}$$
(52)

follow the most exothermic pathway exclusively; (2) an unencumbered cation undergoes product formation with less inversion [see Eqs. (51) (Wiberg, 1950) and (52) (Weiner and Sneen, 1965; Streitweiser and Waiss, 1962) for example]; (3) unencumbered cations better reflect the conformational composition of the starting material; and (4) unencumbered cations are more apt to fragment. The basic explanation for this behavior is that unencumbered cations are formed in exothermic processes which do not require solvent participation, whereas encumbered cations are generated in endothermic processes requiring strong solvent participation. The unencumbered cation is generated "free" in a solvent cavity which has not relaxed (reoriented its dipoles) to solvate the cation. The realignment of solvent dipoles or neighboring anions eventually produces a highly solvated (encumbered) cation, but many intramolecular changes prior to this relaxation account for the differences in behavior. These ideas have been reviewed in detail (Keating and Skell, 1970).

The use of isotopes provides an important and powerful means of indicating the presence of carbonium ions by product analysis. This technique has reached its most advanced state in the work of Collins (1964, 1968), where it has been used to distinguish between two or more pathways, to reject one or more of the conceivable intermediates, and to yield information on the kind of transition state encountered (through isotopic rate effects). An

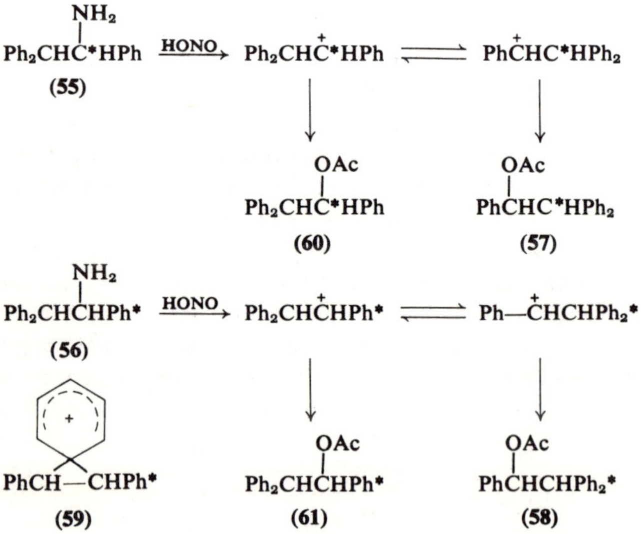

$$\underset{(55)}{Ph_2CH\overset{NH_2}{C}{}^*HPh} \xrightarrow{HONO} Ph_2CH\overset{+}{C}{}^*HPh \rightleftharpoons Ph\overset{+}{C}HC^*HPh_2$$

$$\downarrow \qquad \qquad \downarrow$$

$$\underset{(60)}{Ph_2CH\underset{OAc}{C}{}^*HPh} \qquad \underset{(57)}{PhCH\underset{OAc}{C}{}^*HPh_2}$$
(53)

$$\underset{(56)}{Ph_2CHCHPh^*}\overset{NH_2}{\underset{|}{}} \xrightarrow{HONO} Ph_2CH\overset{+}{C}HPh^* \rightleftharpoons Ph-\overset{+}{C}HCHPh_2{}^*$$

$$\underset{(59)}{PhCH-CHPh^*} \qquad \underset{(61)}{Ph_2CHCHPh^*\atop OAc} \qquad \underset{(58)}{PhCHCHPh_2{}^*\atop OAc}$$
(54)

instructive example is found in the deaminative reactions of 1,2,2-triphenyl derivatives (Collins and Bonner, 1955; Bonner and Collins, 1956). The ^{14}C chain-labeled (**55**) and ring-labeled (**56**) isotope position isomers of 1,2,2-triphenylethyl amine were prepared and subjected to deamination [Eqs. (53) and (54)]. The yields of the rearranged C-1 labeled product (**57**) and the phenyl-labeled product (**58**) were compared. Product **58** was always found in greater yield than **57**. This excess of **58** over **57** was well described by two limiting equations which had been derived using only classic open carbonium ions (instead of bridged ions, such as **59**). It is now recognized that bridging by an adjacent phenyl group to a carbonium ion center is unlikely to occur when that carbonium ion is tertiary or benzylic. Thus, the product analyses seem in agreement. Next, the deamination of optically active 1,2,2-triphenylethyl amine was undertaken (Collins *et al.*, 1959, 1961), and it was shown that 1,2,2-triphenylethyl acetate (**60** or **61**) was produced with 70% retention of configuration. When retention of optical configuration is observed in S_N1 reactions, it is usually associated with bridged ions or some form of neighboring group participation. On the other hand, the formation of open classic ions from optically active centers results in racemization or inversion (see Section II,B). Thus, the double-labeling experiments and the stereochemical experiments seemed at first to be contradictory, especially since the stoichiometry of the double-labeling experiments was not compatible with a significant fraction of bridged ion formation.

To resolve this dilemma, optically active (+)-1,2,2-triphenylethyl amine (**62**), which was ^{14}C labeled in the 1-position, was prepared and deaminated [Eq. (55)] to give four products {[(+)-**63**], [(−)-**63**], [(+)-**63′**], and [(−)-**63′**]},

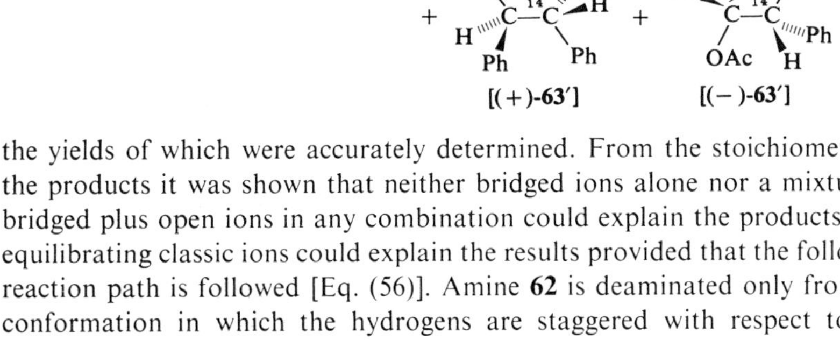

the yields of which were accurately determined. From the stoichiometry of the products it was shown that neither bridged ions alone nor a mixture of bridged plus open ions in any combination could explain the products. Only equilibrating classic ions could explain the results provided that the following reaction path is followed [Eq. (56)]. Amine **62** is deaminated only from the conformation in which the hydrogens are staggered with respect to one another. Second, the phenyls on adjacent carbons in ions **64**, **65**, and **66** tend to remain as far apart as possible. Third, only those rotations in which phenyls and hydrogens eclipse one another occur. Finally, ion **64** reacts with acetic acid preferentially only from one direction.

The predominant retention of configuration is then possible if the k_r/k_ϕ ratio is not too large. Using this reaction mechanism in its complete detail, Collins derived all the equations describing the process. Substituting the observed product yields into these equations (see Collins, 1968) allowed the ratios $k_r/k_\phi = 1.81$ and $k_2/k_\phi = 1.31$ to be calculated. Thus, there seems little doubt that open ions are involved in the deaminations.

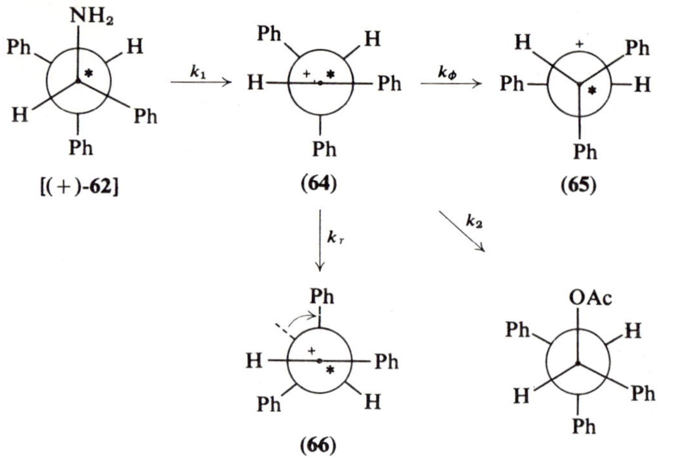

(56)

The use of ^{14}C-labeled cyclopropylcarbinyl amines in the study of carbon scrambling in cyclopropylcarbinyl cationic rearrangements represents another excellent example of the use of radioactive labels in studying carbonium reactions. This experiment is treated in the section on bridged ions (see Section VI,C,2).

III. Methods of Formation

Any attempt to comprehensively treat the entire range of methods of forming carbonium ions would require a chapter in itself. Thus, only a brief survey of some of the major methods will be presented.

A. Electron Removal from an Electrically Neutral Molecule or Radical

The removal of an electron from a neutral molecule is the normal mode of forming the molecular radical cation in mass spectroscopy. For example,

4. CARBONIUM IONS

under the impact of an electron beam the radical cation of 2,2-dimethyl butane can be formed. This in turn fragments readily to the *tert*-butyl cation and the ethyl radical. The presence of the unpaired electron sets radical

$$\underset{\underset{CH_3}{|}}{\overset{\overset{CH_3}{|}}{CH_3-C-CH_2CH_3}} + e^- \longrightarrow \left[\underset{\underset{CH_3}{|}}{\overset{\overset{CH_3}{|}}{CH_3-C-CH_2CH_3}} \right]^{+\cdot} + 2\,e^- \quad (57)$$

$$\downarrow$$

$$(CH_3)_3C^+ + \cdot CH_2CH_3$$

cations apart from the diamagnetic carbonium ions considered thus far. Since radical cations are discussed in Chapter 6, no further treatment is presented here.

The loss of an electron by a free radical results in a diamagnetic carbonium ion. Electron impact in the mass spectrometer has been used to generate carbonium ions from radicals (see Section IV,B,2). This method is not a preparative technique but has been used to provide information on the energetics of carbonium ion formation. In solution, many electrochemical studies (see Section II,D,2) have generated carbonium ions by the one-electron oxidation of free radicals. This is not a frequently used preparative technique since the radicals are often more difficult to generate than carbonium ions. Radicals can be oxidized by other routes. The chemical oxidation of triphenylmethyl radicals to cations with $KMnO_4$ in acetone (Schlenk, 1912) or with nitrobenzene in the absence of air (Goldschmidt and Christmann, 1925) have been claimed. One important preparative reaction which involves the oxidation of radicals to cations is the Kolbe reaction. Anodic oxidation of carboxylate anions leads to carbonium ions by one of two mechanisms (see Eqs. (58) (Corey *et al.*, 1960) and (59) (Koehl, 1964).

$$RCO_2^- \longrightarrow RCO_2\cdot + e^-$$
$$RCO_2\cdot \longrightarrow R\cdot + CO_2 \quad (58)$$
$$R\cdot \longrightarrow R^+ + e^-$$

$$RCO_2^- \longrightarrow RCO_2\cdot \longrightarrow RCO_2^+ \quad (59)$$
$$RCO_2^+ \longrightarrow R^+ + CO_2$$

Similar oxidations of radicals to carbonium ions appear to occur in the lead tetraacetate reaction with carboxylate ions (Corey and Casanova, 1963) and the decomposition of peresters and alkyl hydroperoxides in the presence of cupric ions (Kochi, 1962a,b; Jenkins and Kochi, 1972). Reactions of transition metal salts with radicals generated *in situ* are becoming increasingly important. Electron-transfer oxidations of alkyl radicals by copper(II) acetate, which proceeds via a metastable alkyl copper species, are capable of generating highly unencumbered carbonium ions. Replacing the acetate ligand by the weakly coordinating perchlorate or trifluoromethane sulfonate ligands gives a dramatic improvement in oxidative solvolysis, and even primary radicals such as *n*-butyl and methyl radicals are converted to carbonium ions, resembling those generated by solvolysis in highly ionizing and poorly nucleophilic solvents (Jenkins and Kochi, 1972). Table XV summarizes some representative results for the *n*-butyl system in Eqs. (60) and (61).

$$n\text{-Bu}-\text{CO}_2\text{O}_2\text{C}-n\text{-Bu} + \text{Cu(II)} \longrightarrow n\text{-BuCO}_2\text{Cu(II)} + n\text{-Bu}\cdot + \text{CO}_2 \quad (60)$$

$$n\text{-Bu}\cdot + \text{Cu(II)} \longrightarrow n\text{-Bu}^+ + \text{Cu(I)} \quad (61)$$

Table XV
Reaction of *n*-Butyl Radicals with Copper(II) Salts

Cu(II) Y$_2$	CO$_2$	⌐⌐	⌐⌐	⌐⌐	⌐⌐	OAc (OH)[b]	OAc (OH)[b]
Cu(ClO$_4$)$_2$[a]	93	0	22	7.7	4.4	25	23
Cu(O$_3$S—CF$_3$)$_2$[a]	93	0	24	6.5	3.6	36	19
Cu(ClO$_4$)$_2$[b]	91	0	42.5	—	—	27.0	14.0
Cu(O$_3$S—CF$_3$)$_2$[b]	100	0	54.0	—	—	32.8	11.0

[a] In HOAc.
[b] In H$_2$SO$_4$.

B. Heterolytic Fission of a Larger Molecule

This is the most common method of generating carbonium ions. The solvolysis reactions of alkyl, aryl, and aralkyl halides, tosylates, brosylates, and other derivatives [Eq. (62)] have been dealt with extensively in Sections II,B and II,E. Thus, these processes will not be greatly elaborated on here. As the anion becomes a weaker base, its reactivity as a leaving group im-

$$\text{R}-\text{X} \longrightarrow \text{R}^+ + \text{X}^-$$
$$\text{X} = \text{F, Cl, Br, I, OTs, OBs, OAc, }p\text{NB, CF}_3\text{COO}^-, \text{CF}_3\text{SO}_3^- \quad (62)$$

proves. Perhaps the best leaving group is the trifluoromethyl sulfonate (triflate) group which is the anion of the very strong acid CF$_3$SO$_3$H. Triflates are known to be at least 10^4 to 10^5 times more reactive under solvolytic conditions than the arenesulfonates (Hansen, 1965). Thus, triflates are superior leaving groups in the S$_N$1 generation of vinyl cations [e.g., Eq. (63)

$$\underset{\underset{\text{C}_6\text{H}_5}{|}}{\overset{\overset{\text{C}_6\text{H}_5}{|}}{\text{C}}}=\underset{\underset{\text{C}_6\text{H}_5}{|}}{\overset{\overset{\text{OSO}_2\text{CF}_3}{|}}{\text{C}}} \longrightarrow \underset{\underset{\text{C}_6\text{H}_5}{|}}{\overset{\overset{\text{C}_6\text{H}_5}{|}}{\text{C}}}=\overset{+}{\text{C}}-\text{C}_6\text{H}_5 + \text{CF}_3\text{SO}_3^- \quad (63)$$

4. CARBONIUM IONS

(Jones and Maness, 1969)]. The reactions of halides are greatly facilitated by the use of Ag^+ to assist (to "pull") the halide's departure [Eq. (64)]. Al-

$$R-X + Ag^+ \longrightarrow R^+ + AgX \tag{64}$$

ternatively, Lewis acids such as $AlCl_3$, SbF_6, BF_3, $SbCl_5$, $FeCl_3$, $SbCl_3$, and PCl_3 have been extensively used to assist in the removal of halides and to remove the halide ion from the system [Eq. (65)] (see Olah and Olah, 1970).

$$R-X + SbF_5 \longrightarrow R^+ + SbF_5X^- \tag{65}$$

The decomposition of onium ions constitutes a very important class of carbonium ion generating reactions. For example, the protonation of an alcohol or ester generates an oxonium ion which becomes a good leaving group [Eq. (66)]. High acidity and low water activity promote this reaction.

$$R-OR' + H^+ \longrightarrow R-\overset{+}{O}\!\!\begin{smallmatrix}H\\R'\end{smallmatrix} \longrightarrow R^+ + R'OH \tag{66}$$

R = alkyl, acyl; R' = alkyl, aryl, H

This class of reactions has been thoroughly treated elsewhere in this chapter. The loss of OR' from ethers is promoted by protonation [Eq. (67)] and the use of BF_3 where the formation of B–O bonds favor the reaction (Meerwein

$$R-O-R' \xrightarrow{H^+} R\overset{\overset{H}{|}}{\underset{}{O^+}}R' \longrightarrow R^+ + HOR' \tag{67}$$

$$3\,R-O-Et + 4\,BF_3 \longrightarrow 3\,R^+ + 3\,BF_4^- + B(OEt)_3 \tag{68}$$

et al., 1960a) shown in Eq. (68). Similarly, thiols are protonated and lose H_2S to generate carbonium ions (Olah et al., 1967), whereas protonated sulfides are cleaved to carbonium ions in strongly acid media [Eqs. (69) and (70)]. Sulfonium and ammonium ions are excellent precursors of carbocations by

$$CH_3CH_2\underset{\underset{CH_3}{|}}{\overset{\overset{CH_3}{|}}{C}}-SH \xrightarrow{H^+} CH_3-CH_2-\underset{\underset{CH_3}{|}}{\overset{\overset{CH_3}{|}}{C}}-\overset{+}{S}\!\!\begin{smallmatrix}H\\H\end{smallmatrix} \xrightarrow{+H^+} CH_3CH_2\overset{+}{C}\!\!\begin{smallmatrix}CH_3\\CH_3\end{smallmatrix} + H_3S^+ \tag{69}$$

$$CH_3-\underset{\underset{CH_3}{|}}{\overset{\overset{CH_3}{|}}{C}}-S-CH_3 \xrightarrow{H^+} CH_3-\underset{\underset{CH_3}{|}}{\overset{\overset{CH_3}{|}}{C}}-\overset{+}{S}\!\!\begin{smallmatrix}H\\CH_3\end{smallmatrix} \xrightarrow{+H^+} CH_3-\underset{\underset{CH_3}{|}}{\overset{\overset{CH_3}{|}}{C^+}} + CH_3\overset{+}{S}H_2 \tag{70}$$

virtue of the stability of the leaving groups and the low stability of these precursors [Eqs. (71) and (72)]. Since sulfonium and ammonium ions are

$$R-\overset{+}{\underset{R'}{S}}\overset{R'}{} \longrightarrow R^+ + R'_2S \tag{71}$$

$$R-\overset{+}{\underset{R'}{N}}-R' \longrightarrow R^+ + R'_3N \tag{72}$$

available from sulfides and amines, respectively, this represents a large potential class of cation precursors.

Olah et al. (1970d) demonstrated that alkyl sulfites undergo protonation followed by cleavage to protonated alcohols, sulfur dioxide, and alkyl carbonium ions in powerfully acidic media [Eq. (73)]. Alkyl sulfates decompose similarly, except H_2SO_4 (not SO_2) is the by-product [Eq. (74)].

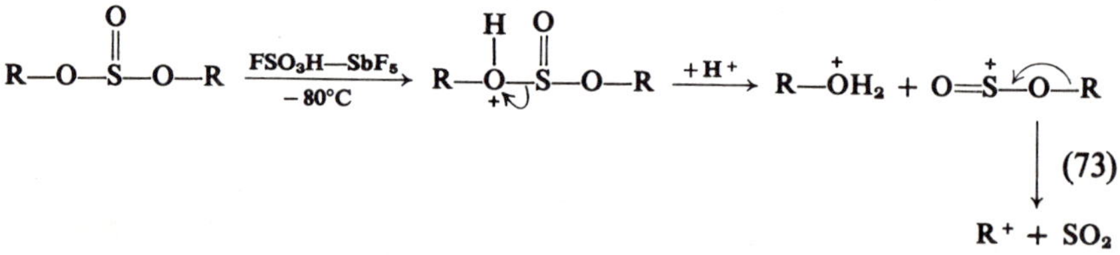

$$R-O-\overset{O}{\underset{\|}{S}}-O-R \xrightarrow[-80°C]{FSO_3H-SbF_5} R-\overset{H}{\underset{+}{O}}-\overset{O}{\underset{\|}{S}}-O-R \xrightarrow{+H^+} R-\overset{+}{O}H_2 + O=\overset{+}{S}-O-R \tag{73}$$

$$\downarrow$$

$$R^+ + SO_2$$

$$\underset{R-O}{\overset{R-O}{\diagdown}}\overset{O}{\underset{O}{\diagup}}S \xrightarrow[-80°C]{FSO_3H-SbF_5} \underset{R-O}{\overset{R-\overset{H}{O^+}}{\diagdown}}\overset{O}{\underset{O}{\diagup}}S \longrightarrow R^+ + \underset{R-O}{\overset{HO}{\diagdown}}\overset{O}{\underset{O}{\diagup}}S \tag{74}$$

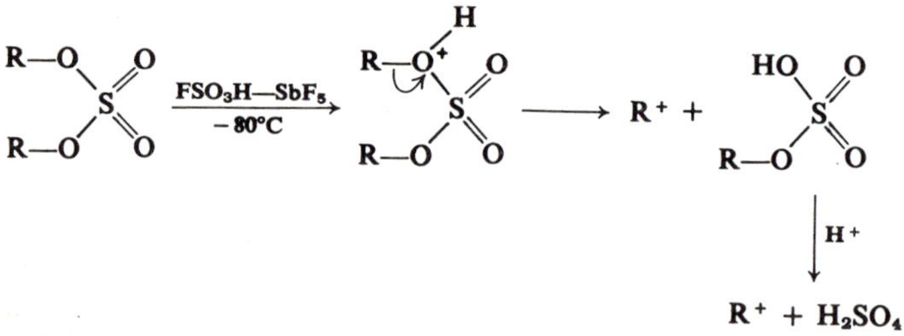

$$\downarrow H^+$$

$$R^+ + H_2SO_4$$

Chlorosulfinates decompose on treatment with Lewis acids to generate cations via Eq. (75).

$$R-O-\overset{O}{\underset{\|}{S}}-Cl + SbF_5 \longrightarrow R^+ + SO_2 + SbF_5Cl^- \tag{75}$$

Tertiary and secondary nitroalkanes cleave after protonation in $FSO_3H-SbF_5-SO_2$ at $-80°C$ to give a carbonium ion NO^+, and H_3O^+. For example, nitrocyclohexane gives the 1-methylcyclopentenyl cation [Eq.

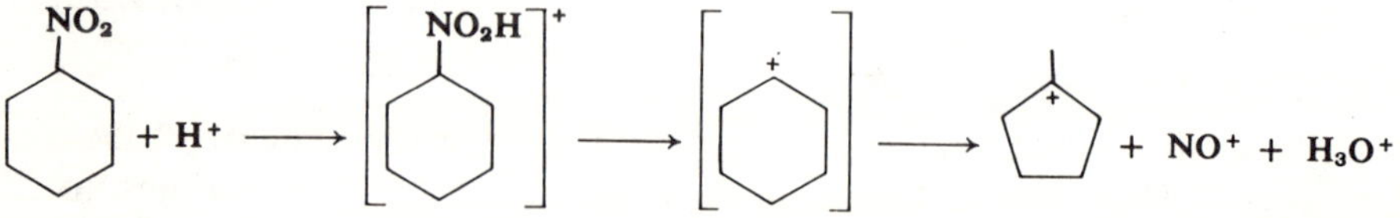

$$\underset{NO_2}{\bigcirc} + H^+ \longrightarrow \left[\underset{NO_2H}{\bigcirc}\right]^+ \longrightarrow \left[\overset{+}{\bigcirc}\right] \longrightarrow \overset{+}{\bigtriangleup} + NO^+ + H_3O^+ \tag{76}$$

(76)]. Primary nitroalkanes must be warmed before appreciable cleavage occurs (Olah and Kiovsky, 1967).

Diazonium ions, with the exception of aryl examples, readily undergo the loss of a nitrogen molecule to generate carbonium ions [Eq. (77)]. This topic is covered in Sections II,F and IV,C,3 and will not be elaborated on here. Two excellent reviews of this area are available (Moss, 1971; Friedman, 1970).

$$R-NH_2 \xrightarrow{HNO_2} R-\overset{+}{N}\equiv N \longrightarrow R^+ + N_2 \qquad (77)$$

Several reactions are functionally related to deamination via diazonium ions. For example, amines can be converted to amides which can then be nitrosated followed by decomposition of the nitrosoamide in suitable solvents [Eq. (78)]. In the rate-determining step, the nitrosoamide rearranges to an alkyl diazo-

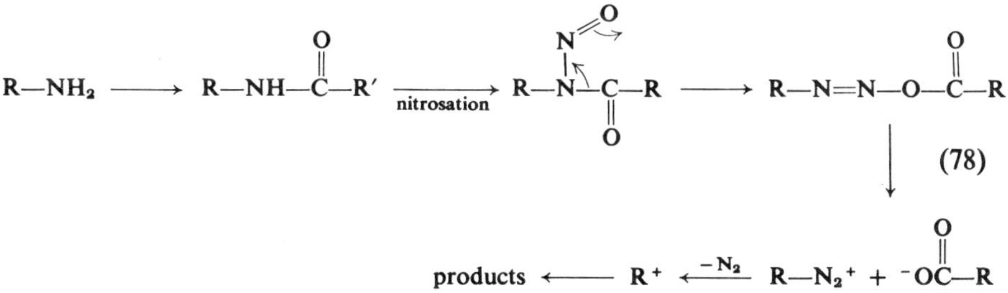

$$R-NH_2 \longrightarrow R-NH-\overset{O}{\underset{\|}{C}}-R' \xrightarrow{\text{nitrosation}} R-\overset{N\overset{O}{\nearrow}}{\underset{\|}{N}}\overset{}{\underset{O}{C}}-R \longrightarrow R-N=N-O-\overset{O}{\underset{\|}{C}}-R$$

$$\text{products} \longleftarrow R^+ \xleftarrow{-N_2} R-N_2{}^+ + {}^-O\overset{O}{\underset{\|}{C}}-R \qquad (78)$$

ester which decomposes to an ion pair with subsequent nitrogen loss leading to the cation. Similarly, the decomposition of N-nitrourethanes appears to proceed via carbonium ions (White and Woodcock, 1968). Aryl diazonium ions react with amines to give triazenes. Triazenes are unstable in acid solution and undergo tautomerization followed by loss of nitrogen to give ion pairs [Eq. (79)]. Diazoalkanes also appear to be a source of carbonium ions.

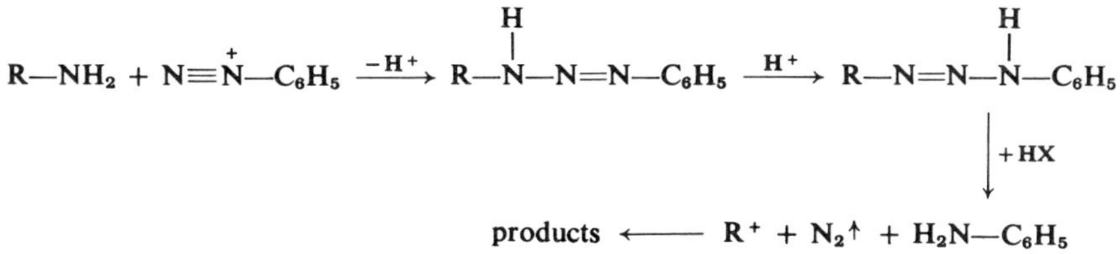

$$R-NH_2 + N\equiv\overset{+}{N}-C_6H_5 \xrightarrow{-H^+} R-\overset{H}{\underset{|}{N}}-N=N-C_6H_5 \xrightarrow{H^+} R-N=N-\overset{H}{\underset{|}{N}}-C_6H_5$$

$$\downarrow +HX$$

$$\text{products} \longleftarrow R^+ + N_2\uparrow + H_2N-C_6H_5$$

$$(79)$$

A diazoalkane is the conjugate base of an alkyl diazonium ion; thus, on protonation, diazoalkanes undergo decompositions that resemble nitrous acid deaminations [Eq. (80)]. Finally, deaminations via alkane diazotate

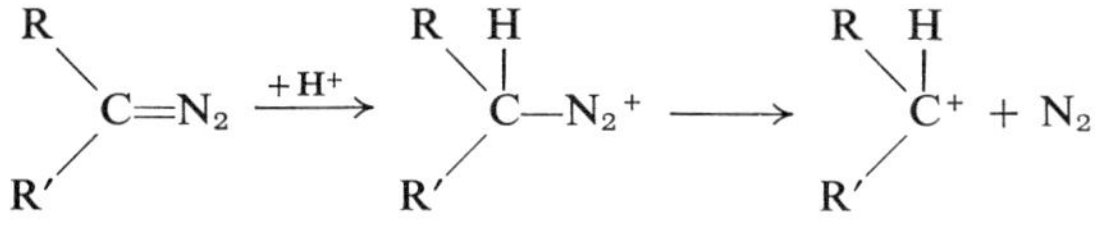

$$\underset{R'}{\overset{R}{\diagdown}}C=N_2 \xrightarrow{+H^+} \underset{R'}{\overset{R\ H}{\diagdown}}C-N_2{}^+ \longrightarrow \underset{R'}{\overset{R\ H}{\diagdown}}C^+ + N_2 \qquad (80)$$

salts are known to lead to carbonium ions. Diazotates are conjugate bases of diazotic acids (which are thought to be intermediates in the nitrous acid deaminations of amines), and protonation should lead to deamination (Moss, 1971) as shown in Eq. (81).

$$R\!-\!N\!=\!N\!-\!O^-K^+ \xrightarrow{H^+} [R\!-\!N\!=\!N\!-\!OH] \xrightarrow{-N_2} R^+ \longrightarrow \text{products} \qquad (81)$$

The reaction of alkyl azides with nitrosonium ion provides a convenient route to cations without concomitant formation of by-products which would interfere with this process [Eq. (82)] (Doyle and Wierenga, 1970). In general,

$$R\!-\!N_3 + NO^+SbF_6^- \longrightarrow \begin{bmatrix} R\!-\!N\!=\!N\!-\!N\!=\!\overset{+}{N}\!=\!O \xrightarrow{-N_2O} R\!-\!N_2{}^+ \\ \text{or} \\ R\!-\!N\!\begin{smallmatrix}\overset{+}{N}\!\equiv\!N\\ \\ N\!=\!O\end{smallmatrix} \xrightarrow{-N_2} R\!-\!\overset{..}{N}\!=\!\overset{+}{N}\!=\!O \end{bmatrix} \begin{smallmatrix}-N_2\\ \searrow\\ \nearrow\\ -N_2O\end{smallmatrix} R^+$$

(82)

N-aryl derivatives of benzylideneimine react with nitrosonium salts to give aryl diazonium ions, whereas N-alkyl derivatives produce carbonium ions, nitrogen, and benzaldehyde [Eq. (83)].

$$Ph\!-\!CH\!=\!NR + NO^+X^- \longrightarrow PhCHO + N_2 + R^+X^- \qquad (83)$$

Recently, Olah *et al.* (1966c) reported that alkyl and acyl sulfinyl amines [Eq. (84)], isocyanates [Eq. (85)], and isothiocyanates [Eq. (86)] react with nitrosonium salts to produce carbonium and oxocarbonium ions. The

$$R\!-\!N\!=\!S\!=\!O + NO^+SbF_6^- \longrightarrow R^+SbF_6^- + N_2\uparrow + SO_2\uparrow \qquad (84)$$

$$R\!-\!N\!=\!C\!=\!O + NO^+SbF_6^- \longrightarrow R^+SbF_6^- + N_2\uparrow + CO_2\uparrow \qquad (85)$$

$$R\!-\!N\!=\!C\!=\!S + NO^+SbF_6^- \longrightarrow R^+SbF_6^- + N_2\uparrow + COS\uparrow \qquad (86)$$

driving force in these three reactions is the great stability of the molecules generated on fragmentation. The products, other than the carbonium ion

(87)

salt, are gaseous and escape, driving the reaction to completion. These reactions are related to the decarbonylation of oxocarbonium ions. Whenever an oxocarbonium ion can lose CO to generate a more stable carbonium ion [e.g., Eq. (87) (Dewar and Ganellin, 1959)], this reaction may be used as a source of carbonium ions from the corresponding acids or acid halides. Since CO escapes, this reaction is not subject to the usual equilibrium consideration. However, this reaction is reversible in the presence of added CO and can be used as a method to form carboxylic acids from cations (Koch and Haaf, 1960; Pincock et al., 1959; Stork and Bersohn, 1960).

Another class of heterolytic fission reactions is hydride transfer. For example, one cation can remove a hydride to generate a second cation [Eq. (88)] (Dauben et al., 1957). Many oxidizing agents, such as NO^+ (Olah and

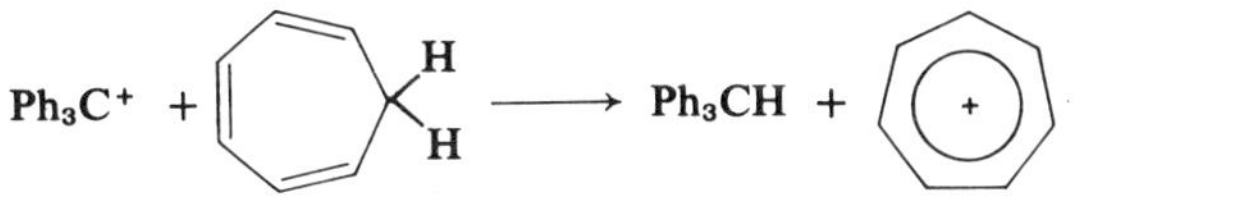

$$Ph_3C^+ + \bigcirc\!\!\!<^H_H \longrightarrow Ph_3CH + \bigcirc^+ \quad (88)$$

Friedman, 1966), bromine and acetic acid-phosphoric oxide (Deno et al., 1962b), molecular oxygen in H_2SO_4 (Deno et al., 1962a), and chromic acid (Necsoiu and Nenitzescu, 1960) are able to remove hydride ions from activated hydrocarbon substrates.

The decomposition of mercury derivatives can result in the formation of carbonium ions. Equation (89) portrays in simplified fashion the decomposition of an organomercuric acetate to a carbonium ion (Jensen and Rickborn, 1968). Finally, the enzyme-induced carbonium ion-type polymerization of

$$R\text{---}Hg^+ \; {}^-OAc \longrightarrow R^+ \; {}^-OAc + Hg \quad (89)$$

isopentenyl pyrophosphate to geranyl pyrophosphate [Eq. (90)] is basic to the construction of the entire terpene family of compounds. In this reaction the pyrophosphate may be thought of as a good leaving group.

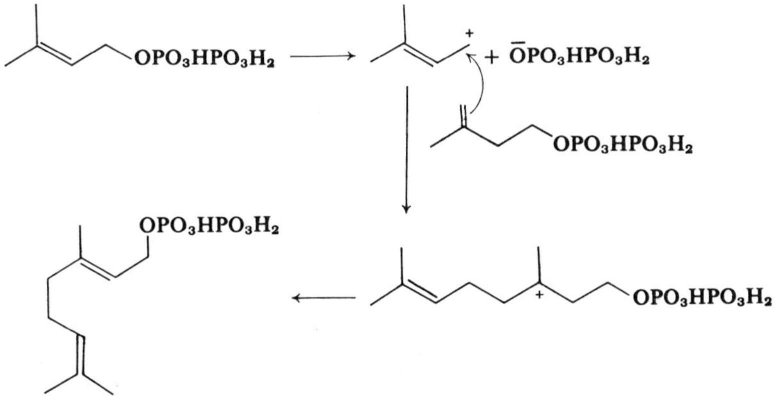

(90)

Geranyl pyrophosphate

C. Addition of a Cation to a Neutral Species

Carbonium ions are generated by the addition of a proton to an olefin or alkyne [Eqs. (91) and (92)] in a manner which produces the more stable ion. These reactions are usually equilibria except in the case of very strongly

$$\underset{R}{\overset{R}{\diagdown}}C=CR_2' \underset{-H^+}{\overset{H^+}{\rightleftarrows}} \underset{R}{\overset{R}{\diagdown}}\overset{+}{C}-CR_2'H \tag{91}$$

$$R-C\equiv C-R' \underset{-H^+}{\overset{H^+}{\rightleftarrows}} R-\overset{+}{C}=C\underset{R'}{\overset{H}{\diagup}} \tag{92}$$

stabilizing R groups or in super-acid media. The protonation of aromatic systems is an extension of this reaction. For example, azulene is easily protonated on the five-membered ring (Plattner *et al.*, 1950). During gas-

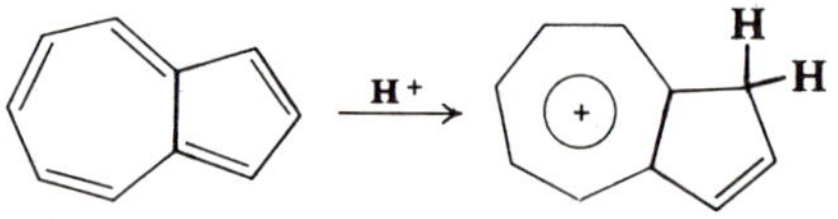

$$\text{[azulene]} \xrightarrow{H^+} \text{[protonated azulene]} \tag{93}$$

phase γ radiolysis of organic compounds in the presence of H_2, CH_4, or the rare gases, highly acidic species such as H_3^+ or CH_5^+ are formed which transfer protons to olefins, cyclopropane, and cyclobutane (Ausloos and Lias, 1965).

Cation transfer agents are useful in the synthesis of carbonium ions. For example, CH_3Cl—$AlCl_3$ is capable of transferring CH_3^+ to hexamethyl benzene to produce the heptamethylcyclohexadienyl cation (Doering *et al.*, 1958). Similarly, $CH_3F + BF_3$ can transfer CH_3^+ to dimethylether, giving the trimethyloxonium ion (Meerwein *et al.*, 1939; Perst, 1971). The rearrangement of one ion to another by cyclizations, hydride shifts, alkyl or aryl shifts,

$$\text{[hexamethylbenzene]} \xrightarrow{\underset{AlCl_3}{CH_3Cl}} \text{[heptamethylcyclohexadienyl cation]} \tag{94}$$

$$\underset{R}{\overset{R}{\diagdown}}O + CH_3F + BF_3 \longrightarrow \underset{R}{\overset{R}{\diagdown}}\overset{+}{O}-CH_3 + BF_4^- \tag{95}$$

and generation of cations by π and σ routes are considered elsewhere but represent intramolecular additions of cations to neutral species.

In a series of recent papers, Olah (see Olah and Olah, 1971, and references therein) has argued that unequivocal experimental evidence has been obtained for penta- and tetracoordinated carbonium ions of which CH_5^+ is the parent member. The protonation of alkanes, such as methane [Eq. (96)], is envisioned as taking place in super-acid media (SbF_5—FSO_3H) to give CH_5^+. The bonding is considered to involve three, two-electron covalent bonds with the fourth bond being a two-electron three-center bond. The use of structure **67**

$$CH_4 + H^+ \rightleftharpoons \underset{(68)}{\text{H}_2\text{C}\overset{+}{\underset{\text{H}}{\langle}}\text{H}_2} \equiv \underset{(67)}{H_3C\overset{+}{\cdots}\overset{H}{\underset{H}{\langle}}} \quad (96)$$

two electron
three centered
bond

has been advanced to best portray structure **68**, and the similarity of this structure to the norbornyl cation has been pointed out. The attacking proton is envisioned as adding to the main lobe of one of the C–H bonds at the point of high electron density and not from the "relatively unimportant back lobes." Thus, Olah attributes a variety of electrophilic aliphatic substitution reactions of alkanes and cycloalkanes to involving carbonium ion analogs of CH_5^+. The key to these reactions is the ability of a two-electron single bond to share their electron pairs with a strong electrophile (H^+, NO_2^+, or R^+). Both C–H and C–C bonds are σ donors to such powerful electrophiles (provided no other nucleophilic center is available). Using these concepts, the generation of CH_3^+, the incorporation of D into H_2, the protolysis reactions of hydrocarbons, the alkylation, and the nitration of alkanes [Eqs. (97–102)] have been explained. This concept of generating pentacoordinated carbonium ions

$$CH_4 + H^+ \rightleftharpoons \left[CH_3\text{---}\overset{H}{\underset{H}{\langle}} \right]^+ \rightleftharpoons CH_3^+ + H_2 \uparrow \quad (97)$$

$$\underset{H}{\overset{H}{|}} + D^+ \rightleftharpoons \left[\underset{H'}{\overset{H}{\rangle}}\text{---}D \right]^+ \rightleftharpoons \underset{D}{\overset{H}{|}} + H^+ \quad (98)$$

$$\underset{\underset{CH_3}{|}}{\overset{CH_3}{\underset{|}{CH_3-C-CH_3}}} \xrightarrow[\text{C-H}]{\overset{\text{C-C}}{H^+}} \begin{array}{l} \left[(CH_3)_3C\text{---}\overset{H}{\underset{CH_3}{\langle}} \right]^+ \longrightarrow (CH_3)_3C^+ + CH_4 \\ \\ \left[(CH_3)_3C-CH_2\text{---}\overset{H}{\underset{H}{\langle}} \right]^+ \longrightarrow \underset{\underset{CH_3}{|}}{\overset{CH_3}{\underset{|}{CH_3-\overset{+}{C}-\overset{+}{C}H_2}}} \end{array} \quad (99)$$

$$\downarrow$$

$$\underset{\underset{CH_3}{|}}{\overset{}{CH_3-\overset{+}{C}-CH_2CH_3}} + H_2$$

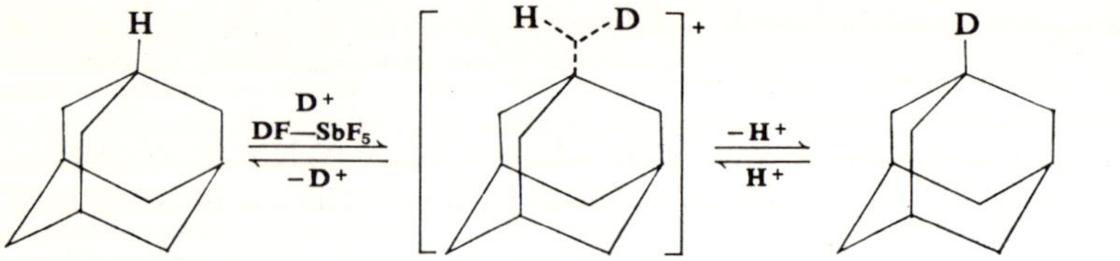

$$\text{[adamantane-H]} \xrightarrow[-D^+]{\underset{DF-SbF_5}{D^+}} \text{[adamantane-H···D]}^+ \xrightarrow[H^+]{-H^+} \text{[adamantane-D]} \quad (100)$$

$$(CH_3)_3CH + {}^+C(CH_3)_3 \xrightarrow[78°C]{SOClF} \left[(CH_3)_3C\cdots\overset{H}{\underset{C(CH_3)_3}{<}}\right]^+ \quad (101)$$
$$\searrow$$
$$(CH_3)_3C-C(CH_3)_3 + H^+$$

$$CH_4 \xrightarrow[\substack{CH_2Cl_2-\text{sulfolane}\\ \text{dark, }250°C}]{NO_2^+ PF_6^-} CH_3\cdots\overset{H}{\underset{NO_2}{<}} \longrightarrow CH_3NO_2 + H^+ \quad (102)$$

via electron sharing of single σ bonds with electrophilic reagents is a bold one and a major contribution to our understanding of a large class of reactions. More recent results from Olah's group include alkylation, chlorination, and nitration of single bonds (see Olah and Mo, 1972, and references therein).

D. Interaction of Neutral Molecules

The formation of carbonium ions by dissociative electron transfer is observed when tetranitromethane reacts with an olefin (Penczek *et al.*, 1968; Patterson, 1955). For example, the reaction of tetranitromethane with 1,1-diphenylethylene gives 1,1-diphenyl-2-nitroethylene and trinitromethane [Eq. (103)]. This reaction proceeds via an initial charge-transfer complex which reacts by donating NO_2^+ to 1,1-diphenylethylene.

$$Ph_2C=CH_2 + C(NO_2)_4 \rightleftharpoons \underset{\text{complex}}{\text{charge-transfer}} \longrightarrow$$
$$Ph_2C^+-CH_2NO_2 + \bar{C}(NO_2)_3 \longrightarrow Ph_2C=CHNO_2 + HC(NO_2)_3 \quad (103)$$

IV. Factors Affecting the Stability of Carbonium Ions

The "stability" of an ion is a relative term. In the gas phase the energy required to form an ion can often be determined quantitatively (see Section IV,B,2). In solution, one measures a change in energy of a reaction which generates an ion. Thus, the assigned ΔH_f or ΔG_f of an ion is assigned as relative to an arbitrary standard. Thus, carbonium ion stabilities are often defined relative to that ion's precursor. Second, in solution many ions which are thermodynamically stable are chemically quite unstable and react to form

4. CARBONIUM IONS

other products in a particular medium. The medium containing the carbonium ion is then of primary importance in considering the stability of an ion.

A. Medium Effects; Kinetic and Thermodynamic Considerations

When an ion is taken from the gas phase into a solution, energy is evolved due to solvation. This is conveniently represented by the free energy cycle in Eq. (104) where $\Delta G^0 = \Delta G_v^0 + \Delta G_g^0 + \Delta G_{solv}$. Heats for the formation of

$$RY_{(g)} \xrightarrow{\Delta G_g^0} R^+_{(g)} Y^-_{(g)}$$
$$\Delta G_v^0 \uparrow \qquad \qquad \downarrow \Delta G_{solv} \qquad (104)$$
$$R-Y \xrightarrow{\Delta G^0} R^+_{(s)} + Y^-_{(s)}$$

carbonium ions are now being obtained using solution calorimetry (Arnett and Larsen, 1968). In this work the heat of ionization of the reaction

$$RY \xrightarrow[HA]{} R^+ + HAY^-$$

is obtained indirectly from the ΔH_s of RY in CCl_4, and the ΔH of reaction measured when RY in CCl_4 is stirred with FSO_3H—SbF_5 to generate R^+. Several assumptions are made, however. At present, no completely satisfactory method is available to accurately determine solvation energies. All methods which have been used to date (Bernal and Fowler, 1933; Evans, 1946; Latimer et al., 1939; Franklin, 1952; Stokes, 1964) rest on the Born equation for the energy required to charge a spherical dielectric [Eq. (105)], where e is the charge on the electron, D is the solvent dielectric constant, r is

$$-\Delta G = (e^2/2r)(1 - 1/D) \qquad (105)$$

the cavity radius, and ΔG is the change in free energy on transferring the ion from the gas phase into the solution. By ignoring dielectric saturation effects about the ion and the free energy change owing to reorganization of the solvent structure by the ion (ΔG_{c^+}), comparisons of the free energies of solvation (ΔG_{solv}) for a series of ions of similar size and gross structures can

$$\Delta G_{solv} = -(e^2/2r)(1 - 1/D) + \Delta G_{c^+} \qquad (106)$$

be made. It is usually thought that the major factor determining the solvation energy is electrostatic. However, a hole must be created in the solvent, and solvent shells are formed about the ion. Thus, it might be expected that large contributions owing to entropy changes will occur. In fact, those solvents with very high dielectric constants often have a high order which is perturbed when a hole is generated. This is followed by extensive ordering of a solvent shell. In nonpolar solvents which have little order to start with, the introduction of a carbonium ion might cause a large increase in order in the region close to the ion. The net effects cannot be predicted accurately. Usually, the smaller the ion (R is smaller) the greater ΔG_{solv} becomes. Thus, ΔG_{solv} should be much larger for the isopropyl cation than for the triphenylcarbonium ion or the tri-*tert*-butyl carbonium ion.

$$CH_3\overset{+}{C}HCH_3 \qquad Ph_3C^+ \qquad \overset{X}{\underset{+}{\overset{+}{C}}}X$$

(67)   (18)   (68)

The solvation of a carbonium ion also depends on steric hindrance. In the tri-*tert*-butyl carbonium ion, it would be more difficult for solvent molecules (especially large ones) to effectively solvate the cationic center. Since solvation would also depend on the specific conformation of the molecule, entropy effects would be different in **18, 67,** and **68** on a steric basis alone. Furthermore, the distinction between electrostatic and partial covalent solvation is not always clear. The "vacant" *p* orbital of the carbonium ion could overlap with filled orbitals of the solvent molecules (see **69**) (Streitwieser,

(69)

1962). Another complication is the counterion. Its effect on the "stability" of the cation varies with (a) the structure of the anion, (b) the solvent, and (c) whether a tight ion pair, solvent-separated ion, or free ions are present.

A very useful qualitative theory of solvent effects has been systematized by Ingold (see Ingold, 1969). Founded on the assumption that charge–solvent interactions are the dominant influences on the rate (via $\Delta H^{\ddagger}$), he concludes: (1) When charge is created going from the reactants to the transition state, the rate of the reaction is sharply enhanced by increases in solvent polarity. (2) When charge is only dispersed during this process, a slight decrease in rate occurs when solvent polarity is raised. (3) When charge is destroyed, large rate decreases occur with increases in the solvent polarity. For example, in an S_N1 reaction, where $RX \rightarrow R^+ + X^-$, charge is generated on going to the transition state, and rate increases of from 3 to 6 powers of ten are expected as the solvent is varied from ethanol to water. This principle is illustrated in Table XVI. As the solvent is varied from ethanol to water, the rate of the S_N1 solvolysis of *tert*-butyl chloride increases by a factor of 2.7×10^5. The values of $\Delta S^{\ddagger}$ vary only from -3.8 to -1.9 eu going from ethanol to formic acid in agreement with the $\Delta H^{\ddagger}$ effects dominating the solvent effect. Ingold's treatment exhibits its limitations in this reaction when H_2O is the solvent. The large $+12.2$ eu $\Delta S^{\ddagger}$ value in water demonstrates that $\Delta S^{\ddagger}$ can be very important. The change in "order" on going to the transition state is little effected by the nature of the solvent in ethanol through formic acid, whereas the substrate–solvent interaction ($\Delta H^\ddagger$) varies systematically. In water the huge entropy effect is due in part to the highly ordered water structure. Thus, the Ingold generalizations do not allow enough understanding to permit detailed depictions of mechanisms going via carbonium ions.

Table XVI

Rate Constants and Activation Parameters for the S_N1 Solvolysis of *tert*-Butyl Chloride at 25°C[a]

Solvent	Dielectric constant	Rate $10^7 K$ (sec^{-1})	$\Delta S^\ddagger$ (cal mole^{-1} deg^{-1})	$\Delta H^\ddagger$ (kcal mole^{-1})
Ethanol	24.3	1.0	−3.2	26.1
Acetic acid	6.2	2.2	−2.5	25.8
Methanol	32.6	7.2	−3.1	24.9
Formamide	109.5	3.7×10^2	−3.8	22.4
Formic acid	74	1.1×10^4	−1.9	21.0
Water	78.4	2.7×10^5	+12.2	23.2

[a] Winstein and Fainberg (1957).

Defining solvent polarity is also difficult as illustrated by Table XVI. Although the reaction does go faster in solvents with higher dielectric constants, there is no direct parallel (i.e., see ethanol versus acetic acid and formamide versus formic acid and water).

Winstein and Fainberg (1957) first demonstrated that a large portion of the rate enhancement in *tert*-butyl chloride solvolyses, going from methanol to water as the solvent, was due to the free energy of solution of *tert*-butyl chloride. Figure 8 illustrates this effect. The rate enhancement (7.2 to 2.7 ×

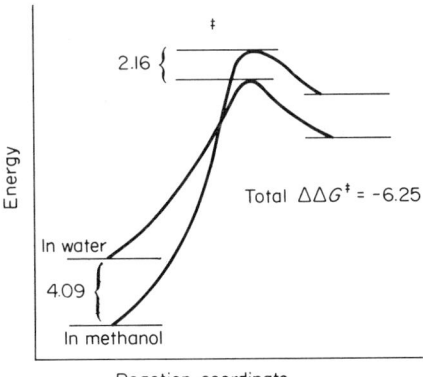

Fig. 8. Solvolyses of *tert*-butyl chloride in water.

10^5, Table XVI) corresponded to a difference of -6.25 kcal/mole in the free energies of activation between the two solvents (i.e., $\Delta\Delta G^{\ddagger} = -6.25$ kcal/mole). Only -2.16 kcal/mole was due to differences in solvation of the transition state, whereas -4.09 kcal/mole was due to differences in the starting solutions. The effect on the rate of interaction of the *tert*-butyl chloride with solvent, going from CH_3OH to H_2O, was larger and in an opposite direction to the respective transition state interactions. Arnett and co-workers (1965) concluded, from partial molal heat of solution measurements, that the importance of the initial state is due to the effect of the solute on the local structure of the solution.

Since the transition state of an S_N1 reaction resembles the carbonium ion formed, kinetic studies do provide a valid means of studying the effect of solvent medium on carbonium ion stability (see Section II,E). As we have seen, simple generalizations do not provide a detailed understanding of solvent effects. Grunwald and Winstein (1948) chose to use the solvolysis of *tert*-butyl chloride to generate an emperical definition or scale of solvent ionizing power. They defined the parameter Y as a measure of this ionizing power. The Y values were derived from the rates of S_N1 reactions of *tert*-butyl chloride in various solvents versus 80% EtOH(aq.) as a standard solvent, where Y was defined as

$$Y = \log k_{\textit{tert}\text{-butyl chloride in solvent}} - \log k_{\textit{tert}\text{-butyl chloride in 80% EtOH(aq.)}}$$

The more usual form of the Grunwald–Winstein relationship is

$$\log k/k_0 = mY \qquad k_0 = k_{\textit{tert}\text{-butyl chloride in 80% EtOH(aq.)}}$$

The more ionizing a solvent is, the faster the solvolysis will proceed regardless of whether the effect stems from the transition state solvation or initial state effects. Thus, Y increases as the value of $\log k - \log k_0$ increases. In other words, the larger Y is, the more ionizing the solvent is. The value of Y for 80% EtOH–20% H_2O is zero, and when Y is greater than zero the solvent is more ionizing than this arbitrary standard. Once the Y values for a series of solvents are determined, a plot of $\log k/k_0$ versus Y should give a straight line with a slope of m as long as the relationship holds. Thus, m is a measure of the sensitivity of a particularly alkyl halide solvolysis to solvent ionizing power. Where $m > 1$ that halide solvolysis is more solvent dependent than that of *tert*-butyl chloride. Table XVIII lists the Y values of a few sample solvents while Table XVII summarizes the m values for a few examples of halides. The failure of all points, from a given substrate in different solvent media, to fall on the same line illustrates the limitations of this classification of solvents.

In most organic solvents, kinetic techniques are used to study carbonium ion stability because the short-lived ions cannot be directly detected. In more

Table XVII
m Values for Alkyl Halide Solvolysis[a]

Compound	Solvent mixtures[b]	m
1-Phenylethyl chloride	10–100% EtOH—H_2O	0.966[c]
	50–100% MeOH—H_2O	0.912[c]
	0–100% CH_3COOH—HCOOH	1.194[c]
tert-Butyl bromide	40–100% EtOH—H_2O	0.941[d]
	1–100% CH_3COOH—HCOOH	0.946[d]
Benzhydryl chloride	80–100% EtOH—H_2O	0.740[e]

[a] At 25°C.
[b] Percentages by volume of first-named component.
[c] Winstein and Fainberg (1957).
[d] Fainberg and Winstein (1957).
[e] Winstein et al. (1957).

Table XVIII
Y Values of Selected Solvents

Solvent	Y
H_2O	3.493
$HCOOH/H_2O$, 50/50	2.546
HCOOH	2.054
$H-C(=O)NH_2$	0.604
CH_3OH	−1.090
CH_3OH/H_2O, 90/10	−0.301
CH_3OH/H_2O, 60/40	1.492
EtOH	−2.033
$EtOH/H_2O$, 90/10	−0.747
$EtOH/H_2O$, 60/40	1.124
$EtOH/H_2O$, 40/60	2.196
$Acetone/H_2O$	−1.856

strongly acidic media such as H_2SO_4, HF, HNO_3, or $HClO_4$, many ions are stable enough to be detected. Thus, their equilibria with their precursors may be observed. For example, the protonation equilibria of dienes **70** and **71** with carbonium ions **35** and **37**, respectively, can be compared. Diene **71** is half-protonated at the lower acidity. Thus, ion **37** is the more stable carbonium ion (relative to its starting diene). Since sulfuric acid may be continually

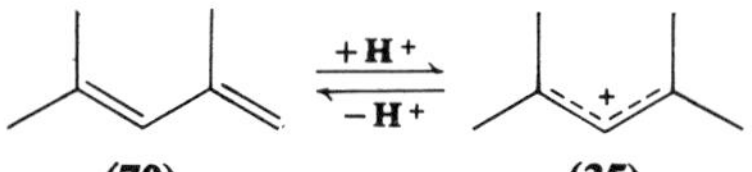

$$(70) \xrightleftharpoons[-H^+]{+H^+} (35) \quad \text{acid concentration for half-protonation } 73\% \quad (107)$$

			acid concentration	
	$\xrightarrow{+H^+}$		for half-protonation	(108)
	$\xleftarrow{-H^+}$		35%	
(71)		(37)		

diluted from concentrated fuming oleum solutions to pure water, it is convenient for the comparison of olefin basisities (carbonium ion stabilities) (for a review, see Liler, 1971).

The activity of an acid cannot be divorced from the reaction in which it participates. Thus, attempts have been made to put relative equilibrium measurements in strong acids on a true thermodynamic basis. Hammett's pioneering work in protonation equilibria (Hammett and Deyrup, 1932) led to the development of the H_0 acidity function from which the relative protonating powers of various acid media could be determined. Thus, for the reaction

$$B + H^+ \rightleftharpoons BH^+ \qquad K_{BH^+} = \frac{[B][H^+]f_B f_{H^+}}{[BH^+]f_{BH^+}} \qquad (109)$$

where f = activity coefficient. By showing that the ratio of the activity coefficients of the base and its conjugate acid (for a series of related bases) is independent of the exact nature of the base, Hammett proposed that the acidity of any strong acid medium should be defined by

$$h_0 = a_{H^+}\left(\frac{f_B}{f_{BH^+}}\right) = K_{BH^+}\left(\frac{[BH^+]}{[B]}\right) \qquad (110)$$

Expressing this as the negative logarithm, the H_0 acidity function is defined as

$$H_0 = pK_{BH^+} - \log\frac{[BH^+]}{[B]} = -\log h_0 \qquad (111)$$

where $[BH^+]$ and $[B]$ are directly measurable by spectroscopic means.

This relation is not applicable to the dehydration equilibria of alcohols to carbonium ions in acids. Deno et al. (1955) and Gold (1955) developed the H_R acidity function to describe these equilibria. For

$$ROH + H^+ \rightleftharpoons R^+ + H_2O$$

the equilibrium constant is

$$K_R = \frac{[ROH]f_{ROH}a_{H^+}}{[R^+]f_{R^+}a_{H_2O}} \qquad (112)$$

Assuming f_{ROH}/f_{R^+} will be about the same for a series of similar indicators, the acidity function H_R can be defined as follows:

$$h_R = \frac{a_{H^+}f_{ROH}}{a_{H_2O}f_{R^+}} = K_R\frac{[R^+]}{[ROH]} \qquad (113)$$

$$H_R = -\log h_R = pK_R - \log\frac{[R^+]}{[ROH]} \qquad (114)$$

Again, $[R^+]$ and $[ROH]$ are measurable. Table XIX lists the H_0 and H_R values of selected concentrations of H_2SO_4 from which one can calculate pK_{BH^+} or pK_{R^+} by making spectroscopic measurements of the position of reactions **109** or **112**. These definitions will be used in discussing substituent effects in the following sections. Many excellent reviews of the acidity function theory are available (Paul and Long, 1957; Arnett, 1963; Boyd, 1969; Rochester, 1970).

Table XIX
The H_0 and H_R Acidity Functions of Sulfuric Acid/Water Solutions

wt% H_2SO_4	$H_0{}^a$	$H_R{}^b$
1	+0.84	−0.92
10	−0.43	0.72
20	−1.10	1.92
30	−1.82	3.22
50	−3.41	6.60
60	−4.51	8.92
70	−5.92	11.52
80	−7.52	14.12
90	−9.03	16.72
98	−10.27	19.64

^{a} Ryabova *et al.* (1966).
^{b} Deno *et al.* (1955).

Fluorosulfonic acid (FSO_3H) is an even stronger protonating acid than H_2SO_4. Its Hammett acidity function (H_0) has a value of −13.9 as compared with −12.1 for sulfuric acid (Gillespie, 1968). Furthermore, its low freezing point (−88.98°C) has enabled it to be used to study carbonium ions which would be unstable at room temperature or even 0°C. In fluorosulfonic acid, even such strong acids as HF, HS_2O_6F, and $HClO_4$ cannot enhance its acidity. However, antimony pentafluoride is capable of increasing the acidity of HSO_3F. When SbF_5 or SO_3 and FSO_3H are mixed an even stronger acid medium is produced. These acid systems truly classify as "super acids." They are estimated to be from 10^3 to 10^5 times stronger than pure HSO_3F ($H_0 \simeq -16$ to -18) and protonate such weak bases as fluorobenzene (Olah and Kiovsky, 1967) and trinitrobenzene completely. They may be diluted with SO_2, $SOCl_2$, SOF_2, or $SOClF$ to give solutions of carbonium ions in which temperatures of −150°C may be reached without freezing. The enormous stability of carbonium ions in these media is unparalleled, as yet, in any other solvent systems. This is due to two factors. First, the huge protonating power prevents the equilibration of the ion and its precursor. Thus, second-order reactions between these two species cannot occur. Second, there are no bases of reasonable strength present. The strongest base might be $FSO_3{}^-$, HSO_3F, or some complex such as $(FSO_3-SbF_5)^-$ or $Sb_2F_{11}{}^-$. One might

$$\text{\Large /\!\!=\!\!\textbackslash} \xrightleftharpoons[\text{SOClF}]{\text{FSO}_3\text{H—SbF}_5} \text{\Large /\!\!\overset{+}{\cdot}\!\!=\!\!\textbackslash} \qquad (115)$$

suspect that covalent solvation of carbonium ions is reduced (thus a certain resemblance to the gas phase exists) but that electrostatic solvation might be considerable (the dielectric constants of these acid systems are not yet available).

B. Inductive and Resonance Effects

1. *In Solution*

As a carbonium ion becomes increasingly able to disperse its charge by resonance, it becomes increasingly stable. This effect is clearly illustrated in

Table XX

The Free Energies and pK_R Constants of Forming Carbonium Ions from Alcohols

No.		Cation	ΔG_R (kcal/mole)	pK_R+	Solvent
	72	Tri-(p-dimethylaminophenyl)methyl	−12.75	+9.36	H_2SO_4[a]
	73	Tri(p-aminophenyl)methyl	−10.3	+7.57	H_2SO_4[a]
	74	Tripropylcyclopropenyl	—	+7.2	50% CH_3CN/H_2O[b]
	75	Tri-p-anisylcyclopropenyl	—	+6.5	50% CH_3CN/H_2O[b]
	76	Triphenylcyclopropenyl	—	+3.1	50% CH_3CN/H_2O[b]
	77	Heptaphenyltropylium	—	+7	50% CH_3CN/H_2O[c]
	20	Tropylium	—	+4.2	H_2O[d]
	78	Benzotropylium	—	+1.7	CH_3COOH[e]
	79	Tri-p-anisylmethyl	−1.12	+0.82	H_2SO_4[a,f]
	80	Tri-p-tolylmethyl	+4.85	−3.56	H_2SO_4[a]
	81	1,1-Di-p-anisyl-1-propyl	7.53	−5.5	H_2SO_4[a]
	82	Di-p-anisylmethyl	+7.78	−5.71	H_2SO_4[a,f]
	83	Tri(*tert*-butylphenyl)methyl	8.85	−6.5	H_2SO_4[a]
	84	Dimesitylmethyl	9.0	−6.6	H_2SO_4[a]
	18	Triphenylmethyl	9.03	−6.63	H_2SO_4[a]
	85	Tri-(p-chlorophenyl)methyl	11.52	−7.74	H_2SO_4[a]
	86	$(C_6H_5)_2C^+(C_6H_4$—p—$NO_2)$	12.45	−9.15	H_2SO_4[a]
	87	$C_6H_5C^+(C_6H_4$—p—$NO_2)_2$	17.55[a] 18.35[f]	−12.90	H_2SO_4[a]
	88	Tri-(p-nitrophenyl)methyl	22.3[a] 24.6[f]	−16.27	H_2SO_4[a,f]
	89	Di-(p-chlorophenyl)methyl	19.0	−13.96	H_2SO_4[a]
	90	Di-(m-chlorophenyl)methyl	23.5	−17.3	—
	44	Diphenylmethyl	18.1	−13.3	H_2SO_4[a]

[a] Deno and Schriesheim (1955).
[b] Breslow *et al.* (1962).
[c] Battiste (1961).
[d] Doering and Knox (1954).
[e] Naville *et al.* (1960).
[f] Arnett and Bushick (1964).

Table XIII (Section II,E), where solvolysis rates vary by about 13 powers of ten as resonance delocalization is continuously increased. These kinetic orders of carbonium ion stabilities are in good agreement with the equilibrium stabilities presented in Tables XX and XXI.

In Tables XX and XXI an enormous range of stabilities from the tri-dimethylaminophenylcarbonium ion (**72**, $pK_{R^+} = +9.36$) to the *tert*-butyl cation (**6**, $pK = -15.5$) is represented. The stabilities of cyclopropenyl cations **74–76** (Table XX) and cycloheptatrienyl cations **20**, **77**, and **78** (Table XX) are due to their aromaticity and charge delocalization over the aromatic system. Intuitively, one would expect cycloheptatrienyl cations to be more stable than their cyclopropenyl analogs (i.e., compare **77** to **76**) since one expects less strain in the seven-membered ring. Cation **74** with its inductive stabilizing propyl groups might seem somewhat out of line. However, cyclopropenyl

Table XXI
The Stabilities of Some Allylic and Dienylic Cations

No.	Cation	% H_2SO_4 for 50% protonation	pK^a	Reference
91	1,3,4,4,5,5-Hexamethylcyclopentenyl	33	−1.9	b
37	1,3-Dimethylcyclopentenyl	35	−2.1	b
92	1-Cyclopropyl-3-methylcyclopentenyl	12	−0.46	c
93	1-Methyl-3-phenylcyclopentenyl	50	−3.4	c
35	2,4-Dimethyl-3-penten-2-yl	73	−6.3	b
94	2,3,4-Trimethyl-3-penten-2-yl	82	−7.7	b
38	1,3,5,5-Tetramethylcyclohexenyl	50	−3.4	b
95	1,5,5-Trimethylcyclohexenyl	80	−7.3	b
96	1,5,5-Trimethyl-3-phenylcyclohexenyl	52	−3.6	c
97	1,3-Dicyclopropyl-5,5-dimethylcyclohexenyl	1.2	+0.6	c
98	2,6-Dimethyl-3,5-heptadien-2-yl	~4	0	b
99	1,1-Diphenylethyl	71	−6.0	d
100	1,3,5-Trimethyl-2,4-cyclohexadien-1-yl (mesitylenonium)	94.5	−9.6	e
101	1,3,5-Trimethoxy-2,4-cyclohexadien-1-yl	7 M $HClO_4$	−5.1	f, g
102	1,5-Dimethyl-2,4-cyclohexadien-1-yl	$HF—BF_3$	−14	h
6	*tert*-Butyl	$FSO_3H—SbF$	Estimate −15.5	i

^a The pK values were calculated from $H_0 = pK - \log [BH^+]/[B]$.
^b Deno *et al.* (1963).
^c Deno *et al.* (1965a).
^d Deno *et al.* (1959a).
^e Kilpatrick and Hyman (1958).
^f Long and Schulze (1964).
^g Kresge and Chiang (1961).
^h Mackor *et al.* (1957).
ⁱ Deno (1964).

cations are in equilibrium with cyclopropenols which may not be directly comparable to cycloheptatrienols.

As expected, the pK_{R^+} values of diaryl carbonium ions are considerably more negative than those for triaryl carbonium ions (see **79** versus **82**, **85** versus **89**, and **18** versus **44**). This is due in part to relief of back strain and the increased size and polarizability of the triaryl systems. The difference in free energy of forming the triphenylmethyl (**18**) and diphenylmethyl (**44**) cations in H_2SO_4 is about 9 kcal/mole which is quite close to the difference in forming these ions from the hydrocarbons in the gas phase.

Table XXII

Effect of Substituents on Carbonium Ion Stability as a Function of Cation Stability

Substituent	$\Delta(\Delta F_{R^+}) = \Delta F_{\text{subst R}^+} - \Delta F_{R^+}$	
	$R^+ = (Ph)_2CH^+$	$R^+ = Ph_3C^+$
4-CH$_3$	−2.3	−1.9
4,4′-CH$_3$	−3.9	3.0
4-OCH$_3$	−7.4	−4.4
4,4′-OCH$_3$	−10.4	−7.3

The effect of a substituent on cation stability should decrease as the ion in question becomes increasingly stable. This follows from the expectation that the more unstable the ion is, the greater the demand on that substitutent for electron density becomes. This effect is illustrated in Table XXII, where the free energy change for some similarly substituted diaryl and triaryl carbonium ions are compared. The addition of an electron-donating substituent increases the stability of the more stable triaryl carbonium ions less than it increases the stability of less stable diaryl carbonium ions.

Inductive and resonance effects are clearly separated in ions **18**, **85**, **89**, **90**, and **44** (Table XX). Cation **85** is 1.1 pK_{R^+} units less stable than the trityl cation (**18**). Moving all three chloro substituents to the meta position [tri-(m-chloro)case, not shown in Table XX] results in a decrease in stability of 4.4 pK_{R^+} units. This reflects an increase in the $-I$ effect because chlorine is brought closer to the charged center. It also reflects the removal of the $+R$ effect which cannot operate when the substituent is meta. In the diphenyl series the ΔpK_{R^+} between ions **44** and **89** is -0.66 pK_{R^+} unit, whereas that between ions **44** and **90** is 4.0 pK_{R^+} units. The ΔpK_{R^+} observed when three p-dimethylamino groups are changed to three p-nitro groups is -25.63. This is a tremendous effect considering the great inherent stability of the triaryl carbonium ion series.

The resonance energy of the allylic cation from HMO theory is $2\sqrt{2}\beta$

versus 2β units for ethylene. This difference is 0.83β units or approximately 45 kcal/mole. By adding distortion energies of one-third that of benzene (12 kcal/mole), one crudely estimates the allyl cation to be 33 kcal/mole more stable because of conjugation. The pK value of the unsubstituted allylic cation is not known. However, the pK values of many substituted allylic cations have been obtained, and a few are listed in Table XXI. The 1,3,5,5-tetramethylcyclohexenyl cation (**38**) is 2.9 pK units more stable than its trialkylated analog **95**. This ΔpK value from equilibrium measurements in H_2SO_4 agrees very well with the ΔpK values calculated from the solvolysis rates of allylic halides in formic acid (Vernon, 1954) shown below.

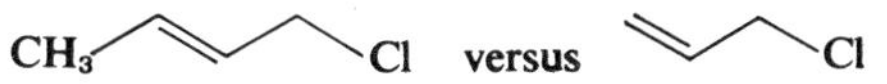

CH_3‒CH=CH‒Cl versus CH$_2$=CH‒CH$_2$‒Cl ΔpK = 3.55

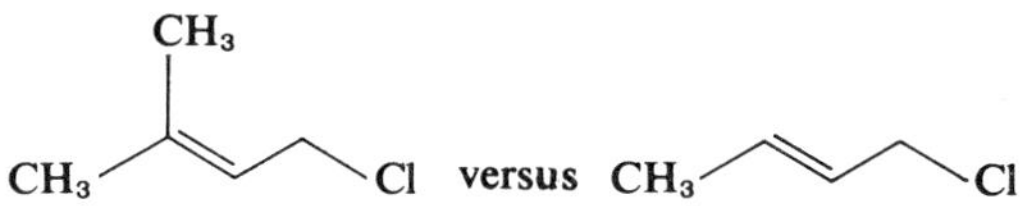

(CH$_3$)$_2$C=CH‒CH$_2$‒Cl versus CH$_3$‒CH=CH‒CH$_2$‒Cl ΔpK = 3.62

The effect of adding terminal methyl groups is additive and results in an increase in stability of the transition state (relative to the chloride) of about 3.6 pK units. Using a value of about 3.75 pK units per methyl group, one estimates the allyl cation would be about 15 pK units less stable than tetrasubstituted allylic cations. Using the 1,3-dimethylcyclopentenyl cation (**37**, Table XXI) as a model, one would roughly estimate the pK of the allyl cation as -16 to -17, which is somewhat less than that estimated for the *tert*-butyl cation. The effectiveness of alkyl groups in stabilizing allylic cations decreases in the order methyl > ethyl > isopropyl > *tert*-butyl.

cyclopentenyl$^+$ (-2.1)   cyclohexenyl$^+$ (-3.1) ΔpK = 1

Alternatively, cyclopropyl groups are about 1.6 pK units more stabilizing than methyl groups (compare ions **37** and **92**, **38** and **97**, Table XXI). Surprisingly, phenyl substituents destabilize both cyclopentenyl and cyclohexenyl cations relative to methyl groups.

The effects of substituents on the solvolysis of cumyl chlorides has been quantitatively assessed by Brown and co-workers (Brown and Chloupek, 1963; Brown and Okamoto, 1957, 1958; Brown and Ichikawa, 1957) in the form of σ^+ constants. Using Eq. (116), solvolysis studies in many ionizing

$$\log(k_X/k_H) = \rho\sigma^+ \qquad (116)$$

solvents led to a consistent set of σ^+ values. These reflect the effect that substituents exert on the transition state of limiting S_N1 solvolyses of the cumyl chlorides. The value of the constant ρ (-4.54 in 90% acetone–water at 25°C) was chosen in order to directly relate σ^+ values to the σ values developed by Hammett. The σ^+ values represent the set of substituent parameters which are most applicable for correlating equilibrium and rate constants in carbonium ion reactions.

For example, σ^+ values predict the positions of arylmethanol–arylmethyl cation equilibria listed in Table XVIII (and in the references in Table XVIII) with good accuracy. Good $\rho\sigma^+$ correlations with large negative values of ρ have been found in reactions such as the acid-catalyzed hydration of styrenes ($\rho = -3.42$; Schubert et al., 1964) and the cis–trans isomerization via double-bond hydration ($\rho = -4.3$; Noyce et al., 1962).

Table XXIII

Substituent Effects on Solvolysis Rates of tert-Cumyl Chlorides

Substituent	σ_{para}^+	σ_{meta}^+
—OCH$_3$	-0.778	$+0.047$
—SCH$_3$	-0.604	$+0.158$
—CH$_3$	-0.311	-0.066
—CH$_2$CH$_3$	-0.298	-0.064
—CH(CH$_3$)$_2$	-0.280	-0.060
—C(CH$_3$)$_3$	-0.256	-0.59
—H	0	0
—F	-0.073	$+0.352$
—Cl	$+0.114$	$+0.399$
—Br	$+0.150$	$+0.405$
—CO$_2$CH$_3$	$+0.489$	$+0.368$
—CF$_3$	$+0.612$	$+0.520$
—CN	$+0.659$	$+0.562$
—NO$_2$	$+0.790$	$+0.674$

Table XXIII lists the σ^+ values of some representative substituents. A negative σ^+ value means the substituent is electron donating relative to hydrogen. The normal order of electron-donating effects (Me > tert-butyl) is observed in alkyl groups. The methoxy and fluorine substituents merit brief mention. The methoxy group is a strong donor ($\sigma^+ = -0.778$) in the para position owing to a large $+R$ effect which more than compensates for its $-I$ effect. However, in the meta position where resonance is not possible the $-I$ effect is clearly exhibited ($\sigma^+ = +0.047$). Fluorine, unlike the other halogens, shows the same behavior. This is expected since the short C–F bond distance allows more effective p_π–p_π overlap with the para carbon.

2. In the Gas Phase

The effects which substituents exhibit in stabilizing or destabilizing carbonium ions are much larger in the gas phase than in solution. The ionization of benzyl radicals to benzyl cations illustrates this principle. Using the ionization potentials of Harrison *et al.* (1961) from Table XXIV, the huge ρ value of -19.1 is obtained by plotting the ionization potentials (kcal/mole) divided

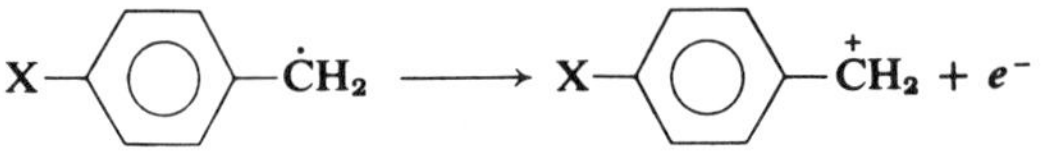

$$\text{X}-\langle\bigcirc\rangle-\dot{\text{C}}\text{H}_2 \longrightarrow \text{X}-\langle\bigcirc\rangle-\overset{+}{\text{C}}\text{H}_2 + e^- \tag{117}$$

Table XXIV
Ionization Potentials of Para-Substituted Benzyl Radicals

para Substituent	Ionization potential (kcal/mole)
—CN	193.0
—F	179.6
—H	178.8
—CH$_3$	172.0
—OCH$_3$	157.2

by 2.303 RT versus σ^+. The electron demand by the bare carbonium ion center is very high, and delocalization into the ring and the substituents is more intense than in solution where a ρ value of about 4 would not be exceeded (in fact, ρ is -2.18 for the hydrolysis of benzyl chlorides, but some solvent participation is probably occurring).

The highest substituent sensitivity ever observed is that reported by Taft *et al.* (1965) for the ionization of substituted methanes to their corresponding cations [according to Eq. (118)] in the gas phase.

$$\text{CH}_3\text{X}_{(g)} + e^- \longrightarrow {}^+\text{CH}_2\text{X} + 2e^- + \text{H} \cdot \tag{118}$$

In this case the substituent is attached directly to the charged center, creating an enormous demand on the substituent in the absence of solvation. There is no ordinary substituent constant which serves to correlate all the results, but the rough correlation which can be made with σ^+ results in a ρ value of -45! The amounts by which these substituents stabilize or destabilize the cation relative to hydrogen are summarized in Table XXV. Surprisingly, the nitro

Table XXV

Relative Stabilization Energies for Monosubstituted Methyl Cations[a]

Ion	ΔA^a (eV)	Stabilization energy (kcal/mole)
$^+CH_2CN$	+0.4	−10
$^+CH_3$	0	0
$^+CH_2F$	−1.1	26
$^+CH_2Cl$	−0.14	32
$^+CH_2CH_3$	−1.5	35
$^+CH_2SCN$	−1.8	42
$^+CH_2Br$	−2.2	51
$^+CH_2I$	−2.3	53
$^+CH_2-NO_2$	−2.35	54
$^+CH_2-C=CH_2$	−2.35	54
$^+CH_2-C\equiv CH$	−2.4	55
$^+CH_2-C_6H_5$	−2.4	55
$^+CH_2OH$	−2.6	60
$^+CH_2SH$	−2.8	64
$^+CH_2OCH_3$	−3.0	69
$^+CH_2SCH_3$	−3.2	74
$^+CH_2P(CH_3)_2$	−3.4	79
$^+CH_2NH_2$	−4.1	95
$^+CH_2NHCH_3$	−4.3	99
$^+CH_2N(CH_3)_2$	−4.6	106

[a] Relative to $^+CH_3$.

group stabilizes the cation relative to hydrogen. This is quite unexpected. Franklin (1968) has suggested that resonance hybrid **103b** might account for this behavior. This result is worthy of further investigation.

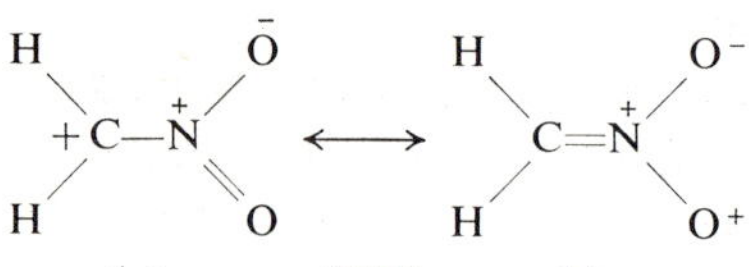

$$\begin{array}{c} H \\ \diagdown \\ ^+C-N \\ \diagup \\ H \end{array} \begin{array}{c} \bar{O} \\ \diagup \\ \diagdown \\ O \end{array} \longleftrightarrow \begin{array}{c} H \\ \diagdown \\ C=N \\ \diagup \\ H \end{array} \begin{array}{c} O^- \\ \diagup \\ \diagdown \\ O^+ \end{array} \qquad (119)$$

(a)   (103)   (b)

The heats of formation of CH_3^+ and $C_3H_7^+$ have both been determined from a large number of starting molecules in the gas phase, and there is general agreement that $\Delta H_f(CH_3^+)$ is close to 260 kcal/mole and $\Delta H_f(C_3H_7^+)$ is 190–195 kcal/mole (see Franklin, 1968). Substituting alkyl groups for hydrogen on CH_3^+ results in a lowering of both the heat of formation and the ionization potential of the corresponding hydrocarbon. As expected, the stability order is 3° > 2° > 1° > CH_3^+ (see Table XXVI). Comparing the allylic, benzenonium (cyclohexadienyl), and tropylium ions, Franklin and

Field (1953) found that the differences in appearance potentials (ΔA_p) agreed closely with the differences in resonance energy (ΔE_R) calculated by HMO

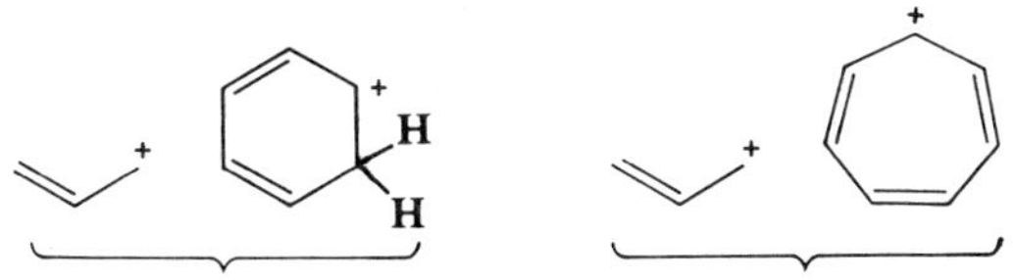

ΔE_R (HMO)	12 kcal/mole	43 kcal/mole
ΔA_P	14 kcal/mole	43 kcal/mole

Table XXVI

Heats of Formation of Carbonium Ions and Ionization Potentials of Their Corresponding Hydrocarbons

Cation, R$^+$	ΔH^f of R$^+$ (kcal/mole)	I_p of RH (kcal/mole)
Methyl	258	227
Ethyl	225	200
1-Propyl	218	200
2-Propyl (*i*-propyl)	194	182
2-Methylpropyl (*tert*-butyl)	174	171
2-Methyl-2-butyl	162	164
Vinyl	285	—

theory. The vinyl cation appears to be very unstable, and it has a high heat of formation (285 kcal/mole) as does the phenyl cation (297 kcal/mole). Adding resonance-stabilizing phenyl groups to CH_3^+ progressively stabilizes the resulting ion and lowers the ionization potential (Ph–CH_2^+, 7.76; Ph_2CH^+, 7.32; Ph_3C^+, 7.26). As expected, these differences become progressively smaller as the ion becomes larger and more stable.

C. Steric Effects

Steric effects play an important role in a variety of reactions involving carbonium ions. Steric effects are most often studied by kinetic techniques, although direct spectroscopic observations have become increasingly important.

1. *Steric Assistance in Carbonium Ion Formation*

There are many reactions which generate carbonium ions in which the steric requirements of the ground state are greater than those of the transition state. This situation has been introduced in Section II,E, and Fig. 6 gives an example of rate enhancement in an S_N1 reaction caused by release of ground-state strain. Such "steric assistance" (Brown, 1946) is found in the solvolysis of the isomeric 4-*tert*-butylcyclohexyl tosylates (Winstein and Holness, 1955). The bulky tosylate substituent experiences greater crowding in the axial position. Therefore, $\Delta G_A^\ddagger$ is smaller than $\Delta G_E^\ddagger$ since the same carbonium ion (and presumably transition states of similar energy) is formed in the two reactions. To the extent that the transition states resemble ion **104**, the

initial difference between the energy levels of the trans and cis tosylate has disappeared, and the less stable cis isomer reacts three to four times faster.

The solvolysis of tertiary ring derivatives shows kinetic effects which can be predicted by examining the steric effects involved in converting the ring carbon from sp^3 to sp^2. Cyclopropyl and cyclobutyl derivatives resist solvolysis because the introduction of an sp^2 ring carbon compounds the strain. On the other hand, this change in hybridization on a cyclopentane ring carbon results in a relief of four bond oppositions (eclipsing interactions) which decreases strain by about 4 kcal/mole. This enhances solvolysis rates of cyclopentyl derivatives. Cyclohexane rings are strain free; thus, conversion of one carbon from sp^3 to sp^2 introduces strain and slows solvolysis reactions. Medium sized rings (from 8 to 11 carbons) have angle deformations, eclipsing interactions, and transannular nonbonded hydrogen–hydrogen repulsions. Thus, converting a ring carbon from sp^3 to sp^2 reduces these repulsions (Prelog, 1960), and this relief of strain enhances S_N1 reactions. Illustrative examples are summarized in Table XXVII.

The cyclization of dienyl cations are remarkably sensitive to steric effects (Deno and Pittman, 1964b). The 2,6-dimethylheptadienyl cation (**39**) cyclizes only slowly ($t_{1/2}$ = 3 days, 25°C) to the 1,2,3,4-tetramethylcyclopentenyl cation [Eq. (121)], but the corresponding cyclization of the 2,4,6-trimethylheptadienyl cation (**105**) in H_2SO_4 is very fast ($t_{1/3}$ < 20 sec, 25°C). Even at −30°C in FSO_3H, **105** cyclizes in 20 sec, whereas **39** has a half-life of years (Sorensen, 1964). This remarkable rate difference appears to be due to the methyl–methyl nonbonded repulsions in **105** which destabilize the trans,trans

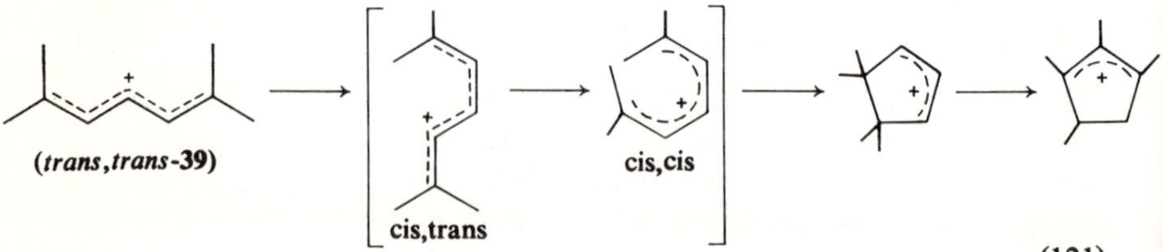

Table XXVII
Solvolytic Studies of Ring Compounds[a]

Ring size	Rates CH₂–C(CH₃)₂–Cl	Rates CH₂–C(CH₃)₂–OTs	Predicted difficulty for $sp^3 \to sp^2$
Acyclic	1.00	1.00	—
3	—	2×10^{-5}	Difficult
4	0.97[b]	8.5[b]	Difficult
5	43.7	10.5	Easy
6	0.35	0.75	Difficult
7	38.0	19.0	Easy
8	100.	144	Easy
9	15.4	129	Easy
10	6.22	286	Easy
11	4.21	30.8	Easy
12	—	2.44	Uncertain
15	0.64	1.65	[c]
17	0.67	1.63	[c]

[a] Brown (1956a,b).
[b] Solvolysis of cyclobutyl derivatives involves "nonclassical ions."
[c] No appreciable strain is predicted in 14-membered or larger rings.

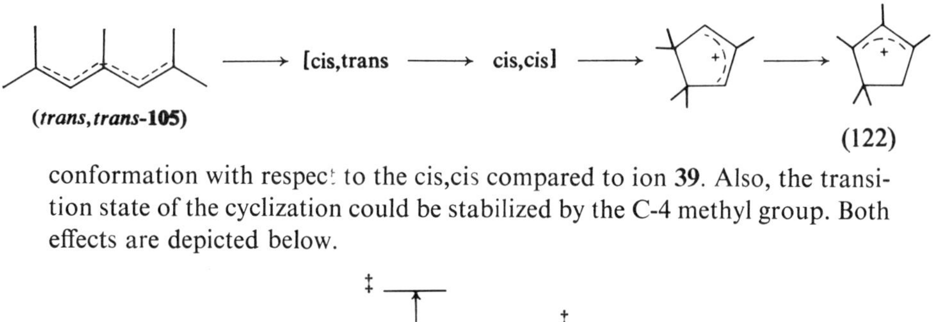

(*trans,trans*-**105**) ⟶ [cis,trans ⟶ cis,cis] ⟶ ⟶   (122)

conformation with respect to the cis,cis compared to ion **39**. Also, the transition state of the cyclization could be stabilized by the C-4 methyl group. Both effects are depicted below.

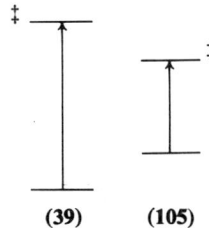

(**39**)   (**105**)

An increase in solvolysis rates was first systematically recognized in the classic work of Brown and Fletcher (1949, 1951) and Bartlett (1951; Bartlett and Stiles, 1955; Bartlett and Swain, 1955) for derivatives of *tert*-butyl halides (or *p*-nitrobenzoates) as steric crowding was systematically increased (see Table XXVIII). These rate enhancements were first attributed to "B strain," a repulsive interaction between bulky groups in the starting materials which was relieved on going to the transition state. However, it was noted that the greatest rate accelerations were in systems where the reaction gave prevalent or total molecular rearrangement. Thus, it was also possible to interpret the rate enhancements to neighboring group participation, especially since the systems with the highest migratory aptitudes also produced the largest reaction rates (Stiles and Mayer, 1959). Brown and Kornblum (1954) maintained that the high occurrence of rearrangements was conceivably due to steric hindrance encountered by nucleophiles attacking the carbonium ion, thus increasing the lifetime of the ion so that it could rearrange. Much later, Shiner and Meier (1966) found that 2-chloro-2,3,3,4,4-pentamethylpentane (**106**) solvolyzed in 1:2 water–dioxane exclusively to unrearranged olefin (**107**) 27,900 times faster than *tert*-butyl chloride. Thus, in this case, participation of neighboring alkyl groups was ruled out as a factor contributing to the

$$\underset{(\mathbf{106})}{\overset{\text{Cl}}{\diagup}\!\!\!\!\diagdown\!\!+\!\!+\!\!+} \longrightarrow \underset{(\mathbf{107})}{\diagdown\!\!\!\diagdown\!\!\!+\!\!+\!\!+} \quad (123)$$

rate enhancement. In a paper completing this classic series, Bartlett and Tidwell (1968) determined the first-order solvolysis rates for the *p*-nitro-

Table XXVIII

Relative Rate Constants for Solvolysis of Highly Branched Tertiary Derivatives versus *tert*-Butyl Chloride Solvolysed under the Same Conditions[a]

Compound	k_{rel}	Solvent
Me$_2$-*n*-BuCCl	1.4	80% aq. EtOH 25°C
Me$_2$-*t*-BuCCl	1.21	80% aq. EtOH 25°C
MeEt$_2$CCl	2.6	80% aq. EtOH 25°C
Me-*i*-Pr$_2$CCl	13.6	80% aq. EtOH 25°C
Me-*t*-Bu$_2$CCl	18.4	80% aq. EtOH 25°C
i-Pr$_3$C—O—*p*NB	403	60% aq. dioxane 40°C
(Np)$_3$C—O—*p*NB	560	60% aq. dioxane 40°C
i-Pr-*t*-BuC—O—*p*NB	3,440	60% aq. dioxane 40°C
i-Pr-*t*-Bu$_2$C—O—*p*NB	3,330	60% aq. dioxane 40°C
t-Bu$_3$C—O—*p*NB	13,000	60% aq. dioxane 40°C
t-Bu$_2$NpC—O—*p*NB	19,000	60% aq. dioxane 40°C
RC(Me)$_2$Cl	27,900	1:2 water-dioxane 25°C
t-Bu(Np)$_2$C—O—*p*NB	68,000	60% aq. dioxane 40°C

[a] Np = neopentyl and R = 1,1,2,2-tetramethylpropyl group.

4. CARBONIUM IONS

benzoates of tri-*tert*-butylcarbinol (**108**), di-*tert*-butylneopentylcarbinol, (**109**), *tert*-butyldineopentylcarbinol (**110**), and trineopentylcarbinol (**111**) relative to *tert*-butyl *p*-nitrobenzoate (in 60% aqueous dioxane at 40°C). The rate enhancements were 13,000, 19,000, 68,000, and 560, respectively. The products in the solvolyses of **110** and **111** were predominantly of unrearranged carbon skeletons. Since the system, **111**, giving the greatest rate enhancement did not give skeletal rearrangement (and in view of studies on **106**), neighboring group participation can be ruled out as the origin of the rate en-

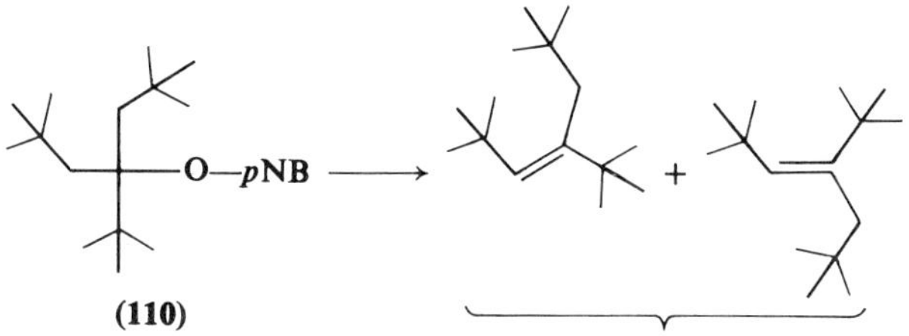

hancement. Fisher–Hirschfelder–Taylor models of **108**, **109**, and **110** could not be constructed without fission of one or more bonds. The model of the *tert*-butyldineopentylcarbonium ion shows that the central carbon is buried deep in the model and overlaid by methyls of the two neopentyl groups. The only *p*-nitrobenzoate model which could be constructed was that of less reactive **111**.

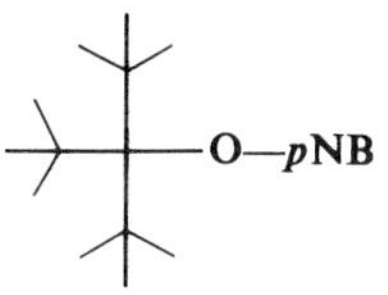

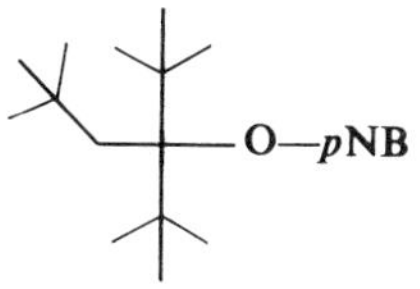

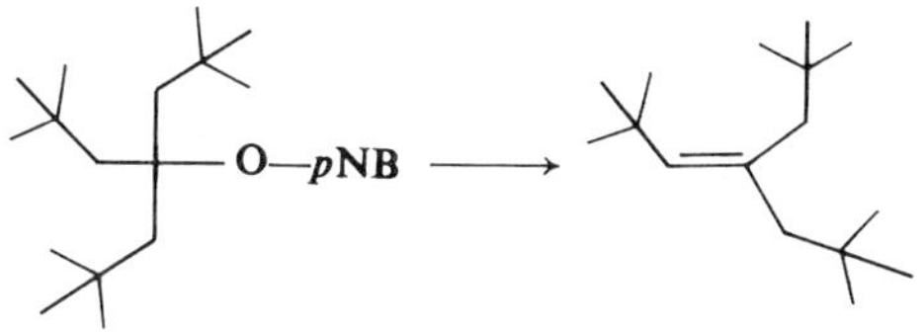

2. Steric Hindrance to Ionization

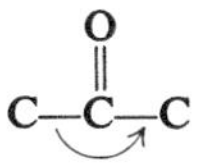

As the C–C–C bond angle in a secondary system is reduced, the stretching frequency of a carbonyl group at the central carbon is increased. The size of the C–C–C angle also affects the ease of ionization of C—CHX—C → C—C$^+$H—C. As this angle is reduced (in a rigid system) it becomes more difficult to convert an sp^3 carbon to sp^2. This results in a lower rate of ionization (Schleyer and Nicholas, 1961a). Foote (1964) showed that the rate constants for secondary alkyl tosylate solvolyses can be estimated from the carbonyl stretching frequencies which depend on the angle. Schleyer (1964; Fry et al., 1970) developed a more elaborate version of this approach which corrects the rate, predicted from the carbonyl frequency, for (1) torsional strain about the reaction center, (2) polar effects, and (3) changes in nonbonded interactions going to the transition state (TS). Table XXIX compares the Foote–Schleyer predictions with experimental results, and the general correlation is good. As the stretching frequency of the carbonyl increases, the rate of solvolysis of the corresponding tosylate decreases.

Table XXIX
Relative Rate Constants for Acetolysis of Secondary Alkyl Tosylates Calculated by the Foote–Schleyer Method

Compound	$v_c = 0$	log k_{rel} Observed	log k_{rel} Calculated	log (k_{obs}/k_{cal})
Cyclohexyl	1716	0.00	−0.1	0.1
3-Methyl-2-butyl	1718	0.93	0.6	0.3
endo-2-Norbornenyl	1745	−1.48	−1.0	−0.5
endo-2-Norbornyl	1751	0.18	−0.2	0.4
7-Norbornyl	1773	−7.0	−7.0	0.0

Solvolysis studies (Brown, 1966; Brown and Hammar, 1967; Brown et al., 1966; Schleyer et al., 1965) have indicated that the ionization of endo tosylates **112**, **113**, and **114** are retarded by *steric effects which destabilize the solvolytic transition state*. These all solvolyze *slower* than *endo*-2-norbornyltosylate

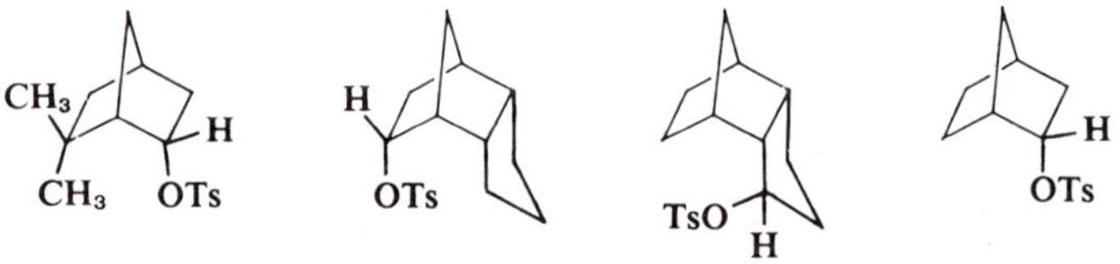

	(112)	(113)	(114)	(115)
relative rate	0.054	0.10	0.19	1

despite the fact that the ground state in **112**, **113**, and **114** is far more crowded than **115**. Thus, if one assumes that the TS strain is $\simeq 0$ for the leaving group, tosylates **112**, **113**, and **114** should relieve the ground-state strain going to the TS, and a rate acceleration versus **115** would be expected; for example, compare **113** to **115**. Their corresponding ketones exhibit stretching frequencies of 1743 and 1751 cm^{-1}, respectively. From the Foote–Schleyer correlation scheme, this predicts an increase in acetolysis rate of **113** over **115** by a factor of 10. The steric strain in **115** versus **113** experienced by the leaving group is 1.3 versus 4.0 kcal/mole. Thus, another factor of 100 should be added to the acetolysis rate of **113** versus **115**. Thus, one expects (assuming no TS strain is

4. CARBONIUM IONS

experienced by the leaving group) that **113** should undergo acetolysis 1000 times faster than **115**. However, the rate ratio $k_{113}/k_{115} = 0.1$. Thus, a discrepancy of 10,000 exists. This means that the leaving group is experiencing an increase in strain in going to the TS which is large enough to retard the rate by 10,000! (see Fig. 9).

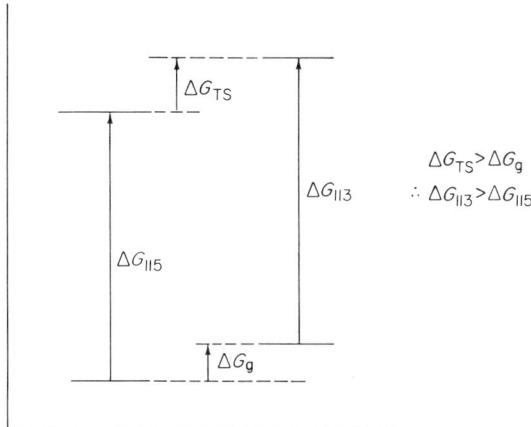

Fig. 9. Comparison of the free energy changes between the ground state and the transition state in **113** and **115**.

The origin of the TS destabilization is not well understood. Destabilization could be due to a physical barrier to the departure of the anion, or it could arise from steric hindrance which interferes with the anion's solvation as it approaches the TS.

3. *Steric Hindrance to Solvent Assistance*

Assistance by solvent during the formation of the S_N1 transition state should be especially important in the solvolysis of isopropyl chloride (**116a**). Schleyer and co-workers (1970) demonstrated that the first-order solvolysis of **116** is dependent on the nucleophilicity of the solvent. Thus, in 80% ethanol, $k_s/k_c \sqrt{} 1$. The ratio k_s/k_c progressively decreases as the solvent

$$CH_3-CHX-CH_3 \xrightarrow[\text{no solvent assistance}]{k_c} CH_3\overset{+}{C}HCH_3 \quad X^- \quad \begin{array}{l}(\textbf{116a}) \ X = Cl \\ (\textbf{116b}) \ X = OTs\end{array} \quad (126)$$

(**116**) with solvent assistance k_s

becomes progressively less nucleophilic in the solvent series: 80% ethanol, acetic acid, formic acid, 97% trifluoroethanol, trifluoroacetic acid. Highly branched derivatives such as di-*tert*-butylcarbinyl chloride (**117a**) or tosylate (**117b**) might sterically prevent backside solvent participation.

Liggero et al. (1970) have shown that **117a** actually solvolyzes *more slowly* than **116a** despite the fact that the Foote–Schleyer correlation above predicts

>−CH−<
   |
   X   (117a) X = Cl
 (117)   (117b) X = OTs

117 to solvolyze $10^{5.3}$ times *faster* than **116**. A rate enhancement owing to the relief of B strain in **117** should be added to this. Thus, **117** must be solvolyzing without solvent assistance, k_c, whereas **116** is proceeding with assistance, k_s. This interpretation nicely explains the solvolysis rate ratios of **117b** versus **116b** as a function of solvent nucleophilicity. This ratio ($k_{117b\ obs}/k_{116b\ obs}$) rapidly increases with a decrease in solvent nucleophilicity as follows: 8 in 80% ethanol, 71 in acetic acid, 320 in formic acid, 630 in 97% trifluoroethanol, and approximately $10^{5.1}$ in trifluoroacetic acid. This is excellent evidence for steric hindrance to solvent participation in system **117**.

In bridgehead derivatives, such as adamantyl bromide, their very construction precludes solvent participation from the rear as the leaving group departs. Solvolyses of these derivatives are pure k_c processes, and they may be used as models to compare with acyclic tertiary compounds. A large body of evidence has now been compiled showing that many acyclic tertiary derivatives exhibit about the same sensitivity to solvent structure and nucleophilicity as do bridgehead derivatives (Doering and Schoenewaldt, 1951; Doering et al., 1953; Finkelstein, 1955; Fort and Schleyer, 1966a). For example, the m values from the Grunwald–Winstein equation (see Section IV, A) measure the sensitivity of a given compound to changes in the ionizing power (Y) of the solvent. If the solvent plays a different role in the transition states for bridgehead and tertiary acyclic solvolyses, one would expect a large difference in the m values. As indicated below, the m values for the *tert*-butyl, bicyclooctyl, and adamantyl derivatives are about the same. Thus, covalent solvation of the TS of tertiary derivatives is probably small.

	$(CH_3)_3CBr$	bicyclic-Br	bicyclic-Br	adamantyl-Br, R	R = H, alkyl
$m =$	0.94	1.13	1.13	0.89–1.0	

Krapcho and Horn (1966) showed that increasing methyl substitution resulted in a decrease in the solvolysis rates of secondary cyclopentyl tosylates. This methyl substitution retards nucleophilic participation of solvent (k_s). Just the opposite trend is found in the solvolysis of tertiary cyclopentyl chlorides. Brown and Chloupek (1963) showed that introducing methyl substituents into the ring markedly increases the solvolysis rate. In these tertiary derivatives, where nucleophilic solvent participation (k_s) is not important, the relief of steric strain going to the TS becomes the dominant effect. These results are summarized in Table XXX.

Table XXX
Solvolysis Results for Secondary Cyclopentyl Tosylates and Tertiary Cyclopentyl Chlorides with Increasing Methyl Substitution

Compound	Relative rate	Compound	Relative rate
		tert-Butyl chloride	1.0
 cyclopentyl-OTs	1.0	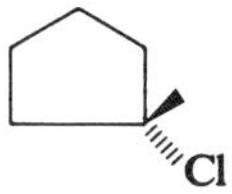 1-methylcyclopentyl-Cl	66
 2-methylcyclopentyl-OTs	0.86	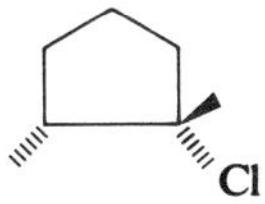 1,2-dimethylcyclopentyl-Cl	171
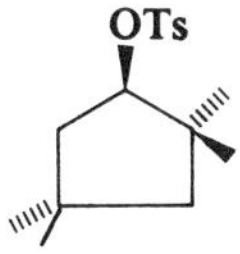 2,5-dimethylcyclopentyl-OTs	0.15	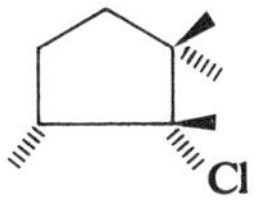 1,2,5-trimethylcyclopentyl-Cl	2380
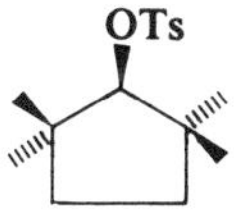 2,2,5-trimethylcyclopentyl-OTs	0.07	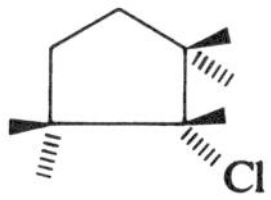 1,2,2,5-tetramethylcyclopentyl-Cl	5390

4. *Bridgehead Carbonium Ions*

Evidence for the preferred planarity of carbonium ions was presented earlier. In bridgehead derivatives the carbonium ion cannot achieve planarity. Thus, questions arise: (1) Can nonplanar carbonium ions be generated, and (2) if so, how will their stability compare to similar planar ions? In a classic paper, Bartlett and Knox (1939) showed that apocamphyl chloride (**118**) could not be solvolyzed. Even when boiled with aqueous alcoholic silver nitrate for 48 hours, no precipitate of AgCl was formed! In contrast, *tert*-butyl chloride reacts with silver nitrate rapidly in the cold, and the rate ratio between these systems is well over 10^{11}. 1-Bromonorbornane is equally unreactive (Doering and Schoenewaldt, 1951). Similar solvolytic inactivity was demonstrated for the bridgehead triptycyl halides (i.e., **119**) with $AgNO_3$ and $SnCl_4$ (Bartlett and Lewis, 1950). Thus, it looked as if bridgehead carbonium ions might not exist. However, Doering *et al.* (1953) showed that bridgehead carbonium ions could be formed in the bicyclo[2.2.2]octane series by demonstrating that **120** solvolyzes in aqueous ethanol in the presence of silver nitrate in 4 hr. In the absence of silver nitrate in aqueous dioxane the solvolysis rate of **120** was very slow (about 10^{-6} versus 1 for *tert*-butyl bromide).

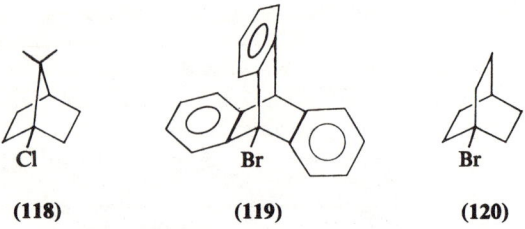

(118)   (119)   (120)

The unreactivity of **119** (versus **120**) may be attributed to the adverse inductive effects of the three phenyl groups. These phenyl groups cannot stabilize the developing carbonium ion by resonance since the *p* orbital forming at the bridgehead is orthogonal to the π clouds of the benzene rings. This is an excellent example of steric inhibition of resonance.

The finding that bridgehead bicyclo[2.2.2]octyl derivatives are solvolyzed much faster than their bridgehead bicyclo[2.2.1]heptyl counterparts are agreed with the concept that the cations become progressively less stable as they are progressively deformed from a plane. However, 1-bromoadamentane solvolyzes readily [only 800 times slower than *tert*-butyl bromide (Stetter *et al.*, 1959, 1960; Schleyer and Nicholas, 1961b)], and β-caryophyllene chloride (**121**) is readily solvolyzed to the acetate in boiling acetic acid (Henderson *et al.*, 1926). Since **121** contains the [4.3.1] system, it should be possible to distort the ring system to allow the carbonium ion to become planar without an excessive increase in strain.

(121)

1-Adamantyl cations are quite stable in FSO_3H—SbF_5—SO_2 (Olah *et al.*, 1966a) and SbF_5—SO_2 (Schleyer *et al.*, 1964a). Combining this information with the solvolysis data, it becomes clear that carbonium ions, although preferring to be planar, do not experience serious destabilization until the C–C$^+$–C angles are decreased below 109 deg. If only the energies of the carbonium ions themselves or the energies of the precursors are considered, the kinetic data cannot be explained. However, the solvolysis rates are adequately explained by calculating the strain in the precursor and then calculating the strain in the cation (assumed to be close to that of the TS). *The increase in strain going from the precursor to the TS is the appropriate factor which correlates with the solvolysis rates* (Schleyer and Nicholas, 1961a; Fort and

4. CARBONIUM IONS

Schleyer, 1966a,b). Table XXXI summarizes the results of these calculations and illustrates the correlation for the series; *tert*-butyl bromide, homoadamantyl bromide, adamantyl bromide, bicyclo[2.2.1]octyl bromide, and 1-norbornyl bromide.

Table XXXI
The Effects of Strain on Bridgehead Reactivity

System	$(CH_3)_3CBr$	homoadamantyl-Br	adamantyl-Br	bicyclo[2.2.1]octyl-Br	1-norbornyl-Br
Ground state strain (kcal)	0	10.6	0	9.3	—
Transition state strain (kcal)	0	11.1	3.9	16.1	—
Increase in strain upon ionization (kcal)	0	0.5	3.9	6.8	—
Relative solvolysis rate	1	0.5	10^{-3}	10^{-6}	10^{-13}

The general form of these calculations can be indicated with 1-adamantyl bromide. As the bridgehead position flattens during the ionization process, the gain in stability that results from this move toward planarity is opposed by the distortion of other angles in the molecule. This process is illustrated in Fig. 10. As the bridgehead moves toward planarity ($\theta \to 120$ deg) the strain in θ is decreased because the bridgehead carbon would like to achieve sp^2 hybridization. However, as this process occurs, ϕ is progressively decreased until it reaches 90 deg when $\theta = 120$ deg. This large distortion of ϕ is

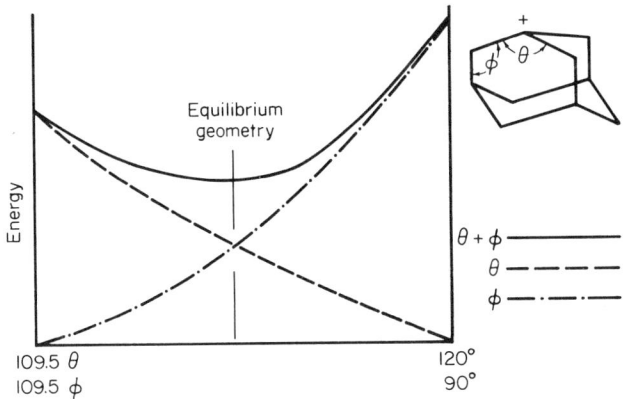

Fig. 10. Angle strain in the 1-adamantyl cation.

energetically unfavorable. Thus, θ will only deform a fraction of this amount. Quantitative calculations for the adamantyl system have been made (Fort and Schleyer, 1966b) using C–C–C bending force constant of 0.8×10^{-4} erg radian^{-2}, and $E(\text{kcal}) = 0.0175X^2$ [where X is the deviation of θ or ϕ from its preferred value (Westheimer, 1956; Pitzer and Donath, 1959)]. Application of this force constant to the bicyclo[2.2.1]heptyl system, however, is known to overestimate strain because distortions become very large in this system.

Deamination of bridgehead amines, unlike their corresponding solvolyses reactions, proceeds with great ease. Examples include the deamination of apocamphyl amine with nitrous acid or nitrosyl chloride (Bartlett and Knox, 1939) [Eq. (127)], the formation of 1-norbornanol from its amine (Whelan, 1952), and the deamination of such highly strained bridgehead amines as **122** (Wilhelm and Curtin, 1957) and **123** (Hart and Martin, 1960). Streitwieser (1957) proposed that the great stability of the leaving nitrogen molecule

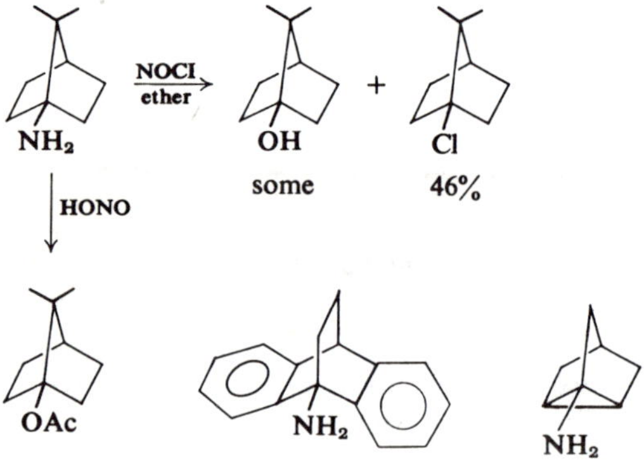

(127)

resulted in a very low activation energy for deamination. If E_{act} is low, according to the Hammond postulate, the transition state leading to the bridgehead carbonium ion should resemble the starting diazonium ion and not the resulting ion. Since the change in geometry going to the TS is small, the increase in strain in this process should also be small. Therefore, deamina-

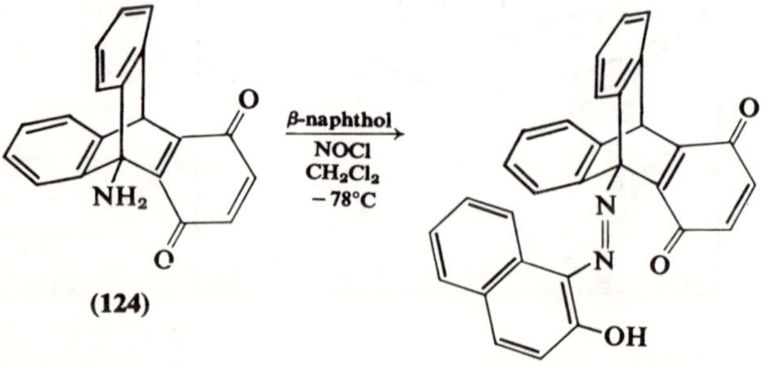

(128)

4. CARBONIUM IONS

tions should not experience nearly as much resistance as solvolysis reactions do in bridgehead systems. In one case (124) where the resulting carbonium ion is predicted to be very unstable, Curtin *et al.* (1962) trapped the bridgehead diazonium ion before decomposition [Eq. (128)].

5. Steric Influences and the Curtin–Hammett Principle

The Curtin–Hammett principle states that the proportion of the products formed in a reaction depends only on the relative activation energies of the processes leading to these products (Curtin, 1954). It follows that, provided the barrier to rotation is low compared to the energy of activation of two competing reactions, the proportion of the products in no way reflects the relative population of the ground-state conformations A and B but reflects only on the values of $\Delta G_1^{\ddagger}$ and $\Delta G_2^{\ddagger}$. Thus, for the reaction depicted in Fig. 11, the ratio of products 1 and 2 does not depend on the population of

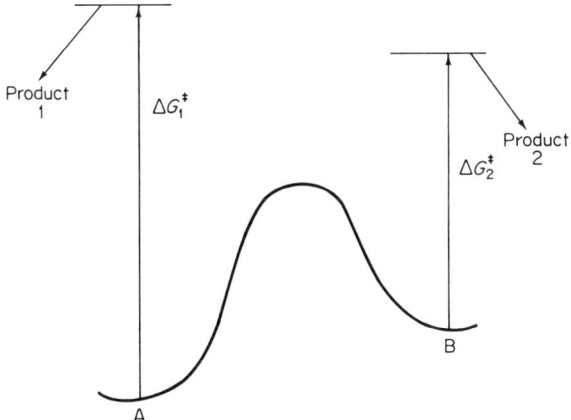

Fig. 11. The Curtin–Hammett principle.

conformers A and B. In fact, product 2, which arises from the stable conformer B, would be the major product in this case (for a detailed discussion, see Eliel, 1962). When the activation energies for a set of reactions is very low, the barrier to rotation (or other conformational changes) may be as high or higher. In these cases the Curtin–Hammett principle no longer will hold. If the activation energy of the reaction is significantly lower than the rotational barrier, it follows that the ratio of products is equal to the population of the respective starting states. This situation is represented in Fig. 12. Deamination reactions have such low activation energies that the product distribution might reflect the original conformer populations. The previously described

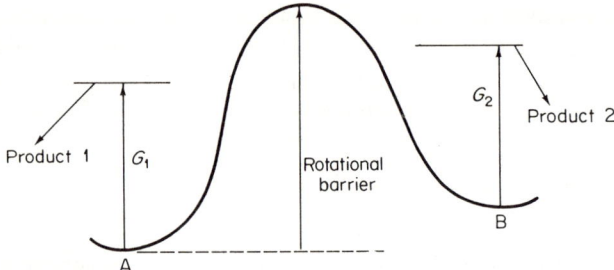

Fig. 12. An example of a situation where the Curtin–Hammett principle is inoperative.

deamination of optically active, stereospecifically labeled 1-phenyl-1-phenyl-^{14}C-2-aminopropanol-1 (Section II,F) is an example where rotation about the single bond and aryl migration proceed at comparable rates. In these cases the rotational barrier is approximately equal to the free energies of activation of the processes (intermediate situation between Figs. 11 and 12). Thus, the product distribution in that reaction partially reflects the population of the ground-state conformations.

6. *Miscellaneous Steric Effects*

The steric effects on nucleophilic reactivity toward carbonium ions is little studied. Bunton *et al.* (1954) found that fluoride ion was more reactive than bromide toward the *tert*-butyl cation in SO_2, and attributed this phenomenon to steric effects. Bethell and Howard (1966) showed that the reactivity of a series of aliphatic alcohols toward diarylmethyl cations in acetonitrile appeared to follow a steric order. As more steric compressions were predicted in the TS, the rate of reaction was reduced. However, both nucleophiles and ions are always surrounded by a solvent shell, and steric effects are difficult to separate from solvent effects. Intermolecular hydride transfer reactions do, however, provide an interesting example of steric factors at work. Hydride transfer is a reversible reaction [Eq. (129)], and hydride transfer equilibria can be used to provide quantitative information on the relative stabilities of carbonium ions (Stewart, 1957; Grinter and Mason, 1964; Deno *et al.*, 1962a; Bartlett and McCollum, 1956). If $^+CR_3'$ is the more stable ion with respect

$$R_3C^+ + H-CR_3' \rightleftharpoons R_3CH + {}^+CR_3' \tag{129}$$

to $CR_3{}^+$, then the equilibrium in Eq. (129) is shifted to the right. However, when the rates of hydride transfer are compared, steric as well as electronic factors become important. The log k for hydride transfer from xanthene (**125**) to a series of triaryl carbonium ions is a linear function of ΔpK_{R^+}. The same type of log k verses ΔpK_{R^+} relation is found on replacing triaryl with diaryl carbonium ions. However, the rate of hydride transfer to the diaryl carbonium ions is much faster (Deno et al., 1962a). This could be due to a buildup of

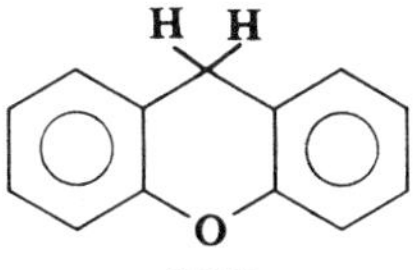

(125)

front strain in the transition state. The TS for hydride transfer to triaryl carbonium ions is more crowded than that for transfer to diaryl carbonium ions.

V. Reactions of Carbonium Ions

Many reaction pathways are open to carbonium ions. First, we must stress again that these reactions are affected by many factors; among these are (1) the source of the cation, (2) the temperature and nature of the medium, (3) the nature of the nucleophilic species present, and (4) factors affecting the stability of the cation. These factors have been discussed earlier, and examples were given to illustrate the specific effect on the reactivity of the cation.

Since carbonium ions are electrophilic species, their reactions are those generally expected of electrophiles. Other pathways arise as the cation becomes increasingly complex. For example, the methyl cation can only *react* as an electrophile in reactions such as Eqs. (130)–(134).

i. addition of a nucleophile

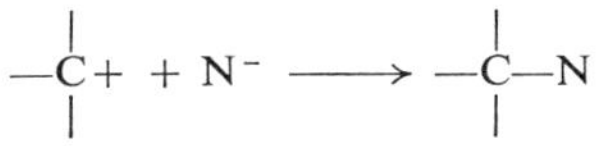

$$-\overset{|}{\underset{|}{C}}{}^+ + N^- \longrightarrow -\overset{|}{\underset{|}{C}}-N \quad (130)$$

ii. addition to an alkene to form a new cation

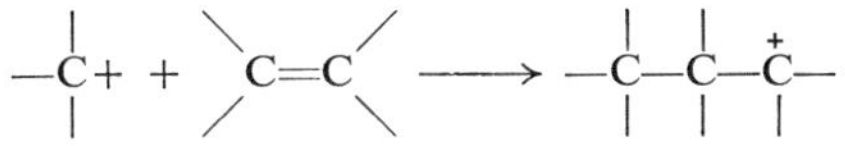

$$-\overset{|}{\underset{|}{C}}{}^+ + \overset{\diagdown}{\underset{\diagup}{C}}=\overset{\diagup}{\underset{\diagdown}{C}} \longrightarrow -\overset{|}{\underset{|}{C}}-\overset{|}{\underset{|}{C}}-\overset{|}{\underset{|}{C}}{}^+ \quad (131)$$

iii. abstraction of a hydride ion to form an alkane and a new cation

$$-\overset{|}{\underset{|}{C}}{}^+ + H-C\overset{R}{\underset{\diagdown}{\diagup}} \longrightarrow \overset{\diagdown}{\underset{\diagup}{C}}-H + {}^+C\overset{R}{\underset{|}{\diagup}} \quad (132)$$

iv. addition to an aromatic compound to yield, after expulsion of a hydrogen, a substituted aromatic species

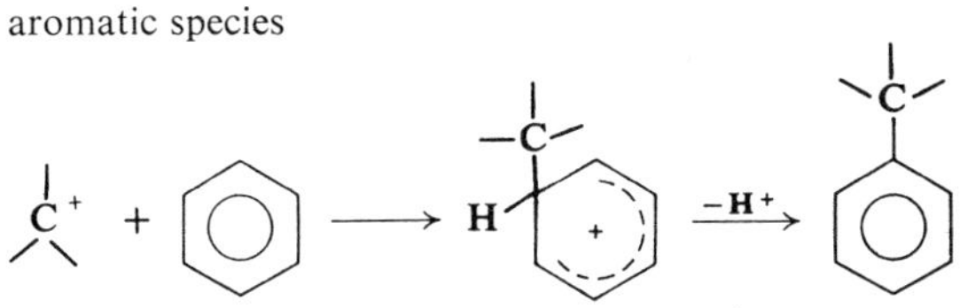

(133)

v. elimination of an α proton to form a carbene remains a possibility

$$CH_3^+ \xrightarrow[B:]{} :CH_2 + BH^+ \quad (134)$$

The ethyl cation can undergo each of these reactions, and, in addition, it can:

vi. undergo elimination of a β proton to form ethylene

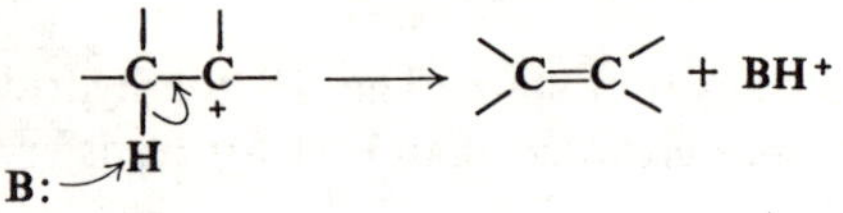

$$-\overset{|}{\underset{B:\nearrow H}{C}}-\overset{|}{\underset{+}{C}}- \longrightarrow {>}C=C{<} + BH^+ \quad (135)$$

vii. rearrange by a 1,2-hydride shift

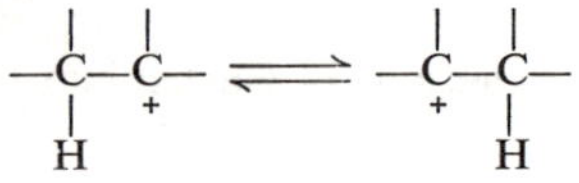

$$-\overset{|}{\underset{+}{C}}-\overset{|}{\underset{H}{C}}- \rightleftharpoons -\overset{|}{\underset{H}{C}}-\overset{|}{\underset{+}{C}}- \quad (136)$$

As the cation becomes more complex it may undergo other modes of reaction such as:

viii. 1,2-alkyl and aryl shifts

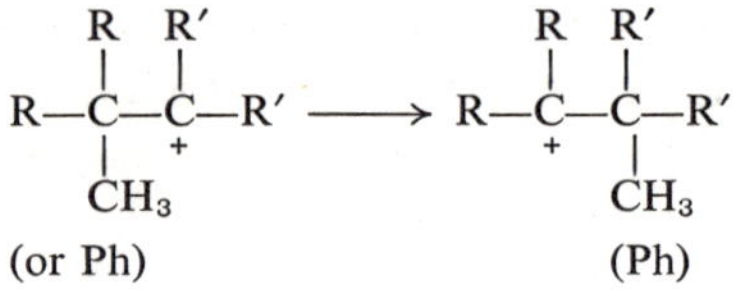

$$\begin{array}{c} R R' \\ | | \\ R-C-C-R' \\ | + \\ CH_3 \\ (\text{or Ph}) \end{array} \longrightarrow \begin{array}{c} R R' \\ | | \\ R-C-C-R' \\ + | \\ CH_3 \\ (\text{Ph}) \end{array} \quad (137)$$

ix. formation of bridged ions

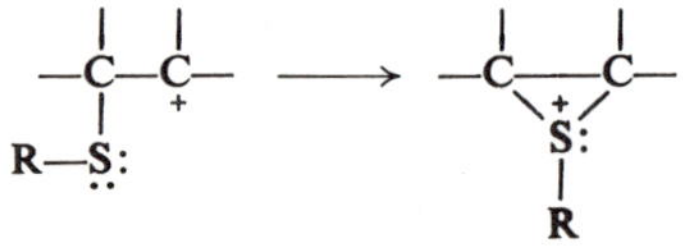

$$-\overset{|}{\underset{R-S:}{C}}-\overset{|}{\underset{+}{C}}- \longrightarrow -\overset{|}{\underset{\underset{R}{\overset{|}{S:}}}{C}}\overset{+}{\cdots}\overset{|}{C}- \quad (138)$$

x. fragmentation to an alkene and a new cation

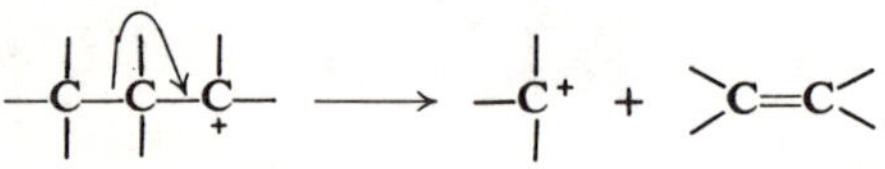

$$-\overset{|}{\underset{|}{C}}-\overset{|}{\underset{|}{C}}-\overset{|}{\underset{+}{C}}- \longrightarrow -\overset{|}{C^+} + {>}C=C{<} \quad (139)$$

xi. rearrange through a protonated cyclopropane intermediate

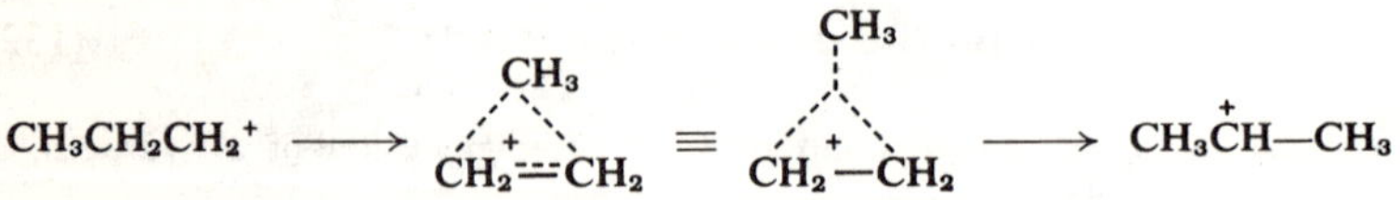

$$CH_3CH_2CH_2^+ \longrightarrow \underset{CH_2\cdots CH_2}{\overset{CH_3}{\triangle^+}} \equiv \underset{CH_2-CH_2}{\overset{CH_3}{\triangle^+}} \longrightarrow CH_3\overset{+}{C}H-CH_3 \quad (140)$$

xii. internal alkylation of a double bond or aryl group

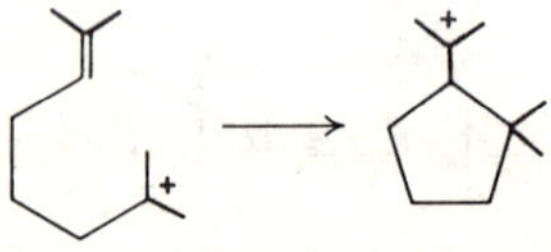

$$\quad (141)$$

xiii. 1,3 and higher order hydride transfers (see Section V,C,1)

A. Elimination of a β Proton

Proton elimination from a carbonium ion is the reverse of protonation. It has been argued that the transition state for the protonation of aliphatic olefins resembles the resulting cation (Gold and Kessick, 1965a,b). Thus, loss of a proton, according to the principle of microscopic reversibility, should proceed through the same transition state, and this should resemble the cation. This agrees with the observation that the addition of bases usually has only a minor effect on the ratio of substitution to elimination in E1 solvolysis reactions and the observation of very low deuterium isotope effects in these processes (Silver, 1961).

The elimination/substitution ratio increases as the cation becomes increasingly branched within the tertiary series (Brown and Fletcher, 1950; Hughes et al., 1953). For example, in the solvolysis of tert-alkyl chlorides in 80% EtOH–H$_2$O (v/v) at 25°C the tert-butyl derivative gives only 16% elimination, whereas 3-ethyl-3-pentenyl and 3-tert-butyl-3-pentenyl cations gave 40 and 90% elimination, respectively. This could be due to steric hindrance to the nucleophile's approach (reduces k_{subst}) or greater stabilization of the transition state to proton loss by alkyl substituents on the incipient double bond. If an alkyl cation can produce two or more olefin products, the one which is most rapidly formed (in highest yield) is usually the more substituted olefin [Saytzeff elimination product Eq. (142)]. As the size of the

$$\begin{array}{c} CH_3 \\ \diagdown \\ CH{-}\overset{+}{C}H{-}CH_3 \\ \diagup \\ CH_3 \end{array} \longrightarrow \begin{array}{c} CH_3 \\ \diagdown \\ C{=}C \\ \diagup \\ CH_3 \end{array}\begin{array}{c} H \\ \diagup \\ \diagdown \\ CH_3 \end{array} + H^+ \qquad (142)$$

Saytzeff product

β-alkyl substituents increase significantly, a greater yield of terminal olefin (Hofmann product) is observed, and the cis/trans ratio dramatically decreases. This effect has been explained in terms of the populations of the conformations of the cations and the rates of proton loss from them and eclipsing strain (steric effects, see Brown and Nakagawa, 1955).

β Proton elimination is much more common than α proton elimination. Drowning strong acid solutions of carbonium ions into rapidly stirred solutions of base [Eq. (143)] usually results in β elimination although hydration competes.

$$\begin{array}{c}\text{Ph}\\ \diagdown\\ \diagup\\ \text{OH}\end{array} \xrightarrow[\text{H}_2\text{O}]{\text{H}_2\text{SO}_4} \begin{array}{c}\text{Ph}\\ \diagdown\\ \overset{+}{}\\ \diagup\end{array} \xrightarrow[\text{H}_2\text{SO}_4]{\text{aq. OH}^-} \begin{array}{c}\\ \diagdown\\ \diagup\end{array}\!\!\text{-Ph} \qquad (143)$$

Elimination is often discussed in connection with solvolytic studies since S_N1 and E1 processes compete. We may make the following generalizations about E1 reactions (see Ingold, 1969): (1) The elimination/substitution ratio and the olefin A/olefin B ratio (if more than one can be formed) are unaffected by the leaving group unless the media is poorly solvating (ion pairing is important, Cocivera and Winstein, 1963); (2) the elimination/substitution ratio for alkyl cations is usually less than unity; and (3) elimination becomes more important as the cation varies from 1° to 3°.

B. Elimination of an α Proton

The formation of carbenes from carbonium ions by elimination of a proton is rare. An example of a reaction that probably proceeds via both cation and carbene intermediates is the aprotic deamination of the vinyl amine **126** (Curtin et al., 1965).

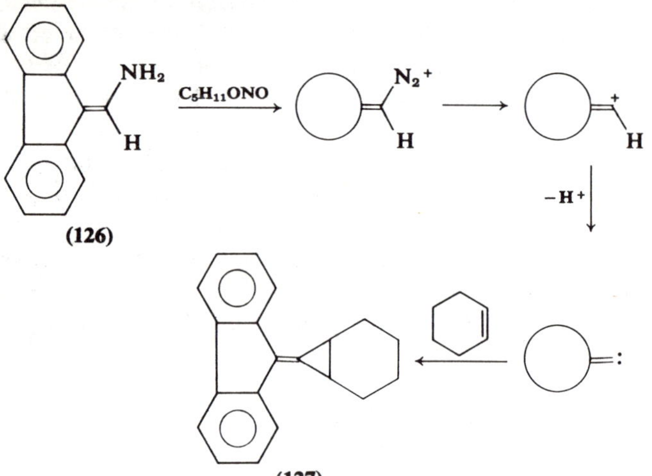

C. Rearrangement to Form Another Cation

Cations can rearrange by alkyl, aryl, and hydride shifts. We will first treat intramolecular hydride shifts (for a review, see Fry and Karabatsos, 1970).

1. Intramolecular Hydride Shifts

Hydride shifts are not involved in tertiary to primary or secondary to primary carbonium ion rearrangements. This is expected since tertiary cations are estimated to be about 33 kcal/mole more stable than primary ions, and the difference is about 22 kcal/mole between 2° and 1° cations. Thus, the rearrangement of the methylcyclopentyl cation to the cyclohexyl cation [Eq. (145)] does not proceed via a 1,2-hydride shift to the primary cation followed by a 1,2-alkyl shift. Instead, a 1,2- (tertiary to secondary) hydride shift would take place followed by formation of a protonated cyclopropane intermediate which opens to the cyclohexyl cation. The tertiary to secondary shift is permitted since the energy difference is only about 11 kcal/mole going from 3° to 2°.

$$\text{cyclohexyl}^+ \rightleftharpoons \text{cyclopentyl-CH}_2^+ \not\rightleftharpoons \text{cyclopentyl}^+ \rightleftharpoons \text{cyclopentyl}^+ \rightleftharpoons$$

$$\text{methylcyclopentyl cation} \longrightarrow \text{cyclohexyl}^+ \quad (145)$$

Secondary to secondary and tertiary to tertiary cation rearrangements can proceed via 1,2-hydride shifts. For example, the 12% rearrangement observed (Karabatsos et al., 1966) in the deamination of 2-butylamine-2-d proceeds by a 2° to 2° hydride (deuteride) shift in the intermediate cation. The interconversion by 1,2-hydride shifts of the 2,3-dimethyl-2-butyl cations (**128**)

$$\underset{\underset{NH_2}{|}}{CH_3CH_2CDCH_3} \xrightarrow[H_2O]{HNO_2} \underset{\underset{OH}{|}}{CH_3CH_2CDCH_3} + \underset{\underset{OH}{|}}{CH_3CHCHDCH_3} \quad (146)$$
$$ 88\% 12\%$$

was shown to proceed with a rate constant greater than 10^4 sec^{-1} at $-85°$C by Brouwer et al. (1965). Hydride shifts which produce more stable cations

$$\underset{\overset{H}{|}}{\overset{Me\;Me}{\underset{|}{|\;|}}}\!\!Me\!-\!\overset{+}{C}\!-\!C\!-\!Me \rightleftharpoons \underset{\overset{H}{|}}{\overset{Me\;Me}{\underset{|}{|\;|}}}\!\!Me\!-\!C\!-\!\overset{+}{C}\!-\!Me \quad (147)$$
$$\text{(a)} (\mathbf{128}) \text{(b)}$$

are well known. For example, the introduction of 1,2,2-triphenyl-1-d-ethanol into SO_2—SbF_5—FSO_3H generates the 1,1,2-triphenyl-2-d-1-ethyl cation quantitatively (Olah and Pittman, 1965) via a 1,2-hydride shift.

$$\underset{\underset{Ph}{|}}{\overset{\overset{OH}{|}}{Ph\!-\!CH\!-\!CD\!-\!Ph}} \longrightarrow \underset{\underset{Ph}{|}}{Ph\!-\!CH\!-\!\overset{+}{C}D\!-\!Ph} \longrightarrow \underset{\underset{Ph}{|}}{Ph\!-\!\overset{+}{C}\!-\!CHDPh} \quad (148)$$
$$(\mathbf{129}) (\mathbf{130})$$

1,3-Hydride shifts and higher order hydride shifts are well known and have been summarized (Fry and Karabatsos, 1970). The transannular hydride shifts discussed in Section II,E are probably the most common examples in this class.

2. Alkyl and Aryl Shifts

Alkyl and aryl groups show a high propensity to migrate if a more stable cation results. The Wagner–Meerwein and pinacolic rearrangements are examples which have been reviewed (Pocker, 1963). To emphasize the facility with which alkyl groups can migrate, consider the 2,3,3-trimethyl-2-butyl cation (**22**) which undergoes fast equilibration of all five methyl groups. In FSO_3H—SbF_5—SO_2ClF at $-160°C$ the PMR spectrum is still a singlet! Thus, the shift rate at this temperature is exceeding 5×10^3 sec^{-1}, and the energy barrier must be less than 5 kcal/mole (Olah and White, 1969). A combination of rapid reversible methyl and hydride shifts shown in Eq. (150) accounts for a detailed NMR variable temperature ($-60°$ to $180°C$) lineshape

$$CH_3-\underset{\underset{CH_3}{|}}{\overset{\overset{CH_3}{|}}{C}}-\overset{+}{\underset{\underset{CH_3}{|}}{C}}-CH_3 \rightleftharpoons CH_3-\overset{+}{\underset{\underset{CH_3}{|}}{C}}-\underset{\underset{CH_3}{|}}{\overset{\overset{CH_3}{|}}{C}}-CH_3 \quad (149)$$

(a)   (**22**)   (b)

analysis and spin-echo study by Saunders and Hagen (1968).

$$CH_3-\underset{\underset{+}{|}}{\overset{\overset{CH_3}{|}}{C}}-CH_2CH_2 \rightleftharpoons CH_3-\underset{\underset{H}{|}}{\overset{\overset{CH_3}{|}}{C}}-\overset{+}{C}HCH_3 \rightleftharpoons$$

$$CH_3-\overset{+}{\underset{\underset{H}{|}}{C}}-\underset{}{\overset{\overset{CH_3}{|}}{C}}HCH_3 \rightleftharpoons CH_3CH_2\overset{}{\underset{+}{\overset{\overset{CH_3}{|}}{C}}}-CH_3 \quad (150)$$

The cations produced from several precursors [Eq. (151)] all partition under similar conditions to give pinacol and pinacolone (Pocker, 1959). The similar product ratios for reactions proceeding at vastly different rates argues strongly for a cation intermediate.

$$\begin{array}{c}
(CH_3)_2C\text{——}C(CH_3)_2 \\
\diagdown\diagup \\
\overset{+}{O} \\
| \\
H
\end{array}$$

$$(CH_3)_2C\text{——}C(CH_3)_2$$
$$\overset{|}{OH} \quad \overset{|}{Cl}$$

$$(CH_3)_2C\text{——}C(CH_3)_2$$
$$\overset{|}{OH} \quad \overset{|}{N_2^+}$$

$$(CH_3)_2C\text{——}C(CH_3)_2$$
$$\overset{|}{OH} \quad \overset{|}{{}^+OH_2}$$

$\xrightarrow[\text{or Ag}^+\text{catalyzed}]{-Cl^-}$ uncatalyzed solvolysis, $-N_2$, $-H_2O$

$$(CH_3)_2C\text{——}\overset{+}{C}(CH_3)_2$$
$$\overset{|}{OH}$$

1,2-Me Shift ↓

$$CH_3-\overset{\overset{CH_3}{|}}{\underset{\underset{OH}{|}}{C}}-\overset{+}{\underset{\underset{CH_3}{|}}{C}}-CH_3 \xrightarrow{-H^+} CH_3-\underset{\underset{O}{||}}{C}-C(CH_3)_3$$

$$\xrightarrow{H_2O, -H^+} (CH_3)_2C\text{——}C(CH_3)_2$$
$$\overset{|}{OH} \quad \overset{|}{OH} \quad (151)$$

4. CARBONIUM IONS

The ability of a substituent to migrate to a cation center is called its "migratory aptitude." These aptitudes have been discussed by Cram (1956). In general, the migrating ability decreases in the series H > aryl > alkyl, but there are many exceptions because several effects are involved. They are solvent perturbations, the configuration and conformation of the starting compound, steric effects at the migration origin, eclipsing effects in the TS of the rearrangement, and the electron demand at the terminus. One example is the rearrangement of cation **129** to **130**. Does hydride transfer occur before phenyl equilibration or can the phenyl rings equilibrate (**129** ⇌ **131**) prior to

$$Ph_2-CH-\overset{+}{C}D-Ph \rightleftharpoons Ph_2-\overset{+}{C}-CD-Ph \qquad (152)$$
$$\text{\textbf{129}} \qquad\qquad \underset{\text{stable}}{\overset{|}{H}}$$
$$\updownarrow \qquad\qquad \text{\textbf{130}}$$
$$Ph-\overset{+}{C}H-CD-Ph_2$$
$$\text{\textbf{(131)}}$$

hydride transfer? Using methyl group labeling experiments, Olah *et al.* (1966a) demonstrated that phenyl equilibration did not occur in strong acid media. However, in certain deamination reactions, cation **129** can undergo phenyl equilibration prior to or in competition with the hydride shift (Collins, 1964).

The pinacol rearrangement has often been used to study migratory aptitudes. First, let us consider which hydroxyl group will be lost. The one which generates the more stable cation invariably is lost first. 1,1-Diphenylethanediol (**132**) gives diphenylacetaldehyde (**133**) and not phenylacetophenone (**134**). This case, then, does not represent an example of the preference of H migration over phenyl migration. Since the preference of losing one hydroxyl group versus another is not always so clear, mixtures may result.

$$Ph_2-\underset{\underset{\text{OH}}{|}}{\overset{\overset{\text{OH}}{|}}{C}}-CH_2 \xrightarrow[-H_2O]{H^+} Ph_2-\underset{+}{\overset{|}{C}}-\underset{\overset{|}{OH}}{CH_2} \xrightarrow{H \text{ shift}} Ph_2-\underset{\overset{|}{H}}{\overset{|}{C}}-\underset{\overset{|}{OH}}{\overset{+}{C}}-H$$
$$\text{\textbf{(132)}}$$
$$\qquad\qquad\qquad\qquad\qquad\qquad\qquad\qquad\qquad\qquad\qquad (153)$$
$$\downarrow \qquad\qquad\qquad\qquad\qquad\qquad\qquad\qquad \downarrow -H^+$$
$$Ph_2-\underset{\overset{|}{OH}}{\overset{|}{C}}-\underset{+}{CH_2} \longrightarrow Ph-\underset{\overset{||}{O}}{C}-CH_2Ph \qquad Ph_2-\underset{\overset{|}{H}}{\overset{|}{C}}-\underset{\overset{||}{O}}{C}-H$$
$$\qquad\qquad\qquad \text{\textbf{(134)}} \qquad\qquad \text{\textbf{(133)}}$$

If symmetrical pinacols are used, we may compare migratory aptitudes, providing proper considerations are given to stereochemical factors. In *threo*-1,2-diphenylethanediol the preferred conformer is probably **135** rather

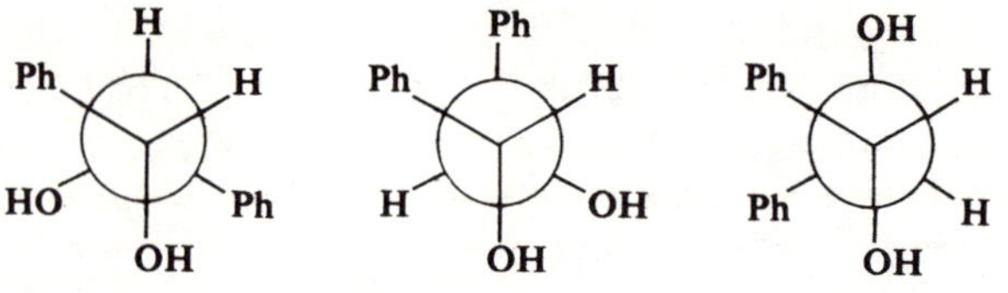

(154)

(135) (136) (137)

than **136** or **137**. Thus, no matter which hydroxyl group is lost the hydrogen is in the preferred position to migrate since no rotation is required for hydride migration from its antiperiplanar position. An antiperiplanar conformation with its favored orbital alignment is preferred.

If the erythro (meso) isomer is considered, one observes that conformer **138** is probably preferred. However, the hydroxyl group does not migrate.

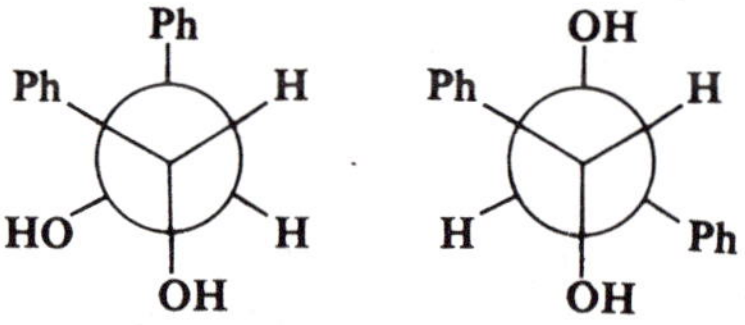

(139) (138)

If a hydroxyl group is lost in conformation **138**, then a rotation is required to satisfy orbital considerations. In conformer **139**, depending on which hydroxy group is lost, either a phenyl or hydrogen is antiperiplanar. Thus, arguments about true migratory aptitudes occasionally get complex.

3. *Cyclization to a New Carbonium Ion*

If a neighboring multiple bond is arranged so that ring formation will result in a five- or six-membered ring, cyclization occurs readily. The best

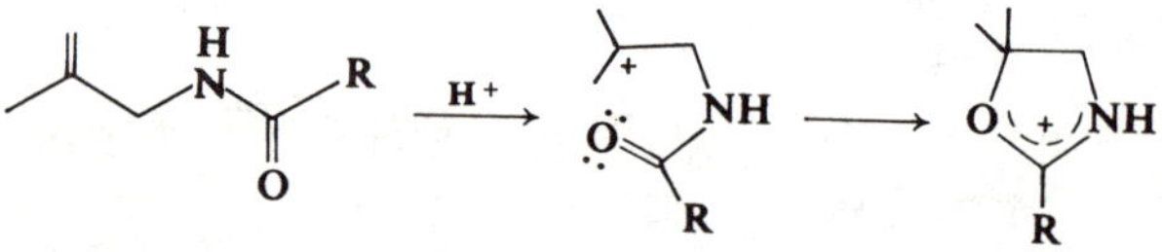

(140) (142) (155)

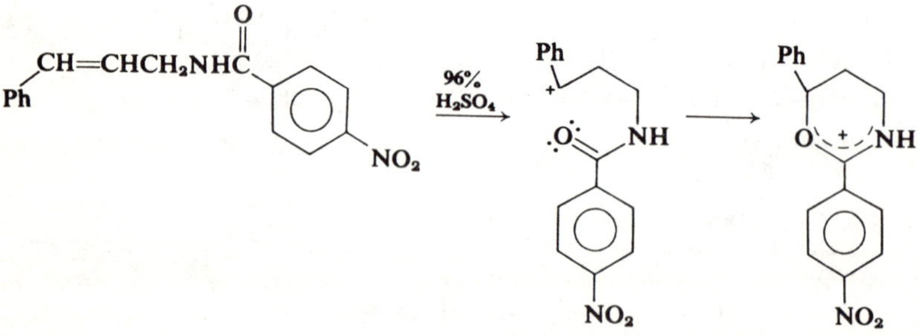

(143) (156)

4. CARBONIUM IONS

known examples in this class involve carbonyl groups (see Pittman *et al.*, 1972b, for a review). For example, the amide group intercepts the carbonium ion intermediate in the reactions of **140** and **141** to produce the heteronuclear stabilized cations **142** and **143** (McManus *et al.*, 1970, 1972).

The participation of olefinic and acetylenic groups occurs as well. Such examples may be considered as special cases of the reaction discussed in Section V,D.

D. ADDITION TO AN ALKENE (ALKYLATION)

This reaction can be illustrated by commercially important reactions of carbonium ions. A well-known process for making "isooctane" (2,2,4-trimethylpentane), the octane standard of 100, and an important high performance fuel involves the acid-catalyzed dimerization of isobutylene [shown earlier in Eq. (47)] to give trimethylpentenes. The alkylation process will only take place when the acidity is strong enough to allow a sufficient cation lifetime (or concentration) to exist so that this second-order reaction can occur. In dilute acids only olefin–alcohol hydration equilibrium exists. In 55 to 80% H_2SO_4 the alkylation shown in Eq. (47) generates dimer, trimer, and some traces of higher boiling olefins.

Thomas and Sparks (1944) found that Lewis acids were acceptable catalysts for the cationic copolymerization of mixtures of isobutylene and dienes. Butyl rubber, an extremely important elastomer, is prepared by polymerizing isobutylene with about 1.5% of isoprene [Eq. (157)]. The isoprene introduces a small amount of residual double bonds to allow vulcanization (cross-linking). Note that the intermediate dimer **144** is essentially the same as the

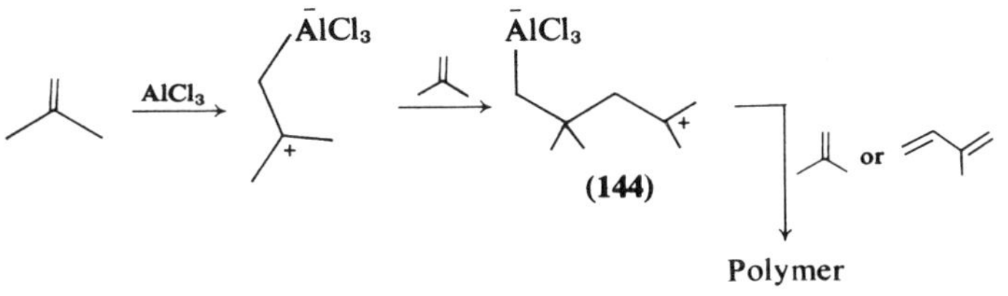

intermediate **51** in Eq. (47). The different end result illustrates the importance of conditions on the reaction.

E. INTERMOLECULAR HYDRIDE TRANSFER

Saturated alkanes can be produced from the olefin alkylation reaction [Eq. (158)] in one step by adding alkanes such as isobutane to the reaction mixture. Bartlett *et al.* (1944) first proposed an intermolecular hydride shift to account for this behavior when they observed that the reaction of *tert*-butyl chloride with $AlBr_3$ in the presence of isopentane gave isobutane and

$$(CH_3)_3CCl \xrightarrow{AlBr_3} (CH_3)_3C^+ + H-C(CH_3)_2C_2H_5 \longrightarrow (CH_3)_3CH + C_2H_5\overset{+}{C}(CH_3)_2 \quad (158)$$

tert-pentyl bromide. Extrapolation of these results to the alkylation of olefins makes it apparent that isobutane becomes the hydride donor to the dimer and trimer alkyl cations formed in the alkylation process. The new *tert*-butyl

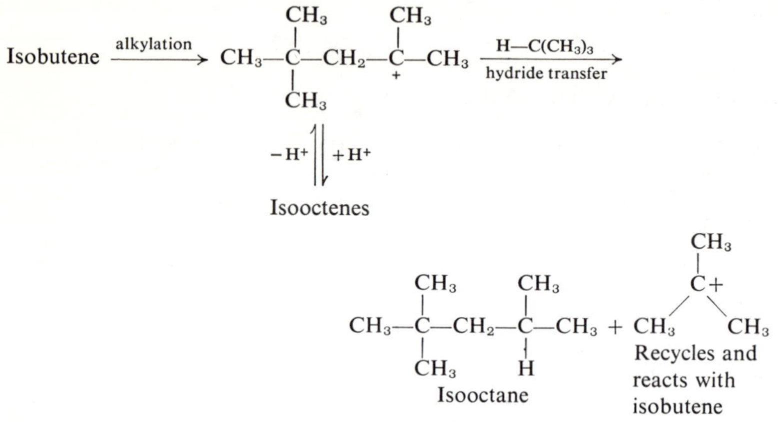

$$\text{Isobutene} \xrightarrow{\text{alkylation}} CH_3-\underset{\underset{CH_3}{|}}{\overset{\overset{CH_3}{|}}{C}}-CH_2-\overset{\overset{CH_3}{|}}{\underset{+}{C}}-CH_3 \xrightarrow[\text{hydride transfer}]{H-C(CH_3)_3} \quad (159)$$

$$\updownarrow \; -H^+ \;\; +H^+$$

Isooctenes

$$CH_3-\underset{\underset{CH_3}{|}}{\overset{\overset{CH_3}{|}}{C}}-CH_2-\underset{\underset{H}{|}}{\overset{\overset{CH_3}{|}}{C}}-CH_3 \;+\; CH_3\overset{\overset{CH_3}{|}}{\underset{CH_3}{C+}}$$

Isooctane — Recycles and reacts with isobutene

cation generated by this hydride transfer then propagates a chain reaction by alkylating another isobutene molecule. Usually, intermolecular hydride transfer gives the most stable cation. Thus, the highly stabilized 1,3-dioxolan-2-ylium cation (**145**) is readily formed when **146** is reacted with the stable

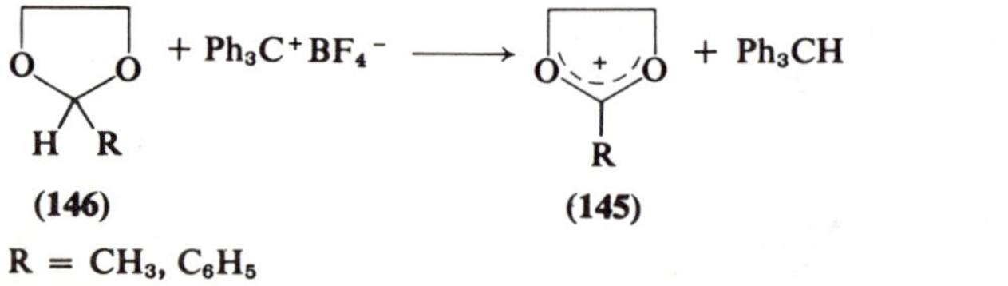

$$\underset{\underset{H \quad R}{(146)}}{\overset{O\frown O}{\diagdown\diagup}} + Ph_3C^+BF_4^- \longrightarrow \underset{\underset{R}{(145)}}{\overset{O\cdots\overset{+}{\cdots}O}{\diagdown\diagup}} + Ph_3CH \quad (160)$$

R = CH₃, C₆H₅

triphenylmethyl cation (see Pittman *et al.*, 1972b). Another example is the hydride transfer of an allylic hydrogen to an alkyl cation to generate an

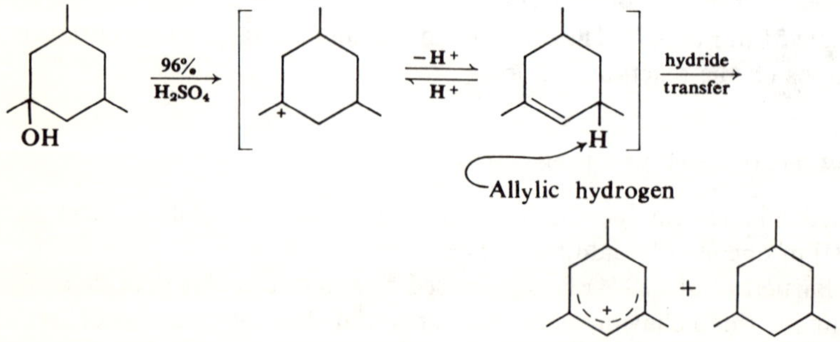

(161)

Allylic hydrogen

4. CARBONIUM IONS

alkane plus an allylic cation. For example, Deno and Pittman (1964b) showed that 1,3,5-trimethylcyclohexanol when dissolved into 96% H_2SO_4 undergoes disproportionation to the 1,3,5-trimethylcyclohexenyl cation and 1,3,5-trimethylcyclohexane by hydride transfer [Eq. (161)] before other rearrangements can take place.

F. Fragmentation

This process is formally the reverse of alkylation and is usually only observed when (1) a more stable cation results during cleavage or (2) when the lifetime of the ions are sufficient to allow fragmentation to proceed in competition with proton loss, rearrangement, and reaction with a nucleophile. An example of fragmentation generating a more stable cation is illustrated in Eq. (162). Alcohol **147** fragments to isobutylene and the very stable tropylium ion (**20**) (Conrow, 1959); however, $Ph_3CCH_2CH_2OH$ does not fragment (Deno and Sacher, 1965).

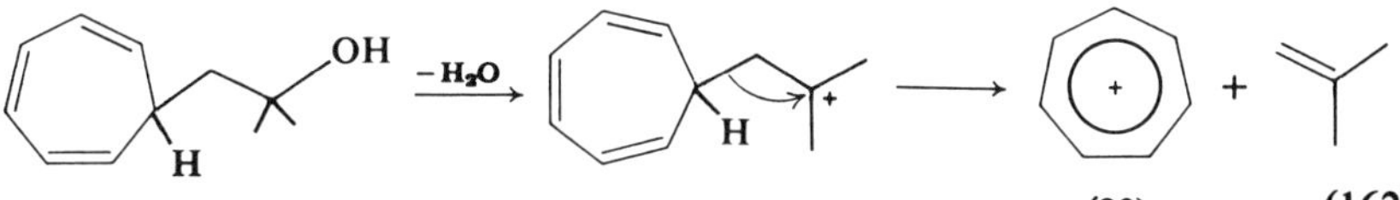

Fragmentation has been shown to be dependent on the arrangement of the participating orbitals. The required stereochemistry for maximum effect in **148** is obtained when the lone pair of the nitrogen and the plane of the C–X bond are both antiperiplanar to the fragmenting C_β–C_α bond. This arrangement is present in both 3-bromoadamantan-1-yl amine (**149**) and 4-bromo-

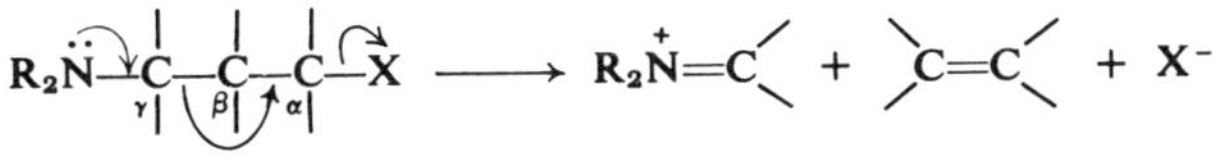

$$R_2\ddot{N}-\underset{\gamma}{C}-\underset{\beta}{C}-\underset{\alpha}{C}-X \longrightarrow R_2\overset{+}{N}=C\diagup + \diagup C=C\diagup + X^- \quad (163)$$

quinuclidine (**150**). Solvolysis of **149** and **150** in aqueous organic solvents occurs at rates which are, respectively, 30–500 times and 50,000 times greater than those of 1-alkyl-3-bromoadamantanes and 1-bromobicyclo[2.2.2]octane, respectively (Grob and Schwartz, 1964; Brenneisen et al., 1965).

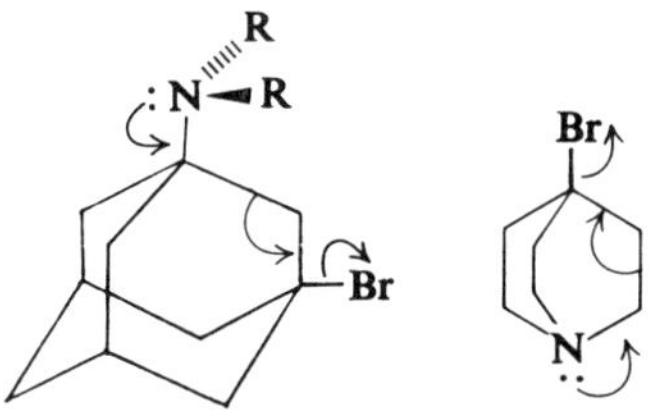

The stereoelectronic requirements are refined even further by comparing the reactions of 3β-chlorotropane (**151**) with those of 3α-chlorotropane (**152**). Since only one isomer undergoes the fragmentation process and since it reacts faster, this is strong evidence for a concerted process with the stereochemical requirements discussed above (Grob and Schiess, 1967).

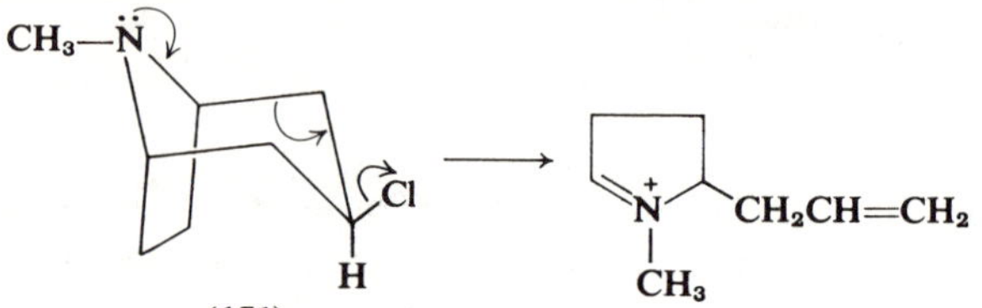

$$\text{(151)} \longrightarrow \overset{+}{N}(CH_3)\text{-}CH_2CH=CH_2 \quad (164)$$

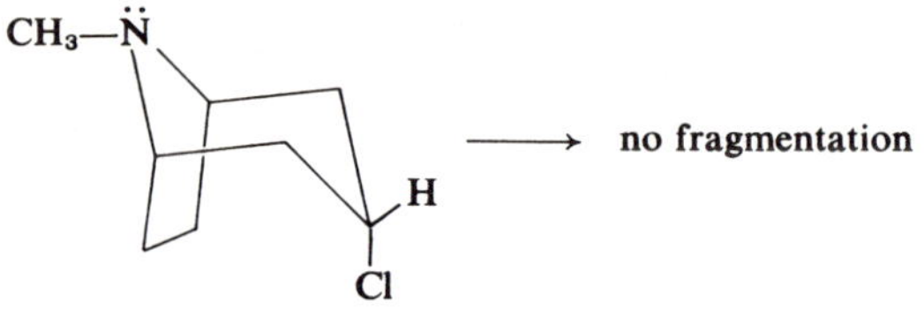

$$\text{(152)} \longrightarrow \text{no fragmentation} \quad (165)$$

We have seen that in 50 to 80% H_2SO_4 olefin alkylation reactions take place to give dimeric, trimeric, and some higher alkenes, but no fragmentation to intermediate species takes place. Thus, isobutylene (C_4) produces C_8, C_{12}, and C_{16} olefins but no C_5, C_6, C_7, C_9, C_{10}, or C_{11} products. This lack of fragmentation has been demonstrated for coalkylation of secondary and tertiary butyl alcohols and for mixed amyl and butyl alcohols (Whitmore *et al.*, 1941; Whitmore and Mixon, 1941). Alternatively, in 80 to 100% H_2SO_4, aliphatic alcohols or olefins disproportionate to produce cycloalkenyl cations in the acid layer and saturated, linear, highly branched alkanes which form a separate phase. The redox nature of this reaction was recognized by Ipatieff and co-workers (1953; Ipatieff and Linn, 1947) but only extensively elucidated by Deno *et al.* (1964). In this more strongly acidic media the lifetime of alkyl carbonium ions is sufficient to permit fragmentation to compete with other reaction paths.

G. Friedel–Crafts Alkylation of an Aromatic Ring

On examining the monumental treatise on the Friedel–Crafts reaction (Olah, 1963), one concludes that Friedel–Crafts alkylation of aromatic compounds is an important, but not necessarily a simple, process. Specifically, the Friedel–Crafts alkylation occurs under the influence of acid catalysts. Typically, the reaction can be represented as in Eqs. (166) and (167). The kinetics are in agreement with the rate-determining step being the formation of the σ complex (**153**).

4. CARBONIUM IONS

The formation of a carbonium ion as an intermediate is strongly indicated by several facts: (1) Many alkyl cations have been observed spectroscopically

$$(CH_3)CCl + AlCl_3 \longrightarrow (CH_3)_3C^+ + \bar{A}lCl_4 \qquad (166)$$

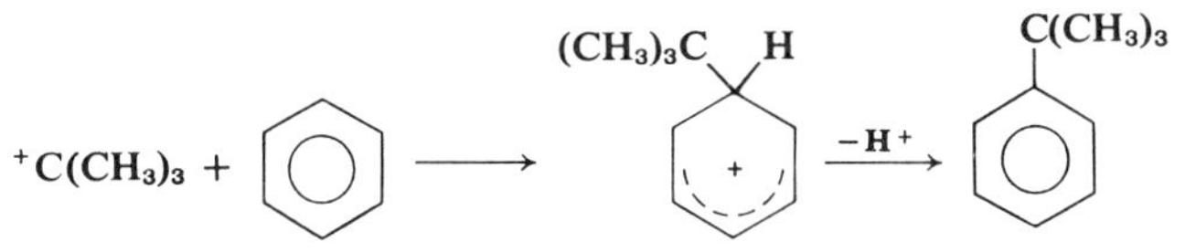

$$^+C(CH_3)_3 + \underset{}{\bigcirc} \longrightarrow \underset{(153)}{\overset{(CH_3)_3C\;\;H}{\bigodot_+}} \xrightarrow{-H^+} \underset{}{\overset{C(CH_3)_3}{\bigcirc}} \qquad (167)$$

and chemically under conditions similar to those of the Friedel–Crafts reaction (Olah and Olah, 1970); (2) some alkyl halides, especially primary, give skeletal rearrangement products [Eq. (168) (Gustavson, 1878)]; and (3)

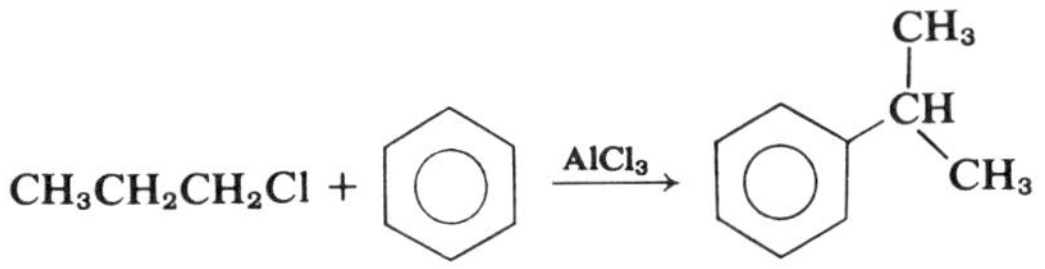

$$CH_3CH_2CH_2Cl + \underset{}{\bigcirc} \xrightarrow{AlCl_3} \underset{}{\bigcirc}\overset{\overset{CH_3}{|}\;\;\;\;\;}{\underset{CH_3}{\overset{CH}{\diagdown}}} \qquad (168)$$

alkylating agents with chirality at the potentially cationic center generally give predominantly racemization with a slight amount of inversion of configuration (Streitwieser and Stang, 1965).

This representation is an oversimplification, however, since there is much evidence to indicate that an open carbonium ion is not involved in all cases (see Bethell and Gold, 1967). The transition state in those examples (normally primary alkyl halides) has been suggested as that represented by **154** (Jungk *et al.*, 1956). Thus, we may consider the carbonium ion mechanism as the

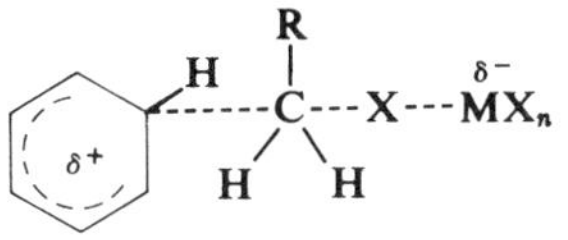

$$\underset{\delta^+}{\bigcirc}\overset{H}{\underset{}{\cdots}}\overset{\overset{R}{|}}{\underset{H\;\;H}{\overset{}{C}}}\cdots X \cdots \overset{\delta^-}{MX_n}$$

(154)

typical case for most secondary and tertiary systems and the limiting case for primary compounds.

H. COMBINATION WITH A NUCLEOPHILE

This pathway is one of the most common for carbocations since it is the one observed in unimolecular nucleophilic substitution reactions and unimolecular solvolytic reactions in aliphatic systems. Many examples of this reaction have been included throughout this chapter. Specifically, Section II,B has many examples.

The ability of a cation to combine with a nucleophile depends to a large extent on the nucleophilicity of the nucleophile. Swain and Scott (1953) have attempted to quantify nucleophilic character especially with respect to S_N2 reactions. Although the application here is not exact, a partial list of nucleophilicity constants relative to water as zero is included in Table XXXII. In particular, note that the nucleophilic character is not always a function of the ability of the anion to stabilize a negative charge.

Table XXXII
Nucleophilicity Constants for Some Common Nucleophiles

Nucleophile	Nucleophilicity constant	Nucleophile	Nucleophilicity constant
SH^-	5.1	OAc^-	2.72
CN^-	5.1	SO_4^{2-}	2.5
I^-	5.04	F^-	2.5
SCN^-	4.77	$2,4,5\text{-}(NO_2)_3C_6H_2O^-$	1.9
OH^-	4.20	NO_3^-	1.03
N_3^-	4.00	TsO^-	< 1.0
Br^-	3.89	H_2O	0.0
Cl^-	3.04	ClO_4^-	< 0

Nucleophilicity can sometimes be related to the base strength of an anion, but only when the same attacking atom is involved; for example, $CH_3O^- > HO^- > H_2O > HCO_3^-$.

I. Photochemical Reactions

The photochemical reactions of carbonium ions have been almost entirely neglected, and this represents an area where exciting research progress is still to be made. Only a few very stable cations such as the tropylium, triphenylcyclopropenyl, and triphenylmethyl have been photolyzed. Dauben (1960) isolated 9-phenylfluorene from tritylperchlorate which had lain on a desk top for 15 days. Van Tamelen and co-workers (1968) photolyzed the triphenylmethyl cation in a variety of solvents and got a variety of both monomeric (9-phenylfluoren-9-ol, triphenylmethane, 9-phenylfluorene, and

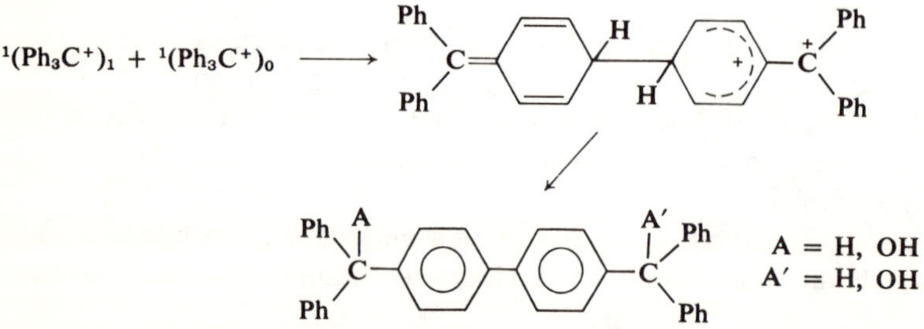

$$^1(Ph_3C^+)_1 + {}^1(Ph_3C^+)_0 \longrightarrow \text{[structures]} \quad A = H, OH; \quad A' = H, OH \tag{169}$$

4. CARBONIUM IONS

benzophenone) and dimeric products. The dimeric products arose from an initial electrophilic attack of an excited singlet carbonium ion on the ground-state ion [Eq. (169)] to give the para product exclusively.

Irradiation of the tropylium ion in 5% aq. H_2SO_4 gave a 58% yield of bicyclo[3.2.0]hepta-3,6-dien-2-ol (**155**) and its corresponding ether (**156**) via the intermediate valence-bond isomeric cation (**157**). Childs and Taguchi

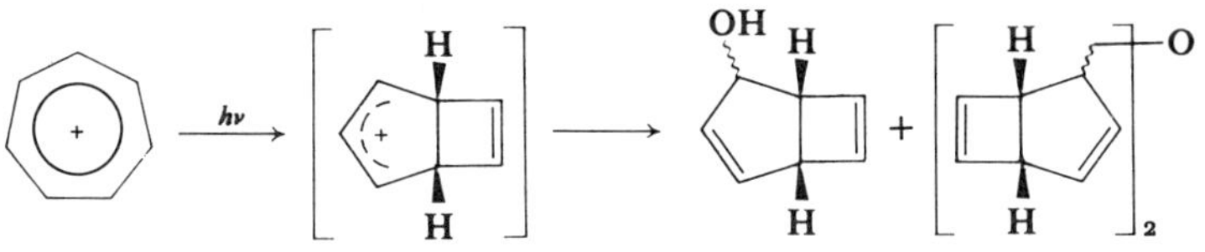

$$\text{(157)} \quad \text{(155)} \quad \text{(156)} \tag{170}$$

(1970) irradiated tropylium tetrafluoroborate in FSO_3H at $-60°C$ in order to observe intermediate cations which would be unstable in more nucleophilic solvents. They observed the norbornadien-7-yl cation (**158**) as the reaction product which presumably arose via rearrangement of **157**. Irradiation of the

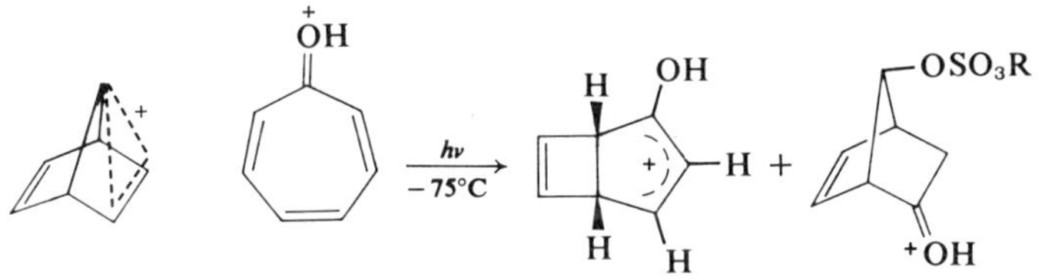

$$\text{(158)} \quad \text{(159)} \quad \text{(160)} \tag{171}$$

methyltropylium cation gave the 2-methylnorbornadienyl cation. That **157** was the primary photoproduct in these reactions was supported by the

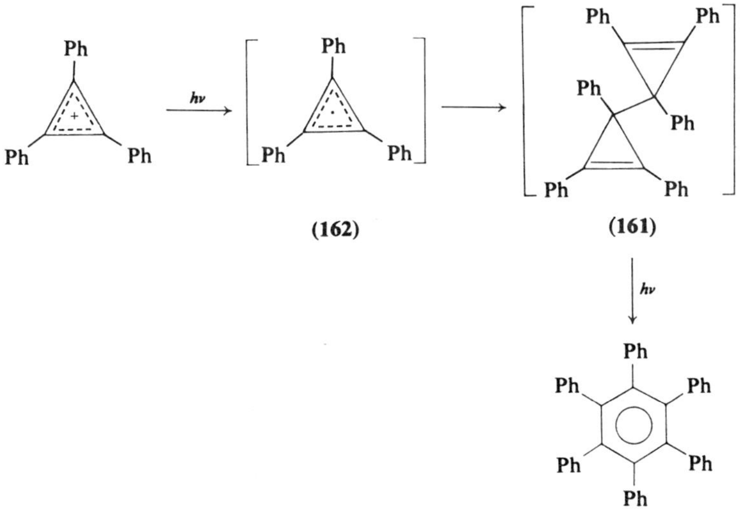

$$\text{(162)} \quad \text{(161)} \tag{172}$$

observation that protonated tropone ($-75°C$, FSO_3H) gave protonated bicyclo[3,2,0]hepta-3,6-dienone (**159**) and **160** as the observed products.

Irradiation of the triphenylcyclopropenyl cation (van Tamelen et al., 1968) gave hexaphenylbenzene in 49% yield. Since bis(triphenylcyclopropenyl) (**161**) is known to give hexamethylbenzene, it seems likely that the cation, on irradiation, undergoes charge transfer to radical **162**. The photochemical reactions of alkyl, allylic, dienylic, monoaryl, hetrocyclic, and other classes of cations, now readily available in strong acids, have not been reported.

VI. Bridged Carbonium Ions

A. Evidence for Bridged Ions

The concept of bridged ions is embraced by the larger concept of neighboring group participation. To account for the stereochemistry of addition of bromine to alkenes, Roberts and Kimball (1937) suggested an intermediate containing a halogen bridge. The extension (Winstein and Lucas, 1939) of the

$$\begin{array}{c} R_1 \quad\quad R_2 \\ \diagdown \quad \diagup \\ C - C \\ \diagup \ \diagdown_+\diagup\ \diagdown \\ R_3 \quad X \quad R_4 \end{array}$$

halonium ion concept to nucleophilic substitution reactions at saturated carbon was the starting point of a series of investigations of *neighboring group effects* in reactions of compounds containing oxygen, nitrogen sulfur, and halogens. Reviews by Bunton (1963), Capon (1964), Goodman (1967), Lwowski (1958), Pittman et al. (1972b), and Streitwieser (1962) are available. Participation by neighboring carbon ultimately resulted in studies of systems which have attracted widespread attention.

The basic concept of neighboring group participation in a solvolysis reaction is shown in Eq. (173). A pair of electrons is donated either from a single atom (lone pair of nonbonding electrons) or from a group of atoms ($C=O$, $C=C$, aryl, etc.). Where G is participating in the bond-breaking stage, the neighboring group is said to lend "anchimeric assistance" (Winstein and Grunwald, 1948). Orbital symmetry and steric considerations require that participation by the neighboring group and solvent attack occur from the direction opposite the leaving group or bridge. Thus, path (*a*) leads to overall retention of stereochemistry at the carbon undergoing solvolysis, and path (*b*) leads to rearranged product. Path (*c*) is an alternative resulting from the ambident nature (Pittman et al., 1972b; Hünig, 1964) of the bridged ion.

The participation of neighboring groups is experimentally observed in several ways. Anchimeric assistance generally is reflected by a large increase

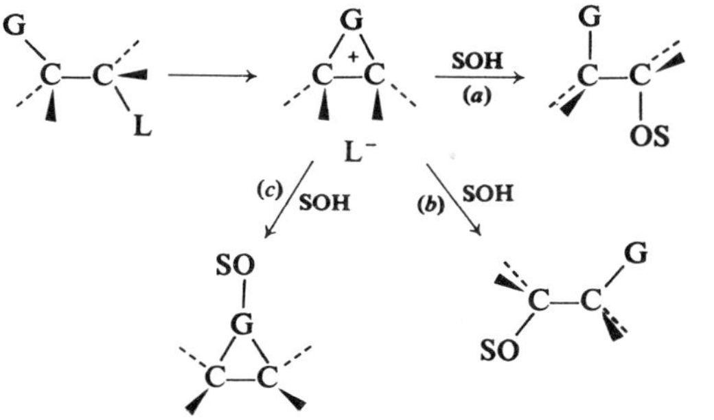

[Scheme (173) showing carbonium ion intermediate reactions with SOH pathways (a), (b), (c)]

(173)

in the reaction rate. For example, the rate data of Oae (1956) and Pasto and Serve (1965) for the solvolysis of some halo ketones (Table XXXIII) can be

Table XXXIII
Relative Rates of Solvolysis of Halo Ketones

Halide	Relative rate	
	Cl^a	Br^b
n-Butyl halide	1.0	1.0
Phenacyl halide	1.3	0.055
3-Halopropriophene	7.9	0.31
4-Halobutyrophenone	759.0	71.1
5-Halovalerophenone	21.3	c
6-Halocapriophenone	2.7	c

^{a} Reaction with $AgClO_4$ in 80% aqueous ethanol at 56.2°C (Pasto and Serve, 1965).
^{b} Reaction with mercuric nitrate in weakly acidic dioxane at 40.05°C (Oae, 1956).
^{c} Bromide not determined.

taken as evidence for participation by the carbonyl oxygen in the solvolysis of certain halides as shown in Eq. (174). This proposal leads to three other

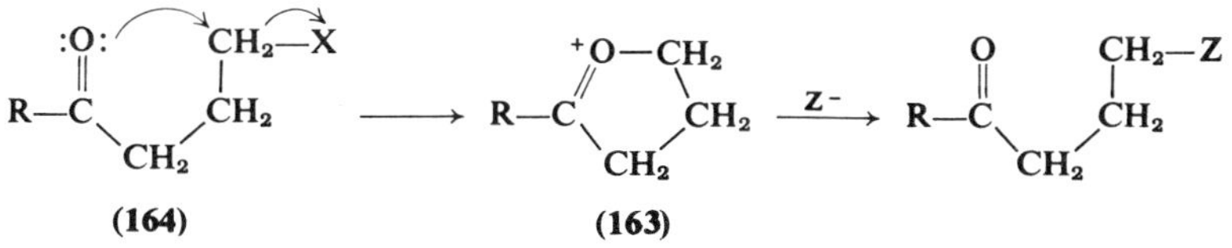

[Scheme (174) showing conversion of (164) through oxonium ion (163) to product]

(174)

methods of observing neighboring group participation: (1) observation of the intermediate; (2) isolation of the intermediate; and (3) determination of the stereochemical fate of the reaction. Ward and Sherman's (1968) follow-up to the work of Pasto and Serve (1965) illustrates two methods. First, they observed the presence of the oxonium ion intermediate in several cases using NMR techniques. This spectroscopic observation of the intermediate then led to the successful isolation of a salt of the intermediate **163** (R = phenyl) from the solvolysis of **164** (R = phenyl, X = *p*-bromobenzene sulfonate). This, of course, supports the original suggestion that neighboring group participation occurs.

Since solvolysis should lead to inversion of configuration or racemization through either an S_N1 or S_N2 mechanistic pathway (Ingold, 1969), the study of a reaction involving solvolysis at an asymmetric carbon should yield stereochemical evidence of participation. The classic Walden inversion [Eq. (175)] typifies the use of this kind of evidence (Winstein and Lucas, 1939). Ionization of the bromide ion in a weakly basic solution is assisted by the carboxylate group. The α-lactone is formed by inversion. A subsequent

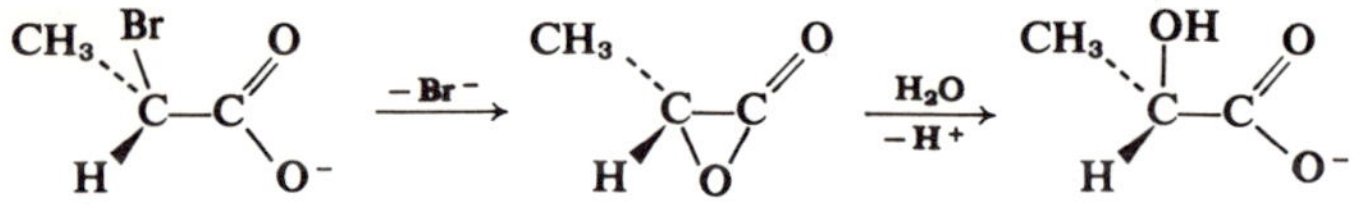

$$\underset{H}{\overset{CH_3}{\underset{|}{\overset{Br}{C}}}}-\overset{O}{\underset{O^-}{C}} \xrightarrow{-Br^-} \underset{H}{\overset{CH_3}{C}}-\overset{O}{\underset{O}{C}} \xrightarrow[-H^+]{H_2O} \underset{H}{\overset{CH_3}{\underset{|}{\overset{OH}{C}}}}-\overset{O}{\underset{O^-}{C}} \quad (175)$$

inversion (on water attack at C-2) gives the final product with overall retention.

A final way that neighboring group participation or anchimeric assistance is manifested is through a molecular rearrangement. This is exemplified in the reactions of cholesteryl tosylate (**165**). Shoppee (1946) observed that cholesteryl chloride undergoes nucleophilic substitutions with complete retention of configuration which resulted from participation of the double bond as in **166**. Similarly, Winstein and Adams (1948) observed that **165** undergoes methanolysis to give a product of retention (**167**). When the same reaction is carried out in the presence of acetate ion, the *i*-cholesteryl ether (**168**) is formed.

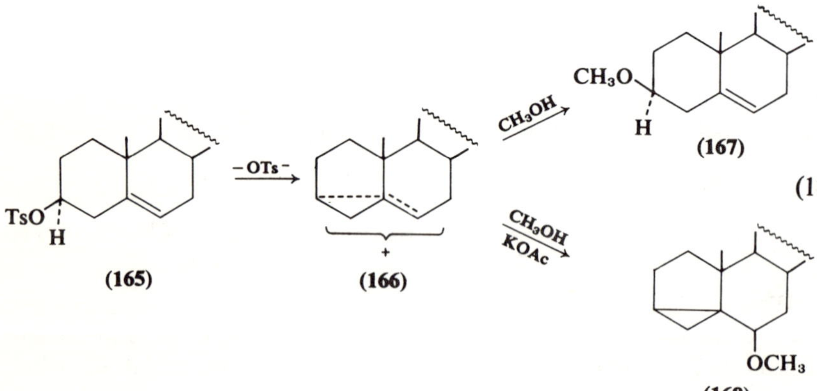

(176)

The foregoing techniques can be placed into four classifications: (i) kinetic methods; (ii) application of spectroscopic methods; (iii) stereochemical observations; and (iv) product analysis. There are, of course, hazards in interpreting the results of a single investigation of some of the above classes in terms of bridged ions. For example, a comparison of the reaction rates reflects only the differences in the free energies of activation and, thus, provides no information beyond the transition state along the reaction coordinate. Spectroscopic techniques, such as proton NMR, may only give "average" structures owing to the measurement time and, therefore, may not yield definitive data on the existence of a bridged ion intermediate [as

$$\overset{X}{\underset{\diagup}{C}}\overset{+}{-}\overset{}{\underset{\diagdown}{C}}\diagdown \qquad \overset{X}{\underset{\diagup}{C}}\overset{}{-}\overset{}{\underset{+}{C}}\diagdown \rightleftarrows \overset{X}{\underset{\diagup}{C}}\overset{+}{-}\overset{}{\underset{\diagdown}{C}}\diagdown \qquad (177)$$

bridged ion   **equilibrating ions**

opposed to equilibrating ions, i.e., Eq. (177)]. One also must recognize that structural factors inherent to a system may have the major influence on its

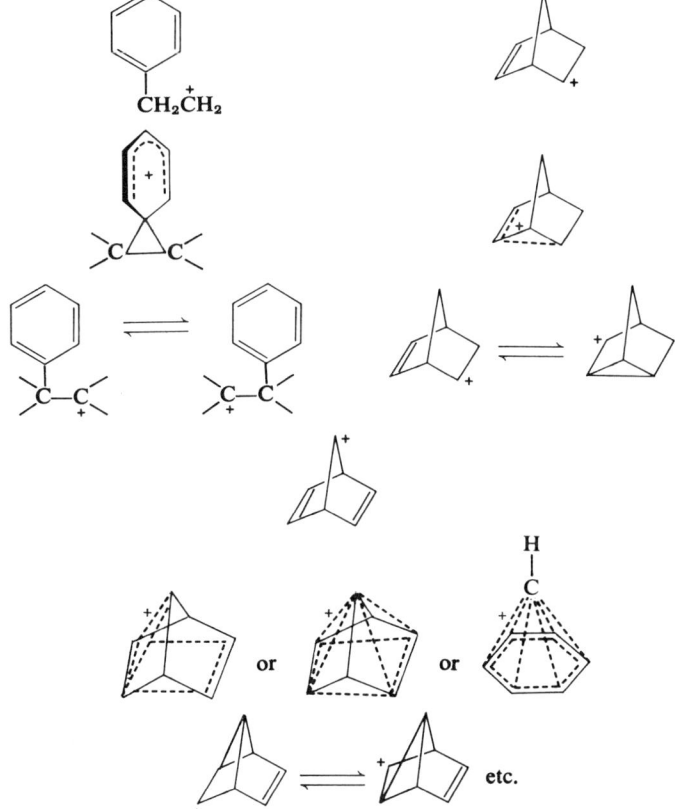

Fig. 13. Some carbonium ions for which electron-sufficient nonclassic structures have been considered.

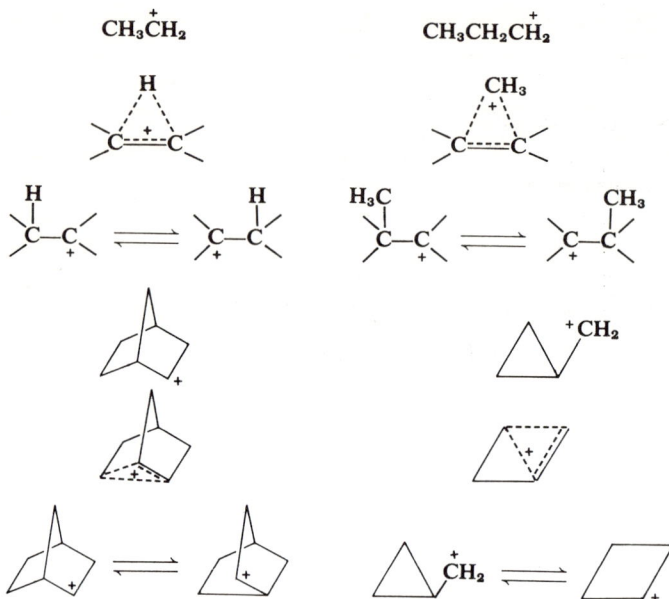

Fig. 14. Some carbonium ions for which electron-deficient nonclassic structures have been considered.

reactions. Thus, care should be exercised in drawing conclusions based on fragile experimental evidence.

At the 139th National Meeting of the American Chemical Society in St. Louis, Missouri in 1961, at the symposium on "The Transition State" (Brown, 1962) at Sheffield, England in 1962, and at several conferences since then, Brown has argued that several carbonium ions popularly regarded as nonclassic species are actually pairs of equilibrating ions. Concerning the experimental basis for such proposals, Brown stated "but the Emperor is naked" (Brown, 1967). Among those ions which Brown defines as equilibrating classic ions are those in Figs. 13 and 14. The ions are separated into classes called electron-sufficient or electron-deficient bridged ions and are treated in those classifications here.

In a footnote in the preface to his book "Non-classical Ions—Reprints and Commentary," Bartlett (1965) quotes from *Chemical Abstracts*: "Only a dark, undistillable resin remained upon removal of the solvent. This suggests that a nonclassical carbonium ion intermediate is involved in the mechanism." Such an interpretation of data is an extreme example of the sort of nonobjectivity that apparently became too common for Brown to accept. In the following sections, some of the existing data for selected controversial ions will be treated.

B. Electron-Sufficient Bridged Ions

Roberts and Lee (1951) introduced the term "non-classical carbonium ion" to describe a structure for the norbornyl cation originally proposed by Winstein and Trifan (1949). Although the word "non-classical" has since been widely used to describe all kinds of bridged carbonium ions, Bartlett (1965) has redefined an ion to be "non-classical if its ground state has delocalized bonding σ electrons." Electron-sufficient bridged carbonium ions clearly do not qualify as "non-classical" ions according to Bartlett's definition. Despite their nomenclature, the controversy over the structure of many of these intermediates places them in a class apart from simple carbonium ions.

1. Aryl Bridged Ions

In 1949, Cram presented stereochemical evidence which led to his postulation of phenyl group assistance in the acetolysis of optically active *erythro*- and *threo*-3-phenyl-2-butyl tosylates (Cram, 1949). The "phenonium ion" (Cram, 1952) intermediate from the *erythro*-tosylate can only lead to the

$$\underset{\text{L-}threo}{\overset{\text{H}}{\underset{\text{C}_6\text{H}_5}{\text{H}_3\text{C}}}\!\!\!\!\!\!\!\!\overset{*}{\text{C}}\!\!-\!\!\overset{*}{\underset{\text{H}}{\text{C}}}\!\!\!\!\!\!\!\!\overset{\text{OTs}}{\underset{\text{CH}_3}{}}} \xrightarrow{-\text{OTs}^-} \underset{\substack{\text{Symmetrical}\\\text{phenonium ion (Cram, 1949)}}}{\overset{\text{H}}{\underset{\text{H}_3\text{C}}{}}\!\!\!\!\!\!\!\!\text{C}\!\!=\!\!\underset{\text{CH}_3}{\overset{\text{H}}{\text{C}}}\!\underset{\overset{+}{\text{C}_6\text{H}_5}}{}}$$

$$\downarrow \text{AcOH}$$

$$\underset{96\% \text{ Racemic } threo\text{-acetate}}{\left[\underset{\text{C}_6\text{H}_5}{\overset{\text{H}}{\text{H}_3\text{C}}}\!\!\!\!\!\!\!\!\text{C}\!\!-\!\!\underset{\text{H}}{\overset{\text{OAc}}{\text{C}}}\!\!\!\!\!\!\!\!\overset{}{\underset{\text{CH}_3}{}} + \underset{\text{H}}{\overset{\text{AcO}}{\text{H}_3\text{C}}}\!\!\!\!\!\!\!\!\text{C}\!\!-\!\!\underset{\text{C}_6\text{H}_5}{\overset{\text{H}}{\text{C}}}\!\!\!\!\!\!\!\!\overset{\text{CH}_3}{}\right]}$$

(178)

optically active *erythro*-acetate. The *threo*-tosylate, however, leads to a symmetric phenonium ion and, thus, a racemic product.

The phenonium ion is actually a special example of the σ complex found in electrophilic aromatic substitution (Berliner, 1964). This is evident in Cram's representation of the phenonium ion [Eq. (180, Cram, 1964]. Brown *et al.* (1965) considered the phenonium ion concept as essentially untenable in light of the small rate accelerations present in such reactions. Brown *et al.* (1965) chose instead to account for the stereochemical results of such reactions in terms of rapidly equilibrating ions. Such steric control was descriptively named the "windshield wiper effect." In the case of certain activated

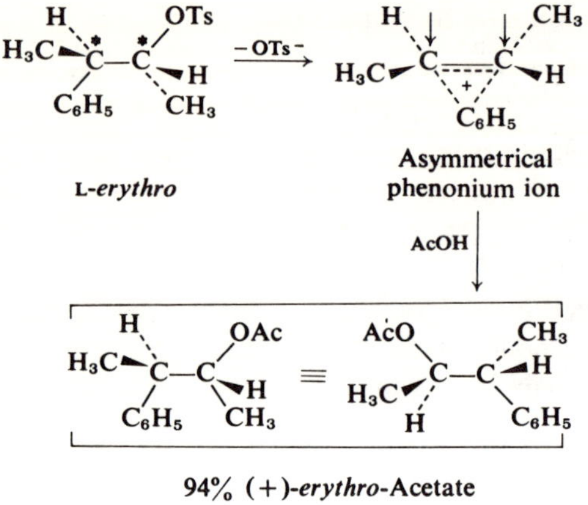

$$\text{L-}erythro \xrightarrow{-\text{OTs}^-} \text{Asymmetrical phenonium ion} \xrightarrow{\text{AcOH}} 94\% \ (+)\text{-}erythro\text{-Acetate} \tag{179}$$

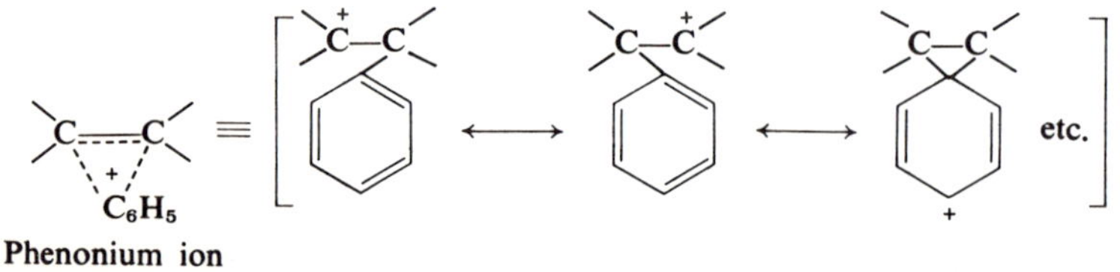

$$\text{Phenonium ion} \equiv \left[\ \leftrightarrow \ \leftrightarrow \ \text{etc.} \right] \tag{180}$$

β-arylethyl derivatives such as β-anisylethyl derivatives, Brown proposed to explain the kinetic acceleration of such systems in terms of rapidly equilibrating π bridged intermediates, although he recognized the phenonium ion as

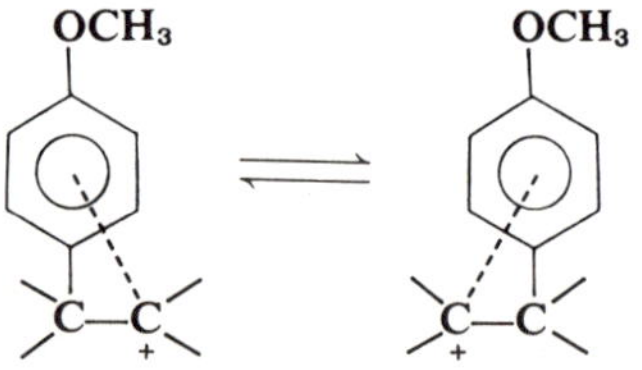

$$\tag{181}$$

an extreme possibility [Brown et al., 1965; Eq. (181), Brown and Kim, 1968].

Sufficient evidence has now been reported which demonstrates that two discrete, strongly assisted pathways k_s (solvent assisted) and k_Δ (aryl assisted) compete without crossover in the solvolysis of primary (Bentley and Dewar, 1970; Harris et al., 1969; Jablonski and Snyder, 1969; Jones and Coke, 1969; Coke et al., 1969; Diaz et al., 1968; Nordlander and Deadman, 1968) and secondary (Brown et al., 1970; Diaz and Winstein, 1969; Lancelot and Schleyer, 1969; Schleyer and Lancelot, 1969; Lancelot et al., 1969; Brown and Kim, 1971; Thompson and Cram, 1969; Nordlander and Kelly, 1969; Nordlander and Deadman, 1968) β-arylalkyl arenesulfonates. The terms can be treated quantitatively by the use of the Winstein equation (182) (Eberson et al., 1965), where k_t is the total titrimetric rate or the total solvolysis rate

$$k_t = k_s + Fk_\Delta \tag{182}$$

depending on the treatment, and F is the fraction of aryl-assisted reaction which leads to product (i.e., a correction for ion-pair return). A Hammett plot (Harris et al., 1969) (Fig. 15) is sufficient to show that such systems may

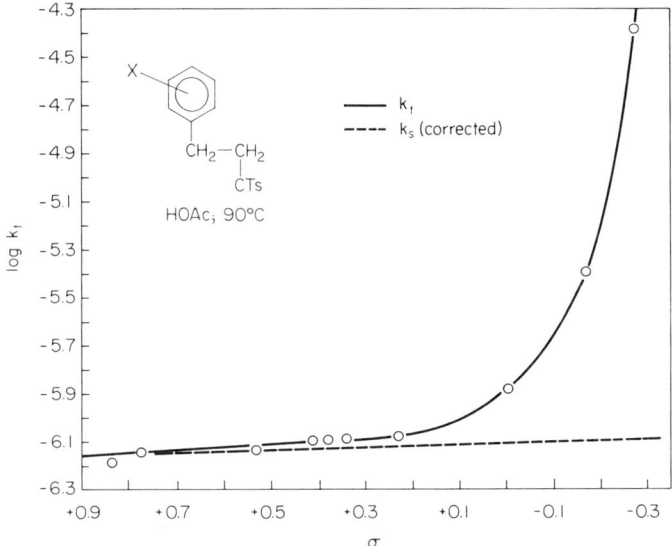

Fig. 15. Log k_t versus σ; β-arylethyl tosylate acetolysis. Reprinted from J. Amer. Chem. Soc. 91, 7508 (1969). Copyright (1969) by the American Chemical Society. Reprinted by permission of the copyright owner.

essentially follow k_s or k_Δ pathways. The conclusion that the k_s and k_Δ pathways are in fact separate and independent comes from the close agreement between the rate and product studies (Harris et al., 1969).

The lack of a marked increase in rate results from an already efficient S_N2 solvent-assisted process with which aryl assistance must compete (Schleyer and Lancelot, 1969). In the aryl participation process the high degree of charge delocalization into the participating aryl nucleus is revealed by the ρ, -2.4 at 115°C, from the plot of σ^+ versus log k_Δ (Harris et al., 1969).

The presence of both favorable rate and product studies offers strong evidence for a phenonium ion transition state (Schleyer and Lancelot, 1969; Harris et al., 1969) for the aryl-assisted fraction.

There is direct evidence for the existence of phenonium ion intermediates from other studies. For example, Baird and Winstein (1957, 1963) isolated the spirodienone (**169**) from the methanolysis of **170**. The reaction of **169** with methanol is acid catalyzed.

Olah *et al.* (1969) found that ionization of 3-*p*-X-phenyl-2,3-dimethyl-2-butyl chloride (**171**) in SbF_5—SO_2 at $-78°C$ yields the most thermodynamically stable cation or mixture of cations. For X = H, the tetramethylethylene phenonium ion (**172**) was found to be most stable. For

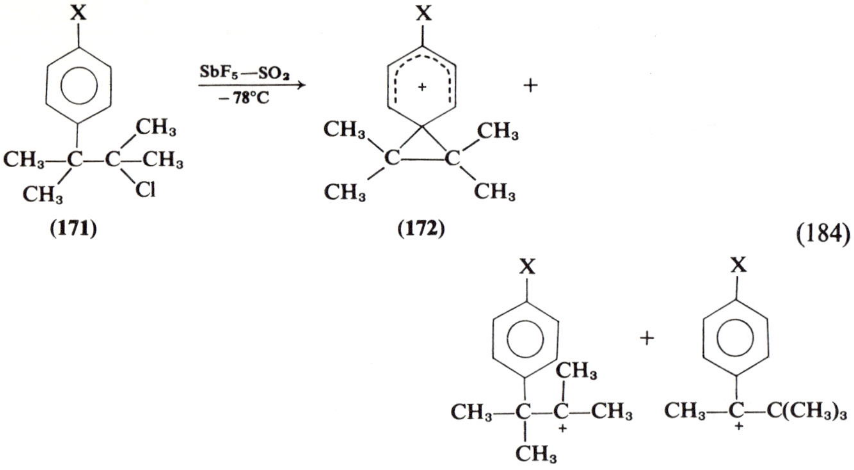

X = CH_3, a near-equal mixture of the phenonium ion and the rearranged classic ion **173** was formed. With X = OCH_3, the rearranged ion **173** was most stable, and with X = CF_3 the classic cation **174** was most stable.

Olah and Porter (1970, 1971) have now reported observing the stable ethylenephenonium ion (**175**) in mixture with the styryl cation in SbF_5—SO_2ClF at $-78°C$. Since the solvent is quite different from that found in solvolytic reactions, a one-to-one comparison is not possible. Nevertheless, Olah and Porter have demonstrated that delocalization of charge in the benzene ring shows a charge distribution similar to protonated benzene (i.e., a cyclohexadienyl cation), and the delocalization of charge into the spirocyclopropane portion resembles charge delocalization into secondary cyclopropylcarbinyl cations. Thus, this ion is actually a spiro[2.5]octadienyl cation. The ^1H and ^{13}C NMR data summarized below elegantly point out these conclusions and confirms the classic nature of **175**.

According to Hammond's postulate (Hammond, 1955), one should predict that the transition state for the solvolysis of β-arylethyl derivatives resembles the first intermediate (Brown and Kim 1968). Thus, for that fraction of

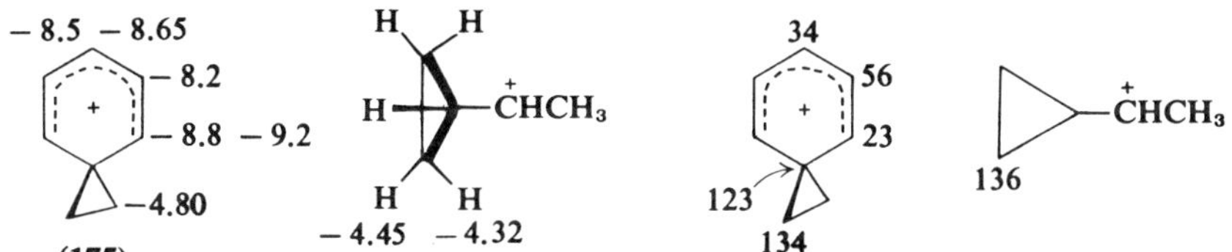

4. CARBONIUM IONS

−8.5 −8.65
−8.2
−8.8 −9.2
−4.80
(175)

^{1}H Chemical shifts
(relative to TMS)

H, H, H, CHCH$_3$, H, H
−4.45 −4.32

34, 56, 23
123, 134

136 —CHCH$_3$

^{13}C Chemical shifts
(relative to ^{13}CS$_2$)

solvolysis reactions proceeding with aryl assistance, the first *intermediate* is best described as a phenonium ion (i.e., a spiro[2.5]octadienyl cation).

2. Homoallylic Cations

An earlier example was given to illustrate the first well-studied cation system of this class, the cholesteryl → *i*-cholesteryl rearrangement. Although stereochemical factors may influence specific systems of this class to behave differently, this system can actually be considered as a π route to the cyclo-

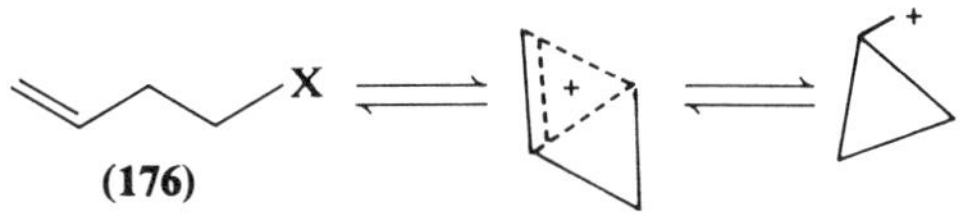

$$\text{(176)} \qquad \rightleftharpoons \qquad \rightleftharpoons \qquad \qquad (185)$$

propylcarbinyl nonclassic cation system [Eq. (185)]. Except for the differences which can be accounted for by solvent assistance (k_s pathway), the reaction products from the formolysis of but-3-enyl tosylate (**176**, X = OTs) are similar to those obtained from solvolytic reactions of cyclopropylmethyl and cyclobutyl compounds. Since the latter compounds are discussed in the next section and since a common intermediate may explain the results of the solvolysis of these homallylic cations, these ions will be treated in the next section.

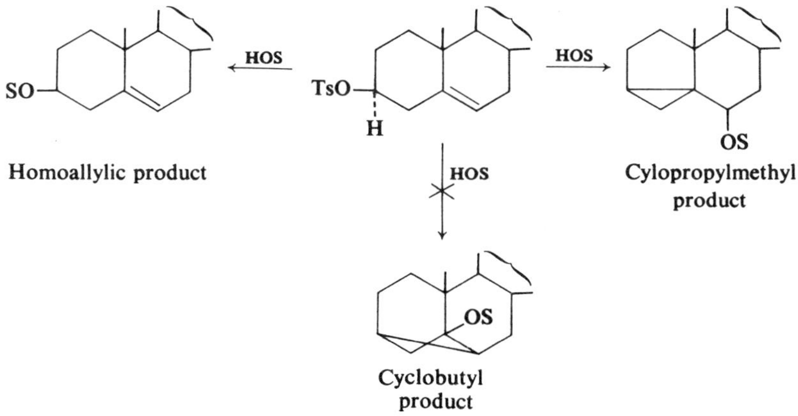

Homoallylic product ←HOS— TsO—H —HOS→ Cylopropylmethyl product   (186)

↓ HOS

Cyclobutyl product

Special steric factors in cyclic alkenyl systems set them apart from simple homoallylic systems. This results in reaction products unlike those expected from the symmetrically bridged homoallylic cation. The cholesteryl system already mentioned is one such example. The cyclobutyl product is not observed presumably because of an unfavorable transition state leading to it. Geometric factors in the starting material are also important in cyclic compounds. For example, a marked rate acceleration occurs only with the leaving group in the β configuration in cholesteryl derivatives.

endo-5-Norbornenyl derivatives show a depressed reaction rate although anchimeric assistance appears to be present in the corresponding exo derivatives. A rate comparison for several norbornyl derivatives is given in Fig. 16.

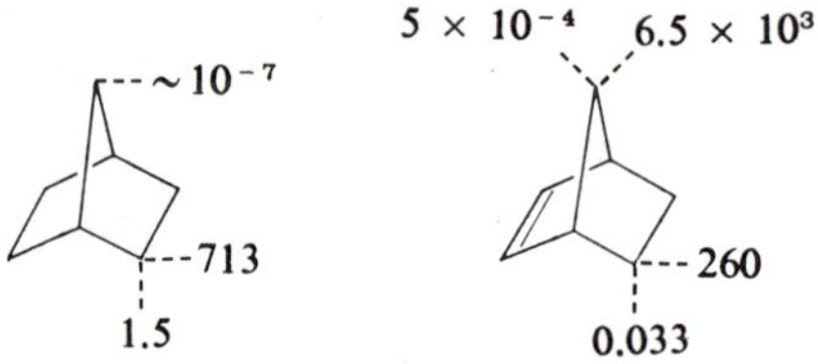

Fig. 16. Reactivities (acetolysis at 25°C, relative to cyclohexyl *p*-bromobenzene sulfonate = 1) of various bicyclic *p*-bromobenzene sulfonates. Data compiled by Berson (1963).

The kinetic evidence clearly suggests that the alkenyl derivatives with anti 7-substituents are greatly assisted during the ionization stage. These systems have near optimum stereochemical arrangement (see **177**) for π orbital

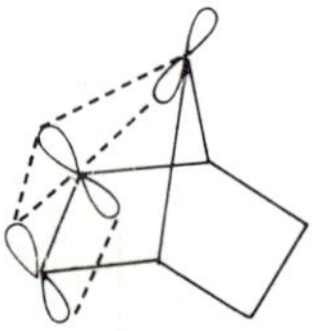

(**177**)

interaction with the back lobe of the σ bonding orbital which is becoming a p orbital.

Goldstein and Hoffmann (1971) have predicted from theory that the 7-norbornadienyl cation (**178**) is among several longicyclic systems that

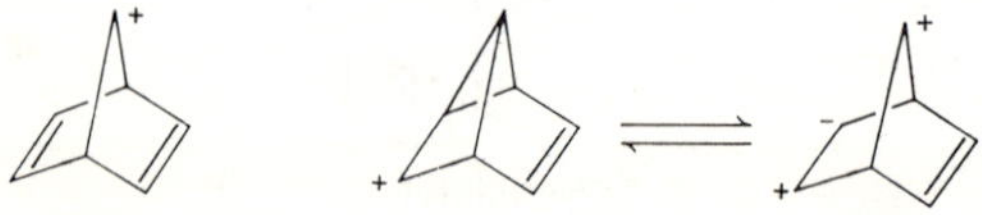

(**178**)   (**179**)   (**187**)

4. CARBONIUM IONS

should be stabilized by π electron overlap. Interestingly, the alternate ionic structure (equilibrating ions) preferred by Brown (1967) is also potentially stabilized since it is the collapsed form of **179** which is predicted to be stabilized.

In a search for "bridge flipping" (interconversion of **178a** to **178b**) in the 7-norbornadienyl cation, Lustgarten et al. (1967) observed that the NMR

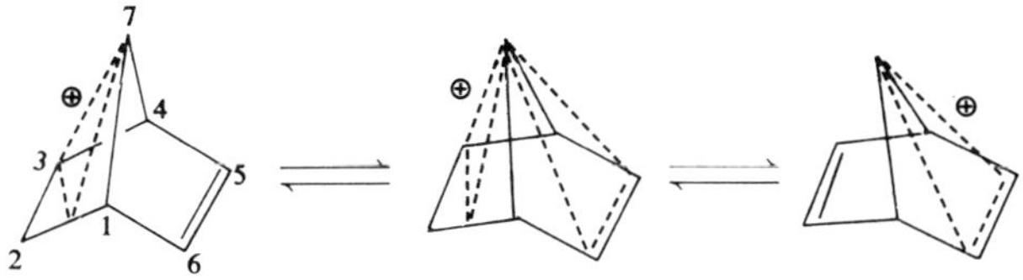

$$\text{(178a)} \rightleftharpoons \rightleftharpoons \text{(178b)} \quad (188)$$

spectrum at $-77°C$ in FSO_3H showed a broadening of the signals for the protons on C-7, C-1, and C-4, C-5, and C-6, but not for those on C-2 or C-3. The observation of partial carbon degeneracy with the use of the labeled precursors **180**, **181**, and **182** led Lustgarten et al. (1967) to propose a 5-carbon degenerate rearrangement (Eq. 189) involving ring contraction to the bicyclo-

(180) 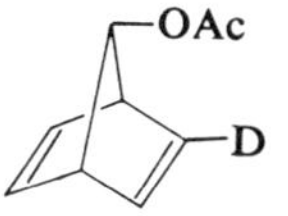 **(181)** 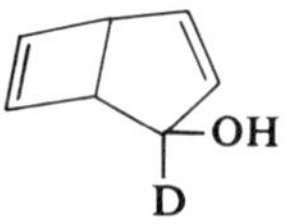 **(182)**

heptadienyl cation followed by ring expansion back to the norbornadienyl cation. Ring expansion and contraction apparently does not occur under

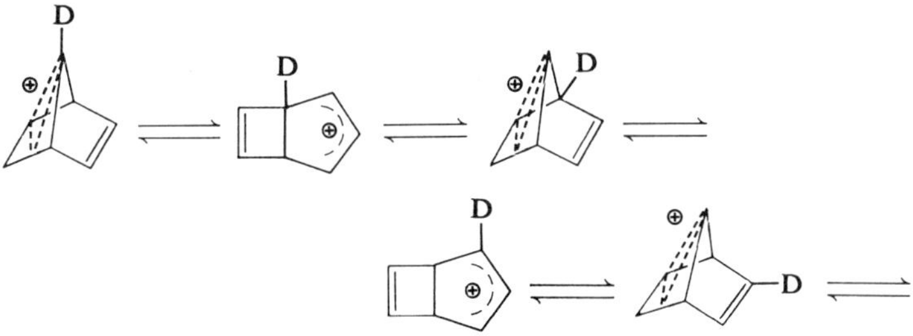

$$\rightleftharpoons \rightleftharpoons \rightleftharpoons$$
$$\rightleftharpoons \rightleftharpoons \text{etc.} \quad (189)$$

conditions of short carbonium ion lifetimes such as in solvolytic media (Story and Cooke, 1968).

Variable temperature studies of the 7-methylnorbornadienyl cation (**183**) in FSO_3H revealed evidence supporting the unsymmetrical bridged cation **183** at $-45°C$ and rapid "bridge flipping" near $0°C$ (Brookhart et al., 1967). A good treatment of the 7-norbornadienyl cation is contained in a recent

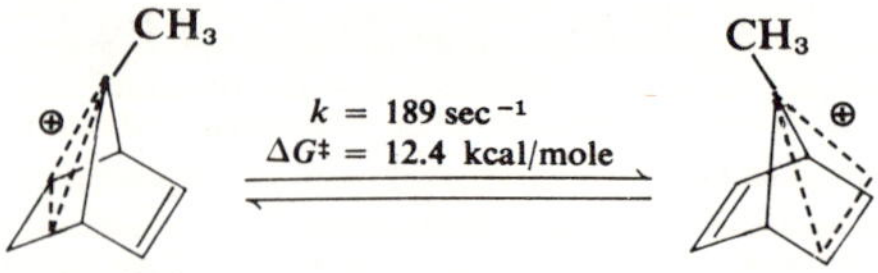

$$k = 189 \text{ sec}^{-1}$$
$$\Delta G^\ddagger = 12.4 \text{ kcal/mole}$$

(183a) ⇌ (183b)　　(190)

review on degenerate carbonium ions (Leone and Schleyer, 1970).

Like the 7-norbornadienyl cation, available data on the 5-norbornenyl cation do not allow its representation by a simple static bridged ion. Labeling experiments with ^{14}C (Roberts et al., 1955) and deuterium (Cristol et al.,

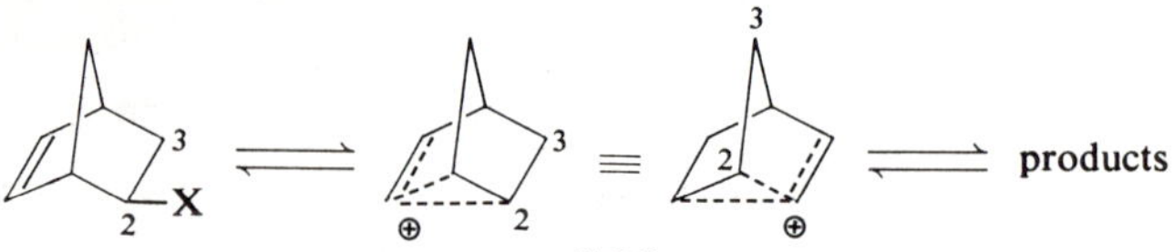

⇌ ≡ ⇌ products　　(191)

(184)

1966) led to evidence of scrambling of the label. The latter data were consistent with either the bridged ion **184** or with rapidly equilibrating classic ions and were inconsistent with the process shown in Eq. (192).

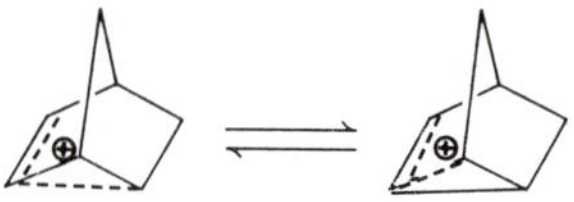

⇌ 　　(192)

C. Electron-Deficient Bridged Ions

1. Bridging by Hydrogen and Alkyl Groups

One of the most elementary rearrangements discussed in undergraduate textbooks is intramolecular hydride transfer. This phenomenon is the topic of a review by Fry and Karabatsos (1970). Although the literature yields an abundance of data to fully document that intramolecular hydride shifts occur, there is little or no evidence on which to argue for a bridged intermediate. The strongest evidence for bridging, perhaps, comes from hydrogen participation studies such as those of Shiner and Jewett (1965), who obtained a k_H/k_D of 2.2 for one neighboring axial hydrogen in the solvolysis of cis-4-tert-butylcyclohexyl brosylate. The product mixture contains 86% tert-butylcyclohexene formed by loss of an axial β hydrogen atom. Shiner and

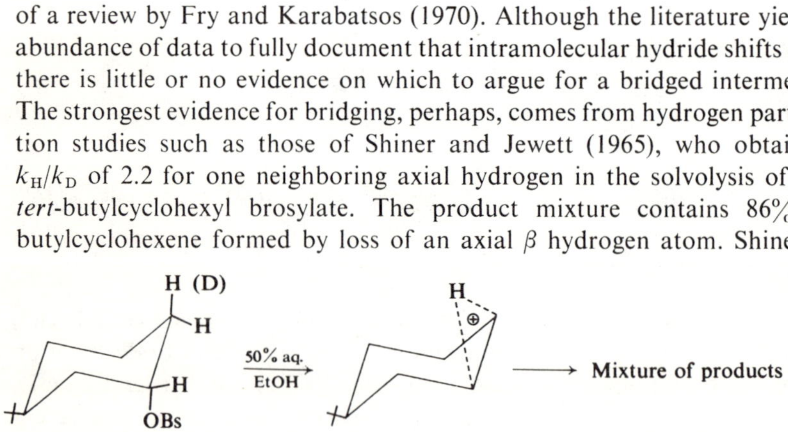

→ 50% aq. EtOH → Mixture of products　　(193)

Jewett stated: "Since the participation is from a secondary hydrogen to an adjacent secondary solvolytic center it is difficult to conceive of any driving force for it other than the formation of a stabilized, bridged, nonclassical intermediate carbonium ion."

There is good evidence against bridging in some tertiary cases (Fry and Karabatsos, 1970). This is in line with the lower expectation of a bridged ion in cases where a relatively stable classic ion can be formed. At the other extreme, bridging might be expected in the case of the ethyl cation. Despite expectations, *ab initio* calculations show the classical cation to be 11.4 (Williams *et al.*, 1970) and 12.4 kcal/mole (Pfeiffer and Jewett, 1970) more

$$CH_3\overset{\oplus}{-}CH_2 \rightleftharpoons \underset{(185)}{CH_2\overset{\overset{H}{\underset{\oplus}{\cdots}}}{-}CH_2} \rightleftharpoons \overset{\oplus}{CH_2}-CH_3 \qquad (194)$$

stable than the ethylene protonium ion (**185**). Additional calculations failed to provide any support for **185** as an intermediate in the conversion of one primary cation into another. Thus, the gas-phase activation energy for the hydride shift in an ethyl cation (through **185** as a transition state) can be assumed to be 11.4–12.4 kcal/mole.

Protonated cyclopropanes, a class of controversial intermediates with much evidence in their favor, can be represented either as a 1,3-proton bridged cation (edge-protonated, **186**) or as a 1,2-methyl bridged cation (corner protonated, **187**). Both types have been postulated to occur as

$$\underset{(186)}{\overset{H_2C\cdots H}{\underset{CH_2-CH_2}{\overset{\oplus}{\diagup}}}} \qquad \underset{(187)}{\overset{CH_3}{\underset{CH_2\cdots CH_2}{\overset{\oplus}{\diagup}}}} \equiv \overset{CH_3}{\underset{CH_2-CH_2}{\overset{|}{\diagup}\overset{+}{\diagdown}}}$$

intermediates in chemical reactions (Collins, 1969; Leone and Schleyer, 1970; Lee, 1970). Notably, deuterium and ^{14}C scrambling results in simple propyl systems, and some rearrangements are best explained by invoking protonated cyclopropane intermediates (the norbornyl cation discussed below is also best described as a special case of a protonated cyclopropane derivative). As is true with the other controversial ions discussed here, there are documented examples of likely cations which fail to show any desire to exist as stable protonated cyclopropane derivatives. The 2,3,3-trimethylbut-2-yl cation is one such case (see Section II,C,1,b).

2. *Cyclopropylcarbinyl Cations*

Spurred by the observations that cyclopropylcarbinyl derivatives solvolyzed at rates far greater than ordinary primary systems and that there was considerable interconversion between cyclobutyl, allylcarbinyl, and cyclopropylcarbinyl derivatives (Roberts and Mazur, 1951a), J. D. Roberts and his co-workers investigated the cyclopropylcarbinyl system over a period of years. These research efforts have led to a rather widely accepted proposal of a "bicyclobutonium" ion intermediate (**188**) (Mazur et al., 1959). Simpler intermediates failed to adequately account for the near degeneracy of such cations. For example, cyclopropylcarbinyl amine-α-^{14}C undergoes nitrous

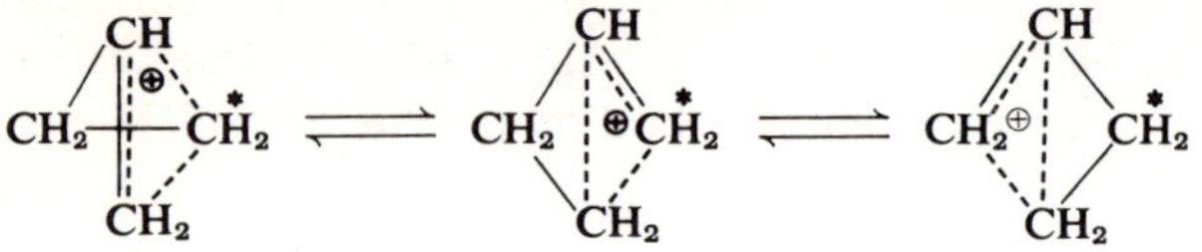

$$\underset{(a)}{\underset{CH_2}{\overset{CH}{\underset{|}{\bigoplus}}}\!\!\!\!\!\!\!\!-\!\!\!\!\!\!\!\!\overset{*}{CH_2}} \rightleftharpoons \underset{(b)}{\underset{CH_2}{\overset{CH}{\underset{|}{}}}\overset{*}{\bigoplus}\!CH_2} \rightleftharpoons \underset{(c)}{\underset{CH_2}{\overset{CH}{\underset{|}{}}}\overset{*}{CH_2}^{\oplus}} \qquad (195)$$

(**188**)

acid deamination to give products **189**, **190**, and **191** with the ^{14}C tag distributed as shown (Mazur et al., 1959). Wiberg et al. (1972) have recently reviewed the work on allylcarbinyl cations.

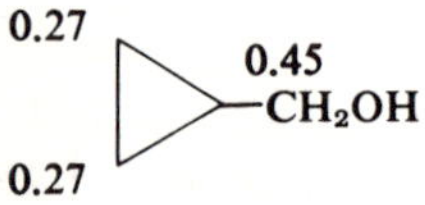

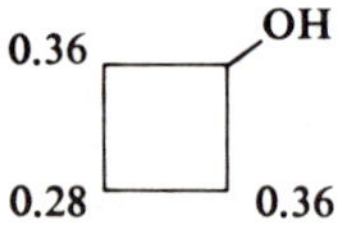

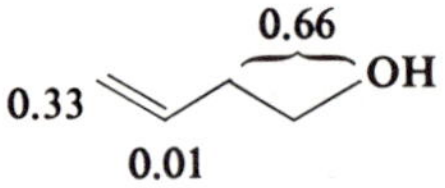

(**189**) 48%   (**190**) 47%   (**191**) 5%

Majerski and Schleyer (1971) have published their studies which confirm that (1) hydride shifts do not occur in these systems, (2) scrambling among the carbons is extensive but not complete, and (3) that under solvolytic conditions, cyclopropylcarbinyl rearrangements of the initially formed intermediate occur at a rate comparable to solvent capture. They further found that the products of the cyclopropylcarbinyl → cyclopropylcarbinyl, the cyclopropylcarbinyl → cyclobutyl, and the cyclopropylcarbinyl → allylcarbinyl rearrangements were remarkably stereospecific. The latter evidence strongly supports some sort of bridged intermediate.

The results of ^{1}H and ^{13}C NMR spectroscopic studies (Olah et al., 1970c, 1972) on the long-lived cyclopropylcarbinyl cation (**192**) provide additional evidence for the nonclassical nature of this ion. Comparison of the spectra of

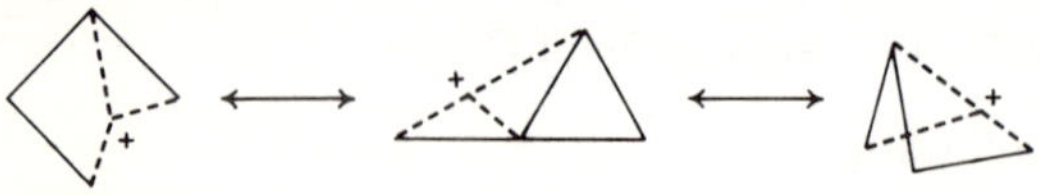

(a) (b) (c)
(**192**) (**193**)

192 with that of secondary and tertiary derivatives of 192 and with other model cations led to the conclusion that 192 is a three-center bonded carbonium ion. The tricyclobutonium ion (193) geometry is excluded by theoretical considerations (Pittman et al., 1972a).

3. The 2-Norbornyl Cation

The norbornyl cation and its derivatives have received so much attention that mention of the terms bridged or nonclassical carbonium ion may conjure up the picture of the norbornyl cation in the minds of organic chemists trained in the past decade. In this brief treatment, only a few facts can be presented. This ought to be sufficient to develop events leading to the current level of understanding in this field. The treatment cannot, however, capture the occasionally emotional drama which has characterized the study of bridged carbonium ions, and especially the norbornyl cation, during most of the past decade. The reader is referred to the treatments by Bartlett (1965), Sargent (1972), Brown (1967), and to a special report of the 21st National Organic Chemistry Symposium (Bernhard, 1969). In the latter report, Brown was said to be on the University of California at Los Angeles faculty along with Cram and Winstein. In a letter to the Editor, Brown replied (1969) that he was located "at Purdue University, in Layfayette, Indiana, where the air is still sufficiently clear that we see carbon–carbon single bonds sharply, without the fuzziness such bonds appear to acquire in smoggy atmospheres."

Winstein and Trifan (1949, 1952) first suggested a bridged norbornyl ion to account for several important phenomena associated with the solvolysis of norbornyl arenesulfonates. They observed (1) the exclusive formation of exo product, (2) total racemization attending solvolysis of *exo*-norbornyl arenesulfonates and extensive racemization (92%) from the endo isomer, (3) internal return of optically active *exo*-norbornyl arenesulfonate with racemization attending solvolysis in solvents of high ionizing strength but low nucleophilic reactivity, and (4) k_{exo}/k_{endo} of about 350 in solvolysis under identical conditions. Winstein and Trifan (1952) considered and eliminated the simple alternatives to the nonclassical structure proposed.

Winstein and Trifan (1952) accounted for the major results as follows. The *exo*-norbornyl arenesulfonate ionizes with anchimeric assistance (enhanced rate) to give the norbornyl cation [Eq. (196)]. The norbornyl cation is symmetrical, therefore internal return would lead to racemization. The endo isomer is said to proceed without anchimeric assistance, but possibly with solvent assistance, to a classical cation which can be trapped by solvent (k_s process). The majority of the classical cation, however, goes to the more stable, symmetrical norbornyl cation through leakage.

Howe et al. (1965) later showed that exo substitution in these reactions

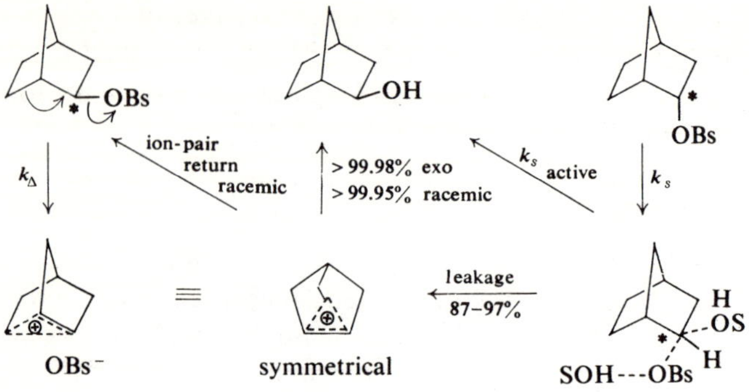

(196)

[Eq. (197)] results from other than steric preference for exo attack since borohydride attacks apocamphor predominantly on the endo side. The equilibrium mixture favors endo alcohol (63%).

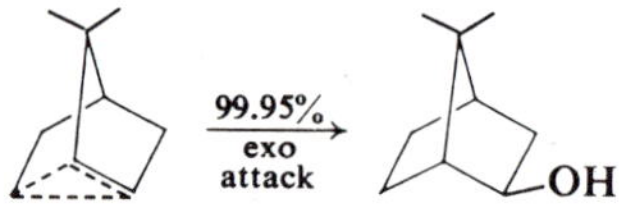

(197)

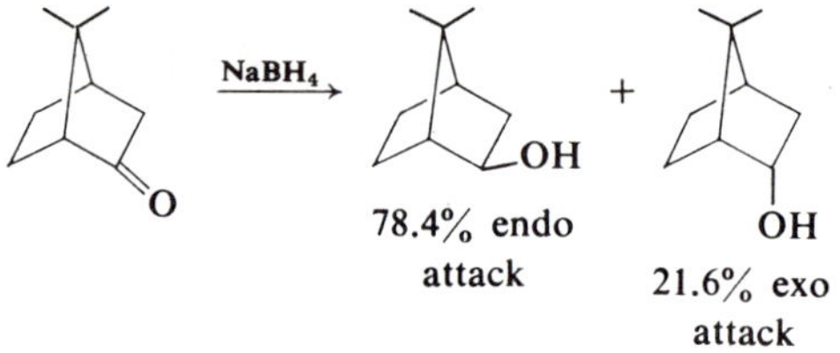

(198)

Roberts et al. (1954) used ^{14}C-labeled norbornyl derivatives to show that the structure proposed by Winstein and Trifan (1949, 1952) was not fully adequate. Their results could be explained if 2,6- or 1,6-hydride shifts were occurring. A "nortricyclonium" ion intermediate (**194–196**) or its equivalent was proposed.

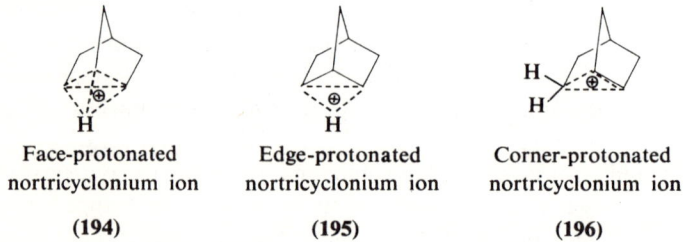

Face-protonated nortricyclonium ion
(**194**)

Edge-protonated nortricyclonium ion
(**195**)

Corner-protonated nortricyclonium ion
(**196**)

NMR studies of the norbornyl cation at different temperatures clearly indicate that the three processes shown in Eq. (199) are occurring (Schleyer *et al.*, 1964b; Saunders *et al.*, 1964; Jensen and Beck, 1966; Olah *et al.*, 1968b). A Raman spectral analysis (Olah *et al.*, 1968b) of the norbornyl

(199)

cation was used as further support for either **195** or **196** as the species present at low temperature. The observed relative high energy barrier for the 3,2-hydride shift in the norbornyl cation was explained as being the result of the necessary transition to the classic cation from the more stable nonclassic ion prior to the 3,2-hydride shift.

Olah *et al.* (1970b) have now employed lower temperatures, Raman spectroscopy, and ^{13}C NMR studies to allow a more refined interpretation of the structural aspects of the norbornyl cation (see Section II, C). At $-120°C$, the 3,2-hydride shift is completely frozen out, whereas the 6,2-hydride shift is frozen out at $-156°C$. The latter spectrum is relatively simple and is best explained in terms of the corner-protonated ion **31**. The various edge-protonated species **195** (a degenerate cation) are considered to be transition states between the various corner-protonated species.

Brown's (1967) argument for rapidly equilibrating classical ions for the norbornyl cation is not well accepted in light of the large amount of support for the bridged ion. Brown, however, was instrumental in pointing out that not all norbornyl derivatives produce nonclassical ions (Bernhard, 1969). Tertiary carbonium ions in this system, as in most systems studied, prefer a classical structure (Winstein *et al.*, 1952).

4. *Homoaromatic Cations*

Winstein (1969) has reviewed much of his work on the class of novel cations which he labeled homoaromatic compounds. Since the cyclopropenyl cation is the smallest aromatic compound, the homocyclopropenyl cations are the simplest homoaromatic species. *cis*-3-Bicyclo[3.1.0]hexyl tosylate

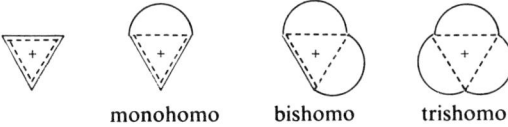

monohomo   bishomo   trishomo

(197), its trans isomer (198), and its *d*-3 derivative (199) give evidence for the simplest trishomo species (200). The *cis* compound, 197, undergoes acetolysis about nine times faster (k_Δ) than 198 (k_s plus leakage) and yields exclusively

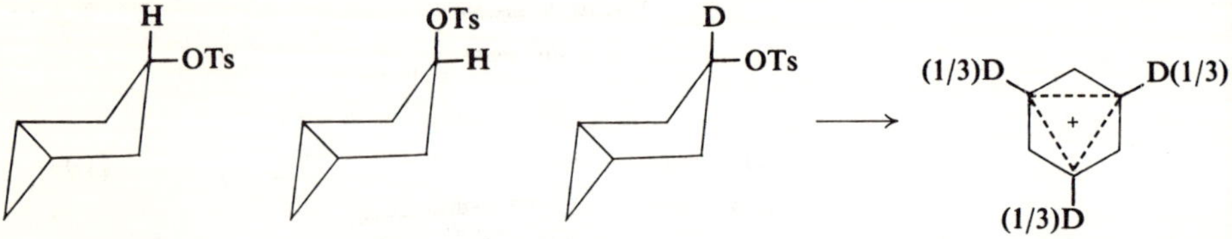

(197) (198) (199) (200) (200)

cis-acetate. The trans isomer yields a mixture of *cis*-acetate and olefin (2:1). The deuterium label in 199 is scrambled as would be expected from formation of the degenerate ion, 200.

A more dramatic rate effect is seen in the hydrolysis of ester 201. The endo-fused cyclopropane ring (edge participation) is effective in accelerating hydrolysis whereas the exo-fused cyclopropane ring (face participation) is not (Tanida *et al.*, 1967; Battiste *et al.*, 1967).

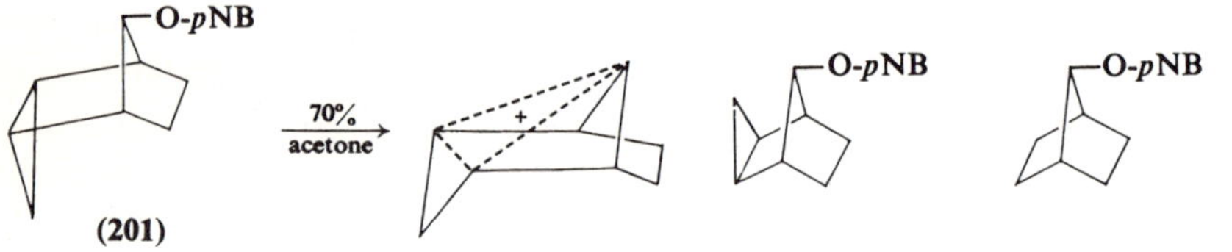

(201)

relative rate = 10^{14} relative rate = 1   relative rate = 3

(201)

5. Heteronuclear Bridged Species

No attempt will be made to give examples of these bridged species since the variety is so great. For the interested reader, however, recent reference to important articles or reviews which discuss bridged ions are given. Olah and Porter (1971) and Yates and McDonald (1971) have spectral and thermochemical data indicative of the bridged bromonium ion (see also Hegarty *et al.*, 1972, and references therein for a discussion of nonbridged β-bromocarbonium ions), while Peterson (1971) has reviewed cyclic halonium ions with five-membered rings. Olah and Clifford (1971) have presented convincing evidence for stable mercurinium ions. Finally, a more extensive review of certain bridged ions with O, S, and N (but principally dioxolan-2-ylium or acetoxonium ions) has been written by Pittman *et al.* (1972b).

D. Summary

Since many bridged ion species have been discussed here, we would be remiss not to remind the reader that bridged ions exist solely because other arrangements are less stable. Therefore, one should not consider that a bridged ion is automatically formed just because a bridged structure can be easily drawn. To quote from the commentary (Bartlett, 1965, p. 77) on the paper of Winstein *et al.* (1952), concerning driving forces in the Wagner–Meerwein rearrangement: "... noting the dwindling participational driving force as the α-carbon atom is changed from primary to secondary, one forecasts from [Winstein's] paper that bridged ions will not provide any important driving force in the benzpinacol rearrangement or other cases where the charge delocalization in a classical ion is overwhelmingly efficient. These anticipations were elegantly verified in tracer work by Collins and his co-workers."

VII. Related Species

The foregoing discussion contains countless examples where the presence of trivalent carbocations (carbenium ions*) either as stable entities or reaction intermediates is indubitable. The evidence for related species is not nearly as abundant. Pauling electronegativities (Table XXXIV) and our current information about carbocations suggest that there might be some difficulty in

Table XXXIV
Pauling Electronegativities of Some Elements

B	C	N	O	F
2.0	2.5	3.0	3.5	4.0
Al	Si	P	S	Cl
1.5	1.8	2.1	2.5	3.0

obtaining solutions of stable organic cations of first row elements such as the isoelectronic derivatives **202** and **203**. On the other hand, a stable trivalent

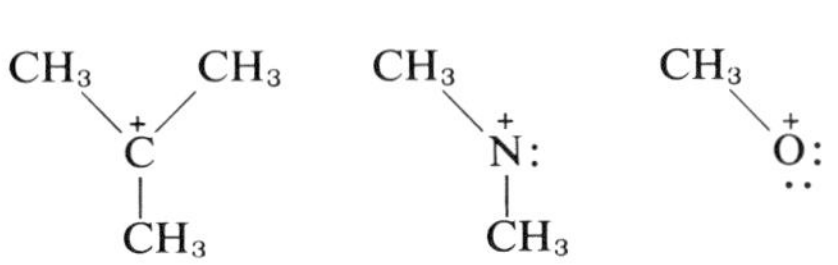

$$(CH_3)_3 \overset{+}{C} \qquad (CH_3)_3 \overset{+}{N}: \qquad (CH_3)_3 \overset{+}{O}:$$

(202) (203)

stable in strong acid media

unknown as stable species in solution

silicocation would be predicted to be more stable than an analogous carbenium ion, other things being equal. We will present some evidence to show that

* In this section alone, Olah's carbenium and carbonium nomenclature (see Section I,D) will be used so that the proper analogy can be seen in the names.

second row elements such as silicon do not readily form cations isoelectronic with carbenium ions (i.e., nitrenium and oxenium ions may be capable of isolation as stable intermediates under special circumstances).

A. Nitrenium and Oxenium Ions

There are some acid-catalyzed rearrangements of carbon, nitrogen, and oxygen derivatives that show a distinct similarity. In the case of the carbocation

$$Ph_3-C-CH_2OH \xrightarrow[-H_2O]{acid} [Ph_3-C-\overset{+}{C}H_2] \longrightarrow Ph_2-\overset{+}{C}-CH_2Ph \longrightarrow products \quad (202)$$

$$Ph_3-C-NHOH \xrightarrow[-H_2O]{acid} [Ph_3-C-\overset{+}{N}H] \longrightarrow Ph_2-\overset{+}{C}-NHPh \xrightarrow{H_2O} products \quad (203)$$

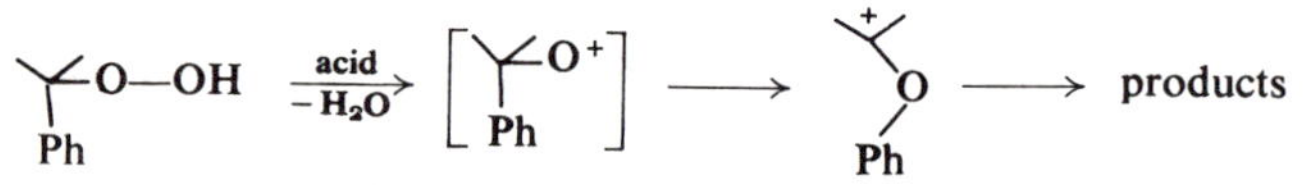

$$\underset{Ph}{\diagup}\!\!-O-OH \xrightarrow[-H_2O]{acid} \left[\underset{Ph}{\diagup}\!\!-O^+\right] \longrightarrow \underset{Ph}{\overset{+}{\diagup}}\!\!-O \longrightarrow products \quad (204)$$

reaction [Eq. (202)], the initially formed cationic carbon shown is primary and would be an intermediate not expected to exist for any length of time under the conditions of the reactions. Initial loss of water concurrent with migration (k_Δ) is also a pathway which is open to the alcohol. The hydroxylamine and the hydroperoxide have similar options, and the existence of the cationic oxygen and nitrogen intermediates in the reactions shown would have to be proved by a more thorough and systematic search. Although there is strong evidence against the intermediacy of oxenium and nitrenium species in many reactions (Smith, 1963), there is recent evidence for the existence of these species as intermediates. For example, strong evidence for nitrenes as intermediates in several processes has now been presented (Gassman, 1970; Gassman and Nakai, 1971; also see Chapter 3, Section VI). The isolation of stable salts such as **204** (Dimroth *et al.*, 1967) enhances the probability of

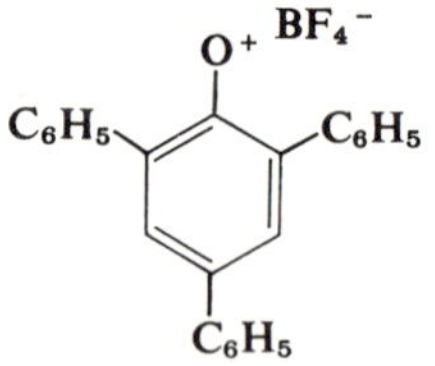

$$\underset{(204)}{\underset{C_6H_5}{\diagup}\!\!\!\overset{O^+ \ BF_4^-}{\underset{C_6H_5}{\diagdown}\!\!\!C_6H_5}}$$

(204)

gaining evidence to support the existence of oxenium ion intermediates in organic reactions.

B. Silicenium Ions

As predicted from electronegativity values, Hess et al. (1964) found from electron impact experiments that slightly less energy is required to ionize covalently bound silicon as compared to carbon. Despite their apparent greater stability, trivalent silicocations (silicenium ions*) have been elusive species in solution.

Among the most stable carbenium ions are the triaryl derivatives. Stable triphenylmethyl cations are characterized by their deeply colored solutions. Typical of highly resonance-stabilized carbenium ions is crystal violet (**205**).

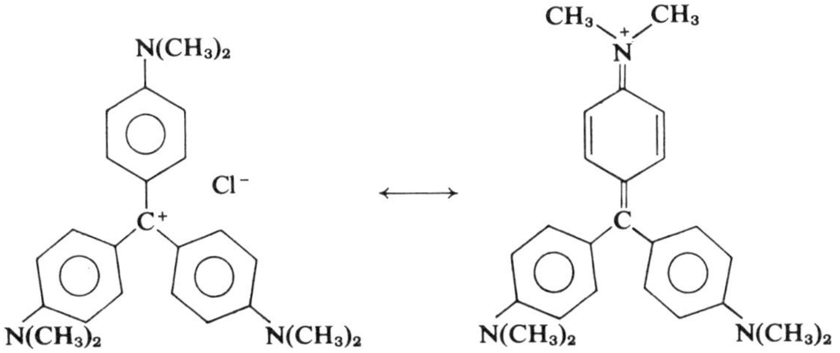

(**205**)

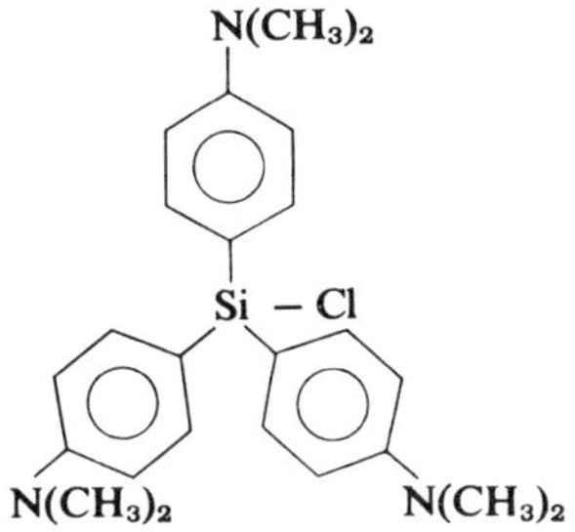

(**206**)

Silico-crystal violet (**206**) does not form colored solutions in polar organic solvents even in the presence of Lewis acids. These and other studies of triarylsilicon compounds (see Sommer, 1965) provide no evidence for the formation of silicenium ions. Sommer (1965) attributes the failure of silico-crystal violet to ionize to the inability of silicon to form stable multiple bonds ($3p_\pi-2p_\pi$). Because of this bonding difference, the aryl groups may destabilize (inductively) rather than stabilize (resonance) the hypothetical silicenium ion from **206**.

* In analogy with carbo*nium* ion terminology, these ions are often called silico*nium* ions; for reasons given earlier, we advocate the use of the name silic*enium* ions.

Although there are examples where silicenium ions are indicated as reaction intermediates in solution, such examples are few, and the evidence is generally less compelling than in analogous carbenium ion reactions (for references to the literature, see Bethell and Gold, 1967; Olah and Mo, 1971). Recently, aryl substituents have been avoided in systematic attempts to obtain evidence of stable silicenium ions. Olah and Mo (1971) used fluorine substituents in some of their unsuccessful attempts to obtain stable silicenium ions. Again, fluorine probably stabilizes carbenium ions by $p_\pi-p_\pi$ overlap; thus, as in the case with phenyl substitution, one would not expect enhanced stability in fluorine-substituted silicenium cations. The substituents most likely to stabilize trivalent silicocations are alkyl groups. Recent attempts to prepare the trimethylsilicenium ion in super-acid media, in which the *tert*-butyl cation is extremely stable, failed (Olah and Mo, 1971). The availability of the empty $3d$ orbitals of the silicon atom were thought responsible for the inability to observe the trimethylsilicenium ion. The empty $3d$ orbital, according to that argument, is capable of changing the tetrahedral sp^3 hybridization to the trigonal pyramidal sp^3d hybridization. Olah and Mo (1971) concluded that the species that they observed when trimethylsilyl fluoride was dissolved in SbF_5/SO_2ClF at $-78°C$ was the pentacoordinate species shown in Eq. (206).

$$(CH_3)_3SiF + SbF_5/SO_2ClF \longrightarrow \begin{array}{c} CH_3 \\ | \\ CH_3-Si \\ | \\ CH_3 \end{array} \begin{array}{c} F \\ \diagup \\ \diagdown \\ F \end{array} \begin{array}{c} F \\ | \\ Sb \\ | \\ F \end{array} \begin{array}{c} F \\ \diagup \\ \diagdown \\ F \end{array} \quad (206)$$

O'Brien and Harbordt (1970, 1971) and Brook and Pannell (1970) have also failed in attempts to observe stable silicenium ions in solution.

References

Arndt, F., and Lorenz, L. (1930). *Chem. Ber.* **63**, 3121.
Arnett, E. M. (1963). *Progr. Phys. Org. Chem.* **1**, 223.
Arnett, E. M., and Bushick, R. D. (1964). *J. Amer. Chem. Soc.* **86**, 1564.
Arnett, E. M., and Larsen, J. W. (1968). *In* "Carbonium Ions" (G. A. Olah and P. von R. Schleyer, eds.), Vol. I, pp. 441–456. Wiley (Interscience), New York.
Arnett, E. M., and McKelvy, D. R. (1965). *Rec. Chem. Progr.* **26**, 185.
Arnett, E. M., Bentrude, W. G., Burke, J. J., and Duggleby, P. M. (1965). *J. Amer. Chem. Soc.* **87**, 1541.
Ausloos, P., and Lias, S. G. (1965). *Discuss. Faraday Soc.* **39**, 36.
Baeyer, A. (1905). *Chem. Ber.* **38**, 569.
Baeyer, A., and Villiger, V. (1902). *Chem. Ber.* **35**, 1189 and 3013.
Baird, R. L., and Winstein, S. (1957). *J. Amer. Chem. Soc.* **79**, 4238.
Baird, R. L., and Winstein, S. (1963). *J. Amer. Chem. Soc.* **85**, 567.
Bartlett, P. D. (1951). *Bull. Soc. Chim. Fra.* [5] **18**, 101C.
Bartlett, P. D. (1965). "Nonclassical Ions." Benjamin, New York.

Bartlett, P. D., and Knox, L. H. (1939). *J. Amer. Chem. Soc.* **61**, 3184.
Bartlett, P. D., and Lewis, E. S. (1950). *J. Amer. Chem. Soc.* **72**, 1005.
Bartlett, P. D., and McCollum, J. D. (1956). *J. Amer. Chem. Soc.* **78**, 1441.
Bartlett, P. D., and Stiles, R. M. (1955). *J. Amer. Chem. Soc.* **77**, 2806.
Bartlett, P. D., and Swain, M. S. (1955). *J. Amer. Chem. Soc.* **77**, 2801.
Bartlett, P. D., and Tidwell, T. T. (1968). *J. Amer. Chem. Soc.* **90**, 4421.
Bartlett, P. D., Condon, F. E., and Schneider, A. (1944). *J. Amer. Chem. Soc.* **66**, 1531.
Bateman, L. C., Hughes, E. D., and Ingold, C. K. (1940). *J. Chem. Soc., London* p. 960.
Battiste, M. A. (1961). *J. Amer. Chem. Soc.* **83**, 4101.
Battiste, M. A., Deyrup, C. L., Pincock, R. E., and Haygood-Farmer, J. (1967). *J. Amer. Chem. Soc.* **89**, 1954.
Bayles, J. W., Evans, A. G., and Ross, J. R. (1955). *J. Chem. Soc., London* p. 206.
Benfey, O. T., Hughes, E. D., and Ingold, C. K. (1952). *J. Chem. Soc., London* p. 2488.
Bentley, M. D., and Dewar, M. J. S. (1970). *J. Amer. Chem. Soc.* **92**, 3996.
Bernal, J. D., and Fowler, R. H. (1933). *J. Chem. Phys.* **1**, 515.
Bernhard, R. (1969). *Sci. Res.* Aug. 18, pp. 26–28.
Berliner, E. (1964). *Progr. Phys. Org. Chem.* **2**, 253.
Berson, J. A. (1963). *In* "Molecular Rearrangements" (P. deMayo, ed.) Part One, p. 111. Wiley (Interscience), New York.
Bethell, D., and Gold, V. (1967). "Carbonium Ions: An Introduction." Academic Press, New York.
Bethell, D., and Howard, R. D. (1966). *Chem. Commun.* p. 94.
Betteridge, D., and Baker, A. D. (1970). *Anal. Chem.* **42**, 43A.
Boer, F. P. (1966). *J. Amer. Chem. Soc.* **88**, 1572.
Bonner, W. A., and Collins, C. J. (1956). *J. Amer. Chem. Soc.* **78**, 5587.
Boyd, R. H. (1969). *In* "Solute-Solvent Interactions" (J. F. Coetzee and C. D. Ritchie, eds.), pp. 98–218. Dekker, New York.
Brenneisen, P., Grob, C. A., Jackson, R. A., and Ohta, M. (1965). *Helv. Chim. Acta* **48**, 146.
Breslow, R., and Chu, W. (1970). *J. Amer. Chem. Soc.* **92**, 2169.
Breslow, R., and Yuan, C. (1958). *J. Amer. Chem. Soc.* **80**, 5991.
Breslow, R., Bahary, W., and Reinmuth, W. (1961). *J. Amer. Chem. Soc.* **83**, 1763.
Breslow, R., Hover, H., and Chang, H. W. (1962). *J. Amer. Chem. Soc.* **84**, 3168.
Brook, A. G., and Pannell, K. H. (1970). *Can. J. Chem.* **48**, 3679.
Brookhart, M., Lustgarten, R. K., and Winstein, S. (1967). *J. Amer. Chem. Soc.* **89**, 6354.
Brouwer, D. M., MacLean, C., and Mackor, E. L. (1965). *Discuss. Faraday Soc.* **39**, 129.
Brown, H. C. (1946). *Science* **103**, 385.
Brown, H. C. (1956a). *Bull. Soc. Chim. Fra.* [5] p. 980.
Brown, H. C. (1956b). *J. Chem. Soc., London* p. 1248.
Brown, H. C. (1962). *Chem. Soc., Spec. Publ.* **16**, 140.
Brown, H. C. (1966). *Chem. Brit.* p. 199.
Brown, H. C. (1967). *Chem. Eng. News* **45**, No. 7, 86.
Brown, H. C. (1969). *Sci. Res.* Dec. 22, p. 5.
Brown, H. C., and Chloupek, F. J. (1963). *J. Amer. Chem. Soc.* **85**, 2322.
Brown, H. C., and Fletcher, R. S. (1949). *J. Amer. Chem. Soc.* **71**, 1845.
Brown, H. C., and Fletcher, R. S. (1950). *J. Amer. Chem. Soc.* **72**, 1223.
Brown, H. C., and Fletcher, R. S. (1951). *J. Amer. Chem. Soc.* **73**, 1317.
Brown, H. C., and Hammar, W. J. (1967). *J. Amer. Chem. Soc.* **89**, 6378.
Brown, H. C., and Ichikawa, K. (1957). *Tetrahedron* **1**, 221.

Brown, H. C., and Kim, C. J. (1968). *J. Amer. Chem. Soc.* **90**, 2082.
Brown, H. C., and Kim, C. J. (1971). *J. Amer. Chem. Soc.* **93**, 5764.
Brown, H. C., and Kornblum, R. B. (1954). *J. Amer. Chem. Soc.* **76**, 4510.
Brown, H. C., and Nakagawa, M. (1955). *J. Amer. Chem. Soc.* **77**, 3610 and 3614.
Brown, H. C., and Okamoto, Y. (1957). *J. Amer. Chem. Soc.* **79**, 1913.
Brown, H. C., and Okamoto, Y. (1958). *J. Amer. Chem. Soc.* **80**, 4979.
Brown, H. C., Morgan, K. J., and Chloupek, F. J. (1965). *J. Amer. Chem. Soc.* **87**, 2137.
Brown, H. C., Rothberg, I., Schleyer, P. von R., Donaldson, M. M., and Harper, J. J. (1966). *Proc. Nat. Acad. Sci. U.S.* **56**, 1653.
Brown, H. C., Kim, C. J., Lancelot, C. J., and Schleyer, P. von R. (1970). *J. Amer. Chem. Soc.* **92**, 5244.
Bryan, R. F. (1964). *J. Amer. Chem. Soc.* **86**, 733.
Bunton, C. A. (1963). "Nucleophilic Substitution at Saturated Carbon." Elsevier, Amsterdam.
Bunton, C. A., Greenstreet, C. H., Hughes, E. D., and Ingold, C. K. (1954). *J. Chem. Soc., London* pp. 642 and 647.
Capon, B. (1964). *Quart. Rev., Chem. Soc.* **18**, 45.
Carney, J., and Pittman, C. U., Jr. (1971). Unpublished results.
Childs, R. F., and Taguchi, V. (1970). *Chem. Commun.* p. 695.
Cocivera, M., and Winstein, S. (1963). *J. Amer. Chem. Soc.* **85**, 1702.
Coke, J. L., McFarlane, F. W., Mourning, M. C., and Jones, M. G. (1969). *J. Amer. Chem. Soc.* **91**, 1154.
Collins, C. J. (1964). *Advan. Phys. Org. Chem.* **2**, 1.
Collins, C. J. (1968). *In* "Carbonium Ions" (G. A. Olah and P. von R. Schleyer, eds.), Vol. I, pp. 307–352. Wiley (Interscience), New York.
Collins, C. J. (1969). *Chem. Rev.* **69**, 543.
Collins, C. J., and Bonner, W. A. (1955). *J. Amer. Chem. Soc.* **77**, 92.
Collins, C. J., Bonner, W. A., and Lester, C. T. (1959). *J. Amer. Chem. Soc.* **81**, 466.
Collins, C. J., Christie, J. B., and Raaen, V. F. (1961). *J. Amer. Chem. Soc.* **83**, 4267.
Conant, J. B., and Chow, B. F. (1933). *J. Amer. Chem. Soc.* **55**, 3752.
Conant, J. B., Small, L. F., and Taylor, B. S. (1925). *J. Amer. Chem. Soc.* **47**, 1959.
Conrow, K. (1959). *J. Amer. Chem. Soc.* **81**, 5461.
Cook, D. (1959). *Can. J. Chem.* **37**, 48.
Cook, D. (1963). *In* "Friedel-Crafts and Related Reactions" (G. A. Olah, ed.), Vol. I, p. 767. Wiley, New York.
Cooke, I., Suaz, B. P., and Herschmann, C. (1954). *Helv. Chim. Acta* **37**, 1280.
Cope, A. C., Fenton, S. W., and Spencer, C. F. (1952). *J. Amer. Chem. Soc.* **74**, 5884.
Corey, E. J., and Casanova, J. (1963). *J. Amer. Chem. Soc.* **85**, 165.
Corey, E. J., Bauld, N. L., LaLonde, R. T., Casanova, J., and Kaiser, E. T. (1960). *J. Amer. Chem. Soc.* **82**, 2645.
Cox, L. E., Jack, J. J., and Hercules, D. M. (1972). *J. Amer. Chem. Soc.* **94**, 6575.
Cram, D. J. (1949). *J. Amer. Chem. Soc.* **71**, 3863 and 3875.
Cram, D. J. (1952). *J. Amer. Chem. Soc.* **74**, 2129 and 2137.
Cram, D. J. (1956). *In* "Steric Effects in Organic Chemistry" (M. S. Newman, ed.), pp. 249–303. Wiley, New York.
Cram, D. J. (1964). *J. Amer. Chem. Soc.* **86**, 3767.
Cristol, S. J., Morrill, T. C., and Sanchez, R. A. (1966). *J. Amer. Chem. Soc.* **88**, 3087.
Curtin, D. Y. (1954). *Rec. Chem. Progr.* **15**, 111.
Curtin, D. Y., Klanderman, B. H. and Tavares, D. F. (1962). *J. Org. Chem.* **27**, 2709.
Curtin, D. Y., Kampmeier, J. A., and O'Connor, B. R. (1965). *J. Amer. Chem. Soc.* **87**, 863.

Dauben, H. J., Jr. (1960). *J. Org. Chem.* **25**, 1442.
Dauben, H. J., Jr., Gadecki, F. A., Harmon, K. M., and Pearson, D. L. (1957). *J. Amer. Chem. Soc.* **79**, 4557.
Deno, N. C. (1964). *Progr. Phys. Org. Chem.* **2**, 129.
Deno, N. C., and Pittman, C. U., Jr. (1964a). *J. Amer. Chem. Soc.* **86**, 1744.
Deno, N. C., and Pittman, C. U., Jr. (1964b). *J. Amer. Chem. Soc.* **86**, 1871.
Deno, N. C., and Sacher, E. (1965). *J. Amer. Chem. Soc.*, **87**, 5120.
Deno, N. C., and Schriesheim, A. (1955). *J. Amer. Chem. Soc.* **77**, 3051.
Deno, N. C., Jaruzelski, J. J., and Schriesheim, A. (1955). *J. Amer. Chem. Soc.* **77**, 3044.
Deno, N. C., Groves, P. J., and Saines, G. (1959a). *J. Amer. Chem. Soc.* **81**, 5790.
Deno, N. C., Berkheimer, H. E., Evans, W. L., and Peterson, H. J. (1959b). *J. Amer. Chem. Soc.* **81**, 2344.
Deno, N. C., Saines, G., and Spangler, M. S. (1962a). *J. Amer. Chem. Soc.* **84**, 3295.
Deno, N. C., Friedman, N., Hodge, J. D., MacKay, F. F., and Saines, G. (1962b). *J. Amer. Chem. Soc.* **84**, 4713.
Deno, N. C., Richey, H. G., Jr., Hodge, J. D., and Wisotsky, M. J. (1962c). *J. Amer. Chem. Soc.* **84**, 1498.
Deno, N. C., Richey, H. G., Jr., Liu, J. S., Hodge, J. D., Houser, J. J., and Wisotsky, M. J. (1962d). *J. Amer. Chem. Soc.* **84**, 2016.
Deno, N. C., Bollinger, J., Hafer, K., Hodge, J. D., and Houser, J. J. (1963). *J. Amer. Chem. Soc.* **85**, 2998.
Deno, N. C., Boyd, J. D., Hodge, J. D., Pittman, C. U., Jr., ana Turner, J. O. (1964). *J. Amer. Chem. Soc.* **86**, 1745.
Deno, N. C., Richey, H. G., Jr., Liu, J. S., Lincoln, D. N., and Turner, J. O. (1965a). *J. Amer. Chem. Soc.* **87**, 4533.
Deno, N. C., Pittman, C. U., Jr., and Turner, J. O. (1965b). *J. Amer. Chem. Soc.* **87**, 2153.
Dewar, M. J. S. (1962). "Hyperconjugation." Ronald Press, New York.
Dewar, M. J. S., and Ganellin, C. R. (1959). *J. Amer. Chem. Soc.* **81**, 2438.
Diaz, A., and Winstein, S. (1969). *J. Amer. Chem. Soc.* **91**, 4300.
Diaz, A., Lazdins, I., and Winstein, S. (1968). *J. Amer. Chem. Soc.* **90**, 6546.
Dilthey, W., and Dinklage, R. (1929). *Chem. Ber.* **62**, 1834.
Dimroth, K., Umbach, W., and Thomas, H. (1967). *Chem. Ber.* **100**, 132.
Doering, W. von E., and Knox, L. H. (1954). *J. Amer. Chem. Soc.* **76**, 3203.
Doering, W. von E., and Schoenewaldt, E. F. (1951). *J. Amer. Chem. Soc.* **73**, 2333.
Doering, W. von E., Levitz, M., Sayigh, A., Sprecher, M., and Whelan, W. P. (1953). *J. Amer. Chem. Soc.* **75**, 1008.
Doering, W. von E., Saunders, M., Boyton, H. G., Earhart, H. W., Wadley, E. F., Edwards, W. R., and Laber, G. (1958). *Tetrahedron* **4**, 178.
Doyle, M. P., and Wierenga, W. (1970). *J. Amer. Chem. Soc.* **92**, 4999.
Eberson, L., Petrovich, J. P., Baird, R., Dyckes, D., and Winstein, S. (1965). *J. Amer. Chem. Soc.* **87**, 3504.
Eliel, E. L. (1962). "Stereochemistry of Carbon Compounds," pp. 149–156. McGraw-Hill. New York.
Eriks, K., and Koh, L. L. (1963). *Petrol. Res. Fund Rep.* No. 8, p. 5.
Evans, A. G. (1946). *Trans. Faraday Soc.* **42**, 719.
Evans, J. C. (1968). *In* "Carbonium Ions" (G. Olah and P. von R. Schleyer, eds.), Vol. I, pp. 223–236. Wiley (Interscience), New York.
Fainberg, A. H., and Winstein, S. (1957). *J. Amer. Chem. Soc.* **79**, 1602.
Fateley, W. G., Curnutte, B., and Lippincott, E. R. (1957). *J. Chem. Phys.* **26**, 1471.
Feldman, M., and Flythe, W. C. (1969). *J. Amer. Chem. Soc.* **91**, 4577.

Feldman, M., and Flythe, W. C. (1971). *J. Amer. Chem. Soc.* **93**, 1547.
Field, F. H., and Franklin, J. L. (1957). "Electron Impact Phenomena and the Properties of Gaseous Ions." Academic Press, New York.
Finkelstein, M. (1955). Ph.D. Dissertation, Yale University, New Haven, Connecticut.
Foote, C. S. (1964). *J. Amer. Chem. Soc.* **86**, 1853.
Fort, R. C., Jr., and Schleyer, P. von R. (1966a). *Advan. Alicyclic Chem.* **1**, 283.
Fort, R. C., Jr., and Schleyer, P. von R. (1966b). *Chem. Rev.* **64**, 277.
Fraenkel, G., and Farnum, D. G. (1968). *In* "Carbonium Ions" (G. Olah and P. von R. Schleyer, eds.), Vol. I, pp. 237–256. Wiley (Interscience), New York.
Franklin, J. L. (1952). *Trans. Faraday Soc.* **48**, 443.
Franklin, J. L. (1968). *In* "Carbonium Ions" (G. A. Olah and P. von R. Schleyer, eds.), Vol. I, pp. 77–110. Wiley (Interscience), New York.
Franklin, J. L., and Field, F. H. (1953). *J. Amer. Chem. Soc.* **75**, 2819.
Freedman, H. H., and Frantz, A. M. (1962). *J. Amer. Chem. Soc.* **84**, 4165.
Friedman, L. (1970). *In* "Carbonium Ions" (G. A. Olah and P. von R. Schelyer, eds.), Vol. II, pp. 655–714. Wiley (Interscience), New York.
Fry, J. L., and Karabatsos, G. J. (1970). *In* "Carbonium Ions" (G. Olah and P. von R. Schleyer, eds.), Vol. II, pp. 521–572. Wiley (Interscience), New York.
Fry, J. L., Lancelot, C. J., Lam, L. K. M., Harris, J. M., Bingham, R. C., Raber, R. J., Hall, R. E., and Schleyer, P. von R. (1970). *J. Amer. Chem. Soc.* **92**, 2538.
Gassman, P. G. (1970). *Accounts Chem. Res.* **3**, 26.
Gassman, P. G., and Nakai, T. (1971). *J. Amer. Chem. Soc.* **93**, 5897.
Gillespie, R. J. (1968). *Accounts Chem. Res.* **1**, 202.
Gillespie, R. J., and Robinson, E. A. (1968). *In* "Carbonium Ions" (G. Olah and P. von R. Schleyer, eds.), Vol. I, pp. 111–134. Wiley (Interscience), New York.
Gleave, J. L., Hughes, E. D., and Ingold, C. K. (1935). *J. Chem. Soc., London* p. 236.
Goering, H. L., and Chang, S. (1965). *Tetrahedron Lett.* p. 3607.
Goering, H. L., and Levy, J. F. (1962). *J. Amer. Chem. Soc.* **84**, 3853.
Goering, H. L., and Levy, J. F. (1964). *J. Amer. Chem. Soc.* **86**, 120.
Goering, H. L., and Linsay, E. C. (1969). *J. Amer. Chem. Soc.* **91**, 7435.
Goering, H. L., Briody, R. G., and Sandrock, G. (1970). *J. Amer. Chem. Soc.* **92**, 7401.
Gold, V. (1955). *J. Chem. Soc., London* p. 1263.
Gold, V., and Kessick, M. A. (1965a). *J. Chem. Soc., London* p. 6718.
Gold, V., and Kessick, M. A. (1965b). *Discuss. Faraday Soc.* **39**, 84.
Gold, V., Howes, B. W. V., and Tye, F. L. (1952). *J. Chem. Soc., London* p. 2167.
Goldschmidt, S., and Christmann, F. (1925). *Justus Liebigs Ann. Chem.* **442**, 246.
Goldstein, M. J., and Hoffmann, R. (1971). *J. Amer. Chem. Soc.* **93**, 6193.
Gomberg, M. (1902). *Chem. Ber.* **35**, 2397 and 2405.
Gomes de Mesquita, A. H., MacGillavry, C. H., and Eriks, K. (1965). *Acta Crystallogr.* **18**, 437.
Goodman, L. (1967). *Advan. Carbohyd. Chem.* **22**, 109.
Gould, E. S. (1959). "Mechanism and Structure in Organic Chemistry," pp. 220–230. Holt, New York.
Grant, D. M. (1964). *Annu. Rev. Phys. Chem.* **15**, 489.
Grant, D. M., and Litchman, W. M. (1965). *J. Chem. Phys.* **87**, 3994.
Grinter, R., and Mason, S. F. (1964). *Trans. Faraday Soc.* **60**, 889.
Grob, C. A., and Schiess, P. W. (1967). *Angew. Chem., Int. Ed. Engl.* **6**, 1.
Grob, C. A., and Schwartz, W. (1964). *Helv. Chim. Acta* **47**, 1870.
Grunwald, E., and Winstein, S. (1948). *J. Amer. Chem. Soc.* **70**, 846.
Gustavson, G. (1878). See Wagner (1878).

Hafner, K., and Pelster, H. (1961). *Angew. Chem.* **73**, 342.
Hammett, L. P. (1940). "Physical Organic Chemistry." McGraw Hill, New York.
Hammett, L. P., and Deyrup, A. J. (1932). *J. Amer. Chem. Soc.* **54**, 2721.
Hammett, L. P., and Deyrup, A. J. (1933). *J. Amer. Chem. Soc.* **55**, 1900.
Hammond, G. S. (1955). *J. Amer. Chem. Soc.* **77**, 334.
Hansen, R. L. (1965). *J. Org. Chem.* **30**, 4322.
Hantzsch, A. (1907). *Z. Phys. Chem. (Leipzig)* **61**, 257.
Hantzsch, A. (1921). *Chem. Ber.* **54**, 2573 and 2578.
Harned, H. S., and Owen, B. B. (1958). "The Physical Chemistry of Electrolytic Solutions," 3rd ed., pp. 70–75. Van Nostrand-Reinhold, Princeton, New Jersey.
Harris, J. M., Schadt, F. L., Schleyer, P. von R., and Lancelot, C. J. (1969). *J. Amer. Chem. Soc.* **91**, 7508.
Harrison, A. G., Kenarle, P., and Lossing, F. P. (1961). *J. Amer. Chem. Soc.* **3**, 777.
Hart, H., and Martin, R. A. (1960). *J. Amer. Chem. Soc.* **82**, 6362.
Hegerty, A. F., Lomas, J. S., Wright, W. V., Bergmann, E. D., and Dubois, J. E. (1972). *J. Org. Chem.* **37**, 2222.
Henderson, G. G., Robertson, J. M., and Kerr, C. A. (1926). *J. Chem. Soc., London* p. 62.
Hercules, D. M. (1970). *Anal. Chem.* **42**, 20A.
Hess, C. G., Lampe, F. W., and Sommer, L. H. (1964). *J. Amer. Chem. Soc.* **86**, 3174.
Hollander, J. M., and Jolly, W. L. (1970). *Accounts Chem. Res.* **3**, 193.
Howe, R., Friedrich, E. C., and Winstein, S. (1965). *J. Amer. Chem. Soc.* **87**, 379.
Hughes, E. D., Ingold, C. K., and Shiner, V. J. (1953). *J. Chem. Soc., London* p. 3827.
Hünig, S. (1964). *Angew. Chem., Int. Ed. Engl.* **3**, 548.
Ingold, C. K. (1927). *Annu. Rep. Progr. Chem.* **24**, 156.
Ingold, C. K. (1969). "Structure and Mechanism in Organic Chemistry," 2nd ed. Cornell Univ. Press, Ithaca, New York.
Ingold, C. K., and Rothstein, E. (1928). *J. Chem. Soc., London* p. 1217.
Ipatieff, V. N., and Linn, C. B. (1947). U.S. Pat. 2,421,946; *Chem. Abstr.* **41**, 5296 (1947).
Ipatieff, V. N., Schaad, R. E., and Shanley, W. B. (1953). In "The Science of Petroleum," Vol. V, Part II, pp. 14–15. Oxford Univ. Press, London and New York.
Jablonski, R. J., and Snyder, E. I. (1969). *J. Amer. Chem. Soc.* **91**, 4445.
Jackman, L. M., and Sternhell, S. (1969). "Applications of NMR Spectroscopy in Organic Chemistry," 2nd ed. Pergamon, Oxford.
Jaffe, H. (1953). *Chem. Rev.* **53**, 191.
Jenkins, C. L., and Kochi, J. K. (1972). *J. Amer. Chem. Soc.* **94**, 843.
Jennen, J. J. (1966). *Chimia* **20**, 309.
Jensen, E. D., and Taft, R. W. (1964). *J. Amer. Chem. Soc.* **86**, 116.
Jensen, F. R., and Beck, B. H. (1966). *Tetrahedron Lett.* p. 4287.
Jensen, F. R., and Rickborn, B. (1968). "Electrophilic Substitution of Organomercurials." McGraw-Hill, New York.
Jones, M. G., and Coke, J. L. (1969). *J. Amer. Chem. Soc.* **91**, 4284.
Jones, W. M., and Maness, D. D. (1969). *J. Amer. Chem. Soc.* **91**, 4314.
Jungk, H., Smoot, C. R., and Brown, H. C. (1956). *J. Amer. Chem. Soc.* **78**, 2185.
Karabatsos, G. J., Mount, R. A., Rickter, D. O., and Meyerson, S. (1966). *J. Amer. Chem. Soc.* **88**, 5651.
Karplus, M., and Grant, D. (1959). *Proc. Nat. Acad. Sci. U.S.* **45**, 1269.
Katz, T. J., and Gold, E. H. (1964). *J. Amer. Chem. Soc.* **86**, 1600.
Keating, J. T., and Skell, P. S. (1970). In "Carbonium Ions" (G. Olah and P. von R. Schleyer, eds.), Vol. II, pp. 573–654. Wiley (Interscience), New York.

Kehrmann, F., and Wentzel, F. (1901). *Chem. Ber.* **34**, 3815.
Kilpatrick, M., and Hyman, H. H. (1958). *J. Amer. Chem. Soc.* **80**, 77.
Kispert, L. D., Dyas, C., Engelman, K., and Pittman, C. U., Jr. (1971). *J. Amer. Chem. Soc.* **93**, 6948.
Kitaigorodskii, A. I., Struchkov, V. T., Khotsyanova, T. L., Volpin, M. E., and Kursanov, D. N. (1960). *Izve. Akad. Nauk SSSR* p. 45. (*Bull. Acad. Sci. U.S.S.R.* 32).
Koch, H., and Haaf, W. (1960). *Justus Liebigs Ann. Chem.* **638**, 111 and 122.
Kochi, J. K. (1962a). *J. Amer. Chem. Soc.* **84**, 3271.
Kochi, J. K. (1962b). *Tetrahedron* **18**, 483.
Koehl, W. J. (1964). *J. Amer. Chem. Soc.* **86**, 4686.
Krapcho, A. P. and Horn D. E. (1966). *Tetrahedron Lett.* p. 6107.
Kresge, A. J., and Chiang, Y. (1961). *Proc. Chem. Soc., London* p. 81.
Kresge, A. J., Lichtin, N. N., Rao, K. N., and Weston, R. E., Jr. (1965). *J. Amer. Chem. Soc.* **87**, 437.
Lancelot, C. J., and Schleyer, P. von R. (1969). *J. Amer. Chem. Soc.* **91**, 4291 and 4296.
Lancelot, C. J., Harper, J. J., and Schleyer, P. von R. (1969). *J. Amer. Chem. Soc.* **91**, 4294.
Latimer, W. M., Pitzer, K. S., and Slansky, C. M. (1939). *J. Chem. Phys.* **7**, 108.
Lauterbur, P. C. (1957). *J. Chem. Phys.* **26**, 217.
Lauterbur, P. C. (1962). In "Determination of Organic Structures by Physical Methods" (F. C. Nachod and W. D. Phillips, eds.), Vol. II, p. 465. Academic Press, New York.
Lee, C. C. (1970). *Progr. Phys. Org. Chem.* **7**, 129.
Leftin, H. P. (1968). In "Carbonium Ions" (G. A. Olah and P. von R. Schleyer, eds.), Vol. I, p. 353. Wiley (Interscience), New York.
Leone, R. E., and Schleyer, P. von R. (1970). *Angew. Chem., Int. Ed. Engl.* **9**, 860.
Levy, H. A., and Brockway, L. O. (1937). *J. Amer. Chem. Soc.* **59**, 2085.
Lichtin, N. N. (1968). In "Carbonium Ions" (G. Olah and P. von R. Schleyer, eds.), Vol. I, pp. 135–152. Wiley (Interscience), New York.
Lichtin, N. N., and Vignale, M. J. (1957). *J. Amer. Chem. Soc.* **79**, 579.
Liggero, S. H., Harper, J. J., Schleyer, P. von R., Krapcho, A. P., and Horn, D. E. (1970). *J. Amer. Chem. Soc.* **92**, 3789.
Liler, M. (1971). "Reaction Mechanisms in Sulphuric Acid." Academic Press, New York.
Long, F. A., and Schulze, J. (1964). *J. Amer. Chem. Soc.* **86**, 327.
Lustgarten, R. K., Brookhart, M., and Winstein, S. (1967). *J. Amer. Chem. Soc.* **89**, 6350.
Lwowski, W. (1958). *Angew. Chem.* **70**, 483.
McCubbin, R. J., and Adkins, H. (1930). *J. Amer. Chem. Soc.* **52**, 2547.
Mackor, E. L., Hofstra, A., and van der Waals, J. H. (1957). *Trans. Faraday Soc.* **53**, 1309.
McLafferty, F. W., ed. (1963). "Mass Spectrometry of Organic Ions." Academic Press, New York.
McManus, S. P., Carroll, J. T., and Pittman, C. U., Jr. (1970). *J. Org. Chem.* **35**, 3768.
McManus, S. P., Pittman, C. U., Jr., and Fanta, P. E. (1972). *J. Org. Chem.* **37**, 2353.
Mahler, J. E., Jones, D. A. K., and Pettit, R. (1964). *J. Amer. Chem. Soc.* **86**, 3589.
Majerski, Z., and Schleyer, P. von R. (1971). *J. Amer. Chem. Soc.* **93**, 665.
Mazur, R. H., White, W. N., Semanov, D. A., Lee, C. C., Silver, M. S., and Roberts, J. D. (1959). *J. Amer. Chem. Soc.* **81**, 4390.
Meerwein, H., and Van Emster, K. (1922) *Chem. Ber.* **55**, 2500.

Meerwein, H., Hammel, O., Serini, A., and Vorster, J. (1927). *Justus Liebigs Ann. Chem.* **453**, 16.
Meerwein, H., Battenberg, E., Gold, H., Pfeil, E., and Willfang, G. (1939). *J. Prakt. Chem.* [N.S.] **154**, 83.
Meerwein, H., Bodenbanner, K., Borner, P., Kunert, F., and Wunderlich, K. (1960a). *Justus Liebigs Ann. Chem.* **632**, 38.
Meerwein, H., Hederich, V., Morschel, H., and Wunderlich, K. (1960b). *Justus Liebigs Ann. Chem.* **635**, 1.
Moss, R. A. (1971). *Chem. Eng. News* **49**, (48) (November 22), p. 28.
Muller, N. (1962). *J. Chem. Phys.* **36**, 359.
Muller, N., and Pritchard, D. E. (1959). *J. Chem. Phys.* **31**, 768 and 1471.
Mulliken, R. S. (1958). *Can. J. Chem.* **36**, 10.
Naville, G., Strauss, H., and Heilbronner, E. (1960). *Helv. Chim. Acta* **43**, 1221.
Necsoiu, I., and Nenitzescu, C. D. (1960). *Chem. Ind. (London)* p. 337.
Nenitzescu, C. D. (1968). In "Carbonium Ions" (G. Olah and P. von R. Schleyer, eds.), Vol. I, pp. 1–75. Wiley (Interscience), New York.
Nordlander, J. E., and Deadman, W. G. (1968). *J. Amer. Chem. Soc.* **90**, 1590.
Nordlander, J. E., and Kelly, W. J. (1969). *J. Amer. Chem. Soc.* **91**, 996.
Norris, J. F. (1901). *Amer. Chem. J.* **25**, 117.
Noyce, D. S., King, P. S., Lane, C. A., and Reed, W. L. (1962). *J. Amer. Chem. Soc.* **84**, 1638.

Oae, S. (1956). *J. Amer. Chem. Soc.* **78**, 4030.
O'Brien, D. H., and Harbordt, C. M. (1970). *J. Organometal. Chem.* **21**, 321.
O'Brien, D. H., and Harbordt, C. M. (1971). *J. Organometal. Chem.* **24**, 327.
Offer, R. R., and Shriner, R. L. (1951). *J. Amer. Chem. Soc.* **73**, 887.
Olah, G., ed. (1963). "Friedel-Crafts and Related Reactions," Vols. I–IV. Wiley, New York.
Olah, G. A. (1971). *Chem. Technol.* **1**, 566.
Olah, G. A. (1972). *J. Amer. Chem. Soc.* **94**, 808.
Olah, G. A., and Clifford, P. R. (1971). *J. Amer. Chem. Soc.* **93**, 1260 and 2320.
Olah, G. A., and Comisarow, M. B. (1966). *J. Amer. Chem. Soc.* **88**, 1818.
Olah, G. A., and Comisarow, M. B. (1969). *J. Amer. Chem. Soc.* **91**, 2955.
Olah, G. A., and Friedman, N. (1966). *J. Amer. Chem. Soc.* **88**, 5350.
Olah, G. A., and Kiovsky, T. E. (1967). *J. Amer. Chem. Soc.* **89**, 5692.
Olah, G. A., and Lukas, J. (1967). *J. Amer. Chem. Soc.* **89**, 2227 and 4739.
Olah, G. A., and Meyer, M. W. (1963). In "Friedel-Crafts and Related Reactions" (G. A. Olah, ed.), Vol. I, p. 623. Wiley, New York.
Olah, G. A., and Mo, Y. K. (1971). *J. Amer. Chem. Soc.* **93**, 4942.
Olah, G. A., and Mo, Y. K. (1972). *J. Amer. Chem. Soc.* **94**, 6864.
Olah, G. A., and Olah, J. A. (1970). In "Carbonium Ions" (G. Olah and P. von R. Schleyer, eds.), Vol. II, pp. 715–782. Wiley (Interscience), New York.
Olah, G. A., and Olah, J. A. (1971). *J. Amer. Chem. Soc.* **93**, 1256.
Olah, G. A., and Pittman, C. U., Jr. (1965). *J. Amer. Chem. Soc.* **87**, 3597.
Olah, G. A., and Pittman, C. U., Jr. (1966). *Advan. Phys. Org. Chem.* **4**, 305.
Olah, G. A., and Porter, R. D. (1970). *J. Amer. Chem. Soc.* **92**, 7627.
Olah, G. A., and Porter, R. D. (1971). *J. Amer. Chem. Soc.* **93**, 6877.
Olah, G., and Schleyer, P. von R., eds. (1968–1972). "Carbonium Ions," Vols. I–IV. Wiley (Interscience), New York.
Olah, G. A., and White, A. M. (1968). *J. Amer. Chem. Soc.* **90**, 1884.

Olah, G. A., and White, A. M. (1969). *J. Amer. Chem. Soc.* **91**, 3954 and 5801.
Olah, G. A., Kuhn, S. J., Tolgyesi, W. S., and Baker, E. B. (1962). *J. Amer. Chem. Soc.* **84**, 2733.
Olah, G. A., Tolgyesi, W. S., Kuhn, S. J., Moffett, M. E., Bastien, I. J., and Baker, E. B. (1963). *J. Amer. Chem. Soc.* **85**, 1328.
Olah, G. A., Baker, E. B., Evans, T. C., Tolgyesi, W. S., McIntyre, J. S., and Bastien, I. J. (1964). *J. Amer. Chem. Soc.* **86**, 1360.
Olah, G. A., Pittman, C. U., Jr., Namanworth, E., and Comisarow, M. B. (1966a). *J. Amer. Chem. Soc.* **88**, 5571.
Olah, G. A., Pittman, C. U., Jr., Waack, R., and Doran, M. (1966b). *J. Amer. Chem. Soc.* **88**, 1488.
Olah, G. A., Friedman, N., Bollinger, J. M., and Lukas, J. (1966c). *J. Amer. Chem. Soc.* **88**, 5328.
Olah, G. A., O'Brien, D. H., and Pittman, C. U., Jr. (1967). *J. Amer. Chem. Soc.* **89**, 2996.
Olah, G. A., Pittman, C. U., Jr., and Symons, M. C. R. (1968a). In "Carbonium Ions" (G. Olah and P. von R. Schleyer, eds.), Vol. I, pp. 153–222. Wiley (Interscience), New York.
Olah, G. A., Commeyras, A., and Liu, C. Y. (1968b). *J. Amer. Chem. Soc.* **90**, 3882.
Olah, G. A., Comisarow, M. B., and Kim, C. J. (1969). *J. Amer. Chem. Soc.* **91**, 1458.
Olah, G. A., Mateescu, G. D., Wilson, L. A., and Gross, M. H. (1970a). *J. Amer. Chem. Soc.* **92**, 7231.
Olah, G. A., White, A. M., DeMember, J. R., Commeyras, A., and Liu, C. Y. (1970b). *J. Amer. Chem. Soc.* **92**, 4627.
Olah, G. A., Kelly, D. P., Jeuell, C. L., and Porter, R. D. (1970c). *J. Amer. Chem. Soc.* **92**, 2546.
Olah, G. A., Ku, A. T., and Olah, J. A. (1970d). *J. Org. Chem.* **35**, 3929.
Olah, G. A., DeMember, J. R., Commeyras, J. R., and LucBribes, J. (1971). *J. Amer. Chem. Soc.* **93**, 459.
Olah, G. A., Jeuell, C. L., Kelly, D. P., and Porter, R. D. (1972). *J. Amer. Chem. Soc.* **94**, 146.
Oulevey, G., and Susz, B. P. (1965). *Helv. Chim. Acta* **48**, 630.
Pasto, D. J., and Serve, M. P. (1965). *J. Amer. Chem. Soc.* **87**, 1515.
Patterson, J. M. (1955). *J. Org. Chem.* **20**, 1277.
Paul, M. A., and Long, F. A. (1957). *Chem. Rev.* **57**, 1.
Penczek, S., Jagar-Grodzinski, J., and Szwarc, M. (1968). *J. Amer. Chem. Soc.* **90**, 2174.
Pepper, D. C. (1964). In "Friedel-Crafts and Related Reactions" (G. A. Olah, ed.), Vol. II, p. 1293. Wiley (Interscience), New York.
Perst, H. (1971). "Oxonium Ions in Organic Chemistry." Verlag Chemie-Academic Press, New York.
Peterson, P. E. (1971). *Accounts Chem. Res.* **4**, 407.
Pfeiffer, G. V., and Jewett, J. G. (1970). *J. Amer. Chem. Soc.* **92**, 2143.
Pfeiffer, P. (1927). "Organishe Molekulverbundungen." Enke, Stuttgart.
Pincock, R. E., Grigat, E., and Bartlett, P. D. (1959). *J. Amer. Chem. Soc.* **81**, 6332.
Pittman, C. U., Jr., and Miller, W. (1972). To be published.
Pittman, C. U., Jr., and Olah, G. A. (1965). *J. Amer. Chem. Soc.* **87**, 2998 and 5123.
Pittman, C. U., Jr., Dyas, C., Engleman, C., and Kispert, L. D. (1972a). *Trans. Faraday Soc.* **68**, 345.
Pittman, C. U., Jr., McManus, S. P., and Larson, J. W. (1972b). *Chem. Rev.* **72**, 357.
Pitzer, K. S., and Donath, W. E. (1959). *J. Amer. Chem. Soc.* **81**, 3213.
Plattner, P. A., Heilbronner, E., and Weber, S. (1950). *Helv. Chim. Acta* **33**, 1663.

Pocker, Y. (1959). *Chem Ind. (London)*, p. 332.
Pocker, Y. (1963). *In* "Molecular Rearrangements" (P. deMayo, ed.), Part One, pp. 1–26. Wiley (Interscience), New York.
Prelog, V. (1957). *Rec. Chem. Prog.* **18**, 247.
Prelog, V. (1960). *Bull. Soc. Chim. Fra.* [5] p. 1433.
Prelog, V., and Traynham, J. G. (1963). *In* "Molecular Rearrangements" (P. deMayo, ed.), Part One, pp. 593–616. Wiley (Interscience), New York.
Roberts, I., and Kimball, G. E. (1937). *J. Amer. Chem. Soc.* **59**, 947.
Roberts, J. D., and Lee, C. C. (1951). *J. Amer. Chem. Soc.* **13**, 5009.
Roberts, J. D., and Mazur, R. H. (1951a). *J. Amer. Chem. Soc.* **73**, 3542.
Roberts, J. D., and Mazur, R. H. (1951b). *J. Amer. Chem. Soc.* **73**, 2509.
Roberts, J. D., Lee, C. C., and Saunders, W. H. (1954). *J. Amer. Chem. Soc.* **76**, 4501.
Roberts, J. D., Lee, C. C., and Saunders, W. H. (1955). *J. Amer. Chem. Soc.* **77**, 3034.
Rochester, C. H. (1970). "Acidity Functions." Academic Press, New York.
Ryabova, R. S., Medvetskaya, I. M., and Vinnik, M. I. (1966). *Zh. Fiz. Khim.* **40**, 339.
Sargent, G. D. (1966). *Quart. Rev., Chem. Soc.* **20**, 299.
Sargent, G. D. (1972). *In* "Carbonium Ions" (G. Olah and P. von R. Schleyer, eds.), Vol. III, Chap. 24. Wiley (Interscience), New York.
Saunders, M., and Hagen, E. L. (1968). *J. Amer. Chem. Soc.* **90**, 6881.
Saunders, M., Schleyer, P. von R., and Olah, G. A. (1964). *J. Amer. Chem. Soc.* **86**, 5680.
Schlenk, W. (1912). *Justus Liebigs Ann. Chem.* **394**, 178.
Schleyer, P. von R. (1964). *J. Amer. Chem. Soc.* **86**, 1854.
Schleyer, P. von R., and Lancelot, C. J. (1969). *J. Amer. Chem. Soc.* **91**, 4297.
Schleyer, P. von R., and Nicholas, R. D. (1961a). *J. Amer. Chem. Soc.* **83**, 182.
Schleyer, P. von R., and Nicholas, R. D. (1961b). *J. Amer. Chem. Soc.* **83**, 2700.
Schleyer, P. von R., Fort, R. C., Watts, W. E., Comisarow, M. B., and Olah, G. A., (1964a). *J. Amer. Chem. Soc.* **86**, 4195.
Schleyer, P. von R., Watts, W. E., Fort, R. C., Jr., Comisarow, M. B., and Olah, G. A. (1964b). *J. Amer. Chem. Soc.* **86**, 5679.
Schleyer, P. von R., Donaldson, M. M., and Watts, W. E. (1965). *J. Amer. Chem. Soc.* **87**, 375.
Schleyer, P. von R., Fry, J. L., Lam, L. K. M., and Lancelot, C. J. (1970). *J. Amer. Chem. Soc.* **92**, 2542.
Schubert, W. M., Lamm, B., and Keeffe, J. R. (1964). *J. Amer. Chem. Soc.* **86**, 4727.
Schuster, I. I., Cotter, A. K., and Kurland, R. J. (1968). *J. Amer. Chem. Soc.* **90**, 4679.
Seel, F. (1943). *Z. Anorg. Allg. Chem.* **252**, 24.
Sharp, D. W. A., and Sheppard, N. (1957). *J. Chem. Soc., London* p. 674.
Shiner, V. J., and Jewett, J. G. (1965). *J. Amer. Chem. Soc.* **87**, 1382.
Shiner, V. J., and Meier, G. F. (1966). *J. Org. Chem.* **31**, 137.
Shoppee, C. W. (1946). *J. Chem. Soc., London* p. 1138.
Silver, M. S. (1961). *J. Amer. Chem. Soc.* **83**, 3487.
Silver, M. S. (1963). *J. Org. Chem.* **28**, 1686.
Skell, P. S., and Maxwell, R. J. (1962). *J. Amer. Chem. Soc.* **84**, 3963.
Smith, P. A. S. (1963). *In* "Molecular Rearrangements" (P. de Mayo, ed.), Part One, pp. 457–592. Wiley (Interscience), New York.
Sommer, L. H. (1965). "Stereochemistry, Mechanism and Silicon." McGraw-Hill, New York.
Sorensen, T. S. (1964). *Can. J. Chem.* **42**, 2768.
Sorensen, T. S. (1965a). *J. Amer. Chem. Soc.* **87**, 5075.
Sorensen, T. S. (1965b). *Can. J. Chem.* **43**, 2744.
Stetter, H., Schwarz, M., and Hirschhorn, H. H. (1959). *Chem. Ber.* **92**, 1629.

Stetter, H., Mayer, J., Schwarz, M., and Wulff, K. (1960). *Chem. Ber.* **93**, 226.
Stewart, R. (1957). *Can. J. Chem.* **35**, 766.
Stiles, R. M., and Mayer, R. P. (1959). *J. Amer. Chem. Soc.* **81**, 1497.
Stokes, R. S. (1964). *J. Amer. Chem. Soc.* **86**, 979 and 982.
Stork, G., and Bersohn, M. (1960). *J. Amer. Chem. Soc.* **82**, 1261.
Story, P. R., and Cooke, B. J. A. (1968). *Chem. Commun.* p. 1080.
Stothers, J. B. (1965). *Quart. Rev., Chem. Soc.* **19**, 144.
Streitwieser, A., Jr. (1957). *J. Org. Chem.* **22**, 861.
Streitwieser, A., Jr. (1962). "Solvolytic Displacement Reactions." McGraw-Hill, New York.
Streitwieser, A., Jr., and Stang, P. J. (1965). *J. Amer. Chem. Soc.* **87**, 4953.
Streitwieser, A., Jr., and Waiss, A. C. (1962). *J. Org. Chem.* **27**, 290.
Sundaralingam, M., and Jensen, L. H. (1966). *J. Amer. Chem. Soc.* **88**, 198.
Susz, B. P., and Cassimatis, D. (1961). *Helv. Chim. Acta* **44**, 395.
Susz, B. P., and Wuhrmann, J. J. (1957). *Helv. Chim. Acta* **40**, 722 and 971.
Swain, C. G., and Scott, C. B. (1953). *J. Amer. Chem. Soc.* **75**, 141.
Taft, R. W. (1956). *In* "Steric Effects in Organic Chemistry" (M. S. Newman, ed.), pp. 556–675. Wiley, New York.
Taft, R. W., Martin, R. H., and Lampe, F. W. (1965). *J. Amer. Chem. Soc.* **87**, 2490.
Tanida, H., Tsuji, T., and Irie, T. (1967). *J. Amer. Chem. Soc.* **89**, 1953.
Thomas, R. M., and Sparks, W. J. (1944). U.S. Pat. 2,356,128.
Thompson, J. A., and Cram, D. J. (1969). *J. Amer. Chem. Soc.* **91**, 1778.
Tobey, S. W., and West, R. (1964). *J. Amer. Chem. Soc.* **86**, 1459.
van Tamelen, E. E., Cole, T. M., Greeley, R., and Schumacher, H. (1968). *J. Amer. Chem. Soc.* **90**, 1372.
Velthorst, N. V., and Hoijtink, G. J. (1965). *J. Amer. Chem. Soc.* **87**, 4529.
Vernon, C. A. (1954). *J. Chem. Soc., London* p. 423.
Volz, H., and Lotsch, W. (1969). *Tetrahedron Lett.* p. 2275.
Waack, R., and Doran, M. A. (1963). *J. Amer. Chem. Soc.* **85**, 1651.
Wagner, G. (1878). *Chem. Ber.* **11**, 1251.
Wagner, G. (1899). *Chem. Ber.* **32**, 2302.
Walden, P. (1902). *Chem. Ber.* **35**, 2018.
Walsh, A. D. (1953). *J. Chem. Soc., London* p. 2266.
Ward, A. M. (1927). *J. Chem. Soc., London* pp. 445 and 2285.
Ward, H. R., and Sherman, P. D. (1968). *J. Amer. Chem. Soc.* **90**, 3812.
Wawzonek, S., Berkey, R. A., and Thomson, D. T. (1956). *J. Electrochem. Soc.* **103**, 513.
Weiner, H., and Sneen, R. A. (1965). *J. Amer. Chem. Soc.* **87**, 292.
West, R., Sado, A., and Tobey, S. W. (1966). *J. Amer. Chem. Soc.* **88**, 2488.
Westheimer, F. H. (1956). *In* "Steric Effects in Organic Chemistry" (M. S. Newman, ed.), p. 523. Wiley, New York.
Whelan, W. P. (1952). Ph.D. Dissertation, Columbia University, New York.
White, E. H., and Woodcock, D. J. (1968). *In* "Chemistry of the Amino Group" (S. Patai, ed.), p. 440. Wiley (Interscience), New York.
Whitmore, F. C. (1932). *J. Amer. Chem. Soc.* **54**, 3274.
Whitmore, F. C. (1934). *Ind. Eng. Chem.* **26**, 94.
Whitmore, F. C. (1948). *Chem. Eng. News* **26**, 688.
Whitmore, F. C., and Mixon, L. W. (1941). *J. Amer. Chem. Soc.* **63**, 1460.
Whitmore, F. C., Laughlin, K. C., Matuszesky, F., and Surmatis, J. D. (1941). *J. Amer. Chem. Soc.* **63**, 736.
Wiberg, K. B. (1950). Ph.D. Dissertation, Columbia University, New York.

Wiberg, K. B., Andes, B. A., Jr., and Ashe, A. J. (1972). *In* "Carbonium Ions" (G. Olah and P. von R. Schleyer, eds.), Vol. III, Chap. 26. Wiley (Interscience), New York.
Wilhelm, M., and Curtin, D. Y. (1957). *Helv. Chim. Acta* **40**, 2129.
Williams, J. E., Buss, V., Allen, L. C., Schleyer, P. von R., Latham, W. A., Hehre, W. J., and Pople, J. A. (1970). *J. Amer. Chem. Soc.* **92**, 2141.
Winstein, S. (1969). *Quart. Rev., Chem. Soc.*, **23**, 141.
Winstein, S., and Adams, R. (1948). *J. Amer. Chem. Soc.* **70**, 838.
Winstein, S., and Fainberg, A. H. (1957). *J. Amer. Chem. Soc.* **79**, 5937.
Winstein, S., and Grunwald, E. (1948). *J. Amer. Chem. Soc.* **70**, 828.
Winstein, S., and Holness, N. J. (1955). *J. Amer. Chem. Soc.* **77**, 5562.
Winstein, S., and Lucas, H. J. (1939). *J. Amer. Chem. Soc.* **61**, 1576 and 2845.
Winstein, S., and Takahashi, J. (1958). *Tetrahedron* **2**, 316.
Winstein, S., and Trifan, D. S. (1949). *J. Amer. Chem. Soc.* **71**, 2953.
Winstein, S., and Trifan, D. S. (1952). *J. Amer. Chem. Soc.* **74**, 1147 and 1154.
Winstein, S., Morse, B. K., Grunwald, E., Schreiber, K. C., and Corse, J. (1952). *J. Amer. Chem. Soc.* **74**, 1113.
Winstein, S., Clippinger, E., Fainberg, A. H., Heck, R., and Robinson, G. C. (1956). *J. Amer. Chem. Soc.* **78**, 328.
Winstein, S., Fainberg, A. H., and Grunwald, E. (1957). *J. Amer. Chem. Soc.* **79**, 4146.
Winstein, S., Appel, B., Baker, R., and Diaz, A. (1965). *Chem. Soc., Spec. Publ.* **19**, 109.
Yates, K., and McDonald, R. S. (1971). *J. Amer. Chem. Soc.* **93**, 6297.

5

CARBANIONS

E. M. Kaiser and D. W. Slocum

I. General Aspects	.	337
II. Methods of Investigation of Carbanions	.	339
III. Preparation of Carbanions	.	341
IV. Factors Affecting the Stability of Carbanions	.	342
V. Reactions of Carbanions	.	344
A. Chemistry of Monocarbanions	.	344
B. Multiple Anion Chemistry	.	365
C. Anion Rearrangements	.	381
D. Metalation of Aromatic Compounds	.	403
References	.	415

I. General Aspects

Perhaps the earliest question examined in carbanion chemistry was that of the relative acidities of C–H compounds as acids. Two early scales of such acidities, each derived from equilibration studies, were proposed in the 1930's. The first of these, that of Conant and Wheeler, involved a colorimetric method for the estimation of the relative concentrations of the ionized and un-ionized species, while the second, proposed by McEwen, involved measurement of the differences in optical rotation of l-menthol and its sodio derivative as indicative of the extent of ionization of a particular carbon acid. Reexamination of these methods has suggested that dissociation to ion pairs rather than true dissociation was measured. A more recent method established by Streitwieser consisted of the equilibration of various carbon acids with lithium or cesium cyclohexylamide in cyclohexylamine. Other methods involving M–M exchange and X–M exchange have also provided an ordering of the acidity of hydrocarbons. A further discussion of these methods including an assessment of the relative merits of each can be found in the noted monograph on carbanions by Cram (1965) and in a review by J. R. Jones (1971).

The above methods which measure thermodynamic acidities are to be contrasted with techniques that measure rates of ionization or exchange of carbon acids, thereby affording kinetic acidities. Thus, dissociation of weak acids can be detected, and the rate determined by studying the exchange between hydrogen isotopes. For example, potassium amide-catalyzed deuterium exchange has established the following order of acidities (Cram, 1965):

$$H\text{—}H > H_3C\text{—}H > H_3C \cdot CH_2\text{—}H > \text{cyclohexyl-H} > (CH_3)_3C\text{—}H$$

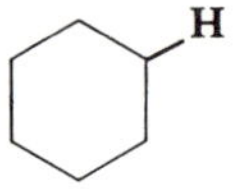

Similarly, the rates of exchange of cyclopropyl, allylic, and benzylic systems were found to place relatively high on this scale. Also, the relative rates of cyclohexylamide salt-catalyzed proton exchange of the very weak acids (**1–3**) were found to be in the relative order of 1, 10^{-3}, and 10^{-8}, respectively (Streitwieser and Ziegler, 1969).

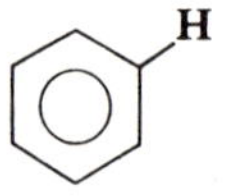

(1) phenyl-H **(2)** 3,5-di-tert-butylphenyl-H **(3)** thiacyclohexyl-H (S)

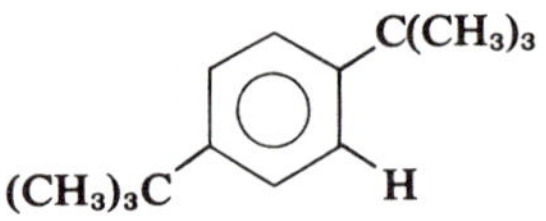

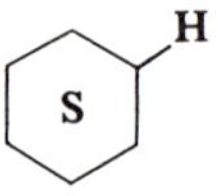

It should be pointed out that the relative rates of ionization of a pair of carbon acids (kinetic acidity) need not be indicative of their relative thermodynamic acidities. The ionization data in Table I illustrate this point (Ayres, 1966).

Table I
Kinetic and Thermodynamic Acidities of Two Carbon Acids

Compound	k_1	k_{-1}	$K = k_1/k_{-1}$
$CH_2(CO_2C_2H_5)_2$	2.5×10^{-5}	5×10^8	5×10^{-14}
$CH_3CH_2NO_2$	3.7×10^{-8}	1.5×10^1	2.5×10^{-9}

In addition to the McEwen and Streitwieser acidity scales, two other approaches have been reported. These have resulted in the scales proposed by Applequist and by Dessy which involve, respectively, X–M and M–M equilibration. Cram (1965) has correlated the data of McEwen, Streitwieser, Applequist, and Dessy to the extent that a scale of carbon acidities ranging from a pK_a of 11 to 45 has been set up (Table II).

Table II

McEwen–Streitwieser–Applequist–Dessy (MSAD) pK_a Scale

Compound	pK_a	Compound	pK_a
Fluoradene	11	Ethylene	36.5
Cyclopentadiene	15	Benzene	37
9-Phenylfluorene	18.5	Cumene (α-position)	37
Indene	18.5	Triptycene (α-position)	38
Phenylacetylene	18.5	Cyclopropane	39
Fluorene	22.9	Methane	40
Acetylene	25	Ethane	42
1,3,3-Triphenylpropene	26.5	Cyclobutane	43
Triphenylmethane	32.5	Neopentane	44
Toluene (α-position)	35	Propane (s position)	44
Propene (α-position)	35.5	Cyclopentane	44
Cycloheptatriene	36	Cyclohexane	45

II. Methods of Investigation of Carbanions

Carbanion-containing species can be conveniently divided into three relatively distinct classes (Coates and Wade, 1967). First, organolithium and organomagnesium compounds are usually considered separately from other organoalkali or alkaline earth derivatives since these C–M bonds are more covalent than the others; of course, many transition M–C bonds are quite covalent. This relatively high covalent character for the C–Li and C–Mg bonds imparts special properties to such carbanions, notably a high solubility in ether and hydrocarbon solvents. Second, organoalkali derivatives of the other alkali metals whose anionic charges are not delocalized are, in contrast to the above, essentially ionic. These compounds exhibit a more intense reactivity than organolithium reagents because of a higher negative charge on carbon. Based on electronegativities of the metal atoms, C–Cs compounds should be among the most ionic of all compounds containing a C–M bond. Finally, organoalkali compounds whose anionic charges are delocalized as in benzylpotassium or sodiocyclopentadienide are brightly colored, less reactive, and more stable species.

A number of studies have appeared in recent years dealing with the ion-pair nature of carbanions. Fluorenyl derivatives have been thoroughly investigated. For example, solvent effects on the equilibrium between contact and solvent-separated ion pairs in fluorenyllithium were most pronounced (Chan and Smid, 1968). In the series tetrahydrofuran, 2-methyltetrahydrofuran, and 2,5-dimethyltetrahydrofuran, the relative percentage of solvent-separated ion pairs decreased from 80 to 25 to 2%. Only solvent-separated ion pairs were observed in hexamethylphosphoric triamide at 25°C; in 3-oxacyclohexene, only contact species were observed. In each of these solvents, 9-(2-hexyl)fluorenyllithium (4) is much more highly dissociated than fluorenyllithium itself, presumably for steric reasons. Use of cyclic chelating polyethers for complexation of the alkali cation (Pedersen, 1967, 1968)

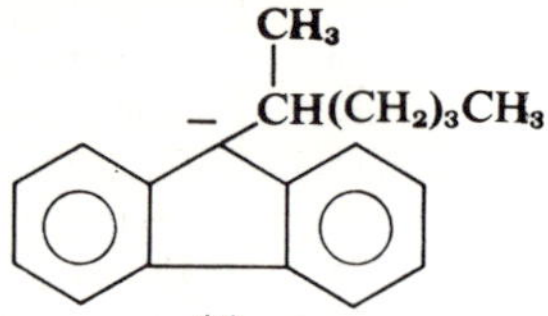

(4)

significantly increased the extent of formation of separated ion pairs (Wong et al., 1970).

Various spectroscopic methods have been useful in the study of the structural aspects of carbanions in solution. Electronic spectra of undissociated species and intimate and solvent-separated ion pairs differ significantly, a fact which is of great utility for relatively quantitative work. An NMR spectrum of a carbanion is complicated by the ion-pair or covalent nature of the particular derivative which, in turn, is influenced by solvent, temperature, counterion, and extent of electron delocalization (Jackman and Sternhell, 1969). All these will, of course, influence the chemical shift. As expected, the extent of negative charge localized at a particular carbon atom can be related to the degree of upfield shift experienced by the neighboring protons.

Certain carbanions have been shown to be capable of preserving their stereochemical integrity; among others, these include cyclopropyl carbanions (Impasto and Walborsky, 1962), vinyllithium reagents (Casey and Boggs, 1971), and certain α-sulfonyl carbanions. A relatively recent technique for deciphering the stereochemical course of a carbanionic reaction has been conceived by Cram and his co-workers (Cram, 1968). This method involves the measurement of the rate of isotopic exchange (k_e) versus the rate of loss of optical activity (k_α) of a particular C–H isotope bond. The scale shown in the following tabulation has been elucidated to define four distinct steric courses by which a C–H isotope bond can be exchanged:

k_e/k_α	Steric course
∞	100% Retention
1	100% Racemization
0.5	100% Inversion
0	100% Racemization with exchange (isoracemization)

5. CARBANIONS

No loss of optical activity, i.e., $k_e/k_\alpha = \infty$, signifies retention of configuration. Total racemization is indicated if the ratio of k_e/k_α is unity, since for every two acts of exchange only two molecules of racemic product are produced. If the rate of exchange is one-half the rate of loss of optical activity, inversion is indicated since each act of exchange produces two molecules of racemic product. Ratios intermediate between 0.5 and 1.0 indicate various amounts of racemization and inversion. When the ratio of k_e/k_α approaches zero, isoracemization is occurring. This is a process whereby the rate of isotopic exchange of a C–H bond lags far behind the rate of loss of optical activity. Isoracemization presumably proceeds via an intramolecular transfer of a proton from one side of the carbon atom to the other. This process is now called the "conducted tour mechanism."

III. Preparation of Carbanions

Carbanions can usually be prepared by any of six different methods. Metalation (M–H exchange) [Eq. (1)]:

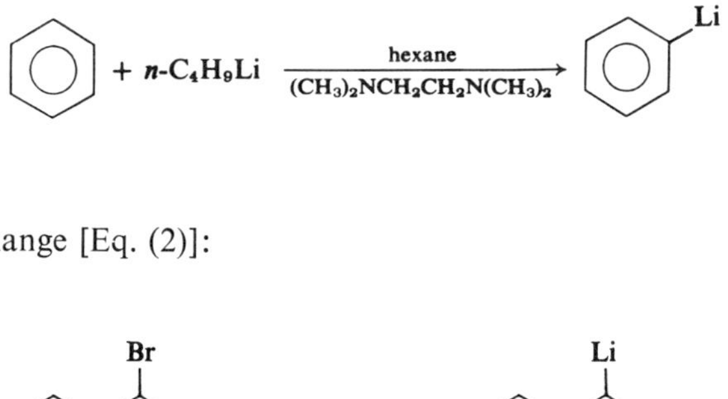

$$\text{C}_6\text{H}_6 + n\text{-C}_4\text{H}_9\text{Li} \xrightarrow[(CH_3)_2NCH_2CH_2N(CH_3)_2]{\text{hexane}} \text{C}_6\text{H}_5\text{Li} \quad (1)$$

(Langer, 1965)

M–X exchange [Eq. (2)]:

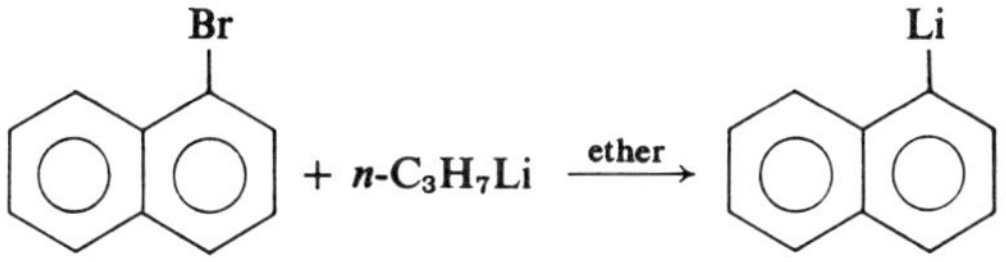

$$\text{1-Br-naphthalene} + n\text{-C}_3\text{H}_7\text{Li} \xrightarrow{\text{ether}} \text{1-Li-naphthalene} \quad (2)$$

(Gilman and Jones, 1951)

Reduction of alkyl or aryl halides by metal [Eq. (3)]:

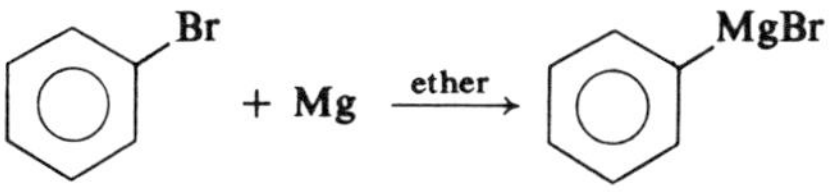

$$\text{C}_6\text{H}_5\text{Br} + \text{Mg} \xrightarrow{\text{ether}} \text{C}_6\text{H}_5\text{MgBr} \quad (3)$$

(Kharasch and Reinmuth, 1954)

Addition of metals to unsaturated systems [Eq. (4)]:

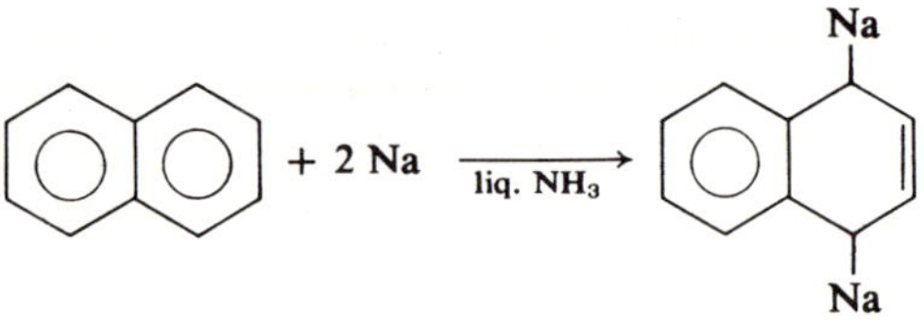

$$\text{naphthalene} + 2\,\text{Na} \xrightarrow{\text{liq. NH}_3} \text{1,4-dihydronaphthalene-1,4-diyl-disodium} \qquad (4)$$

(H. Smith, 1963)

Reduction of certain ethers [Eq. (5)]:

$$C_6H_5CH_2OCH_3 + 2\,Li \xrightarrow[\text{below 0°C}]{\text{THF}} C_6H_5CH_2Li \qquad (5)$$

(Gilman and McNinch, 1961)

Transmetalation [Eq. (6)]:

$$4\,C_6H_5Li + (CH_2{=}CH)_4Sn \xrightarrow{\text{ether}} 4\,CH_2{=}CHLi + (C_6H_5)_4Sn \qquad (6)$$

(Seyferth and Weiner, 1961)

Carbanions are usually prepared with the intent of performing reactions with them. Section V (on carbanionic reactions) lists a much more extensive collection of examples for each of the above preparatory methods.

IV. Factors Affecting the Stability of Carbanions

Relative stabilities of certain carbanions have been discussed above with regard to exchange studies. It has also been mentioned that the extent of covalent or ionic character of the various C–M bonds can influence the nucleophilicity of any carbanion. Such a situation exists for the α-carbanion of ethyl phenyl ketone. The rates of reaction of the lithium, sodium, and potassium salts of this compound with ethyl iodide have been found to be different, the potassium salt being the fastest and the lithium salt the slowest (Zook and Gumby, 1960). This can be attributed to a relative difference in ionic character of the bonding, resulting in a difference in solvation and, hence, in the intimacy of the ion pairs involved. A somewhat different solvent effect is to be found in the report that the rate of racemization with potassium methoxide of 2-methyl-3-phenylpropionitrile (5) in 98.5% dimethyl sulfoxide/ methanol exceeds the rate of racemization in methanol by a factor of about 10^8 (Cram et al., 1961). Thus, the reactivity of a carbanion can be seen to be dependent on the ionic/covalent character of its bonding and the ionizing power of the medium, these two factors, of course, being quite dependent on one another.

5. CARBANIONS

Electronic factors obviously can affect the stabilities and, hence, the reactivities of carbanions. Thus, resonance and conjugative effects serve to delocalize a negative charge on a carbon atom, thereby diluting its nucleophilicity. For example, for the multiple anion **6**, the carbanion which is delocalized to the lesser extent, namely, the primary one, is the more nucleophilic. An extreme example of the extent of delocalization possible for a

$C_6H_5CH_2CHCH_3$
$\qquad\quad\;\;|$
$\qquad\quad\;\;CN$

(5) 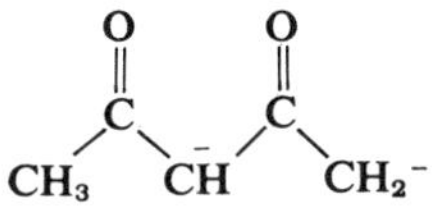 (6) 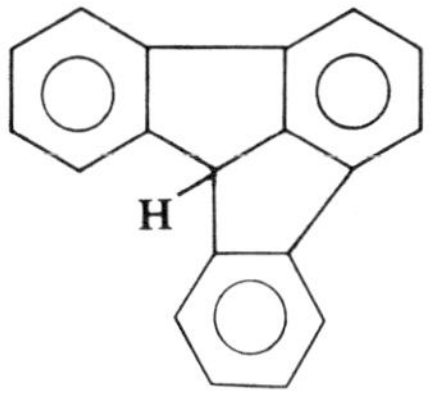 (7) 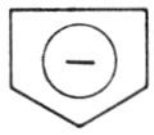 (8)

carbon acid anion is the pK_a of fluoradene (**7**) which has been found to be 10.5 in dimethyl sulfoxide (DMSO) and 17 in methanol (Ritchie and Uschold, 1968). These values are of the order of magnitude of oxygen acids. Cyclopentadienide anion (**8**) is also extraordinarily stable, a fact which can be attributed to delocalization of the anionic charge over an aromatic sextet.

The hybridization of a carbanion plays an important role in its stability. For example, acetylene is 11.5 pK_a units more acidic than ethylene which in turn is 5.5 pK_a units more acidic than ethane (Table II). The order of hybridized carbanion reactivity falls off in the order $sp^3 > sp^2 > sp$, a fact attributable to the increasing s character of the orbital. Resonance delocalization of carbanions into vacant d orbitals of such elements as sulfur has been recognized with the relative acidity of the protons in the 2-position of thiophene

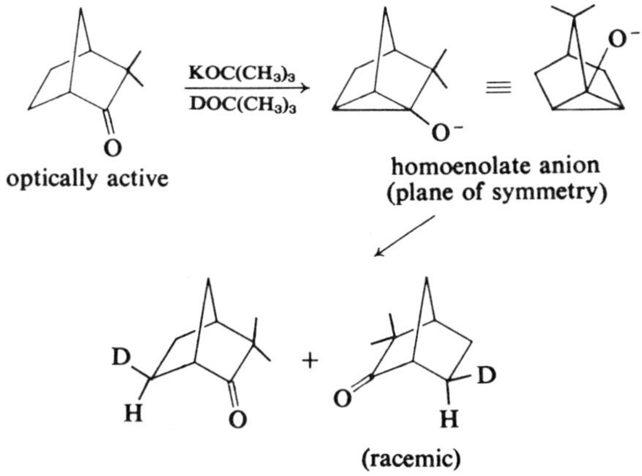

optically active $\xrightarrow[DOC(CH_3)_3]{KOC(CH_3)_3}$ homoenolate anion (plane of symmetry)

(racemic)

being attributed to this. Carbanion homoconjugative effects, initially observed by Nickon and Lambert (1966), have added a new dimension to conjugative effects possible for carbanion delocalization.

V. Reactions of Carbanions

A. Chemistry of Monocarbanions

Regardless of their mode of formation, carbanions enter into a wide variety of condensation reactions which can be conveniently codified according to simple mechanistic classifications. The following subdivisions are appropriate and will be employed in this section: (1) nucleophilic additions, (2) nucleophilic acyl substitutions, and (3) nucleophilic substitutions on aliphatic and aromatic systems. The carbanions used will often be pictured as ionic compounds even though C–M bonds can be mostly covalent as in Grignard reagents or mostly ionic as in organopotassium compounds.

1. Nucleophilic Additions

In general, most nucleophilic addition reactions involve the conversion of a carbonyl group to a substituted alcohol or its dehydrated product (Nielsen and Houlihan, 1968). The mechanism of such additions is pictured simply as an attack by the electron pair of the carbanion on the relatively positive carbonyl carbon of an aldehyde or ketone [Eq. (7)]. Prior complexation of the carbonyl oxygen atom with the metallic cation of the organometallic may or may not be important and depends on the cation, magnesium complexing strongly and potassium hardly at all.

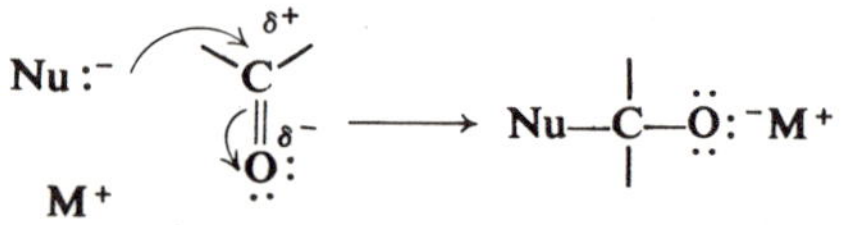

$$Nu:^- \overset{\delta+}{C} \longrightarrow Nu-C-\ddot{O}:^- M^+ \tag{7}$$

More specifically, aldehydes or ketones with α-hydrogens on treatment with bases (or with acids) combine with themselves or with other aldehydes and ketones to give β-hydroxycarbonyl compounds or aldols (Nielsen and Houlihan, 1968). Often, such aldols are dehydrated to give α,β-unsaturated carbonyl compounds either spontaneously or on treatment with acid or heat. Such aldol condensations may be realized with two molecules of the same

$$2\ CH_3\overset{O}{\overset{\|}{C}}H \xrightarrow[H_2O]{KCN} CH_3\overset{OH}{\overset{|}{C}}HCH_2\overset{O}{\overset{\|}{C}}H \tag{8}$$

(Royals, 1954)

aldehyde [Eq. (8)], two molecules of the same ketone [Eq. (9)], two different

$$2\ CH_3\overset{O}{\overset{\|}{C}}CH_3 \xrightarrow[\substack{\text{Soxhlet}\\100°C}]{Ba(OH)_2} (CH_3)_2\overset{OH}{\overset{|}{C}}CH_2\overset{O}{\overset{\|}{C}}CH_3 \quad (9)$$

(Conant and Tuttle, 1941)

aldehydes, though a mixture of products may be obtained unless one of the

$$C_6H_5\overset{O}{\overset{\|}{C}}H + C_6H_5CH_2\overset{O}{\overset{\|}{C}}H \xrightarrow[CH_3OH]{NaOCH_3} C_6H_5\overset{OH}{\overset{|}{C}}H-CH\overset{C_6H_5}{\underset{\underset{O}{\overset{\|}{CH}}}{\diagup}} \quad (10)$$

(Royals, 1954)

aldehydes has no α-hydrogens [Eq. (10)], and two different ketones [Eq. (11)].

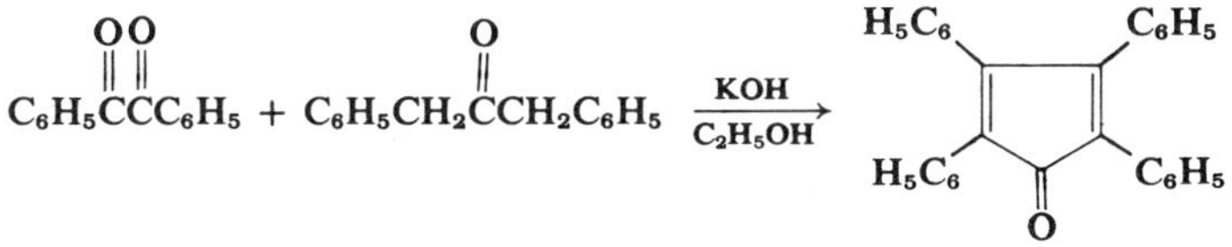

$$C_6H_5\overset{OO}{\overset{\|\,\|}{CC}}C_6H_5 + C_6H_5CH_2\overset{O}{\overset{\|}{C}}CH_2C_6H_5 \xrightarrow[C_2H_5OH]{KOH} \begin{matrix}H_5C_6 \diagdown \diagup C_6H_5 \\ \\ H_5C_6 \diagup \underset{O}{\diagdown} C_6H_5\end{matrix} \quad (11)$$

(J. R. Johnson and Grummitt, 1955)

In addition, ketones with α-hydrogens can be condensed with aldehydes usually without α-hydrogens in the Claisen–Schmidt reaction [Eq. (12)].

$$C_6H_5\overset{O}{\overset{\|}{C}}H + C_6H_5\overset{O}{\overset{\|}{C}}CH_3 \xrightarrow[\substack{H_2O,\ C_2H_5OH\\15°-30°C}]{NaOH} C_6H_5CH=CH\overset{O}{\overset{\|}{C}}C_6H_5 \quad (12)$$

(Kohler and Chadwell, 1941)

Active hydrogen compounds other than aldehydes and ketones can similarly be caused to condense with aldehydes and ketones in the Knoevenagel condensation (G. Jones, 1967). Such reactions, effected by bases, usually lead to olefins rather than alcohols. This can be illustrated by the following general equation (13), where G is $-CO_2R$, $-CN$, $-NO_2$, $-SOR$, $-SO_2R$, etc.

$$R-\overset{O}{\overset{\|}{C}}-R + G-CH_2-G' \xrightarrow{base} \begin{matrix}R \diagdown \diagup G \\ C=C \\ R \diagup \diagdown G'\end{matrix} \quad (13)$$

A specific example further illustrates the reaction [Eq. (14)].

$$NCCH_2CO_2C_2H_5 + (C_2H_5)_2C=O \xrightarrow[\substack{CH_3CO_2H\\C_6H_6\\\Delta}]{NH_4O\overset{O}{\overset{\|}{C}}-CH_3} \begin{matrix}C_2H_5 \diagdown \diagup CN \\ C=C \\ C_2H_5 \diagup \diagdown CO_2C_2H_5\end{matrix} \quad (14)$$

(Cope and Hancock, 1955)

A very large number of Knoevenagel-like reactions [Eq. (15)] are also known which differ from Eq. (14) in two respects. First, the active hydrogen compounds are weaker acids than those employed in the Knoevenagel reaction, thus requiring the use of stronger bases to afford the intermediate carbanion. Second, alcohols rather than olefins are normally obtained as products.

$$CH_3CN \xrightarrow[NH_3]{NaNH_2} NaCH_2CN \xrightarrow[\text{2. acid}]{\text{1. } (C_6H_5)_2CO} (C_6H_5)_2\overset{\overset{OH}{|}}{C}CH_2CN \quad (15)$$

(Kaiser and Hauser, 1968)

Similarly, the reaction of Grignard reagents with aldehydes or ketones usually affords alcohols [Eq. (16)] (Kharasch and Reinmuth, 1954). Certain side reactions, particularly reduction [Eq. (17)] or enolization [Eq. (18)] of the carbonyl compound, can be realized, however, if the reagents are bulky or sterically hindered. The proposed mechanisms for addition, reduction, and enolization have been published but may have to be changed, depending on the outcome of the study of the structures of Grignard reagents (Ashby and Parris, 1971).

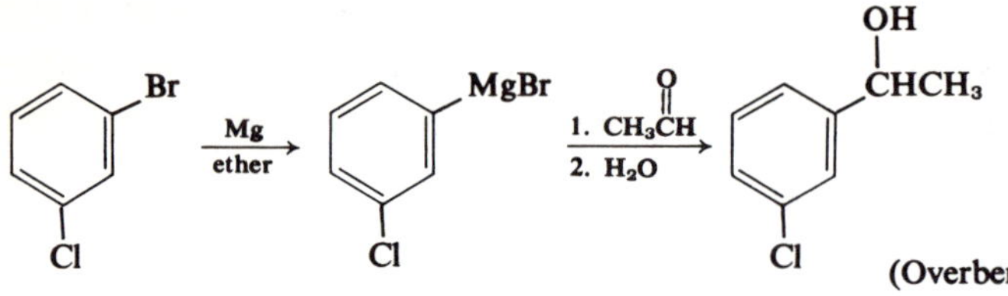

$$\underset{Cl}{\underset{|}{C_6H_4}}-Br \xrightarrow[\text{ether}]{Mg} \underset{Cl}{\underset{|}{C_6H_4}}-MgBr \xrightarrow[\text{2. } H_2O]{\text{1. } CH_3CHO} \underset{Cl}{\underset{|}{C_6H_4}}-\overset{\overset{OH}{|}}{CH}CH_3 \quad (16)$$

(Overberger et al., 1955)

$$(CH_3)_2CHMgBr + (CH_3)_2CHCOCH(CH_3)_2 \xrightarrow{\text{ether}} \xrightarrow{\text{acid}} (CH_3)_2CH\overset{\overset{OH}{|}}{CH}CH(CH_3)_2 \quad (17)$$

(Whitmore and George, 1942)

$$(CH_3)_3CCH_2MgCl + (CH_3)_2CHCOCH(CH_3)_2 \xrightarrow{\text{ether}} \underset{CH_3}{\overset{CH_3}{\diagdown}}C=C\underset{CH(CH_3)_2}{\overset{OMgCl}{\diagup}} \quad (18)$$

(Whitmore and George, 1942)

In a related acyl addition reaction, Grignard reagents and other organometallics react with carbon dioxide under a variety of conditions to give carboxylic acids [Eqs. (19) and (20)] (Kharasch and Reinmuth, 1954).

$$CH_3CH_2\underset{CH_3}{\underset{|}{CH}}MgCl \xrightarrow[\text{2. } H_2SO_4]{\text{1. gaseous } CO_2} CH_3CH_2\underset{CH_3}{\underset{|}{CH}}CO_2H \quad (19)$$

(Gilman and Catlin, 1941)

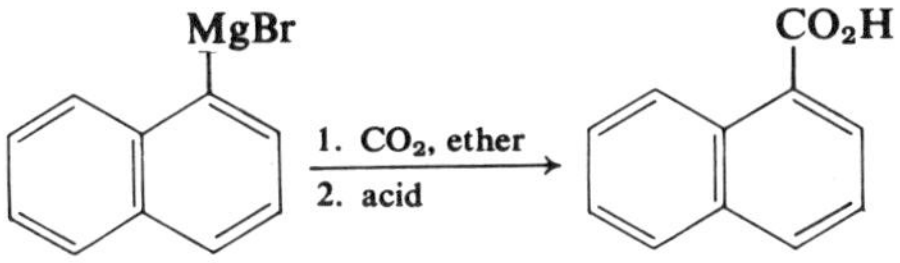

5. CARBANIONS

$$\text{1-Naphthyl-MgBr} \xrightarrow[\text{2. acid}]{\text{1. CO}_2\text{, ether}} \text{1-Naphthyl-CO}_2\text{H} \quad (20)$$

(Gilman et al., 1943)

Similar addition reactions are observed with other electrophiles such as nitriles, azo compounds, anils, and thio ketones [Eqs. (21)–(23)].

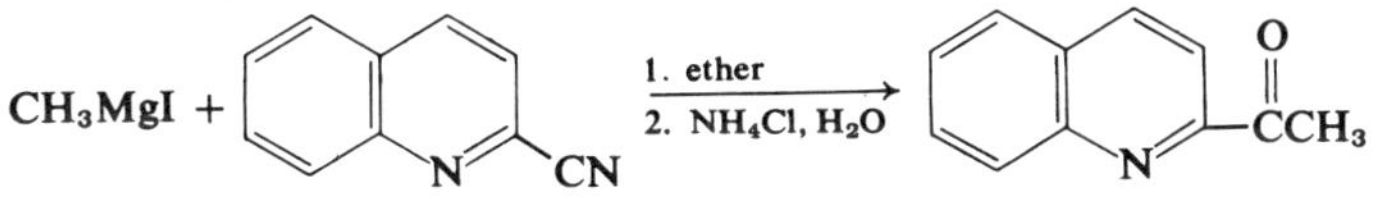

$$\text{CH}_3\text{MgI} + \text{quinoline-2-CN} \xrightarrow[\text{2. NH}_4\text{Cl, H}_2\text{O}]{\text{1. ether}} \text{quinoline-2-C(O)CH}_3 \quad (21)$$

(Kharasch and Reinmuth, 1954)

$$(C_6H_5)_2\text{CHK} + C_6H_5N=NC_6H_5 \xrightarrow{\text{NH}_3} \xrightarrow{\text{acid}} (C_6H_5)_2\text{CH}-\underset{\underset{C_6H_5}{|}}{N}-\text{NHC}_6H_5 \quad (22)$$

(Kaiser and Bartling, 1972)

$$C_6H_5\text{Li} + (C_6H_5)_2C=NC_6H_5 \xrightarrow[\text{2. acid}]{\text{1. ether}} (C_6H_5)_3C-\text{NHC}_6H_5 \quad (23)$$

(Gilman and Kirby, 1933)

Other active hydrogen compounds likewise can be condensed with aldehydes and ketones often under rather specific conditions. For example, the Reformatsky reaction involves the reaction of aldehydes or ketones with bromo esters, α-halo nitriles, and α-halo N,N-disubstituted amides to give the corresponding β-hydroxy compounds [Eqs. (24) and (25)] (Shriner, 1942). Such reactions, effected by zinc metal, are thought to involve organozinc reagents, compounds similar to, but less reactive than Grignard reagents.

$$C_6H_5\overset{\text{O}}{\underset{\|}{\text{CH}}} + \text{BrCH}_2\text{CO}_2\text{C}_2\text{H}_5 \xrightarrow[\text{2. H}_2\text{SO}_4]{\text{1. Zn, C}_6\text{H}_6, \Delta} C_6H_5\overset{\text{OH}}{\underset{|}{\text{CH}}}\text{CH}_2\text{CO}_2\text{C}_2\text{H}_5 \quad (24)$$

(Hauser and Breslow, 1955)

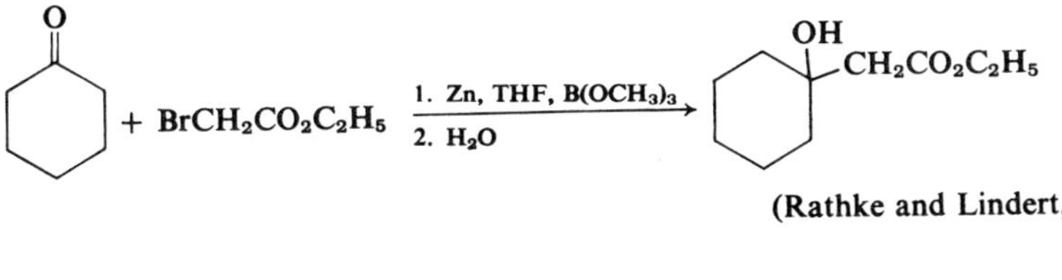

$$\text{cyclohexanone} + \text{BrCH}_2\text{CO}_2\text{C}_2\text{H}_5 \xrightarrow[\text{2. H}_2\text{O}]{\text{1. Zn, THF, B(OCH}_3)_3} \text{1-hydroxy-1-(CH}_2\text{CO}_2\text{C}_2\text{H}_5\text{)cyclohexane} \quad (25)$$

(Rathke and Lindert, 1970)

Reformatsky products can also be obtained by interaction of the conjugate bases of esters with aldehydes and ketones provided excess base is present to convert the ester completely to its carbanion [Eq. (26)].

$$CH_3CO_2C_2H_5 \xrightarrow[NH_3]{excess\ LiNH_2} LiCH_2CO_2C_2H_5 \xrightarrow[2.\ NH_4Cl]{1.\ (C_6H_5)_2CO} (C_6H_5)_2\overset{OH}{\underset{|}{C}}CH_2CO_2C_2H_5 \quad (26)$$

(Dunnavant and Hauser, 1960)

In the Stobbe condensation (W. S. Johnson and Daub, 1960), diethyl succinate is condensed with aldehydes or ketones to afford α,β-unsaturated half-acid esters [Eq. (27)]. The mechanism has been elucidated.

$$C_6H_5\overset{O}{\overset{\|}{C}}CH_3 + \underset{\underset{CO_2C_2H_5}{|}}{(CH_2)_2}\overset{CO_2C_2H_5}{} \xrightarrow[2.\ CH_3CO_2H,\ H_2O]{1.\ NaH,\ C_6H_6} \underset{CH_3}{\overset{C_6H_5}{}}C=C\underset{CH_2CO_2H}{\overset{CO_2C_2H_5}{}} \quad (27)$$

Acid anhydrides are condensed with aldehydes to give α,β-unsaturated acids in the Perkin reaction [Eqs. (28) and (29)] (J. R. Johnson, 1942).

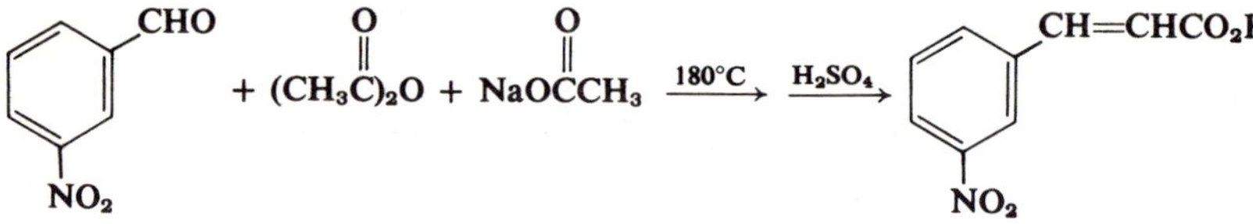

$$\underset{NO_2}{C_6H_4}\text{-CHO} + (CH_3\overset{O}{\overset{\|}{C}})_2O + NaO\overset{O}{\overset{\|}{C}}CH_3 \xrightarrow[H_2SO_4]{180°C} \underset{NO_2}{C_6H_4}\text{-CH=CHCO}_2H \quad (28)$$

(Thayer, 1941)

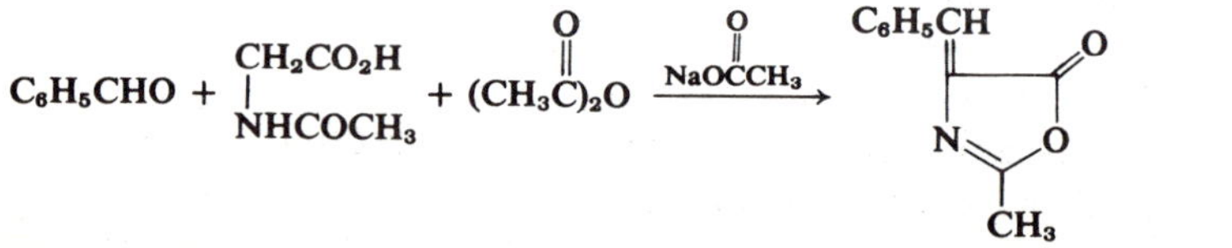

$$C_6H_5CHO + \underset{NHCOCH_3}{\overset{CH_2CO_2H}{\underset{|}{}}} + (CH_3\overset{O}{\overset{\|}{C}})_2O \xrightarrow{NaO\overset{O}{\overset{\|}{C}}CH_3} \text{oxazolone product} \quad (29)$$

(Herbst and Shemin, 1943)

Interaction of α-halo esters, ketones, nitriles, or N,N-disubstituted amides with aldehydes or ketones in the presence of base constitutes the Darzens glycidic ester synthesis (Newman and Magerlein, 1949; Ballester, 1955). The glycidic esters so obtained [Eq. (30)] can be easily decarboxylated by acid to give aldehydes or ketones.

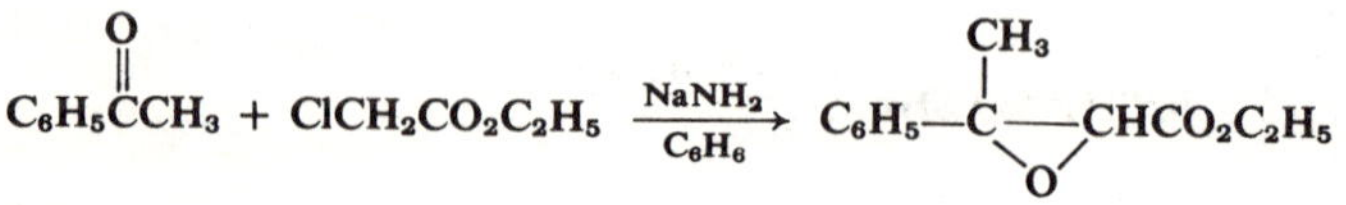

$$C_6H_5\overset{O}{\overset{\|}{C}}CH_3 + ClCH_2CO_2C_2H_5 \xrightarrow[C_6H_6]{NaNH_2} C_6H_5-\underset{\underset{O}{\diagdown\ \diagup}}{\overset{CH_3}{\underset{|}{C}}}-CHCO_2C_2H_5 \quad (30)$$

(Allen and Van Allan, 1955)

5. CARBANIONS

In the Thorpe or Thorpe–Ziegler reactions (Schaefer and Bloomfield, 1967), nitriles with active α-hydrogens can be self-condensed by base to afford ultimately ketones. The Thorpe–Ziegler method has found use in the synthesis of ring compounds provided a large amount of solvent is employed (high dilution technique) [Eq. (31)].

$$\text{cyclobutane-1,1-bis(CH}_2)_2\text{CN} + \text{Ph--N(Li)CH}_3 \xrightarrow[\text{mix over 48 hours}]{\text{ethyl ether}} \text{bicyclic ketone} \quad (31)$$

The coupling of two molecules of aromatic aldehydes, the benzoin condensation (Ide and Buck, 1948), gives α-hydroxy ketones or benzoins. The success of such reactions depends on the fact that aldehydic protons can be made sufficiently acidic by prior addition of cyanide ion (or other nucleophiles) and that a carbanion can be formed which adds to a second equivalent of the aldehyde [Eq. (32)].

$$2\ C_6H_5\overset{O}{\underset{\|}{C}}H + \text{NaCN} \xrightarrow{H_2O,\ C_2H_5OH} \left[\ C_6H_5\underset{CN}{\overset{OH}{\underset{|}{C}}}:^-\ \right] \longrightarrow C_6H_5\overset{HO}{\underset{|}{C}}H\overset{O}{\underset{\|}{C}}C_6H_5 \quad (32)$$

(Adams and Marvel, 1932)

Another condensation involving aldehydes is the Mannich reaction (Blicke, 1942). In this reaction, an aldehyde (usually formaldehyde), an amine, and an active hydrogen compound are combined resulting in aminomethylation of the active hydrogen compound [Eq. (33)].

$$\diagdown\!\!\text{N}\!\!\diagup + \overset{O}{\underset{\diagdown}{\underset{\|}{C}}\diagup} + \text{CH}_3\text{G} \xrightarrow{\text{base or acid}} \diagdown\!\!\text{N--CH}_2\text{--CH}_2\text{--G} \quad (33)$$

The group G has been $-\overset{O}{\underset{\|}{C}}-\text{R(Ar)}$, $-\overset{O}{\underset{\|}{C}}-\text{H}$, $-\overset{O}{\underset{\|}{C}}\text{OR}$, $-\text{CN}$, $-\text{NO}_2$, pyridine, and others. A base-catalyzed Mannich reaction is shown in Eq. (34) (Cope et al., 1963).

$$\text{H}\overset{O}{\underset{\|}{C}}(\text{CH}_2)_3\overset{O}{\underset{\|}{C}}\text{H} \xrightarrow[\text{NaOH, H}_2\text{O}]{\underset{\text{HOCCH}_2\text{CCH}_2\text{COH}}{\overset{\text{CH}_3\text{NH}_2}{}}} \xrightarrow{\underset{\Delta}{\text{HCl}}} \text{bicyclic N--CH}_3 \text{ ketone} \quad (34)$$

with $\text{HOCCH}_2\text{CCH}_2\text{COH}$ (each C=O)

Various phosphorus-, arsenic-, and sulfur-containing compounds have also been found to undergo interesting and useful reactions with aldehydes and ketones. For example, in the Wittig reaction (Maercker, 1965), ylides of phosphines [Eq. (35)] or phosphonate esters [Eq. (36)] combine with such carbonyl compounds to give olefins. Certain phosphonamide and sulfona-

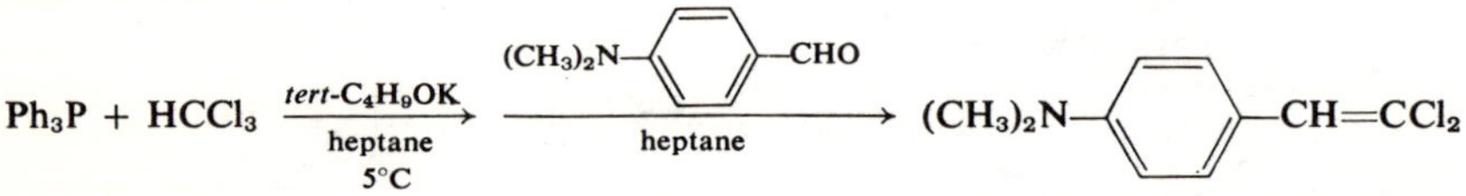

$$Ph_3P + HCCl_3 \xrightarrow[\text{heptane}]{\text{tert-}C_4H_9OK} \xrightarrow[\text{heptane}]{(CH_3)_2N-\text{C}_6H_4-CHO} (CH_3)_2N-\text{C}_6H_4-CH=CCl_2$$

(Speziale et al., 1965) (35)

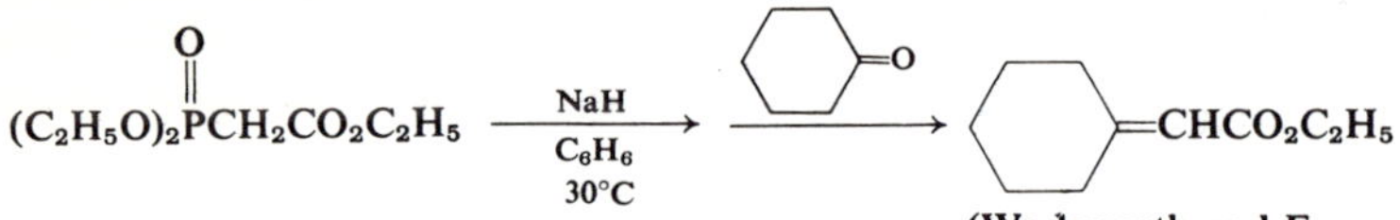

$$(C_2H_5O)_2\overset{O}{\overset{\|}{P}}CH_2CO_2C_2H_5 \xrightarrow[\substack{C_6H_6 \\ 30°C}]{NaH} + \text{cyclohexanone} \longrightarrow \text{cyclohexylidene}=CHCO_2C_2H_5 \quad (36)$$

(Wadsworth and Emmons, 1965)

mide ylides behave similarly (Corey and Kwiatkowski, 1966; Corey and Durst, 1966).

In contrast, sulfonium ylides ($R_2S=CXY \leftrightarrow R_2\overset{+}{S}-\overset{-}{C}XY$) and sulfoxonium ylides ($R_2SO=CXY \leftrightarrow R_2S^+O-C^-XY$) have been found to act as methylene transfer reagents toward aldehydes and ketones (Corey and Chaykovsky, 1965). The former ylide is a more powerful reagent with less steric requirements than the latter one [Eqs. (37) and (38)]. Similar reactions

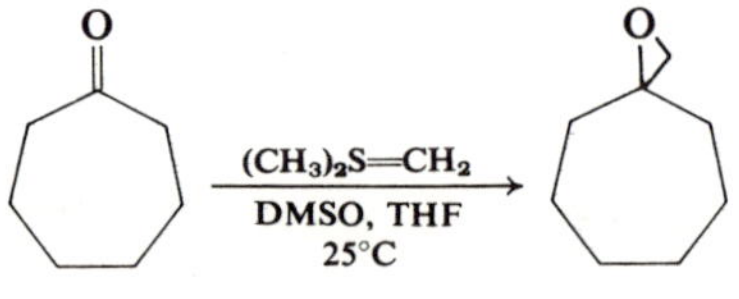

$$\text{cycloheptanone} \xrightarrow[\substack{\text{DMSO, THF} \\ 25°C}]{(CH_3)_2S=CH_2} \text{epoxide} \quad (37)$$

$$\text{4-t-butylcyclohexanone} \xrightarrow[\substack{\text{THF} \\ 25°C}]{(CH_3)_2S=CH_2} \text{axial epoxide} \quad (38)$$

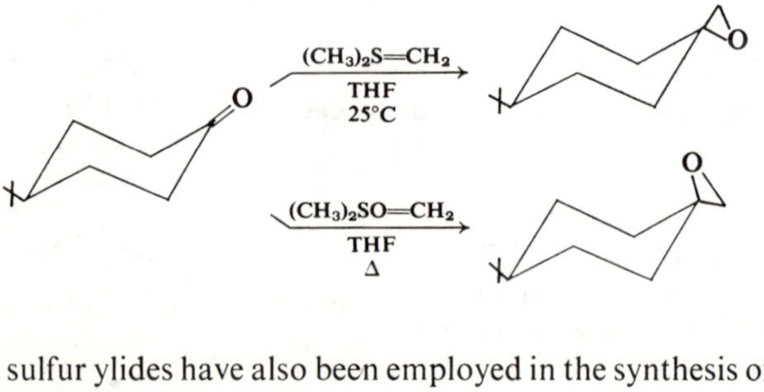

$$\xrightarrow[\Delta]{(CH_3)_2SO=CH_2 \atop \text{THF}} \text{equatorial epoxide}$$

of the above sulfur ylides have also been employed in the synthesis of aziridines and episulfides [Eqs. (39) and (40)].

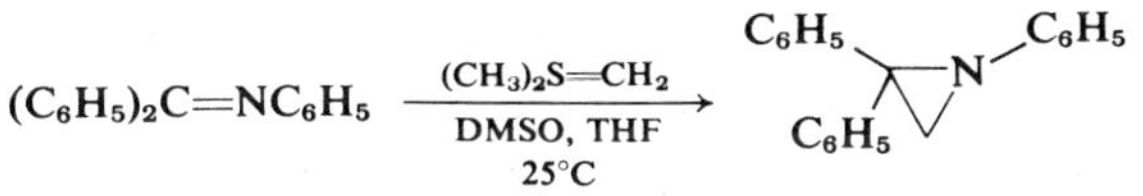

$$(C_6H_5)_2C=NC_6H_5 \xrightarrow[\substack{\text{DMSO, THF} \\ 25°C}]{(CH_3)_2S=CH_2} \underset{C_6H_5}{\overset{C_6H_5}{\triangle}}\!\!\!\!N\text{—}C_6H_5 \quad (39)$$

$$(C_6H_5)_2C=S \xrightarrow[\substack{\text{THF} \\ 25°C}]{(CH_3)_2SO=CH_2} \underset{C_6H_5}{\overset{C_6H_5}{\triangle}}\!\!\!\!S \quad (40)$$

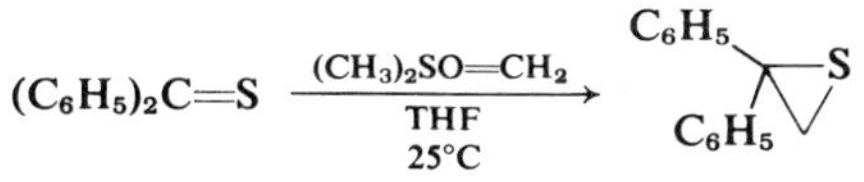

In condensations related to those with the various carbanions described above, a wide variety of α,β-unsaturated aldehydes, ketones, esters, nitriles, and similar compounds undergo reactions with nucleophiles, not at the carbonyl or nitrile carbon, but at the carbon beta to such functional groups. Such reactions, called conjugate or 1,4-additions, can also be pictured as giving alkoxides as intermediates which, by virtue of their being enolate anions, give back the original carbonyl or nitrile functional group on protonation

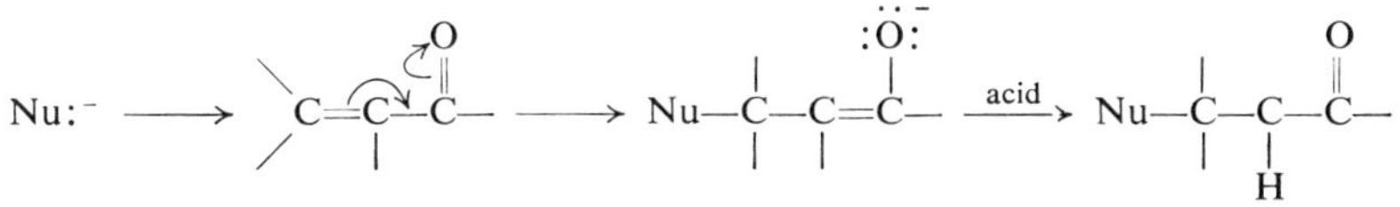

$$\text{Nu:}^- \longrightarrow \overset{\displaystyle \diagdown}{\underset{\displaystyle \diagup}{C}}=C\text{—}\overset{O}{\overset{\|}{C}}\text{—} \longrightarrow \text{Nu}\text{—}\overset{|}{\underset{|}{C}}\text{—}\overset{:\ddot{O}:^-}{\underset{|}{C}}=C\text{—} \xrightarrow{\text{acid}} \text{Nu}\text{—}\overset{|}{\underset{|}{C}}\text{—}\overset{|}{\underset{H}{C}}\text{—}\overset{O}{\overset{\|}{C}}\text{—} \quad (41)$$

(Kaiser *et al.*, 1970)

[Eq. (41)]. Conjugate addition reactions can be subdivided into two classes (Kaiser *et al.*, 1970): (1) those that form a neutral adduct which is more thermodynamically stable than either of the two neutral starting materials and (2) those that form an adduct whose carbanion (enolate anion) is more stable (more weakly basic) than the initial nucleophile.

The best known example of the first class of reactions is the Michael condensation which is effected by only a catalytic amount of base (Bergmann *et al.*, 1959; Brunson, 1949). Only a small amount of base need be employed in these reactions since the resulting enolate anion is converted to final product by reaction with a protonic solvent or an un-ionized active hydrogen

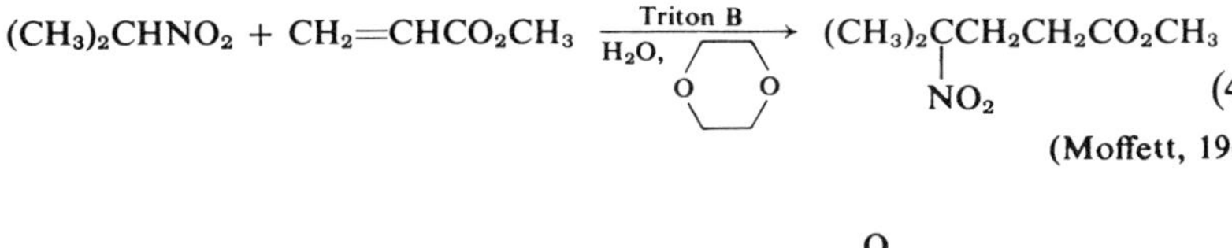

$$(CH_3)_2CHNO_2 + CH_2=CHCO_2CH_3 \xrightarrow[H_2O,]{\text{Triton B}} (CH_3)_2\underset{NO_2}{\overset{|}{C}}CH_2CH_2CO_2CH_3 \quad (42)$$

(Moffett, 1963)

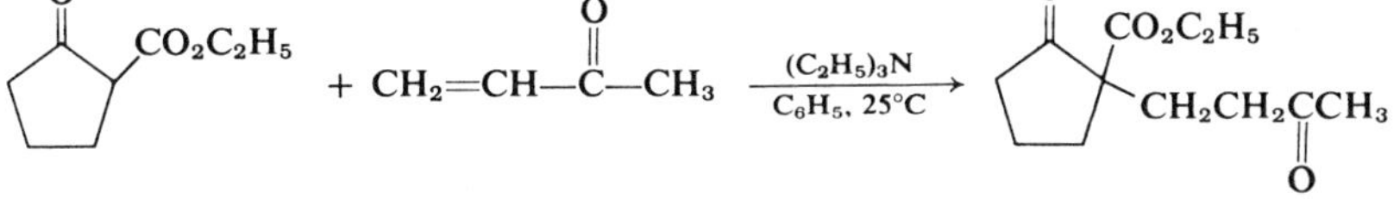

(Dauben and McFarland, 1960)

compound [Eqs. (42) and (43)]. Michael condensations involving acrylonitrile as the acceptor molecule belong to a special class of reactions called cyanoethylations (Brunson, 1949) [Eq. (44)]. When the active hydrogen compound

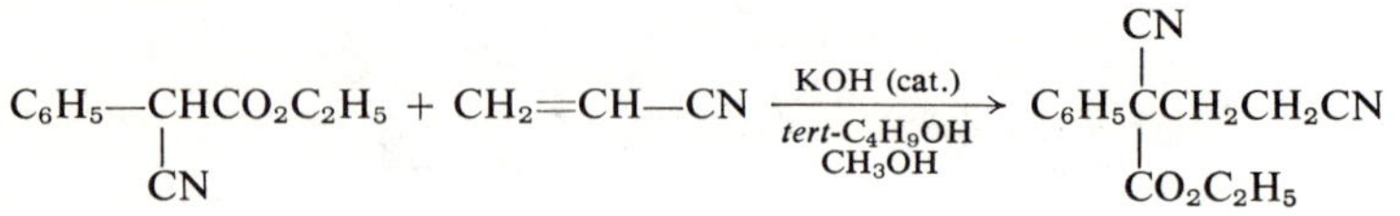

$$C_6H_5-\underset{CN}{\underset{|}{CH}}CO_2C_2H_5 + CH_2=CH-CN \xrightarrow[\substack{tert-C_4H_9OH \\ CH_3OH}]{KOH\ (cat.)} C_6H_5\underset{CO_2C_2H_5}{\underset{|}{\overset{CN}{\overset{|}{C}}}}CH_2CH_2CN \quad (44)$$

(Horning and Finelli, 1963)

initially has more than one acidic hydrogen atom, multiple Michael condensations can be realized [Eq. (45)].

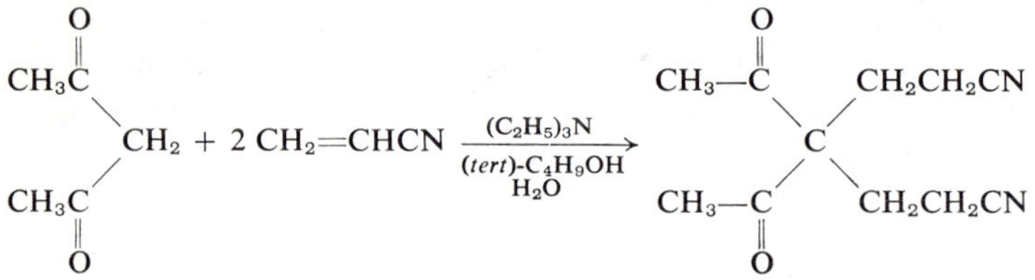

$$\underset{CH_3\overset{O}{\overset{||}{C}}}{\overset{CH_3\overset{O}{\overset{||}{C}}}{\diagdown}}CH_2 + 2\ CH_2=CHCN \xrightarrow[\substack{(tert)-C_4H_9OH \\ H_2O}]{(C_2H_5)_3N} \underset{CH_3-\overset{O}{\overset{||}{C}}}{\overset{CH_3-\overset{O}{\overset{||}{C}}}{\diagdown}}\underset{CH_2CH_2CN}{\overset{CH_2CH_2CN}{\diagup}}C \quad (45)$$

(Adamcik and Miklasiewicz, 1963)

Especially active and unstable α,β-unsaturated carbonyl compounds can be employed as Michael acceptors without appreciable side reactions by employing Robinson's modification. This method uses a suitable precursor of the carbonyl compound which apparently is converted to the acceptor *in situ* under the reaction conditions [Eq. (46)].

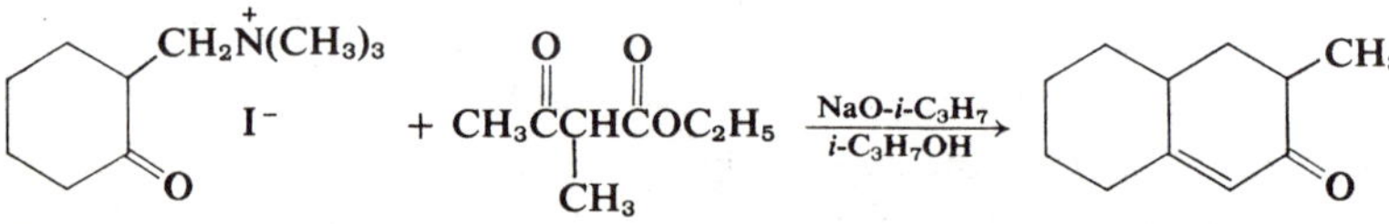

$$\text{[cyclohexanone-CH}_2\overset{+}{N}(CH_3)_3\ I^-] + CH_3\overset{O}{\overset{||}{C}}\underset{CH_3}{\underset{|}{CH}}\overset{O}{\overset{||}{C}}OC_2H_5 \xrightarrow[i-C_3H_7OH]{NaO-i-C_3H_7} \text{[octahydronaphthalenone-CH}_3\text{]} \quad (46)$$

(Logan et al., 1954)

Michael adducts have also been obtained from the interaction of certain enamines with Michael acceptors [Eq. (47)].

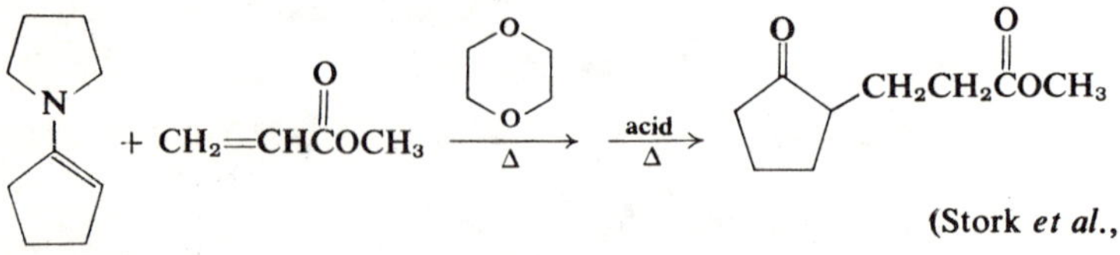

$$\text{[pyrrolidinyl-cyclopentene]} + CH_2=CH\overset{O}{\overset{||}{C}}OCH_3 \xrightarrow[\Delta]{} \text{[morpholine intermediate]} \xrightarrow[\Delta]{acid} \text{[cyclopentanone-CH}_2CH_2\overset{O}{\overset{||}{C}}OCH_3\text{]} \quad (47)$$

(Stork et al., 1963)

More strongly basic, often relatively nonresonance-stabilized carbanions also undergo conjugate additions to α,β-unsaturated systems, but these belong to class 2 since the resulting adduct is present as its resonance-stabilized weakly basic conjugate base until an acid is added [Eq. (48)].

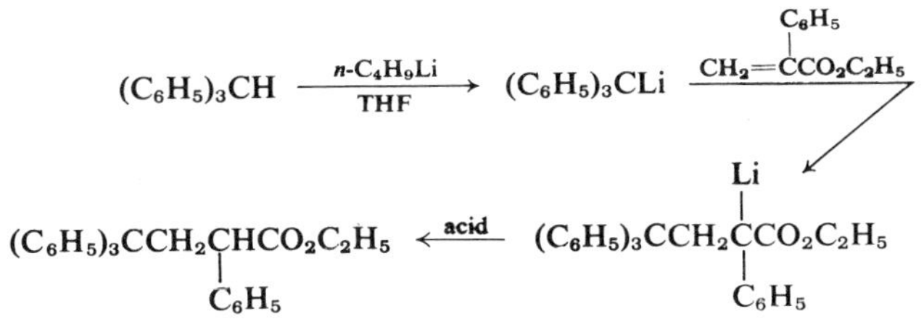

$$(C_6H_5)_3CH \xrightarrow[THF]{n\text{-}C_4H_9Li} (C_6H_5)_3CLi \xrightarrow{CH_2=\overset{C_6H_5}{\underset{}{C}}CO_2C_2H_5}$$

$$(C_6H_5)_3CCH_2\underset{C_6H_5}{\overset{}{C}}HCO_2C_2H_5 \xleftarrow{\text{acid}} (C_6H_5)_3CCH_2\underset{C_6H_5}{\overset{Li}{\underset{}{C}}}CO_2C_2H_5 \quad (48)$$

(Kaiser et al., 1970)

Such reactions can be accompanied or even dominated by 1,2-addition of the nucleophile across the unsaturated linkage of the functional group. Thus, for example, α,β-unsaturated aldehydes and Grignard reagents invariably give β,γ-unsaturated alcohols [Eq. (49)]. In addition, lithio reagents have a

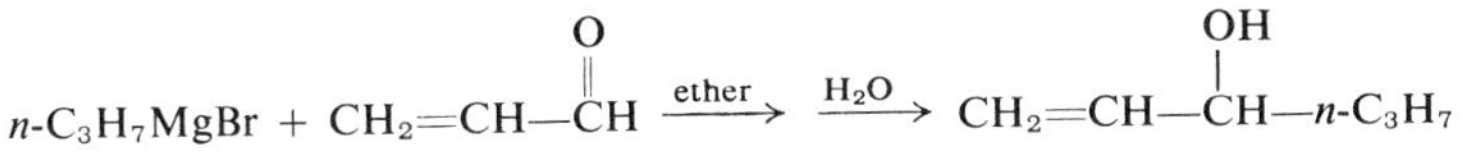

$$n\text{-}C_3H_7MgBr + CH_2=CH-\overset{O}{\overset{\|}{C}}H \xrightarrow{\text{ether}} \xrightarrow{H_2O} CH_2=CH-\overset{OH}{\underset{}{C}}H-n\text{-}C_3H_7 \quad (49)$$

(Kharasch and Reinmuth, 1954)

greater tendency to undergo 1,2-addition on even α,β-unsaturated ketones. A similar interaction of Grignard reagents with α,β-unsaturated ketones can afford either 1,4- or 1,2-addition products depending on the structure of the ketone. Thus, steric hindrance at the carbonyl carbon atom enhances the amount of 1,4-adduct, whereas similar hindrance at the β-carbon atom tends to give more 1,2-adduct. Table III illustrates this steric effect (Royals, 1954).

The interaction of Grignards with α,β-unsaturated carbonyl compounds is dramatically affected by the presence of copper salts even in catalytic amounts. For example, isophorone and methylmagnesium bromide give the following alcohol but not the ketone [Eq. (50)]. In contrast, in the presence of only

Table III
Reaction of Phenylmagnesium Bromide with α,β-Unsaturated Ketones

Ketone	1,4-Addition with C_6H_5MgBr (%)
$C_6H_5CH=CHCOCH_3$	12
$C_6H_5CH=CHCOC_2H_5$	40
$C_6H_5CH=CHCO-i\text{-}C_3H_7$	88
$C_6H_5CH=CHCO-tert\text{-}C_4H_9$	100
$C_6H_5CH=CHCOC_6H_5$	94
$(C_6H_5)_2C=CHCOC_6H_5$	0
$C_6H_5CH=C(C_6H_5)COC_6H_5$	100

1 mole% of cuprous bromide, the same reaction affords only the ketone; indeed, the alcohol is absent (Kharasch and Reinmuth, 1954). More recent

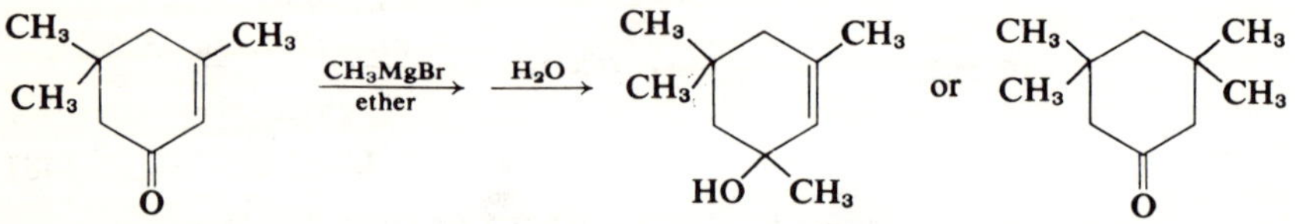

(50)

work dealing with copper-catalyzed conjugate addition reactions is of interest [Eqs. (51)–(53)]. Copper–ate complexes and electron-transfer reactions appear to be involved in these transformations.

$$CH_3CH=CH-\overset{O}{\underset{\|}{C}}-CH_3 \xrightarrow[\text{ether}]{\underset{(n-C_4H_9)_3PCuI}{CH_3Li}} \xrightarrow{\underset{H_2O}{NH_4Cl}} (CH_3)_2CHCH_2\overset{O}{\underset{\|}{C}}CH_3 \quad (51)$$

(House et al., 1966)

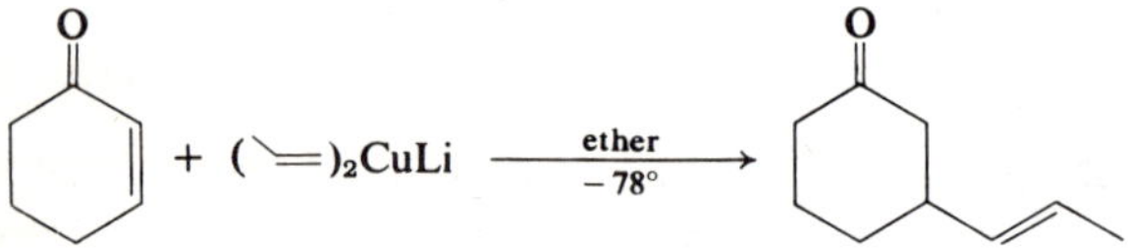

$$\text{cyclohexenone} + (\!=\!)_2\text{CuLi} \xrightarrow[-78°]{\text{ether}} \text{product} \quad (52)$$

(Casey and Boggs, 1971)

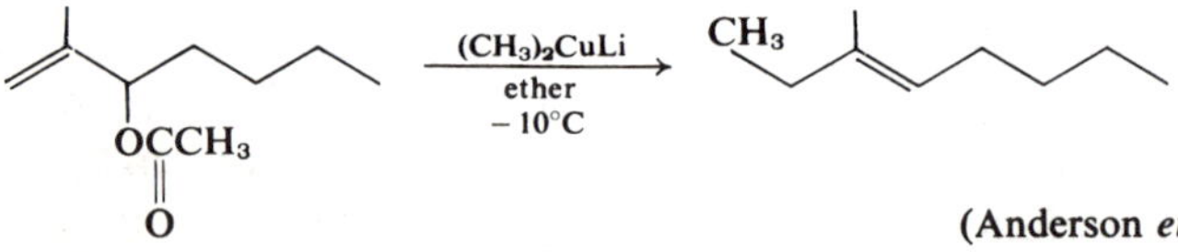

$$\xrightarrow[\text{ether} \\ -10°C]{(CH_3)_2CuLi} \quad (53)$$

(Anderson et al., 1970)

2. Nucleophilic Acyl Substitution

In general, this section will be characterized by the conversion of carboxylic acids or their derivatives to alcohols or ketones by a mechanism whose first step, similar to that in the previous section, involves the addition of a nucleophile across the carbonyl carbon to give an intermediate alkoxide-like ion [Eq. (54)].

$$R-\overset{O}{\underset{\|}{C}}-G + Nu:^-M^+ \rightleftharpoons R-\overset{\overset{..}{O}:^-M^+}{\underset{Nu}{\underset{|}{C}}}-G \quad G = Cl, OR, OCCH_3, \text{etc.} \quad (54)$$

5. CARBANIONS

However, in contrast to the behavior cited in the previous section, such ions derived from carboxylic acids and their derivatives undergo a subsequent substitution or elimination reaction because the group G initially bonded to the carbonyl carbon is a good leaving group. It should be noted that the elimination pathway apparently proceeds by a metal-complexed or uncomplexed ketone intermediate which, depending on the reagents, can often be isolated after hydrolysis [Eq. (55)].

$$\begin{array}{c}
\ddot{\text{O}}{:}^-\text{M}^+ \\
\text{R—C—G} \\
\text{Nu}
\end{array} \rightleftharpoons \text{R—C(=O)—Nu} \xrightarrow{\text{Nu}{:}^-\text{M}^+} \begin{array}{c}
\ddot{\text{O}}{:}^-\text{M}^+ \\
\text{R—C—Nu} \\
\text{Nu}
\end{array} \xrightarrow{\text{acid}} \begin{array}{c}
\text{OH} \\
\text{R—C—Nu} \\
\text{Nu}
\end{array} \quad (55)$$

More specifically, the interaction of Grignard reagents with acid derivatives, particularly esters, affords tertiary alcohols [Eq. (56)];

$$2\ \text{C}_6\text{H}_5\text{MgBr} \xrightarrow[\text{ether, benzene}]{\text{C}_6\text{H}_5\text{CO}_2\text{C}_2\text{H}_5} \xrightarrow{\text{NH}_4\text{Cl}} (\text{C}_6\text{H}_5)_3\text{COH} \quad (56)$$

(Bachmann and Hetzner, 1955)

the use of acyl halides or anhydrides is usually not as suitable for these preparations since such acid derivatives have a propensity toward more side reactions, e.g., reduction and enolization. Certain esters like formates [Eq. (57)] or carbonates [Eq. (58)] lead to secondary or tertiary alcohols

$$2\ n\text{-C}_4\text{H}_9\text{MgBr} \xrightarrow[\text{ether}]{\text{HCO}_2\text{C}_2\text{H}_5} \xrightarrow[\text{H}_2\text{O}]{\text{H}_2\text{SO}_4} (n\text{-C}_4\text{H}_9)_2\text{CHOH} \quad (57)$$

(Coleman and Craig, 1943)

$$3\ \text{C}_2\text{H}_5\text{MgBr} \xrightarrow[\text{ether}]{\text{H}_5\text{C}_2\text{OCOC}_2\text{H}_5} \xrightarrow[\text{H}_2\text{O}]{\text{NH}_4\text{Cl}} (\text{C}_2\text{H}_5)_3\text{COH} \quad (58)$$

(Moyer and Marvel, 1943)

which contain two or three of the same alkyl or aryl groups, respectively; the use of lactones usually affords diols [Eq. (59)].

$$2\ \text{CH}_3\text{MgBr} + \underset{n\text{-C}_4\text{H}_9}{\text{(lactone)}} \xrightarrow{\text{ether}} \xrightarrow[\text{H}_2\text{O}]{\text{HCl}} n\text{-C}_4\text{H}_9\text{CH}(\text{CH}_2)_2\text{C}(\text{CH}_3)_2 \quad (59)$$
$$\qquad\qquad\qquad\qquad\qquad\qquad\qquad\qquad \overset{|}{\text{OH}} \quad \overset{|}{\text{OH}}$$

(Colonge and Marey, 1963)

On the other hand, the interaction of organolithium reagents with carboxylic acids affords ketones in synthetically useful yields [Eq. (60)]. Interest-

$$\underset{}{C_6H_{11}CO_2H} \xrightarrow[DME]{LiH} C_6H_{11}CO_2Li \xrightarrow[DME]{CH_3Li} \xrightarrow[H_2O]{HCl} C_6H_{11}C(O)CH_3 \quad (60)$$

(Jorgenson, 1970)

ingly, ketones are not usually prepared by the reaction of Grignard or organolithium compounds with other carboxylic acid derivatives, a notable exception being the reaction of Grignards with acid anhydrides at low temperature [Eq. (61)]. The less reactive organozinc or organocadmium

$$n\text{-}C_4H_9MgBr \xrightarrow[\substack{\text{ether}\\-70°C}]{CH_3COCCH_3(O)(O)} \xrightarrow[H_2O]{NH_4Cl} n\text{-}C_4H_9CCH_3 \quad (61)$$

(Newman and Booth, 1945)

reagents, however, do combine with acid chlorides to afford ketones [Eq. (62)] (Shirley, 1954).

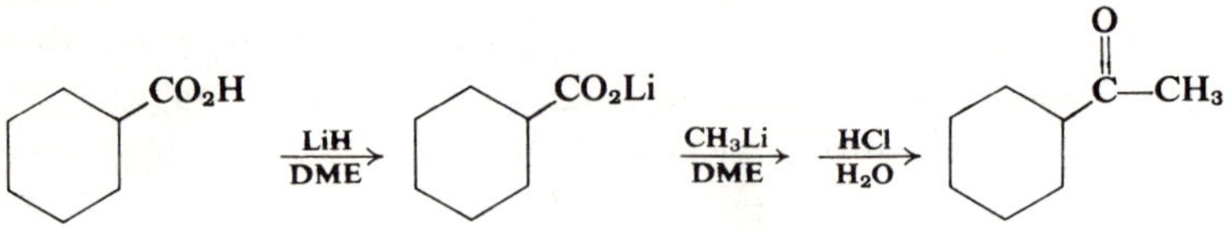

$$(i\text{-}C_3H_7CH_2CH_2)_2Cd \xrightarrow[\substack{1.\ ClC(CH_2)_2COCH_3\\ 2.\ H_2O}]{} i\text{-}C_3H_7CH_2CH_2C(CH_2)_2COCH_3 \quad (62)$$

Regardless of the metallic cation, carbanions that are resonance delocalized, for example, G_2CHM or GCH_2M, are easily converted to ketones on reaction with acid derivatives, particularly esters. One of the better known examples of such acylation reactions is the Claisen condensation which involves the self-condensation of esters to give β-keto esters [Eq. (63)] (Hauser and Hudson, 1947).

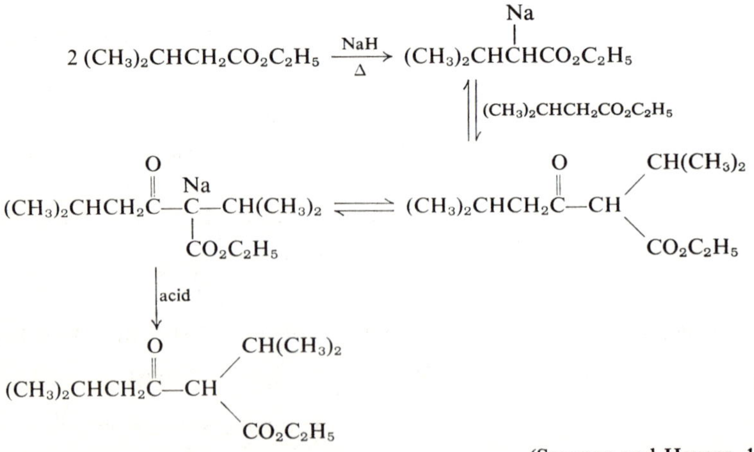

$$2\,(CH_3)_2CHCH_2CO_2C_2H_5 \xrightarrow[\Delta]{NaH} (CH_3)_2CH\overset{Na}{C}HCO_2C_2H_5$$

$$\Big\Updownarrow (CH_3)_2CHCH_2CO_2C_2H_5$$

$$(CH_3)_2CHCH_2\overset{O}{\overset{\|}{C}}-\overset{Na}{\underset{CO_2C_2H_5}{C}}-CH(CH_3)_2 \rightleftharpoons (CH_3)_2CHCH_2\overset{O}{\overset{\|}{C}}-CH\Big\langle\begin{smallmatrix}CH(CH_3)_2\\ CO_2C_2H_5\end{smallmatrix}$$

$$\Big\downarrow \text{acid}$$

$$(CH_3)_2CHCH_2\overset{O}{\overset{\|}{C}}-CH\Big\langle\begin{smallmatrix}CH(CH_3)_2\\ CO_2C_2H_5\end{smallmatrix} \quad (63)$$

(Swamer and Hauser, 1946)

The success of the Claisen condensation is ascribed to ionization of the more strongly acidic β-keto ester product which counteracts unfavorable equilibria toward un-ionized product. Interestingly, the failure of ethyl isobutyrate to undergo the Claisen condensation is ascribed to the inability of the product β-keto ester to undergo such an ionization [Eq. (64)].

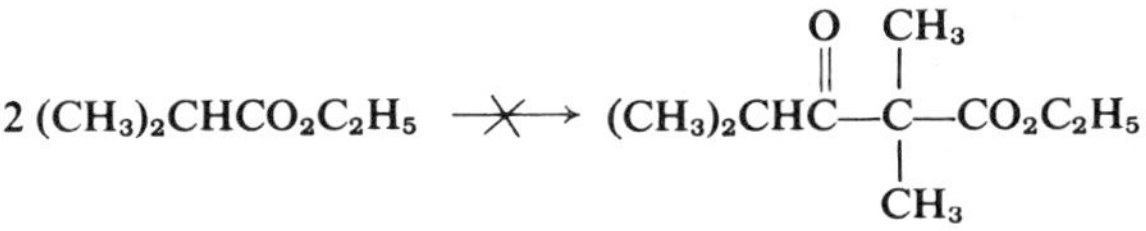

$$2\,(CH_3)_2CHCO_2C_2H_5 \xrightarrow{\times} (CH_3)_2CHC(=O)-C(CH_3)_2-CO_2C_2H_5 \tag{64}$$

Similar synthetically useful ester condensations are realized by employing one ester with and one without α-hydrogens (the crossed Claisen) [Eq. (65)]

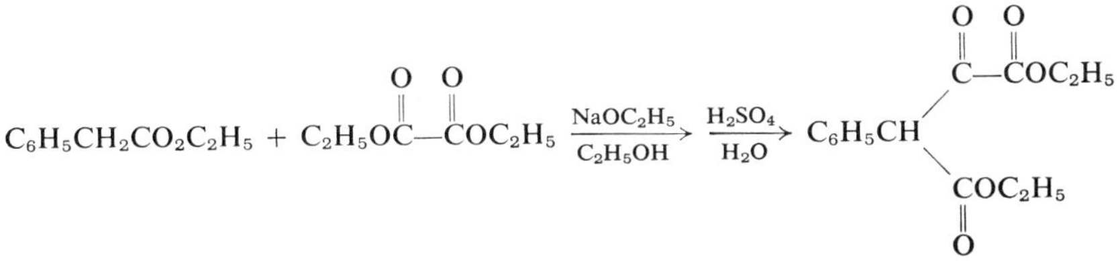

$$C_6H_5CH_2CO_2C_2H_5 + C_2H_5OC(=O)-C(=O)OC_2H_5 \xrightarrow[C_2H_5OH]{NaOC_2H_5} \xrightarrow[H_2O]{H_2SO_4} C_6H_5CH(COC_2H_5)(COC_2H_5) \tag{65}$$

(Levene and Meyer, 1943)

and by cyclizing diesters (the Dieckmann condensation) [Eq. (66)] (Schaefer and Bloomfield, 1967).

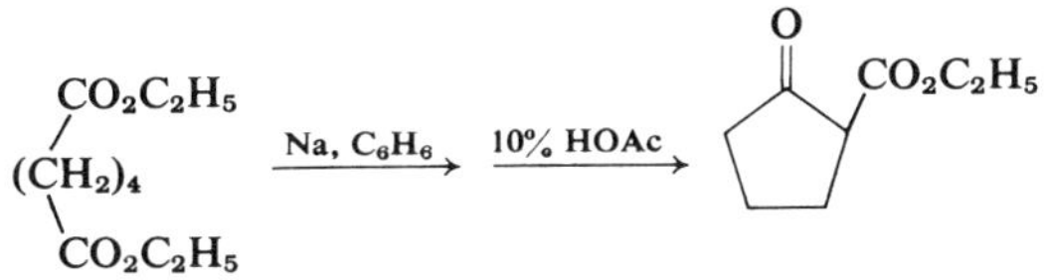

$$(CH_2)_4(CO_2C_2H_5)_2 \xrightarrow{Na,\,C_6H_6} \xrightarrow{10\%\,HOAc} \text{2-(ethoxycarbonyl)cyclopentanone} \tag{66}$$

(Pinkney, 1943)

Ketones, nitriles, sulfones, and other active hydrogen compounds whose conjugate carbanions are also delocalized are similarly acylated [Eqs. (67)–(69)] (Hauser et al., 1962).

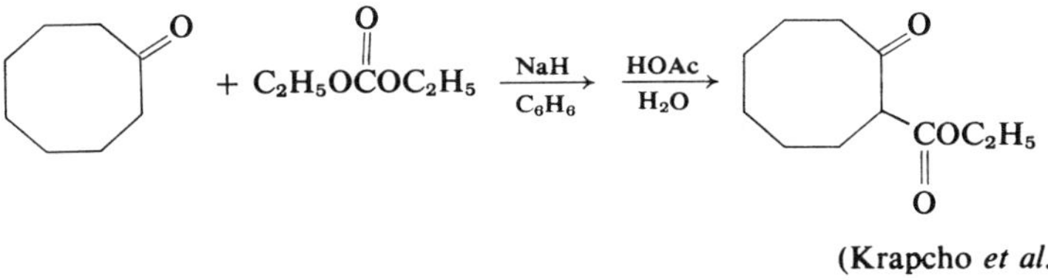

$$\text{cyclooctanone} + C_2H_5OCOC_2H_5 \xrightarrow{NaH}_{C_6H_6} \xrightarrow{HOAc}_{H_2O} \text{2-(ethoxycarbonyl)cyclooctanone} \tag{67}$$

(Krapcho et al., 1967)

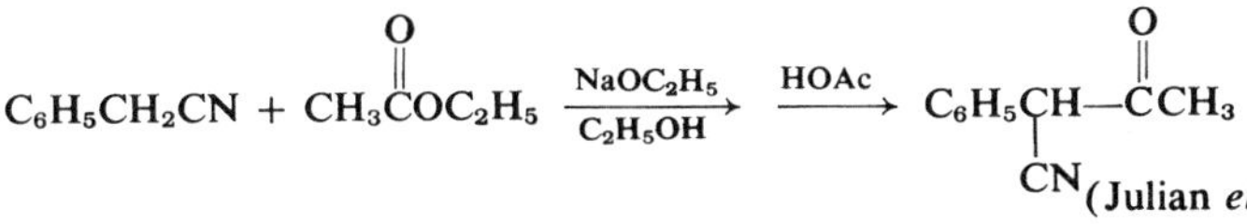

$$C_6H_5CH_2CN + CH_3COC_2H_5 \xrightarrow[C_2H_5OH]{NaOC_2H_5} \xrightarrow{HOAc} C_6H_5CH(CN)-COCH_3 \tag{68}$$

(Julian et al., 1943)

$$CH_3SO_2CH_3 + C_6H_5COCH_3 \xrightarrow[\text{glyme}]{\text{NaH}} \xrightarrow{H_2O} (C_6H_5CCH_2)_2SO_2 \quad (69)$$

(Miles and Hauser, 1964)

Selective acylations have recently been shown to occur on organothallium derivatives of certain β-dicarbonyl compounds (Taylor and McKillop, 1970). The use of acetyl fluoride at 25°C or acetyl chloride at −78°C in such reactions gives rise to C-acetyl and O-acetyl derivatives, respectively [Eq. (70)].

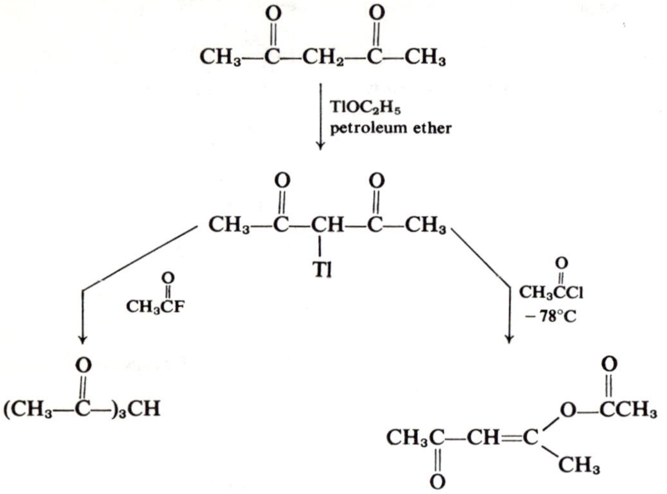

$$CH_3-C-CH_2-C-CH_3 \xrightarrow{\text{TlOC}_2H_5, \text{petroleum ether}} CH_3-C-CH(Tl)-C-CH_3$$

$$\xrightarrow{CH_3CF} (CH_3-CO-)_3CH \qquad \xrightarrow[-78°C]{CH_3CCl} CH_3C(=O)-CH=C(OCCH_3)(CH_3) \quad (70)$$

3. Nucleophilic Substitution

Nucleophilic substitution reactions that do not involve a carbonyl group can be conveniently divided into aliphatic systems which undergo S_N1 and S_N2 reactions and aryl systems which undergo reaction by an addition–elimination or by a benzyne mechanism.

a. Aliphatic Systems. Most carbanions condense with aliphatic halides and related compounds, though the nature of the metallic cation of the organometallic is often of great importance. For example, synthetically useful coupling of Grignard reagents, including those derived from acetylenes (Jacobs, 1949), seems limited to rather reactive halides [Eqs. (71) and (72)]

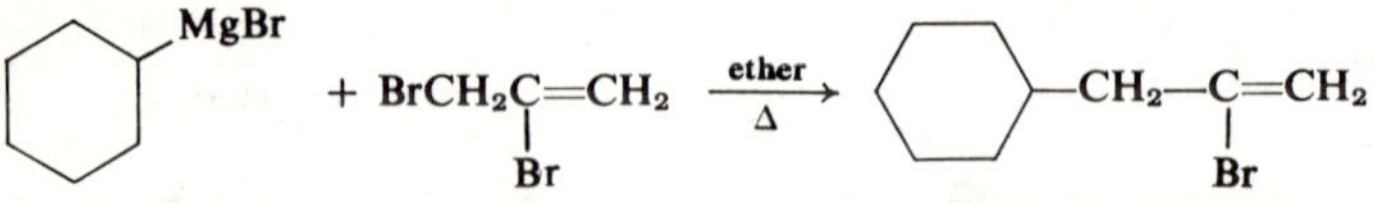

$$\text{C}_6\text{H}_{11}\text{MgBr} + \text{BrCH}_2\text{C}(\text{Br})=\text{CH}_2 \xrightarrow[\Delta]{\text{ether}} \text{C}_6\text{H}_{11}\text{-CH}_2\text{-C}(\text{Br})=\text{CH}_2 \quad (71)$$

(Lespieau and Bourguel, 1941)

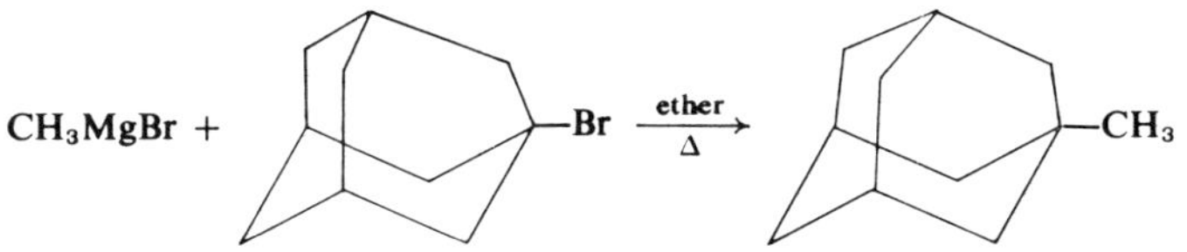

$$\text{CH}_3\text{MgBr} + \text{[adamantyl]-Br} \xrightarrow[\Delta]{\text{ether}} \text{[adamantyl]-CH}_3 \qquad (72)$$

(Osawa et al., 1971)

or other special compounds which exhibit a high reactivity toward nucleophiles (Kharasch and Reinmuth, 1954). Examples of the latter class of compounds are α-halo esters [Eq. (73)], esters of sulfuric and sulfonic acids

$$n\text{-C}_4\text{H}_9\text{MgBr} \xrightarrow[\text{ether}]{\text{ClCH}_2\text{OCH}_3} \text{CH}_3(\text{CH}_2)_4\text{OCH}_3 \qquad (73)$$

(Kharasch and Reinmuth, 1954)

[Eq. (74)], and epoxides. The latter reagents, however, appear to afford

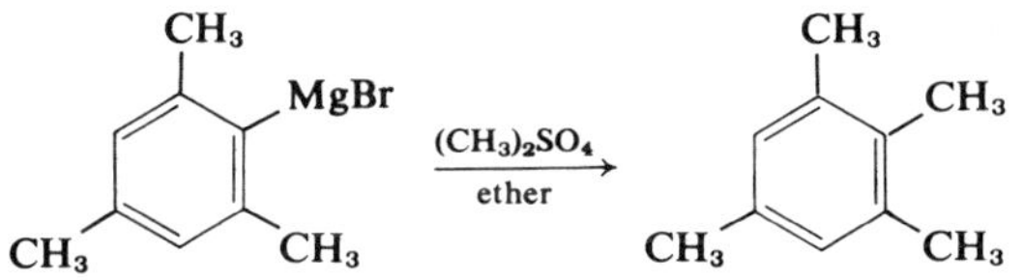

$$\text{[2,4,6-trimethylphenyl]MgBr} \xrightarrow[\text{ether}]{(\text{CH}_3)_2\text{SO}_4} \text{[1,2,3,5-tetramethylbenzene]} \qquad (74)$$

(L. I. Smith, 1943)

cleaner reaction mixtures with organocopper than with Grignard reagents [Eq. (75)].

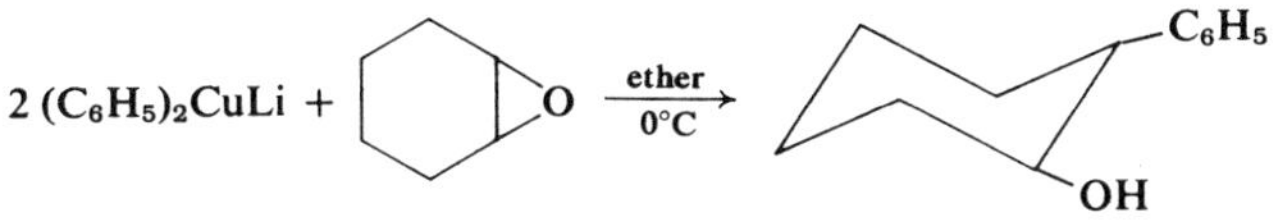

$$2\,(\text{C}_6\text{H}_5)_2\text{CuLi} + \text{[cyclohexene oxide]} \xrightarrow[0°\text{C}]{\text{ether}} \text{[trans-2-phenylcyclohexanol]} \qquad (75)$$

(Herr et al., 1970)

Organoalkali derivatives seem to be more reactive toward alkyl halides than Grignards, and, in fact, the scope of such reactions is larger. For example, alkylations of the conjugate bases of esters and nitriles have been thoroughly studied (Cope et al., 1957). Thus, mono- or dialkylation of diethyl malonate followed by hydrolysis and decarboxylation conveniently gives rise to mono- or dialkylated acetic acids, respectively [Eq. (76)].

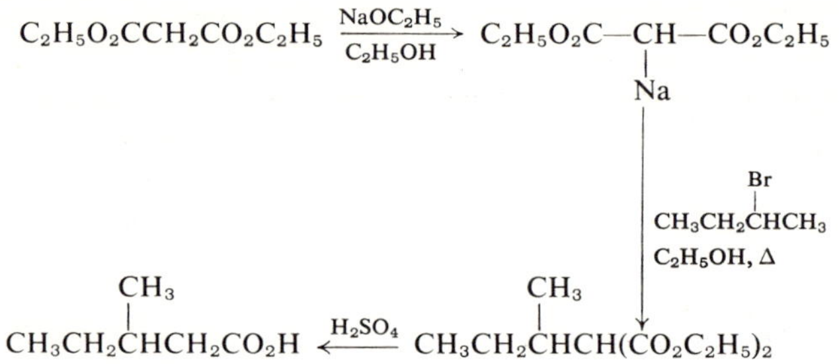

$$C_2H_5O_2CCH_2CO_2C_2H_5 \xrightarrow[C_2H_5OH]{NaOC_2H_5} C_2H_5O_2C-\underset{Na}{CH}-CO_2C_2H_5$$

$$\downarrow \begin{array}{c} \overset{Br}{|} \\ CH_3CH_2CHCH_3 \\ C_2H_5OH, \Delta \end{array}$$

$$CH_3CH_2\underset{CH_3}{\overset{|}{C}H}CH_2CO_2H \xleftarrow{H_2SO_4} CH_3CH_2\underset{CH_3}{\overset{|}{C}H}CH(CO_2C_2H_5)_2 \qquad (76)$$

(Marvel, 1955)

Even monoesters and nitriles can be alkylated in this manner, but care often must be taken to avoid dialkylation [Eq. (77)]. A method has been developed to afford monoalkylated derivatives of phenylacetonitrile involving phenylcyanoacetic acid [Eq. (78)].

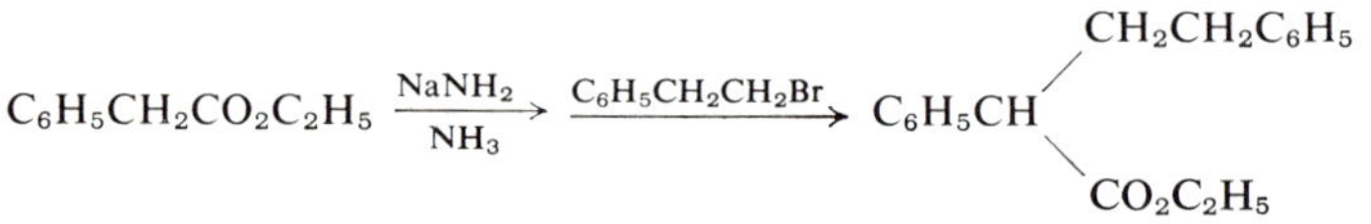

$$C_6H_5CH_2CO_2C_2H_5 \xrightarrow[NH_3]{NaNH_2} \xrightarrow{C_6H_5CH_2CH_2Br} C_6H_5CH \begin{array}{c} \diagup CH_2CH_2C_6H_5 \\ \diagdown CO_2C_2H_5 \end{array} \qquad (77)$$

(Kaiser et al., 1967a)

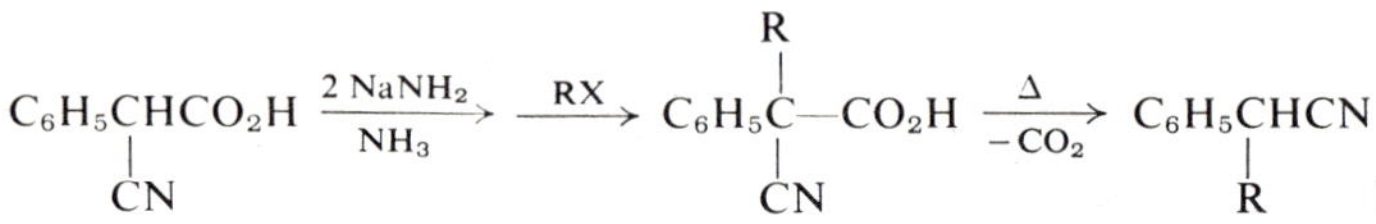

$$C_6H_5\underset{CN}{\overset{|}{C}H}CO_2H \xrightarrow[NH_3]{2\,NaNH_2} \xrightarrow{RX} C_6H_5\underset{CN}{\overset{R}{\underset{|}{\overset{|}{C}}}}-CO_2H \xrightarrow[-CO_2]{\Delta} C_6H_5\underset{R}{\overset{|}{C}H}CN \qquad (78)$$

(Kaiser and Hauser, 1966a)

The conjugate bases of ketones also undergo alkylation, but not always cleanly. Thus, in the case of unsymmetrical ketones, two alkylated products are possible because the α-hydrogens on either side of the carbonyl are acidic, and two enolate anion intermediates can thus be formed [Eq. (79)] (House, 1972).

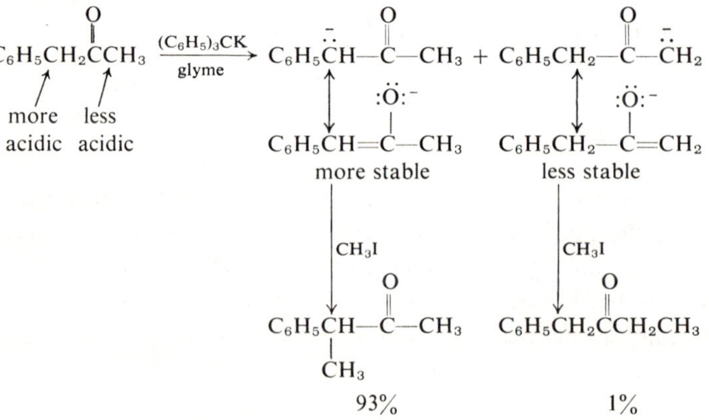

$$\underset{\underset{\text{acidic acidic}}{\text{more less}}}{C_6H_5CH_2\overset{O}{\overset{\|}{C}}CH_3} \xrightarrow[\text{glyme}]{(C_6H_5)_3CK} C_6H_5\overset{-}{C}H-\overset{O}{\overset{\|}{C}}-CH_3 + C_6H_5CH_2-\overset{O}{\overset{\|}{C}}-\overset{-}{C}H_2$$

$$\downarrow :\ddot{O}:^- \qquad\qquad \downarrow :\ddot{O}:^-$$

$$C_6H_5CH=\underset{\text{more stable}}{C}-CH_3 \qquad C_6H_5CH_2-\underset{\text{less stable}}{C}=CH_2$$

$$\downarrow CH_3I \qquad\qquad \downarrow CH_3I$$

$$\underset{\underset{CH_3}{|}}{C_6H_5\overset{|}{C}H}-\overset{O}{\overset{\|}{C}}-CH_3 \qquad C_6H_5CH_2\overset{O}{\overset{\|}{C}}CH_2CH_3 \qquad (79)$$

$$93\% \qquad\qquad\qquad 1\%$$

5. CARBANIONS

Sometimes, though, ionization of the most acidic α-hydrogen does not give rise to the most stable enolate anion; thus, caution must be exercised in predicting the products of such reactions [Eq. (80)]. Clean, predictable

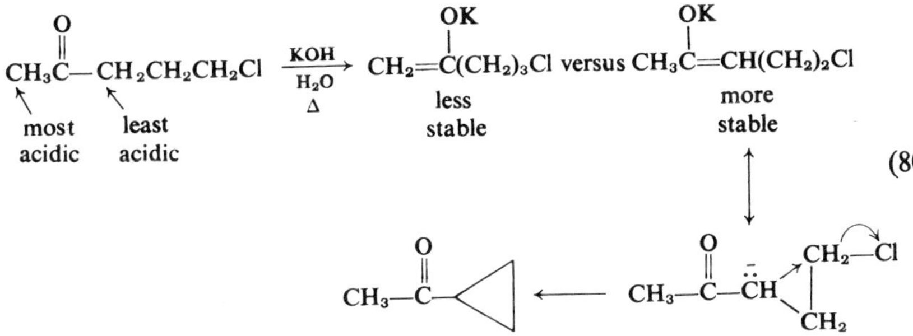

$$\underset{\underset{\text{acidic}}{\text{most}}}{CH_3C}-\underset{\underset{\text{acidic}}{\text{least}}}{\overset{O}{\underset{\|}{C}}}H_2CH_2CH_2Cl \xrightarrow[\Delta]{KOH\atop H_2O} \underset{\text{less stable}}{CH_2=\overset{OK}{\underset{|}{C}}(CH_2)_3Cl} \text{ versus } \underset{\text{more stable}}{CH_3\overset{OK}{\underset{|}{C}}=CH(CH_2)_2Cl} \qquad (80)$$

$$CH_3-\overset{O}{\underset{\|}{C}}-\triangleleft \;\longleftarrow\; CH_3-\overset{O}{\underset{\|}{C}}-\overset{..}{\underset{\diagdown}{CH}}\overset{\overset{\frown}{CH_2-Cl}}{\underset{CH_2}{\diagup}}$$

(Cannon et al., 1963)

alkylations of ketones can be realized by reaction of enamines with alkylating agents [Eq. (81)] (Stork et al., 1963).

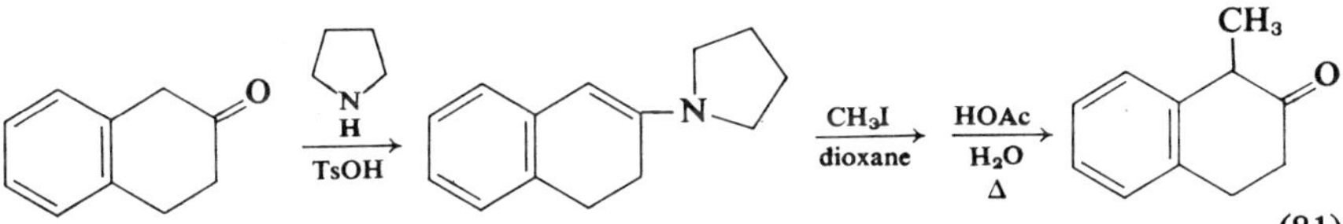

(81)

β-Keto esters and β-diketones also are alkylated conveniently. One of the best known of the β-keto esters is ethyl acetoacetate which, after alkylation, hydrolysis, and decarboxylation, affords substituted acetones [Eq. (82)].

$$\underset{}{CH_3\overset{O}{\underset{\|}{C}}CH_2\overset{O}{\underset{\|}{C}}OC_2H_5} \xrightarrow[\underset{\Delta}{C_2H_5OH}]{NaOC_2H_5} \xrightarrow[\Delta]{BrCH_2\overset{CH_3}{\underset{|}{C}}HOC_6H_5} \xrightarrow[-CO_2]{H_2O\atop \Delta} CH_3\overset{O}{\underset{\|}{C}}(CH_2)_2\underset{\underset{CH_3}{|}}{C}HOC_6H_5 \qquad (82)$$

(Brown and Partridge, 1945)

The use of organothallium derivatives of β-diketones and β-keto esters in alkylations has been shown to be most useful [Eq. (83)].

$$CH_3\overset{O}{\underset{\|}{C}}CH_2\overset{O}{\underset{\|}{C}}CH_3 \xrightarrow[\text{petroleum ether}]{TlOC_2H_5} \xrightarrow[\Delta]{CH_3I} CH_3-\overset{O}{\underset{\|}{C}}-\underset{\underset{CH_3}{|}}{CH}-\overset{O}{\underset{\|}{C}}-CH_3 \qquad (83)$$

(Taylor and McKillop, 1970)

Dialkylations of active hydrogen compounds with the same alkyl halide can be conveniently accomplished by using a one-pot reaction mixture containing the compound to be alkylated and an excess of the base and alkyl halide [Eq. (84)]. In a few cases, *gem*-dicarbanion intermediates can be formed which are then dialkylated (see Section V, B, 6).

$$CH_2(CO_2C_2H_5)_2 \xrightarrow[\substack{3\ C_6H_5CH_2Cl \\ THF \\ \Delta}]{3\ NaH} (C_6H_5CH_2)_2C(CO_2C_2H_5)_2 \qquad (84)$$

(Kaiser *et al.*, 1971)

Alkylations of two rather novel carbanionic systems have proved interesting in the quest for syntheses of aldehydes and ketones. In the first, lithio derivatives of dithianes are alkylated and then hydrolyzed to give aldehydes or ketones (Corey and Seebach, 1966) [Eqs. (85) and (86)].

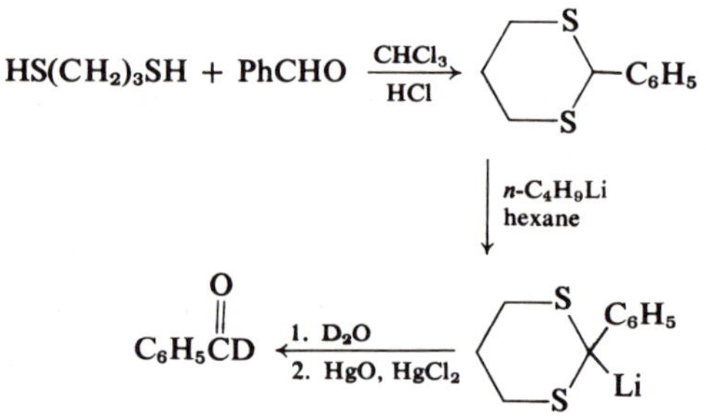

$$HS(CH_2)_3SH + PhCHO \xrightarrow[HCl]{CHCl_3} \text{[dithiane-}C_6H_5\text{]}$$

$$\downarrow n\text{-}C_4H_9Li,\ hexane$$

$$C_6H_5\overset{O}{\underset{\|}{C}}D \xleftarrow[2.\ HgO,\ HgCl_2]{1.\ D_2O} \text{[dithiane-}C_6H_5,\ Li\text{]} \qquad (85)$$

(Seebach *et al.*, 1966)

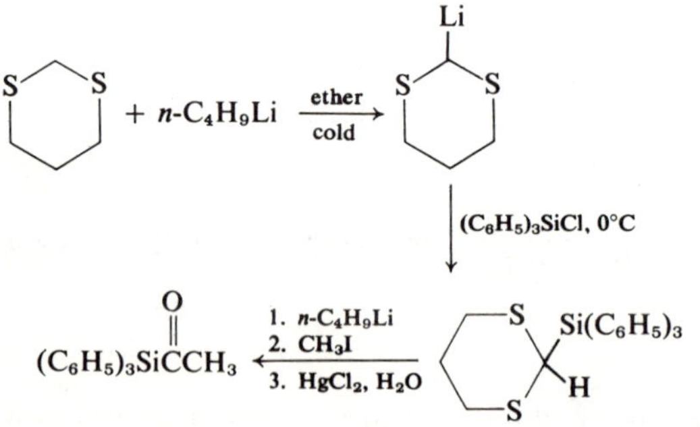

$$\text{[dithiane]} + n\text{-}C_4H_9Li \xrightarrow[cold]{ether} \text{[dithiane-Li]}$$

$$\downarrow (C_6H_5)_3SiCl,\ 0°C$$

$$(C_6H_5)_3Si\overset{O}{\underset{\|}{C}}CH_3 \xleftarrow[\substack{1.\ n\text{-}C_4H_9Li \\ 2.\ CH_3I \\ 3.\ HgCl_2,\ H_2O}]{} \text{[dithiane-Si}(C_6H_5)_3,\ H\text{]} \qquad (86)$$

(Brook *et al.*, 1967)

In the second, lithio salts of oxazines and related compounds likewise undergo alkylation. Subsequent reduction of the products by borohydride followed by hydrolysis gives aldehydes [Eq. (87)]. Dialkylation of the oxazines ultimately gives ketones.

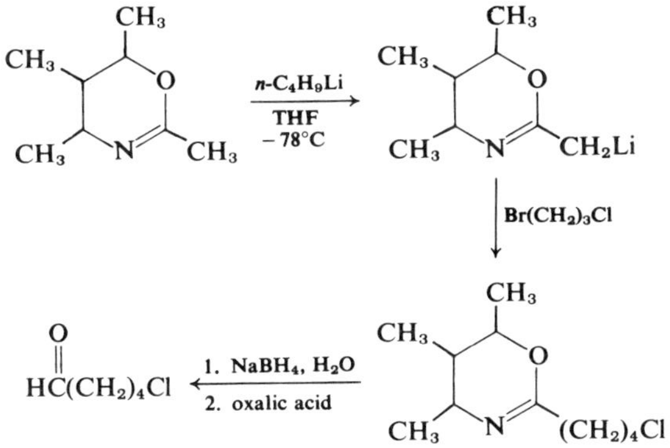

5. CARBANIONS

$$\text{dihydrooxazine} \xrightarrow[\text{THF} \\ -78°C]{n\text{-}C_4H_9Li} \text{lithiated species} \quad (87)$$

$$\downarrow \text{Br(CH}_2)_3\text{Cl}$$

$$HC(CH_2)_4Cl \xleftarrow[\text{2. oxalic acid}]{\text{1. NaBH}_4,\ H_2O} \text{alkylated oxazine}$$

(A. I. Meyers *et al.*, 1969)

Finally, certain nucleophilic substitutions on alkyl halides are interesting because the carbanion attacks or displaces on the halogen atom, not the α-carbon atom, to give a different carbanion or products derived from it [Eqs. (88) and (89)].

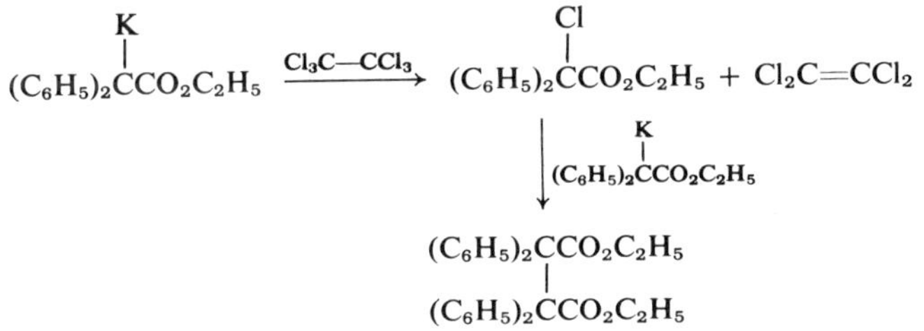

$$(C_6H_5)_2\overset{K}{C}CO_2C_2H_5 \xrightarrow{Cl_3C-CCl_3} (C_6H_5)_2\overset{Cl}{C}CO_2C_2H_5 + Cl_2C{=}CCl_2 \quad (88)$$

$$\downarrow (C_6H_5)_2\overset{K}{C}CO_2C_2H_5$$

$$(C_6H_5)_2CCO_2C_2H_5$$
$$|$$
$$(C_6H_5)_2CCO_2C_2H_5$$

(Kofron and Hauser, 1970)

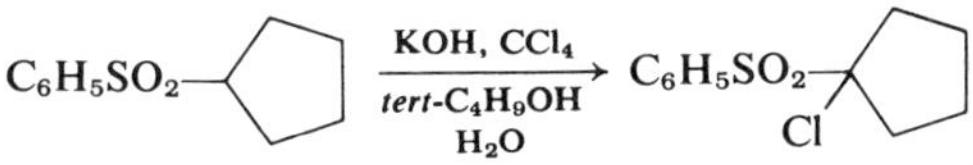

$$C_6H_5SO_2\text{-cyclopentyl} \xrightarrow[\text{tert-}C_4H_9OH \\ H_2O]{KOH, CCl_4} C_6H_5SO_2\text{-(Cl)cyclopentyl} \quad (89)$$

(C. Y. Meyers *et al.*, 1969b)

b. Aromatic Systems. Aromatic halides and related compounds likewise undergo nucleophilic substitution reactions. At first glance, such reactions appear to resemble S_N2 reactions since their net result is the replacement of a leaving group by some nucleophile. However, one of two rather special conditions must be met to effect these reactions which commonly occur by two different mechanisms. The aromatic compound must possess electron-withdrawing groups ortho and/or para to the leaving group (addition–elimination mechanism) since the resulting intermediate carbanion must be capable of extensive delocalization [Eq. (90)] (Bunnett and Zahler, 1951).

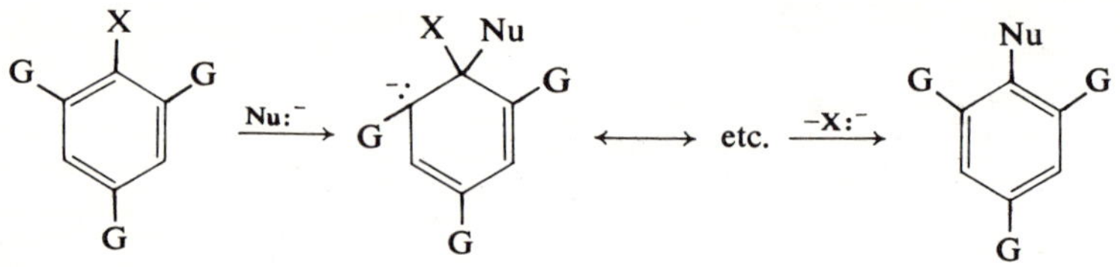

[Structures for Eq. (90) showing nucleophilic aromatic substitution via addition-elimination] (90)

Alternatively, the aromatic compound must possess a hydrogen atom ortho to a leaving group which is removed ultimately giving a benzyne intermediate. Attack of the benzyne by the nucleophile at either of the carbon atoms that comprise the aryne then gives the substitution product [Eq. (91)] (see Chapter 7).

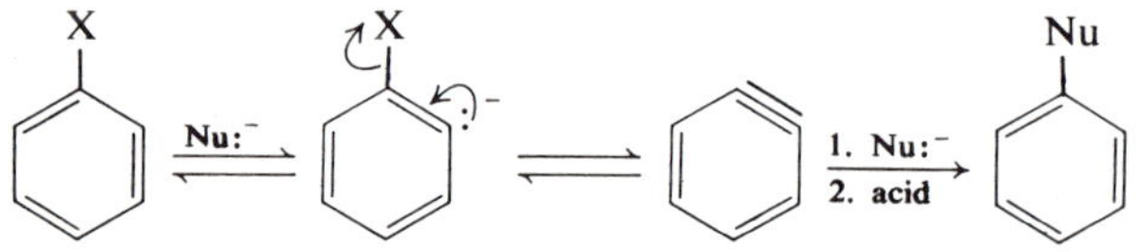

[Structures for Eq. (91) showing benzyne mechanism] (91)

Although the number of carbanions that have been reacted with aromatic systems via the addition–elimination mechanism is relatively small, those that are known seem to be synthetically useful [Eq. (92)]. Even more useful

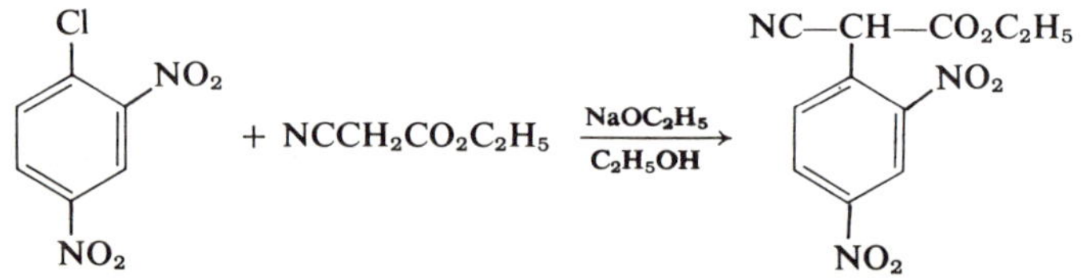

[Structures for Eq. (92): 1-chloro-2,4-dinitrobenzene + $NCCH_2CO_2C_2H_5$ with $NaOC_2H_5/C_2H_5OH$ giving substituted product] (92)

(Fairbourne and Fawson, 1927)

are similar reactions effected on pyridine, quinoline, and related compounds (Paquette, 1968), compounds which act like deactivated benzene rings [Eqs. (93)–(95)].

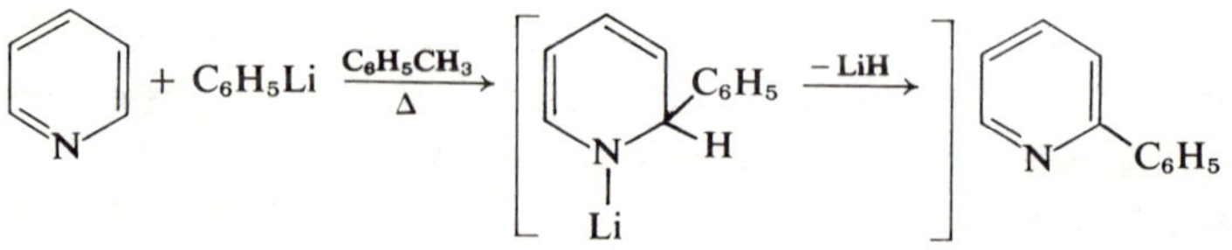

[Structures for Eq. (93): pyridine + C_6H_5Li in $C_6H_5CH_3$, Δ giving dihydropyridyl lithium intermediate, then $-LiH$ to 2-phenylpyridine] (93)

(Evans and Allen, 1943)

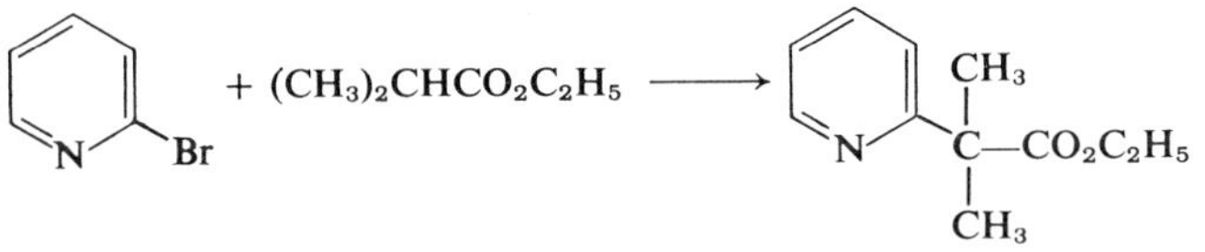

5. CARBANIONS

$$\text{Pyridine-Br} + (CH_3)_2CHCO_2C_2H_5 \longrightarrow \text{Pyridine-}\underset{CH_3}{\underset{|}{\overset{CH_3}{\overset{|}{C}}}}-CO_2C_2H_5 \quad (94)$$

(Doering and Pasternak, 1950)

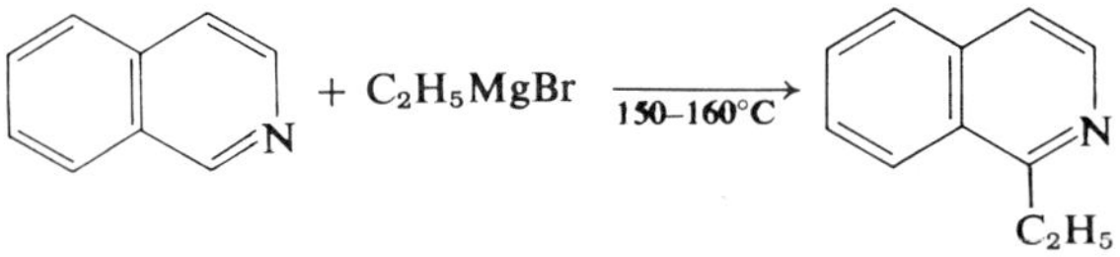

$$\text{Isoquinoline} + C_2H_5MgBr \xrightarrow[150-160°C]{} \text{1-Ethylisoquinoline (}C_2H_5\text{)} \quad (95)$$

(Paquette, 1968)

Related reactions of aryne intermediates are discussed in Chapter 7.

B. MULTIPLE ANION CHEMISTRY

Although multiple anion chemistry dates back to 1904 when Grignard himself found that phenylacetic acid could be converted to a dianion [Eq. (96)] (Grignard, 1904), little further work was done in this area until the late 1950's when Hauser and his students began their investigation of organic

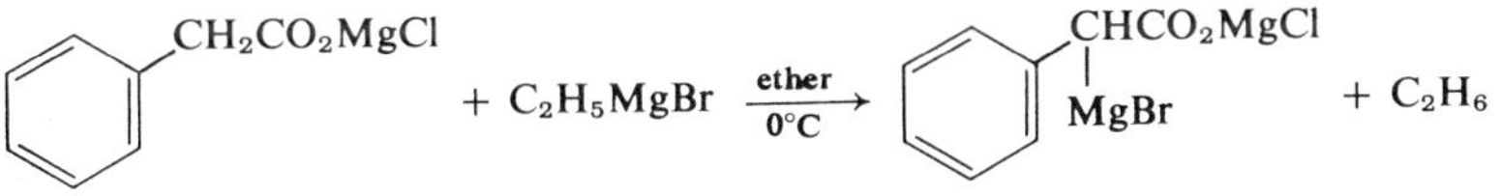

$$\text{PhCH}_2CO_2MgCl + C_2H_5MgBr \xrightarrow[0°C]{ether} \text{PhCH(MgBr)}CO_2MgCl + C_2H_6 \quad (96)$$

active hydrogen compounds containing more than one type of acidic hydrogen. The resulting studies have led to the discovery of very useful synthetic approaches to a wide variety of compounds described in well over 100 papers.

By definition, multiple anion systems are organic compounds which contain at least two anionic sites, one of which is usually on a carbon atom. Such multiple anions prepared to date generally have fallen into one of the classes which are illustrated below. In these examples, G is an electron-withdrawing group by resonance. By far the most extensive work has been performed on 1,3-dianions though specific examples of the other types will also be presented.

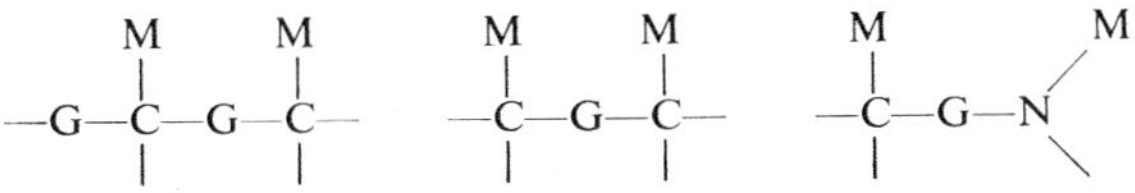

$$\underset{(1,3\text{-Dianions})}{-G-\overset{M}{\underset{|}{C}}-G-\overset{M}{\underset{|}{C}}-\qquad -\overset{M}{\underset{|}{C}}-G-\overset{M}{\underset{|}{C}}-\qquad -\overset{M}{\underset{|}{C}}-G-N\overset{M}{\diagdown}}$$

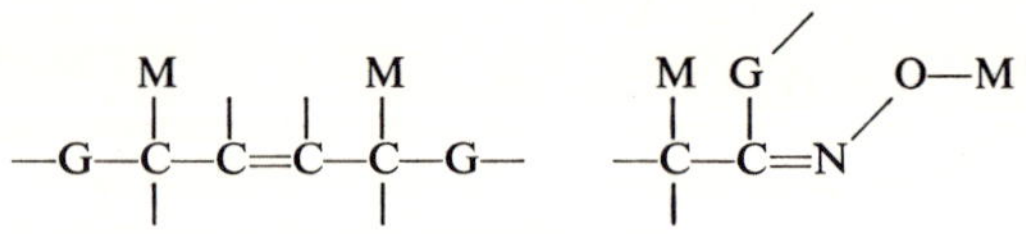

$$\begin{array}{cc} \text{M} \quad \text{M} \quad \text{M} & \text{M} \quad \text{G} \quad \text{O—M} \\ |||| & ||/ \\ \text{—G—C—C=C—C—G—} & \text{—C—C=N} \\ || & | \end{array}$$

(1,4-Dianions)

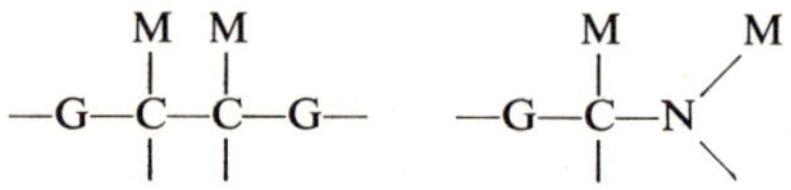

$$\begin{array}{cc} \text{M M} & \text{M} \quad \text{M} \\ || & |/ \\ \text{—G—C—C—G—} & \text{—G—C—N} \\ || & |\backslash \end{array}$$

(1,2-Dianions)

$$\begin{array}{c} \text{M} \\ | \\ \text{—G—C—G—} \\ | \\ \text{M} \end{array}$$

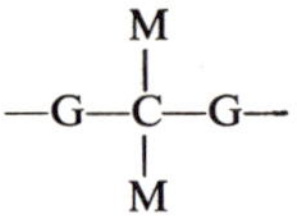

(1,1-Dianion)

Before becoming more specific, several general points about 1,3- and certain 1,4-dianions can be made. First, the procedure most often employed for the preparation of these multiple anions involves addition of the active hydrogen compound to two or more molecular equivalents of an alkali metal amide in ammonia or *n*-butyllithium in ether or THF followed, after several minutes, by an electrophile. Second, the most acidic proton, usually one which is on a carbon atom alpha to two electron-withdrawing groups or on a nitrogen or oxygen atom, is ionized to afford a relatively nonnucleophilic anion. This ionization is followed by ionization of the second proton to give an anion which is invariably the more nucleophilic of the two anions formed. With but very few exceptions, subsequent condensations of such multiple anions occur on the latter anion exclusively. Third, even though the multiple anions prepared to date are resonance delocalized (Miles *et al.*, 1966), only the canonical forms that lead directly to products are usually drawn. Fourth, once multiple anions are formed, they readily enter into single condensations with electrophiles such as alkyl halides, aldehydes, ketones, esters, and α,β-unsaturated carbonyl compounds.

1. *β-Diketones, β-Keto Esters, and Related Compounds*

The conversion of non-nitrogen-containing β-dicarbonyl compounds to their corresponding 1,3-dicarbanions by means of alkali metal amides in liquid ammonia has been reviewed (Harris and Harris, 1969). A general example is shown in Eq. (97). Just 8 of the over 36 β-dicarbonyl multiple

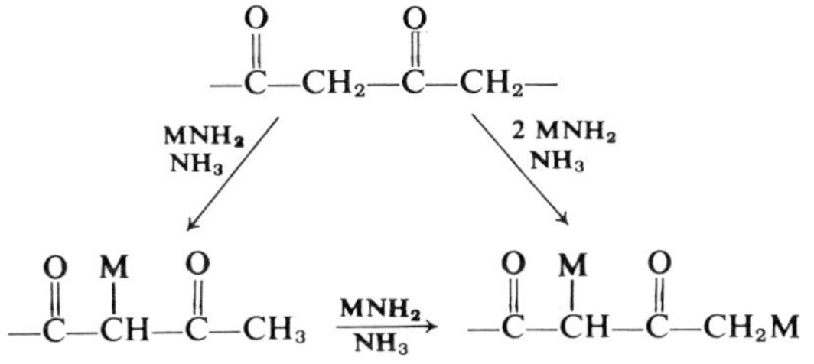

5. CARBANIONS

$$\underset{MNH_2/NH_3}{\swarrow} \overset{O \quad\quad O}{-\overset{\|}{C}-CH_2-\overset{\|}{C}-CH_2-} \underset{2\,MNH_2/NH_3}{\searrow}$$

$$-\overset{O}{\overset{\|}{C}}-\overset{M}{\underset{|}{C}H}-\overset{O}{\overset{\|}{C}}-CH_3 \xrightarrow{MNH_2/NH_3} -\overset{O}{\overset{\|}{C}}-\overset{M}{\underset{|}{C}H}-\overset{O}{\overset{\|}{C}}-CH_2M \quad\quad (97)$$

anions that have been prepared (Harris and Harris, 1969) are shown below, where M is usually sodium or potassium though it can also be lithium. The carbanion underlined is the one that undergoes reaction with electrophiles.

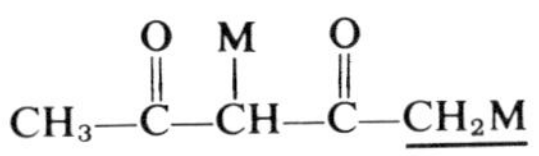

$CH_3-\overset{O}{\overset{\|}{C}}-\overset{M}{\underset{|}{C}H}-\overset{O}{\overset{\|}{C}}-\underline{CH_2M}$
(Hampton et al., 1965a)

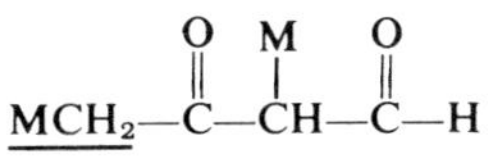

$\underline{MCH_2}-\overset{O}{\overset{\|}{C}}-\overset{M}{\underset{|}{C}H}-\overset{O}{\overset{\|}{C}}-H$
(Harris et al., 1963)

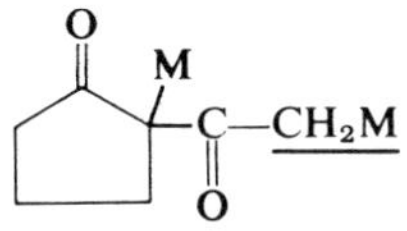

(Harris and Hauser, 1959)

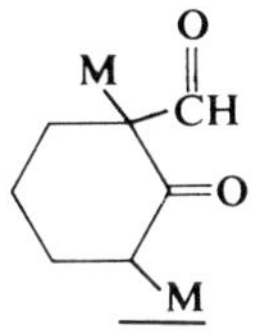

(Boatman et al., 1965a)

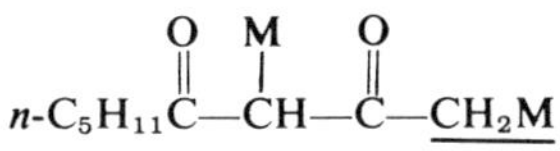

$n\text{-}C_5H_{11}\overset{O}{\overset{\|}{C}}-\overset{M}{\underset{|}{C}H}-\overset{O}{\overset{\|}{C}}-\underline{CH_2M}$
(Hampton et al., 1965a)

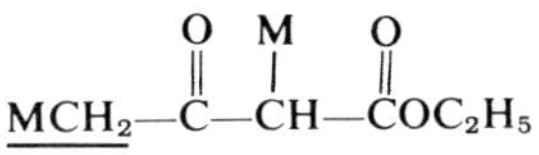

$\underline{MCH_2}-\overset{O}{\overset{\|}{C}}-\overset{M}{\underset{|}{C}H}-\overset{O}{\overset{\|}{C}}OC_2H_5$
(Wolfe et al., 1964)

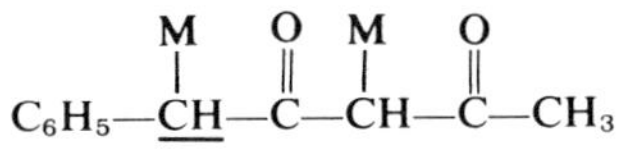

$C_6H_5-\overset{M}{\underset{|}{\underline{C}H}}-\overset{O}{\overset{\|}{C}}-\overset{M}{\underset{|}{C}H}-\overset{O}{\overset{\|}{C}}-CH_3$
(Hampton et al., 1966)

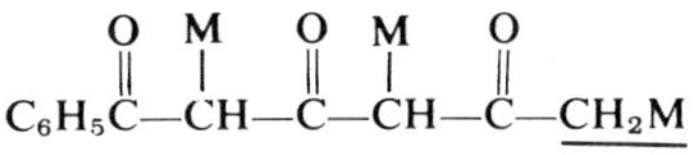

$C_6H_5\overset{O}{\overset{\|}{C}}-\overset{M}{\underset{|}{C}H}-\overset{O}{\overset{\|}{C}}-\overset{M}{\underset{|}{C}H}-\overset{O}{\overset{\|}{C}}-\underline{CH_2M}$
(Hampton et al., 1965b)

Representative transformations of all the β-dicarbonyl multiple anions are illustrated with benzoylacetone [Eq. (98)] in Table IV.

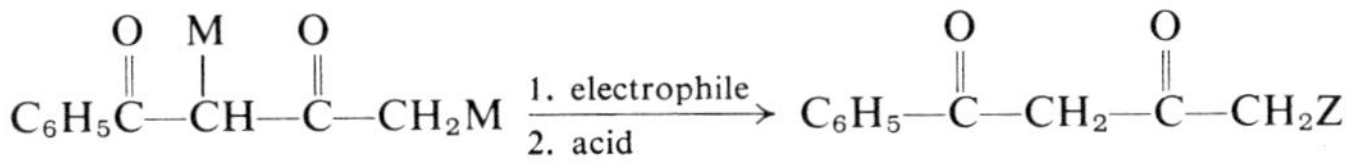

$$C_6H_5\overset{O}{\overset{\|}{C}}-\overset{M}{\underset{|}{C}H}-\overset{O}{\overset{\|}{C}}-CH_2M \xrightarrow[\text{2. acid}]{\text{1. electrophile}} C_6H_5-\overset{O}{\overset{\|}{C}}-CH_2-\overset{O}{\overset{\|}{C}}-CH_2Z \quad\quad (98)$$

In contrast to β-dicarbonyl compounds, only one monocarbonyl compound, dibenzyl ketone, had been converted to a dicarbanion prior to 1967 [Eq. (99)].

Table IV
Conversions of Dialkalibenzoylacetone

Electrophile	Z	Yield (%)	Reference
$C_6H_5CH_2Cl$	$-CH_2C_6H_5$	77	Hauser and Harris (1958)
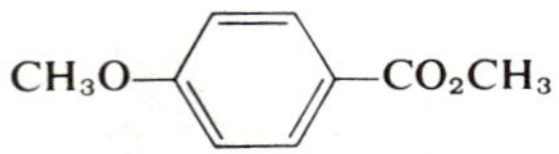 4-CH_3O-C_6H_4-CO_2CH_3	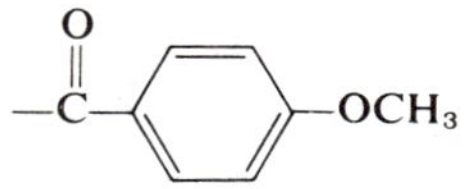 4-CH_3O-C_6H_4-$C(=O)-$	61	Light and Hauser (1960)
CO_2	$-CO_2H$	58	Hauser and Harris (1958)
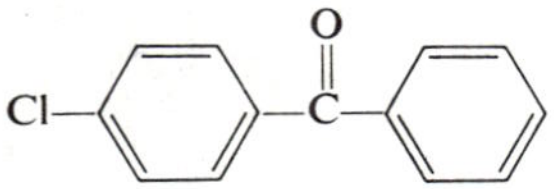 4-Cl-C_6H_4-C(=O)-C_6H_5 (benzophenone)	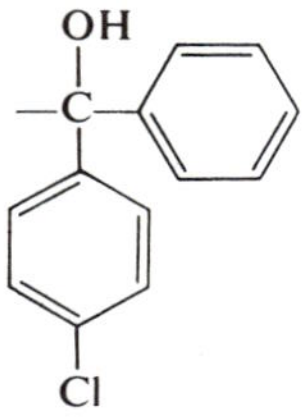 $-C(OH)(C_6H_5)(4$-Cl-$C_6H_4)$	69	Light and Hauser (1961)
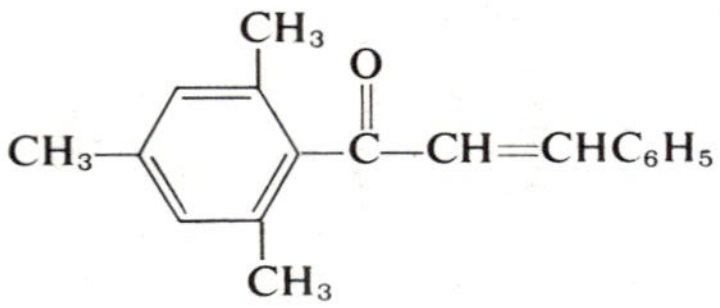 2,4,6-(CH$_3$)$_3$-C_6H_2-C(=O)-CH=CHC$_6$H$_5$	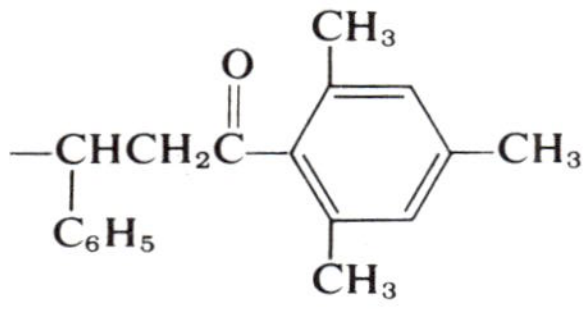 2,4,6-(CH$_3$)$_3$-C_6H_2-C(=O)-CH(C$_6$H$_5$)-CH$_2$-	84	Light et al. (1961)
$(C_6H_5)_2ICl$	$-C_6H_5$	61	Hampton et al. (1964)

5. CARBANIONS

$$C_6H_5CH_2COCH_2C_6H_5 \xrightarrow[NH_3]{2\ KNH_2} C_6H_5\overset{K}{C}HCO\overset{K}{C}HC_6H_5$$

$$\xrightarrow[\text{2. acid}]{1.\ C_6H_5CH_2Cl} C_6H_5\overset{CH_2C_6H_5}{\underset{|}{C}}HCOCH_2C_6H_5 \qquad (99)$$

(Hauser and Harris, 1959)

Since that time a new two-base technique has greatly expanded the number of dicarbanions derived from such monoketones (Mao et al., 1967). The method simply involves monoionization of the ketone by an alkali amide in ammonia, replacement of the ammonia by THF, and then subsequent secondary ionization of the enolate anion by n-butyllithium [Eq. (100)]. Prior single

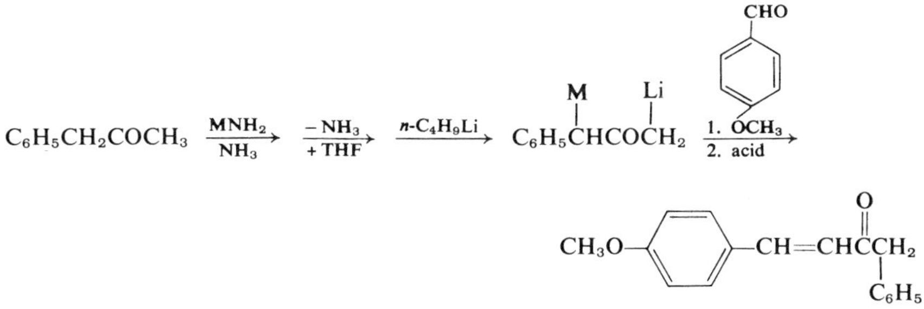

$$C_6H_5CH_2COCH_3 \xrightarrow[NH_3]{MNH_2} \xrightarrow[+THF]{-NH_3} \xrightarrow{n\text{-}C_4H_9Li} C_6H_5\overset{M}{C}HCO\overset{Li}{C}H_2 \xrightarrow[\text{2. acid}]{1.\ \overset{CHO}{\underset{}{C_6H_4\text{-}OCH_3}}}$$

$$CH_3O-\!\!\bigcirc\!\!-CH=CH\overset{O}{\underset{|}{C}}CH_2$$
$$\qquad\qquad\qquad\qquad C_6H_5$$

(100)

ionization by alkali amide is necessary since n-butyllithium adds to the carbonyl group of ketones but not to that of their enolate anions.

2. Carboxylic Acids

Phenylacetic acid has been converted to its dialkali derivatives which react with electrophiles only at the α-carbon atom. Some examples of these condensations [Eq. (101)] are listed in Table V. Both the preparation and reactions of such dialkali salts seem more convenient than the corresponding dimagnesium or sodium–magnesium salts (Ivanoff reagents) (Ivanoff et al., 1932).

$$C_6H_5\overset{M}{\underset{|}{C}}HCO_2M \xrightarrow[\text{2. acid}]{1.\ \text{electrophile}} C_6H_5\overset{Z}{\underset{|}{C}}HCO_2H \qquad (101)$$

Two results in Table V deserve special mention, viz., the condensations with cyclohexanone. In such reactions of carbanions with enolizable ketones,

Table V
Reactions of Dialkaliphenylacetate

M	Reagent	Z	Yield (%)	Reference
Na	$C_6H_5\overset{Cl}{\underset{\|}{C}}HCH_3$	$C_6H_5\overset{\|}{\underset{\|}{C}}HCH_3$ (erythro)	84	Meyer and Hauser (1961b)
Li	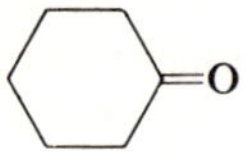 cyclohexanone	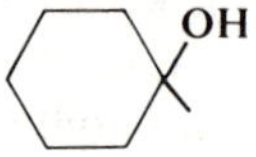 1-phenylcyclohexanol (OH)	93	Hamrick and Hauser (1960)
Na	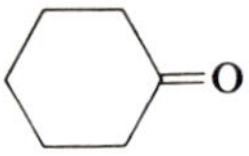 cyclohexanone	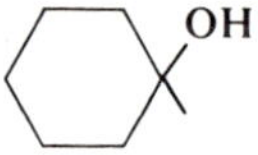 1-phenylcyclohexanol (OH)	64	Hamrick and Hauser (1960)
K	$C_6H_5CO_2CH_3$	$C_6H_5\overset{O}{\underset{\|}{C}}-$	22	Work et al. (1964)
Na	$C_6H_5CH=CHCO_2C_2H_5$	$C_6H_5CHCH_2CO_2C_2H_5$	96	Meyer and Hauser (1961a)

lithium rather than sodium or potassium cations generally affords higher yields of the desired alcohols since the latter metallic cations have a greater propensity to enolize rather than add to the ketone (O'Sullivan et al., 1961).

More recently, it has been demonstrated that various aliphatic and toluic acids can be converted to dilithio salts by means of lithium diisopropylamide in THF or hexamethylphosphoramide (Creger, 1967, 1970). Two examples follow, the first of which illustrates a viable alternative to the malonic ester synthesis of substituted acetic acids [Eqs. (102) and (103)].

$$(CH_3)_2CHCO_2H \xrightarrow[\text{THF-hexane, 0°C}]{2\ \text{LiN}(i\text{-}C_3H_7)_2} (CH_3)_2\overset{Li}{\underset{\|}{C}}CO_2Li$$

$$\xrightarrow[\text{2. acid}]{1.\ \frac{1}{2}\ Br(CH_2)_4Br} \begin{array}{c} (CH_3)_2\overset{\|}{C}CO_2H \\ | \\ (CH_2)_4 \\ | \\ (CH_3)_2\overset{\|}{C}CO_2H \end{array} \quad (102)$$

$$\underset{CH_3}{\underset{}{C_6H_4}}\text{-}CO_2H \xrightarrow[\text{THF-hexane, 0°C}]{2\ \text{LiN}(i\text{-}C_3H_7)_2} \underset{CH_2Li}{\underset{}{C_6H_4}}\text{-}CO_2Li \xrightarrow[\text{2. acid}]{1.\ n\text{-}C_4H_9Br} \underset{C_5H_{11}}{\underset{}{C_6H_4}}\text{-}CO_2H \quad (103)$$

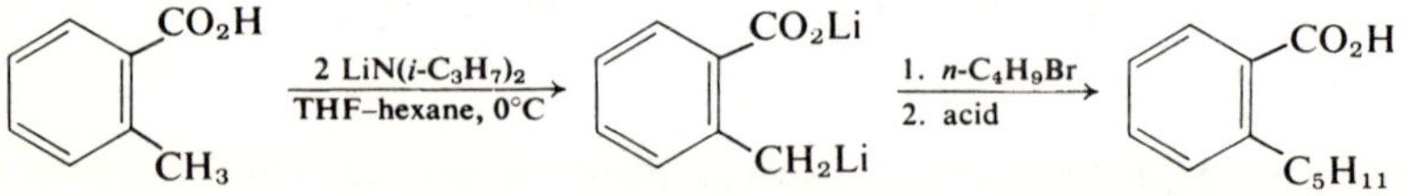

Such dilithio acids have also been condensed with aldehydes and ketones to give 19 different Reformatsky-type products conveniently (Moersch and Burkett, 1971) [Eq. (104)].

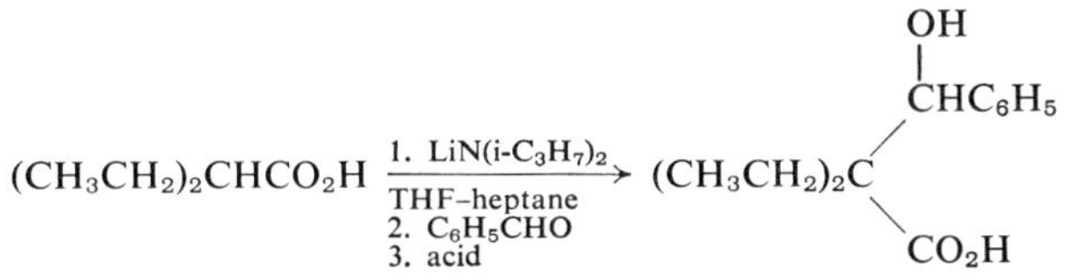

$$(CH_3CH_2)_2CHCO_2H \xrightarrow[\substack{\text{THF-heptane} \\ \text{2. } C_6H_5CHO \\ \text{3. acid}}]{\text{1. } LiN(i\text{-}C_3H_7)_2} (CH_3CH_2)_2C \genfrac{}{}{0pt}{}{\overset{\overset{\displaystyle OH}{|}}{CHC_6H_5}}{CO_2H} \quad (104)$$

3. Carboxamides and Related Nitrogen-Containing Compounds

Over 48 amides, imides, pyridones, and similar compounds have been converted to multiple anions by means of alkali amides in ammonia or *n*-butyllithium in ether–hexane. Such anions can be arbitrarily divided into various types as shown below. As shown above, the C–M bond underlined in each example is the one that reacts with electrophiles.

$$\overset{M}{\underset{|}{C_6H_5\underline{C}H}}CO\overset{M}{\underset{|}{N}}C_6H_5$$
(M = Li, Na, or K)
(Work et al., 1964)

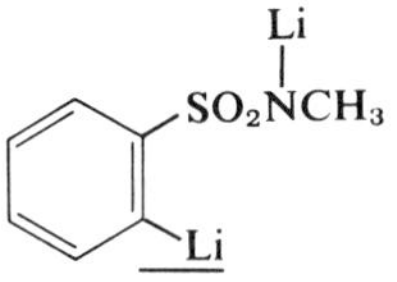

[o-Li-C₆H₄-SO₂N(Li)CH₃]
(Watanabe et al., 1968)

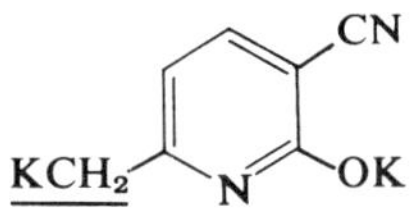

[pyridine with KCH₂ at 2-position, OK at 6-position, CN at 3-position]
(Boatman et al., 1965b)

$C_6H_5CHLiCONLi_2$
(Kaiser et al., 1967b)

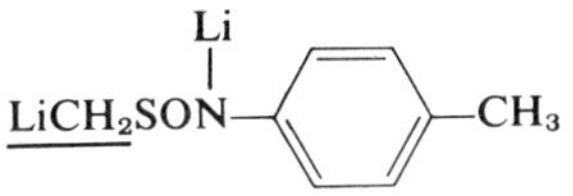

$LiCH_2SON(Li)$—C₆H₄—CH₃
(Corey and Durst, 1968)

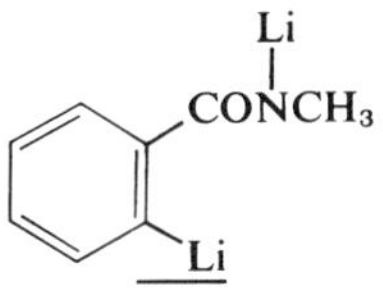

[o-Li-C₆H₄-CON(Li)CH₃]
(Puterbaugh and Hauser, 1964a)

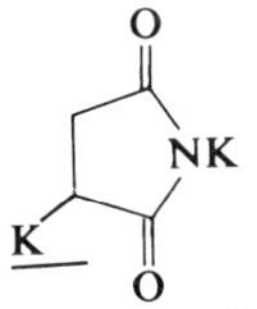

[succinimide with NK, α-K]
(Bryant and Hauser, 1961)

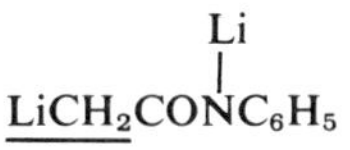

$Li\underline{CH_2}CON(Li)C_6H_5$
(Gay et al., 1965)

$\overset{K}{\underset{|}{C_6H_5\underline{C}H}}CON\overset{K}{\underset{|}{CO}}\overset{K}{\underset{|}{C}}HC_6H_5$
(Wolfe et al., 1966)

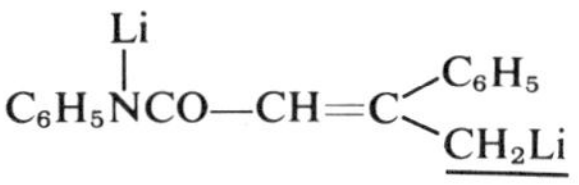

$C_6H_5\overset{Li}{\underset{|}{N}}CO-CH=C\genfrac{}{}{0pt}{}{C_6H_5}{CH_2Li}$
(Wolfe and Mao, 1967)

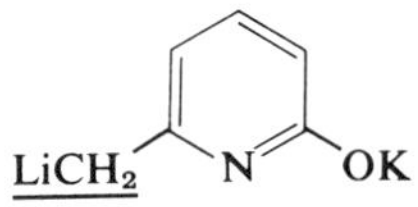

[pyridine with LiCH₂ at 2-position, OK at 6-position]
(Gay et al., 1965)

Most of the multiple anions of amides and related compounds undergo condensations with electrophiles as illustrated with dialkaliphenylacetamide [Eq. (105)] in Table VI.

$$C_6H_5\overset{M}{\underset{|}{C}}HCONHM \xrightarrow[\text{2. acid}]{\text{1. electrophile}} C_6H_5-\overset{Z}{\underset{|}{C}}HCONH_2 \qquad (105)$$

Table VI

Reactions of $C_6H_5\overset{M}{\underset{|}{C}}HCONHM$

M	Electrophile	Z	Yield (%)	Reference
Na	n-C_4H_9Br	n-C_4H_9—	86	Meyer and Hauser (1961b)
Na	$(C_6H_5)_2CO$	$(C_6H_5)_2\overset{OH}{\underset{|}{C}}$—	83	Kaiser and Hauser (1966b)
K	$C_6H_5CO_2CH_3$	$C_6H_5\overset{O}{\overset{\|}{C}}$—	50	Work et al. (1964)
K	$C_6H_5CH=CHCO_2C_2H_5$	$C_6H_5\underset{|}{C}HCH_2CO_2H$[a]	89	Hauser and Tetenbaum (1958)

[a] After hydrolysis.

A large number of other electrophiles have been reacted with a variety of the types of multiple anions listed above. These include carbon dioxide to give amido amides (Gay and Hauser, 1967a), benzonitrile to afford imino amides (Barnish et al., 1968a), cyclohexene oxide to yield lactones (Puterbaugh and Hauser, 1964a), isocyanates to give diamides (Watanabe et al., 1969a), aldehydes and ketones followed by heat to yield olefins (Corey and Durst, 1968), azobenzene to give hydrazines (Kaiser and Bartling, 1972), and aliphatic diazo compounds to afford hydrazones (Kaiser and Warner, 1971). Multiple anions of certain amides have been decarboxamidated (H. A. Smith and Hauser, 1969).

4. Sulfur-Containing Compounds

Dibenzyl sulfone and dibenzyl sulfoxide have been converted to their corresponding 1,3-dialkali salts in both ammonia and THF–hexane. The former salt undergoes either mono- or, surprisingly, dibenzylation, depending on the amount of benzyl chloride employed [Eq. (106)].

5. CARBANIONS

$$\underset{\underset{C_6H_5}{|}}{\overset{\overset{K}{|}}{C}}HSO_2\underset{\underset{C_6H_5}{|}}{\overset{\overset{K}{|}}{C}}HC_6H_5 \begin{array}{c} \xrightarrow[\text{2. acid}]{\text{1. } C_6H_5CH_2Cl} \\ \text{NH}_3 \\ \xrightarrow[\text{2. acid}]{\text{1. 2 } C_6H_5CH_2Cl} \\ \text{NH}_3 \end{array} \begin{array}{c} C_6H_5CH_2\underset{\underset{C_6H_5}{|}}{C}HSO_2CH_2C_6H_5 \\ \\ C_6H_5CH_2\underset{\underset{C_6H_5}{|}}{C}HSO_2\underset{\underset{C_6H_5}{|}}{C}HCH_2C_6H_5 \end{array} \quad (106)$$

(Hauser and Harris, 1959)

This salt has also been condensed with ethyl cinnamate to form the monoadduct [Eq. (107)], but not with benzophenone unless magnesium or aluminum

$$\underset{\underset{C_6H_5}{|}}{\overset{\overset{K}{|}}{C}}HSO_2\underset{\underset{C_6H_5}{|}}{\overset{\overset{K}{|}}{C}}HC_6H_5 \xrightarrow[\text{2. acid}]{\text{1. } C_6H_5CH=CHCOC_2H_5} \underset{\underset{C_6H_5\overset{|}{C}HSO_2CH_2C_6H_5}{|}}{C_6H_5CHCH_2CO_2C_2H_5} \quad (107)$$

(Hauser and Harris, 1959)

cations are used [Eq. (108)]. To date, dialkalidibenzyl sulfoxides have only

$$\underset{\underset{C_6H_5}{|}}{\overset{\overset{MgBr}{|}}{C}}HSO_2\underset{\underset{C_6H_5}{|}}{\overset{\overset{MgBr}{|}}{C}}HC_6H_5 \xrightarrow[\text{2. acid}]{\text{1. }(C_6H_5)_2CO} \underset{\underset{C_6H_5\overset{|}{C}HSO_2CH_2C_6H_5}{}}{\overset{(C_6H_5)_2COH}{|}} \quad (108)$$

(Kaiser and Hauser, 1967a)

been dideuterated [Eq. (109)].

$$\underset{}{\overset{\overset{Li}{|}}{C_6H_5}\overset{\overset{Li}{|}}{C}HSO\overset{}{C}HC_6H_5} \xrightarrow[\text{THF}]{D_2O} C_6H_5CHDSOCHDC_6H_5 \quad (109)$$

(Beard, 1970)

The corresponding dialkali derivatives of dimethyl sulfone and dimethyl sulfoxide not only undergo bis-alkylations similar to dibenzyl sulfone (Kaiser and Hauser, 1967b; Moskowitz et al., 1968) but even enter into twofold aldol-type condensations with benzophenone in ammonia to give the diols shown [Eqs. (110) and (111)].

$$MCH_2SO_2CH_2M \xrightarrow[\text{2. acid}]{\text{1. }(C_6H_5)_2CO} \left[(C_6H_5)_2\overset{\overset{OH}{|}}{C}CH_2\right]_2SO_2 \quad (110)$$

(Kaiser and Hauser, 1967b)

$$MCH_2SOCH_2M \xrightarrow[\text{2. acid}]{\text{1. }(C_6H_5)_2CO} \left[(C_6H_5)_2\overset{\overset{OH}{|}}{C}CH_2\right]_2SO \quad (111)$$

(Kaiser and Beard, 1968)

Tetrahydrothiophene-1,1-dioxide has also been converted to its 1,3-dialkali salts in ammonia. The results of subsequent condensations with benzophenone are a function of the metal cation, however, since only the monohydroxy sulfone is obtained with lithium, whereas a dihydroxy sulfone is obtained with sodium and potassium [Eq. (112)].

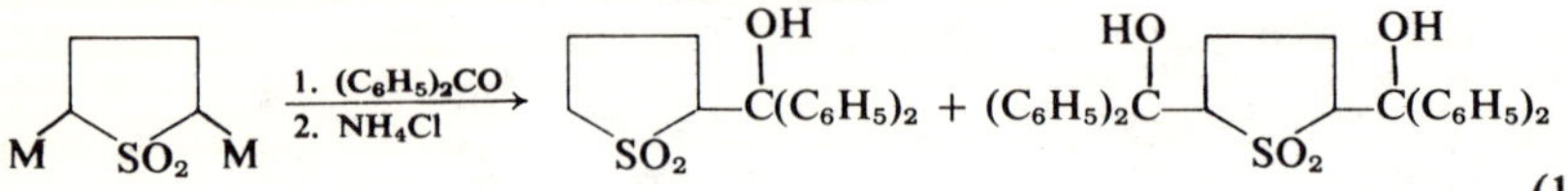

$$\underset{M}{\overset{}{\bigcap}}\underset{SO_2}{\overset{}{\bigcap}}_M \xrightarrow[2.\ NH_4Cl]{1.\ (C_6H_5)_2CO} \underset{SO_2}{\overset{OH}{\bigcap}}-C(C_6H_5)_2 + (C_6H_5)_2C-\underset{SO_2}{\overset{HO}{\bigcap}}-C(C_6H_5)_2 \qquad (112)$$

(Kaiser and Hauser, 1967b)

Benzyl phenyl sulfone has been converted to a *gem*-dicarbanion by *n*-butyllithium; subsequent alkylations give α,α-dialkyl derivatives in high yield [Eq. (113)] (Kaiser et al., 1971).

$$C_6H_5CH_2SO_2C_6H_5 \xrightarrow[THF-hexane]{2\ n-C_4H_9Li} C_6H_5\overset{Li}{\underset{Li}{C}}SO_2C_6H_5$$

$$\xrightarrow{C_6H_5CH_2Cl} C_6H_5-\overset{CH_2C_6H_5}{\underset{CH_2C_6H_5}{C}}-SO_2C_6H_5 \qquad (113)$$

Benzyl phenyl sulfoxide and *N*,*N*-dimethyl-α-phenylmethane sulfonamide have similarly been converted to 1,1-dianions as evidenced by deuteration and alkylation, respectively [Eqs. (114) and (115)].

$$C_6H_5\overset{Li}{\underset{Li}{C}}SOC_6H_5 \xrightarrow[THF]{D_2O} C_6H_5CD_2SOC_6H_5 \qquad (114)$$

(Beard, 1970)

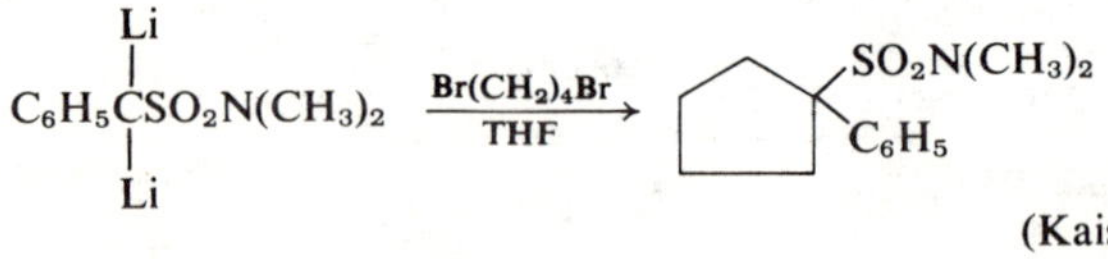

$$C_6H_5\overset{Li}{\underset{Li}{C}}SO_2N(CH_3)_2 \xrightarrow[THF]{Br(CH_2)_4Br} \underset{C_6H_5}{\overset{SO_2N(CH_3)_2}{\bigcirc}} \qquad (115)$$

(Kaiser et al., 1971)

5. Multiple Anions Derived from Hydrazones, Oximes, Azines, and Osazones

Several derivatives of carbonyl compounds have been converted to 1,4- or 1,6-multiple anions by the usual bases and subsequently condensed with electrophiles to afford useful and practical syntheses of compounds which often are difficult to obtain otherwise. For example, certain phenylhydrazones have been alkylated [Eq. (116)], condensed with ketones to give hydroxy-

$$\underset{\underset{K}{|}}{\overset{\overset{CH_2K}{|}}{C_6H_5C}}=NNC_6H_5 \xrightarrow[\text{2. acid}]{\text{1. } C_6H_5CH_2Cl} \underset{}{\overset{\overset{CH_2CH_2C_6H_5}{|}}{C_6H_5C}}=NNHC_6H_5 \qquad (116)$$

(Henoch et al., 1969)

hydrazones [Eq. (117)], and aroylated and subsequently cyclized by acid to

$$\underset{}{\overset{\overset{K}{|}}{C_6H_5CH}}\underset{}{\overset{\overset{K}{|}}{CH}}=NNC_6H_5 \xrightarrow[\text{2. acid}]{\text{1. } (C_6H_5)_2CO} \underset{}{\overset{\overset{(C_6H_5)_2C-OH}{|}}{C_6H_5-CHCH}}=NNHC_6H_5 \qquad (117)$$

(Henoch et al., 1967)

give N-substituted pyrazoles [Eq. (118)].

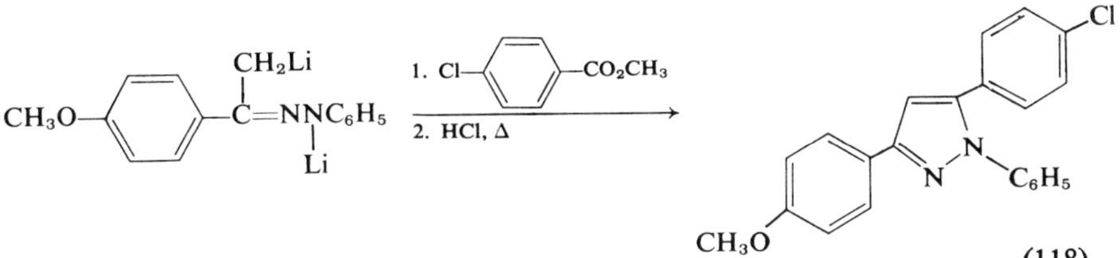

$$CH_3O-\bigcirc-\underset{\underset{Li}{|}}{\overset{\overset{CH_2Li}{|}}{C}}=NNC_6H_5 \xrightarrow[\text{2. HCl, }\Delta]{\text{1. Cl}-\bigcirc-CO_2CH_3} \qquad (118)$$

(Foote et al., 1970)

N-Unsubstituted hydrazones have likewise been converted to 1,1- and 1,1,4-multiple anions. For example, Wolff–Kishner reductions which proceed via *gem*-dicarbanions [Eq. (119)] can now be realized in ethers, and N-

$$(C_6H_5)_2CH_2 \xrightarrow[\substack{NH_3 \text{ or} \\ \text{ethers}}]{\text{base}} (C_6H_5)_2C=NNK_2$$
$$\qquad\qquad\qquad\qquad\qquad \downarrow \Delta \atop -N_2 \qquad (119)$$
$$(C_6H_5)_2CH_2 \xleftarrow{\text{acid}} (C_6H_5)_2CK_2$$

(Kaiser et al., 1968)

unsubstituted pyrazoles can be conveniently obtained on aroylation and cyclization [Eq. (120)].

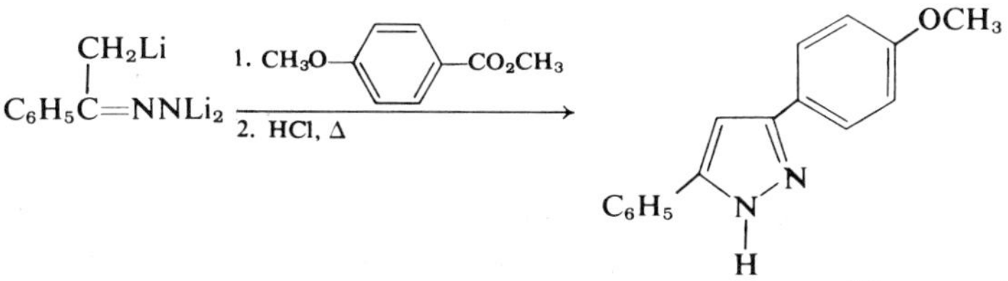

$$\underset{}{\overset{\overset{CH_2Li}{|}}{C_6H_5C}}=NNLi_2 \xrightarrow[\text{2. HCl, }\Delta]{\text{1. } CH_3O-\bigcirc-CO_2CH_3} \qquad (120)$$

(Beam et al., 1971)

Similarly, multiple anions of oximes can be alkylated followed by hydrolysis to give ketones [Eq. (121)], aroylated followed by cyclization to give

$$C_6H_5\overset{NOH}{\underset{\|}{C}}CH_3 \xrightarrow[\text{THF-hexane}]{2\,n\text{-}C_4H_9Li} C_6H_5\overset{NOLi}{\underset{\|}{C}}CH_2Li \xrightarrow[\text{2. H}_2\text{O, HCl}]{1.\;\tfrac{1}{2}\,Br(CH_2)_4Br} C_6H_5\overset{O}{\underset{\|}{C}}(CH_2)_6\overset{O}{\underset{\|}{C}}C_6H_5 \quad (121)$$

(Henoch et al., 1969)

isoxazoles [Eq. (122)], carbonated, then cyclized to give Δ^2-isoxazolin-5-ones

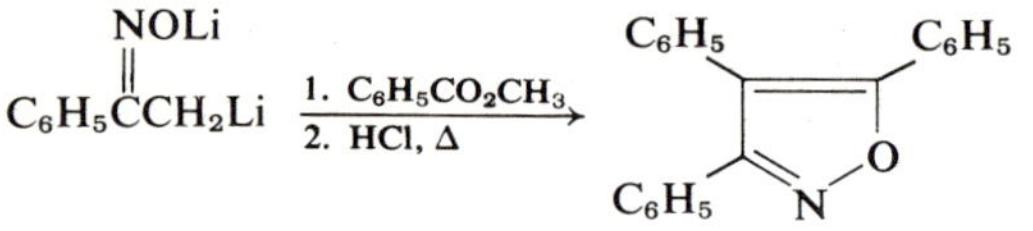

$$C_6H_5\overset{NOLi}{\underset{\|}{C}}CH_2Li \xrightarrow[\text{2. HCl, }\Delta]{1.\;C_6H_5CO_2CH_3} \quad\text{[isoxazole ring with }C_6H_5\text{ substituents]}\quad (122)$$

(Beam et al., 1970)

[Eq. (123)], and reacted with aldehydes and ketones to give hydroxy oximes

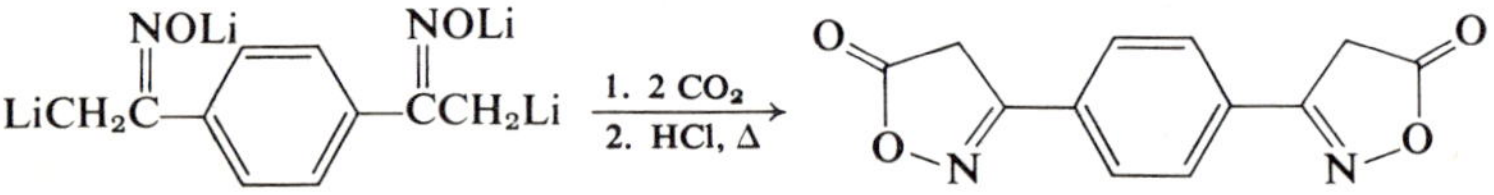

$$LiCH_2\overset{NOLi}{\underset{\|}{C}}{-}\!\!\left\langle\;\right\rangle\!\!{-}\overset{NOLi}{\underset{\|}{C}}CH_2Li \xrightarrow[\text{2. HCl, }\Delta]{1.\;2\,CO_2} \quad\text{[bis-isoxazolinone]} \quad (123)$$

(Griffiths et al., 1971)

which on cyclization give isoxazolines [Eq. (124)].

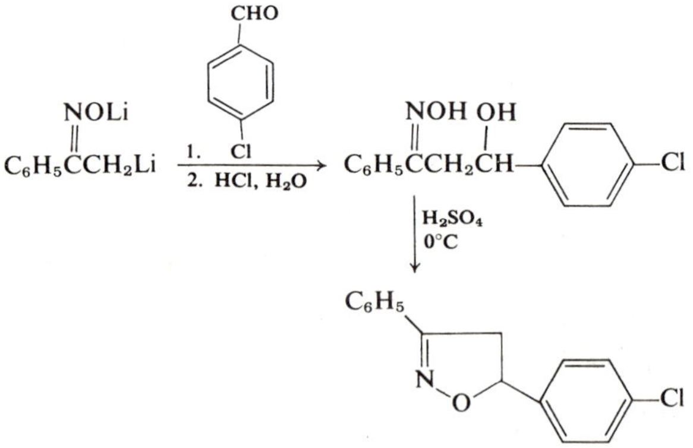

$$C_6H_5\overset{NOLi}{\underset{\|}{C}}CH_2Li \xrightarrow[\text{2. HCl, H}_2\text{O}]{1.\;\text{4-Cl-C}_6H_4\text{CHO}} C_6H_5\overset{NOH}{\underset{\|}{C}}CH_2\overset{OH}{\underset{|}{C}}H{-}\!\!\left\langle\;\right\rangle\!\!{-}Cl$$

$$\xrightarrow[0°C]{H_2SO_4}\quad\text{[isoxazoline with }C_6H_5\text{ and }p\text{-Cl-C}_6H_4\text{]}\quad (124)$$

(Kaiser and Kaufman, 1972)

Finally, considering 1,6-multiple anions, the dilithio salt of acetophenone azine has been cyclized by a displacement reaction on an alkyl halide to give a dihydropyridazine [Eq. (125)], and the trilithio salt of a simple osazone has

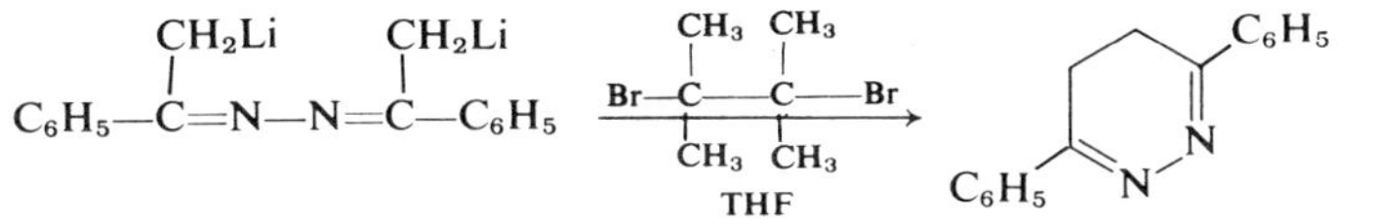

5. CARBANIONS

$$\underset{\underset{C_6H_5}{|}}{\overset{\overset{CH_2Li}{|}}{C}}=N-N=\underset{\underset{C_6H_5}{|}}{\overset{\overset{CH_2Li}{|}}{C}} \quad \xrightarrow[\text{THF}]{\underset{\underset{CH_3\ CH_3}{|}}{\overset{\overset{CH_3\ CH_3}{|\ \ \ |}}{Br-C-C-Br}}} \quad$$ (125)

(Henoch et al., 1969)

been converted to a hydroxyosazone [Eq. (126)].

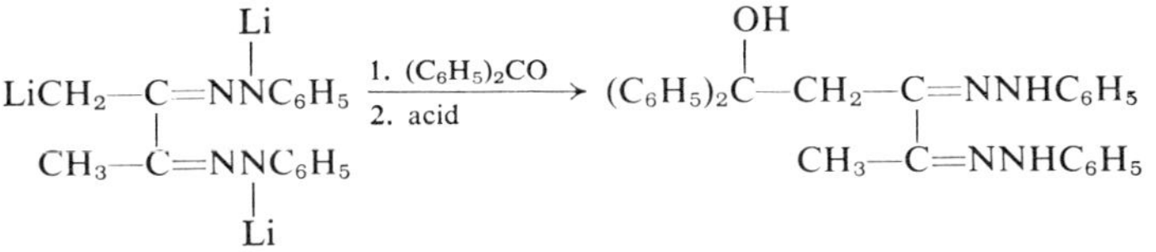

$$\begin{array}{c}\overset{Li}{|}\\ LiCH_2-C=NNC_6H_5\\ |\\ CH_3-C=NNC_6H_5\\ |\\ Li\end{array} \xrightarrow[2.\ acid]{1.\ (C_6H_5)_2CO} \begin{array}{c}\overset{OH}{|}\\ (C_6H_5)_2C-CH_2-C=NNHC_6H_5\\ |\\ CH_3-C=NNHC_6H_5\end{array}$$ (126)

(Henoch et al., 1969)

6. Other 1,1-, 1,2-, and 1,4-Multiple Anions

In the last few years, there has arisen a surge of interest concerning the development of 1,1- or *gem*-dicarbanions. For example, on treatment with *n*-butyllithium in hexane, propyne is perlithiated to give C_3Li_4, a compound which is converted to a polysilane with chlorotrimethylsilane [Eq. (127)].

$$CH_3C\equiv CH \xrightarrow[\text{hexane}]{n\text{-}C_4H_9Li} Li_2C=C=CLi_2$$

$$\xrightarrow[-70°C,\ \text{then}\ 25°C]{ClSi(CH_3)_3} [(CH_3)_3Si]_2C=C=C[Si(CH_3)_3]_2 \quad (127)$$

(West et al., 1965; West and Jones, 1969)

Other polylithioacetylenes similarly prepared and subsequently silylated or deuterated include $CH_3C_3Li_3$, $C_3H_7C_3Li_3$, and $n\text{-}C_5H_{11}C_3Li_3$ (West et al., 1965; West and Jones, 1969); $C_6H_5C_3Li_3$ (Mulvaney et al., 1967); $C_6H_2Li_3\text{-}C_3Li_3$ and $C_6H_3Li_2C_3Li_3$ (West and Gornowicz, 1971); and $ArC\equiv C\text{-}CLi_2C(R)=CHR$ (Klein and Brenner, 1969).

Certain nitriles have likewise been converted to *gem*-dicarbanions. Thus, treatment of phenylacetonitrile with *n*-butyllithium in THF–hexane gives the corresponding dilithio salt which has been alkylated and acylated (Kaiser et al., 1971) [Eq. (128)]. Dilithioacetonitrile has similarly been silylated

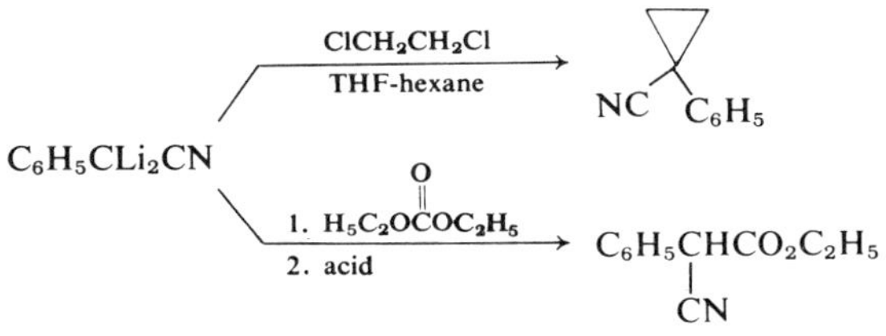

$$C_6H_5CLi_2CN \begin{array}{c}\xrightarrow[\text{THF-hexane}]{ClCH_2CH_2Cl} NC\ \ C_6H_5\\ \\ \xrightarrow[2.\ acid]{1.\ H_5C_2O\overset{O}{\overset{\|}{C}}OC_2H_5} C_6H_5CHCO_2C_2H_5\\ |\\ CN\end{array}$$ (128)

[Eq. (129)] or has been reacted with carbonyl compounds followed by

$$Li_2C_2HN \xrightarrow[THF, -30°C]{ClSi(CH_3)_3} [(CH_3)_3Si]_2C=C=NSi(CH_3)_3 + (CH_3)_3SiC\equiv C-N[Si(CH_3)_3]_2 \quad (129)$$

(Gornowicz and West, 1971)

chlorotrimethylsilane [Eq. (130)].

$$Li_2C_2HN \xrightarrow[THF]{2\ CH_3CHO} \xrightarrow{ClSi(CH_3)_3} (CH_3CH)_2\overset{\overset{OSi(CH_3)_3}{|}}{C}HCN \quad (130)$$

(Gornowicz and West, 1971)

Finally, even toluene has been polymetalated by n-butyllithium–TMEDA in hexane as evidenced by silylation and deuteration experiments [Eq. (131)].

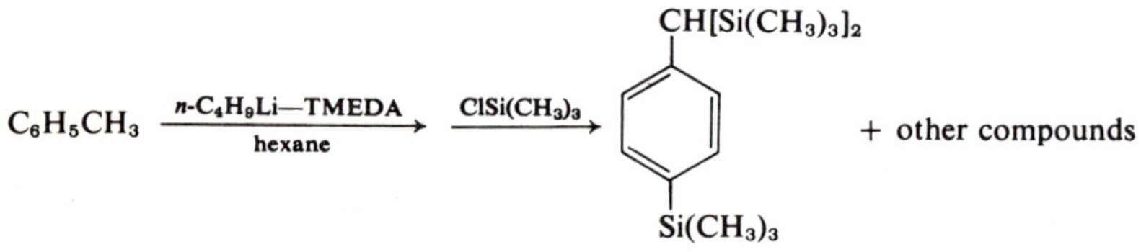

$$C_6H_5CH_3 \xrightarrow[hexane]{n\text{-}C_4H_9Li\text{--}TMEDA} \xrightarrow{ClSi(CH_3)_3} \text{[1,4-disubstituted benzene with } CH[Si(CH_3)_3]_2 \text{ and } Si(CH_3)_3\text{]} + \text{other compounds} \quad (131)$$

(West and Jones, 1968)

1,2-Dianions, though rare, have also proved to be synthetically useful. For example, benzophenone is converted rapidly to its 1,2-disodio salt in ammonia, which can be condensed with alkyl halides, aldehydes [Eq. (132)], and other

$$(C_6H_5)_2CO \xrightarrow[NH_3]{2\ Na} (C_6H_5)_2\overset{\overset{Na}{|}}{C}-ONa \xrightarrow[2.\ acid]{1.\ C_6H_5CHO} (C_6H_5)_2\overset{\overset{HO}{|}}{C}-\overset{\overset{OH}{|}}{C}HC_6H_5 \quad (132)$$

(Hamrick and Hauser, 1959)

electrophiles. Certain anils have similarly been reduced to their 1,2-dialkali salts which alkylate on carbon but which, surprisingly, acylate on nitrogen [Eq. (133)].

$$(C_6H_5)_2\overset{\overset{Na}{|}}{C}-\overset{\overset{Na}{|}}{N}C_6H_5 \begin{array}{c} \xrightarrow[2.\ acid]{1.\ n\text{-}C_3H_7Br} (C_6H_5)_2\overset{\overset{n\text{-}C_3H_7}{|}}{C}NHC_6H_5 \\ \xrightarrow[2.\ acid]{1.\ C_6H_5CO_2CH_3} (C_6H_5)_2CH\overset{}{N}\overset{\overset{O}{\|}}{C}C_6H_5 \\ \underset{C_6H_5}{|} \end{array} \quad (133)$$

(J. G. Smith and Veach, 1967)

5. CARBANIONS

The related 1,2-dicarbanions are known particularly when both negatively charged carbon atoms are flanked by electron-withdrawing groups [e.g.,

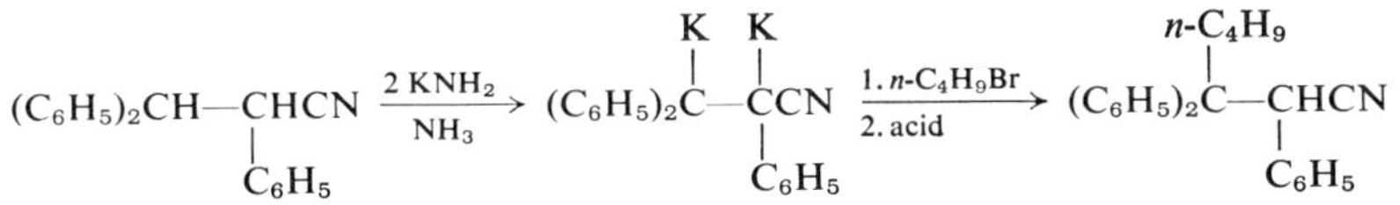

$$(C_6H_5)_2CH\text{---}\underset{\underset{C_6H_5}{|}}{CH}CN \xrightarrow[NH_3]{2\ KNH_2} (C_6H_5)_2\underset{\underset{C_6H_5}{|}}{\overset{\overset{K}{|}}{C}}\text{---}\overset{\overset{K}{|}}{C}CN \xrightarrow[2.\ acid]{1.\ n\text{-}C_4H_9Br} (C_6H_5)_2\underset{\underset{C_6H_5}{|}}{\overset{\overset{n\text{-}C_4H_9}{|}}{C}}\text{---}CHCN \quad (134)$$

(Kofron *et al.*, 1962)

Eq. (134)]. Attempts to prepare similar dicarbanions by reduction of olefins has sometimes been successful but, more often than not, only a radical anion is formed which dimerizes to give a 1,4-dianion. Such a dianion undergoes carbonation and other related reactions [Eq. (135)].

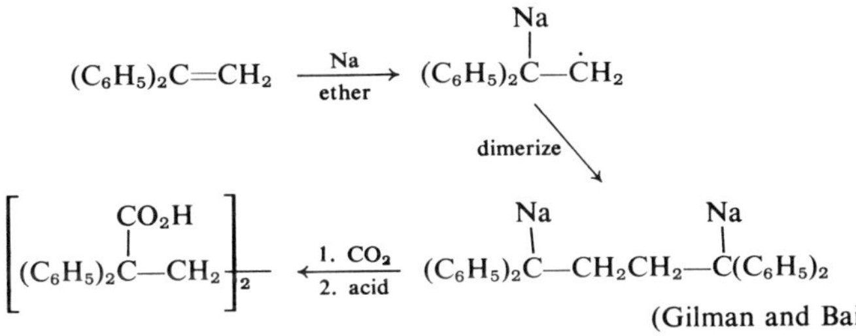

$$(C_6H_5)_2C{=}CH_2 \xrightarrow[ether]{Na} (C_6H_5)_2\overset{\overset{Na}{|}}{C}\text{---}\dot{C}H_2$$

$$\left[(C_6H_5)_2\overset{\overset{CO_2H}{|}}{C}\text{---}CH_2\right]_2 \xleftarrow[2.\ acid]{1.\ CO_2} (C_6H_5)_2\overset{\overset{Na}{|}}{C}\text{---}CH_2CH_2\text{---}\overset{\overset{Na}{|}}{C}(C_6H_5)_2 \quad (135)$$

(Gilman and Bailie, 1943)

Similar reductive dimerizations of certain acetylenes have resulted in an excellent method for the preparation of novel heterocycles [Eq. (136)].

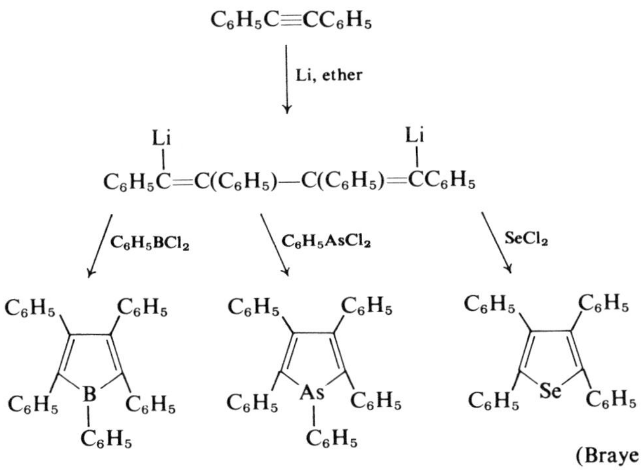

$$C_6H_5C{\equiv}CC_6H_5 \xrightarrow{Li,\ ether} C_6H_5\overset{\overset{Li}{|}}{C}{=}C(C_6H_5)\text{---}C(C_6H_5){=}\overset{\overset{Li}{|}}{C}C_6H_5$$

[C₆H₅BCl₂ → boron heterocycle with C₆H₅ groups; C₆H₅AsCl₂ → arsenic heterocycle; SeCl₂ → selenium heterocycle] (136)

(Braye *et al.*, 1961)

Related reductive cleavages of disubstituted acetylenes followed by hydrolysis give allenes [Eq. (137)].

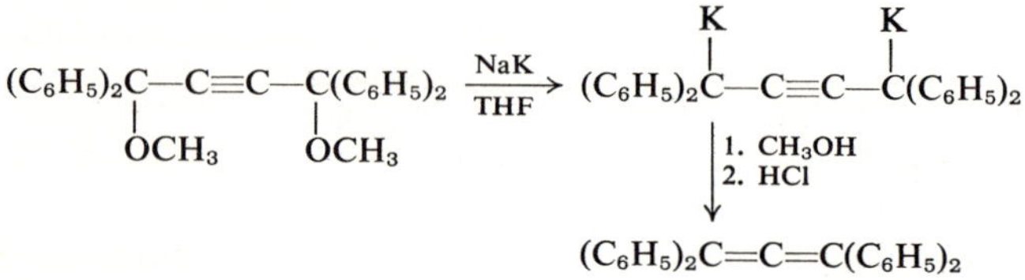

$$(C_6H_5)_2C-C\equiv C-C(C_6H_5)_2 \xrightarrow{\text{NaK}}_{\text{THF}} (C_6H_5)_2\overset{K}{C}-C\equiv C-\overset{K}{C}(C_6H_5)_2$$
$$\overset{|}{OCH_3} \quad \overset{|}{OCH_3}$$

$$\downarrow \begin{array}{l} 1.\ CH_3OH \\ 2.\ HCl \end{array}$$

$$(C_6H_5)_2C=C=C(C_6H_5)_2 \tag{137}$$

(Edinger and Day, 1971)

Finally, the interaction of tolan with various organolithium reagents results in the formation of a novel 1,4-dianion which has been carbonated

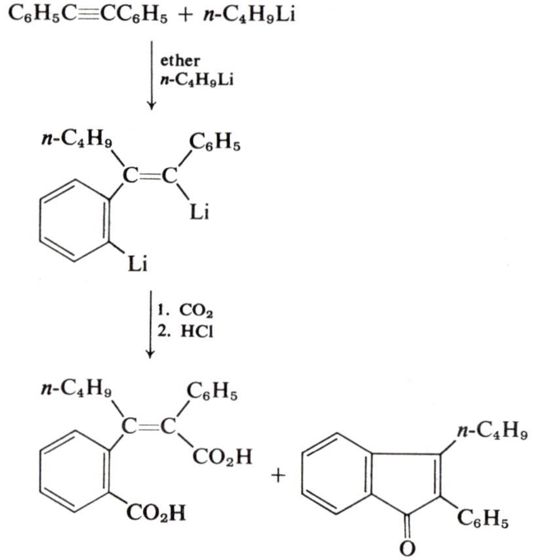

$$C_6H_5C\equiv CC_6H_5 + n\text{-}C_4H_9Li$$

$$\downarrow \begin{array}{l} \text{ether} \\ n\text{-}C_4H_9Li \end{array}$$

[structure: ortho-lithiated phenyl with C=C bearing n-C$_4$H$_9$ and C$_6$H$_5$, and Li]

$$\downarrow \begin{array}{l} 1.\ CO_2 \\ 2.\ HCl \end{array} \tag{138}$$

[product: diacid structure + indanone with n-C$_4$H$_9$ and C$_6$H$_5$]

(Mulvaney et al., 1966)

[Eq. (138)] and has been reacted with a variety of metallic and pseudo-metallic halides to give metallo-indene systems [Eq. (139)].

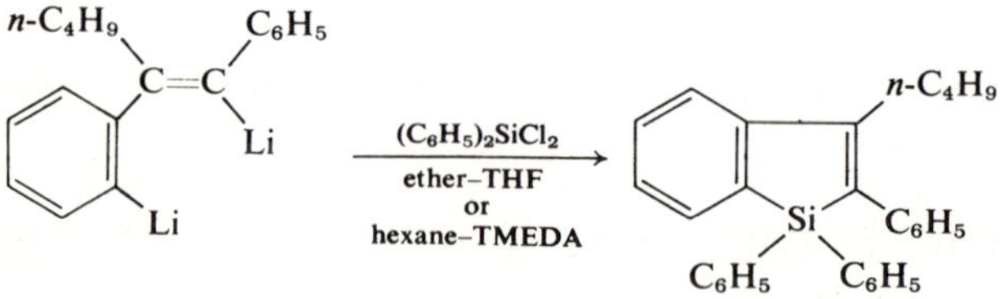

[structure → silaindene product]

$$\xrightarrow[\substack{\text{ether-THF} \\ \text{or} \\ \text{hexane-TMEDA}}]{(C_6H_5)_2SiCl_2} \tag{139}$$

(Rausch and Klemann, 1967)

C. Anion Rearrangements

1. 1,2-Rearrangements

a. Stevens Rearrangement. Interest in the Stevens and other anion rearrangements was revived in the late 1960's. A number of recent reviews are available (Pine, 1971; Lepley and Giumanini, 1971; Musker, 1970).

Early examples of this rearrangement were provided by Stevens and coworkers from whom the reaction got its name [Eqs. (140) and (141)]. Later

$$\text{PhCH}_2\overset{+}{\text{N}}(\text{CH}_3)(\text{CH}_2\text{Ph})\text{CH}_3 \xrightarrow[\text{NaNH}_2]{\text{fused NaOCH}_3} \text{PhCH}(\text{CH}_2\text{Ph})\text{N}(\text{CH}_3)_2 \quad (140)$$

(Thompson and Stevens, 1932)

$$\text{PhCOCH}_2\overset{+}{\text{N}}(\text{CH}_3)_2(\text{CH}_2\text{Ph}) \xrightarrow{\text{OH}^-} \text{PhCOCH}(\text{CH}_2\text{Ph})\text{N}(\text{CH}_3)_2 \quad (141)$$

(Dunn and Stevens, 1934)

examples have involved the treatment of quaternary ammonium salts with much stronger bases, namely, organolithium reagents [Eqs. (142) and (143)].

$$\text{PhCH}_2\text{N}(\text{CH}_3)_3{}^+\text{Br}^- \xrightarrow{\text{PhLi}} \text{PhCH}(\text{CH}_3)\text{N}(\text{CH}_3)_2 \quad (142)$$

(Wittig et al., 1948)

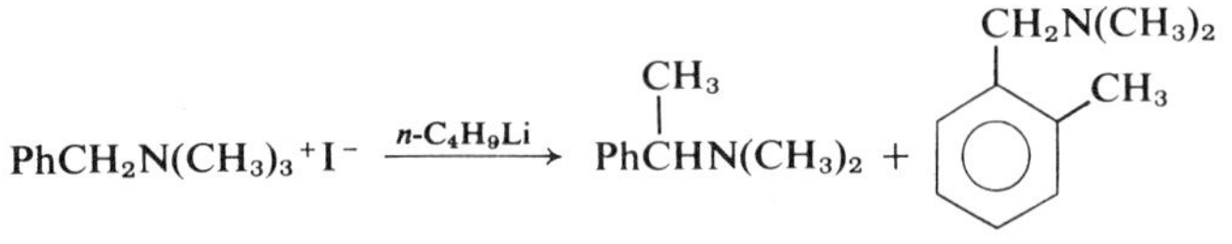

$$\text{PhCH}_2\text{N}(\text{CH}_3)_3{}^+\text{I}^- \xrightarrow{n\text{-C}_4\text{H}_9\text{Li}} \text{PhCH}(\text{CH}_3)\text{N}(\text{CH}_3)_2 + \text{[o-CH}_3\text{C}_6\text{H}_4\text{CH}_2\text{N}(\text{CH}_3)_2\text{]} \quad (143)$$

(Lepley and Becker, 1965a)

The last reaction, that of Lepley and Becker, is interesting in that it illustrates the simultaneous generation of product from the Sommelet–Hauser rearrangement as well as from the Stevens rearrangement. About a 4:1 ratio of the Sommelet–Hauser versus Stevens product is obtained when butyllithium is the base, but only the Stevens product is obtained when phenyllithium is the base [Eq. (144)]. Various products of the Stevens rearrangement can be written depending on which C–H bond is ionized and, in some cases, which carbon undergoes substitution. Examples of the attainment of mixtures of Stevens rearrangement products are known.

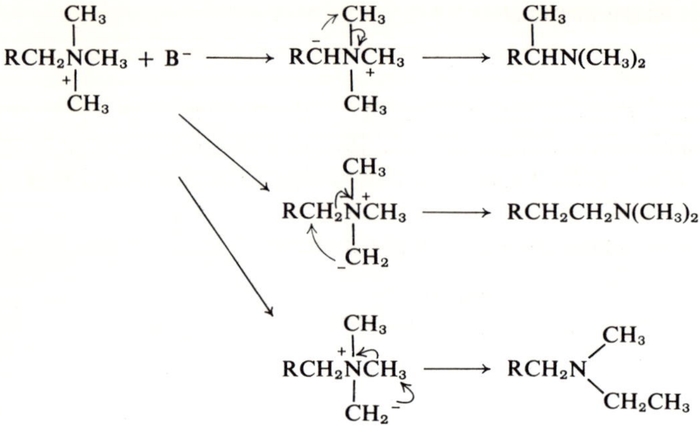

$$\begin{array}{c}
\text{CH}_3 \\
| \\
\text{RCH}_2\text{NCH}_3 + \text{B}^- \\
+| \\
\text{CH}_3
\end{array}
\longrightarrow
\begin{array}{c}
\text{CH}_3 \\
| \\
\text{RCHNCH}_3 \\
|+ \\
\text{CH}_3
\end{array}
\longrightarrow
\begin{array}{c}
\text{CH}_3 \\
| \\
\text{RCHN(CH}_3)_2
\end{array}$$

$$\begin{array}{c}
\text{CH}_3 \\
| \\
\text{RCH}_2\text{NCH}_3 \\
| \\
\text{CH}_2
\end{array}
\longrightarrow \text{RCH}_2\text{CH}_2\text{N(CH}_3)_2 \quad (144)$$

$$\begin{array}{c}
\text{CH}_3 \\
+| \\
\text{RCH}_2\text{NCH}_3 \\
| \\
\text{CH}_2^-
\end{array}
\longrightarrow \text{RCH}_2\text{N}\begin{array}{c}\text{CH}_3 \\ \text{CH}_2\text{CH}_3\end{array}$$

Significant evidence has been amassed in support of an S_Ni mechanism. For instance, the lack of cross-over products in a mixed Stevens rearrangement of two different quaternary ammonium salts is taken as indicative of a concerted internal substitution. An S_Ni mechanism would also demand retention as well as stereospecificity. This is now known to be the case for this rearrangement [Eq. (145)] (A. Campbell *et al.*, 1947; Brewster and

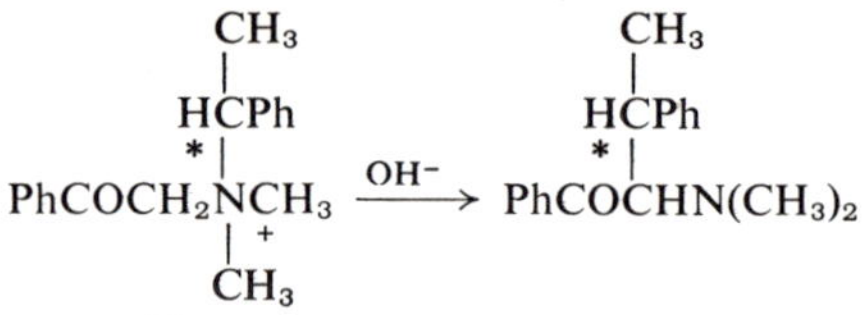

$$\begin{array}{c}
\text{CH}_3 \\
| \\
\text{HCPh} \\
*| \\
\text{PhCOCH}_2\text{NCH}_3 \\
|+ \\
\text{CH}_3 \\
(9)
\end{array}
\xrightarrow{\text{OH}^-}
\begin{array}{c}
\text{CH}_3 \\
| \\
\text{HCPh} \\
*| \\
\text{PhCOCHN(CH}_3)_2
\end{array}
\quad (145)$$

* = Enantiomeric carbon

Kline, 1952). Essentially no racemization is observed in the course of this transformation.

Incorporation of electron-withdrawing substituents at the para position of the benzoyl group in compounds similar to **9** are known to retard the reaction. These same withdrawing substituents facilitate the reaction when located at the para or meta position of the migrating group. This electron withdrawal at the migration origin and donation at the migration terminus serve to facilitate the reaction.

Interesting anomalies of the Stevens rearrangement being retarded by electron-donating groups at the migration origin are provided by the quaternary ammonium ferrocene salts **10** and **11**. In **10**, ionization of a methylene adjacent to either a ferrocenyl or a phenyl group can be realized. The well-known electron-donor properties of the ferrocenyl group suggest that the

5. CARBANIONS

rearrangement should take place such that the ferrocenylmethylene becomes the terminus. In actuality, the ferrocenylmethyl system is the migrating group [Eq. (146)]. More to the point, ferrocenylmethyltrimethylammonium iodide

$$\begin{bmatrix} C_5H_5FcC_5H_4CH_2 \diagdown \diagup CH_3 \\ N \\ PhCH_2 \diagup \diagdown CH_3 \end{bmatrix}^+ I^- \xrightarrow{n\text{-}C_4H_9Li} C_5H_5FcC_5H_4CH_2\overset{N(CH_3)_2}{\underset{|}{C}}HPh \quad (146)$$

(10)   (Ustynyuk and Perevalova, 1964)

(**11**) has been rearranged to give a single Stevens product in 50% yield [Eq. (147)]. This product is comparable in structure to neither the Stevens

$$FcCH_2N(CH_3)_3^+ I^- + NaNH_2 \longrightarrow FcCH_2CH_2N(CH_3)_2 \quad (147)$$

(11)   (Lindsay and Hauser, 1957)

nor Sommelet–Hauser product from the corresponding benzyl methiodide [Eqs. (142) and (143)]. Again, migration of a group with strong electron-donation properties, namely, the ferrocenylmethyl group, is realized. An explanation for this apparent reversal is that ionization of a ferrocenyl-methylene proton may be so slight that reaction can proceed only via ionization of a benzyl proton.

A recent stereochemical result deserves comment. Hill and Chan (1966) have demonstrated a transfer of asymmetry from nitrogen to carbon in this reaction by employing an optically active ammonium salt as the starting

$$\underset{\underset{CH_2Ph}{\overset{+|}{|}}}{\overset{\overset{CH_3}{\underset{|}{|}}}{Ph\overset{*}{-}N}}-CH_2CH=CH_2 \xrightarrow[DMSO]{tert\text{-}C_4H_9OK} PhCH_2\overset{\overset{H}{\overset{*|}{|}}}{\underset{CH_3NPh}{C}}CH=CH_2 \quad (148)$$

* = Enantiomeric atom

material [Eq. (148)]. This result also favors an S_Ni mechanism for the Stevens rearrangement or, at the very least, a tight ion pair.

One last mechanistic qualification should be mentioned. Recently, several communications have appeared in which a CIDNP (chemically induced dynamic nuclear polarization) effect has been detected during the first few minutes of a Stevens rearrangement. For example, the quaternary salt (**12**) on treatment with aqueous alkali gives the isolable nitrogen ylide (**13**) which with heating rearranges smoothly to the Stevens product (**14**) [Eq. (149)]. A CIDNP effect has been noted during this rearrangement and is interpreted as signifying a homolytic cleavage taking place during the migration. Since a CIDNP effect has not been noted during all Stevens rearrangements, it is difficult to assess the actual significance of this observation.

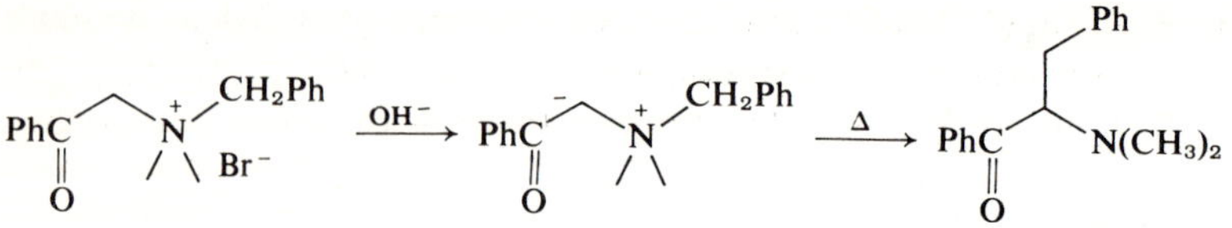

$$\text{PhC(=O)}\overset{+}{\text{N}}(\text{CH}_2\text{Ph})_2 \text{Br}^- \xrightarrow{\text{OH}^-} \text{PhC(=O)}\overset{+}{\text{N}}^-(\text{CH}_2\text{Ph})(\text{CH}_2\text{Ph}) \xrightarrow{\Delta} \text{PhC(=O)CH(Ph)N(CH}_3)_2 \quad (149)$$

(12)   (13)   (14)

(Jemison et al., 1970)

Homologs of the Stevens rearrangement are known. Rearrangement of benzyl allyl methiodide (15) under a variety of conditions produces various amounts of the 1,4-rearrangement product in addition to the expected 1,2-product [Eq. (150)]. Again, this favors the S_{Ni} or tight ion-pair mechan-

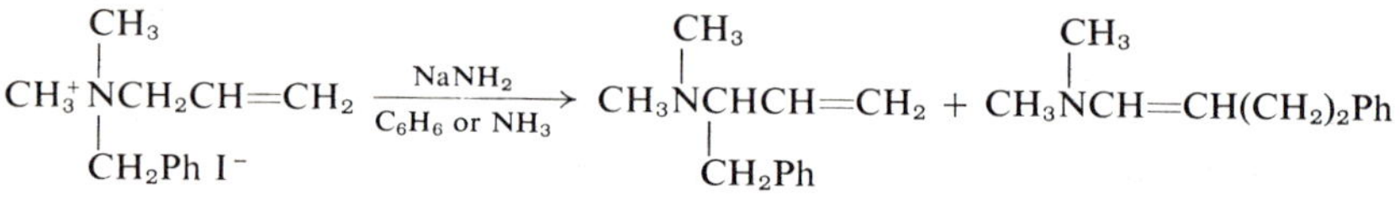

$$\underset{(15)}{\text{CH}_3\overset{+}{\text{N}}(\text{CH}_3)(\text{CH}_2\text{Ph})\text{CH}_2\text{CH}=\text{CH}_2 \; \text{I}^-} \xrightarrow[\text{C}_6\text{H}_6 \text{ or NH}_3]{\text{NaNH}_2} \underset{1,2\text{-}}{\text{CH}_3\text{N}(\text{CH}_3)(\text{CH}_2\text{Ph})\text{CHCH}=\text{CH}_2} + \underset{1,4\text{-}}{\text{CH}_3\text{N}(\text{CH}_3)\text{CH}=\text{CH}(\text{CH}_2)_2\text{Ph}} \quad (150)$$

(Jenny and Druey, 1962)

ism. More recently, Jenny and Melzer have reported an example of a 1,3-Stevens rearrangement [Eq. (151)].

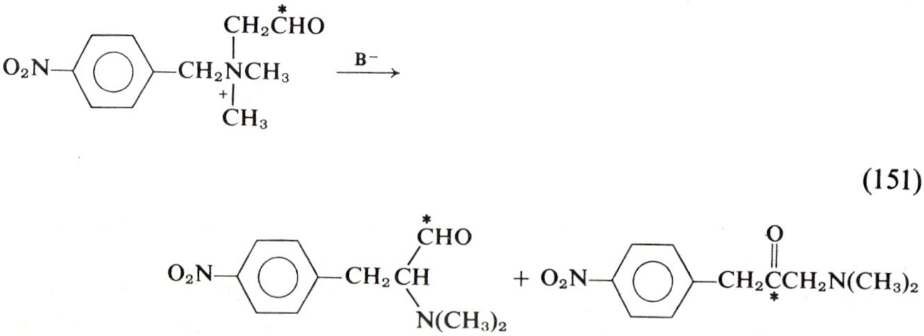

$$\text{O}_2\text{N-C}_6\text{H}_4\text{-CH}_2\overset{+}{\text{N}}(\text{CH}_3)_2\text{CH}_2\overset{*}{\text{C}}\text{HO} \xrightarrow{\text{B}^-}$$

(151)

$$\underset{1,2\text{-}}{\text{O}_2\text{N-C}_6\text{H}_4\text{-CH}_2\text{CH}(\overset{*}{\text{CHO}})\text{N}(\text{CH}_3)_2} + \underset{1,3\text{-}}{\text{O}_2\text{N-C}_6\text{H}_4\text{-CH}_2\text{C}(=\text{O})\overset{*}{\text{C}}\text{H}_2\text{N}(\text{CH}_3)_2}$$

* = ^{14}C-labeled carbon   (Jenny and Melzer, 1966)

Stevens rearrangements have been observed with certain sulfonium salts, but examples are rare. For the most part, sulfonium salts tend to undergo the Sommelet–Hauser rearrangement. Recently, it was observed that the chiral ylide (16) could be isolated and thermally converted into the Stevens product (17) [Eq. (152)]. A CIDNP effect and retention of configuration with significant racemization was noted during the course of this reaction. A loose ion-pair or radical-pair mechanism is indicated.

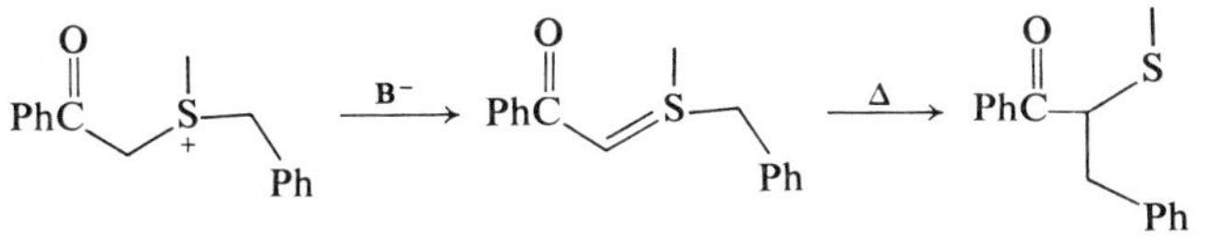

$$\underset{(16)}{\underset{Ph}{PhC(O)-S^+-CH_2}} \xrightarrow{B^-} \underset{Ph}{\underset{(17)}{PhC(O)-S=CH}} \xrightarrow{\Delta} PhC(O)-CH(S)-CH_2Ph \quad (152)$$

(Baldwin et al., 1970)

b. Sommelet–Hauser Rearrangement. Sommelet initially reported that the quaternary ammonium hydroxide (**18**) could be rearranged to a disubstituted aromatic compound as illustrated in Eq. (153). In the 1950's and 1960's,

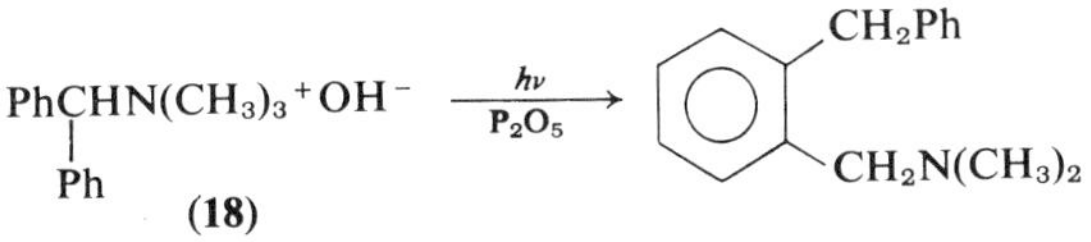

$$\underset{(18)}{PhCH(Ph)N(CH_3)_3{}^+OH^-} \xrightarrow[P_2O_5]{h\nu} \text{o-}C_6H_4(CH_2Ph)(CH_2N(CH_3)_2) \quad (153)$$

(Sommelet, 1937)

Hauser and co-workers extensively investigated this rearrangement such that Hauser's name is also linked to the reaction. The recent reviews that are available for the Stevens rearrangement also cover the Sommelet–Hauser rearrangement. These are Musker (1970), Lepley and Giumanini (1971), and Pine (1971).

Kantor and Hauser (1951) observed that in strong base the salt (**19**) can be almost quantitatively rearranged to the ortho derivative (**20**) [Eq. (154)].

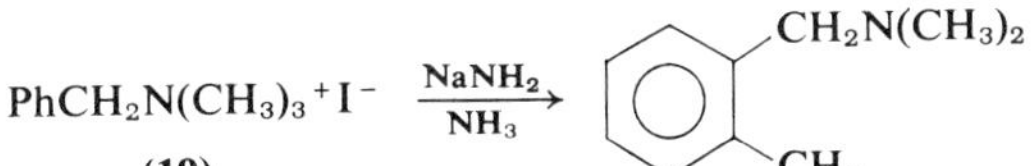

$$\underset{(19)}{PhCH_2N(CH_3)_3{}^+I^-} \xrightarrow[NH_3]{NaNH_2} \underset{(20)}{\text{o-}C_6H_4(CH_2N(CH_3)_2)(CH_3)} \quad (154)$$

In a novel example of the pervasiveness of the reaction the ortho product was converted to its methiodide which was rearranged. This process was repeated until the ultimate product pentamethyldimethylbenzylamine was obtained.

Note should be taken that the quaternary ammonium salts in Eqs. (153) and (154) can, in principle, undergo the Stevens rearrangement. Lepley and Becker (1965b) observed a product from the rearrangement of benzyltrimethylammonium iodide which consisted of a mixture of the Sommelet–

Hauser product and a Stevens rearrangement product, i.e., both rearrangements evidently took place [Eq. (143)]. A similar mixture of the products of these two rearrangements has been noted in the rearrangement of the methiodide of 2-thenylamine [Eq. (155)].

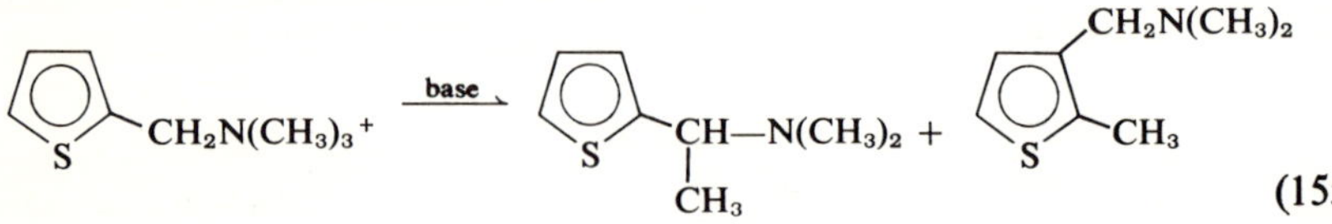

$$\text{2-thienyl-}CH_2N(CH_3)_3^+ \xrightarrow{\text{base}} \text{2-thienyl-}CH(CH_3)-N(CH_3)_2 + \text{3-methyl-2-thienyl-}CH_2N(CH_3)_2 \quad (155)$$

(Slocum and Gierer, 1968)

For a number of years the accepted mechanism for this reaction has involved the nucleophilic attack of an ionized methylene ylide on an aromatic ring. A tautomeric step serves to rearomatize the ring [Eq. (156)]. In the

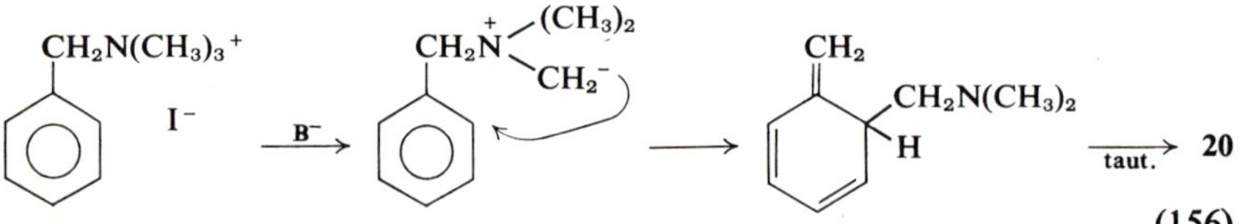

$$PhCH_2N(CH_3)_3^+ I^- \xrightarrow{B^-} PhCH_2\overset{+}{N}(CH_3)_2(CH_2^-) \longrightarrow \text{(cyclohexadienyl-}CH_2N(CH_3)_2\text{, H)} \xrightarrow{\text{taut.}} \mathbf{20} \quad (156)$$

instance of the mesitylene derivative (**21**), such an intermediate has been isolated (Hauser and Van Eenam, 1957), since there is no proton at the ortho

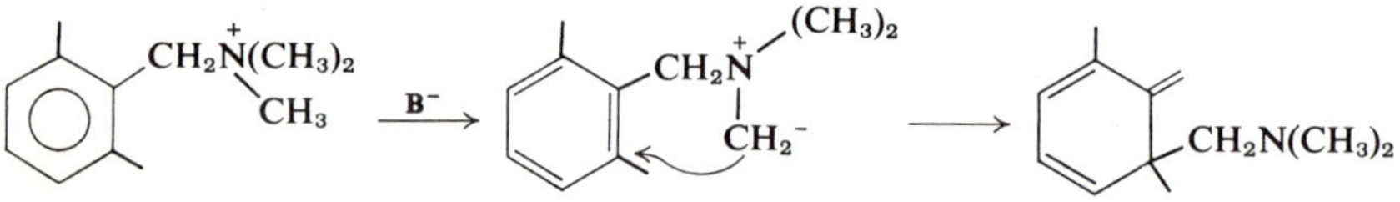

$$(\mathbf{21}) \xrightarrow{B^-} \text{ylide intermediate} \longrightarrow \text{rearranged product with }CH_2N(CH_3)_2 \quad (157)$$

positions which can tautomerize [Eq. (157)]. This mechanism has since been supported by carbon-labeling experiments.

The apparent fact that a benzylic hydrogen is not abstracted by base has resulted in the suggestion that ionization takes place at this position initially and that ionization at the methyl group occurs subsequent to this. Proton transfer may then take place intramolecularly. This benzylic-ionized intermediate has been intercepted by the route described in Eq. (158).

Lepley and Becker (1965b) have studied the influence of the halide anion, solvent, and base on the reaction shown in Eq. (158). Their data show a significant increase in yields of the two products with an increase in base concentration. When the iodide or chloride salt is rearranged in pentane

$$\text{PhCH}_2\text{N}(\text{CH}_3)_3{}^+ \xrightarrow{\text{NaNH}_2} \xrightarrow[\text{2. H}_2\text{O}]{\text{1. Ph}_2\text{CO}} \underset{\substack{|\\ \text{CHCPh}_2\text{OH}\\|\\ \text{N}(\text{CH}_3)_3{}^+\text{I}^-}}{\text{C}_6\text{H}_5} \quad (158)$$

(Puterbaugh and Hauser, 1964b)

under standard conditions, the chloride salt gives significantly more product. This can be attributed to a greater extent of dissolution of the chloride salt in pentane. A more difficult to explain observation is that ether, relative to pentane, permits both a much greater yield of the two products and a significant change in the ratio of the two products. Apparently, ether favors the Sommelet–Hauser rearrangement. From other works coupled with the data in this chapter, it can be concluded that not only is n-butyllithium the best organolithium base to effect this rearrangement, but also that organolithium bases, in general, favor the Stevens rearrangement. On the other hand, amide bases favor the Sommelet–Hauser rearrangement. This may be due in part to a temperature effect. In any case, the Sommelet–Hauser rearrangement appears to be favored by low temperature.

A second mechanism for the Sommelet–Hauser rearrangement has been proposed. This mechanism incorporates the formation of an immonium ion intermediate which electrophilicly attacks the aromatic ring. Although such an ion pair would, in most cases, rearrange itself with a minimum of energy expenditure that ortho attack would be exclusive, under certain conditions other orientations might be observed. A notable instance of this has recently been provided by the observation of a para Sommelet–Hauser product from the rearrangement of the hindered quaternary salt (22) [Eq.

$$\underset{\substack{|\\ \text{N}(\text{CH}_3)_3{}^+\text{Cl}^-\\(\mathbf{22})}}{\text{PhCHC}(\text{CH}_3)_3} \xrightarrow{\text{B}^-} \underset{\text{Sommelet–Hauser}}{\overset{\substack{\text{C}(\text{CH}_3)_3\\|\\\text{CH}_2}}{\bigcirc}\text{-CH}_2\text{N}(\text{CH}_3)_2} + \underset{\text{Stevens}}{(\text{CH}_3)_3\text{C}-\overset{\substack{\text{Ph}\\|}}{\text{C}}\text{HN}(\text{CH}_3)_2}$$

$$+ \; S_N2 \text{ product } + \; \underset{\substack{|\\ \text{CH}_2\text{N}(\text{CH}_3)_2}}{\overset{\substack{\text{CH}_2\text{C}(\text{CH}_3)_3\\|}}{\bigcirc}} \quad (159)$$

(Pine et al., 1971)

(159)]. This immonium ion mechanism can actually be envisioned to encompass both the Stevens and Sommelet–Hauser rearrangements [Eq. (160)].

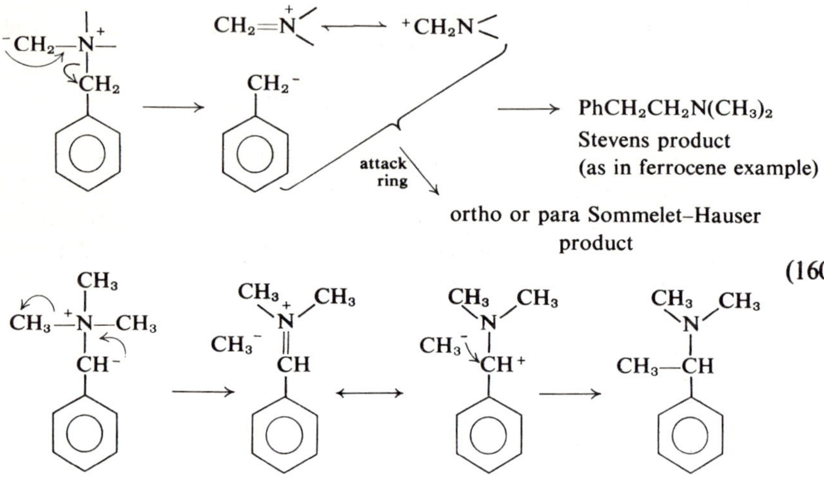

$$\longrightarrow PhCH_2CH_2N(CH_3)_2$$
Stevens product
(as in ferrocene example)

ortho or para Sommelet–Hauser product

(160)

As mentioned, sulfonium salts, when a suitable aryl group is present, undergo the Sommelet–Hauser rearrangement. Some examples of these are

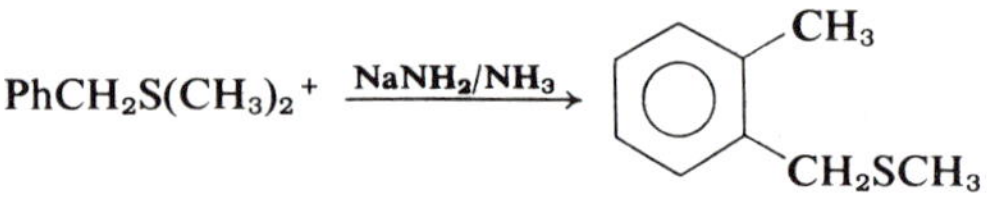

$$PhCH_2S(CH_3)_2{}^+ \xrightarrow{NaNH_2/NH_3}$$ [aryl product with CH$_3$ and CH$_2$SCH$_3$] (161)

(Hauser et al., 1953)

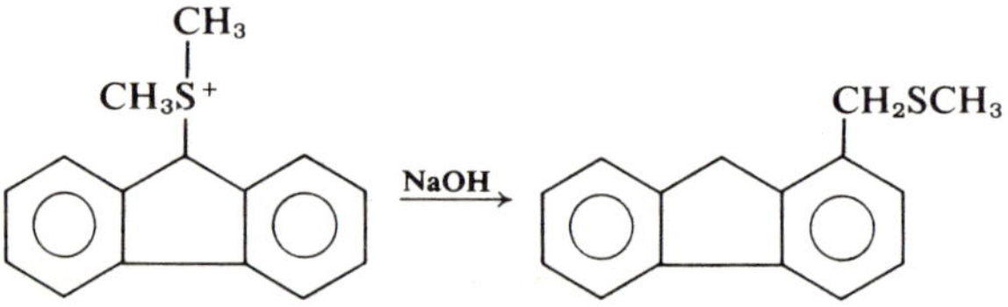

(162)

(A. W. Johnson and LaCount, 1961)

illustrated in Eqs. (161) and (162). An intriguing study was recently performed by Hayashi and Oda (1968). These authors studied the effect of base concentration on the rearrangement of sulfonium salt (23) [Eq. (163)]. As the

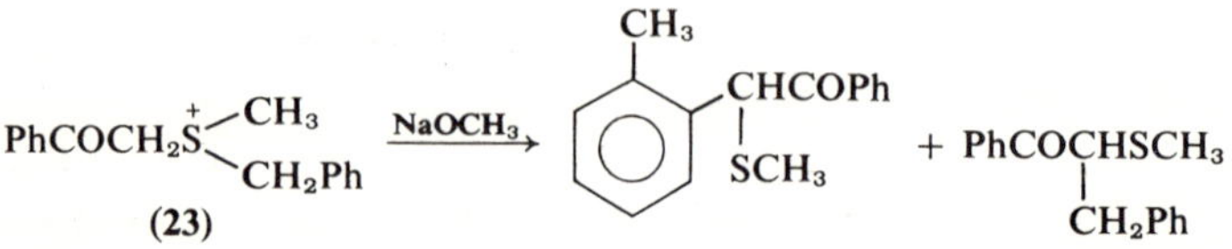

$$PhCOCH_2\overset{+}{S}{\overset{\displaystyle CH_3}{\underset{\displaystyle CH_2Ph}{}}} \xrightarrow{NaOCH_3} [\text{aryl with CH}_3, \text{CHCOPh, SCH}_3] + PhCOCHSCH_3 \quad (163)$$
(23) $\quad\quad\quad\quad\quad\quad\quad\quad\quad\quad\quad\quad\quad\quad\quad\quad\quad\quad\quad\quad\quad\quad\quad$ CH$_2$Ph

concentration of methoxide was raised, a significant shift of Stevens to Sommelet–Hauser product was noted. A similar observation was made for the rearrangement of the cyclic ammonium salt (**24**).

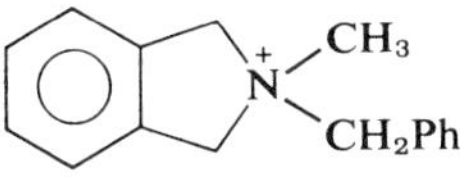

$$\text{(24)} \quad \underset{\text{CH}_2\text{Ph}}{\overset{\text{CH}_3}{\text{isoindolinium}^+}}$$

c. Wittig Rearrangement. Although there are divergences, the Wittig rearrangement appears formally similar to the Stevens rearrangement described earlier. A recent review and assessment of this rearrangement has been published (Schöllkopf, 1970).

An early example of the Wittig rearrangement involved the treatment of benzyl methyl ether with strong base [Eq. (164)]. A recent study has de-

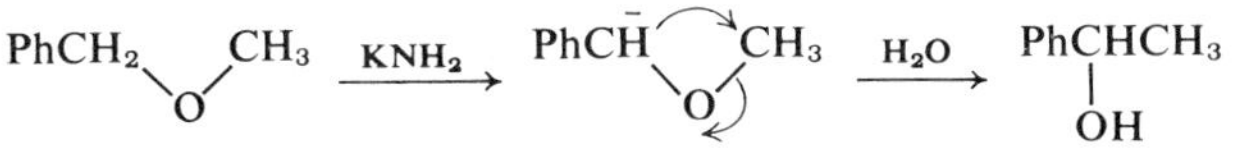

$$\text{PhCH}_2\text{-O-CH}_3 \xrightarrow{\text{KNH}_2} \text{PhCH}^-\text{-O-CH}_3 \xrightarrow{\text{H}_2\text{O}} \underset{\text{OH}}{\text{PhCHCH}_3} \quad (164)$$

(Wittig and Löhman, 1942)

termined the extent of rearrangement for a series of $PhCR^1R^2(CH_2)_nOCH_3$ with butyllithium; only when $R^1 = R^2 = H$ and $R^1 = H$, $R^2 = -CH_3$ are Wittig rearrangement products identified (Finnegan and Altschuld, 1967). Interestingly, ferrocenylmethyl methyl ether, the ferrocene analog of benzyl methyl ether, gives no Wittig rearrangement products when treated with butyllithium; instead, 2-metalation directed by an oxygen atom is observed

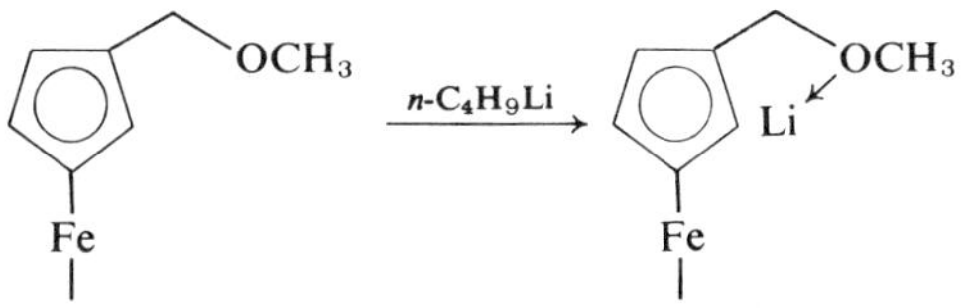

$$\text{FcCH}_2\text{OCH}_3 \xrightarrow{n\text{-C}_4\text{H}_9\text{Li}} \text{Li-FcCH}_2\text{OCH}_3 \quad (165)$$

(Slocum and Koonsvitsky, 1969)

[Eq. (165)]. Ustynyuk *et al.* (1964) have reported the Wittig rearrangement of the ferrocene compound (**25**) [Eq. (166)]. As was the case in the Stevens

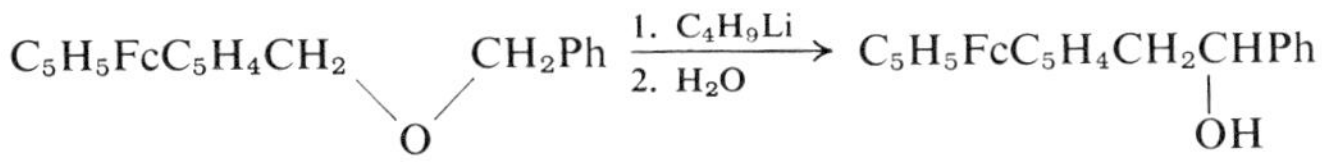

$$\underset{\text{(25)}}{\text{C}_5\text{H}_5\text{FeC}_5\text{H}_4\text{CH}_2\text{-O-CH}_2\text{Ph}} \xrightarrow[\text{2. H}_2\text{O}]{1.\ \text{C}_4\text{H}_9\text{Li}} \underset{\text{OH}}{\text{C}_5\text{H}_5\text{FeC}_5\text{H}_4\text{CH}_2\text{CHPh}} \quad (166)$$

rearrangement, a benzyl proton is ionized in preference to a ferrocenylmethyl proton, and thus the latter is the group that migrates.

In line with current thinking, three mechanisms may be envisioned for the Wittig rearrangement [Eq. (167)].

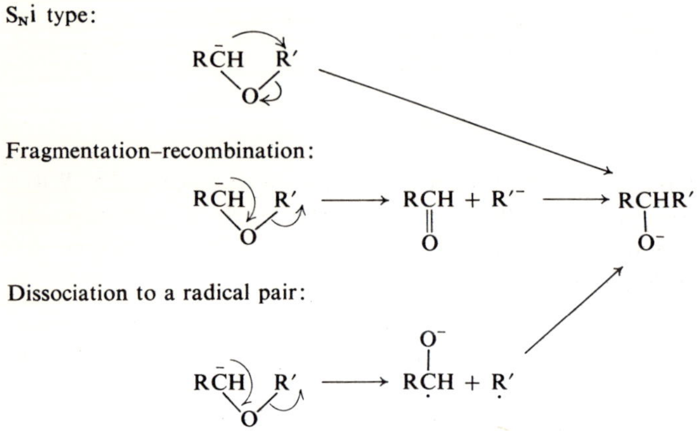

$S_N i$ type:

Fragmentation–recombination:

$$R\bar{C}H\;R' \longrightarrow RCH + R'^{-} \longrightarrow RCHR' \quad (167)$$

Dissociation to a radical pair:

Each mechanism, of course, involves the ionization of a relatively acidic hydrogen. The evidence for a choice among these mechanisms, however, is conflicting. Migratory aptitudes of R' which fall in the order allyl, benzyl > methyl, ethyl > phenyl appear to favor the $S_N i$ mechanism, since electron density localization on methyl and ethyl groups would not be likely. Observations of some racemization of an optically active R' (Schöllkopf and Schäffer, 1963) and the detection of cross-over products in the product mixture from a mixed Wittig rearrangement (Lansbury and Pattison, 1962) are indicative that the fragmentation or radical pathway may be operating (Lansbury et al., 1966). The fact that aldehydes are detected as side products of the reaction (Cast et al., 1960) also is indicative of these last two mechanisms.

d. Benzilic Acid Rearrangement. The benzilic acid rearrangement (Selman and Easthan, 1961) consists of the rearrangement in base of certain α-diketones. For example, benzil, when treated with hydroxide ion, yields benzilic acid [Eq. (168)]. When benzil is treated with methoxide or certain

$$PhCOCOPh \xrightarrow{OH^-} Ph_2CHCO_2H \quad (168)$$

other alkoxide ions, the corresponding esters are obtained. The reaction formally involves a nucleophilic attack at a carbonyl such that the adjacent carbon, the terminus, becomes relatively electron poor while the carbon attacked becomes relatively electron rich. Migration of the aryl group bonded to the attacked carbon is facilitated [Eq. (169)].

$$\underset{(26)}{\underset{\underset{O}{\|}}{\overset{Ar}{C}}-\underset{\underset{O}{\|}}{\overset{Ar}{C}}} \quad \underset{OH^-}{\rightleftharpoons} \quad \underset{(27)}{\underset{\underset{O_-}{\|}}{\overset{Ar}{C}}-\underset{\underset{O-OH}{\ }}{\overset{Ar}{C}}} \longrightarrow Ar_2\underset{\underset{OH}{\ }}{C}-CO_2^- \qquad (169)$$

Evidence for this mechanism consists of the following: (1) A second-order rate law for the rearrangement is observed (Westheimer, 1936a,b); (2) the diketone **26** undergoes ^{18}O exchange at a rate which is rapid compared to the rate of rearrangement, i.e., **27** is an intermediate and is reversibly generated (I. Roberts and Urey, 1938); (3) electron-withdrawing substituents at the meta or para positions of the aryl groups enhance the rate of rearrangement, whereas donating substituents at the same positions retard the rate (Pfeil *et al.*, 1956). This last observation suggests that the rearrangement is controlled by the migratory aptitude of the respective aryl group to delocalize a negative charge.

The benzilic ester rearrangement can be effected only by alkoxides with no α-hydrogens, otherwise abstraction of a hydride at the α-position takes place with the result that a benzoin is produced (Doering and Urban, 1956).

For unsymmetrical benzils, use of isotopically labeled carbonyl carbon has helped to identify the aryl group which migrates [Eq. (170)] (Clark *et al.*, 1955). Analysis of the ratio of the carbon label present at the carboxyl and α-carbon positions gives the ratio of the extent of migration of the two aryl groups.

$$\underset{* \,=\, \text{Carbon label}}{\text{Ph}-\overset{O}{\underset{\|}{C^*}}-\overset{O}{\underset{\|}{C}}-\text{Ar}(X)} \xrightarrow{OH^-} \underset{X}{\text{Ph}\overset{OH}{\underset{|}{C^*}}-CO_2H} + \underset{X}{\text{Ph}\overset{OH}{\underset{|}{C}}-\overset{*}{C}O_2H} \qquad (170)$$

An interesting variation of this rearrangement occurs when benzils are treated with Grignard reagents [Eq. (171)].

$$\text{PhC}-\overset{O}{\underset{\|}{C}}\text{Ph} \xrightarrow{RMgX} \text{Ph}-\underset{\underset{R}{|}}{\overset{OMgX}{C}}-\underset{\underset{O}{\|}}{C}\text{Ph} \xrightarrow{H_2O} R\overset{O}{\underset{\|}{C}}-\underset{\underset{\ }{|}}{\overset{OH}{C}}\text{Ph}_2 \qquad (171)$$

(Selman and Easthan, 1961)

e. *Grovenstein–Zimmermann Rearrangement.* It is often the case that a new reaction will be discovered virtually simultaneously by two different researchers. This is the case of the reaction which involves the migration of an aryl group from a highly arylated carbon atom to a vicinal carbanion

$$Ph_3CCH_2Cl + Na \longrightarrow Ph_3CCH_2Na \longrightarrow Ph_2\overset{\underset{\mid}{Na}}{C}CH_2Ph \qquad (172)$$

(Grovenstein, 1957; Zimmerman and Smentowski, 1957)

[Eq. (172)]. Probably the driving force of this reaction is the enhanced stability of the product anion compared to the initial anion, i.e., reasoning completely analogous to that used in the rationalization of the stability of carbonium ions.

The identity of the alkali metal ion is evidently of importance. In Eq. (172), if the counter ion is lithium, an interesting phenomenon is observed. Halogen metal exchange at $-30°$ to $-65°C$ and treatment with CO_2 or ROH produces only unrearranged product. However, if this same procedure is carried out at $0°C$ only rearranged product is isolated. Further study has demonstrated that the order of extent of rearrangement with respect to the alkali metal ion is RK ~ RNa > RLi > R_2Mg. In fact, a dialkylmagnesium compound has been found which gives no rearrangement whatsoever under any of these conditions.

For the most part, the reaction has been found to be intramolecular with one curious exception. Grovenstein and Wentworth (1963) found that 2,2,3-triphenylpropyllithium at $0°C$ or above and in the presence of labeled benzyllithium rearranges to an intermediate which on carbonation produces two acids, each with a portion of the label in percentages close to that calculated for a completely intermolecular reaction [Eq. (173)].

$$PhCH_2\overset{\underset{\mid}{Ph}}{\underset{\mid}{C}}\text{—}CH_2Li + Ph\overset{*}{C}H_2Li \xrightarrow[0°C]{THF} \left[Ph_2C{=}CH_2 + \begin{cases} Ph\overset{*}{C}H_2Li \\ PhCH_2Li \end{cases} \right]$$

$$\xrightarrow[\text{2. }H_2O]{\text{1. }CO_2} Ph_2\underset{\underset{\mid}{CO_2H}}{C}CH_2\overset{*}{C}H_2Ph + Ph\overset{*}{C}H_2CO_2H \qquad (173)$$

$$\text{50\% label}\text{50\% label}$$

* = Radioactive carbon label

A further study has been defined to differentiate, for the intramolecular migration, whether the mechanism involves an elimination–readdition step or a concerted bond-making–bond-breaking process reminiscent of the migration intermediates of aryl cations (phenonium ion intermediate), i.e., whether a phenyl dissociates to an sp^2 carbanion (**28**) or an sp^3 intermediate with the negative charge delocalized throughout the remainder of the aryl

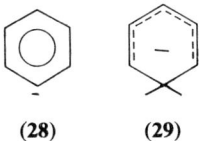

(**28**) (**29**)

ring (**29**). The question has been answered by Grovenstein and Wentworth (1967) in their study of the rearrangement of **30** [Eq. (174)]. Analysis of the

$$Ph-\underset{\underset{(30)}{\overset{\overset{Ph}{|}}{Ph}}}{\overset{\overset{Ph}{|}}{C}}-CH_2Li \xrightarrow{0°C} Ph-\underset{\underset{Li}{|}}{\overset{\overset{Ph}{|}}{C}}-CH_2-\overset{}{\underset{}{}}-Ph \longrightarrow product \quad (174)$$

product showed that the *p*-biphenylyl group migrates preferentially by a factor of at least 50:1. This favors the delocalized intermediate **29** since the phenyl group in the para position would assist in the delocalization of negative charge. In fact, if the other intermediate were involved, a slight preference of *m*-biphenylyl migration would be expected since inductive withdrawal by phenyl would be the only effect operating.

2. *Rearrangements via Three-Membered Ring Intermediates with Subsequent 1,3-Elimination*

a. *The Favorskii Rearrangement.* In 1894, Favorskii discovered the skeletal rearrangement that allows conversion of an α-halo ketone to an ester with alkoxide or to an acid with hydroxide (Kende, 1960) [Eq. (175)]. Two distinct

$$(CH_3)_2CBr\overset{O}{\overset{\|}{C}}CH_3 + OR^- \longrightarrow (CH_3)_3CCO_2R \quad (175)$$
R = H, alkyl

mechanisms obtain: (1) that involving a cyclopropanone intermediate (symmetrical mechanism) and (2) a "semibenzilic acid" mechanism. Ketones with an α-halogen and an α'-hydrogen can undergo rearrangement via an intermediate formed by 1,3-dehydrohalogenation. This is the normal Favorskii rearrangement. α-Halo ketones which do not possess an α'-hydrogen also undergo the rearrangement. Obviously, a 1,3-elimination cannot occur, and a second pathway, that analogous to the benzilic acid rearrangement, has been proposed. Considerable effort has been expended to delineate the normal Favorskii mechanism, whereas the semibenzilic acid process may not have received the attention it deserves.

Labeling experiments by Loftfield (1951) lend credence to the cyclopropanone intermediate idea. The initial label was located at ring carbons 1 and 2 in 2-chlorocyclohexanone as shown in Eq. (176). Reaction via a closing and a random reopening of a three-membered ring intermediate breaks down to

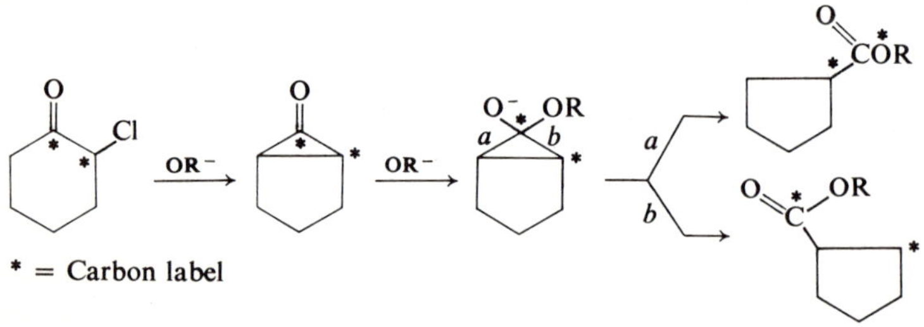

$$(176)$$

* = Carbon label

half the label being located at the carbonyl carbon and half distributed equally between carbons 1 and 2 of the five-membered ring of the product. This prediction has been verified experimentally. This experiment, perhaps more than any other, served to demonstrate the symmetrical nature of the intermediate, thereby documenting the cyclopropanone intermediate concept.

Additional evidence for a cyclopropanone intermediate has been gained from a trapping experiment performed by Fort (1962). A Diels–Alder adduct (**32**) was isolated when the rearrangement of **31** with base was run in the presence of furan [Eq. (177)]. A second interception of a cyclopropanone

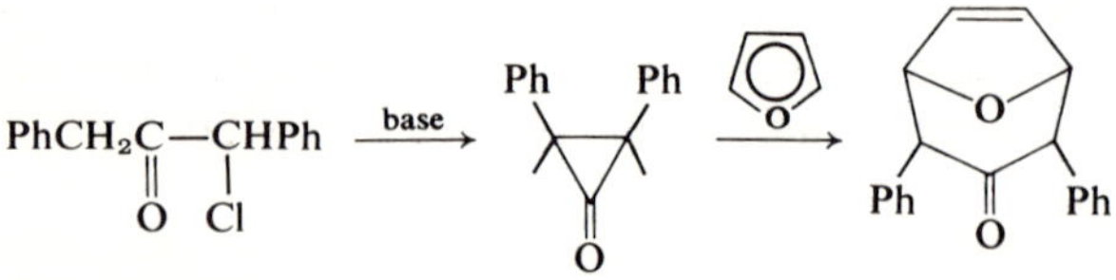

$$(177)$$

(**31**)   (**32**)

intermediate, an example which provides retention of the three-membered ring in an isolable product, has been realized by Breslow et al. (1963). Thus, treatment of either of the diastereoisomers of α,α'-dibromodibenzyl ketone

(33) or a mixture thereof with trimethylamine in methylene chloride has afforded diphenylcyclopropenone [Eq. (178)].

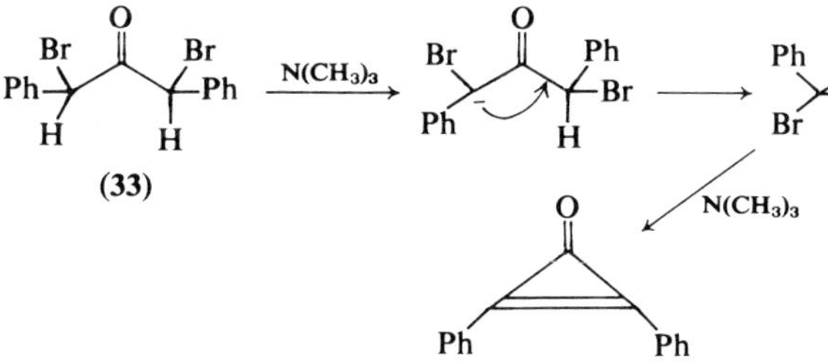

$$\text{(33)} \xrightarrow{\text{N(CH}_3)_3} \cdots \xrightarrow{} \cdots \xrightarrow{\text{N(CH}_3)_3} \text{diphenylcyclopropenone} \quad (178)$$

In a series of elegant studies conducted by Rappe and Knutsson at Uppsala and Turro at Columbia, the reactivity of various cyclopropanones has been demonstrated to be consistent with their intervention as intermediates in the Favorskii reaction. Cyclopropanone 34 or its hemiketal 35 yields, with various bases, the same ratio of ester products as that from the Favorskii rearrangement of α-halo ketones which would give cyclopropanone 34 as an intermediate (Rappe *et al.*, 1970) [Eq. (179)]. For the most part, it was shown that the direction of ring opening could be predicted on the basis of the more stable carbanion being formed, i.e., ester 36 was predominate in the product mixture. However, the percentage of ester 37 is increased if the cyclopropanone intermediate possesses substituent groups larger than methyl. Steric compression in the transition state thus causes a shift toward a cleavage with formation of the less stable carbanion.

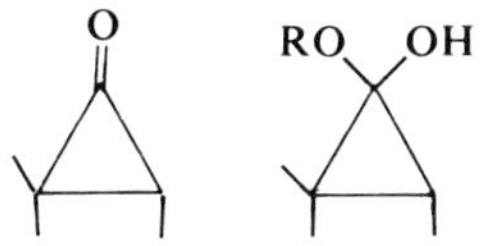

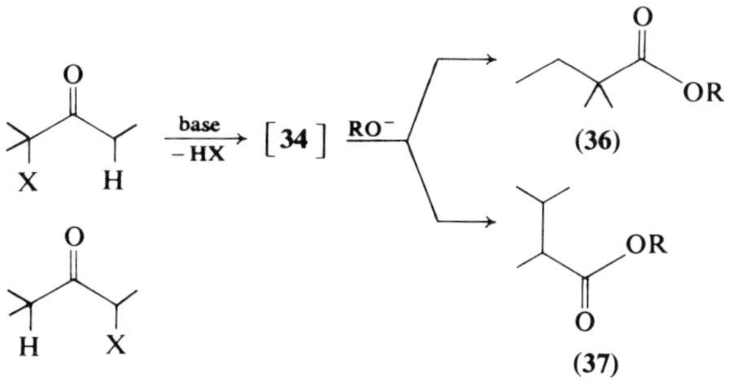

$$\text{α-halo ketones} \xrightarrow[-\text{HX}]{\text{base}} [34] \xrightarrow{\text{RO}^-} \text{(36)} + \text{(37)} \quad (179)$$

Some interesting variations of the Favorskii rearrangement have been discovered. The reaction may be carried out on α,β-epoxy ketones and on α,α-dihalo ketones [Eqs. (180) and (181)]. The latter reaction is stereoselective and gives the *cis*-olefin.

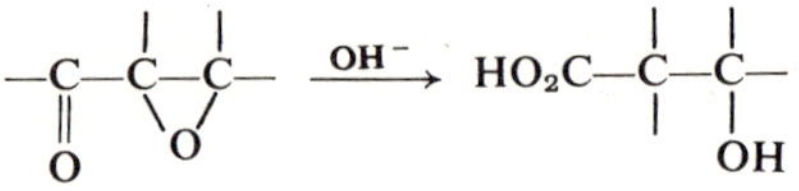

$$-\overset{\underset{\parallel}{O}}{C}-\overset{|}{\underset{\underset{O}{\diagdown\diagup}}{C}}-\overset{|}{C}- \xrightarrow{OH^-} HO_2C-\overset{|}{\underset{|}{C}}-\overset{|}{\underset{OH}{C}}- \quad (180)$$

(House and Gilmore, 1961)

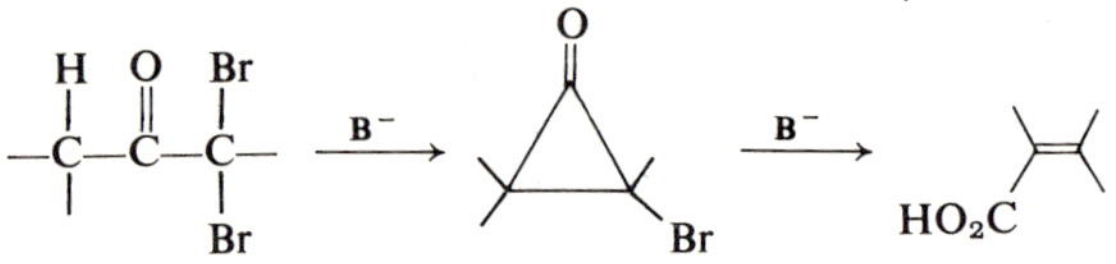

$$-\overset{\underset{|}{H}}{\underset{|}{C}}-\overset{\underset{\parallel}{O}}{\underset{|}{C}}-\overset{\underset{|}{Br}}{\underset{|}{C}}- \xrightarrow{B^-} \underset{Br}{\triangle}^{O} \xrightarrow{B^-} \underset{HO_2C}{\diagup}\!\!=\!\!\diagdown \quad (181)$$

(Kennedy et al., 1964)

Even for α-halo ketones with no α-hydrogens present in the molecule, the Favorskii rearrangement can still be observed to take place. For example, compound **38** undergoes the rearrangement [Eq. (182)]. Esterification of product **39** with ethanol and formation of the hydrochloride produces Demerol (meperidine, a strong analgesic). Treatment of α-bromocyclo-

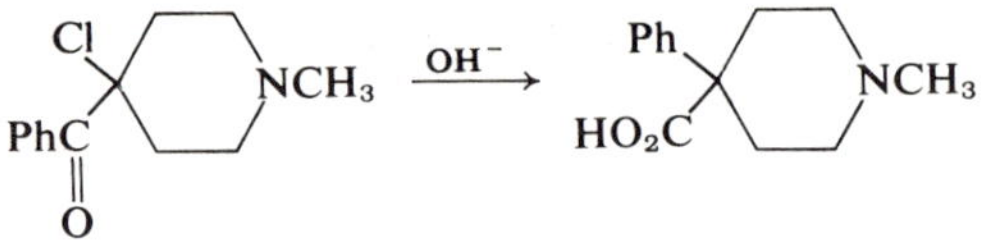

$$\underset{\text{PhC}\underset{\parallel}{}\text{O}}{\overset{Cl}{\diagdown}}\!\!\bigcirc\!\!\text{NCH}_3 \xrightarrow{OH^-} \underset{HO_2C}{\overset{Ph}{\diagdown}}\!\!\bigcirc\!\!\text{NCH}_3 \quad (182)$$

(38)   (39)

(Smissman and Hite, 1959)

butanone with deuteroxide ion gave cyclopropane carboxylic acid, but without incorporation of deuterium at C-2. If the cyclopropanone mechanism were operating, the ring should have opened to form a carbanion which, in turn, would have picked up a deuteron from solvent [Eq. (183)].

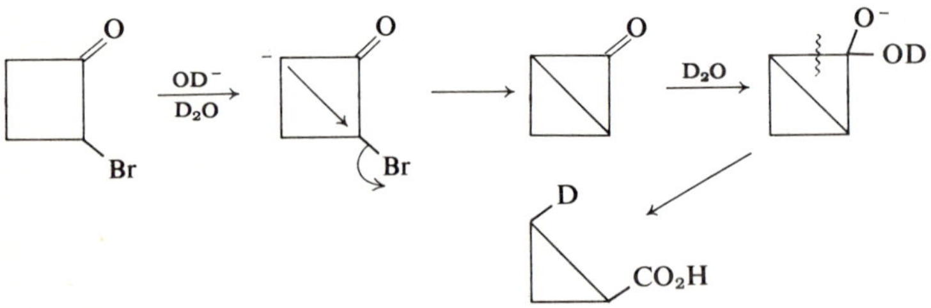

$$\underset{Br}{\square}^{O} \xrightarrow[D_2O]{OD^-} \underset{Br}{\square}^{O} \longrightarrow \underset{D}{\square}^{O} \xrightarrow{D_2O} \underset{}{\square}^{O^-}\!\!-OD \quad (183)$$
$$\searrow \underset{CO_2H}{\triangle}_D$$

Each of these anomalous Favorskii rearrangements have been postulated to proceed via the semibenzilic acid mechanism. This is illustrated for α-bromocyclobutanone [Eq. (184)]. Note that here, although there are

5. CARBANIONS

α-hydrogens present, the benzilic acid mechanism is apparently operating. This mechanism rationalizes the lack of deuterium incorporation and can be attributed to angle strain in the bicyclobutanone intermediate, thereby reversing the normal order of transition states.

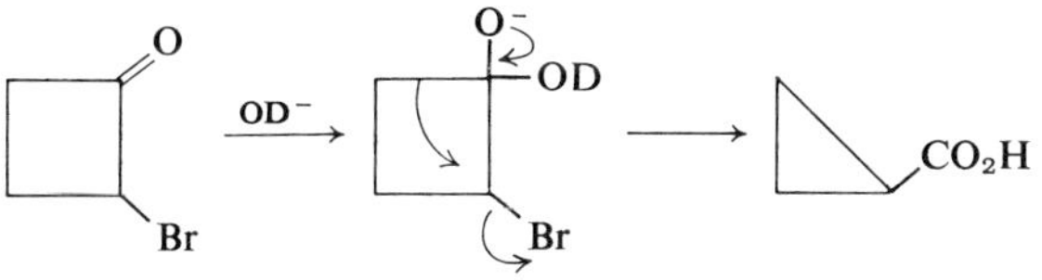

$$\text{cyclobutanone-Br} \xrightarrow{OD^-} \text{intermediate} \longrightarrow \text{cyclopropane-}CO_2H \quad (184)$$

A system in which there is a delicate balance between the two mechanisms has been investigated (Warnhoff et al., 1968). Using the criterion of deuteron incorporation, a decision between the two pathways can be made. For $n = 3$

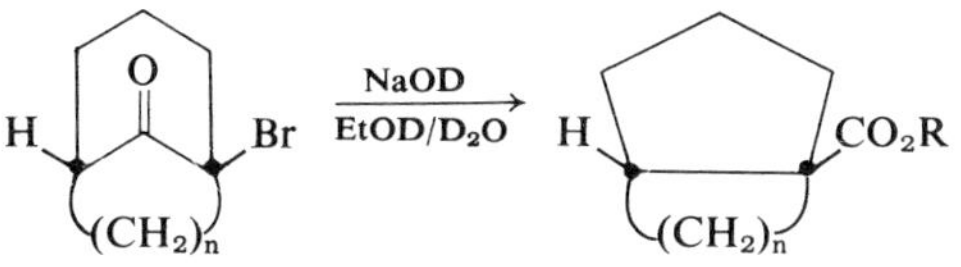

$$\text{bicyclic bromoketone} \xrightarrow[\text{EtOD/D}_2\text{O}]{\text{NaOD}} \text{cyclopentane-}CO_2R \quad (185)$$

and 4 [Eq. (185)], no deuterium is incorporated, and it has been concluded that the semibenzilic acid mechanism is operating. For $n = 5$, 0.90 atom deuterium is incorporated; thus the symmetrical mechanism is extant. If the base is changed to a relatively unusual base for the Favorskii rearrangement, e.g., $AgNO_3$, the rearrangement takes place but with no deuterium incorporation. Although in this study there were no examples of a system where a Favorskii rearrangement occurs by a mixture of the two mechanisms, i.e., 0.5 deuterium incorporation, the authors suggest this possibility. Further study may indeed show that some of the puzzling aspects of this rearrangement may be explained by consideration of a combination of the two mechanisms.

b. The Ramberg–Bäcklund Reaction. This reaction was discovered by the bearers of its name in 1940 and constitutes the 1,3-elimination with rearrangement of α-halo sulfones [Eq. (186)] (Ramberg and Bäcklund, 1940). The reaction very recently has been extensively reviewed (Scott and Meyers, 1972).

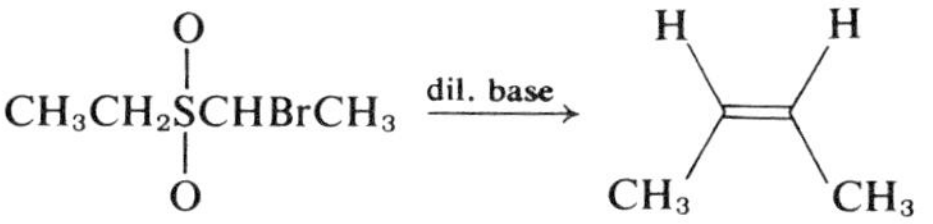

$$CH_3CH_2\overset{\overset{O}{\|}}{\underset{\underset{O}{\|}}{S}}CHBrCH_3 \xrightarrow{\text{dil. base}} \overset{H}{\underset{CH_3}{\diagdown}}C=C\overset{H}{\underset{CH_3}{\diagup}} \quad (186)$$

(Ramberg and Bäcklund, 1940)

It was natural, with the known structure of episulfides and from the nature of the products of the reaction itself, that an episulfone be proposed as an intermediate. The observation that reaction was first order in hydroxide ion as well as first order in sulfone led Bordwell and Cooper (1951) to propose a rapid reversible ionization of an α'-hydrogen followed by a rate-determining nucleophilic displacement of the α-halogen to form the episulfone [Eq. (187)].

$$RCH_2\underset{SO_2}{\diagdown\diagup}CHR\underset{X}{|} + OH^- \rightleftharpoons R\bar{C}H\underset{SO_2}{\diagdown\diagup}CX\underset{H}{\overset{R}{|}} \xrightarrow[slow]{-X^-}$$

$$RCH-CHR\underset{SO_2}{\diagdown\diagup} \xrightarrow{-SO_2} \underset{R}{\overset{H}{\diagdown}}=\underset{R}{\overset{H}{\diagup}} \qquad (187)$$

Reexamination of the reaction later showed that it was not stereospecific as had been thought, but rather a 4:1 mixture of *cis*- to *trans*-2-butene was produced (Neureiter, 1966). In the original paper, Bordwell and Cooper had also found that α-chlorobenzyl benzyl sulfone under similar conditions produced essentially *trans*-stilbene.

Much of the subsequent work on this reaction has involved examination of the episulfone intermediate which can be prepared separately. Surprisingly, an episulfone has never been isolated from Ramberg–Bäcklund reaction conditions.

Exchange with D_2O during the course of the reaction results in deuteration at the 2- and 3-positions [Eq. (188)]. Since decomposition of the *cis*-episulfone

$$CH_3CH_2SO_2CHClCH_3 + OD^- \xrightarrow{D_2O} \underset{D}{\overset{CH_3}{\diagdown}}=\underset{D}{\overset{CH_3}{\diagup}} + \underset{D}{\overset{CH_3}{\diagdown}}=\underset{CH_3}{\overset{D}{\diagup}} \qquad (188)$$
$$\qquad\qquad\qquad\qquad\qquad\qquad 80\% \qquad\qquad 20\%$$

under identical conditions gives no deuteron incorporation, exchange must take place in the acyclic sulfone. Treatment of this acyclic sulfone or of the corresponding *cis*- or *trans*-episulfone with potassium *tert*-butoxide in *tert*-butyl alcohol O–D in each case affords the dideuterated mixture of olefins but containing only 20% of the cis isomer. To explain this it was argued that exchange of the episulfones themselves takes place under these conditions and that as a result equilibration of the episulfone intermediates occurs. Under these relatively strong basic conditions there is greater equilibration toward the more stable *trans*-episulfone, and, hence, more *trans*-2-butene is observed. The fact that essentially completely *trans*-stilbene is produced from α-chlorobenzyl benzyl sulfone can be accounted for by the greater acidity of the α-hydrogens in this sulfone allowing greater equilibration of the episulfone intermediates to be realized.

Paquette and Wittenbrook (1968) have provided a conformational explanation for the formation of *cis*-episulfone in many of these reactions. Because of dipolar interactions, the lowest energy conformer is assumed to be **40** with the sulfonyl oxygens and α-chlorine atom trans to one another. Other arguments account for the greater kinetic acidity of the hydrogen syn to the sulfonyl oxygens in the conformation where the R groups are trans staggered [Eq. (189)]. The α-sulfonyl carbanion does not retain its configuration and

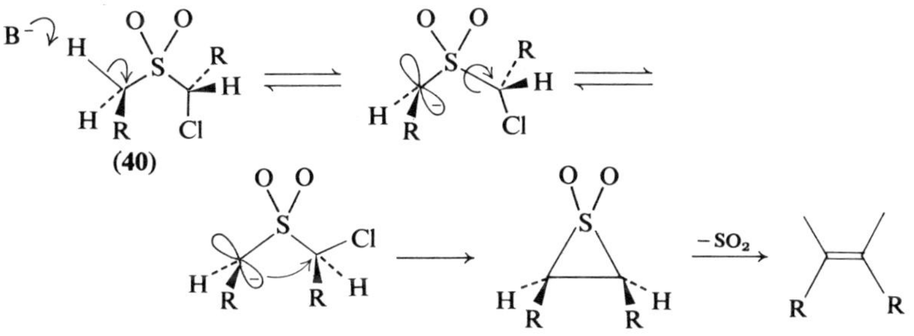

$$\text{(189)}$$

can only react further via its invertomer as shown. Although this view is consistent with the observation that the predominance of *cis*-episulfone falls off somewhat with increasing bulk of R, it is inconsistent with the demonstrated configurational stability of α-sulfonyl carbanions (Corey and Lowry, 1965).

An elegant demonstration of a likely synchronous double inversion in the formation of the episulfone intermediate has been conceived (Bordwell *et al.*, 1968). The diastereomeric bis(α-bromobenzyl) sulfones were each subjected to a Ramberg–Bäcklund reaction and a 1,3-dehalogenation. The results are illustrated for the *dl* isomer [Eqs. (190) and (191)]. The most conceivable course for these transformations is that inversions are effected at each carbon

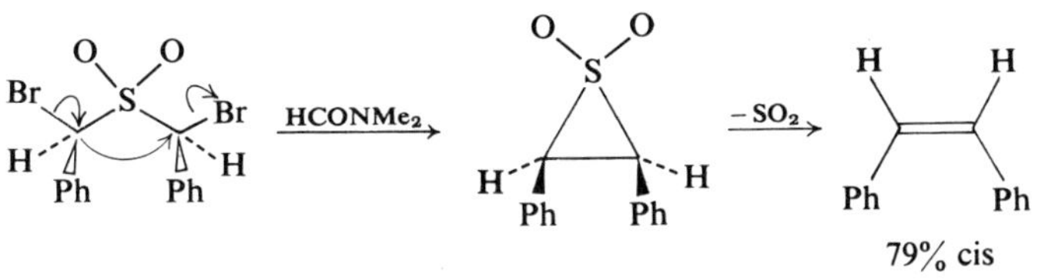

$$\text{(190)}$$

79% cis

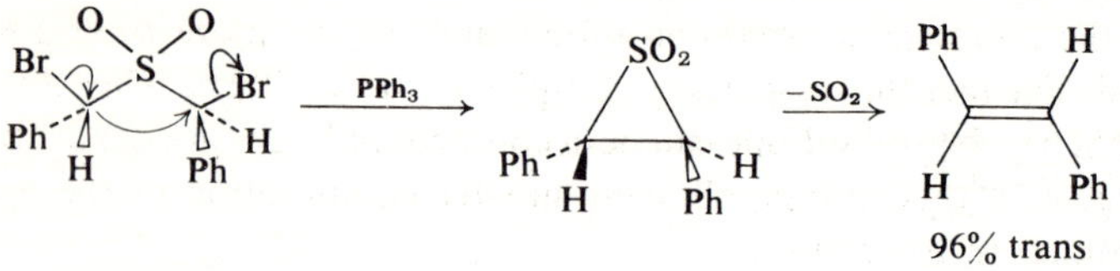

$$\text{(191)}$$

atom. An interesting example in a [3.3.1] bicyclic system must proceed with two inversions [Eq. (192)]. Under Ramberg–Bäcklund conditions, rearrange-

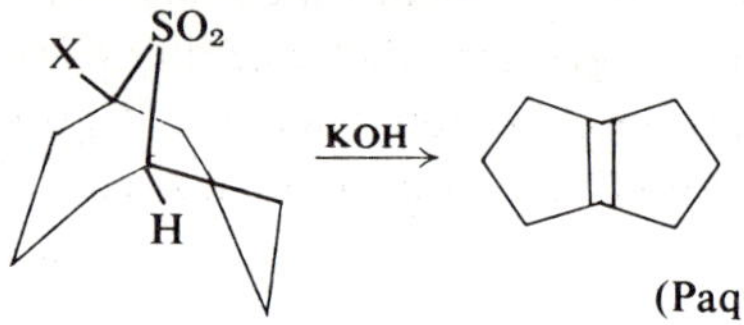

$$\text{(192)}$$

(Paquette and Houser, 1966)

ment of α,α-dihalo and α,α′-dihalo sulfones evidently proceed via an intermediate thiirene-1,1-dioxide. Paquette and Wittenbrook (1967) have examined the behavior of α,α-dichloro-*p*-methylbenzyl benzyl sulfone and similar compounds in aqueous base [Eq. (193)]. The presence of the acetylene product

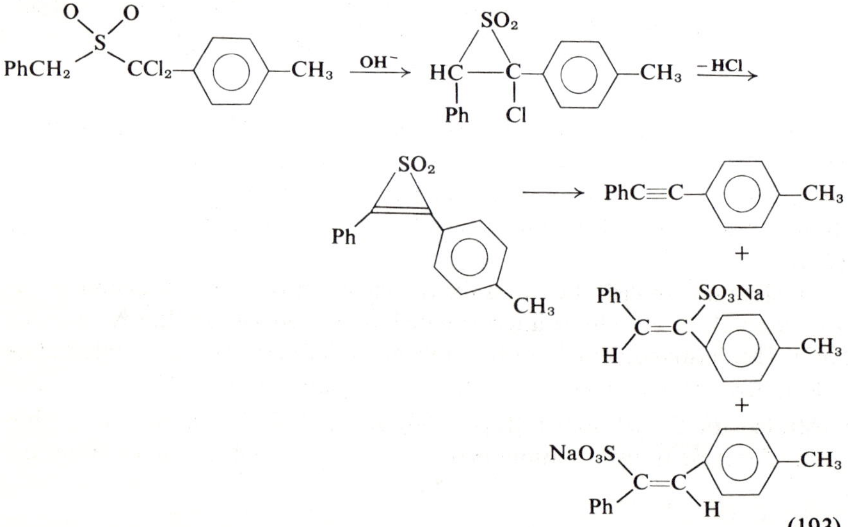

$$\text{(193)}$$

has been interpreted as strongly indicative of a thiirene intermediate.

C. Y. Meyers and co-workers (1969a) have discovered an interesting one-step Ramberg–Bäcklund reaction starting with the sulfone itself. Ionic chlorination using carbon tetrachloride in the presence of strong base allows the generation *in situ* of an α-chloro sulfone which undergoes Ramberg–Bäcklund rearrangement as soon as it is formed. The reaction is illustrated by the 100% conversion of dibenzyl sulfone to *trans*-stilbene. A second chlorination is thought to occur at a rate much slower than that of the rearrangement.

$$(PhCH_2)_2SO_2 \xrightarrow[\textit{tert-BuOH, KOH}]{CCl_4} PhCH_2SO_2CHClPh \xrightarrow[-SO_2]{-HCl} \begin{array}{c} H \\ \diagup \\ Ph \end{array} = \begin{array}{c} Ph \\ \diagdown \\ H \end{array} \qquad (194)$$

c. The Neber Rearrangement. This rearrangement consists of the treatment with base of a ketoxime or aldoxime tosylate (other leaving groups instead of tosylate may be used) which has at least one α-hydrogen to produce an α-amino ketone or aldehyde (Neber and Friedolsheim, 1926). The same ketone is obtained irrespective of which geometrical isomer of the oxime is used as starting material [Eq. (195)]. Of course, if one α-position possesses no

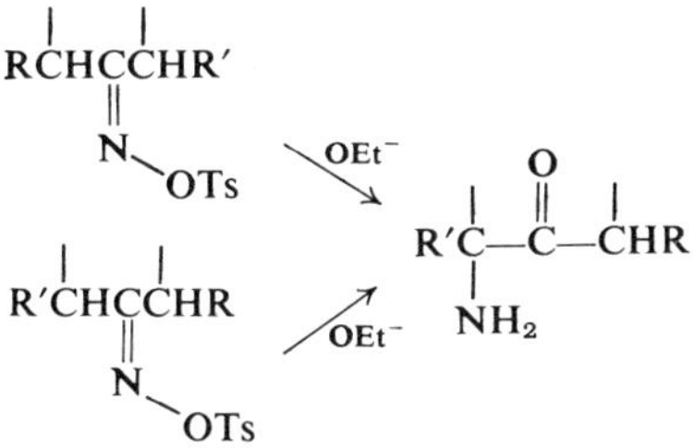

$$\begin{array}{c} RCHCCHR' \\ \parallel \\ N_{\diagdown OTs} \end{array} \xrightarrow[OEt^-]{OEt^-} \begin{array}{c} O \\ \parallel \\ R'C-C-CHR \\ | \\ NH_2 \end{array} \qquad (195)$$
$$\begin{array}{c} R'CHCCHR \\ \parallel \\ N_{\diagdown OTs} \end{array}$$

hydrogens, such as the oxime of an aldehyde or aryl ketone, only one direction for rearrangement is possible. An exhaustive review is available (O'Brien, 1964).

The reaction can most easily be formulated as an equilibrium removal of the most acidic α-proton followed by nucleophilic displacement of tosylate by the carbanion nucleophile [Eq. (196)]. However, this mechanism cannot be

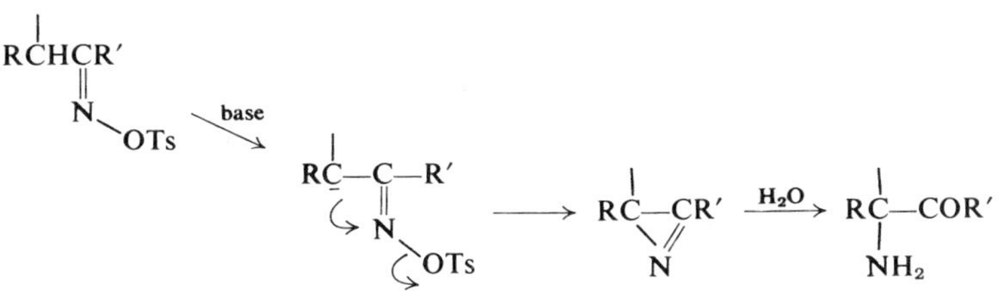

$$\begin{array}{c} RCHCR' \\ \parallel \\ N_{\diagdown OTs} \end{array} \xrightarrow{base} \begin{array}{c} RC-C-R' \\ \parallel \\ N \\ \diagdown OTs \end{array} \longrightarrow \begin{array}{c} RC-CR' \\ \diagdown \diagup \\ N \end{array} \xrightarrow{H_2O} \begin{array}{c} RC-COR' \\ | \\ NH_2 \end{array} \qquad (196)$$

totally correct since the geometric isomer of this ketoxime tosylate will also produce the same product. Such a transformation results in formation of a three-membered azirine ring intermediate which on hydrolysis yields the α-amino ketone product.

Interception or isolation of azirine intermediates from this rearrangement has been achieved. Cram and Hatch (1953) have found that reduction of the reaction mixture allows isolation of the cyclic imine (**41**) [Eq. (197)].

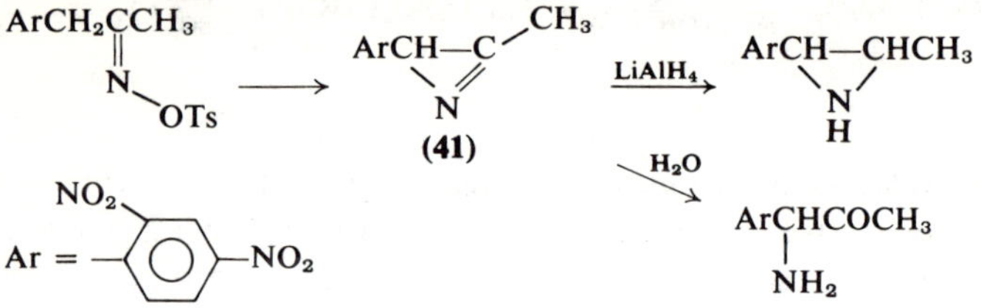

$$\underset{\underset{OTs}{N}}{ArCH_2\overset{\|}{C}CH_3} \longrightarrow \underset{(\mathbf{41})}{ArCH\overset{CH_3}{\underset{N}{-C}}} \xrightarrow{LiAlH_4} \underset{\underset{H}{N}}{ArCH-CHCH_3} \quad (197)$$

$$\xrightarrow{H_2O} \underset{NH_2}{ArCHCOCH_3}$$

$$Ar = -\underset{NO_2}{\overset{NO_2}{\bigcirc}}-NO_2$$

The same starting material gives the α-amino ketone when hydrolyzed in the usual manner. In the case of the rearrangement of the hydrazone methiodide (**42**), an intermediate (**44**) as well as the more usual azirine (**43**) has been isolated [Eq. (198)].

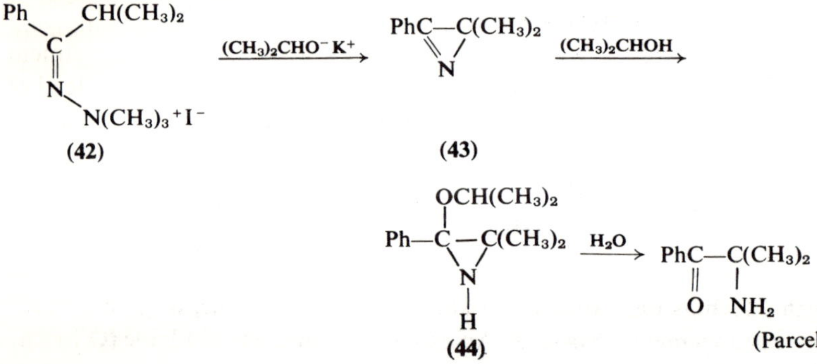

$$\underset{(\mathbf{42})}{\underset{\underset{N(CH_3)_3{}^+I^-}{N}}{\overset{CH(CH_3)_2}{\underset{\|}{Ph\diagdown C\diagup}}}} \xrightarrow{(CH_3)_2CHO^- K^+} \underset{(\mathbf{43})}{\underset{N}{PhC-C(CH_3)_2}} \xrightarrow{(CH_3)_2CHOH}$$

$$\underset{(\mathbf{44})}{\underset{\underset{H}{N}}{Ph-\overset{OCH(CH_3)_2}{\underset{|}{C}}-C(CH_3)_2}} \xrightarrow{H_2O} \underset{O\ \ NH_2}{PhC-C(CH_3)_2} \quad (198)$$

(Parcell, 1963)

Some relatively unusual examples of the Neber rearrangement are shown in Eqs. (199)–(202).

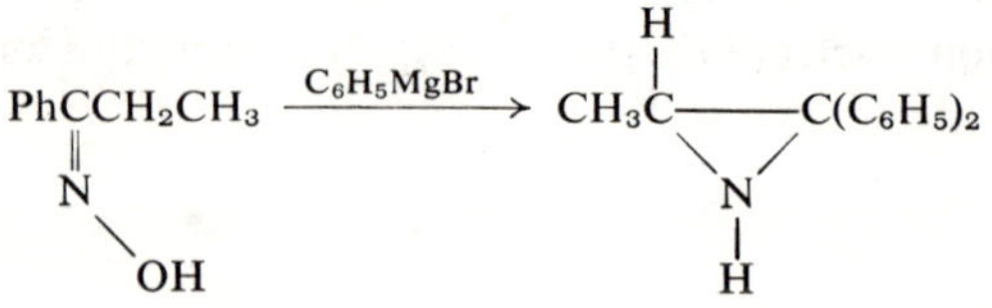

$$\underset{\underset{OH}{\overset{\|}{N}}}{PhCCH_2CH_3} \xrightarrow{C_6H_5MgBr} \underset{\underset{H}{N}}{CH_3\overset{H}{\underset{|}{C}}\!\!-\!\!\!-\!\!C(C_6H_5)_2} \quad (199)$$

(K. N. Campbell et al., 1943)

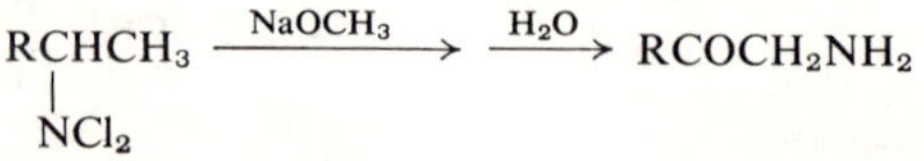

$$\underset{NCl_2}{RCHCH_3} \xrightarrow{NaOCH_3} \xrightarrow{H_2O} RCOCH_2NH_2 \quad (200)$$

(Baumgarten et al., 1960)

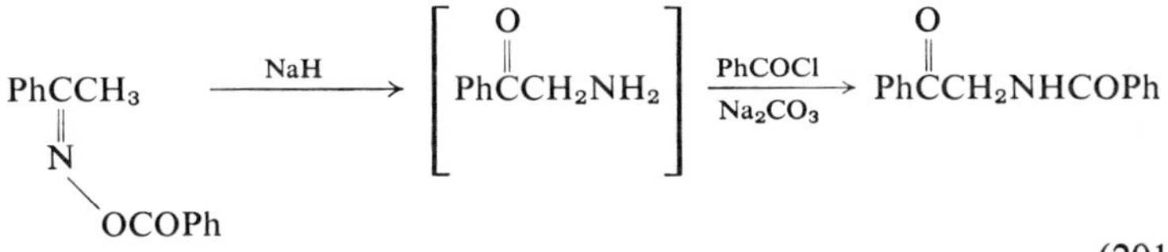

$$\underset{\underset{OCOPh}{\overset{\|}{N}}}{PhCCH_3} \xrightarrow{NaH} \left[Ph\overset{O}{\overset{\|}{C}}CH_2NH_2 \right] \xrightarrow[Na_2CO_3]{PhCOCl} Ph\overset{O}{\overset{\|}{C}}CH_2NHCOPh$$

(201)

(Renfrow et al., 1968)

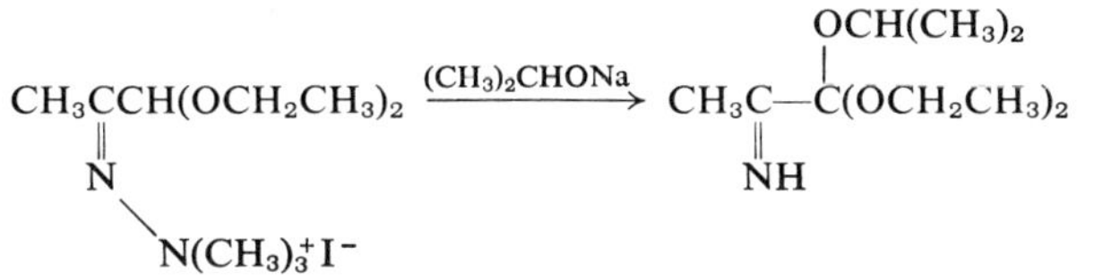

$$\underset{\underset{N(CH_3)_3^+ I^-}{\overset{\|}{N}}}{CH_3CCH(OCH_2CH_3)_2} \xrightarrow{(CH_3)_2CHONa} \underset{NH}{\overset{OCH(CH_3)_2}{\underset{\|}{CH_3C-C(OCH_2CH_3)_2}}}$$

(202)

(Henery-Logan and Fridinger, 1967)

Kinetics and stereochemistry of this rearrangement apparently have not been sufficiently examined. It is also possible that a nitrene intermediate may intervene during the course of this reaction.

D. Metalation of Aromatic Compounds

Although chemistry involving organic anions has been practiced for a number of years (Gilman and Morton, 1954), only recently has significant inquiry been made into the structure and reactions of organoalkali reagents with various organic molecules (Mallan and Bebb, 1969). The reaction of organoalkali reagents with the C–H bond, i.e., metalation, is one such area where some initial discoveries were allowed to lie fallow for a number of years. Utilization of this reaction for the preparation of aromatic compounds complements electrophilic substitution as a method for the introduction of new substituents into an aromatic ring.

1. General Metalation Methods

The metalation of benzene with butyllithium cannot be accomplished without a catalyst. Use of tetramethylethylenediamine (TMEDA) or triethylenediamine (DABCO) to form essentially a TMEDA–butyllithium or DABCO–butyllithium complex produces a reagent which does metalate benzene (Broaddus, 1970). Likewise, use of potassium *tert*-butoxide in conjunction with butyllithium provides metalation of benzene (Schlosser, 1967). Alkyl-substituted benzenes have also been metalated. Metalation of toluene by butylsodium proceeds according to Eq. (203) (Broaddus, 1966). The reaction is kinetically controlled since reaction is rapid under conditions where

$$CH_3CH_2CH_2CH_2Na + C_6H_5CH_3 \longrightarrow C_6H_5CH_2Na + CH_3CH_2CH_2CH_3$$

(203)

p-tolylsodium is only slowly converted to benzylsodium. Treatment of toluene with excess butyllithium–TMEDA complex produces a mixture of polylithiated toluenes [Eq. (131)] (West and Jones, 1968).

The question of the site of metalation of ethylbenzene and cumene (isopropylbenzene) with amylsodium and potassium reagents has been settled (Benkeser *et al.*, 1962, 1963). Cumene is initially metalated in the ring, but a rearrangement of the site of the alkali species to the α-position gradually occurs [Eq. (204)]. Thus, the ring position represents the kinetic site and

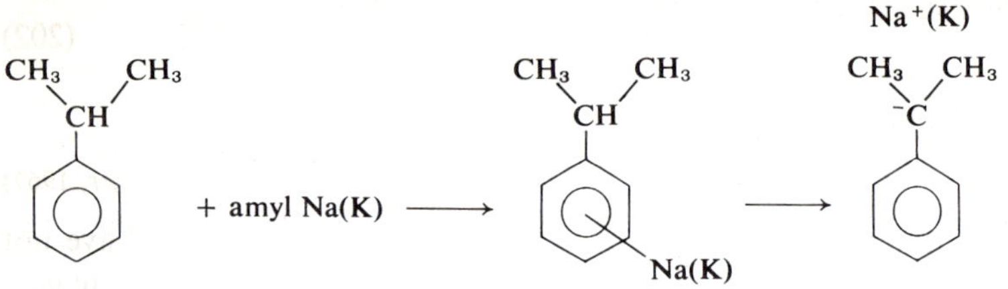

$$\text{Cumene} + \text{amyl Na(K)} \longrightarrow \text{ring-metalated} \longrightarrow \alpha\text{-metalated} \quad (204)$$

α-position the thermodynamic site of metalation. This work served to clear up a number of discrepancies in the literature.

Metalation of other aromatic compounds, notably thiophene, furan, and ferrocene, have also proved fruitful. Both thiophene and furan have been metalated with butyllithium at the 2- (or α-) position, the more reactive position toward both conventional electrophilic substitution and electrophilic substitution involving metalation. A variety of substituents have been introduced at this position via the lithio intermediate **45** [Eq. (205)]. Again, since the

$$\underset{X}{\bigcirc} \xrightarrow{n\text{-BuLi}} \underset{X}{\bigcirc}\text{Li} \xrightarrow{\text{reagent}} \underset{X}{\bigcirc}\text{R} \qquad X = S, O \qquad (205)$$
$$(\mathbf{45})$$

2-position is quite reactive toward electrophilic substitution, the metalation procedure serves to supplement the production of 2-substituted thiophenes and furans.

For the preparation of various substituted metallocenes, metalation has proved to be a much more versatile method than electrophilic substitution. Ferrocene is sufficiently reactive that some care must be exercised that dimetalation does not supercede the production of the monometalated species. The best procedure to use is that of Goldberg *et al.* (1963) where the solvent (ether), the time (5 to 6 hr), and the temperature (25°C), serve to produce a 25% yield of the monometalated ferrocene (Fc) [Eq. (206)]. Since most of the

$$\text{FcH} \xrightarrow{n\text{-BuLi}} \text{FcLi} \xrightarrow{\text{reagent}} \text{FcR} \qquad (206)$$

ferrocene can be recovered, this technique provides a convenient route to a variety of monosubstituted ferrocenes that cannot be prepared by other methods. Previous metalation routes had produced a mixture of mono- and dimetalated ferrocene and, hence, a mixture of the corresponding products which then had to be separated.

Table VII records a number of the monosubstituted ferrocenes that have been synthesized. A greater number of monosubstituted aromatics have been prepared in the ferrocene series using this technique than in any other (Slocum *et al.*, 1969). Of particular interest is the direct synthesis of amino- and nitroferrocene via lithioferrocene, both compounds being inaccessible by more conventional electrophilic substitution techniques.

Table VII
Representative Monosubstituted Ferrocenes Formed by the Metalation of Ferrocene[a]

X in Fc—X	Reactant	Yield (%)
—SiMe$_3$	X—Cl	—
—SiHEt$_2$	X—Cl	—
—SiPh$_3$	X—Cl	27, 50–54
—Si(n-C$_6$H$_{13}$)$_3$	X—Br	—
—NH$_2$	X—OCH$_3$	8
	X—OCH$_2$Ph	25
—NO$_2$	CH$_3$CH$_2$CH$_2$O—X	—
	N$_2$O$_4$	2
—B(OH)$_2$	(n-BuO)$_3$B	—
—COFc	FcCN	80
	FcCO$_2$CH$_3$	—
—N$_3$	p-CH$_3$C$_6$H$_4$SO$_2$N$_3$	—
—α-Pyridyl	Pyridine	32
—CH$_2$CHOHCH$_2$Cl	Epichlorohydrin	—
—C(OH)(CH$_2$)$_3$CH$_2$	Cyclopentanone	—
—N—NFc	N$_2$O	30
—CH$_2$CH$_2$OH	Ethylene oxide	31
—Fc	RX, CoCl$_2$	2, 1.9
—C(=O)NHPh	PhNCO	—

[a] From Slocum *et al.* (1969).

1,1'-Dimetalation of ferrocene with amylsodium and n-BuLi/TMEDA has been accomplished, the latter now being preferred. Carbonation has demonstrated a greater than 90% yield of this intermediate. High yields of the dihalo compounds have been obtained by treatment of this intermediate with the

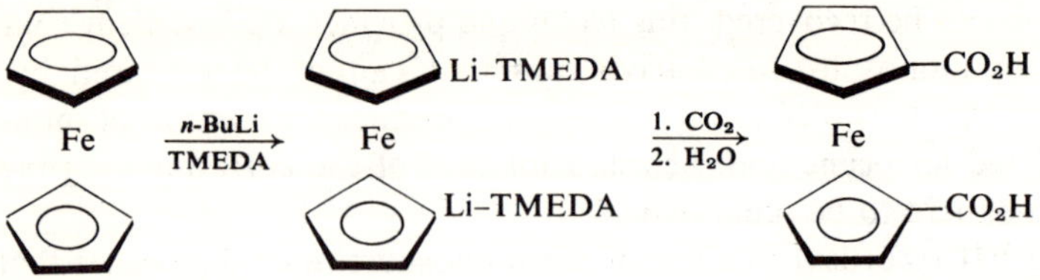

[Ferrocene] —n-BuLi/TMEDA→ [1,1'-di-Li-TMEDA ferrocene] —1. CO₂ / 2. H₂O→ [1,1'-ferrocene dicarboxylic acid]   (207)

(Rausch and Ciappenelli, 1967)

appropriate reagent (Kovar *et al.*, 1971). Ruthenocene also can be 1,1'-dimetalated with butyllithium in 86% yield *without* the use of TMEDA (Rausch *et al.*, 1960). Evidently, ruthenocene undergoes lithiation more readily than does ferrocene.

Bisbenzenechromium undergoes metalation in a manner similar to the metallocenes. The quantitative conversion of bisbenzenechromium to a metalated intermediate by means of *n*-amylsodium has been reported (Fischer and Brunner, 1961). By condensation with carbon dioxide and subsequent methylation with dimethyl sulfate, this intermediate yields a mixture of dimethyl carboxylates. Later it was found that the reaction yields several disubstituted products [Eq. (208)] as well as the monosubstituted product.

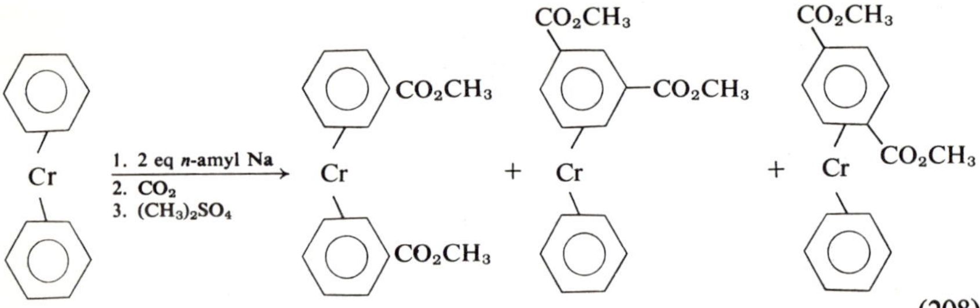

$$\text{bisbenzenechromium} \xrightarrow[\text{2. CO}_2\text{; 3. (CH}_3)_2\text{SO}_4]{\text{1. 2 eq }n\text{-amyl Na}} \text{mono-CO}_2\text{CH}_3 \text{ product} + \text{1,3-disubstituted} + \text{1,4-disubstituted} + \text{1,2-disubstituted}$$

(208)

(Fischer and Brunner, 1962, 1965)

2. Directed Metalation

a. Substituents Which Direct Metalation. In a directed metalation of an aromatic system, a lithium atom replaces a hydrogen atom on a carbon adjacent to the site of the directing substituent. The intermediate in this reaction, to be discussed subsequently, most often involves a 5- or 6-membered coordinated ring containing the lithium atom. At times, inductive stabilization of the lithio anion has also been invoked. The ability of butyllithium or butyllithium–TMEDA to effect metalation of an aromatic compound suggests that many aromatic systems should prove susceptible to metalation. From the preceding discussion it should be clear that any aromatic system

5. CARBANIONS

which will undergo metalation can also be made to undergo a directed metalation, once the proper directing substituent and experimental conditions can be found.

In the benzene series, anisole was the first monosubstituted benzene derivative to provide an example of directed metalation [Eq. (209)] (Gilman and Morton, 1954). In the ensuing years other substituents have also been

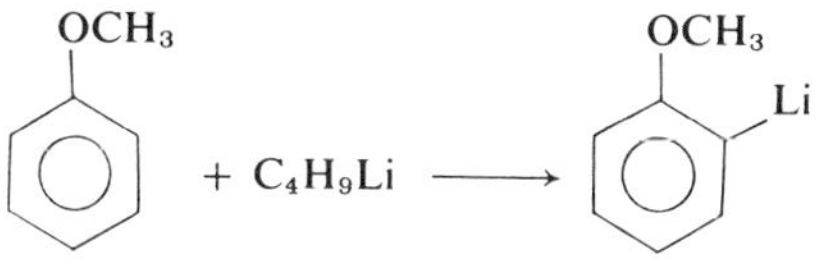

$$\text{PhOCH}_3 + C_4H_9Li \longrightarrow \text{2-Li-PhOCH}_3 \tag{209}$$

found, notably through the efforts of C. R. Hauser and co-workers, to promote 2-lithiation of benzene or ferrocene such that the number has swelled to more than a dozen (Table VIII).

Table VIII
Substituents Which Promote 2-Metalation of Aromatic Substrates

Substituent	Aromatic system	Reference
—F	Benzene, naphthalene	Gilman and Soddy (1957)
—Cl	Ferrocene	Slocum et al. (1972)
—CF_3	Benzene	J. D. Roberts and Curtin (1946)
—OCH_3	Benzene, ferrocene, naphthalene, thiophene	Shirley et al. (1968)
—$N(CH_3)_2$	Benzene	Slocum et al. (1970)
—$CH_2N(CH_3)_2$	Benzene, ferrocene, naphthalene, thiophene	Gay and Hauser (1967b); Slocum and Gierer (1971)
—$CH_2CH_2N(CH_3)_2$	Benzene, ferrocene	Slocum et al. (1971)
—CH_2NHR	Benzene	Ludt and Hauser (1971)
—CONHR	Benzene, ferrocene thiophene	Puterbaugh and Hauser (1964a)
—SO_2NHR	Benzene	Watanabe et al. (1968)
—SO_2NR_2	Benzene, cymantrene	Sutherland and Unni (1970)
—CPh_2OH	Benzene, ferrocene	Benkeser et al. (1961)
—CH_2OCH_3	Ferrocene, thiophene	Slocum and Koonsvitsky (1969)

In most cases, yields of the 2-metalated intermediate were good; however, for the substituents –$N(CH_3)_2$ and –$CH_2CH_2N(CH_3)_2$ very poor yields were obtained when the aromatic systems was benzene. For the former, use of butyllithium–TMEDA significantly increased the yield, but for the latter, use of this reagent reduced the yield to nothing. In all probability, elimination of the elements of dimethylamine from the dimethylaminoethyl moiety competes quite favorably with the 2-lithiation reaction. This was not observed when the dimethylaminoethyl substituent was attached to a ferrocene system; good yields of 2-metalation product were reported for this system.

b. *Aromatic Systems Which Undergo Directed Metalation.* Many aromatic systems should undergo the directed metalation reaction. Prototypes of several systems have been examined, namely, metallocenes and related systems (ferrocene, ruthenocene, and cymantrene), benzene, a polynuclear hydrocarbon (naphthalene), and a heterocycle (thiophene).

i. *Metallocene and related systems.* The directed metalation reaction is of considerable synthetic value in the ferrocene system which is relatively unstable toward strong acids and oxidizing agents. In many cases the conditions for electrophilic substitution are too strong for the ferrocene molecule to survive. Thus, reactions involving metalation and the directed metalation have proved to be the methods of choice for the synthesis of a considerable number of mono- and 1,2-disubstituted ferrocenes.

Substituents, which on a ferrocene ring have been demonstrated to promote 2-metalation, are recorded in Eq. (210). Many of these are, of course, the

$$R = -Cl, -OCH_3, -CH_2OCH_3, -CH_2N(CH_3)_2, -CH_2CH_2N(CH_3)_2, -CONHR, -CPh_2OH \quad (210)$$

same substituents that have been found to promote ortho lithiation on the benzene ring, although there are some notable differences. Interestingly, the $-CH_2OCH_3$ group promotes 2-metalation on ferrocene [Eq. (165)] (Slocum and Koonsvitsky, 1969), whereas on benzene the Wittig rearrangement is effected under similar conditions (Finnigan and Altschuld, 1967). This can probably be attributed to the decreased acidity of the α-methylene hydrogens in the ferrocene molecule owing to the known electron-donating properties of the ferrocene ring. A similar explanation is sufficient for the observation that the $-CH_2CH_2N(CH_3)_2$ group when treated with butyllithium provides mostly polystyrene when attached to benzene with only a small amount of the product of ortho metalation (Slocum et al., 1968), but it provides exclusive metalation when attached to a ferrocene system. Dimethylamino-

5. CARBANIONS

methylruthenocene and the dimethyl sulfonamide of cymantrene have also been successfully 2-lithiated (Hofer and Schlögl, 1968; Sutherland and Unni, 1970).

ii. The benzene system. Directed metalation to yield ortho-disubstituted products may perhaps find the greatest utility in the benzene system since the demand for such isomers is potentially so high. Substituents illustrated in

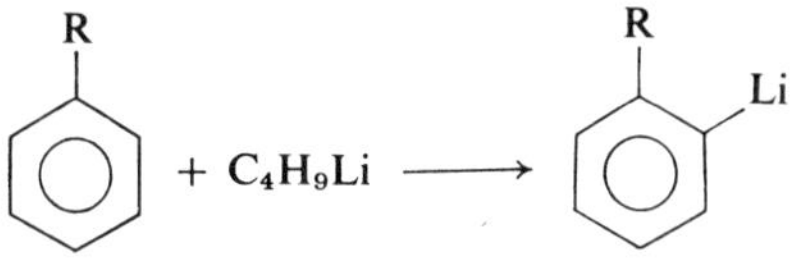

$$\text{PhR} + C_4H_9Li \longrightarrow \text{PhR(Li)} \tag{211}$$

R = —F, —OCH$_3$, —N(CH$_3$)$_2$, —CF$_3$, —CH$_2$NHR, —CH$_2$N(CH$_3$)$_2$,
—CH$_2$CH$_2$N(CH$_3$)$_2$, —CONHR, —SO$_2$N(CH$_3$)$_2$, —SO$_2$NHCH$_3$

Eq. (211) have been demonstrated to provide simple 2-metalation in isolable yield (Table VIII). Ortho-disubstituted benzenes are often difficult or tedious to prepare by other routes. Electrophilic substitution of a monosubstituted benzene of course involves the separation of the ortho from the para and perhaps also the meta isomer, thus involving extra work and a diminished yield of the desired isomer. The directed metalation route in most cases yields a single product in high yield.

iii. The naphthalene system. Three of the known directing substituents have been reported to direct metalation in naphthalene (Table VIII). Perhaps the most interesting aspect of this series is the observation of significant amounts of 8-metalation product in addition to the anticipated product from 2-metalation. A mixture of 8- and 2-metalated intermediates has been postulated [Eq. (212)].

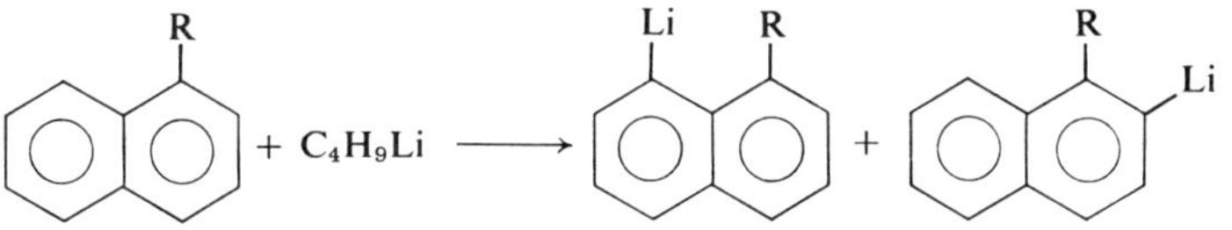

$$\text{Naphthalene-R} + C_4H_9Li \longrightarrow \text{8-Li product} + \text{2-Li product} \tag{212}$$

R = —OCH$_3$, —CH$_2$N(CH$_3$)$_2$, —F

iv. The thiophene system. Recent work has shown that the dimethylaminomethyl group will exert a directive effect in the metalation of thiophene (Slocum and Gierer, 1971). In 5-methyl-2-thenylamine, metalation has been observed at the 3-position [Eq. (213)]; if the 5-position is not blocked, metalation at the more reactive 5-position is found.

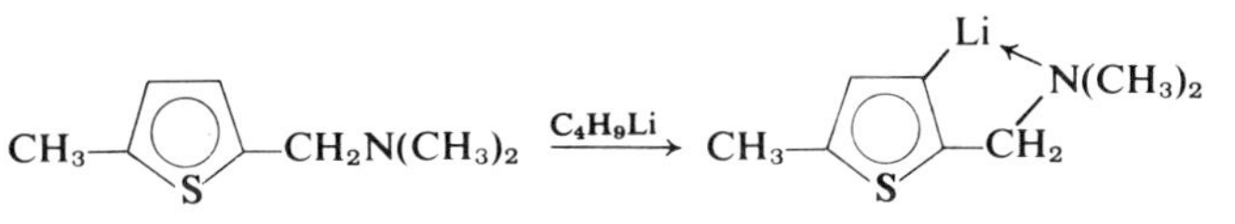

$$CH_3\text{-thiophene-}CH_2N(CH_3)_2 \xrightarrow{C_4H_9Li} CH_3\text{-thiophene(Li)-}CH_2\text{-}N(CH_3)_2 \tag{213}$$

3-Substituted thiophenes in several instances have been found to direct metalation to the 2-position [Eq. (214)] (Table VIII).

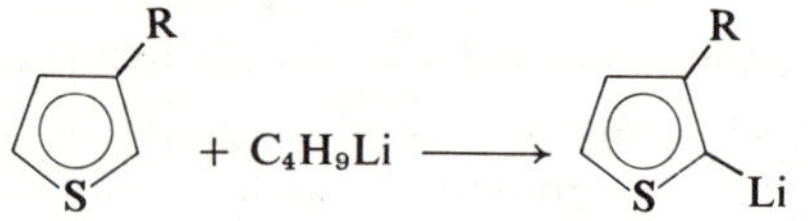

$$\underset{S}{\overset{R}{\bigcirc}} + C_4H_9Li \longrightarrow \underset{S}{\overset{R}{\bigcirc}}{}_{Li} \quad (214)$$

R = —CH$_2$N(CH$_3$)$_2$, —CH$_2$OCH$_3$, —OCH$_3$

3. Competitive Metalation

A group of nine ortho-directing substituents for the benzene ring are known: –CH$_2$N(CH$_3$)$_2$, –CH$_2$CH$_2$N(CH$_3$)$_2$, –N(CH$_3$)$_2$, –CONHR, –OCH$_3$, –CF$_3$, –F, –SO$_2$NHR, and –SO$_2$NR$_2$ (cf. Table VIII). Data are available only for the competitive metalation of the methoxy group versus the eight other directing groups (Slocum and Jennings, 1971; Jennings, 1971). These results are based on the competitive lithiation of the appropriate para-substituted anisoles as well as in certain instances, the corresponding meta and ortho isomers. Utilizing known conditions for the metalation of the appropriate mono-substituted benzene, each member of the series has been treated with butyllithium. It was ascertained that the groups –SO$_2$NR$_2$, –SO$_2$NHR, –CONHR, and –CH$_2$N(CH$_3$)$_2$ all controlled the site of metalation in their respective isomers and that the –N(CH$_3$)$_2$, –CH$_2$CH$_2$N(CH$_3$)$_2$, –F, and –CF$_3$ groups were weaker ortho-directing substituents than the methoxy group [Eqs. (215) and (216)].

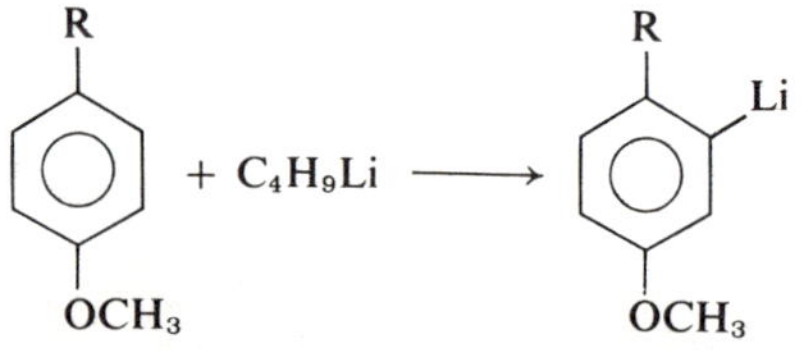

$$\underset{OCH_3}{\overset{R}{\bigcirc}} + C_4H_9Li \longrightarrow \underset{OCH_3}{\overset{R}{\bigcirc}}{}^{Li} \quad (215)$$

R = —SO$_2$N(CH$_3$)$_2$, —SO$_2$NH(CH$_3$),
   —CONHR′, —CH$_2$N(CH$_3$)$_2$

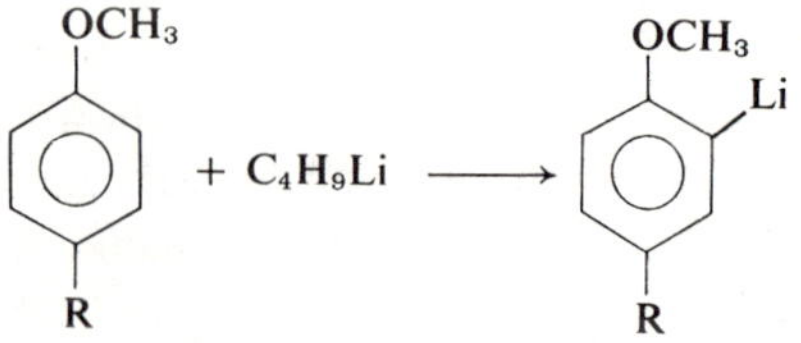

$$\underset{R}{\overset{OCH_3}{\bigcirc}} + C_4H_9Li \longrightarrow \underset{R}{\overset{OCH_3}{\bigcirc}}{}^{Li} \quad (216)$$

R = —CH$_2$CH$_2$N(CH$_3$)$_2$, —N(CH$_3$)$_2$, —F, —CF$_3$

In all cases where a meta isomer has been examined, the site of metalation is ortho to each of the directing groups, that is, the 2-position of a 1,3-disubstituted benzene [Eq. (217)].

5. CARBANIONS

For the metalation of ortho disubstituted benzenes the results indicate the same directive order as have been observed in the corresponding para

$$\underset{Y}{\underset{|}{\overset{X}{\overset{|}{C_6H_4}}}} + C_4H_9Li \longrightarrow \underset{Y}{\underset{|}{\overset{X}{\overset{|}{C_6H_3}}}}\text{-Li} \quad (217)$$

disubstituted compound. Thus, if X were the stronger ortho-directing substituent in the para-disubstituted benzene, it was also found to control the site of metalation in the corresponding ortho-disubstituted benzene [Eq. (218)]. Anticipated steric effects in the metalation of some of the ortho-disubstituted benzenes have not been detected.

$$\underset{}{\overset{X}{\underset{}{\overset{|}{C_6H_4}}}}\text{-Y} + C_4H_9Li \longrightarrow \text{Li-}\underset{}{\overset{X}{\underset{}{\overset{|}{C_6H_3}}}}\text{-Y} \quad (218)$$

Extension of these directive studies to other aromatic systems should provide further insight. The fundamental question is whether these directors will exert the same order of competitive directive effect in ferrocene, naphthalene, or thiophene as they do in benzene.

4. The Directing Mechanism

In all directed metalations studied to date, it has been found that the lithium atom is directed to a proton site adjacent to the directing substituent. No single explanation at present can be proposed to account for all the known examples of the directed metalation reaction; rather, a combination of varying degrees of a coordination mechanism coupled with an inductive effect seems most plausible.

A good example of the intervention of a coordination mechanism is that in the ortho metalation of dimethylbenzylamine (F. N. Jones et al., 1963). The methylene group insulates the ring from essentially any inductive influence of the nitrogen atom. The fact that this molecule can be ortho metalated strongly indicates that some other effect is operating. Such an effect involves the coordinated lithio intermediate depicted in Eq. (219). A

$$\underset{}{\overset{N(CH_3)_2}{\underset{}{\overset{|}{C_6H_5}}}} \xrightarrow{n\text{-}C_4H_9Li} \underset{}{\overset{N(CH_3)_2 \searrow Li}{\underset{}{\overset{|}{C_6H_4}}}} \quad (219)$$

coordination mechanism would also seem to be the most likely directive effect in the case of the $-CH_2CH_2N(CH_3)_2$, $-CONHR$, and $-CH_2NHR$ side chains.

A most intriguing demonstration of the coordinating effect of nitrogen in dimethylbenzylamine is provided by a study of ring versus side-chain metalation with alkylsodium reagents (Puterbaugh and Hauser, 1963). The benzylamine was initially metalated at the ortho position but, after 20 hours, rearrangement to the more stable α-position was complete [Eq. (220)].

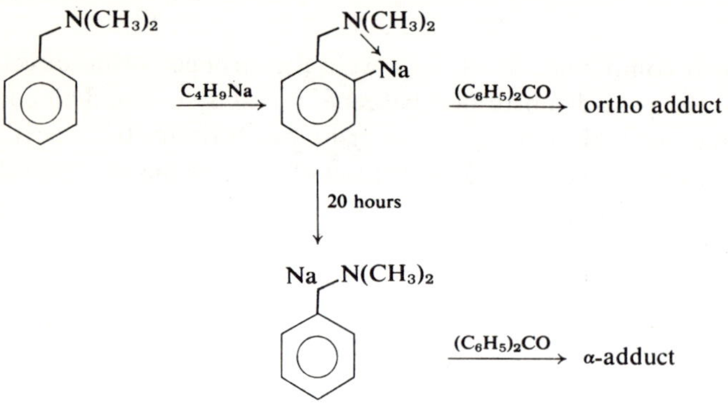

$$\text{PhCH}_2\text{N(CH}_3)_2 \xrightarrow{\text{C}_4\text{H}_9\text{Na}} \text{ortho-Na derivative} \xrightarrow{(\text{C}_6\text{H}_5)_2\text{CO}} \text{ortho adduct}$$

$$\Big\downarrow \text{20 hours} \qquad (220)$$

$$\text{α-Na derivative} \xrightarrow{(\text{C}_6\text{H}_5)_2\text{CO}} \text{α-adduct}$$

Moreover, this rearrangement could be reversed by the addition of lithium bromide to the solution containing the α-metalated species [Eq. (221)]. These

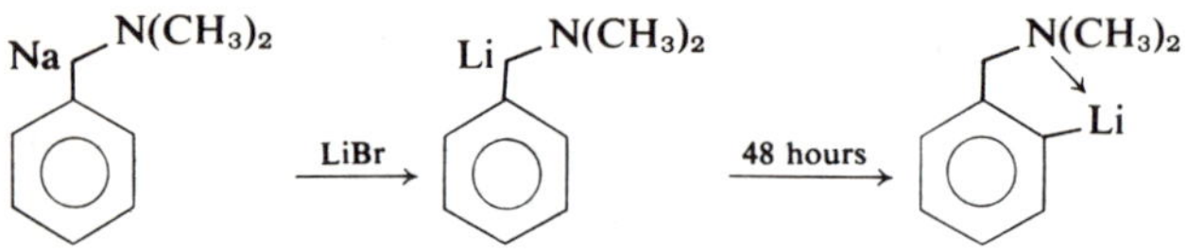

$$\text{α-Na} \xrightarrow{\text{LiBr}} \text{α-Li} \xrightarrow{\text{48 hours}} \text{ortho-Li} \qquad (221)$$

results can be interpreted by assuming that the α-metalated species is more carbanionic in the case of the sodio relative to the lithio derivative and, hence, more conducive to resonance stabilization. In contrast, the ortho-metalated species is greatly stabilized by coordination in the case of the lithio but not the sodio intermediate. Stated simply, thermodynamic and kinetic roles have been reversed in these two cases. No meta or para product was detected in either sequence.

The remaining substituents in Table VIII significantly polarize the aromatic ring and, thus, metalation in these cases may be said to be assisted by an inductive or even a field effect. Certainly the two sulfonamides and probably –CF$_3$ involve such an effect. Some coordination may contribute to the transition states in the 2-lithiation involving the –OMe, –NMe, and –F substituents. A pseudo 5-membered coordinated ring can be invoked for these transition states if a suitable oligomer structure (**46**) (tetramer) (Slocum and Jennings, 1971) for butyllithium is assumed. One anomaly exists however. The metalation of anisole and fluorobenzene can be accomplished in 70 and

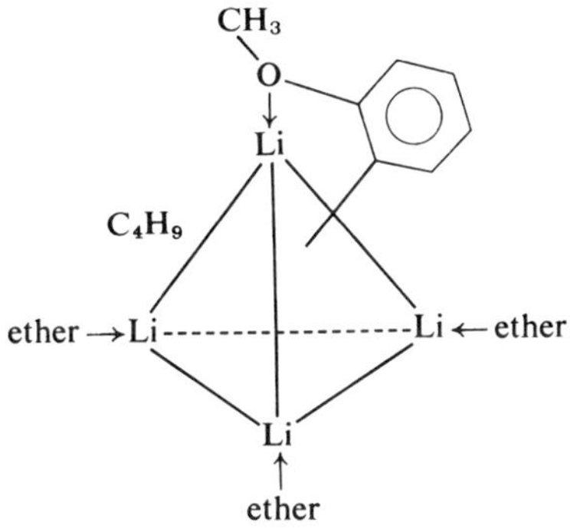

5. CARBANIONS

(46)

36% yield, respectively, but that of dimethylaniline is extremely low. Coordination effects would be expected to fall in the order $-NMe_2$, $-OMe$, $-F$ whereas the inductive order would be $-OMe > NMe_2 > F$. Since neither order is observed, a combination of effects might be suspected.

5. Heterocycles and Natural Products

A most useful synthetic application of the directed metalation reaction is the synthesis of complex heterocyclic systems. In many cases the synthetic route provided by ortho lithiation provides either the only route or at least a more convenient method for syntheses of a few of these compounds. Furthermore, many of the heterocyclic compounds produced via metalation procedures are of extreme interest inasmuch as they are natural products or important derivatives thereof.

Several of the simple heterocyclic systems which can be obtained via the directed metalation reaction are illustrated in Eqs. (222)–(224). In each case, heat or acid catalysis was sufficient to effect cyclization of the ortho-disubstituted benzenes.

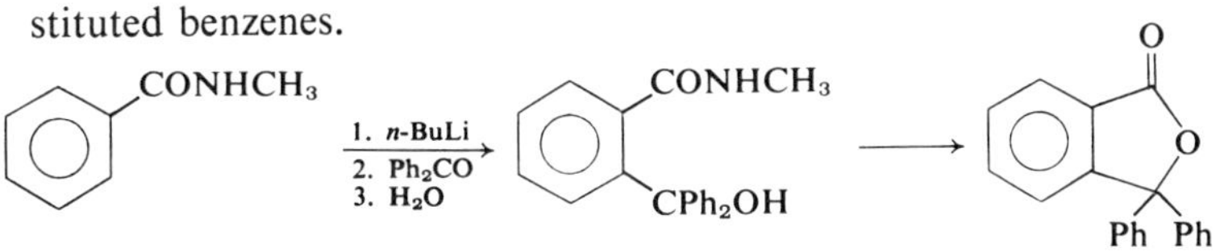

a lactone
(Puterbaugh and Hauser, 1964a)   (222)

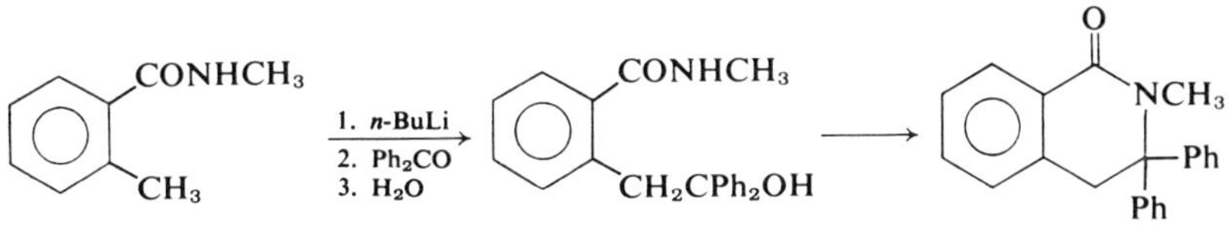

an isocarbostyril
(Barnish et al., 1968b)   (223)

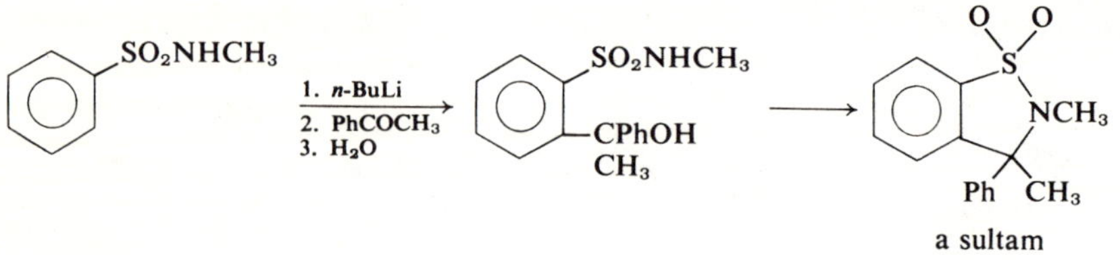

[Reaction scheme 224 — Watanabe et al., 1969b: ArSO₂NHCH₃ → (1. n-BuLi, 2. PhCOCH₃, 3. H₂O) → ortho-substituted ArSO₂NHCH₃ with –CPhOH(CH₃) → a sultam]

(224)

a sultam

(Watanabe et al., 1969b)

Narasimhan and Ranade (1967) have reported the use of the 2-metalation technique in the preparation of isoquinolines. The parent heterocycle can be prepared from dimethylbenzylamine and ethylene oxide [Eq. (225)]. Complex

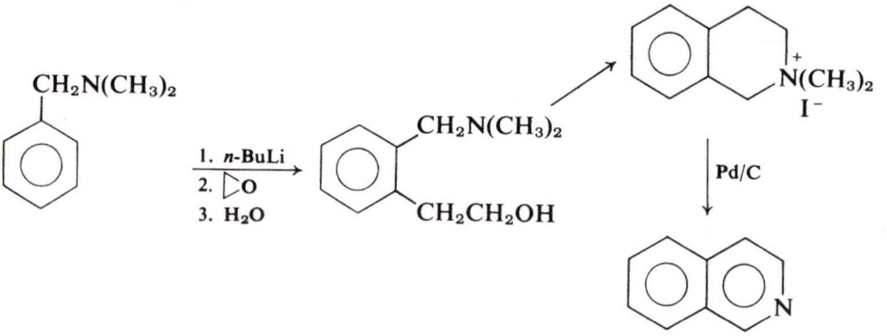

[Reaction scheme 225: PhCH₂N(CH₃)₂ → (1. n-BuLi, 2. ethylene oxide, 3. H₂O) → 2-(2-hydroxyethyl)benzyldimethylamine → cyclization to N-methylisoquinolinium iodide → (Pd/C) → isoquinoline]

(225)

(Narasimhan and Ranade, 1967)

derivatives of quinoline have also been prepared. The syntheses of dictamnine, a quinoline derivative extracted from the leaves of *Skimmia*, an ornamental Asian evergreen, are used here as an example [Eq. (226)].

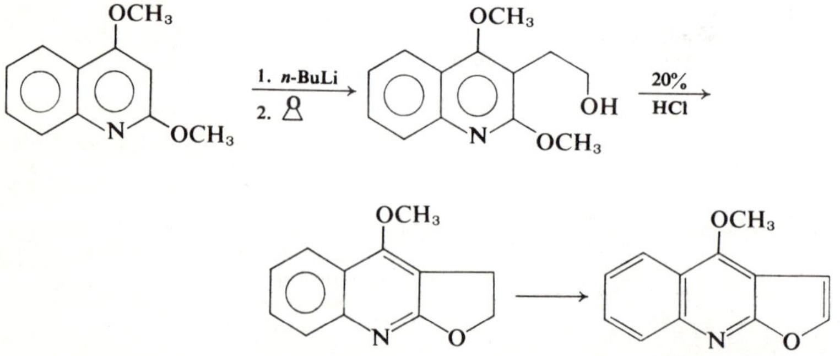

[Reaction scheme 226: 4-methoxy-2-methoxyquinoline → (1. n-BuLi, 2. ethylene oxide) → 3-(2-hydroxyethyl) derivative → (20% HCl) → dihydrofuroquinoline → dictamnine]

(226)

Dictamnine

(Narasimhan et al., 1971)

5. CARBANIONS

By way of the carboxamide group, Narasimhan and Bhide (1970) have synthesized isocoumarins. Among the derivatives of isocoumarin these workers have been able to prepare is the natural product mellein [Eq. (227)].

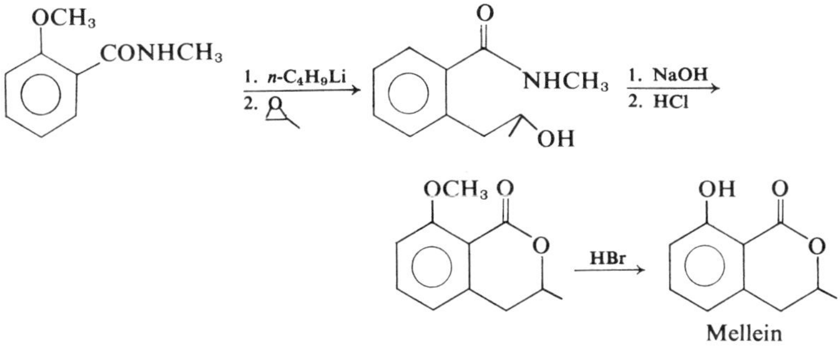

$$\text{(227)}$$

Acknowledgments

The authors thank Professors Scott and Meyers for furnishing us with a prepublication copy of their chapter "The Chemistry of Aliphatic Sulfonyl Compounds and Derivatives."

References

Adamcik, J. A., and Miklasiewicz, E. J. (1963). *J. Org. Chem.* **28**, 336.
Adams, R., and Marvel, C. S. (1932). *Org. Syn.* **I**, 94.
Allen, C. F. H., and Van Allan, J. (1955). *Org. Syn.* **III**, 727.
Anderson, R. J., Henrick, C. A., and Siddall, J. B. (1970). *J. Amer. Chem. Soc.* **92**, 735.
Ashby, E., and Parris, G. (1971). *J. Amer. Chem. Soc.* **93**, 1206.
Ayres, D. C. (1966). "Carbanions in Synthesis." Oldbourne Press, London.
Bachmann, W. E., and Hetzner, H. P. (1955). *Org. Syn.* **III**, 839.
Baldwin, J. E., Erickson, W. F., Hackler, R. E., and Scott, R. M. (1970). *Chem. Commun.* p. 576.
Ballester, M. (1955). *Chem. Rev.* **55**, 283.
Barnish, I. T., Hauser, C. R., and Wolfe, J. F. (1968a). *J. Org. Chem.* **33**, 2116.
Barnish, I. T., Mao, C. L., Gay, R. L., and Hauser, C. R. (1968b). *Chem. Commun.* p. 564.
Baumgarten, H. E., Dirks, J. E., Peterson, J. M., and Wolf, D. C. (1960). *J. Amer. Chem. Soc.* **82**, 4422.
Beam, C. F., Dyer, M. C. D., Schwarz, R. A., and Hauser, C. R. (1970). *J. Org. Chem.* **35**, 1806.
Beam, C. F., Foote, R. S., and Hauser, C. R. (1971). *J. Chem. Soc.*, C p. 1658.
Beard, R. D. (1970). Ph.D. Thesis, University of Missouri, Columbia.
Benkeser, R. A., Fitzgerald, W. P., and Melzer, M. S. (1961). *J. Org. Chem.* **26**, 2569.
Benkeser, R. A., Trevilyan, A. E., and Hooz, J. (1962). *J. Amer. Chem. Soc.* **84**, 4971.
Benkeser, R. A., Hooz, J., Liston, T. V., and Trevilyan, A. E. (1963). *J. Amer. Chem. Soc.* **85**, 3984.

Bergmann, E. D., Ginsburg, D., and Pappo, R. (1959). *Org. React.* **10**, 179.
Blicke, F. F. (1942). *Org. React.* **1**, 303.
Boatman, S., Harris, T. M., and Hauser, C. R. (1965a). *J. Amer. Chem. Soc.* **87**, 82.
Boatman, S., Harris, T. M., and Hauser, C. R. (1965b). *J. Amer. Chem. Soc.* **87**, 5198.
Bordwell, F. G., and Cooper, G. D. (1951). *J. Amer. Chem. Soc.*, **73**, 5187.
Bordwell, F. G., Jarvis, B. B. and Corfield, P. W. R. (1968). *J. Amer. Chem. Soc.*, **90**, 5298.
Braye, E. H., Hubel, W., and Caplier, I. (1961). *J. Amer. Chem. Soc.* **83**, 4406.
Breslow, R., Posner, J., and Krebs, A. (1963). *J. Amer. Chem. Soc.* **85**, 234.
Brewster, J. H., and Kline, M. W. (1952). *J. Amer. Chem. Soc.* **74**, 5179.
Broaddus, C. D. (1966). *J. Amer. Chem. Soc.* **88**, 4174.
Broaddus, C. D. (1970). *J. Org. Chem.* **35**, 10.
Brook, A. G., Duff, J. M., Jones, P. F., and Davis, N. R. (1967). *J. Amer. Chem. Soc.* **89**, 431.
Brown, G. B. and Partridge, C. W. H. (1945). *J. Amer. Chem. Soc.* **67**, 1423.
Brunson, H. A. (1949). *Org. React.* **5**, 79.
Bryant, D. R., and Hauser, C. R. (1961). *J. Amer. Chem. Soc.* **83**, 3468.
Bunnett, J. F., and Zahler, R. E. (1951). *Chem. Rev.* **49**, 273.
Campbell, A., Houston, A. H. J., and Kenyon, J. (1947). *J. Chem. Soc., London.* p. 93.
Campbell, K. N., Campbell, B. K., McKenna, J. F., and Chaput, E. P. (1943). *J. Org. Chem.* **8**, 103.
Cannon, G. W., Ellis, R. C., and Leal, J. R. (1963). *Org. Syn.* **IV**, 597.
Casey, C. P., and Boggs, R. A. (1971). *Tetrahedron Lett.* p. 2455.
Cast, J., Stevens, T. S., and Holmes, J. (1960). *J. Chem. Soc., London* p. 3521.
Chan, L. L., and Smid, J. (1968). *J. Amer. Chem. Soc.* **90**, 4654.
Clark, M. T., Hendley, E. G., and Neville, O. K. (1955). *J. Amer. Chem. Soc.* **77**, 3280.
Coates, G. E., and Wade, K. (1967). "Organometallic Compounds," Vol. I, Chapter 1. Methuen, London.
Coleman, G. H., and Craig, D. (1943). *Org. Syn.* **II**, 179.
Colonge, J., and Marey, R. (1963). *Org. Syn.* **IV**, 601.
Conant, J. B., and Tuttle, N. (1941). *Org. Syn.* **I**, 199.
Cope, A. C., and Hancock, E. M. (1955). *Org. Syn.* **III**, 399.
Cope, A. C., Holmes, H. L., and House, H. O. (1957). *Org. React.* **9**, 107.
Cope, A. C., Aryden, H. L., Jr., and Howell, C. F. (1963). *Org. Syn.* **IV**, 816.
Corey, E. J., and Chaykovsky, M. (1965). *J. Amer. Chem. Soc.* **87**, 1353.
Corey, E. J., and Durst, T. (1966). *J. Amer. Chem. Soc.* **88**, 5656.
Corey, E. J., and Durst, T. (1968). *J. Amer. Chem. Soc.* **90**, 5548.
Corey, E. J., and Kwiatkowski, G. T. (1966). *J. Amer. Chem. Soc.* **88**, 5652.
Corey, E. J., and Lowry, T. H. (1965). *Tetrahedron Lett.* p. 803.
Corey, E. J., and Seebach, D. (1966). *Angew. Chem., Int. Ed. Engl.* **4**, 1075 and 1077.
Cram, D. J. (1965). "Fundamentals of Carbanion Chemistry." Academic Press, New York.
Cram, D. J. (1968). *Surv. Progr. Chem.* **4**, 45–67.
Cram, D. J., and Hatch, M. J. (1953). *J. Amer. Chem. Soc.* **75**, 33 and 38.
Cram, D. J., Rickborn, B., Kingsbury, C. A., and Haberfield, P. (1961). *J. Amer. Chem. Soc.* **83**, 3678.
Creger, P. C. (1967). *J. Amer. Chem. Soc.* **89**, 2500.
Creger, P. C. (1970). *J. Amer. Chem. Soc.* **92**, 1396 and 1397.
Dauben, W. G., and McFarland, J. W. (1960). *J. Amer. Chem. Soc.* **82**, 4245.
Doering, W. von E., and Pasternak, V. Z. (1950). *J. Amer. Chem. Soc.* **72**, 143.

Doering, W. von E., and Urban, R. S. (1956). *J. Amer. Chem. Soc.* **78**, 5938.
Dunn, J. L., and Stevens, T. S. (1934). *J. Chem. Soc., London* p. 1926.
Dunnavant, W. R., and Hauser, C. R. (1960). *J. Org. Chem.* **25**, 503.
Edinger, J. M., and Day, A. R. (1971). *J. Org. Chem.* **36**, 240.
Evans, J. C. W., and Allen, C. F. H. (1943). *Org. Syn.* **II**, 517.
Fairbourne, A., and Fawson, H. R. (1927). *J. Chem. Soc., London* p. 46.
Finnegan, R. A., and Altschuld, J. W. (1967). *J. Organometal. Chem.* **9**, 193.
Fischer, E. O., and Brunner, H. (1961). *Z. Naturforsch. B* **16**, 406.
Fischer, E. O., and Brunner, H. (1962). *Chem. Ber.* **95**, 1999.
Fischer, E. O., and Brunner, H. (1965). *Chem. Ber.* **98**, 175.
Foote, R. S., Beam, C. F., and Hauser, C. R. (1970). *J. Heterocycl. Chem.* **7**, 589.
Fort, A. W. (1962). *J. Amer. Chem. Soc.* **84**, 2620 and 2625.
Gay, R. L., and Hauser, C. R. (1967a). *J. Amer. Chem. Soc.* **89**, 1647.
Gay, R. L., and Hauser, C. R. (1967b). *J. Amer. Chem. Soc.* **89**, 2297.
Gay, R. L., Boatman, S., and Hauser, C. R. (1965). *Chem. Ind. (London)* p. 1789.
Gilman, H., and Bailie, J. C. (1943). *J. Amer. Chem. Soc.* **65**, 267.
Gilman, H., and Catlin, W. E. (1941). *Org. Syn.* **I**, 188.
Gilman, H., and Jones, R. G. (1951). *Org. React.* **6**, 339.
Gilman, H., and Kirby, R. H. (1933). *J. Amer. Chem. Soc.* **55**, 1265.
Gilman, H., and McNinch, H. A. (1961). *J. Org. Chem.* **26**, 3723.
Gilman, H., and Morton, J. W., Jr. (1954). *Org. React.* **8**, 258.
Gilman, H., and Soddy, T. S. (1957). *J. Org. Chem.* **22**, 1957.
Gilman, H., St. John, N. B., and Schulze, F. (1943). *Org. Syn.* **II**, 425.
Goldberg, S. I., Keith, L. H., and Prokopov, T. S. (1963). *J. Org. Chem.* **28**, 850.
Gornowicz, G. A., and West, R. (1971). *J. Amer. Chem. Soc.* **93**, 1714.
Griffiths, J. S., Beam, C. F., and Hauser, C. R. (1971). *J. Chem. Soc., C* p. 974.
Grignard, V. (1904). *Bull. Soc. Chim. Fr.* **31**, 751.
Grovenstein, E., Jr. (1957). *J. Amer. Chem. Soc.* **79**, 4985.
Grovenstein, E., Jr., and Wentworth, G. (1963). *J. Amer. Chem. Soc.* **85**, 3305.
Grovenstein, E., Jr., and Wentworth, G. (1967). *J. Amer. Chem. Soc.* **89**, 2348.
Hampton, K. G., Harris, T. M., and Hauser, C. R. (1964). *J. Org. Chem.* **29**, 3511.
Hampton, K. G., Harris, T. M., and Hauser, C. R. (1965a). *J. Org. Chem.* **30**, 61.
Hampton, K. G., Harris, T. M., and Hauser, C. R. (1965b). *J. Org. Chem.* **30**, 4263.
Hampton, K. G., Harris, T. M., and Hauser, C. R. (1966). *J. Org. Chem.* **31**, 663.
Hamrick, P. J., Jr., and Hauser, C. R. (1959). *J. Amer. Chem. Soc.* **81**, 493.
Hamrick, P. J., Jr., and Hauser, C. R. (1960). *J. Amer. Chem. Soc.* **82**, 1957.
Harris, T. M., and Harris, C. M. (1969). *Org. React.* **17**, 155.
Harris, T. M., and Hauser, C. R. (1959). *J. Amer. Chem. Soc.* **81**, 1160.
Harris, T. M., Boatman, S., and Hauser, C. R. (1963). *J. Amer. Chem. Soc.* **85**, 3273.
Hauser, C. R., and Breslow, D. S. (1955). *Org. Syn.* **III**, 408.
Hauser, C. R., and Harris, T. M. (1958). *J. Amer. Chem. Soc.* **80**, 6360.
Hauser, C. R., and Harris, T. M. (1959). *J. Amer. Chem. Soc.* **81**, 1154.
Hauser, C. R., and Hudson, B. E., Jr. (1947). *Org. React.* **1**, 266.
Hauser, C. R., and Tetenbaum, M. T. (1958). *J. Org. Chem.* **23**, 1146.
Hauser, C. R., and Van Eenam, D. N. (1957). *J. Amer. Chem. Soc.* **79**, 6274, 6277.
Hauser, C. R., Kantor, S. W., and Brasen, W. R. (1953). *J. Amer. Chem. Soc.* **75**, 2660.
Hauser, C. R., Swamer, F. W., and Adams, J. T. (1962). *Org. React.* **8**, 59.
Hayashi, Y., and Oda, R. (1968). *Tetrahedron Lett.* p. 5381.
Henery-Logan, K. R., and Fridinger, T. L. (1967). *J. Amer. Chem. Soc.* **89**, 5724.
Henoch, F. E., Hampton, K. G., and Hauser, C. R. (1967). *J. Amer. Chem. Soc.* **89**, 463.

Henoch, F. E., Hampton, K. G., and Hauser, C. R. (1969). *J. Amer. Chem. Soc.* **91**, 676.
Herbst, R. M., and Shemin, D. (1943). *Org. Syn.* **II**, 1.
Herr, R. W., Wieland, D. M., and Johnson, C. R. (1970). *J. Amer. Chem. Soc.* **92**, 3813.
Hill, R. K., and Chan, T. (1966). *J. Amer. Chem. Soc.* **88**, 866.
Hofer, O., and Schlögl, K. (1968). *J. Organometal. Chem.* **13**, 443.
Horning, E. C., and Finelli, A. F. (1963). *Org. Syn.* **IV**, 776.
House, H. O. (1972). "Modern Synthetic Reactions." Benjamin, New York.
House, H. O., and Gilmore, W. F. (1961). *J. Amer. Chem. Soc.* **83**, 3980.
House, H. O., Respess, W. L., and Whitesides, G. M. (1966). *J. Org. Chem.* **31**, 3128.
Ide, W. S., and Buck, J. S. (1948). *Org. React.* **4**, 269.
Impasto, F. J., and Walborsky, H. M. (1962). *J. Amer. Chem. Soc.* **84**, 4838.
Ivanoff, D., Mihova, M., and Christova, T. (1932). *Bull. Soc. Chim. Fr.* [4] **51**, 1321.
Jackman, L. M., and Sternhell, S. (1969). "Applications of Nuclear Magnetic Resonance Spectroscopy in Organic Chemistry." Pergamon, Oxford.
Jacobs, T. L. (1949). *Org. React.* **5**, 1.
Jemison, R. W., Mageswaran, S., Ollis, W. D., Potter, S. E., Pretty, A. J., Sutherland, I. O., and Thebtaranonth, Y. (1970). *Chem. Commun.* p. 1201.
Jennings, C. A. (1971). Ph.D. Dissertation, Southern Illinois University, Carbondale.
Jenny, E. F., and Druey, J. (1962). *Angew. Chem.*, Int. Ed. Engl. **1**, 155.
Jenny, E. F., and Melzer, A. (1966). *Tetrahedron Lett.* p. 3507.
Johnson, A. W., and LaCount, R. B. (1961). *J. Amer. Chem. Soc.* **83**, 417.
Johnson, J. R. (1942). *Org. React.* **1**, 210.
Johnson, J. R., and Grummitt, O. (1955). *Org. Syn.* **III**, 806.
Johnson, W. S., and Daub, G. H. (1960). *Org. React.* **6**, *1*.
Jones, F. N., Vaulx, R. L., and Hauser, C. R. (1963). *J. Org. Chem.* **28**, 3461.
Jones, G. (1967). *Org. React.* **15**, 204.
Jones, J. R. (1971). *Quart. Rev., Chem. Soc.* **25**, 365.
Jorgenson, M. J. (1970). *Org. React.* **18**, 1.
Julian, P. L., Oliver, J. J., Kimball, R. H., Pike, A. B., and Jefferson, G. D. (1943). *Org. Syn.* **II**, 487.
Kaiser, E. M., and Bartling, G. J. (1972). *J. Org. Chem.* **37**, 490.
Kaiser, E. M., and Beard, R. D. (1968). *Tetrahedron Lett.* p. 2583.
Kaiser, E. M., Fries, J. A., and Simonsen, W. J. (1971). *Org. Prep. Proced. Int.* **3**, 305.
Kaiser, E. M., and Hauser, C. R. (1966a). *J. Org. Chem.* **31**, 3873.
Kaiser, E. M., and Hauser, C. R. (1966b). *J. Org. Chem.* **31**, 3317.
Kaiser, E. M., and Hauser, C. R. (1967a). *J. Amer. Chem. Soc.* **89**, 4566.
Kaiser, E. M., and Hauser, C. R. (1967b). *Tetrahedron Lett.* p. 3341.
Kaiser, E. M., and Hauser, C. R. (1968). *J. Org. Chem.* **33**, 3402.
Kaiser, E. M., and Kaufman, R. J. (1972). Unpublished observations.
Kaiser, E. M., and Warner, C. D. (1971). *J. Organometal. Chem.* **31**, C17.
Kaiser, E. M., Kenyon, W. G., and Hauser, C. R. (1967a). *Org. Syn.* **47**, 72.
Kaiser, E. M., Vaulx, R. L., and Hauser, C. R. (1967b). *J. Org. Chem.* **32**, 3640.
Kaiser, E. M., Henoch, F. E., and Hauser, C. R. (1968). *J. Amer. Chem. Soc.* **90**, 7287.
Kaiser, E. M., Mao, C. L., Hauser, C. F., and Hauser, C. R. (1970). *J. Org. Chem.* **35**, 410.
Kaiser, E. M., Solter, L. E., Schwarz, R., Beard, R. D., and Hauser, C. R. (1971). *J. Amer. Chem. Soc.* **93**, 4237.
Kantor, S. W., and Hauser, C. R. (1951). *J. Amer. Chem. Soc.* **73**, 1437 and 4122.
Kende, A. S. (1960). *Org. React.* **11**, 261–316.

Kennedy, J., McCorkindale, R. A., Scott, W. T., and Zwanenburg, B. (1964). *Proc. Chem. Soc., London* p. 148.
Kharasch, M. S., and Reinmuth, O. (1954). "Grignard Reactions of Nonmetallic Substances." Prentice-Hall, Englewood Cliffs, New Jersey.
Klein, J., and Brenner, S. (1969). *J. Amer. Chem. Soc.* **91**, 3094.
Kofron, W. G., and Hauser, C. R. (1970). *J. Org. Chem.* **35**, 2085.
Kofron, W. G., Dunnavant, W. R., and Hauser, C. R. (1962). *J. Org. Chem.* **27**, 2737.
Kohler, E. P., and Chadwell, H. M. (1941). *Org. Syn.* **I**, 78.
Kovar, R. F., Rausch, M. D., and Rosenberg, H. (1971). *Organometal. Chem. Syn.* **1**, 173.
Krapcho, A. P., Diamanti, J., Cayen, C., and Bingham, R. (1967). *Org. Syn.* **47**, 20.
Langer, A. W., Jr. (1965). *Trans. N.Y. Acad. Sci.* [2] **27**, 741.
Lansbury, P. T., and Pattison, V. A. (1962). *J. Org. Chem.* **27**, 1933.
Lansbury, P. T., Pattison, V. A., Sidler, J. D., and Bieber, J. B. (1966). *J. Amer. Chem. Soc.* **88**, 78.
Lepley, A. R., and Becker, R. H. (1965a). *Tetrahedron* **21**, 2365.
Lepley, A. R., and Becker, R. H. (1965b). *J. Org. Chem.* **30**, 3888.
Lepley, A. R., and Giumanini, A. G. (1971). *In* "Mechanisms of Molecular Migrations" (B. S. Thyagarijan, ed.), Vol. 3, pp. 297–440. Wiley, New York.
Lespieau, R., and Bourguel, M. (1941). *Org. Syn.* **I**, 186.
Levene, P. A., and Meyer, G. M. (1943). *Org. Syn.* **II**, 288.
Light, R. J., and Hauser, C. R. (1960). *J. Org. Chem.* **25**, 538.
Light, R. J., and Hauser, C. R. (1961). *J. Org. Chem.* **26**, 1716.
Light, R. J., Harris, T. M., and Hauser, C. R. (1961). *J. Org. Chem.* **26**, 1344.
Lindsay, J. K., and Hauser, C. R. (1957). *J. Org. Chem.* **22**, 355.
Loftfield, R. B. (1951). *J. Amer. Chem. Soc.* **73**, 4707.
Logan, A. V., Marvell, E. N., La Pore, R., and Bush, D. C. (1954). *J. Amer. Chem. Soc.* **76**, 4127.
Ludt, R. E., and Hauser, C. R. (1971). *J. Org. Chem.* **36**, 1607.
Maercker, A. (1965). *Org. React.* **14**, 270.
Mallan, J. M., and Bebb, R. L. (1969). *Chem. Rev.* **69**, 693.
Mao, C. L., Hauser, C. R., and Miles, M. L. (1967). *J. Amer. Chem. Soc.* **89**, 5303.
Marvel, C. S. (1955). *Org. Syn.* **III**, 495.
Meyer, R. B., and Hauser, C. R. (1961a). *J. Org. Chem.* **26**, 3183.
Meyer, R. B., and Hauser, C. R. (1961b). *J. Org. Chem.* **26**, 3696.
Meyers, A. I., Nabeya, A., Adickes, H. W., and Politzer, I. R. (1969). *J. Amer. Chem. Soc.* **91**, 763.
Meyers, C. Y., Malte, A. M., and Matthews, W. S. (1969a). *J. Amer. Chem. Soc.* **91**, 7510.
Meyers, C. Y., Malte, A. M., and Matthews, W. S. (1969b). *J. Amer. Chem. Soc.* **91**, 7512.
Miles, M. L., and Hauser, C. R. (1964). *J. Org. Chem.* **29**, 2329.
Miles, M. L., Moreland, C. G., von Schriltz, D. M., and Hauser, C. R. (1966). *Chem. Ind (London)* p. 2098.
Moersch, G. W., and Burkett, A. R. (1971). *J. Org. Chem.* **36**, 1149.
Moffett, R. B. (1963). *Org. Syn.* **IV**, 652.
Moskowitz, H., Blanc-Guenee, J., and Miocque, M. (1968). *C. R. Acad. Sci., Ser. C* **267**, 898.
Moyer, W. W., and Marvel, C. S. (1943). *Org. Syn.* **II**, 602.

Mulvaney, J. E., Gardlund, Z. G., Gardlund, S. L., and Newton, D. J. (1966). *J. Amer. Chem. Soc.* **88**, 476.
Mulvaney, J. E., Folk, T. L., and Newton, D. J. (1967). *J. Org. Chem.* **32**, 1674.
Musker, W. K. (1970). *In* "Topics in Current Chemistry," Vol. 14, Part 3, pp. 295–365. Springer-Verlag, Berlin and New York.
Narasimhan, N. S., and Bhide, B. H. (1970). *Chem. Commun.* p. 1552.
Narasimhan, N. S., and Ranade, A. C. (1967). *Chem. Ind. (London)* p. 120.
Narasimhan, N. S., Paradka, M. V., and Alukar, R. H. (1971). *Tetrahedron* **28**, 1351.
Neber, P. W., and Friedolsheim, A. (1926). *Justus Liebigs Ann. Chem.* **449**, 109.
Neureiter, N. P. (1966). *J. Amer. Chem. Soc.* **88**, 558.
Newman, M. S., and Booth, W. T., Jr. (1945). *J. Amer. Chem. Soc.* **67**, 154.
Newman, M. S., and Magerlein, B. J. (1949). *Org. React.* **5**, 413.
Nickon, A., and Lambert, J. L. (1966). *J. Amer. Chem. Soc.* **88**, 1905.
Nielsen, A. T., and Houlihan, W. J. (1968). *Org. React.* **16**, 1.
O'Brien, C. (1964). *Chem. Rev.* **64**, 81.
Osawa, E., Majerski, Z., and Schleyer, P. von R. (1971). *J. Org. Chem.* **36**, 205.
O'Sullivan, W. I., Swamer, F. W., Humphlett, W. J., and Hauser, C. R. (1961). *J. Org. Chem.* **26**, 2306.
Overberger, C. G., Saunders, J. H., Allen, R. E., and Gander, R. (1955). *Org. Syn.* **III**, 200.
Paquette, L. A. (1968). "Principles of Modern Heterocylic Chemistry." Benjamin, New York.
Paquette, L. A., and Houser, R. W. (1966). *J. Amer. Chem. Soc.* **91**, 3870.
Paquette, L. A., and Wittenbrook, L. S. (1967). *J. Amer. Chem. Soc.* **89**, 4483.
Paquette, L. A., and Wittenbrook, L. S. (1968). *J. Amer. Chem. Soc.* **90**, 6783.
Parcell, R. F. (1963). *Chem Ind. (London)* p. 1396.
Paul, R., and Tchelitcheff, S. (1968). *Bull. Soc., Chim. Fr.* [5] p. 2134.
Pedersen, C. J. (1967). *J. Amer. Chem. Soc.* **89**, 7017.
Pedersen, C. J. (1968). *Fed. Proc., Fed. Amer. Soc. Exp. Biol.* **27**, 1305.
Pfeil, E., Geissler, G., Jacqueman, W., and Lomker, F. (1956). *Chem. Ber.* **89**, 1210.
Pine, S. H. (1971). *J. Chem. Educ.* **49**, 99.
Pine, S. H., Munemo, E. M., Phillips, T. R., Bartolini, G., Cotton, W. D., and Andrews, G. C. (1971). *J. Org. Chem.* **36**, 984.
Pinkney, P. S. (1943). *Org. Syn.* **II**, 116.
Puterbaugh, W. H., and Hauser, C. R. (1963). *J. Amer. Chem. Soc.* **85**, 2467.
Puterbaugh, W. H., and Hauser, C. R. (1964a). *J. Org. Chem.* **29**, 853.
Puterbaugh, W. H., and Hauser, C. R. (1964b). *J. Amer. Chem. Soc.* **86**, 1105.
Ramberg, L., and Bäcklund, B. (1940). *Ark. Kemi, Mineral. Geol.* **13A**, No. 27.
Rappe, C., Knutsson, L., Turro, N. J., and Gagosian, R. B. (1970). *J. Amer. Chem. Soc.* **92**, 2032.
Rathke, M. W., and Lindert, A. (1970). *J. Org. Chem.* **35**, 3966.
Rausch, M. D., and Ciappenelli, D. J. (1967). *J. Organometal. Chem.* **10**, 127.
Rausch, M. D., and Klemann, L. P. (1967). *J. Amer. Chem. Soc.* **89**, 5732.
Rausch, M. D., Fischer, E. O., and Grubert, H. (1960). *J. Amer. Chem. Soc.* **82**, 76.
Renfrow, W. B., Witte, J. F., Wolf, R. A., and Bohl, W. R. (1968). *J. Org. Chem.* **33**, 150.
Ritchie, C. D., and Uschold, R. E. (1968). *J. Amer. Chem. Soc.* **90**, 2821.
Roberts, I., and Urey, H. C. (1938). *J. Amer. Chem. Soc.* **60**, 880 (1938).
Roberts, J. D., and Curtin, D. Y. (1946). *J. Amer. Chem. Soc.* **68**, 1658.
Royals, E. E. (1954). "Advanced Organic Chemistry." Prentice-Hall, Englewood Cliffs, New Jersey.

Schaefer, J. P., and Bloomfield, J. J. (1967). *Org. React.* **15**, 28.
Schlosser, M. (1967). *J. Organometal. Chem.* **8**, 9.
Schöllkopf, U. (1970). *Agnew. Chem., Int.* Ed. Engl. **9**, 763.
Schöllkopf, U., and Schäffer, H. (1963). *Justus Liebigs Ann. Chem.* **663**, 22.
Scott, R. B., Jr., and Meyers, C. Y. (1972). "The Chemistry of Aliphatic Sulfonyl Compounds and Derivatives." Pergamon, Oxford.
Seebach, D., Erickson, B. W., and Singh, G. (1966). *J. Org. Chem.* **31**, 4303.
Selman, S., and Easthan, J. F. (1961). *Quart. Rev., Chem. Soc.* **14**, 221.
Seyferth, D., and Weiner, M. D. (1961). *J. Amer. Chem. Soc.* **83**, 3583.
Shirley, D. A. (1954). *Org. React.* **8**, 28.
Shirley, D. A., Johnson, J. R., Jr., and Hendrix, J. P. (1968). *J. Organometal. Chem.* **11**, 209.
Shriner, R. L. (1942). *Org. React.* **1**, 1.
Slocum, D. W., and Gierer, P. (1968). *Abstr., 20th Southeast. Reg. Meet. Amer. Chem. Soc., Tallahassee, Fla.* Pap. No. 102; cf. also Paul and Tchelitcheff (1968).
Slocum, D. W., and Gierer, P. L. (1971). *Chem. Commun.* p. 305.
Slocum, D. W., and Jennings, C. A. (1971). *Abstr., 161st Nat. Meet. Amer. Chem. Soc., Los Angeles* Sect. ORGN, Pap. No. 186.
Slocum, D. W., and Koonsvitsky, B. P. (1969). *Chem. Commun.* p. 846.
Slocum, D. W., Book, G., and Jennings, C. A. (1970). *Tetrahedron Lett.* p. 3443.
Slocum, D. W., Englemann, T. R., and Jennings, C. A. (1968). *Aust. J. Chem.* **21**, 2319.
Slocum, D. W., Englemann, T. R., Ernst, C., Jennings, C. A., Jones, W., Koonsvitsky, B. P., Lewis, J., and Shenkin, P. (1969). *J. Chem. Educ.* **46**, 144.
Slocum, D. W., Jennings, C. A., Englemann, T. R., Rockett, B. W., and Hauser, C. R. (1971). *J. Org. Chem.* **36**, 377.
Slocum, D. W., Koonsvitsky, B. P., and Ernst, C. E. (1972). *J. Organometal. Chem.* **38**, 125.
Smissman, E. E., and Hite, G. (1959). *J. Amer. Chem. Soc.* **81**, 1201.
Smith, H. (1963). "Organic Reactions in Liquid Ammonia." Wiley (Interscience), New York.
Smith, H. D., and Hauser, C. R. (1969). *J. Amer. Chem. Soc.* **91**, 7774.
Smith, J. G., and Veach, C. D. (1967). *Can. J. Chem.* **45**, 1785.
Smith, L. I. (1943). *Org. Syn.* **II**, 360.
Sommelet, M. (1937). *C. R. Acad. Sci.* **205**, 56.
Speziale, A. J., Ratts, K. W., and Bissing, D. E. (1965). *Org. Syn.* **45**, 33.
Stork, G., Brizzolara, A., Landesman, H., Szmuszkovicz, J., and Terrell, R. (1963). *J. Amer. Chem. Soc.* **85**, 207.
Streitwieser, A., Jr., and Ziegler, G. R. (1969). *J. Amer. Chem. Soc.* **91**, 5081.
Sutherland, R., and Unni, A. (1970). *Chem. Commun.* p. 555.
Swamer, F. W., and Hauser, C. R. (1946). *J. Amer. Chem. Soc.* **68**, 2647.
Taylor, E. C., and McKillop, A. (1970). *Accounts Chem. Res.* **3**, 338.
Thayer, F. K. (1941). *Org. Syn.* **I**, 398.
Thompson, T., and Stevens, T. S. (1932). *J. Chem. Soc., London* p. 1932.
Ustynyuk, Y. A., and Perevalova, E. G. (1964). *Izv. Akad. Nauk SSSR, Ser. Khim.* p. 62.
Ustynyuk, Y. A., Perevalova, E. G., and Nesmeyanov, A. N. (1964). *Izv. Akad. Nauk SSSR, Ser. Khim.* p. 70.
Wadsworth, W. S., Jr., and Emmons, W. D. (1965). *Org. Syn.* **45**, 44.
Warnhoff, E. W., Wong, C. M., and Tai, W. T. (1968). *J. Amer. Chem. Soc.* **90**, 514.
Watanabe, H., Gay, R. L., and Hauser, C. R. (1968). *J. Org. Chem.* **33**, 900.
Watanabe, H., Mao, C. L., and Hauser, C. R. (1969a). *J. Org. Chem.* **34**, 1786.

Watanabe, H., Schwarz, R. A., and Hauser, C. R. (1969b). *Chem. Commun.* p. 287.
West, R., and Gornowicz, G. A. (1971). *J. Amer. Chem. Soc.* **93**, 1720.
West, R., and Jones, P. C. (1968). *J. Amer. Chem. Soc.* **90**, 2656.
West, R., and Jones, P. C. (1969). *J. Amer. Chem. Soc.* **91**, 6156.
West, R., Carney, P. A., and Mineo, I. C. (1965). *J. Amer. Chem. Soc.* **87**, 3788.
Westheimer, F. H. (1936a). *J. Amer. Chem. Soc.* **58**, 2209.
Westheimer, F. H. (1936b). *J. Org. Chem.* **1**, 1339.
Whitmore, F. C., and George, R. S. (1942). *J. Amer. Chem. Soc.* **64**, 1239.
Wittig, G., and Löhman, L. (1942). *Justus Liebigs Ann. Chem.* **550**, 260.
Wittig, G., Mangold, R., and Felletschin, G. (1948). *Justus Liebigs Ann. Chem.* **580**, 116.
Wolfe, J. F., and Mao, C. L. (1967). *J. Org. Chem.* **32**, 1977.
Wolfe, J. F., Harris, T. M., and Hauser, C. R. (1964). *J. Org. Chem.* **27**, 3249.
Wolfe, J. F., Mao, C. L., Bryant, D. R., and Hauser, C. R. (1966). *J. Org. Chem.* **31**, 3725.
Wong, K. H., Konizer, G., and Smid, J. (1970). *J. Amer. Chem. Soc.* **92**, 666.
Work, S. D., Bryant, D. R., and Hauser, C. R. (1964). *J. Org. Chem.* **29**, 722.
Zimmerman, H. E., and Smentowski, F. J. (1957). *J. Amer. Chem. Soc.* **79**, 5455.
Zook, H. D., and Gumby, W. L. (1960). *J. Amer. Chem. Soc.* **82**, 1386.

6

RADICAL IONS

Glen A. Russell and Robert K. Norris

I. Introduction	423
A. Scope of the Chapter	423
B. Definitions	424
C. Historical Outline of the Discovery and Recognition of Radical Ions	425
II. Formation of Radical Ions	426
A. Radical Anions	426
B. Radical Cations	429
III. Methods of Investigation of Radical Ions	431
A. Electrochemical Methods	431
B. Magnetic Susceptibility	431
C. Spectrophotometric Methods	432
D. Electron Spin Resonance Spectroscopy	432
IV. Reactions Involving Radical Ions	434
A. Observable Radical Ions	434
B. Transient Radical Ions	437
References	444

I. Introduction

A. SCOPE OF THE CHAPTER

The reactions of radical ions and reactions which involve transient radical ion formation are established as an important part of organic chemistry. This chapter briefly traces the historical discovery and initial investigation of radical ions, mentions the modes of formation and methods of investigating radical ions, and discusses the reactions involving radical ions in some depth. The electron spin resonance description has been minimized because of the large number of excellent reviews available on this subject.

B. Definitions

A radical ion may be defined as a molecule which in addition to having one or more unpaired electrons has a net positive (radical cation) or net negative charge (radical anion).

A formal representation of radical ion formation from a molecule (M) is shown in Eqs. (1) and (2).

$$M + e \longrightarrow M\cdot^- \quad \text{radical anion} \tag{1}$$

$$M - e \longrightarrow M\cdot^+ \quad \text{radical cation} \tag{2}$$

Radical ions may arise from "dismutation" processes (Michaelis and Schubert, 1938), formally symbolized (e.g., Russell et al., 1962b) in Eqs. (3) and (4), where π is a molecule having a conjugated unsaturation.

$$\pi + \pi^{2-} \rightleftharpoons 2\pi\cdot^- \tag{3}$$

$$\pi + \pi^{2+} \rightleftharpoons 2\pi\cdot^+ \tag{4}$$

The terms disproportionation and comproportionation (Anschütz et al., 1944) have been applied to reactions (3) and (4) when proceeding from right to left or left to right, respectively.

Radical ions may be formally subclassified according to the orbitals in which the unpaired electron(s) reside. These orbitals may be σ, π, or "p type" and bonding, antibonding, or nonbonding. σ Radical ions may be bonding-type radical cations such as $[X\cdot Y]^+$, which are normally found only in mass spectrometry and in which the unpaired electron is in the bonding σ orbital, or they may be antibonding-type radical anions such as $[R-I]\cdot^-$, where the unpaired electron is in a σ^* orbital. The nonbonding or "p type" radical ions such as $R_3N\cdot^+$ have the unpaired electron in a nonbonding, nondelocalized orbital (usually a p orbital or a hybrid orbital with high p character). By far the largest and most common type of radical ions are the π type. The electronic configuration of these radical ions is best exemplified by the molecular orbital (MO) diagram for benzene (**1**).

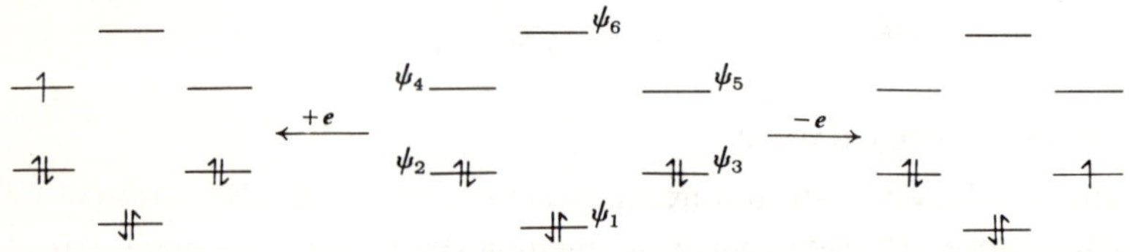

Addition of an electron to benzene will give the benzene radical anion (**1a**) in which the additional unpaired electron goes into the lowest unoccupied MO, which is antibonding in this case. Similarly, removal of an electron leaves the benzene radical cation (**1b**) having an unpaired electron in its highest previously fully occupied MO, which in this case is a bonding orbital. A detailed description of the MO treatment of benzene radical cations and anions together with descriptions of the effect of substitution on the properties of ψ_2 and ψ_3 or ψ_4 and ψ_5 has been presented by Bowers (1968) and also by Vincow (1968).

C. HISTORICAL OUTLINE OF THE DISCOVERY AND RECOGNITION OF RADICAL IONS

In 1836, Laurent observed that on addition of potassium hydroxide solution to benzil, a deep blue coloration developed. This colored species, now recognized as benzil semidione, (**2**; Ar = phenyl) was probably the

$$\text{Ar—C(O·)=C(O}^-\text{)—Ar} \longleftrightarrow \text{Ar—C(O}^-\text{)=C(O·)—Ar}$$
(**2**)

first radical ion observed. Liebermann and Homeyer (1879) further investigated this phenomenon, and in 1885, Bamberger found that a color reaction was given by many α-diketones. Fischer (1882) also observed similar color changes [now ascribed to furil semidione (**2**; R = 2-furyl)] on treatment of either furil or furoin with alcoholic alkali. Scholl (1899) found that whereas heat and/or strongly basic conditions were required for color formation from α-diketones, a mixture of benzil and benzoin gave a color reaction under much milder conditions [now recognized as the type of reaction shown previously in Eq. (3)]. Beckmann and Paul (1891) observed that benzophenone formed deep blue solutions when treated with sodium under an inert atmosphere. They and later workers (e.g., Schlenk and Weickel, 1911) recognized the radical nature of this species, now known as benzophenone ketyl (**3**).

$$\underset{\underset{O^-}{|}}{C_6H_5\text{—}\overset{\cdot}{C}\text{—}C_6H_5} \longleftrightarrow \underset{\underset{O\cdot}{|}}{C_6H_5\text{—}\overset{-}{C}\text{—}C_6H_5}$$
(**3**)

Treatment of tetraarylhydrazines with acids or halogens (Wieland, 1907) and of triarylamines with bromine at low temperatures (Wieland and Wecker, 1910) also was found to give violet or deep blue color reactions now ascribed to hydrazinium (**4**) and triarylaminium radicals (**5**). Schlenk et al. (1914) observed color formation on treatment of many aromatic hydrocarbons, olefins, and azo compounds with sodium in inert solvents. These colorations were again due to radical anion formation.

$$\left[\begin{array}{c} \text{Ar} \\ \diagdown \\ \text{Ar} \end{array} \overset{..}{\text{N}}\text{—}\text{N} \begin{array}{c} \text{Ar} \\ \diagup \\ \text{Ar} \end{array} \right]^{+} \qquad \text{Ar}_3\text{N}\cdot{}^+$$

(**4**)   (**5**)

From these early observations until 1935 there was much controversy as to the structure of these colored species. Willstätter and Piccard (1908) opposed the idea of elements displaying less than normal valence or having unpaired spins. This belief combined with the known dimeric structure of quinhydrone (**6**) (the green solid obtained by mixing equimolar proportions of quinone and hydroquinone) led them to propose "meriquinone" (or dimeric) structures, even in solution. They proposed structure **7**, for Wurster's red, rather than

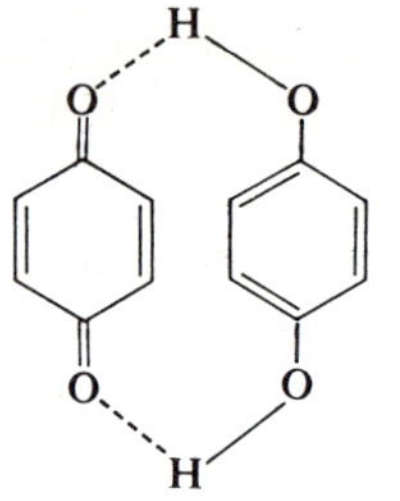

(6) 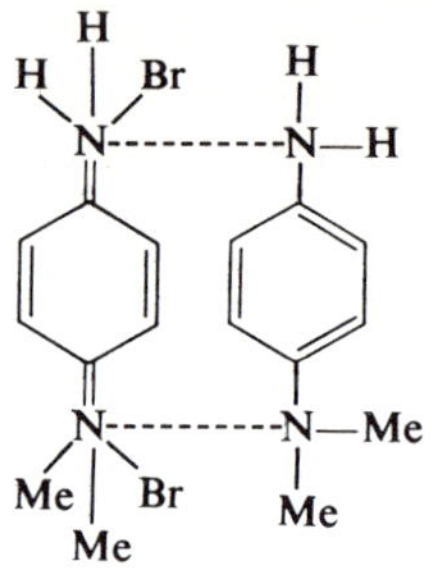   (7)

the monomeric paramagnetic formula. Weitz and Schwechten (1926, 1927), on the other hand, supported monomeric structures for their hydrazinium and triarylamminium radicals. In 1935 Michaelis, in an extensive electrochemical investigation, firmly established the monomeric radical ion structures in solution for many of the radical ions known at that time. Sugden (1934), Müller, and Wiesemann (1936, 1938a,b), and Müller and Janke (1939) also supported this concept by use of magnetic susceptibility measurements. It was also in the period from the 1930's onward that many more radical ions were prepared and identified using these electrochemical and magnetic susceptibility measurements. The application of ESR spectroscopy to the detection and study of radical ions in the late 1950's added renewed interest.

II. Formation of Radical Ions

A. RADICAL ANIONS

1. *Reduction*

   a. Electrolytic Reduction. Michaelis (1935), Michaelis and Schubert (1938), Michaelis and Schwarzenbach, 1938), and Michaelis *et al.* (1939, 1940) had shown by using a combination of electrolytic and spectrophotometric methods that radicals could be produced electrolytically. The radical anions of anthracene, benzophenone, and anthraquinone were prepared and observed by polarographic reduction of the parent molecules followed by transfer of the radical ions in a frozen matrix to the cavity of an ESR spectrometer (Austen et al., 1958). The so-called *intra muros* method, whereby radical ions may be generated and observed in the ESR cavity, was introduced by Maki and Geske in 1959 (1959a,b) and since has developed into the most versatile method of radical ion production. The utility of this method may readily be seen by the wide range of solvents and the variety of types of functional groups which can be employed. The radical anions of nitrobenzene and substituted nitrobenzenes may be formed electrolytically in acetonitrile (Maki and Geske, 1960, 1961), water (Piette et al., 1962), dimethyl sulfoxide, and dimethylformamide (Rieger and Fraenkel, 1963; Bernal and Fraenkel, 1964). In addition to the nitroaromatics already mentioned, some of the wide range of compounds capable of being electrolytically reduced to their radical anions are aromatic hydrocarbons (azulenes, Bernal et al., 1962; phenanthrene, Glarum and Snyder, 1962; and anthracene, Bolton and Fraenkel, 1964); aromatic aldehydes, ketones, and amides (Rieger and Fraenkel, 1962); α-diketones (Dehl and Fraenkel, 1963); fluorinated aliphatic ketones (Janzen and Gerlock, 1967, 1968); unsaturated aliphatic ketones (Russell and Stevenson, 1971); aromatic and aliphatic nitriles (Rieger et al., 1963); 1,3-butadiene (Levy and Meyers, 1964); nitroalkanes (Piette et al., 1962); and *gem*-dinitroalkanes (Shapiro et al., 1968).

b. Metal Reductions. The reduction of molecules using alkali metal has generated some radical anions difficult, if not impossible, to prepare by any other means. The reduction of readily reduced molecules such as benzophenone, stilbene, anthracene, and azobenzene have already been remarked on. The reduction of naphthalene (Scott et al., 1936), benzene, and alkylated benzenes (Tuttle and Weissman, 1958) to their radical anions exemplifies the application of the method to the hydrocarbons. The ketyls derived from hindered aliphatic ketones such as hexamethylacetone (Favorski and Nazarow, 1934; Hiroto and Weissman, 1960) and small ring ketones (Russell et al., 1970b), silyl ketones (P. R. Jones and West, 1968) and even from simple aliphatic ketones such as acetone and methyl ethyl ketone (Bennett et al., 1968) have been reported. Nitrobenzenes are also readily reduced by alkali metals to give radical anions observable by ESR in solution (Weissman et al., 1953) and in some cases isolable in the solid state (Russell and Bemis, 1967). The *o*-semiquinones are also readily prepared from *o*-quinones in ethereal solvents using alkali metals or their amalgams (Müller et al., 1965).

c. Other Reductive Methods. Some of the more stable radical anions can be generated using standard reducing agents. Sodium dithionate has been used in the generation of semiquinones from quinones (Das and Fraenkel, 1962), semidiones from α-diketones (Adams et al., 1958; Russell and Young, 1966), and nitrobenzene radical anions from nitrobenzenes (Kolker and

Waters, 1963). Zinc in alcoholic base has also been used in semidione formation. Glucose in aqueous alkali readily reduces *gem*-dinitroalkanes to the corresponding dinitroalkyl radical dianions (Lagercrantz *et al.*, 1966). Sodium borohydride in basic aqueous or aqueous alcoholic solutions has been found to yield nitroaromatic radical anions from nitroaromatics (Swanwick and Waters, 1970). Hydroxyl radicals produced in the β-irradiation of acqueous solutions (Eiben and Fessenden, 1971) or in the $Ti^{3+}-H_2O_2$ system (Norman and Gilbert, 1967) abstract the α protons from alcohols, giving radicals which readily reduce nitro compounds to their radical ions; for example, with 2-propanol and nitrobenzene the sequence is as shown in Eqs. (5) and (6).

$$\cdot OH + (CH_3)_2CHOH \longrightarrow H_2O + (CH_3)_2\dot{C}OH \qquad (5)$$

$$(CH_3)_2\dot{C}OH + C_6H_5NO_2 \longrightarrow (CH_3)_2CO + H^+ + C_6H_5NO_2\cdot^- \qquad (6)$$

High energy electron beams (2.8 MeV) have been used to reduce acetone, glyoxylic acid, oxalic acid, and nitroalkanes to their respective radical anions (Eiben and Fessenden, 1968, 1971).

2. Oxidation Reactions

Atmospheric oxygen in the presence of base can oxidize some molecules to the radical anions of the corresponding dehydrogenated molecule as shown in Eq. (7).

$$\pi H_2 \xrightarrow{B^-} \pi^{2-} \xrightarrow{O_2} \pi\cdot^- + O_2\cdot^- \qquad (7)$$

This phenomenon was first recognized in the reaction of α-hydroxy ketones in basic solution with oxygen by Weissberger (1932). Hydrazobenzene (Russell and Strom, 1964) and many dihydroaromatics are oxidized in basic solutions via radical anion intermediates (Russell *et al.*, 1962a,b).

3. Disproportionation and Electron-Transfer Processes

Treatment of *o*- and *p*-nitrotoluene in basic solution in the absence of oxygen gives rise to appreciable concentrations of the corresponding nitroaromatic radical anions (Russell and Janzen, 1962). The mode of formation is indicated in Eqs. (8)–(10).

$$p\text{-}NO_2C_6H_4CH_3 \xrightarrow{B^-} p\text{-}NO_2C_6H_4CH_2^- \qquad (8)$$

$$p\text{-}NO_2C_6H_4CH_2^- + p\text{-}NO_2C_6H_4CH_3 \longrightarrow [p\text{-}NO_2C_6H_4CH_3]\cdot^- + p\text{-}NO_2C_6H_4CH_2\cdot \qquad (9)$$

$$2\, p\text{-}NO_2C_6H_4CH_2\cdot \longrightarrow (p\text{-}NO_2C_6H_4CH_2)_2 \qquad (10)$$

6. RADICAL IONS

Another type of disproportionation reaction is the formation of semidiones from α-hydroxy ketones in the absence of oxygen as shown in Eqs. (11)–(13).

$$\text{RCOCH(OH)R} \xrightarrow{B^-} \xrightarrow{B^-} \text{RC(O}^-)\text{=C(O}^-)\text{R} \quad (11)$$

$$\text{RCOCH(OH)R} \xrightarrow{B^-} \text{RCOCOR} + \text{RCH(OH)CH(OH)R} \quad (12)$$

$$\text{RCOCOR} + \text{RC(O}^-)\text{=C(O}^-)\text{R} \rightleftharpoons 2\,\text{RC(O·)=C(O}^-)\text{R} \quad (13)$$

The electronic comproportionation step (13) is but one example of the general process symbolized previously [Eq. (3)]. An extensive list of compounds displaying this phenomenon has been reported by Russell et al. (1962b).

Many carbanions or nitranions may function as electron donors (as well as radical anions themselves, cf. Section IV, A) to unsaturated systems. This process is symbolized in Eqs. (14) and (15) (Strom et al., 1965).

$$R^- + \pi \longrightarrow \pi\cdot^- + R\cdot \quad (14)$$

$$2\,R\cdot \longrightarrow R_2 \quad \text{(or other radical destructive routes)} \quad (15)$$

Examples are given in Eqs. (16) and (17) [see also Eqs. (8)–(10)].

$$2\,C_6H_5COCOC_6H_5 + 2\,(CH_3)_2C=NO_2^- \longrightarrow$$
$$2\,C_6H_5C(O\cdot)=C(O^-)C_6H_5 + (CH_3)_2C(NO_2)C(CH_3)_2NO_2 \quad (16)$$

$$2\,C_6H_5NO_2 + 2\,C_6H_5S^- \longrightarrow 2\,[C_6H_5NO_2]\cdot^- + C_6H_5SSC_6H_5 \quad (17)$$

The radical anions resulting from electron-transfer processes are often insufficiently stable to be observed by the usual procedures (e.g., ESR). These radical anions and their subsequent reactions are discussed fully in Section IV,B.

B. Radical Cations

1. *Oxidation*

a. Electrolytic Oxidation. In contrast to electrolytic reduction, electrolytic oxidation is of limited application and is most commonly used only for the generation of nitrogen-containing radical cations. Examples are Wurster-type radical cations (*p*-phenylenediaminium radical cations), prepared using the electrolytic method by Michaelis et al. (1939), dihydropyrazines and related nitrogen-containing species (Barton and Fraenkel, 1964), triarylaminium radical cations (Seo et al., 1966), tetraalkylhydrazine (Nelson, 1966), and phenothiazine radical cations (Billon et al., 1964).

b. Chemical Oxidation. This procedure is by far the most useful method of preparing radical cations. The most common oxidizing agent is sulfuric acid, which acts both as solvent and oxidant. Following early reports of paramagnetism of sulfuric acid solutions of anthraquinone and thianthrene (e.g., Hisshon *et al.*, 1953), many of the radical cations of aromatic hydrocarbons and their derivatives have been prepared in this way. The first report of the use of sulfuric acid to generate the perylene radical cation (Yokozawa and Miyashita, 1956) was followed by the preparation of the radical cations of anthracene, tetracene, and naphthacene (Weissman *et al.*, 1957; Carrington *et al.*, 1959). Even the unstable benzene radical cation has been prepared by UV irradiation of a sulfuric acid glass containing benzene (Carter and Vincow, 1967). Many other Lewis acid-oxidizing systems have been used, including $BF_3-CF_3CO_2H$, $SbCl_3$, $AlCl_3$, I_2, $K_3Fe(CN)_6-HClO_4$, and the very efficient system $SbCl_5-CH_2Cl_2$ (Lewis and Singer, 1965; these authors also give a comprehensive referenced list of oxidants used up to 1965). Some more recently used oxidants include $AlCl_3-CH_3NO_2$ (Forbes and Sullivan, 1966) and cobalt(III) acetate in trifluoroacetic acid under flow conditions (Dessau *et al.*, 1970). The oxidation of many simple aromatic amines and hydrazines by halogens, lead dioxide, and silver perchlorate–iodine has been reviewed by Weitz (1954). Iodine in tetrahydrofuran has been used to produce 1,4-dihydro-*s*-tetrazine radical cations (Tolles *et al.*, 1969).

2. *Reduction*

The production of radical cations by reductive techniques is effectively limited to molecules containing heteroatoms, particularly nitrogen. Typical examples involve the reduction of dipositive ions to radical cations, formally symbolized in Eq. (18).

$$\pi^{2+} + e \rightleftharpoons \pi^{\cdot +} \tag{18}$$

Zinc dust in organic solvents or acidic media has been used to form the radical cations of dipyridyl (Dimroth and Heene, 1921) and *N*-alkylated dipyridyls (Viologens; Weitz and Ludwig, 1922; Weitz, 1954). Chromium(II) and titanium(II) have also been used to reduce dipyridyls (Dimroth and Frister, 1922) and thiazenes (Ayscough and Thomson, 1962), respectively. Methoxide ion also acts as a reducing agent, reducing the paraquat ion (*N*,*N*-dimethyl-4,4'-bipyridylium dichloride) to the corresponding radical cation (Farrington *et al.*, 1969).

3. *Disproportionation Reactions*

The disproportionation process represented in Eq. (4) has been observed in many instances. For example, tetraphenylxylylene (**8**) reacts with the corresponding dicarbonium ion (**9**) to yield the radical cation (**10**) [Eq. (19)]

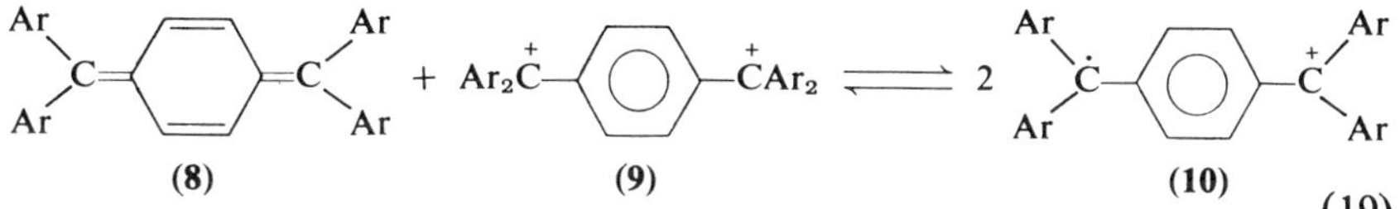

6. RADICAL IONS

$$\text{(8)} + \text{(9)} \rightleftharpoons 2 \text{ (10)} \quad (19)$$

(Weitz, 1954; also Hart et al., 1964). Similarly, the dication of 4,4′-dihydroxybiphenyl reacts with the neutral molecule to give the radical cation (Forbes and Sullivan, 1966). Indolizine dimers (Hünig et al., 1964) and tetrakis(dimethylamino)ethylene (Kuwata and Geske, 1964) also display this phenomenon.

III. Methods of Investigation of Radical Ions

A. Electrochemical Methods

Michaelis (1935), Michaelis and Schubert (1938), and Michaelis and Schwarzenbach (1938) demonstrated radical ion formation potentiometrically and that Eq. (20) was obeyed.

$$E = E_0 + \frac{RT}{nF} \log_e \frac{[\text{oxidant}]}{[\text{reductant}]} \quad (20)$$

The one-electron change ($n = 1$) involved in radical ion production results in a more rapid change in electrode potential than a two-electron change ($n = 2$). Polarography has also been extensively applied to the investigation of radical ion formation (see Heyrovský and Kůta, 1966).

Conductivity measurements have been used not only in the detection of radical ions but also to investigate dimerization and ion-pairing phenomena. Wooster (1934) showed by this procedure that sodium benzophenone ketyl is largely dissociated in liquid ammonia. Buschow et al. (1965) and Slates and Szwarc (1965) observed ion-pairing phenomena in solutions of aromatic radical anions by use of this method.

B. Magnetic Susceptibility

The theory of the magnetic susceptibility method is adequately covered by Hutchison (1955) and, in a more descriptive fashion, by Waters (1948). In summary, it suffices to say that stable paramagnetic species are attracted to a magnetic field and that the magnitude of this attraction (with diamagnetic corrections) can be directly correlated with the number of unpaired spins in a particular sample. Examples of application are equilibria involving ketyls (Müller and Wiesemann, 1936) and, more recently, solid-state measurements of the magnetic susceptibility of potassium nitrobenzenide (Russell and Bemis, 1967).

C. Spectrophotometric Methods

The formation of deep colorations was, as pointed out in the historical introduction (Section I,C), one of the prime reasons for the detection of radical ion formation. More recently, UV-visible spectroscopy has been used to investigate the finer details of the equilibria involved in solutions containing radical ions. The formation of diamagnetic complexes or various types of ion pairs has been investigated for ketyls (Hirota and Weissman, 1964), tetracyanoquinodimethane radical anion (Boyd and Phillips, 1965), sodium naphthenide (Hogen-Esch and Smid, 1966), anthracene, pentacene, and tetracene radical anions (Buschow et al., 1965), and Wurster's blue perchlorate (Hauser and Murrell, 1957).

D. Electron Spin Resonance Spectroscopy

By far the most sensitive and informative tool for investigation of radical anions is electron spin resonance spectroscopy. In a magnetic field the spin transition of an unpaired electron can be observed. The interaction of the electron spin with nuclear spins (^{1}H, ^{13}C, ^{14}N, ^{19}F, etc.) gives hyperfine splitting constants which not only indicate the type and number of nuclei present in the radical ion but also give an indication of spin density (see the review by Bolton, 1968) at these particular nuclei. The application of ESR to radicals, and to radical ions in particular, has been extensively reviewed by many authors. The following references deal generally with applications of ESR to radical ions: Bowers (1965), Ayscough (1967), Forrester et al. (1968), and Scheffler and Stegmann (1970). More specific treatment of the ESR spectra of certain groups of radical ions is also available: ketyls (Hirota, 1968), semidiones (Russell, 1968), aromatic radical anions (M. T. Jones, 1968), radical cations (Vincow, 1968), and sulfur-containing aromatics (Urberg and Kaiser, 1968).

Electron spin resonance is a useful tool in detecting or demonstrating the presence of radical ions as reaction intermediates or reaction products. The value of equilibrium constants for electronic disproportion processes such as reaction (3) can be easily measured in this way, e.g., Eq. (21) (Russell et al.,

$$C_6H_5N{=}NC_6H_5 + [C_6H_5N{-}NC_6H_5]^{-2} \underset{}{\overset{K=6}{\rightleftharpoons}} 2\,[C_6H_5N{=}NC_6H_5]^{\cdot-} \quad (21)$$

1968). Moreover, radical anions can be used as a spin label which will reveal valuable information concerning the structure of the labeled species. Among processes readily investigated in this manner are conformational equilibria, valence isomerization, ion-pairing phenomena, and hydrogen–deuterium exchange studies of ionization equilibria. An example of the use of this technique is the generation of a paramagnetic species by oxidation of either the bicyclic ketone **(11)** or the tricyclic ketone **(12)** (Russell *et al.*, 1970a). Figure 1 requires the formation of a semidione with interaction of the unpaired electron with pairs of equivalent hydrogen atoms (3.20, 2.38, and 0.10 G) and with single hydrogen atoms having a hyperfine interaction of 1.64 and 0.10 G. The multiplicities observed are inconsistent with a planar structure such as **13** and require **14** as the stable valence isomer in this system.

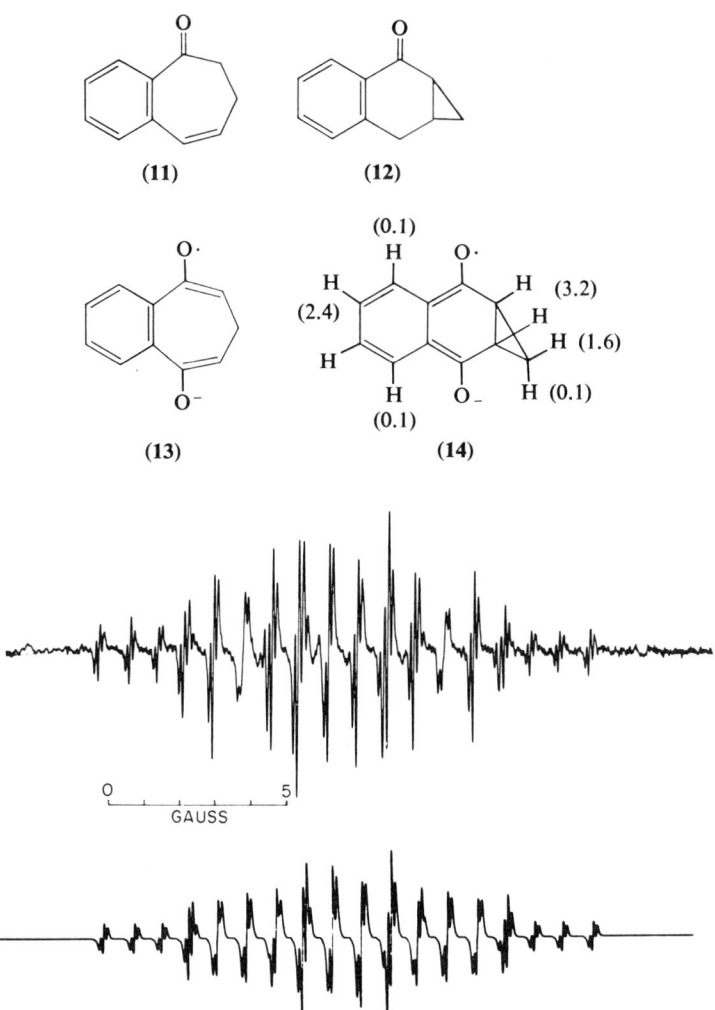

Fig. 1. Top: first derivative ESR spectrum of autooxidation product of **11** or **12** in dimethyl sulfoxide solution in the presence of potassium *tert*-butoxide. Bottom: calculated spectrum with Lorentzian linewidth of 0.1 G and hyperfine splittings by hydrogen atoms of 3.20, 3.20, 2.38, 2.38, 1.64, 0.10, 0.10, and 0.10 G.

IV. Reactions Involving Radical Ions

A. Observable Radical Ions

The radical ions discussed in this section are those which are detectable by ESR and, in some cases, capable of long term existence in suitable solvents or isolable in the solid state. The radical anions of naphthalene and benzophenone will be principally but not exclusively used in the following discussions.

1. Comproportionation and Dimerization Reactions

The comproportionation reactions [symbolized in Eqs. (3) and (4)] have been discussed previously (Sections II,A,5 and II,B,3). However, a further complication may arise when dimerization occurs [Eq. (22)].

$$2\,\pi\cdot^{-} \;\rightleftharpoons\; {}^{-}\pi{-}\pi^{-} \tag{22}$$

Hence, in some of the less polar solvents, ketyls exist predominantly in the form of pinacolates. Naphthalene radical anions and other aromatic radical anions show little tendency to dimerize. Solid-state measurements on many of the more stable radical ions do show some dimerization, however.

2. Hydrolysis of Radical Anions

The protonation of a radical anion (weak base) may proceed via the dianion (strong base) which arises from a comproportionation reaction [see Eq. (3)]. Naphthalene (Nap) radical anion gives a mixture of naphthalene and dihydronaphthalene. Benzophenone ketyl gives benzhydrol and benzophenone in the absence of acid (Schlenk and Weickel, 1911; Wooster, 1928) but tetraphenylpinacol (Bachman, 1933) in the presence of acid. The hydrolysis of nitrobenzene radical ion is pH dependant. Protonation initially gives the radical (**15**) which then disproportionates or reacts with more of the radical anion to give mixtures which include nitrobenzene, nitrosobenzene, phenylhydroxylamine, and azoxybenzene depending on the pH. At pH 10.5, Metcalfe and Waters (1969) propose the sequence in Eqs. (23)–(25).

$$[PhNO_2]\cdot^{-} + H^{+} \;\rightleftharpoons\; [PhNO_2H]\cdot \tag{23}$$
$$(\mathbf{15})$$
$$\mathbf{15} + [PhNO_2]\cdot^{-} \longrightarrow PhNO_2 + [PhNO_2H]^{-} \tag{24}$$
$$[PhNO_2H]^{-} \longrightarrow PhNO + OH^{-} \tag{25}$$

3. Electron-Transfer Reactions

This reaction is by far the most common reaction of radical ions. Nearly all radical anions are destroyed by atmospheric oxygen. This involves the electron-transfer process of Eq. (26).

$$R\cdot^- + O_2 \longrightarrow R + O_2\cdot^- \qquad (26)$$

Iodine may also serve as an electron acceptor as may another π system of higher electron affinity. The intensely colored solutions of the radical anions of naphthalene, benzophenone, and di-*tert*-butyl ketones are destroyed from the air–liquid interface down into the solution clearly demonstrating this reaction. Russell and Bemis (1967) observed almost quantitative formation of potassium superoxide on treatment of potassium nitrobenzenide with oxygen.

Naphthalene radical anion undergoes rapid electron transfer to alkyl halides [Eq. (27); X = Cl, Br, I], forming a variety of products indicated in Eqs. (27)–(30).

$$\text{Nap}\cdot^- + RX \longrightarrow \text{Nap} + [RX]\cdot^- \longrightarrow R\cdot + X^- \qquad (27)$$

$$R\cdot + \text{Nap}\cdot^- \longrightarrow \text{Nap} + R^- \xrightarrow{\text{solvent}} RH \qquad (28)$$

or

$$2R\cdot \longrightarrow R-R \qquad (29)$$

or

$$R\cdot + \text{Nap}\cdot^- \xrightarrow{\text{radical trapping}} \text{alkylated dihydronaphthalenes} \qquad (30)$$

These reactions have been studied in detail by Sargent and Lux (1968) and by Garst *et al.* (1966). Garst and Barton (1969) have investigated the reaction involving alkyl fluorides [Eq. (27); X = F] and find reaction (28) predominates owing to slower reaction and resulting lower concentrations of R·. The formation of alkyl dihydronaphthalenes by radical–radical anion trapping is very similar to the reactions discussed in Section IV,B (see also, Russell, 1970a). The above reactions are of little synthetic utility, but the reactions of sodium naphthenide with 1,3- and 1,4-dihaloalkanes lead to interesting polycyclic molecules (Lipkin *et al.*, 1963). Other synthetic uses of the electron-transfer process are the sodium naphthenide cleavage of toluene sulfonates [Eqs. (31)–(33); Closson *et al.*, 1966] and toluene sulfonamides (Ji *et al.*,

$$\text{Nap}\cdot^- + p\text{-MeC}_6H_4SO_2OR \longrightarrow \text{Nap} + [p\text{-MeC}_6H_4SO_2OR]\cdot^- \qquad (31)$$

$$[p\text{-MeC}_6H_4SO_2OR]\cdot^- \longrightarrow RO^- + p\text{-MeC}_6H_4SO_2\cdot \qquad (32)$$

$$p\text{-MeC}_6H_4SO_2\cdot + \text{Nap}\cdot^- \longrightarrow \text{Nap} + p\text{-MeC}_6H_4SO_2^- \qquad (33)$$

1967), and also benzyl and allyl ethers [Eqs. (34) and (35) and (36)–(38);

Angeleno, 1966]. Sodium naphthenide also readily converts vicinal dihalides

$$\text{EtOCH}_2\text{—CH=CH}_2 + \text{Nap} \cdot ^- \longrightarrow \text{Nap} + \text{EtO}^- + \text{CH}_2\text{=CH—CH}_2 \cdot \quad (34)$$

$$2\,\text{CH}_2\text{=CH—CH}_2 \cdot \longrightarrow (\text{CH}_2\text{=CH—CH}_2)_2 \quad (35)$$

$$\text{PhOCH}_2\text{Ph} + \text{Nap} \cdot ^- \longrightarrow [\text{PhOCH}_2\text{Ph}] \cdot ^- + \text{Nap} \quad (36)$$

$$[\text{PhOCH}_2\text{Ph}] \cdot ^- \longrightarrow \text{PhO}^- + \text{PhCH}_2 \cdot \quad (37)$$

$$\text{PhCH}_2 \cdot + \text{Nap} \cdot ^- \longrightarrow \text{Nap} + \text{PhCH}_2^- \xrightarrow{\text{H}^+} \text{PhCH}_3 \quad (38)$$

to olefins (Garst *et al.*, 1969). Presumably, the intermediate radical is converted to a carbanion [see Eq. (28)] which then eliminates the second chlorine or bromine atom. A similar mechanism has been suggested for the conversion of vicinal dibromides to olefins by the anion of diphylacenitrile (Korzan *et al.*, 1971).

It has been recently shown that chemically induced spin polarization phenomena are associated with the reaction of naphthalene radical anion and alkyl halides, confirming the radical nature of the intermediates (Garst *et al.*, 1970; Garst, 1971). The ketyls of benzophenone (Schlenk and Weickel, 1911; Wooster, 1928; Hirota and Weissman, 1964) and fluorenone (Morantz and Warhurst, 1955) also give C-alkylated derivatives on treatment with alkyl halides.

Benzophenone ketyl initiates polymerization of acrylonitrile and methyl methacrylate (Zilkha *et al.*, 1960), and naphthalene radical anion initiates polymerization of styrene (Szwarc, 1956; Lyssy, 1959), processes which again involve electron transfer [e.g., Eqs. (39) and (40)].

$$\text{Nap} \cdot ^- + \text{Ph—CH=CH}_2 \longrightarrow \text{Nap} + \text{Ph—}\overset{-}{\text{CH}}\text{—}\overset{\cdot}{\text{CH}}_2 \quad (39)$$

$$2\,\text{Ph—}\overset{-}{\text{CH}}\text{—CH}_2 \cdot \longrightarrow$$
$$\text{Ph}\overset{-}{\text{CH}}\text{—CH}_2\text{—CH}_2\text{—}\overset{-}{\text{CH}}\text{Ph} \longrightarrow \text{anionic polymerization} \quad (40)$$

The reaction of nitrosobenzene with phenylhydroxylamine in basic solution in DMSO and alcoholic solutions has been followed by Russell *et al.* (1965, 1967) using ESR. In DMSO the nitrosobenzene radical anion forms and is stable [Eq. (41)]. However, in alcoholic solutions the nitrosobenzene

$$\text{PhNO} + \text{PhNHOH} \xrightarrow{\text{B}^-} 2\,\text{PhNO} \cdot ^- \quad (41)$$

radical anion is only observed under fast flow conditions, decaying when the flow is stopped, and the overall result is a near quantitative yield of azoxybenzene. This result is interpreted in Eqs. (42)–(44).

$$[\text{PhNO}] \cdot ^- \rightleftharpoons \text{PhN}(\text{O}^-)\text{N}(\text{O}^-)\text{Ph} \quad (42)$$

$$\text{PhN}(\text{O}^-)\text{N}(\text{O}^-)\text{Ph} + \text{ROH} \rightleftharpoons \text{PhN}(\text{OH})\text{N}(\text{O}^-)\text{Ph} + \overset{-}{\text{OR}} \quad (43)$$

$$\text{PhN}(\text{OH})\text{N}(\text{O}^-)\text{Ph} \longrightarrow \text{Ph—N=N}^+(\text{O}^-)\text{Ph} + \text{OH}^- \quad (44)$$

6. RADICAL IONS

This result explains the formation of all possible azoxyarenes when nitrosoarenes and arylhydroxylamines having different aryl groups are condensed in basic solution.

The reactions of observable radical cations have been studied but only slightly. The mode of preparation of aromatic hydrocarbon radical cations and their poor stability outside of their preparative conditions (e.g., concentrated sulfuric acid) does not readily allow a study of their chemistry. The chemistry of stable nitrogen-containing radical cations is principally restricted to their redox reactions (e.g., Weitz and Schwechten, 1926, 1927). Seo *et al.* (1966) have observed that triarylaminium salts which do not have para substituents dimerize to tetraarylbenzidines [Eq. (45)].

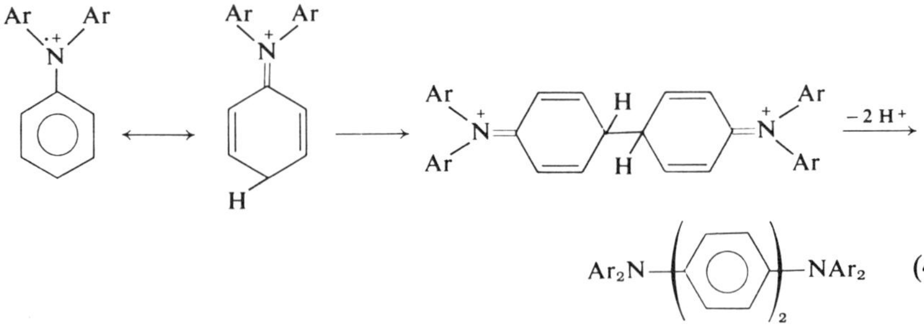

$$Ar_2N-\left(\bigcirc\right)-NAr_2 \quad (45)$$
$$_2$$

The instability of some nitrogen radical cations (e.g., *p*-phenylenediamonium cations without *N*-alkyl groups) may be due to electronic disproportionation to the amine and an unstable dication.

The radical cations of methyl-substituted aromatic hydrocarbons are believed to be involved in oxidations involving transition metal ions, since it has been recently shown by Dessau *et al.* (1970) that under favorable conditions they can be detected by ESR. The reactions of this type of radical cation are discussed in Section IV,B,3.

B. Transient Radical Ions

The radical ions discussed in this section are *not* detectable by ESR or any of the classic methods listed in Section III.

1. *Reactions Involving Nitro Radical Anions*

The first proposal of a nitro radical anion as a reaction intermediate was made in 1954 by Russell to explain the initiation of oxidation by oxygen of 2-nitropropane in deficient base. The initiation step [Eq. (46)] of the chain reaction whose final products are as shown in the overall Eq. (47) must

$$(CH_3)_2{=}NO_2^- + (CH_3)_2CHNO_2 \longrightarrow [(CH_3)_2CHNO_2]\cdot^- + (CH_3)_2\dot{C}NO_2 \quad (46)$$

involve electron transfer to the neutral 2-nitropropane molecule, since in the

$$(CH_3)_2C=NO_2^- + \tfrac{1}{2} O_2 \longrightarrow (CH_3)_2C=O + NO_2^- \qquad (47)$$

presence of excess base [i.e., no un-ionized $(CH_3)_2CHNO_2$ left] the oxidation reaction does not occur. In the presence of excess base and oxygen, however, nitrobenzene acts as a catalytic chain initiator as shown in Eq. (48), and the

$$C_6H_5NO_2 + (CH_3)_2C=NO_2^- \longrightarrow C_6H_5NO_2\cdot^- + (CH_3)_2\dot{C}NO_2 \qquad (48)$$

nitrobenzene radical anion is converted to nitrobenzene and superoxide ion as previously discussed.

Another system whose reactions have been shown to involve radical ions is the p-nitrobenzyl halide system. Hass and Bender (1949) had observed that the treatment of substituted benzyl halides gave predominantly the substituted benzaldehyde, i.e., O alkylation was occurring as shown in Eqs. (49) and (50).

$$ArCH_2X + (CH_3)_2C=NO_2^- \longrightarrow ArCH_2-O\overset{+}{N}(O^-)=C(CH_3)_2 \qquad (49)$$

$$ArCH_2-O\overset{+}{N}(O^-)=CH(CH_3)_2 \longrightarrow ArCHO + (CH_3)_2C=NOH \qquad (50)$$

However, reaction with o- and especially p-nitrobenzyl chloride gave predominantly the C-alkylated compound. Kornblum et al. in 1961, attempted to explain this reaction in terms of a classic S_N2 reaction with different reaction rates for O and C alkylation by the ambident 2-nitro-2-propyl anion on the unreactive benzyl carbon of the nitro compounds, especially since it was observed that the nitrobenzyl bromides and iodides gave successively larger proportions of O alkylation. A radical anion intermediate was later suggested (Kornblum et al., 1964, 1965) as shown in Eqs. (51)–(53) for p-nitrobenzyl

$$p\text{-}NO_2C_6H_4CH_2Cl + (CH_3)_2C=NO_2^- \longrightarrow$$
$$(CH_3)_2\dot{C}NO_2 + [p\text{-}NO_2C_6H_4CH_2Cl]\cdot^- \qquad (51)$$

$$[p\text{-}NO_2C_6H_4CH_2Cl]\cdot^- \longrightarrow p\text{-}NO_2C_6H_4CH_2\cdot + Cl^- \qquad (52)$$

$$p\text{-}NO_2C_6H_4CH_2\cdot + (CH_3)_2\dot{C}NO_2 \longrightarrow p\text{-}NO_2C_6H_4CH_2C(CH_3)_2NO_2 \qquad (53)$$

chloride. This reaction mechanism was subsequently modified to a radical chain mechanism (Russell and Danen, 1966; Kornblum et al., 1966) as shown in Eqs. (51), (52), (54), and (55). The radical anion of the nitrobenzyl chloride formed in Eq. (55) reacts as shown in Eq. (52) to propagate the radical chain.

$$p\text{-}NO_2C_6H_4CH_2\cdot + (CH_3)_2C=NO_2^- \longrightarrow$$
$$[p\text{-}NO_2C_6H_4CH_2C(CH_3)_2NO_2]\cdot^- \qquad (54)$$

$$[p\text{-}NO_2C_6H_4C(CH_3)_3NO_2]\cdot^- + p\text{-}NO_2C_6H_4CH_2Cl \longrightarrow$$
$$p\text{-}NO_2C_6H_4CH_2C(CH_3)_2NO_2 + [p\text{-}NO_2C_6H_4CH_2Cl]\cdot^- \qquad (55)$$

It is instructive to consider some of the factors which make the above mechanism plausible and also some further confirmatory observations. The initial electron-transfer step [Eq. (51)] has been fully substantiated for other nitrobenzenes whose radical anions are stable (Russell et al., 1964). This electron-transfer reaction is light catalyzed as are the above substitution reactions (e.g., Russell and Danen, 1966). The relative rate of the substitution reaction of the *p*-nitrobenzyl chloride with the 2-nitro-2-propyl anion is of the order of 100 times as fast as the rate of substitution of the unsubstituted benzyl chloride, whereas the relative rate of the corresponding benzyl iodides (which give predominantly O alkylation) is only three times as fast. Hence, C- and O-alkylation reactions occur simultaneously, and with the poor leaving groups the S_N2 reaction is slow and the radical chain mechanism (C alkylation) predominates. Oxygen and dinitrobenzenes strongly inhibit the C-alkylation reaction by intercepting the intermediate radical anions. The sterically hindered *p*-nitrocumyl system exhibits all the same reaction characteristics as the *p*-nitrobenzyl system. Hence, a wide variety of reactions are now known [Eq. (56); X = Cl, Y = malonic ester anions, thiophenoxide,

$$\text{CH}_3\underset{\underset{\text{NO}_2}{\bigcirc}}{\overset{\overset{X}{|}}{\underset{|}{C}}}\text{CH}_3 + Y^- \longrightarrow \text{CH}_3\underset{\underset{\text{NO}_2}{\bigcirc}}{\overset{\overset{Y}{|}}{\underset{|}{C}}}\text{CH}_3 + X^- \quad (56)$$

2-nitro-2-propyl, etc. (Kornblum et al., 1967a); X = NO_2, Y as before (Kornblum et al., 1967b); X = NO_2, Cl and Y = various amines (Kornblum and Stuchal, 1970)] in this system and also in entrainment reactions, wherein the 2-nitro-2-propyl anion in small amounts is used to initiate the reaction, and the intermediate *p*-nitrocumyl radical is trapped by anions such as azide or nitrite ion and by amines such as quinuclidine (Kornblum et al., 1970a). Even the *m*-nitrocumyl radical undergoes a radical anion reaction (Kornblum et al., 1968) since the S_N2 reaction here as in other cumyl systems is not feasible.

Many aliphatic nitro compounds are now believed to react via radical anion intermediates. In 1940, Siegle and Hass observed the formation of 2,3-dimethyl-2,3-dinitrobutane on reaction of the 2-nitro-2-propyl anion with the 2-halo-2-nitropropanes. This reaction was shown to proceed via radical anions by Russell and Danen (1966, 1968) as indicated in Eqs. (57)–(60) (X = Cl).

$$(CH_3)_2C\!\!=\!\!NO_2^- + (CH_3)_2C(NO_2)X \longrightarrow$$
$$Me_2\dot{C}NO_2 + [(CH_3)_2C(NO_2)X]\cdot^- \quad (57)$$

$$[(CH_3)_2C(NO_2)X]\cdot^- \longrightarrow (CH_3)_2\dot{C}NO_2 + X^- \quad (58)$$

$$(CH_3)_2\dot{C}NO_2 + (CH_3)_2C{=}NO_2^- \longrightarrow$$
$$[(CH_3)_2C(NO_2)C(CH_3)_2NO_2]\cdot^- \quad (59)$$

$$[(CH_3)_2C(NO_2)C(CH_3)_2NO_2]\cdot^- + (CH_3)_2C(NO_2)X \longrightarrow$$
$$(CH_3)_2C(NO_2)C(CH_3)_2 + [(CH_3)_2C(NO_2)X]\cdot^- \quad (60)$$

The reaction exhibits catalysis by light and inhibition by oxygen, dinitrobenzenes, and "hexaphenylethane." The scope of this reaction has been widened to include α-nitronitriles (X = CN), α-dinitroalkanes (X = NO_2), and α-nitromalonic esters (Russell, 1970a,b; Russell et al., 1971), and also α-nitro esters and α-nitro ketones (Kornblum et al., 1970b; Kornblum and Boyd, 1970) as the electron acceptors in Eq. (57).

Interestingly, in the case of all these nitro aliphatic compounds, except the halo nitro compounds, it is the nitro group which is eliminated. For example, the 2-nitro-2-cyano-2-propyl radical anion reacts as shown in Eqs. (61) and (62).

$$[(CH_3)_2C(NO_2)CN]\cdot^- \longrightarrow (CH_3)_2\dot{C}CN + NO_2^- \quad (61)$$

$$(CH_3)_2\dot{C}CN + (CH_3)_2C{=}NO_2^- \longrightarrow [(CH_3)_2C(CN)C(NO_2)(CH_3)_2]\cdot^- \quad (62)$$

A wide variety of carbanions have now been shown (Russell et al., 1972) to react via radical anion intermediates with 2,2-dinitropropane or 2-bromo-2-nitropropane to give overall reactions as shown in Eqs. (63) and (64).

$$(CH_3)_2C(NO_2)_2 + R{-}\bar{C}(X)(Y) \longrightarrow (CH_3)_2C(NO_2)CR(X)(Y) \quad (63)$$
$$X = CO_2Et, \quad Y = CO_2Et, \quad R = Et$$
$$X = CO_2Et, \quad Y = COMe, \quad R = Et$$
$$X = Y = COMe, \quad R = Me$$

$$(CH_3)_2C(NO_2)Br + R{-}\bar{C}(X)(Y) \longrightarrow (CH_3)_2C(NO_2)CR(X)(Y) \quad (64)$$
$$X = Y = CN, \quad R = Et$$
$$X = (CH_3)_3CCO, \quad Y = CN, \quad R = Me$$
$$X = CO_2Et, \quad Y = CN, \quad R = Me$$

2. Reactions Involving Iodoaryl Radical Anions

Kim and Bunnett (1970) have recently reported that 5-bromo- and 6-bromo- or chloropseudocumenes (**16** or **17**; X = Br or Cl) when treated with potassium amide in liquid ammonia give a constant ratio (aryne ratio) of 5:6-pseudocumidines (**16** and **17**; X = HN_2) as expected of a reaction

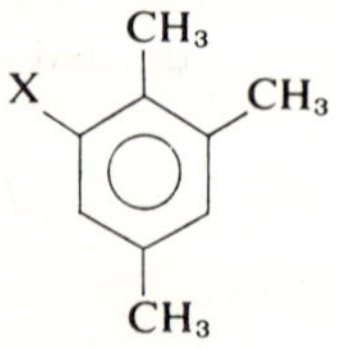

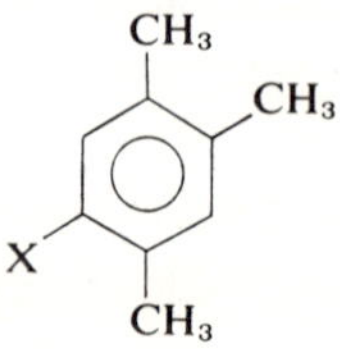

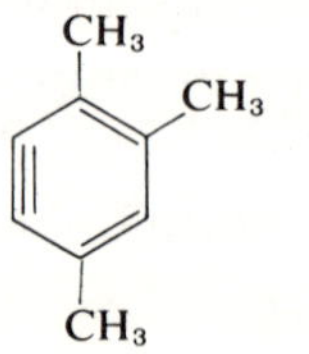

6. RADICAL IONS

involving an aryne intermediate (18). The corresponding 5- and 6-iodopseudocumenes (16 and 17; X = I) react to yield considerably less cine-substituted product. This was interpreted as the result of simultaneous aryne and radical anion reactions. The latter process is given in Eqs. (65)–(68).

$$\text{ArI} + \text{electron donor} \longrightarrow \text{ArI} \cdot^- + \text{residue} \quad (65)$$

$$\text{ArI} \cdot^- \longrightarrow \text{Ar} \cdot + \text{I}^- \quad (66)$$

$$\text{Ar} \cdot + \text{NH}_2^- \longrightarrow \text{ArNH}_2 \cdot^- \quad (67)$$

$$\text{ArNH}_2 \cdot^- + \text{ArI} \longrightarrow \text{ArI} \cdot^- + \text{ArNH}_2 \quad (68)$$

This mechanism was confirmed by noting that 2-methyl-2-nitrosopropane and tetraphenyl hydrazine (both radical trapping agents) brought about substitution ratios for the iodopseudocumenes that closely approached the "aryne ratio." Furthermore, addition of potassium metal (an efficient electron donor) brought about substitution with almost complete absence of rearrangement. A considerable amount of hydrocarbon material was also found using potassium metal, presumably because of hydrogen abstraction from the solvent, liquid ammonia [Eqs. (69) and (70)]. Even more remarkable was

$$\text{Ar} \cdot + \text{RH} \longrightarrow \text{ArH} + \text{R} \cdot \quad (69)$$

$$\text{R} \cdot \longrightarrow \text{nonpropagating steps} \quad (70)$$

the total exclusion of m-anisidine formation when o-halo anisoles were treated with $\text{KNH}_2/\text{K}/\text{NH}_3$ (considerable anisole is formed) compared with appreciable m-anisidine formation in the absence of excess potassium. The reaction of 2-iodo-1,3-xylene, which cannot react via an aryne mechanism, proceeds very smoothly under conditions where excess potassium is present (considerable m-xylene is also formed).

Radical anions derived from a variety of p-nitrohalobenzenes have been detected during nucleophilic aromatic substitution reactions by Shein et al. (1970), and in certain cases it is claimed that the substitution proceeds by a radical mechanism similar to Eqs. (66)–(68) (Shein et al., 1969).

3. Oxidation of Methyl-Substituted Hydrocarbons

Dewar and his co-workers (1966) suggested that oxidation of aromatic hydrocarbons bearing methyl groups probably proceeded via radical cation intermediates as shown in Eqs. (71)–(74). Heiba et al. (1969a) suggested a similar reaction pathway for oxidations involving manganese (III), although

$$\text{ArCH}_3 + \text{Co(III)} \longrightarrow [\text{ArCH}_3] \cdot^+ + \text{Co(II)} \quad (71)$$

$$[\text{ArCH}_3] \cdot^+ \longrightarrow \text{ArCH}_2 \cdot + \text{H}^+ \quad (72)$$

$$\text{ArCH}_2\cdot + \text{Co(III)} \longrightarrow \text{ArCH}_2{}^+ + \text{Co(II)} \quad (73)$$

$$\text{ArCH}_2{}^+ \longrightarrow \text{products} \quad (74)$$

in this case a concurrent free radical process was shown to be taking place, with the radical cation pathway being important for hydrocarbons of lower ionization potential. In this work, Heiba proposed attack on a radical cation by an anion as shown in Eq. (75) as well as loss of the proton from the methyl

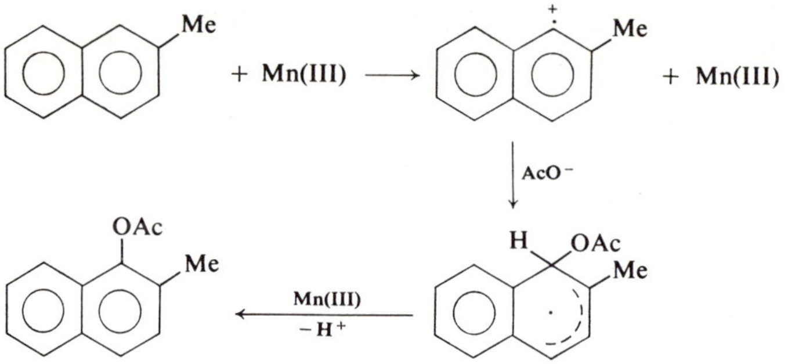

(75)

group. Heiba et al. (1969b) also studied the reaction of cobalt (III) acetate with p-methoxybenzylthiophenyl ether. This oxidation was postulated as proceeding through a radical cation as shown in Eqs. (76)–(79).

$$p\text{-CH}_3\text{OC}_6\text{H}_4\text{CH}_2\text{SC}_6\text{H}_5 + \text{Co(III)} \longrightarrow [p\text{-CH}_3\text{OC}_6\text{H}_4\text{CH}_2\text{SC}_6\text{H}_5]\cdot{}^+ + \text{Co(II)} \quad (76)$$

$$[p\text{-CH}_3\text{OC}_6\text{H}_4\text{CH}_2\text{SC}_6\text{H}_5]\cdot{}^+ \longrightarrow p\text{-CH}_3\text{OC}_6\text{H}_4\text{CH}_2{}^+ + \cdot\text{SC}_6\text{H}_5 \quad (77)$$

$$2\,\text{C}_6\text{H}_5\text{S}\cdot \longrightarrow \text{C}_6\text{H}_5\text{SSC}_6\text{H}_5 \quad (78)$$

$$p\text{-CH}_3\text{OC}_6\text{H}_4\text{CH}_2{}^+ + \text{AcO}^- \longrightarrow \text{CH}_3\text{OC}_6\text{H}_4\text{CH}_2\text{OAc} \quad (79)$$

In solutions containing a high concentration of chloride ion, the radical cation of toluene was found to react principally by two pathways shown in Eqs. (80) and (81).

$$[\text{C}_6\text{H}_5\text{CH}_3]^{\cdot\,+} \xrightarrow{-\text{H}^+} \text{C}_6\text{H}_5\text{CH}_2\cdot \xrightarrow{\text{Co(III)}} \text{C}_6\text{H}_5\text{CH}_2{}^+ \xrightarrow{\text{Cl}^-} \text{C}_6\text{H}_5\text{CH}_2\text{Cl} \quad (80)$$

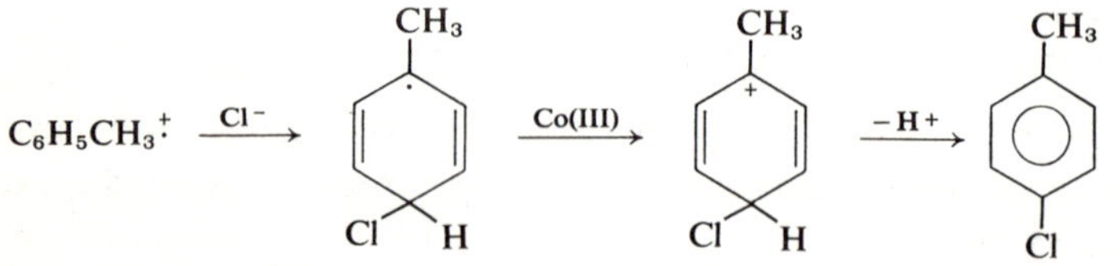

(81)

Electron spin resonance spectra have not been observed in the simple cases discussed above, but by use of a rapid flow system involving cobalt (III) acetate and polyalkylated hydrocarbons in perfluoroacetic acid, the spectra of the radical cations have been observed (Dessau et al., 1970).

6. RADICAL IONS

4. Reactions Involving Transient Heteroatom Cation Radicals

The best-known reaction involving transient nitrogen cation radicals is the Hofmann–Löffler reaction (reviewed by Wolff in 1963) which is generally accepted as proceeding via the route shown in Eqs. (82)–(85).

$$R-(CH_2)_4-\overset{+}{N}H(R')X \xrightarrow{-[X\cdot]} R-(CH_2)_4-\overset{\cdot\,+}{N}HR' \quad X = Br, Cl \quad (82)$$
$$(19) \qquad\qquad\qquad (20)$$

$$R-(CH_2)_4-\overset{\cdot\,+}{N}HR' \longrightarrow R\overset{\cdot}{C}H(CH_2)_3-\overset{+}{N}H_2R' \quad (83)$$

$$19 + R\overset{\cdot}{C}H(CH_2)_3-\overset{+}{N}H_2R' \longrightarrow 20 + RCH(X)(CH_2)_3-\overset{+}{N}H_2R' \quad (84)$$

$$R-CH(X)(CH_2)_3-\overset{+}{N}H_2R' \xrightarrow{OH^-} \underset{\underset{R'}{|}}{\underset{R\quad N}{\bigcirc}} \quad (85)$$

The initial step (82) may be thermal, catalyzed by ultraviolet light or metal ions, e.g., iron (II) or copper (I) [Eq. (86)].

$$R-\overset{+}{N}HR'X + Fe(II) \longrightarrow R\overset{\cdot\,+}{N}HR' + Fe(III) + Cl^- \quad (86)$$

The use of N-chloroamines as chlorinating agents, which has been extensively investigated by Minisci and co-workers (1970; Spanwick and Ingold, 1970), also involves the same radical cation intermediate. The hydrogen atom abstracted by the amino radical cation is an extremely specific process, and in long-chain esters, abstraction of hydrogen (and hence resulting chlorination) occurs predominantly at the penultimate carbon.

A similar type of process involving an oxygen radical cation has been invoked by Deno et al. (1970) to explain selective chlorination of aliphatic esters in sulfuric acid. The proposed scheme is indicated in Eq. (87).

$$CH_3(CH_2)_2CO_2H \xrightarrow{H_2SO_4} CH_3(CH_2)_2\overset{\overset{+}{O}H}{\underset{\|}{C}}-OH \xrightarrow{Cl\cdot} CH_3(CH_2)_2\overset{\overset{\cdot\,+}{O}}{\underset{\|}{C}}-OH \longrightarrow$$

$$\overset{\cdot}{C}H_2(CH_2)_2\overset{O}{\underset{\|}{C}}-OH \xrightarrow{Cl_2} Cl\cdot + CH_2Cl(CH_2)_2COOH \quad (87)$$

The catalysis by iron salts of the conversion of tertiary amine oxides in acid solution to mixtures of tertiary amines, secondary amines, and aldehydes (Ferris et al., 1967) or cyclic products (Smith et al., 1970) has been shown to proceed via nitrogen radical cations as shown in Eqs. (88)–(91) for trimethyl amine N-oxide.

$$(CH_3)_3\overset{+}{N}-O^- \xrightarrow{H^+} (CH_3)_3\overset{+}{N}OH \xrightarrow[H^+]{Fe(II)} (CH_3)_3\overset{\cdot\,+}{N} + Fe(III) + H_2O \quad (88)$$

$$(CH_3)_3\overset{\cdot\,+}{N} \xrightarrow{Fe(II)} (CH_3)_3N + Fe(III) \quad (89)$$

or

$$(CH_3)_3\overset{\cdot\,+}{N} + Fe(III) \longrightarrow (CH_3)_2\overset{+}{N}=CH_2 + H^+ + Fe(II) \qquad (90)$$

$$(CH_3)_2\overset{+}{N}=CH_2 \xrightarrow{H_2O} (CH_3)_2NH + CH_2O \qquad (91)$$

References

Adams, M., Blois, M. S., and Sands, R. H. (1958). *J. Chem. Phys.* **28**, 774.
Angeleno, B. (1966). *Bull. Soc. Chim. Fr.* p. 1091.
Anschütz, L., Broeker, K., and Othneiser, A. (1944). *Chem. Ber.* **77**, 443.
Austen, D. E. G., Given, P. H., Ingram, D. J. E., and Peover, M. E. (1958). *Nature (London)* **182**, 1784.
Ayscough, P. B. (1967). *In* "Electron Spin Resonance in Chemistry," pp. 239–317. Methuen, London.
Ayscough, P. B., and Thomson, D. (1962). *J. Chem. Soc., London* p. 2055.
Bachman, W. E. (1933). *J. Amer. Chem. Soc.* **55**, 1179.
Bamberger, E. (1885). *Chem. Ber.* **18**, 865.
Barton, B. L., and Fraenkel, G. K. (1964). *J. Chem. Phys.* **41**, 1455.
Beckmann, E., and Paul, T. (1891). *Justus Liebigs Ann. Chem.* **266**, 1.
Bennett, J. E., Mile, B., and Thomas, A. (1968). *J. Chem. Soc., A* p. 298.
Bernal, I., and Fraenkel, G. K. (1964). *J. Amer. Chem. Soc.* **86**, 1671.
Bernal, I., Rieger, P., and Fraenkel, G. K. (1962). *J. Chem. Phys.* **37**, 1489.
Billon, J. P., Cauquis, G., and Combrisson, J. (1964). *J. Chim. Phys.* **61**, 374.
Bolton, J. R. (1968). *In* "Radical Ions" (E. T. Kaiser and L. Kevan, eds.), pp. 1–33. Wiley, New York.
Bolton, J. R., and Fraenkel, G. K. (1964). *J. Chem. Phys.* **40**, 3307.
Bowers, K. W. (1965). *Advan. Magn. Resonance* **1**, 317–396.
Bowers, K. W. (1968). *In* "Radical Ions" (E. T. Kaiser and L. Kevan, eds.), pp. 211–244. Wiley, New York.
Boyd, R. H., and Phillips, W. D. (1965). *J. Chem. Phys.* **43**, 2927.
Buschow, K. H. J., Dieleman, J., and Hoijtink, G. J. (1965). *J. Chem. Phys.* **42**, 1993.
Carrington, A., Dravnieks, F., and Symons, M. C. R. (1959). *J. Chem. Soc., London* p. 947.
Carter, M. K., and Vincow, G. (1967). *J. Chem. Phys.* **47**, 292.
Closson, W. D., Wriede, P., and Bank, S. J. (1966). *J. Amer. Chem. Soc.* **88**, 1581.
Das, M. R., and Fraenkel, G. K. (1962). *J. Chem. Phys.* **42**, 1350.
Dehl, R., and Fraenkel, G. K. (1963). *J. Chem. Phys.* **39**, 1793.
Deno, N. C., Fishbein, R., and Wyckoff, J. C. (1970). *J. Amer. Chem. Soc.* **92**, 5274.
Dessau, R. M., Shih, S., and Heiba, E. I. (1970). *J. Amer. Chem. Soc.* **92**, 412.
Dewar, M. J. S., Andrulis, P. J., Dietz, R., and Hunt, R. L. (1966). *J. Amer. Chem. Soc.* **88**, 5473.
Dimroth, O., and Frister, F. (1922). *Chem. Ber.* **55**, 3693.
Dimroth, O., and Heene, R. (1921). *Chem. Ber.* **54**, 2934.
Eiben, K., and Fessenden, R. W. (1968). *J. Phys. Chem.* **72**, 3387.
Eiben, K., and Fessenden, R. W. (1971). *J. Phys. Chem.* **75**, 1186.
Farrington, J. A., Ledwith, A., and Stam, M. F. (1969). *J. Chem. Soc. D* p. 259.
Favorski, A. E., and Nazarow, I. N. (1934). *Bull. Soc. Chim. Fr.* [5] **1**, 46.
Ferris, J. P., Gerwe, R. D., and Gapski, G. R. (1967). *J. Amer. Chem. Soc.* **89**, 5270.

Fischer, E. (1882). *Justus Liebigs Ann. Chem.* **211**, 314.
Forbes, W. F., and Sullivan, P. D. (1966). *J. Amer. Chem. Soc.* **88**, 2862.
Forrester, A. R., May, J. M., and Thomson, R. H., eds. (1968). "Organic Chemistry of Stable Free Radicals," pp. 12–49. Academic Press, New York.
Garst, J. F. (1971). *Accounts Chem. Res.* **4**, 400.
Garst, J. F., and Barton, F. E. (1969). *Tetrahedron Lett.* p. 587.
Garst, J. F., Ayers, P. W., and Lamb, R. C. (1966). *J. Amer. Chem. Soc.* **88**, 4260.
Garst, J. F., Scouten, G. G., Barton, F. E., Burgess, J. R., and Story, P. R. (1969). *J. Chem. Soc., D* p. 78.
Garst, J. F., Cox, R. H., Barbas, J. T., Roberts, R. D., Morris, J. I., and Morrison, R. C. (1970). *J. Amer. Chem. Soc.* **92**, 5761.
Glarum, S. H., and Snyder, L. C. (1962). *J. Chem. Phys.* **36**, 2989.
Hart, H., Fleming, J. S., and Dye, J. L. (1964). *J. Amer. Chem. Soc.* **86**, 2079.
Hass, H. B., and Bender, M. L. (1949). *J. Amer. Chem. Soc.* **71**, 1767, 3482.
Hauser, K. H., and Murrell, J. N. (1957). *J. Chem. Phys.* **27**, 500.
Heiba, E. I., Dessau, R. M., and Koehl, W. J. (1969a). *J. Amer. Chem. Soc.* **91**, 138.
Heiba, E. I., Dessau, R. M., and Koehl, W. J. (1969b). *J. Amer. Chem. Soc.* **91**, 6830.
Heyrovský, J., and Kůta, J. (1966). "Principles of Polarography," pp. 181–187. Academic Press, New York.
Hirota, N. (1968). *In* "Radical Ions" (E. T. Kaiser and L. Kevan, eds.), pp. 38–48. Wiley, New York.
Hirota, N., and Weissman, S. I. (1960). *J. Amer. Chem. Soc.* **82**, 4424.
Hirota, N., and Weissman, S. I. (1964). *J. Amer. Chem. Soc.* **86**, 2538.
Hisshon, J. M., Gardner, D. M., and Fraenkel, G. K. (1953). *J. Amer. Chem. Soc.* **75**, 4115.
Hogen-Esch, T. E., and Smid, J. (1966). *J. Amer. Chem. Soc.* **88**, 307.
Hünig, S., Friedrich, H. J., Scheutzow, D., and Brenninger, W. (1964). *Tetrahedron Lett.* p. 181.
Hutchison, C. A. (1955). *In* "Determination of Organic Structures by Physical Methods" (E. A. Braude and F. C. Ncahod, eds.), vol. 1, pp. 259–321. Academic Press, New York.
Janzen, E. G., and Gerlock, J. L. (1967). *J. Phys. Chem.* **71**, 4577.
Janzen, E. G., and Gerlock, J. L. (1968). *J. Phys. Chem.* **72**, 1832.
Ji, S., Gortler, L. B., Waring, A., Battisti, A., Bank, S., Closson, W. D., and Wriede, P. (1967). *J. Amer. Chem. Soc.* **89**, 5311.
Jones, M. T. (1968). *In* "Radical Ions" (E. T. Kaiser and L. Kevan, eds.), pp. 245–274. Wiley, New York.
Jones, P. R., and West, R. (1968). *J. Amer. Chem. Soc.* **90**, 6978.
Kim, J. K., and Bunnett, J. F. (1970). *J. Amer. Chem. Soc.* **92**, 7463 and 7464.
Kolker, P. L., and Waters, W. A. (1963). *Chem. Commun.* p. 55.
Kornblum, N., and Boyd, S. D. (1970). *J. Amer. Chem. Soc.* **92**, 5784.
Kornblum, N., and Stuchal, F. W. (1970). *J. Amer. Chem. Soc.* **92**, 1804.
Kornblum, N., Pink, P., and Worka, K. V. (1961). *J. Amer. Chem. Soc.* **83**, 2779.
Kornblum, N., Kerber, R. C., and Urry, R. W. (1964). *J. Amer. Chem. Soc.* **86**, 3904.
Kornblum, N., Kerber, R. C., and Urry, R. W. (1965). *J. Amer. Chem. Soc.* **87**, 4520.
Kornblum, N., Michel, R. E., and Kerber, R. C. (1966). *J. Amer. Chem. Soc.* **88**, 5660 and 5662.
Kornblum, N., Davies, T. M., Earl, G. W., Holy, N. L., Kerber, R. C., Musser, M. T., and Snow, D. H. (1967a). *J. Amer. Chem. Soc.* **89**, 725.

Kornblum, N., Davies, T. M., Earl, G. W., Green, G. S., Holy, N. L., Kerber, R. C., Manthey, J. W., Musser, M. T., and Snow, D. H. (1967b). *J. Amer. Chem. Soc.* **89**, 5714.
Kornblum, N., Davies, T. M., Earl, G. W., Holy, N. L., Manthey, J. W., Musser, M. T., and Swiger, R. T. (1968). *J. Amer. Chem. Soc.* **90**, 6219.
Kornblum, N., Swiger, R. T., Earl, G. W., Pinnick, H. W., and Stuchal, F. W. (1970a). *J. Amer. Chem. Soc.* **92**, 5513.
Kornblum, N., Boyd, S. D., and Stuchal, F. W. (1970b). *J. Amer. Chem. Soc.* **92**, 5783.
Korzan, D. G., Chen, F., and Ainsworth, C. (1971). *Chem. Commun.* p. 1053.
Kuwata, K., and Geske, D. H. (1964). *J. Amer. Chem. Soc.* **86**, 2101.
Lagercrantz, C., Torssell, K., and Wold, S. (1966). *Ark. Kemi* **25**, 567.
Laurent, A. (1836). *Ann. Chem.* **17**, 91.
Levy, D. H., and Meyers, R. J. (1964). *J. Chem. Phys.* **41**, 1062.
Lewis, I. C., and Singer, L. S. (1965). *J. Chem. Phys.* **43**, 2712, and references therein.
Liebermann, C., and Homeyer, J. (1879). *Chem. Ber.* **12**, 1971.
Lipkin, D., Galinao, F. R., and Jordan, R. W. (1963). *Chem. Ind. (London)* p. 1657.
Lyssy, T. (1959). *Helv. Chim. Acta* **42**, 2245.
Maki, A. H., and Geske, D. H. (1959a). *Anal. Chem.* **31**, 1450.
Maki, A. H., and Geske, D. H. (1959b). *J. Chem. Phys.* **30**, 1356.
Maki, A. H., and Geske, D. H. (1960). *J. Amer. Chem. Soc.* **82**, 2671.
Maki, A. H., and Geske, D. H. (1961). *J. Amer. Chem. Soc.* **83**, 1852.
Metcalfe, A. R., and Waters, W. A. (1969). *J. Chem. Soc., B* p. 918.
Michaelis, L. (1935). *Chem. Rev.* **16**, 243.
Michaelis, L., and Schubert, M. P. (1938). *Chem. Rev.* **22**, 437.
Michaelis, L., and Schwarzenbach, G. (1938). *J. Amer. Chem. Soc.* **60**, 1668.
Michaelis, L., Schubert, M. P., and Granick, S. (1939). *J. Amer. Chem. Soc.* **61**, 1981.
Michaelis, L., Schubert, M. P., and Granick, S. (1940). *J. Amer. Chem. Soc.* **62**, 204.
Minisci, F., Gardini, G. P., and Bertini, F. (1970). *Can. J. Chem.* **48**, 544, and references therein.
Morantz, D. J., and Warhurst, E. (1955). *Trans. Faraday Soc.* **51**, 1375.
Müller, E., and Janke, W. (1939). *Z. Electrochem.* **45**, 380.
Müller, E., and Wiesemann, W. (1936). *Chem. Ber.* **69**, 2156.
Müller, E., and Wiesemann, W. (1938a). *Justus Liebigs Ann. Chem.* **537**, 86.
Müller, E., and Wiesemann, W. (1938b). *Angew. Chem.* **51**, 657.
Müller, E., Günter, F., Scheffler, K., Ziemek, P., and Rieker, A. (1965). *Justus Liebigs Ann. Chem.* **688**, 134.
Nelson, S. F. (1966). *J. Amer. Chem. Soc.* **88**, 5666.
Norman, R. O. C., and Gilbert, B. C. (1967). *Advan. Phys. Org. Chem.* **5**, 53.
Piette, L. H., Ludwig, P., and Adams, R. N. (1962). *J. Amer. Chem. Soc.* **84**, 4212.
Rieger, P. H., and Fraenkel, G. K. (1962). *J. Chem. Phys.* **37**, 2811.
Rieger, P. H., and Fraenkel, G. K. (1963). *J. Chem. Phys.* **39**, 1793.
Rieger, P. H., Bernal, I., Reinmuth, W. H., and Fraenkel, G. K. (1963). *J. Amer. Chem. Soc.* **85**, 683.
Russell, G. A. (1954). *J. Amer. Chem. Soc.* **76**, 1595.
Russell, G. A. (1968). *In* "Radical Ions" (E. T. Kaiser and L. Kevan, eds.), pp. 87–150. Wiley, New York.
Russell, G. A. (1970a). *Chem. Soc., Spec. Publ.* **24**, 273.
Russell, G. A. (1970b). *Polym. Prepr., Amer. Chem. Soc., Div. Polym. Chem.* **11**, 100.
Russell, G. A., and Bemis, A. G. (1967). *Inorg. Chem.* **6**, 403.
Russell, G. A., and Danen, W. C. (1966). *J. Amer. Chem. Soc.* **88**, 5663.
Russell, G. A., and Danen, W. C. (1968). *J. Amer. Chem. Soc.* **90**, 347.

Russell, G. A., and Geels, E. J. (1965). *J. Amer. Chem. Soc.* **87**, 122.
Russell, G. A., and Janzen, E. G. (1962). *J. Amer. Chem. Soc.* **84**, 4153.
Russell, G. A., and Stevenson, G. R. (1971). *J. Amer. Chem. Soc.* **93**, 2432 (1971).
Russell, G. A., and Strom, E. T. (1964). *J. Amer. Chem. Soc.* **86**, 744.
Russell, G. A., and Young, M. C. (1966). *J. Amer. Chem. Soc.* **88**, 2007.
Russell, G. A., Janzen, E. G., Becker, H.-D., and Smentowski, F. J. (1962a). *J. Amer. Chem. Soc.* **84**, 2652.
Russell, G. A., Janzen, E. G., and Strom, E. T. (1962b). *J. Amer. Chem. Soc.* **84**, 4155.
Russell, G. A., Janzen, E. G., and Strom, E. T. (1964). *J. Amer. Chem. Soc.* **86**, 1807.
Russell, G. A., Geels, E. J., Smentowski, F. J., Chang, K.-Y., Reynolds, J., and Kaupp, G. (1967). *J. Amer. Chem. Soc.* **89**, 3821.
Russell, G. A., Konaka, R., Strom, E. T., Danen, W. C., Chang, K.-Y., and Kaupp, G. (1968). *J. Amer. Chem. Soc.* **90**, 4646.
Russell, G. A., Ku, T., and Lokengard, J. (1970a). *J. Amer. Chem. Soc.* **92**, 3833.
Russell, G. A., Lawson, D. F., and Ochrymowycz, L. A. (1970b). *Tetrahedron* **26**, 4697.
Russell, G. A., Norris, R. K., and Panek, E. J. (1971). *J. Amer. Chem. Soc.* **93**, 5839.
Sargent, G. D., and Lux, G. A. (1968). *J. Amer. Chem. Soc.* **90**, 7160, and references therein.
Scheffler, K., and Stegmann, H. B. (1970). "Electrospinresonanz," pp. 373-480. Springer-Verlag, Berlin and New York.
Schlenk, W., and Weickel, T. (1911). *Chem. Ber.* **44**, 1182.
Schlenk, W., Appenrodt, J., Michael, A., and Thal, A. (1914). *Chem. Ber.* **47**, 473.
Scholl, R. (1899). *Chem. Ber.* **32**, 1809.
Scott, N. D., Walker, J. F., and Hansley, V. L. (1936). *J. Amer. Chem. Soc.* **58**, 2442.
Seo, E. T., Nelson, R. F., Fritsch, J. M., Marcoux, L. S., Leedy, D. W., and Adams, R. N. (1966). *J. Amer. Chem. Soc.* **88**, 3498.
Shapiro, B. I., Kazakova, V. M., and Okhlobystina, L. V. (1968). *Zh. Strukt. Khim.* **8**, 899; *Chem. Abstr.* **70**, 28240 (1969).
Shein, S. M., Bryukhovetskaya, L. V., Khmelinskaya, A. D., Starichenko, V. F., and Ivanova, T. M. (1969). *Reakts Sposobnost Org. Soedin* **6**, 1087; *Chem. Abst.* **73** 24629v. (1970).
Shein, S. M., Bryukhovetskaya, L. V., Pischugin, F. V., Starichenko, V. F., Panfilov, V. N., and Voevodkkii,V. V. (1970). *Zu. Strukt. Khim.* **11**, 243; *Chem. Abstr.* **73**, 55324w (1970).
Siegle, L. W., and Hass, H. B. (1940). *J. Org. Chem.* **5**, 100.
Slates, R. V., and Szwarc, M. (1965). *J. Phys. Chem.* **69**, 4125.
Smith, J. R. L., Norman, R. O. C., and Rowley, A. G. (1970). *J. Chem. Soc., D* p. 1238.
Spanwick, J., and Ingold, K. U. (1970). *Can. J. Chem.* **48**, 546, and references therein.
Strom, E. T., Russell, G. A., and Konaka, R. (1965). *J. Chem. Phys.* **42**, 2033.
Sugden, S. (1934). *Trans. Faraday Soc.* **30**, 18.
Swanwick, M. G., and Waters, W. A. (1970). *J. Chem. Soc., D* p. 63.
Szwarc, M. (1956). *Nature (London)* **178**, 1168.
Tolles, W. M., McBride, W. R., and Thun, W. E. (1969). *J. Amer. Chem. Soc.* **81**, 2443.
Tuttle, T. R., and Weissman, S. I. (1958). *J. Amer. Chem. Soc.* **80**, 5342.
Urberg, M. M., and Kaiser, E. T. (1968). *In* "Radical Ions" (E. T. Kaiser and L. Kevan, eds.), pp. 301-320. Wiley, New York.
Vincow, G. (1968). *In* "Radical Ions" (E. T. Kaiser and L. Kevan, eds.), pp. 151-209. Wiley, New York.
Waters, W. A. (1948). "Chemistry of Free Radicals," 2nd ed., pp. 25-34. Oxford Univ. Press, London and New York.
Weissberger, A. (1932). *Chem. Ber.* **65B**, 1815.

Weissman, S. I., Chu, T. L., Pake, G. E., Paul, D. E., and Townsend, J. (1953). *J. Phys. Chem.* **57**, 504.
Weissman, S. I., DeBoer, E., and Conradi, J. J. (1957). *J. Chem. Phys.* **26**, 963.
Weitz, E. (1954). *Angew. Chem.* **66**, 658.
Weitz, E., and Ludwig, R. (1922). *Chem. Ber.* **55**, 395.
Weitz, E., and Schwechten, H. W. (1926). *Chem. Ber.* **59**, 2307.
Weitz, E., and Schwechten, H. W. (1927). *Chem. Ber.* **60**, 545 and 1203.
Wieland, H. (1907). *Chem. Ber.* **40**, 4263.
Wieland, H., and Wecker, E. (1910). *Chem. Ber.* **43**, 699.
Willstätter, R., and Piccard, J. (1908). *Chem. Ber.* **41**, 1458.
Wolff, M. E. (1963). *Chem. Rev.* **63**, 55.
Wooster, C. B. (1928). *J. Amer. Chem. Soc.* **50**, 1388.
Wooster, C. B. (1934). *J. Amer. Chem. Soc.* **56**, 2436.
Yokozawa, Y., and Miyashita, I. (1956). *J. Chem. Phys.* **25**, 796.
Zilkha, A., Neta, P., and Frankel, M. (1960). *J. Chem. Soc., London*, p. 3357.

7

ARYNES

Ellis K. Fields

I. General Aspects and Historical Background	449
A. Definitions	449
B. Nomenclature	450
C. Structure	451
D. Historical Background	452
II. Methods of Investigating Arynes	455
A. Trapping Arynes	458
B. Spectroscopy	459
C. Metal Complexes of Arynes	459
III. Methods of Formation of Arynes	463
A. In Solution	463
B. In the Gas Phase	466
IV. Factors Affecting the Formation and Stability of Cyclic Acetylenes and Arynes	472
A. Effect of Ring Size	472
B. Effect of Substituents	478
V. Reactions of Arynes	480
A. 2 + 2 Cycloaddition	480
B. 2 + 3 Cycloaddition	481
C. 2 + 4 Cycloaddition	483
D. 2 + 6 Cycloaddition	488
E. 2 + 8 Cycloaddition	489
F. Polymerization	489
G. "Ene" Reaction	491
H. Reaction with Nucleophiles	494
VI. Hetarynes	499
VII. Conclusion	503
References	504

I. General Aspects and Historical Background

A. Definitions

Arynes are aromatic compounds containing a formal C–C triple bond. The first and best known member of the series is benzyne, C_6H_4 (**1**), the aromatic counterpart of acetylene. Formal analogy to acetylene is useful as a means of depicting and dealing with arynes as a class. The analogy also

(1)

extends into the high temperature behavior of aliphatic and aromatic compounds. Thus, just as many aliphatic compounds yield some acetylene on pyrolysis, a great variety of benzene derivatives form benzyne on pyrolysis (Fields and Meyerson, 1968a).

Acetylene and its open-chain homologs have been known for over 100 years (Nieuwland and Vogt, 1945; Viehe, 1969). Cycloalkynes and arynes are of more recent vintage. Cycloalkynes can be isolated if the ring is large enough. Cyclooctyne has been made and characterized, although it polymerizes readily (Blomquist and Lin, 1953); smaller ring acetylenes are even less stable. Cyclohexyne (2) is exceedingly strained and short lived. Benzyne

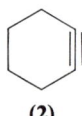

(2)

may be depicted as cyclohexyne with two additional double bonds, therefore, it is a fleeting, highly reactive species.

B. Nomenclature

Chemical Abstracts, usually the final authority on nomenclature, names benzyne as 1,3-cyclohexadiene-5-yne; however, with a total lack of consistency, it terms 1,8-naphthalyne and pyridyne as "perinaphthalyne" and "didehydropyridine," respectively.

The name "aryne" proposed by Roberts *et al.* (1953) has been widely used. Both Wittig (1965) and Hoffmann (1967) prefer dehydrobenzene; however, Wittig uses the structure **1**, whereas Hoffmann prefers **3**.

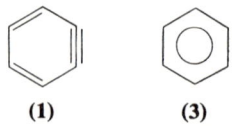

(1)   (3)

In this chapter we use the terms "aryne," "benzyne," "naphthalyne," and generally the ending "yne" to denote aromatic systems with a formal triple bond. For the sake of convenience, we also use the structural formula showing a triple bond (1).

C. Structure

Benzyne is benzene with two hydrogen atoms removed, usually ortho hydrogens. Since benzyne is electrically neutral, two electrons are available for the two orbitals that have nearly trigonal hybridization. These sp^2 orbitals are orthogonal to the π molecular orbitals (MO's) of the ring (Roberts et al., 1956; Coulson, 1958). If the structure is planar, the interaction between

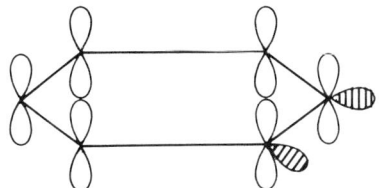

these sets of orbitals will be weak. The sp^2 orbitals can interact to give a singlet or triplet state, but only in the singlet is a new bond formed.

There are three possible ground states of benzyne: a triplet (**4**) and two singlets, one symmetric (**5**), and the other antisymmetric (**6**) with respect to a C_2 axis intraconverting the two sp^2 orbitals containing the free electrons.

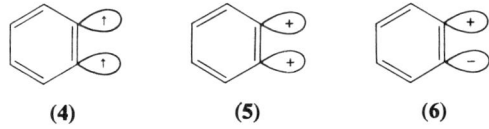

The symmetric benzyne (**5**) should undergo a nonconcerted 2 + 2 cycloaddition; the 2 + 4 reaction should be stereospecific (Woodward and Hoffman, 1968; Vollmer and Servis, 1968; Gill, 1968). The antisymmetric singlet (**6**) should give exactly opposite results. M. Jones and Levin (1969) tested this hypothesis by determining the products of the reactions of benzyne with cis and trans olefins. Benzyne added to the 1,2-dichloroethylenes (2 + 2 cycloaddition) in a nonstereospecific way to give dichlorobenzocyclobutenes with moderate, though not complete, loss of the original stereochemistry. The *cis*-olefin gave 35% of *trans*-benzocyclobutene; *trans*-olefin gave 20% of *cis*-benzocyclobutene. In contrast, the 2 + 4 cycloaddition reaction of benzyne with either *trans,trans*-2,4-hexadiene or the dimethyl ester of *trans*, *trans*-muconic acid was stereospecific. They concluded that the structure of benzyne is, therefore, the symmetric singlet (**5**).

1,8-Perinaphthalyne, however, has been shown as an antisymmetric singlet both by MO calculations (Hoffmann, 1967) and experimentally (C. W. Rees and Storr, 1965, 1969; Hoffmann et al., 1969). That benzyne has a singlet ground state is in line with MO calculations based on a planar benzyne model

(Simmons, 1961; Wilhite and Whitten, 1971; Haselbach, 1971). The new bond in benzyne is $\frac{2}{3}\pi$ and $\frac{1}{3}\sigma$ in nature if no distortion of the ring occurs.

Diradical structure is strongly suggested for *p*-benzyne. Jones and Bergman (1972) found typical radical products from pyrolysis of 1,6-dideuterio-*cis*-1,5-hexadiyn-3-ene (**7**), the *p*-benzyne precursor, with hydrocarbons, carbon tetrachloride, and methanol. In addition, the diiyn-ene rapidly introconverted protium and deuterium through the intermediate *p*-benzyne (**8**).

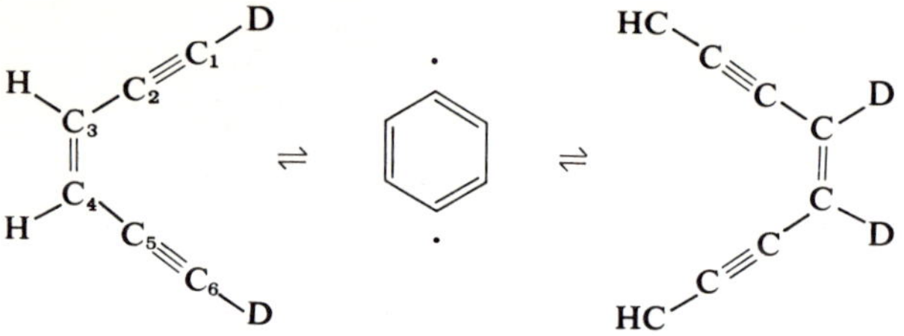

$$(7) \rightleftarrows (8) \rightleftarrows \quad \quad \quad (1)$$

D. Historical Background

Benzyne and products derived from it were probably first formed early in recorded chemical history. Berthelot (1866, 1867a,b) pyrolyzed toluene in a red-hot porcelain tube and obtained methane and naphthalene among the products. Almost 100 years later, Badger and Spotswood (1960) confirmed these results. Neither paper proposed a mechanism for naphthalene formation; however, it appears likely to result from the reaction of benzyne with benzene, formed in much greater amount from toluene at the same time.

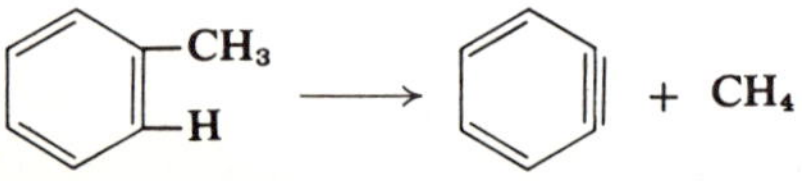

$$C_6H_5\text{-}CH_3 \longrightarrow C_6H_4 + CH_4 \quad (2)$$

$$C_6H_6 + C_6H_4 \longrightarrow \text{[adduct]} \longrightarrow \text{naphthalene} + C_2H_2 \quad (3)$$

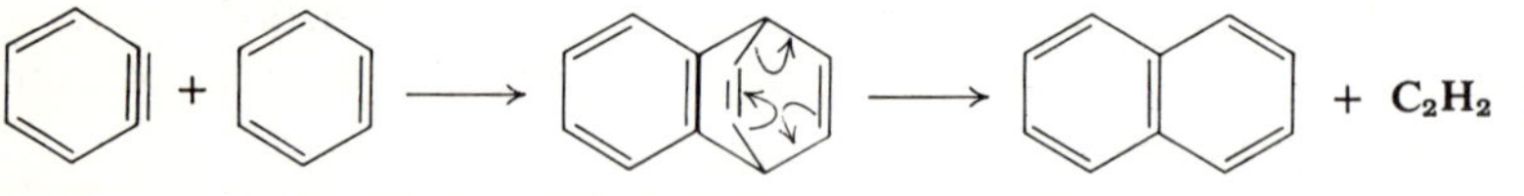

Berthelot undoubtedly also formed benzyne in his pyrolysis of benzene. Fields and Meyerson (1966a) showed that benzene pyrolysis produces naphthalene by intramolecular loss of hydrogen to form benzyne.

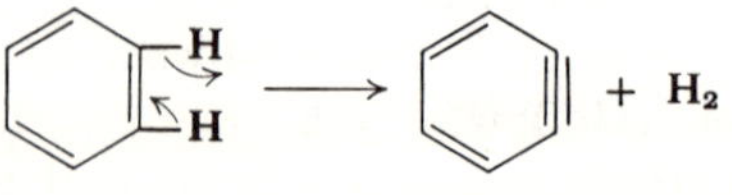

$$C_6H_6 \longrightarrow C_6H_4 + H_2 \quad (4)$$

The amount is small, only 0.2% as much as the biphenyl formed at the same time, and probably was not found by early investigators. The energy required to remove two hydrogen atoms from benzene to give benzyne has been calculated as 157 kcal/mole (Millie et al., 1971).

Kekule (1864) discovered the method of preparing phenols by caustic fusion of benzene sulfonates. Limpricht (1874) obtained resorcinol by alkali fusion of both m- and p-chlorobenzene sulfonates. In both reactions, benzyne was undoubtedly an intermediate.

Meyer and Bergius (1914) observed the formation of appreciable amounts of naphthalene in a phenol process by the reaction of concentrated aqueous sodium hydroxide with chlorobenzene in an autoclave at 300°C. The mechanism may be the following:

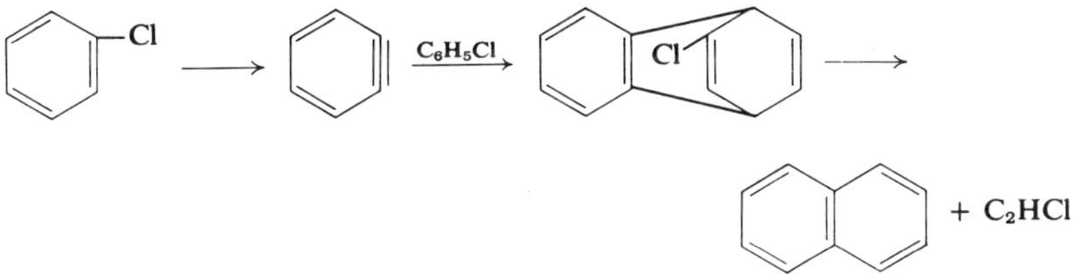

$$\text{PhCl} \longrightarrow \text{benzyne} \xrightarrow{C_6H_5Cl} \text{adduct} \longrightarrow \text{naphthalene} + C_2HCl \quad (5)$$

Chloronaphthalene was most likely formed as well: It was converted to either naphthol or naphthalyne and more highly condensed systems.

Industrial processes for phenol have involved reaction of molten alkali with chlorobenzene or benzenesulfonic acid. This process is almost obsolete now, having been displaced by oxidation of cumene to the hydroperoxide and decomposition of the hydroperoxide by acid to acetone and phenol. However, thousands of tons of phenol have been produced in the past by a reaction involving benzyne, at least in part.

The intermediacy of benzyne in the production of phenol from chlorobenzene and alkali was clearly demonstrated by Bottini and Roberts (1957). Chlorobenzene labeled with ^{14}C at the halogen bond was hydrolyzed at 340°C with 4 N aqueous sodium hydroxide. The resulting phenol had 58% of the labeled carbon atom at the OH bonded carbon and 42% at an ortho position.

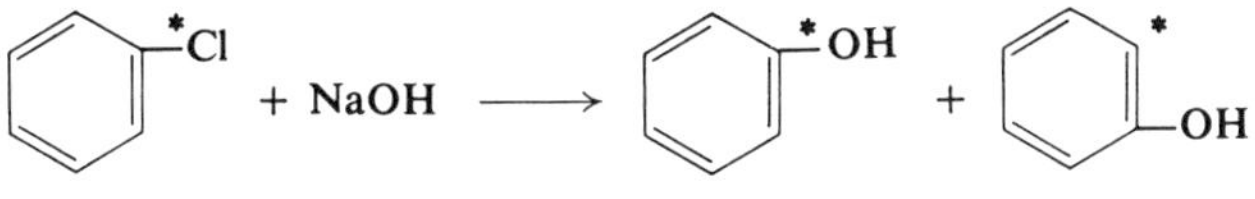

$$\text{Ph}^*\text{Cl} + \text{NaOH} \longrightarrow \text{Ph}^*\text{OH} + \text{Ph}^*(\text{ortho})\text{OH} \quad (6)$$

Rearrangements that occurred during replacement of halogen substituents were also observed by Kym (1895) who reacted p-dibromobenzene with p-toluidine and sodalime and obtained N,N'-di-p-tolyl-m-phenylenediamine. Haeussermann (1900, 1901) found that the three dichlorobenzenes all gave the same product with potassium diphenylamide, N,N,N',N'-tetraphenyl-m-phenylenediamine; some p-diamine also formed from the p-dichlorobenzene.

The first to propose a formal triple bond in a small ring compound were Stoermer and Kahlert (1902). 3-Bromocoumarin on treatment with alcoholic base gave 2-ethoxycoumarin.

A C_6H_4 species was proposed by Luttringhaus and Saaf (1930) as an intermediate in the aryl phenol rearrangement. Morton et al. (1942) suggested that the reaction of amylsodium with chlorobenzene to give amylbenzene proceeded through a benzyne mechanism.

In 1940, Wittig and Fuhrmann firmly established the field of aryne chemistry. Wittig found that fluorobenzene reacted with phenyllithium to yield biphenyl much faster than other halogenobenzenes. Wittig's interpretation was that not biphenyl but 2-biphenylyllithium was the primary product. Phenyl-

$$\text{C}_6\text{H}_4\text{FH} \xrightarrow{\text{PhLi}} \text{C}_6\text{H}_4\text{FLi} \xrightarrow{-\text{LiF}} \text{C}_6\text{H}_4 \xrightarrow{\text{PhLi}} \text{2-biphenylyllithium} \quad (7)$$

lithium, a strong base, removes the ortho hydrogen and generates the ortho fluorophenyl anion and then adds to benzyne as a nucleophile.

Many rearrangements occur in the amination of aryl halides by metal amides in liquid ammonia (reviewed by Bunnett and Zahler, 1951). Although metalation in ortho position relative to the halogen was discussed early by Seibert and Bergstrom (1945) as part of the reaction sequence to give benzyne, the necessity of a hydrogen ortho to the halogen in order to effect amination was established by Benkeser and Buting (1952). Participation of benzyne in these amination reactions was visualized by Roberts et al. (1953) who realized that the amination of [1-^{14}C]chlorobenzene with potassium amide in liquid ammonia via benzyne should lead to equal amounts of [1-^{14}C]aniline and

$$\text{[Ph}^*\text{Cl]} \xrightarrow[\text{NH}_3]{\text{KNH}_2} \text{[Ph}^*\text{]} \xrightarrow{\text{NH}_3} \text{[Ph}^*\text{NH}_2\text{]} + \text{[Ph}^*\text{—NH}_2\text{]} \quad (8)$$

[2-^{14}C]aniline. They obtained both labeled anilines in about a 1:1 ratio and thus confirmed the formation of benzyne.

Supplementary evidence was supplied by Huisgen and Rist (1954) who treated 1- and 2-fluoronaphthalene separately with phenyllithium. After carboxylating the reaction mixture, they obtained 1-phenyl-2-naphthoic acid together with the isomeric 2-phenyl-1-naphthoic acid from each fluoronaphthalene. The reaction evidently proceeded via 1,2-naphthalyne (9).

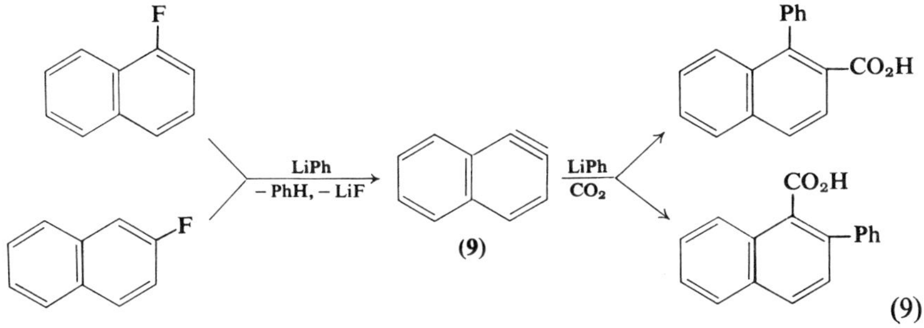

7. ARYNES

[Scheme showing 1-fluoro and 2-fluoronaphthalene treated with LiPh (−PhH, −LiF) to give naphthalyne (9), which with LiPh/CO₂ gives 1-phenyl-2-naphthoic acid and 2-phenyl-1-naphthoic acid.]

(9)

The field of aryne chemistry was widened considerably in its mechanistic and synthetic applications by Wittig and Pohmer (1955). They treated *o*-bromofluorobenzene (10) with lithium amalgam in furan, which as an ether should promote organometallic reactions and as a diene should intercept the benzyne as formed by a Diels–Alder reaction. The product isolated was the endoxide (11) whose structure was confirmed by acid-catalyzed isomerization to α-naphthol, by hydrogenation, and by dehydration to naphthalene.

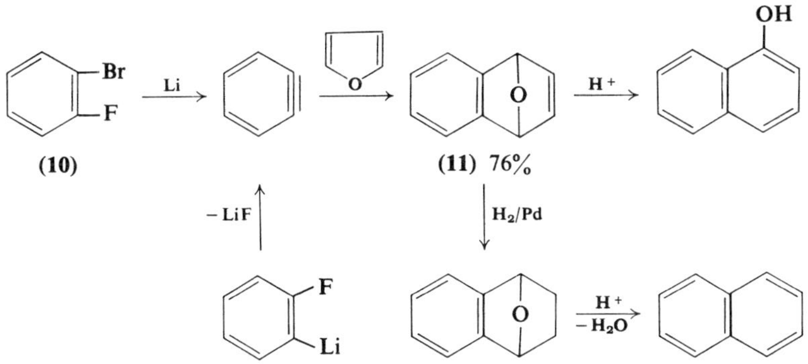

[Scheme: o-bromofluorobenzene (10) → with Li → benzyne → with furan → endoxide (11) 76% → H⁺ → α-naphthol; also via o-fluorophenyllithium (−LiF); and H₂/Pd reduction followed by H⁺/−H₂O giving naphthalene.]

(10)

Realization that benzyne could be trapped and give stable products spurred an ever-increasing amount of research in aryne chemistry. This intensive effort was stimulated still further by the discovery of ways to form arynes in high yields in the absence of nucleophilic solvents and organometallic reagents.

II. Methods of Investigating Arynes

Formation of arynes has been proved by trapping with reagents which yield stable products, by spectroscopic data, and by formation of metal complexes.

Table I
Benzyne Traps and Products

Trap	Product
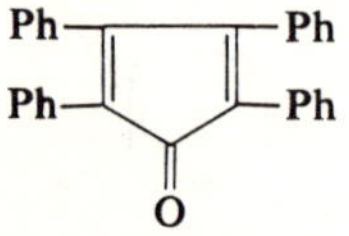 Tetraphenylcyclopentadienone (tetracyclone)	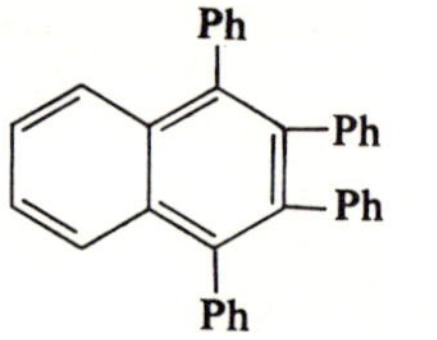 1,2,3,4-Tetraphenylnaphthalene
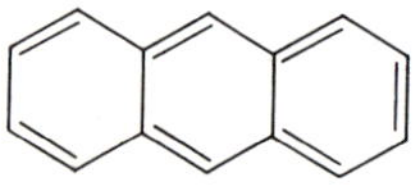 Anthracene	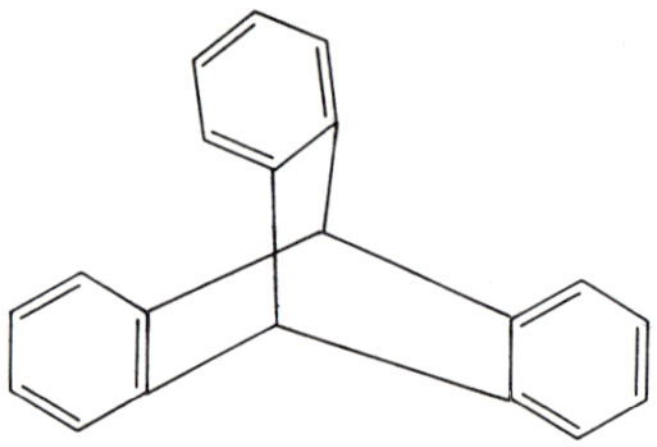 Triptycene
 Furan	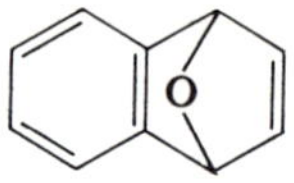 1,4-Dihydronaphthalene endoxide
 Cyclopentadiene	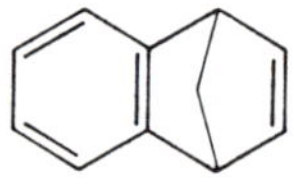 Benzonorbornadiene
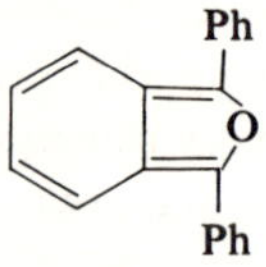 1,3-Diphenylisobenzofuran	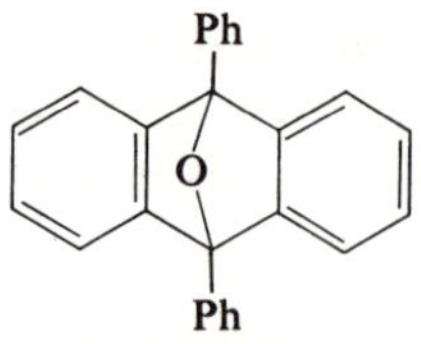 9,10-Dihydro-9,10-diphenylanthracene endoxide
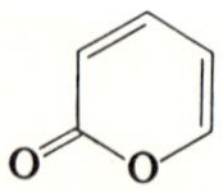 2-Pyrone	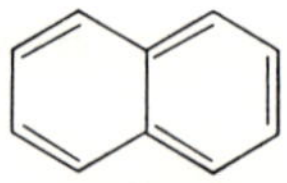 Naphthalene

7. ARYNES

Table I (*continued*)

Trap	Product
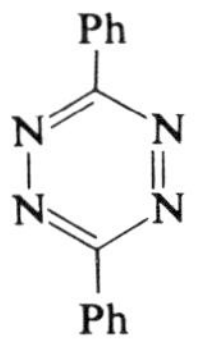 3,6-Diphenyl-*sym*-tetrazene	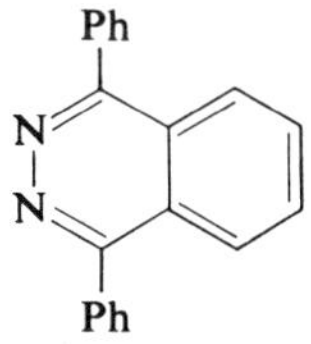 3,8-Diphenylphthalazine
 Benzene	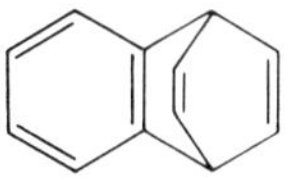 1,4-Adduct
	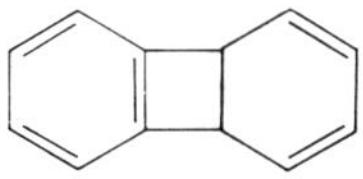 1,2-Adduct
	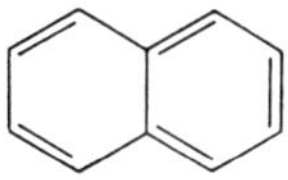 Naphthalene
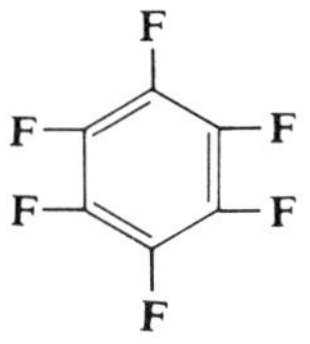 Hexaflurobenzene	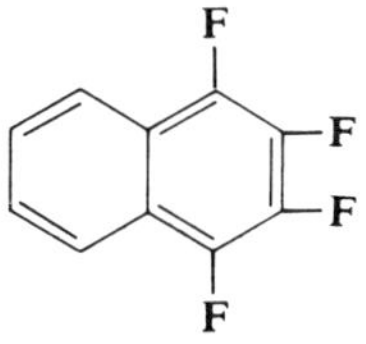 1,2,3,4-Tetrafluoronaphthalene
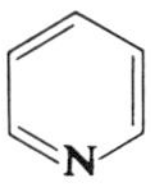 Pyridine	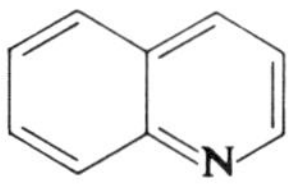 Quinoline
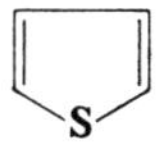 Thiophene	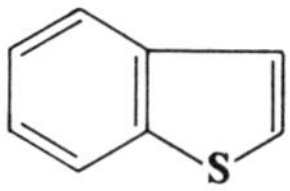 Benzothiophene

A. Trapping Arynes

Reagents often used as traps are tetraphenylcyclopentadienone, anthracene, furan, cyclopentadiene, and diphenylisobenzofuran. Less commonly used but also efficient aryne traps are 2-pyrone, 3,6-diphenyl-*sym*-tetrazine, and dimethyl *sym*-tetrazine-3,6-dicarboxylate. In the gas phase at elevated temperatures, benzene, hexafluorobenzene, pyridine, and thiophene have often been used.

Benzyne traps and their products are shown in Table I. The products from benzyne and the trapping reagents may retain all the atoms of the two reactants or lose some of them. Deciding factors are the temperature and the gains in resonance stability of the product relative to the atom or molecule lost. Thus, the primary adducts of benzyne with tetracyclone (**12**), benzene (**13**), 2-pyrone (**14**), and diphenyltetrazine (**15**) readily aromatize by the loss of the stable molecules CO, C_2H_2, CO_2, and N_2, respectively.

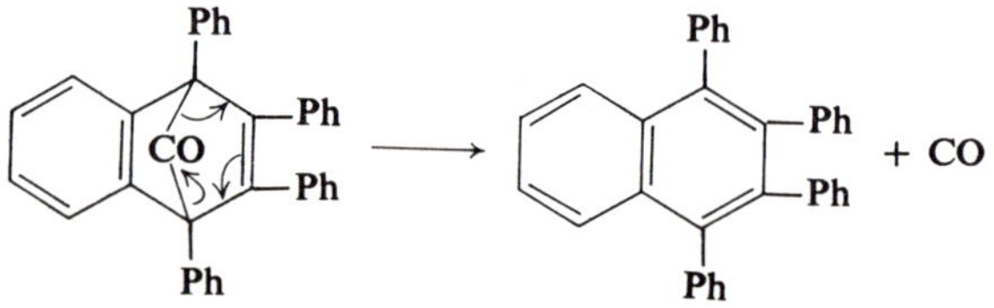

$$\text{(12)} \longrightarrow \text{product} + CO \qquad (11)$$

$$\text{(13)} \longrightarrow \text{product} + C_2H_2 \qquad (12)$$

$$\text{(14)} \longrightarrow \text{product} + CO_2 \qquad (13)$$

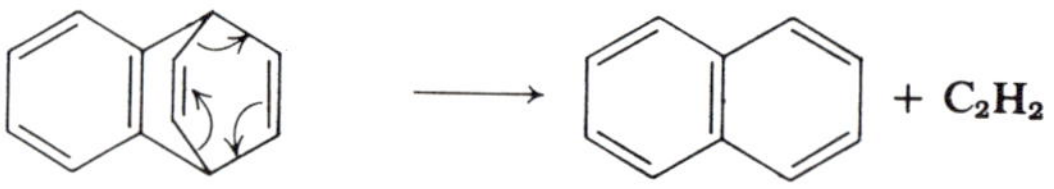

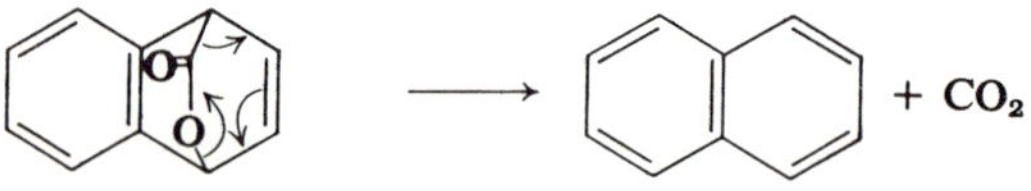

$$\text{(15)} \longrightarrow \text{product} + N_2 \qquad (14)$$

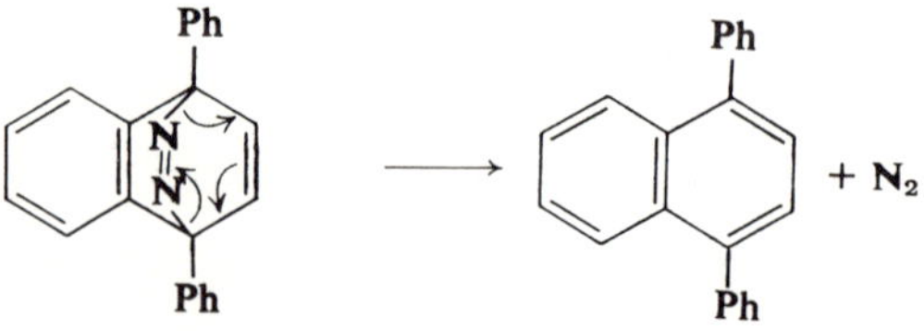

Adducts with furan, 1,3-diphenylisobenzofuran, and cyclopentadiene can aromatize only by the loss of an oxygen or carbon atom and, therefore, are thermally more stable.

B. Spectroscopy

Spectroscopic evidence for the existence of benzyne was obtained by Berry et al. (1964) and by Schafer and Berry (1965). They decomposed o-benzenediazonium carboxylate in an apparatus that coupled flash photolysis with a time-of-flight mass spectrometer and an oscillograph recorder. The species generated by photolysis were recorded at short intervals. At 50 μsec after flash photolysis, the ions of mass 28 (N_2), 44 (CO_2), and 76 (benzyne) were recorded. The intensity of ions of mass 76 decreased and disappeared after 300 μsec; an ion of mass 152 (biphenylene) appeared and reached its maximum at about the same time as m/e 76 disappeared.

In the same apparatus the same authors recorded UV absorption spectra. A series of plots with time showed that a broad absorption band around 2430 nm, attributed to benzyne, developed between 30 and 100 μsec but then was replaced by the much sharper band of biphenylene.

Berry et al. (1965) found the ion of mass 76 was also formed from m- and p-benzenediazonium carboxylate and inferred the formation of m- and p-benzyne. By using a theoretical parameter called "resonance energy per π electron," Hess and Schaad (1971) picture structure **16** for m-benzyne.

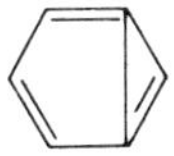

(16)

C. Metal Complexes of Arynes

The first transition metal complex of arynes was prepared by Gowling et al. (1968). Following the lead of Criegee and Schroder (1959) in forming a nickel–π-cyclobutadiene complex, they reacted o-diiodobenzene with tetracarbonyl nickel in cyclohexane and obtained black needles of π-benzyne diiodo-μ-carbonyl nickel dimer (**17**).

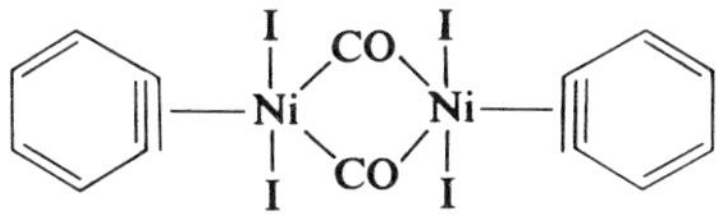

(17)

Elemental analysis gave the formula $[NiC_7H_4I_2O]_n$. Bridging carbonyl groups were indicated by a strong peak at 1790 cm^{-1} in the IR spectrum. The compound was soluble in most organic solvents to gave brown air-sensitive solutions which decolorized instantly when shaken with water and yielded nickel and iodide ions in the aqueous layer. The compound was diamagnetic. The nmr spectrum was of A_2B_2 type ($[J] = 3.5$ cycles/sec and

$\delta 14$ cycles/sec, centered about $\tau 2.33$) in deuteriomethanol. Evidently, the C_6H_4 group retained its aromatic character in the complex.

The mass spectrum gave no parent peak; the heaviest ion was $NiI_2C_6H_4{}^+$. Other ions were $C_6H_5I_2{}^+$ and species derived from it by removal of H or I atoms. Peaks corresponding to $C_{12}H_8{}^+$, perhaps biphenylene$^+$, and related ions with one, two, and four added CO groups as well as $PhCO^+$, $I_2{}^+$, CO^+, and Ph^+ were formed.

Another nickel benzyne derivative (**19**) was prepared by Dobson et al. (1971) by the reaction of trans-chloro(2-bromophenyl)-bis(triethylphosphine)

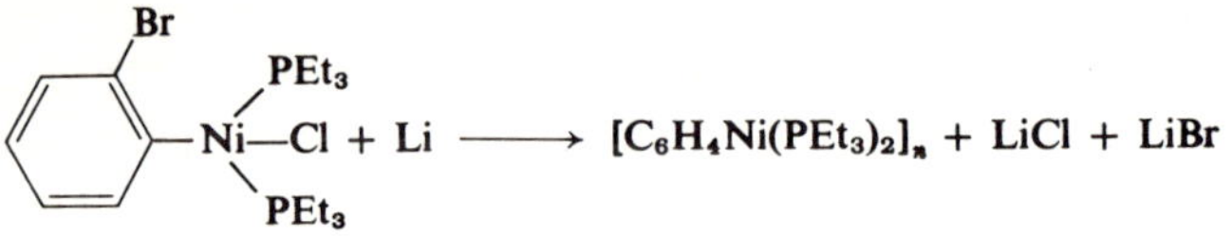

$$\text{(18)} \quad \underset{\text{Br}}{C_6H_4}-\underset{\underset{PEt_3}{|}}{\overset{\overset{PEt_3}{|}}{Ni}}-Cl + Li \longrightarrow [C_6H_4Ni(PEt_3)_2]_n + LiCl + LiBr \quad \text{(15)}$$

(**19**) 100%

nickel (II) (**18**) with lithium metal at $-78°$ to $-40°C$. Compound **19** was a bright yellow crystalline solid, extremely air-sensitive. With iodine it gave trans-iodo(2-iodophenyl) bis(triethylphosphine) nickel(II) (**20**); heating at $92°-96°C$ in tetrachloroethylene transformed it to the nickel compound (**21**) and polymeric hydrocarbons, not further characterized.

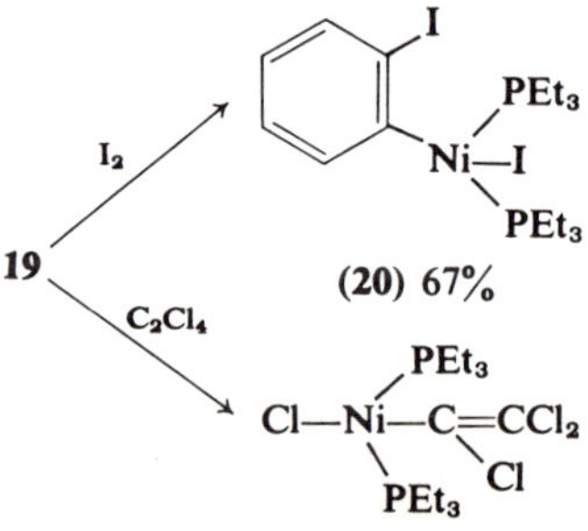

$$\text{19} \xrightarrow{I_2} \underset{(20)\ 67\%}{(2\text{-}IC_6H_4)Ni(I)(PEt_3)_2} \qquad \text{19} \xrightarrow{C_2Cl_4} \underset{(21)\ 36\%}{Cl\text{-}Ni(C=CCl_2)(Cl)(PEt_3)_2} \quad (16)$$

Tetrafluorobenzyne should be one of the best arynes with which to attempt π-type stabilization because back-bonding from the transition metal to antibonding orbitals on the aryne is strongest when highly electronegative groups are present on the π-ligand. Indeed, Roe and Massey (1970) generated tetrafluorobenzyne by thermal decomposition of pentafluorophenyl magnesium bromide in the presence of cobalt carbonyl and obtained a crystalline air-sensitive compound whose elemental analysis corresponded to $C_{16}F_4O_{10}$-Co_4. The mass spectrum contained all the ions of the series $Co_4(CO)_nC_6F_4{}^+$ for $n = 0$ to 10, strongly indicating that the C_6F_4 group was not bonded to CO ligands. Hydrochloric acid generated 1,2,3,4-tetrafluorobenzene, and thermal decomposition *in vacuo* produced cobalt, CO, and octafluorofluorenone but no perfluorotriphenylene or octafluorobiphenylene.

The same authors obtained a complex from tetrafluorobenzyne and triiron dodecacarbonyl (10% yield) whose elemental analysis and IR and mass spectra suggested the following structure:

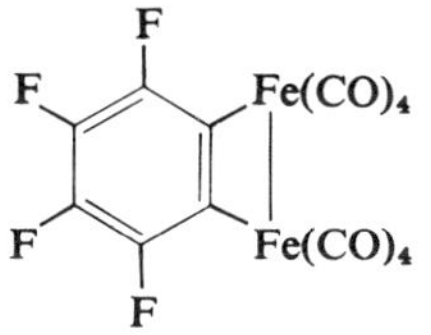

$$\begin{array}{c}\text{F}_4\text{C}_6[\text{Fe(CO)}_4]_2\end{array}$$

In dry dioxane, tetrafluorobenzyne added to nickelocene smoothly to give a 20% yield of $C_5H_5NiC_5H_5C_6F_4$, an approximately 50:50 mixture of two isomers formed by either 1,2- or 1,3-addition of C_6F_4 to one of the cyclopentadienyl rings.

Osmium complexes apparently containing benzyne bound to the metal atom were formed by pyrolysis of osmium carbonyl triphenylphosphine complexes (Bradford *et al.*, 1972; Deeming, *et al.*, 1972; Gainsford *et al.*, 1972).

The complexes described above are stable and can be isolated. Less stable complexes may also play a significant role in product distribution from benzyne reactions. Friedman (1967) found that trace quantities of silver ion greatly affected the product composition in the reaction of benzenediazonium 2-carboxylate with benzene. The yield of the 1,4-adduct (**22**) was decreased and yields of the products of insertion (biphenyl) and 1,2-cycloaddition [benzocyclooctatetraene (**23**)] were increased. Many transition metal

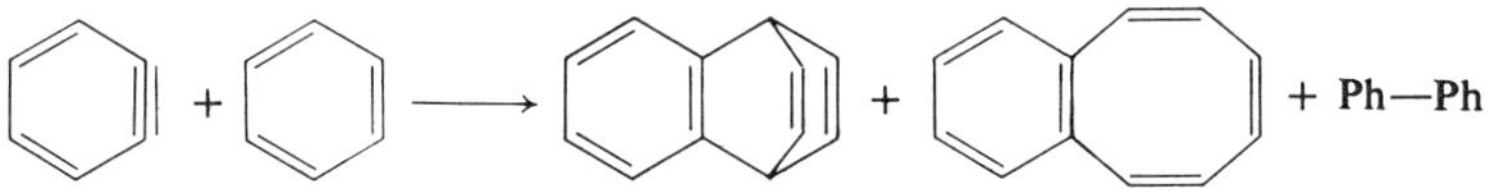

$$\text{benzyne} + \text{benzene} \longrightarrow (\mathbf{22}) + (\mathbf{23}) + \text{Ph}-\text{Ph} \quad (17)$$

compounds such as salts of mercury, tin, copper, lead, chromium, cobalt, and vanadium markedly affected the distribution of products (Tabushi *et al.*, 1969a,b).

Attempts to synthesize the platinum complex of benzyne (**24**) led instead to a complex (**25**) in which the benzyne precursor was trapped before it could decompose to benzyne (Winkler and Wittig, 1963; Cook and Jauhal,

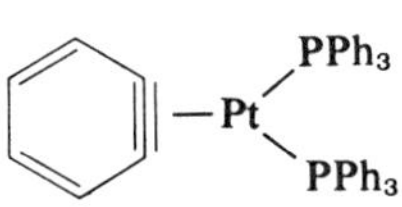

(**24**) benzyne–Pt(PPh₃)₂ (**25**) benzosulfonyl-azo–Pt(PPh₃)₂

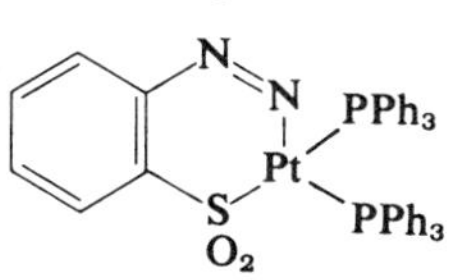

1968; Gilchrist et al., 1968). However, under vigorous conditions the reactions of benzothiodiazole 1,1-dioxide were drastically altered by the presence of zero-valent platinum species. Triphenylene was formed in good yield with no biphenylene; the reverse usually occurs (C. W. Rees et al., 1971).

Vedejs and Shepherd (1970) found a parallel effect in the addition of benzyne to cyclooctatetraene (COT). Decomposition of benzenediazonium 2-carboxylate (26) in COT (27) gave the monoadducts 28, 29, 30, and phenanthrene (Vedejs, 1968). The product ratio was remarkably affected by silver ion catalysis. In the absence of catalyst the reaction between 26 and COT produced 2:1 adduct as a major product. In the presence of 0.5 mole% of silver acetate or silver fluoroborate (based on 26) the relative yields of 30, phenanthrene, and biphenylene were drastically reduced, and 28 became the major hydrocarbon product. The ratio of the 2:1 adduct to 28 was 26:1 for the uncatalyzed reaction and 1:11 for the reaction catalyzed by silver fluoroborate. An equally remarkable Ag ion effect was found by Crews and Loffgren (1971) in the reaction of benzyne with cycloheptatriene.

Friedman (1967) proposed that Ag^+-catalyzed addition of benzyne to benzene involved a benzyne–Ag^+ complex with enhanced electrophilic tendencies. The results of Vedejs and Shepherd (1970) supported this rationale; polar addition of the benzyne–Ag^+ complex to COT would give the homotropylium ion (31), a possible precursor of 28 [Eq. (18)].

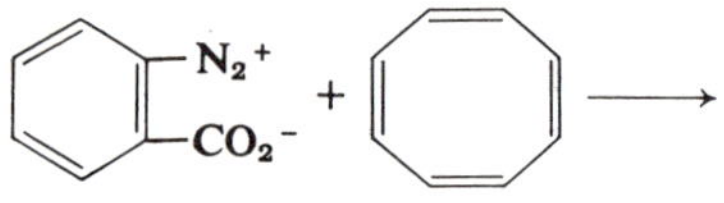

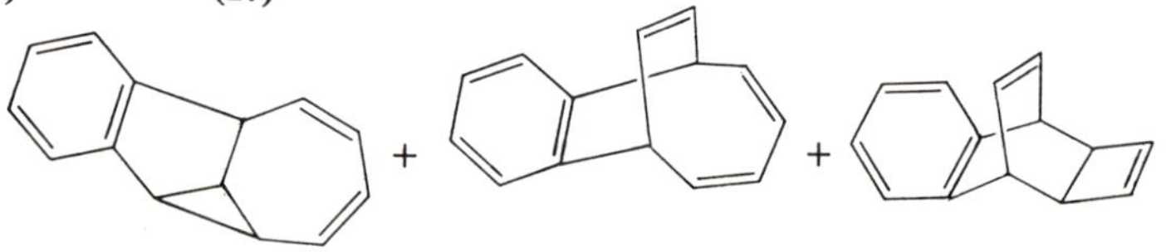

Warner (1971) proposed the complex 31 bound to an additional silver ion since Ag^+ readily complexes dienes but not trienes. This concept was challenged by Paquette (1971) who gave convincing arguments for a mechanism involving only σ-bonded Ag atoms and no extraneous Ag ions.

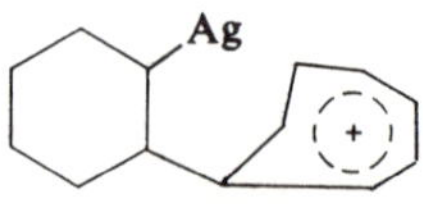

III. Methods of Formation of Arynes

A. IN SOLUTION

In almost all cases, arynes are formed by concerted or stepwise elimination of two ortho substituents in an aromatic ring, leaving the electrons distributed between two formerly bonding orbitals [Eq. (19)]. X and Y may be bonded

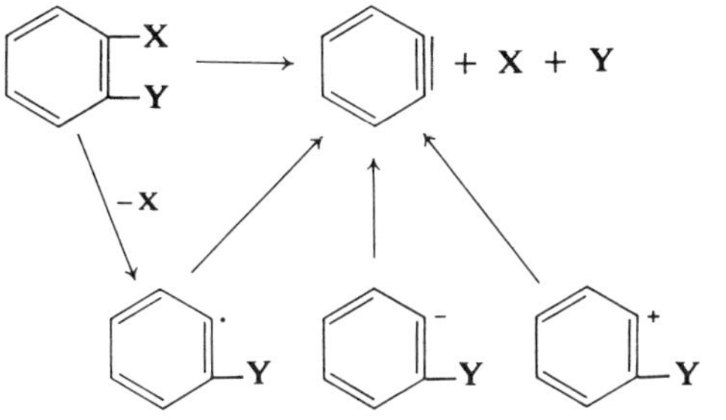

$$\text{[reaction scheme]} \quad (19)$$

directly or with intermediate atoms in a ring system; X and Y may also be in nearby, rather than ortho, positions. Thus, 1,8-naphthalyne, perinaphthalyne (**32**), is formed by the elimination of groups from 1,8-naphthalene derivatives.

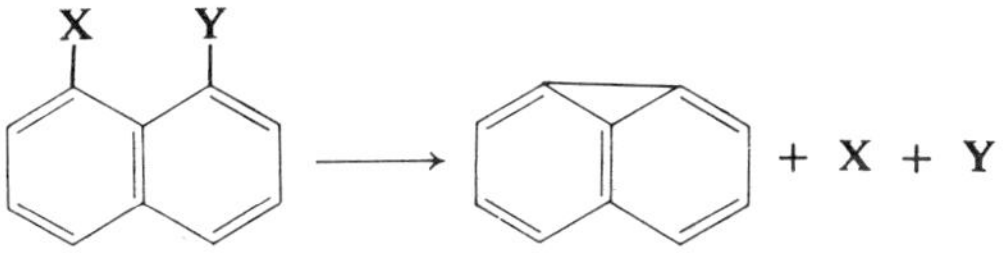

$$\text{[reaction scheme]} \quad (20)$$

An exception is the formation of benzyne from diacetylene and acetylene.

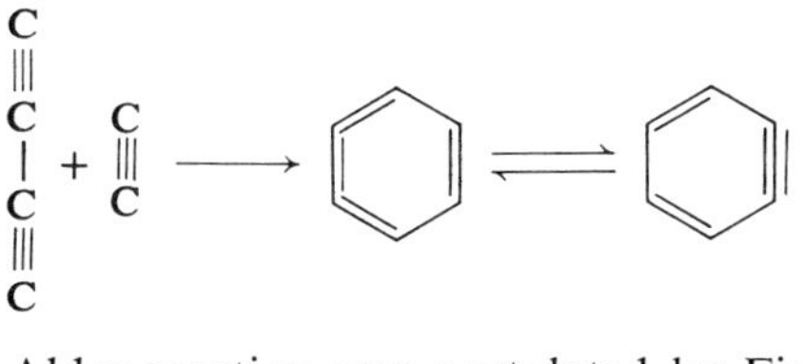

$$\text{[reaction scheme]}$$

This type of Diels–Alder reaction was postulated by Fields and Meyerson (1967b) to account for the similarity in products from a known benzyne precursor, phthalic anhydride, and from acetylene under the same conditions, alone and in their separate reactions with hexafluorobenzene.

In much of the early work, arynes were formed via aryl anions. However, a most significant advance in the discovery of a convenient and easy method of generating benzyne is of comparatively recent origin. Benzenediazonium 2-carboxylate, readily made by diazotization of anthranilic acid, was first prepared by Hantzsch and Davidson (1896) who found it gave salicylic acid in aqueous solution. Hantzsch and Ologauer (1897) ascribed to it the cyclic formula 33.

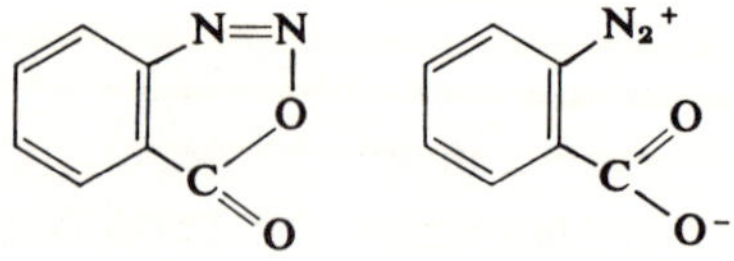

(33)   (34)

Stiles and Miller (1960) determined that benzenediazonium carboxylate had the zwitterion structure **34**, and discovered that in nonaqueous suspensions it lost both CO_2 and N_2 at 39°C and gave adducts derived from benzyne (Stiles et al., 1963; Miller and Stiles, 1963).

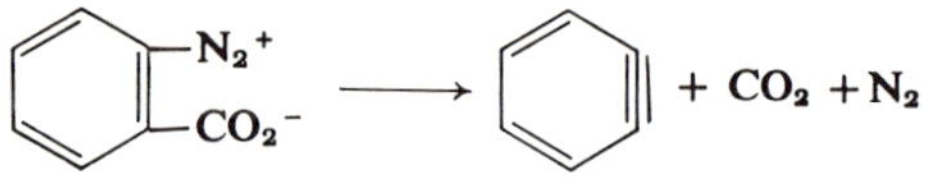

$$\text{(benzenediazonium 2-carboxylate)} \longrightarrow \text{(benzyne)} + CO_2 + N_2 \quad (21)$$

Benzenediazonium 2-carboxylate is explosive and extremely dangerous to store or handle. Friedman and Logullo (1967) developed an ingenious method of forming and using the reagent to generate benzyne by aprotic diazotization with isoamyl nitrite and an even more ingenious way of handling the reagents (Friedman and Logullo, 1965). 2-Carboxybenzene diazonium chloride is prepared by diazotizing anthranilic acid hydrochloride in ethanol with isoamyl nitrite. The diazonium chloride is precipitated with anhydrous ether and may be stored and handled readily. According to the authors, it is non-explosive but flashes on being ignited (Embree, 1971). Benzyne is formed in

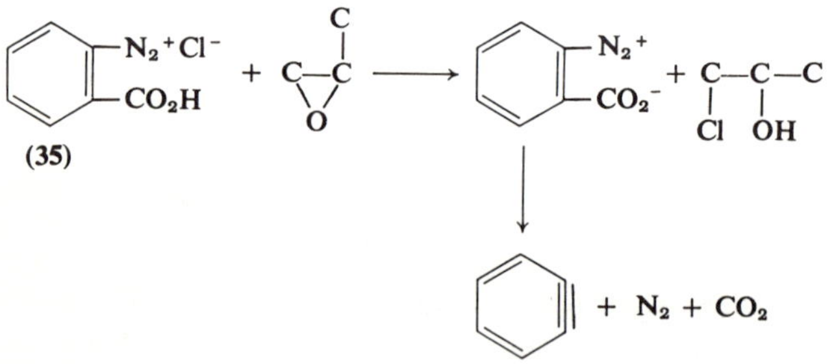

$$(35) + \text{epoxide} \longrightarrow \text{intermediate} + C-C-C \text{ (Cl, OH)} \quad (22)$$

$$\longrightarrow \text{benzyne} + N_2 + CO_2$$

$$(36) + B^- \longrightarrow \text{aryl anion} + HB \quad (23)$$

$$(37) \longrightarrow \text{benzyne} + X^-$$

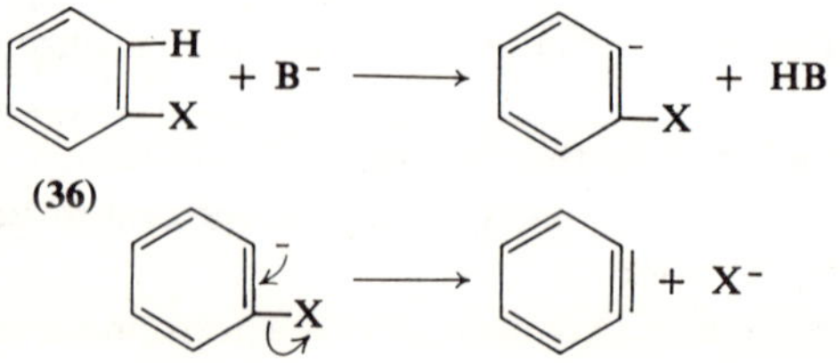

80–90% yields by refluxing a suspension of the diazonium chloride (**35**) in ethylene dichloride with propylene oxide [Eqs. (22) and (23)].

A variety of bases have been used to remove protons from aromatic halogen compounds and generate benzyne via an anion. Bases most commonly used are lithium alkyls and lithium aryls in ether; lithium amides, sodamide, and potassium amide in ether or amine; potassium *tert*-butoxide in dimethyl sulfoxide; and potassium hydroxide in diphenyl ether. Phenyllithium in ether at 20°–25°C gives consistently high yields, especially with bromobenzene (Wittig and Fuhrmann, 1940). Phenyllithium gives arynes with aromatic fluorine derivatives as well. 1,2-Triptycyne (**38**) was prepared

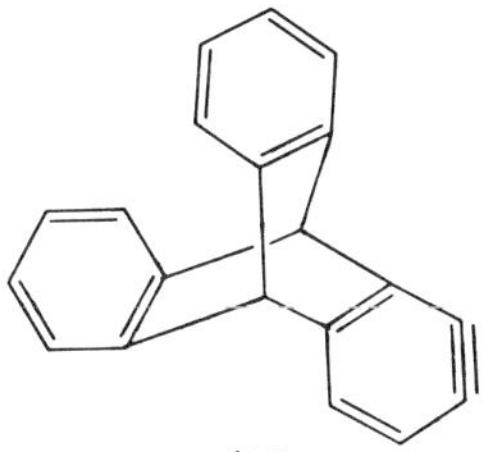

(**38**)

by Petersen and Berg (1971) by the reaction of 2-fluorotriptycene with phenyllithium in ether. Both 1- and 2-phenyltriptycene resulted. Addition of salts such as sodium nitrite, sodium thiocyanate, and potassium thiocyanate increased yields of *N,N*-dimethyl-*o*- and *m*-toluidines from sodamide, chlorobenzene, and dimethyl amine from 5 to 80% (Biehl *et al.*, 1970).

Another general method of forming arynes of considerable utility is from ortho dihalogen-substituted benzenes with lithium or magnesium. Use of strong bases which are good nucleophiles is thus avoided, so the products are other than those from addition of the base to aryne. *o*-Fluorobromobenzene with lithium amalgam in tetrahydrofuran gave biphenylene and triphenylene (Wittig and Pohmer, 1955); in furan, the endoxide was formed (Wittig and Behnisch, 1958).

o-Diiodobenzene and *o*-bromoiodobenzene with magnesium gave benzyne and 3-bromo-4-iodotoluene gave 4-methylbenzyne (Hinton *et al.*, 1958). *o*-Fluorobromopseudocumene with lithium amalgam gave 3,4,6-trimethylbenzyne (Wittig and Härle, 1959). In the same study the authors prepared dimethylbenzadiyne (**40**) from 2,6-difluoro-3,5-dibromo-*p*-xylene (**39**) with

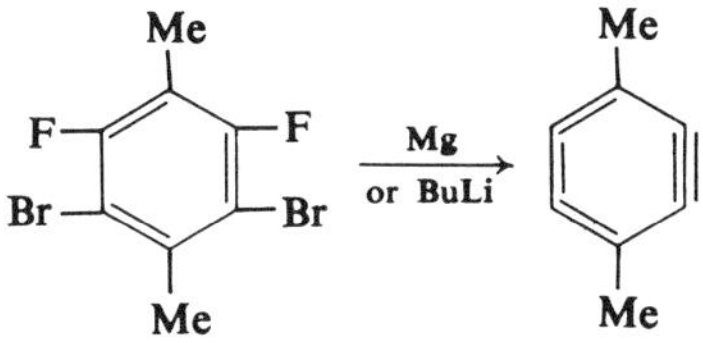

(24)

(**39**) → Mg or BuLi → (**40**)

magnesium in tetrahydrofuran (5% yield) or butyllithium in ether (15% yield) [Eq. (24)].

In the culmination of decades of work by many investigators with *N*-nitrosoacetanilide (**41**) as a free radical precursor, Cadogan *et al.* (1971a) converted aniline to benzyne in one step by reacting it with pentyl nitrite and acetic anhydride in benzene at 80°C [Eq. (25)]. In the presence of tetra-

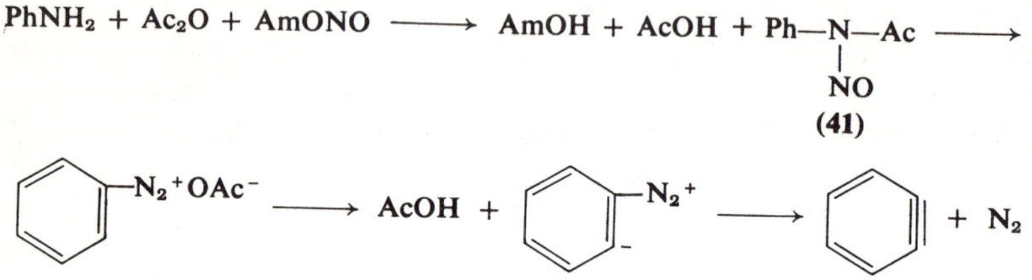

$$PhNH_2 + Ac_2O + AmONO \longrightarrow AmOH + AcOH + Ph-\underset{\underset{NO}{|}}{N}-Ac \longrightarrow$$
(**41**)

$$\text{Ph-}N_2^+OAc^- \longrightarrow AcOH + \text{Ph-}N_2^+ \longrightarrow \text{benzyne} + N_2 \quad (25)$$

phenylcyclopentadienone, aniline gave 1,2,3,4-tetraphenylnaphthalene in 32% yield; in the presence of anthracene, it gave a 10% yield of triptycene (Cadogan, 1971; Cadogan, *et al.*, 1972).

p-Chlorobenzoyl nitrite with acetanilide gave 60–90% yields of benzynes with meta substituents; *p*- and *o*-substituted acetanilides gave much lower yields. The system in which *p*-chlorobenzoyl nitrite furnished the NO group remained virtually aprotic because the *p*-chlorobenzoic acid precipitated as formed.

Rüchardt and Tan (1970) had earlier formed benzyne by decomposition of benzenediazonium chloride or fluoroborate in an acetic acid solution of potassium acetate and tetracyclone [Eq. (26)]. With anthracene the same

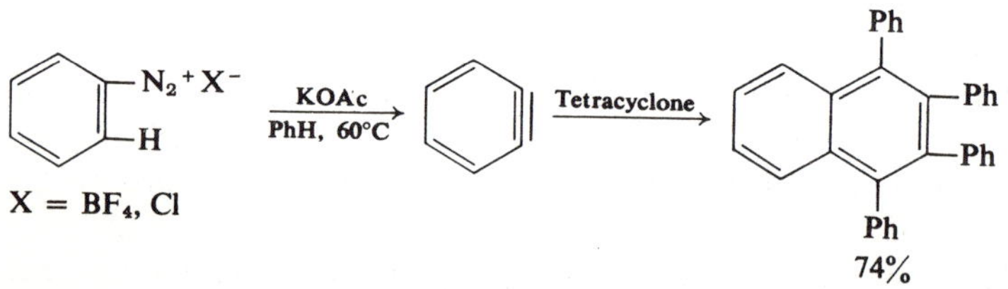

$$\text{Ph-}N_2^+X^- \xrightarrow[\text{PhH, 60°C}]{\text{KOAc}} \text{benzyne} \xrightarrow{\text{Tetracyclone}} \text{tetraphenylnaphthalene} \quad (26)$$
X = BF$_4$, Cl
74%

procedure gave a 41% yield of triptycene.

B. IN THE GAS PHASE

Under electron impact in the mass spectrometer at 10^{-6} Torr, phthalic anhydride decomposes as shown in Table II (McLafferty and Gohlke, 1959; Meyerson, 1965). The large amount of ion of mass 76, $C_6H_4^+$, nominally benzyne, prompted Fields and Meyerson (1965) to duplicate the decomposition of phthalic anhydride together with benzene at 1 atm and 700°C in a flow system. Benzyne was indeed formed and gave the characteristic products biphenylene and triphenylene as well as biphenyl and naphthalene.

Table II
Partial Mass Spectrum of Phthalic Anhydride

Mass	Relative intensity	Tentative structure
148	47	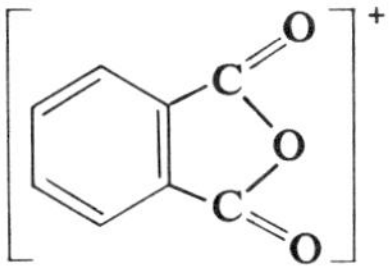 phthalic anhydride cation
$-CO_2 \downarrow$		
104	100	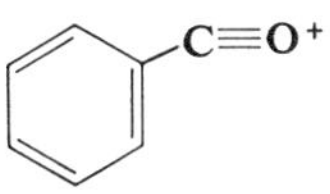 $C_6H_5-C\equiv O^+$
$-CO \downarrow$		
76	85	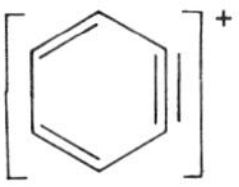 benzyne cation
$-C_2H_2 \downarrow$		
50	40	$HC\equiv C-C\equiv CH$

$$\text{phthalic anhydride} \longrightarrow \text{benzyne} + CO_2 + CO \qquad (27)$$

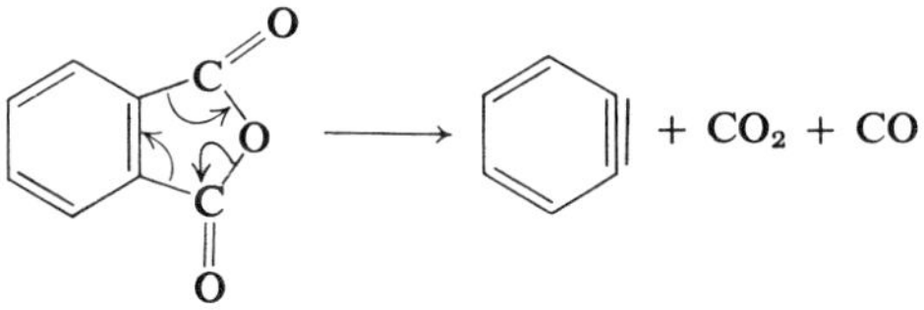

Until the work of Fields and Meyerson (1965) the literature on arynes in the gas phase was limited to a few examples. Wittig (1961) pyrolyzed bis-iodophenyl mercury (42) and phthaloyl peroxide (43) in argon at 600°C and 5 Torr and obtained 54 and 27% yields, respectively, of biphenylene (44).

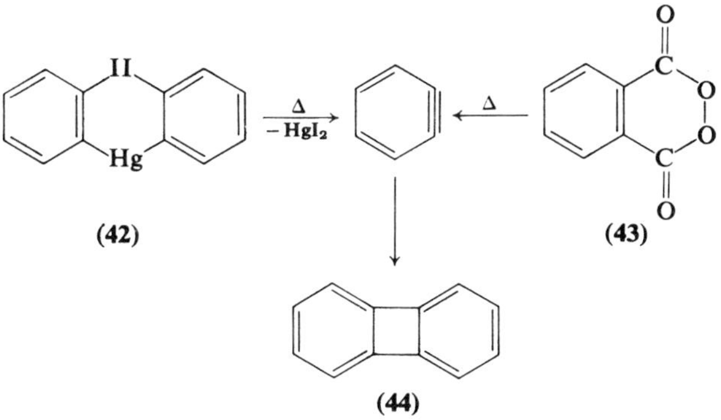

$$(42) \xrightarrow[-HgI_2]{\Delta} \text{benzyne} \xleftarrow{\Delta} (43) \qquad (28)$$

$$\downarrow$$

$$(44)$$

When furan was present in the decomposition zone, α-naphthol resulted from the further reaction of the initially formed 1,4-dihydro-1,4-oxidonaphthalene (**45**). Ebel and Hoffmann (1964) determined a maximum lifetime of 20 msec for benzyne under these conditions.

(**45**)

Fisher and Lossing (1963) decomposed *o*-diiodobenzene at 960°C and 10^{-3} Torr in a reactor coupled to a mass spectrometer and found a product of mass 76, whose vertical ionization potential, 9.75 V, was appreciably higher than that calculated for open-chain C_6H_4 isomers and strongly indicated formation of benzyne. This assignment was supported by the formation of biphenylene, mass 152.

Berry *et al.* (1962, 1964) decomposed solid benzenediazonium 2-carboxylate by flash photolysis; the UV and mass spectra of the products showed that benzyne (76), N_2 (28), and CO_2 (44) had formed in the vapor phase. The mass 76 peak reached a maximum when the UV spectrum showed a broad absorption at 2430 nm assigned to benzyne. A peak at mass 152, biphenylene, appeared as benzyne (76) faded.

Indanetrione at 600°C lost three molecules of CO and also gave benzyne (Brown and Solly, 1965).

Fields and Meyerson (1965), through their discovery of the elimination of CO_2 and CO from aromatic anhydrides at 700°C, furnished a powerful tool for generating a wide variety of arynes from readily available and inexpensive reagents. In all cases the formation of arynes in the gas phase by decomposition of aromatic anhydrides at 600° – 700°C at atmospheric pressure found its counterpart in the decomposition of the identical anhydrides under electron impact in the mass spectrometer at 10^{-6} Torr.

For example, the mass spectra of two naphthalenedicarboxylic anhydrides, the 2,3- (**46**) and 1,8- (**47**) isomers, are qualitatively similar (Fields and Meyerson, 1968a) as shown in Table III.

Pyrolysis of the anhydrides separately in benzene solution at 700°C (Fields and Meyerson, 1968a) gives products derived from 2,3-naphthalyne and perinaphthalyne [(**48**) and (**49**)] which are shown as ions in the mass spectral scheme. 1,8-Naphthalic anhydride gives a much larger parent peak than does the 2,3-anhydride; thermally the 1,8-isomer was also appreciably more stable than the 2,3-anhydride.

Benzadiynes [(**51**) and (**53**)] were formed by pyrolysis of the aromatic tetracarboxylic dianhydrides, pyromellitic (**50**), and mellophanic (**52**); naphthalene-1,4,5,8-tetracarboxylic dianhydride (**54**) gave perinaphthaldiyne (**55**).

7. ARYNES

Table III
Naphthalene Dicarboxylic Anhydride

Mass	Relative intensities		Suggested structures	
	2,3-	1,8-	2,3-	1,8-
198	3.52	53.4	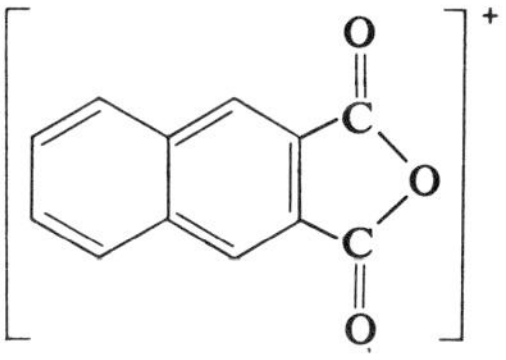 (46)	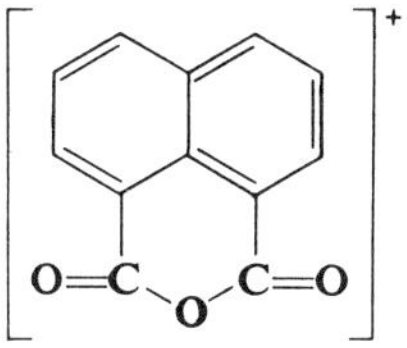 (47)
$-CO_2 \downarrow$				
154	47.5	100.0		
$-CO \downarrow$			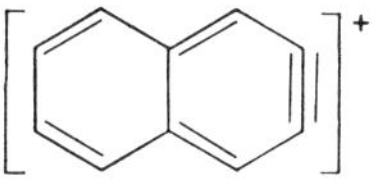	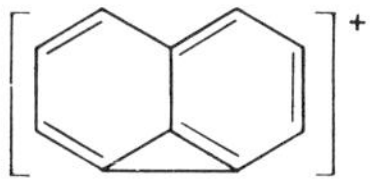
126	100.0	98.8	(48)	(49)

(50) ⟶ (51) + 2 CO$_2$ + 2 CO   (29)

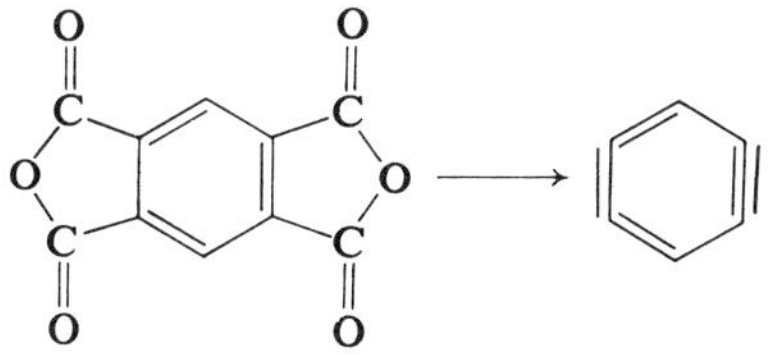

(52) ⟶ (53) + 2 CO$_2$ + 2 CO   (30)

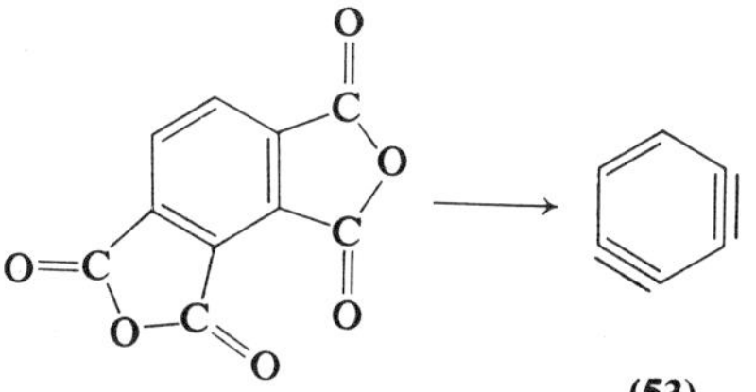

(54) ⟶ (55) + 2 CO$_2$ + 2 CO   (31)

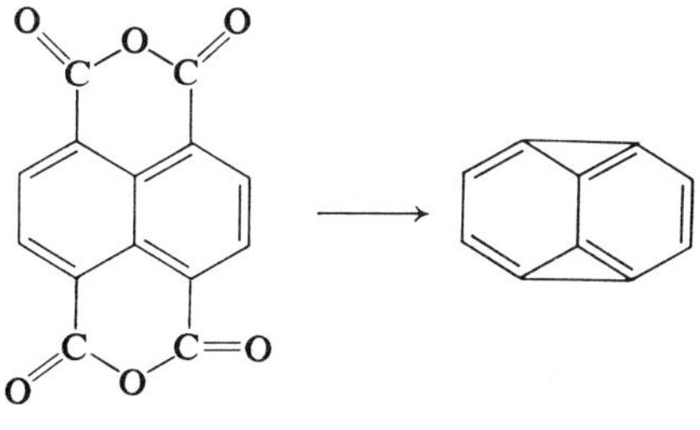

The logical extension of this work by Fields and Meyerson (1968a) would have been the formation of benzatriyne (**57**) from mellitic trianhydride (**56**).

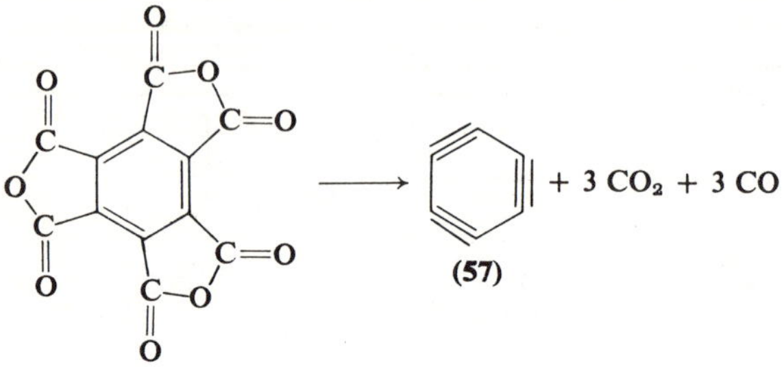

$$\text{(56)} \longrightarrow \text{(57)} + 3\,CO_2 + 3\,CO \qquad (32)$$

Benzatriyne is C_6 or carbon, and carbon was the only product they obtained, although it is doubtful that mellitic trianhydride was in the vapor phase. It decomposes at a relatively low temperature, is extremely insoluble and nonvolatile, and cannot be handled as other anhydrides.

Fields and Meyerson (1971a) decomposed mellitic trianhydride by refluxing its solutions in high-boiling liquids (350°–400°C) for extended periods. The products were pyromellitic and mellophanic anhydrides as well as reaction products of a benzyne tetracarboxylic dianhydride (**58**).

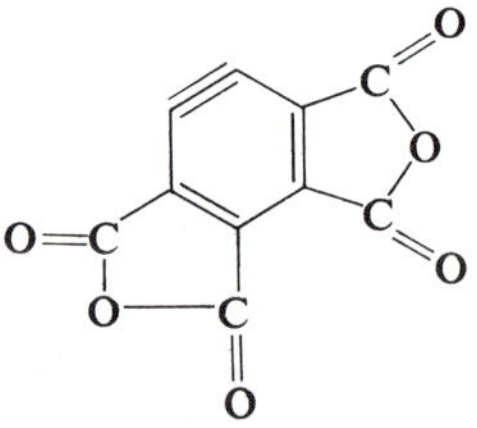

Other anhydrides that lost CO_2 and CO and gave arynes in the vapor phase (Fields and Meyerson, 1969a) were trimellitic anhydride (**59**) to carboxybenzyne (**60**);

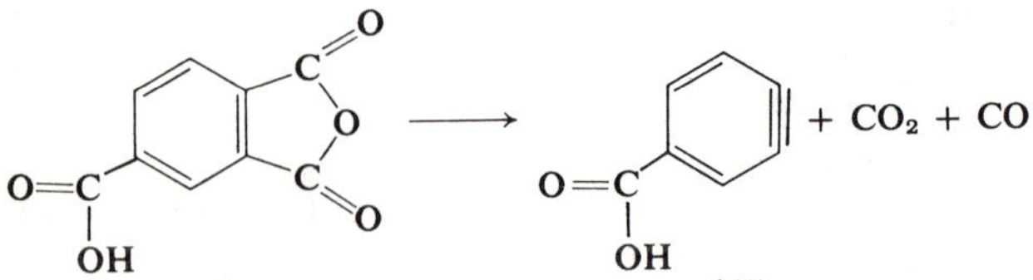

$$\text{(59)} \longrightarrow \text{(60)} + CO_2 + CO \qquad (33)$$

tetrachlorophthalic anhydride (**61**) to tetrachlorobenzyne (**62**);

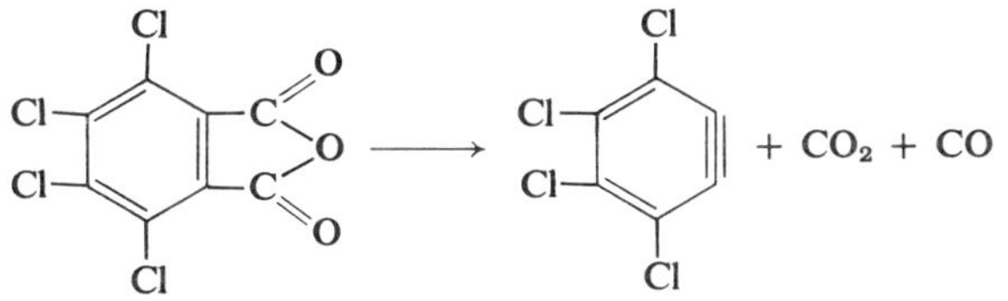

7. ARYNES

tetradeuterophthalic anhydride (63) to tetradeuterobenzyne (64);

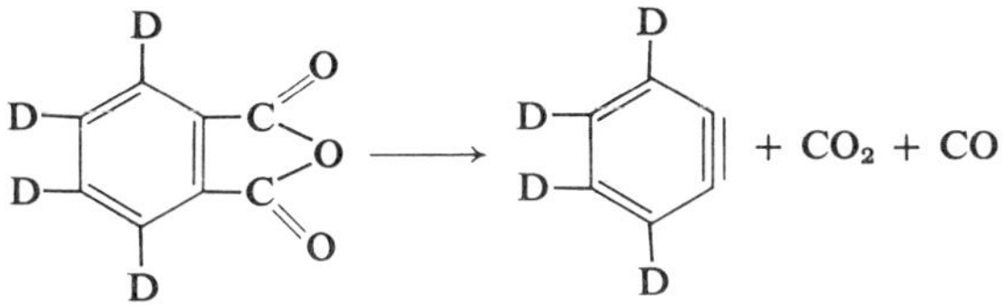

$$\text{(61)} \longrightarrow \text{(62)} + CO_2 + CO \qquad (34)$$

$$\text{(63)} \longrightarrow \text{(64)} + CO_2 + CO \qquad (35)$$

and tetraphenylphthalic anhydride (65) to tetraphenylbenzyne (66).

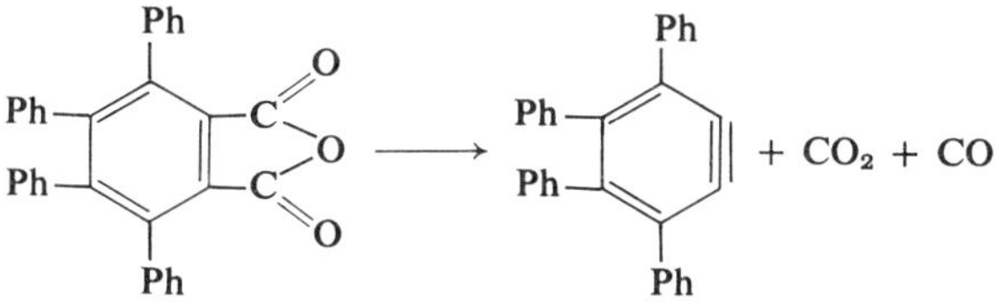

$$\text{(65)} \longrightarrow \text{(66)} + CO_2 + CO \qquad (36)$$

An ion of molecular weight 150, nominally bibenzyne (68), was formed by decomposition of 3,4,3′,4′-biphenyltetracarboxylic anhydride (67) in the mass spectrometer (Martinez de Bertorello et al., 1970).

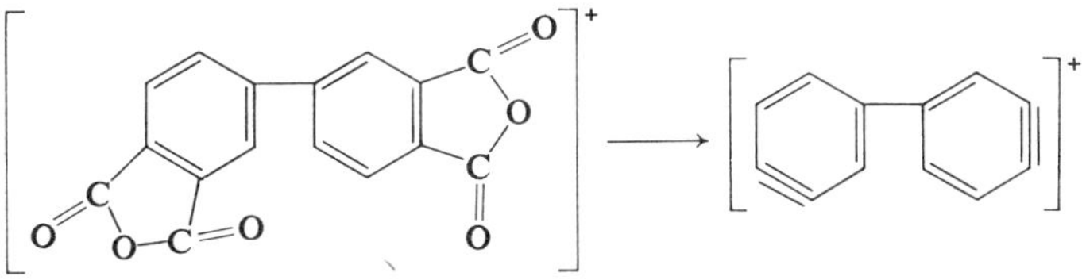

$$[\text{(67)}]^+ \longrightarrow [\text{(68)}]^+ \qquad (37)$$

o-Sulfobenzoic anhydride (69) contains two good leaving groups, SO_2 and

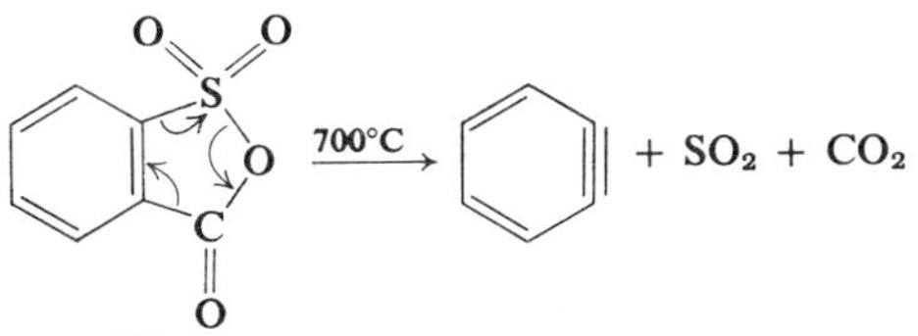

$$\text{(69)} \xrightarrow{700°C} \bigcirc\!\!\!||| + SO_2 + CO_2 \qquad (38)$$

CO_2; Meyerson and Fields (1966) found that it gave benzyne at 700°C in the vapor phase. Under electron impact in the mass spectrometer, sulfobenzoic anhydride gave a $C_6H_4^+$ ion, nominally benzyne ion, by successive loss of SO_3 and CO (Meyerson and Fields, 1966; Brown et al., 1967).

Other aromatic compounds gave benzyne at 600°–700°C; these included chlorobenzene (Fields and Meyerson, 1966b), nitrobenzene (Fields and Meyerson, 1969a), o-chlorofluorobenzene, and benzene itself (Fields and Meyerson, 1968b). Best yields of aryne-derived products are from aromatic anhydrides, whose thermal decomposition is by far the easiest and most versatile way to produce arynes in gas phase. Insoluble anhydrides are best vaporized separately into a nitrogen, helium, or argon stream just before introduction of the mixed stream into the hot tube (Fields and Meyerson, 1971a).

Phthalic anhydride was converted to benzyne by flash photolysis (Porter and Steinfeld, 1968) and in the plasma of a glow discharge (Suhr and Szabo, 1971).

Benzotriazines (**71**), formed from the oxidation of o-aminoaryl ketone hydrazones (**70**) by lead tetraacetate, decomposed at 400°C to give benzyne.

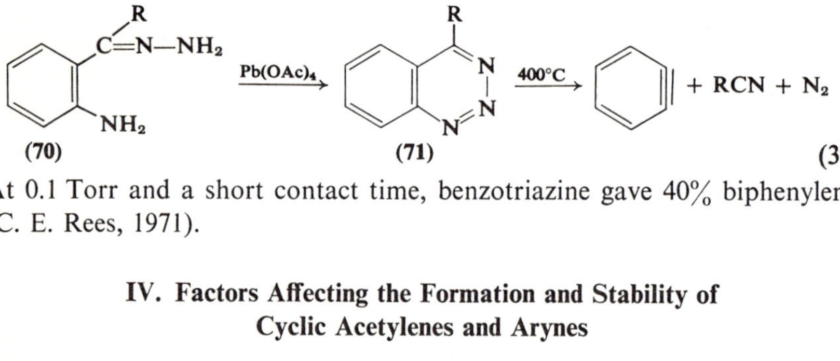

$$\text{(70)} \xrightarrow{\text{Pb(OAc)}_4} \text{(71)} \xrightarrow{400°C} \text{benzyne} + \text{RCN} + N_2 \quad (39)$$

At 0.1 Torr and a short contact time, benzotriazine gave 40% biphenylene (C. E. Rees, 1971).

IV. Factors Affecting the Formation and Stability of Cyclic Acetylenes and Arynes

Cyclic acetylenes are discussed along with arynes in this section since both classes contain a triple bond and are highly reactive organic intermediates. Some of the same methods of formation and proof of existence used for arynes have been applied to ring acetylenes.

A. Effect of Ring Size

Cyclooctyne is the smallest ring acetylene that has been isolated (Domnin, 1938). Ring acetylenes are necessarily disubstituted. The four carbon atoms in the sequence

$$-CH_2-C\equiv C-CH_2-$$

are linear because of sp hybridization of the σ bonds in the acetylene. All ring

7. ARYNES

acetylenes smaller than cyclononyne are, therefore, considerably strained. Baeyer strain is about 8–9 kcal/mole for cyclooctyne, 25–30 kcal/mole for cycloheptyne, and 45–50 kcal/mole for cyclohexyne (Krebs, 1969).

Cyclooctyne (72) is best prepared by the method of Wittig and Dorsch (1968) from 1,2-dibromocyclooctene and magnesium. 1,2-Dibromocyclo-

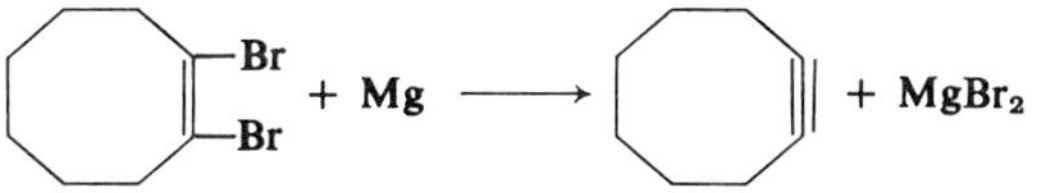

$$\text{[1,2-dibromocyclooctene]} + \text{Mg} \longrightarrow \text{[cyclooctyne (72)]} + \text{MgBr}_2 \quad (40)$$

hexene with magnesium yielded the trimer (73) and tetramer (74) of cyclohexyne (Wittig and Mayer, 1963). 1,3-Diphenylbenzoisofuran proved to be

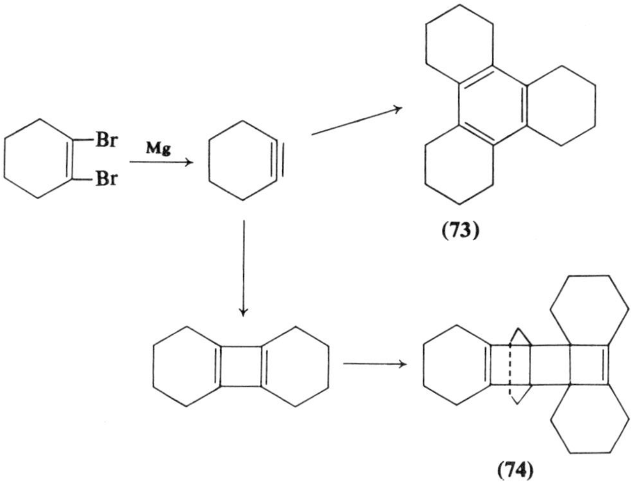

$$\text{[1,2-dibromocyclohexene]} \xrightarrow{\text{Mg}} \text{[cyclohexyne]} \longrightarrow (73) \quad (41)$$

$$\downarrow$$

$$(74)$$

an efficient trap for cyclohexyne, giving the adduct (75) (Wittig and Pohlke,

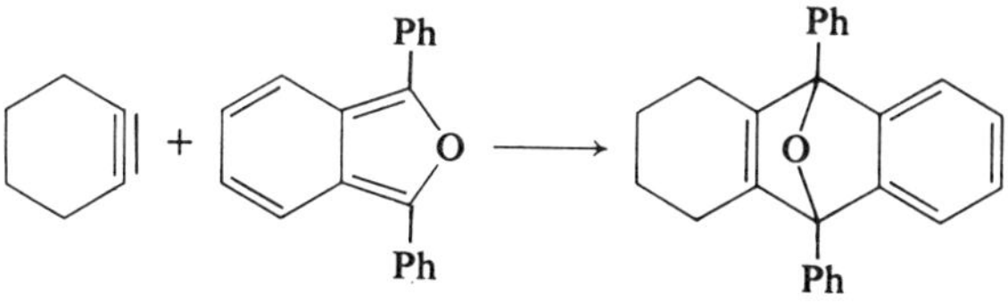

$$\text{[cyclohexyne]} + \text{[1,3-diphenylbenzoisofuran, Ph, O, Ph]} \longrightarrow \text{[adduct 75, Ph, O, Ph]} \quad (42)$$

50.5%
(75)

1961). A somewhat lower yield of the same adduct (75) (35%) resulted when cyclohexyne was generated by the method of Erickson and Wolinsky (1965) from bromomethylene cyclopentane. The yield of cyclopentyne was only 12.3% from bromomethylene cyclobutane, trapped as the 1:2 adduct (76) with diphenylisobenzofuran. α-Bromomethylenecamphene with potassium

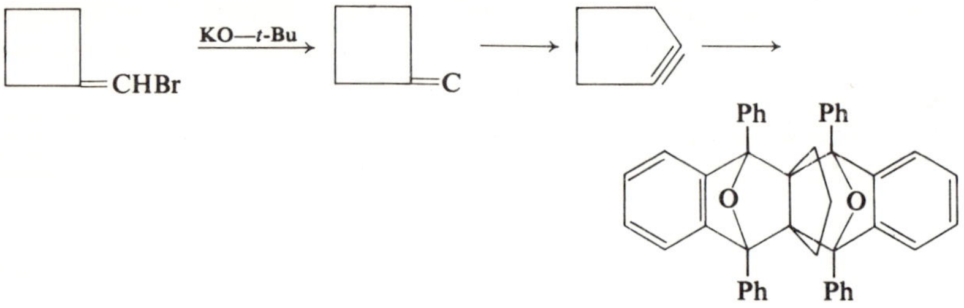

[Reaction scheme (43) showing CHBr bicyclic compound → KO-t-Bu → carbene → cycloalkyne → adduct (76) 12.3% with Ph, Ph, O, Ph, Ph substituents]

butoxide gave a high yield of the bicycloheptyne adduct (77). Cycloheptyne

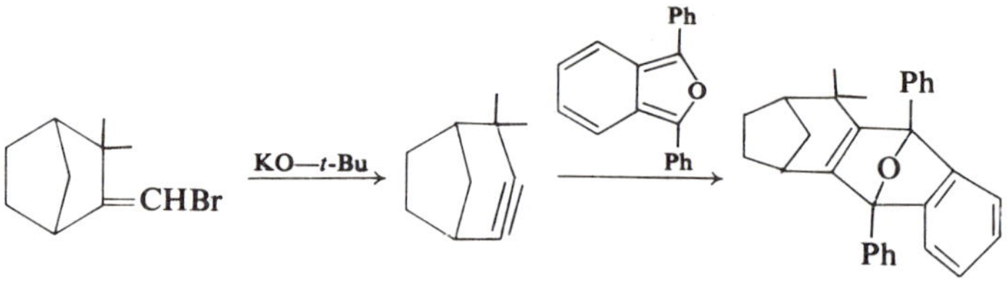

[Reaction scheme (44): bicyclic CHBr compound → KO-t-Bu → cycloalkyne → with Ph/O/Ph diphenylisobenzofuran → adduct (77) 94%]

generated by oxidation of 1-aminocycloheptatriazole (78) with lead tetraacetate at −76°C gave a good yield of adduct (79) with tetracyclone (Wittig and Meske-Schüler, 1968). Nitrogen evolved rapidly from the aminotriazole

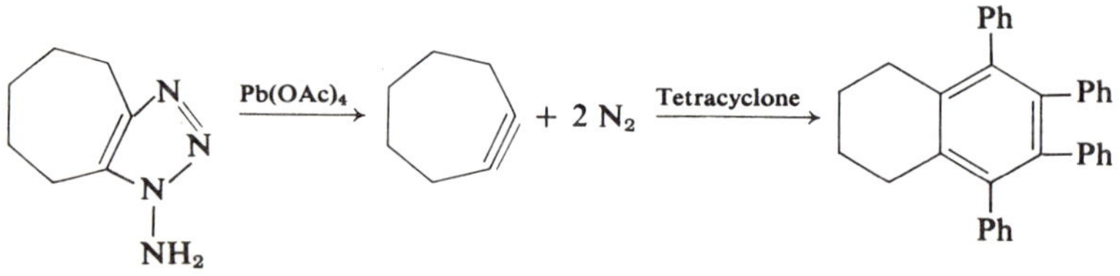

[Reaction scheme (45): (78) aminotriazole → Pb(OAc)₄ → cycloheptyne + 2 N₂ → Tetracyclone → (79) 93%]

(78) on reaction with lead tetraacetate. Adding tetracyclone in methylene chloride at different intervals, the authors obtained the yields of adducts shown in the following tabulation:

Minutes	Yield (%)
0	93
2	74
10	68
60	49
180	35

This semiquantitative kinetic study gave a half-life for cycloheptyne at −76°C of 1 hr. Cyclohexyne has a half-life of a few seconds at −110°C as determined in the same way. 3,3,7,7-Tetramethylcyclophetyne, due to the geminal methyl groups, was about 10^7 times more stable than cycloheptyne (Krebs and Kimling, 1971).

Metal complexes of cyclooctyne have been prepared and characterized. Aqueous silver nitrate gave a crystalline complex (**80**) (Wittig and Dorsch, 1968). Ammonia breaks the complex and releases cyclooctyne.

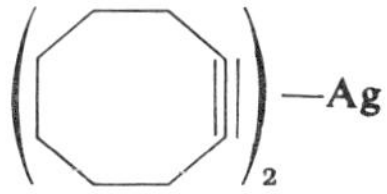

(**80**)

Gilchrist *et al.* (1968) prepared the platinum complex (**81**) from cyclooctyne and tetrakis(triphenylphosphine)platinum.

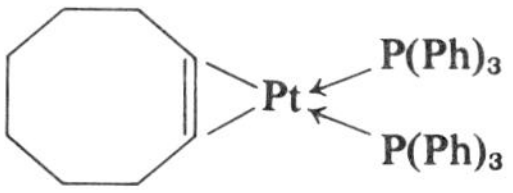

(**81**)

Wittig and Fritze (1968) prepared two nickel complexes from cyclooctyne. Nickel bromide gave the dimeric complex (**82**). Nickel tetracarbonyl con-

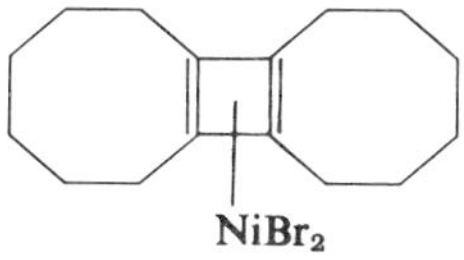

(**82**) 99%

verted cyclooctyne to the cyclopentadienone complex (**83**).

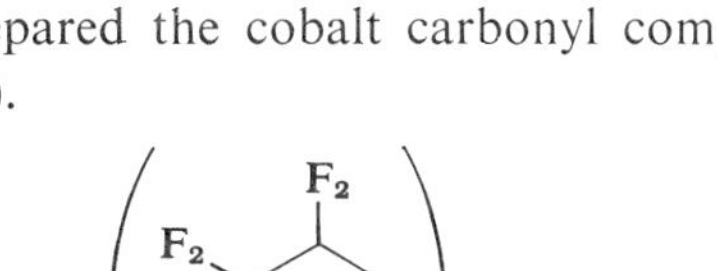

(**83**) 56%

Only one metal complex of a cyclohexyne derivative has been reported. Bailey *et al.* (1964) prepared the cobalt carbonyl complex of hexafluorocyclohex-1-en-3-yne (**84**).

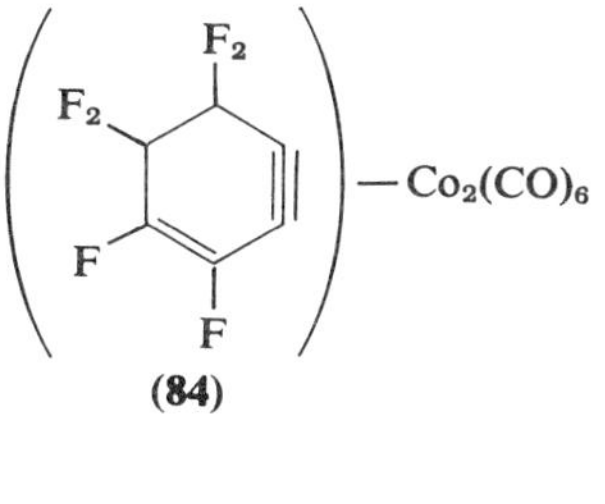

(**84**)

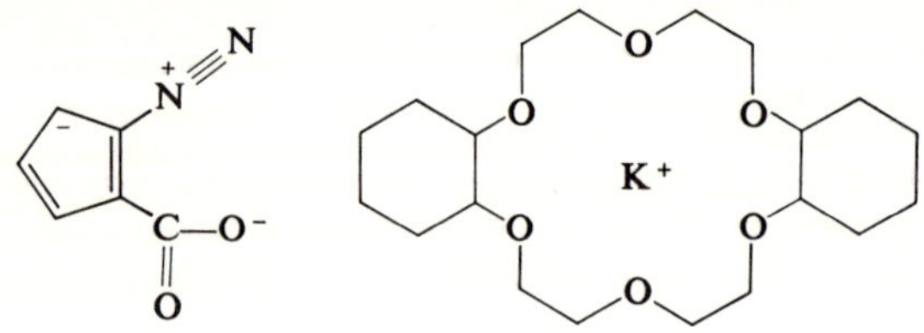

(85)

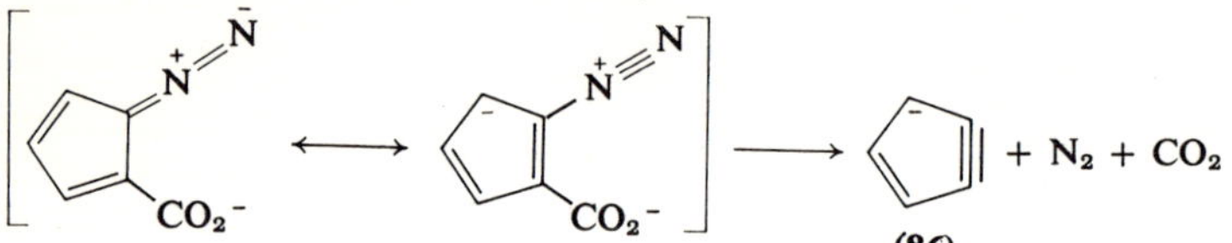

(86)

$$+ N_2 + CO_2 \quad (46)$$

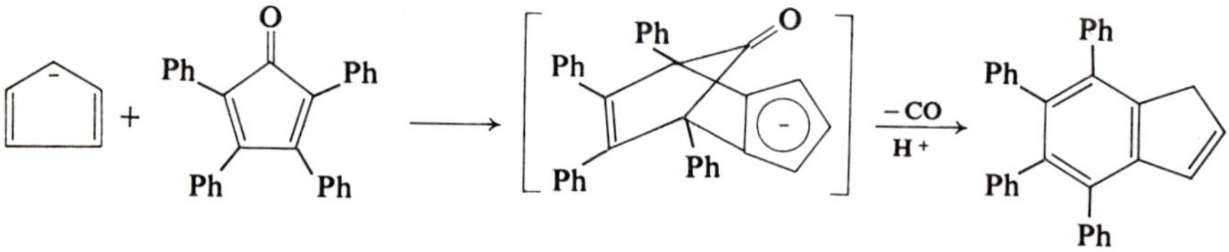

(87) 6.5%

(47)

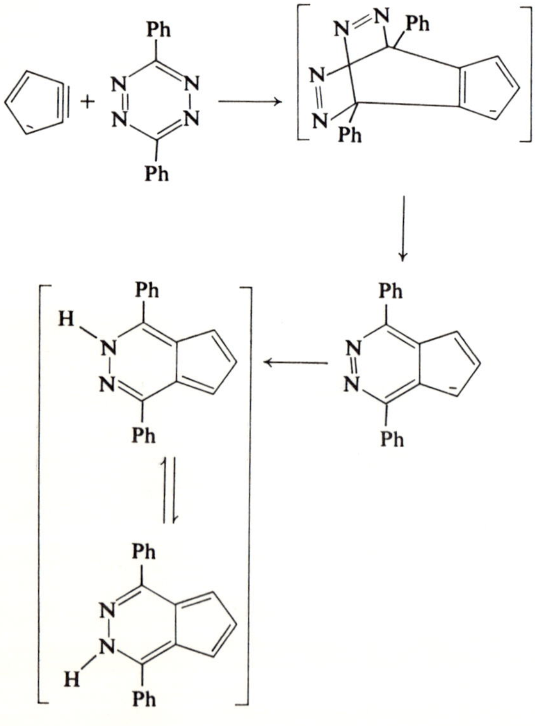

(48)

(88) 8%

7. ARYNES

Dehydrocyclopentadienyl anion (**86**), related to the aromatic 6π electron cyclopentadienyl anion as benzyne is related to benzene, was generated by Martin and Bloch (1971) by pyrolysis of the salt of diazacyclopentadiene carboxylate with potassium dicyclohexyl-18-crown-6 ether (**85**). Adducts of the dehydrocyclopentadienyl anion with tetraphenylcyclopentadienone (**87**) and 3,6-diphenyltetrazine (**88**) were isolated, although the yields were low [Eqs. (47) and (48)].

Cyclic acetylenes containing seven carbon atoms were formed from 3- and 4-bromotropolones by reaction with sodium methylate in dimethyl sulfoxide (Yamatani *et al.*, 1970). The dehydrotropolones were trapped with diphenylisobenzofuran.

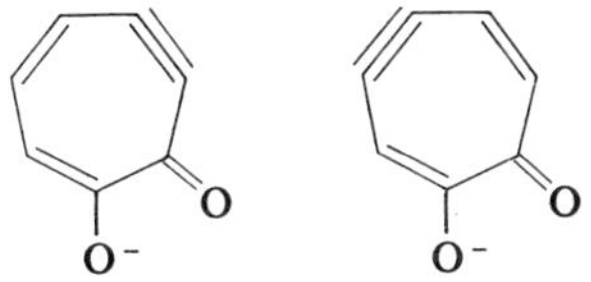

1,2-Dehydrocyclooctatetraene (**89**) prepared from bromocyclooctatetraene and potassium *tert*-butoxide gave 2 + 4 cyclo adducts with conjugated dienes and, thus, has provided a ready synthesis of benzo-substituted cyclooctatetraenes difficult to make otherwise. With butadiene it gave a dihydro adduct which was smoothly dehydrogenated with 2,3-dichloro-5,6-dicyano-*p*-benzoquinone to benzocyclooctatetraene (**90**) (Ellis and Sargent, 1969).

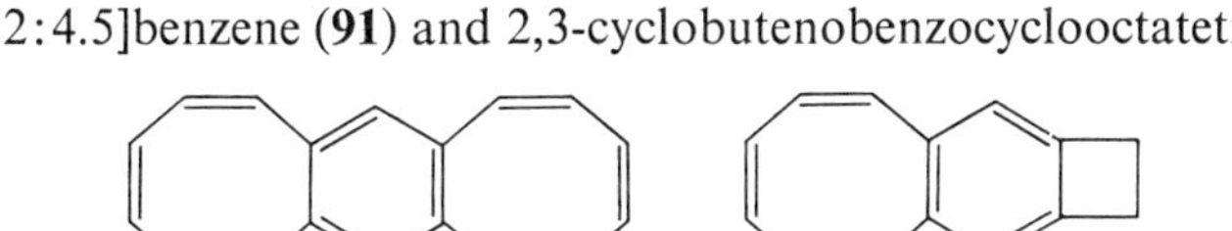

With 7,8-dimethylenecycloocta-1,3,5-triene and 1,2-dimethylenecyclobutane, 1,2-dehydrocyclooctatetraene gave, after dehydrogenation, dicyclooctatetraeno[1.2:4.5]benzene (**91**) and 2,3-cyclobutenobenzocyclooctatetraene (**92**), respectively. With tetracyclone the ring acetylene gave 1,2,3,4-tetraphenylbenzocyclooctatetraene (**93**) (Krebs, 1965).

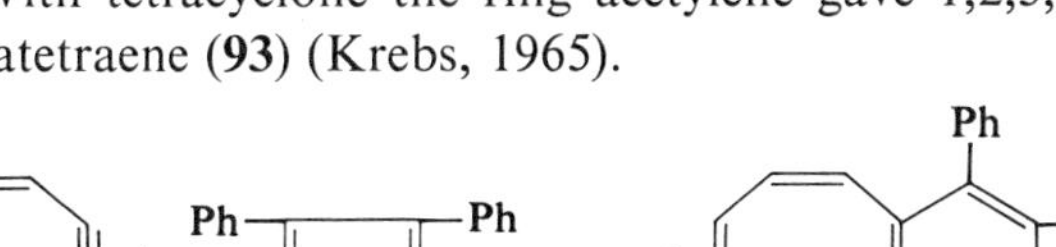

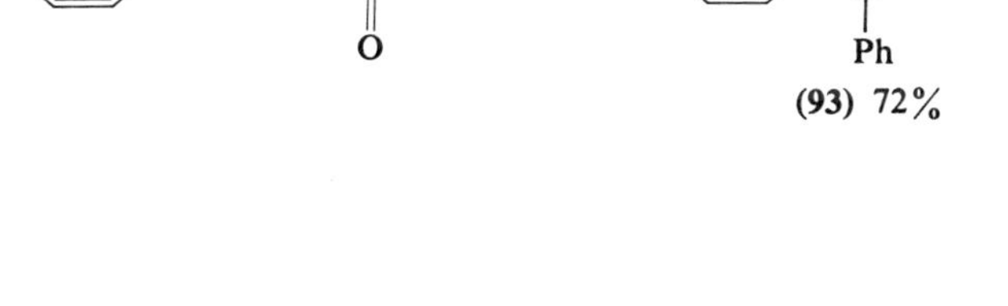

B. EFFECT OF SUBSTITUENTS

The slow step in the formation of benzyne from the halobenzenes may be either the removal of a proton or the loss of halide from the *o*-halogenophenyl anion. Reprotonation of the anion competes with loss of halide; the relative rates depend on the solvent and the halogen. The rates of loss of halide from the anion are in the order Br < Cl < F. With potassium amide in liquid ammonia, the rate of loss of bromide is faster than protonation of the anion, that of chloride is about the same, and that of fluoride is much slower.

From substituted halogenobenzenes, two different arynes may result. Where the substituent is electron withdrawing, more anion **94** than **96** will be formed; if R is electron releasing, **96** is preferred. However, the corresponding arynes are not necessarily also formed preferentially. If loss of halide from the anion is the rate-determining step in the reaction, then the relative rates of halide loss from the two possible anions determine the ratio of arynes actually formed. This is shown by the reactions of *m*-bromo- and *m*-chlorotoluene with potassium amide (Roberts *et al.*, 1956). In the former, the rate-determining step is the formation of the anions; in the latter, it is the loss of chloride from the anions. Since the methyl group is electron releasing, the anion **96** is formed preferentially, but the rate of loss of chloride from **96** is slower than from **94**. Thus, *m*-bromotoluene gives mainly the aryne **97**, but *m*-chlorotoluene gives mainly **95**.

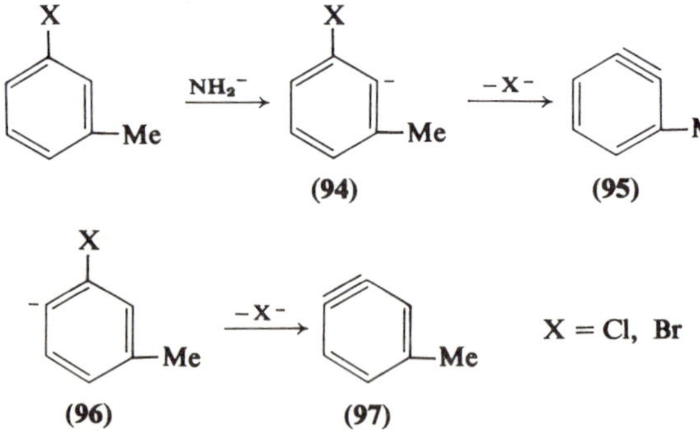

$$
\text{(94)} \xrightarrow{NH_2^-} \text{(94)}^- \xrightarrow{-X^-} \text{(95)} \tag{51}
$$

$$
\text{(96)} \xrightarrow{-X^-} \text{(97)} \quad X = Cl, Br
$$

To avoid isomer formation and troublesome separations and structure proofs, tetrahaloarynes have been frequently used. Heaney (1970) reviews their chemistry, which sometimes is complicated by the susceptibility of fluoroaromatics to nucleophilic attack.

In solution, a nitro group attached to benzyne influences the nature of cycloaddition. Although 4-nitrobenzyne reacts mainly by 2 + 4 cycloaddition, 3-nitrobenzyne gives mainly 2 + 2 cyclo adducts in high yield

7. ARYNES

(Atkin and Claxton, 1970). This was explained on the basis of high electronic charge at the reactive center owing to the 3-nitro group.

The effect of a nitro substituent on the formation of arynes by thermal decomposition of anhydrides is dramatic. Fields and Meyerson (1971b) discovered the formation of an arynyl-free radical by pyrolysis of 4-nitrophthalic anhydride. The arynyl-free radical (98) was trapped with benzene, benzene-d_6, pyridine, and thiophene. The scheme [Eq. (52)] shows the reaction with benzene-d_6:

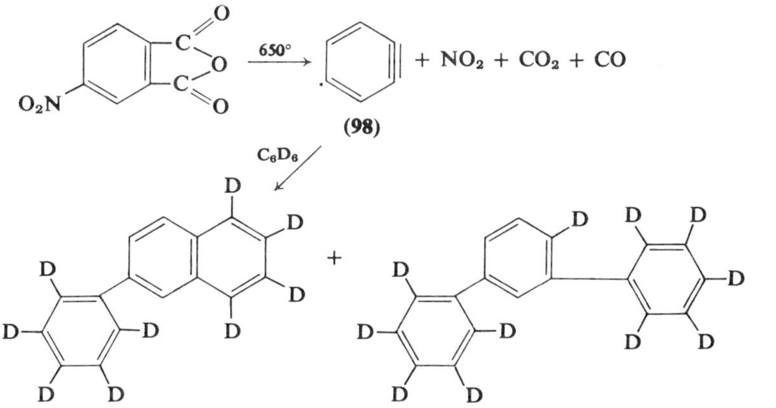

$$\text{4-nitrophthalic anhydride} \xrightarrow{650°} (98) + NO_2 + CO_2 + CO \quad (52)$$

3-Nitrophthalic anhydride and 3-nitronaphthalic anhydride gave the corresponding arynyl-free radicals 99 and 100 at 650°C.

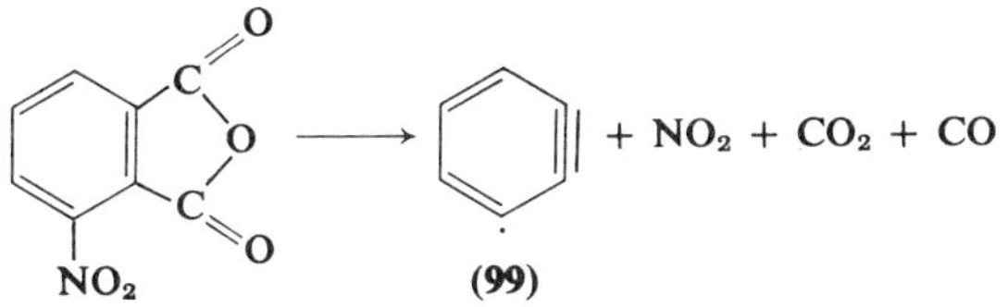

$$\text{3-nitrophthalic anhydride} \longrightarrow (99) + NO_2 + CO_2 + CO \quad (53)$$

$$\text{3-nitronaphthalic anhydride} \longrightarrow (110) + NO_2 + CO_2 + CO \quad (54)$$

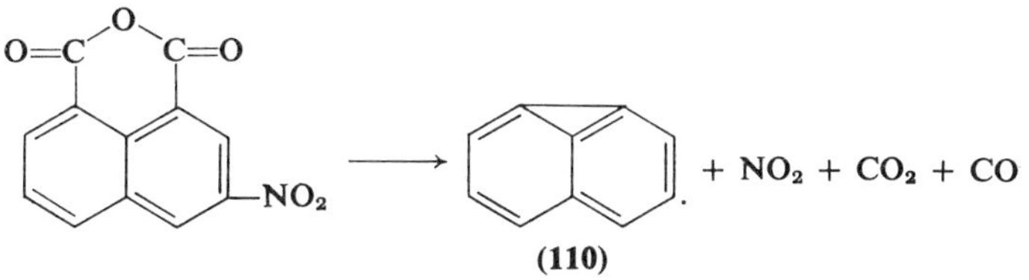

Thermal decomposition paralleled the behavior of the nitro anhydrides under electron impact in the mass spectrometer, as shown by a partial decomposition scheme [Eq. (55)] for 4-nitrophthalic anhydride where the relative intensities of the ions are given in parentheses. The major decomposition path consisted of successive loss of CO_2, NO_2, CO, and H. The ion of mass 75, the second most abundant in the spectrum, has the elemental

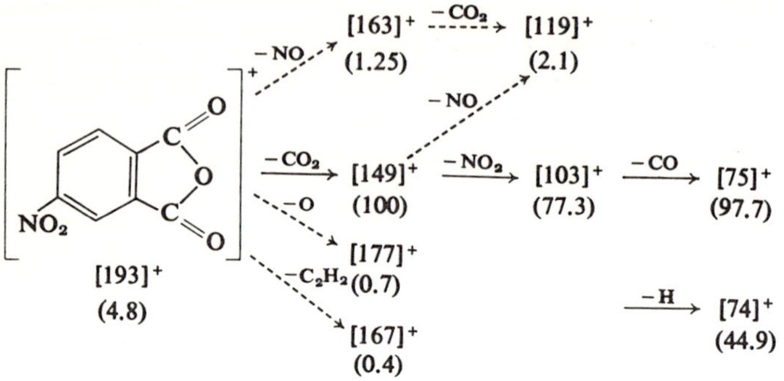

$$[193]^+ \xrightarrow{-NO_2} [163]^+ \xrightarrow{-CO_2} [119]^+$$
(4.8)   (1.25)   (2.1)

$$\xrightarrow{-CO_2} [149]^+ \xrightarrow{-NO_2} [103]^+ \xrightarrow{-CO} [75]^+ \quad (55)$$
(100)   (77.3)   (97.7)

$$[177]^+ \quad (0.7)$$

$$[167]^+ \quad (0.4)$$

$$\xrightarrow{-H} [74]^+ \quad (44.9)$$

composition of a benzynyl free radical and was pictured in terms of the benzynyl structure, differing from that of the benzynyl free radical only by virtue of the charge.

V. Reactions of Arynes

Arynes react with a wide variety of mono- and polyolefins, unsaturated molecules with heteroatoms, and nucleophiles as well as with each other to give dimers and trimers. Cycloaddition to olefins is highly characteristic.

A. 2 + 2 Cycloaddition

Four-membered rings are usually formed by reaction of benzyne with olefins, as shown by the reaction of 2,3-dimethylbutadiene (Wittig and Dürr, 1964).

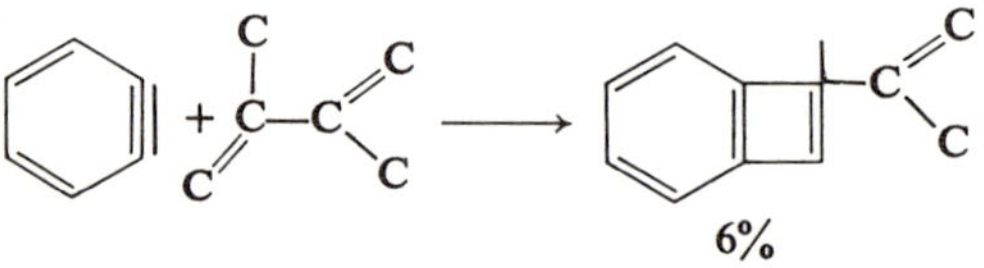

(56)

6%

The benzocyclobutane adduct may undergo ring expansion by valence isomerization and further reaction as Berson and Pomerantz (1964) found for dimethylcyclopropene.

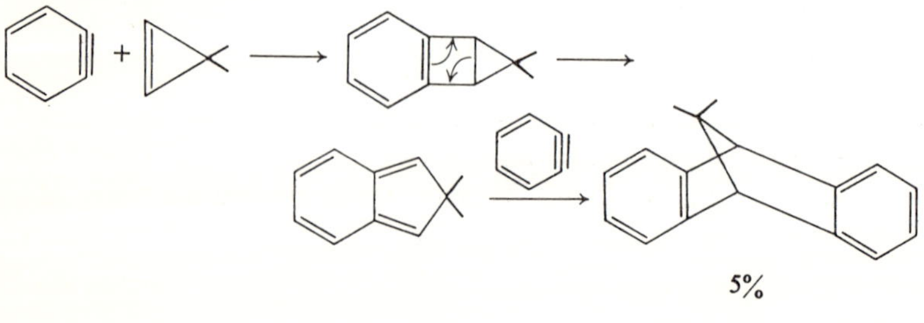

(57)

5%

7. ARYNES

Similarly, after the 1,2-addition of benzyne to benzene, ring expansion led to benzocyclooctatetraene (**101**) (Miller and Stiles, 1963). With norbornene,

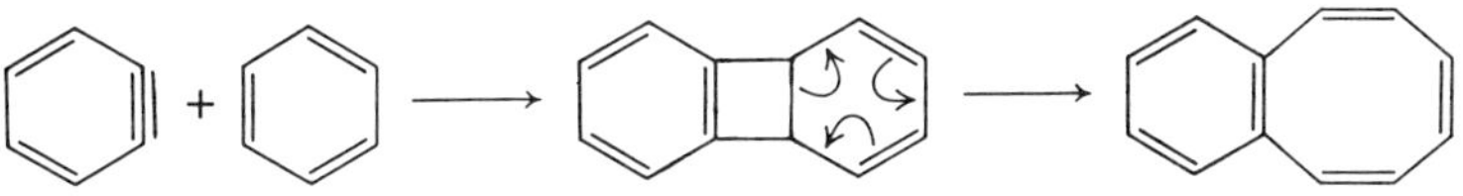

$$\text{benzyne} + \text{benzene} \longrightarrow \text{[bicyclic intermediate]} \longrightarrow \text{(101)} \quad 8\% \tag{58}$$

benzyne formed the exo adduct (**102**) (Simmons, 1961). Cyclobutanes were

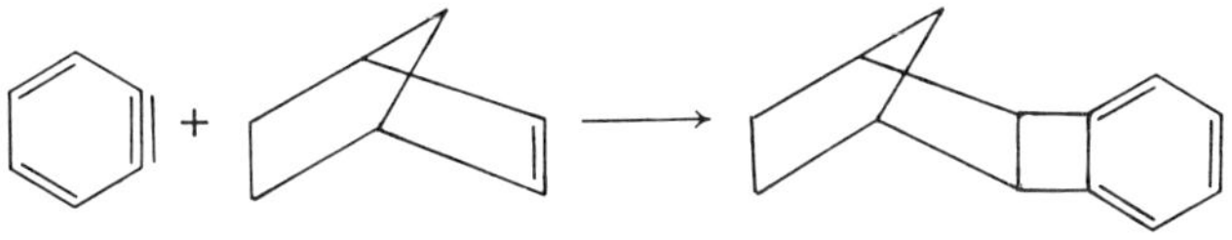

$$\text{benzyne} + \text{norbornene} \longrightarrow \text{(102)} \quad 10\% \tag{59}$$

also obtained from benzyne with acrylonitrile (Matsuda and Mitsuyasu, 1966) and ethyl propenyl ether (Wasserman et al., 1968).

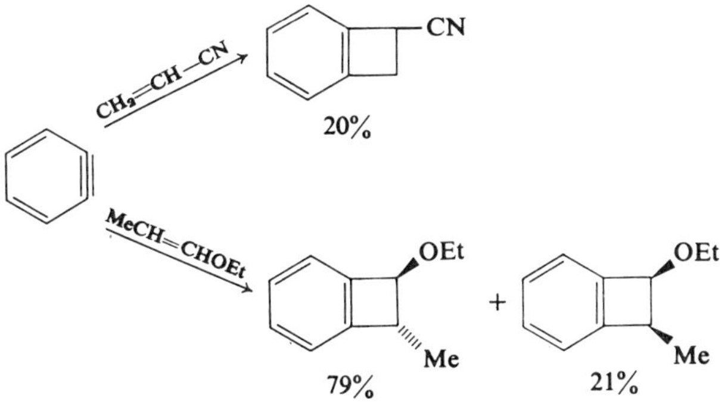

$$\text{benzyne} \xrightarrow{CH_2=CH-CN} \text{[benzocyclobutane-CN]} \quad 20\% \tag{60}$$

$$\xrightarrow{MeCH=CHOEt} \text{[cis-OEt/Me]} \ 79\% + \text{[trans-OEt/Me]} \ 21\%$$

Cyclic ketones, enolized in the presence of sodamide and sodium *tert*-butylate, reacted with benzyne to give cyclobutanols (Caubere et al., 1971), as shown for cyclohexanone.

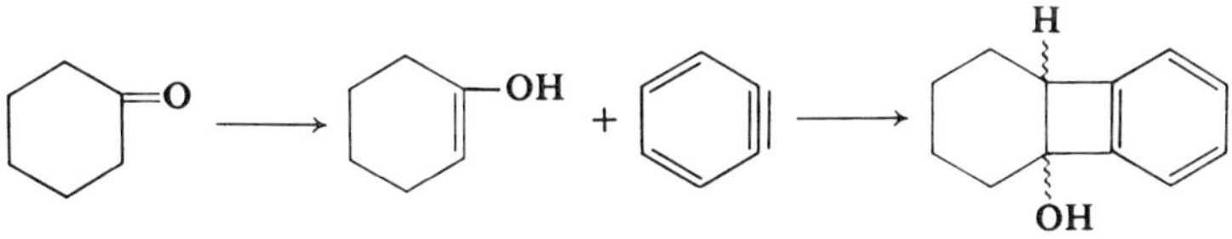

$$\text{cyclohexanone} \longrightarrow \text{enol} + \text{benzyne} \longrightarrow \text{cyclobutanol product} \tag{61}$$

B. 2 + 3 CYCLOADDITION

1,3-Dipolar compounds such as azides, diazo compounds, and nitrile oxides add readily to arynes to give benzo derivatives of five-membered heterocyclic compounds in good yields. With phenyl azide, benzyne gave 1-phenylbenzotriazole (**103**) (Huisgen and Knorr, 1961). Benzoyldiazo-

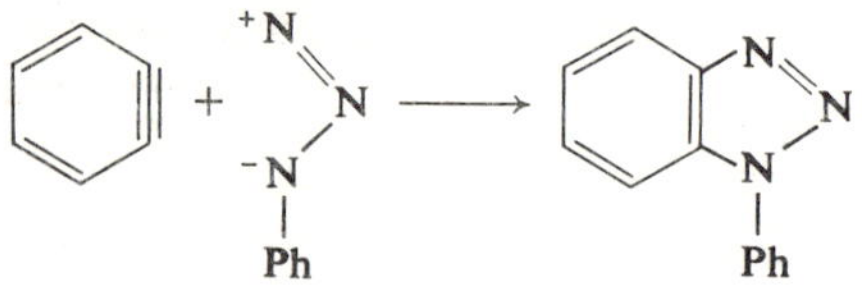

$$\text{benzene} + \text{PhN}_2^+\text{N}^-\text{-N} \longrightarrow \text{benzotriazole-Ph} \quad (62)$$

(103) 88%

methane gave 3-benzoylindazole (104) (Reid and Schön, 1965). Benzonitrile

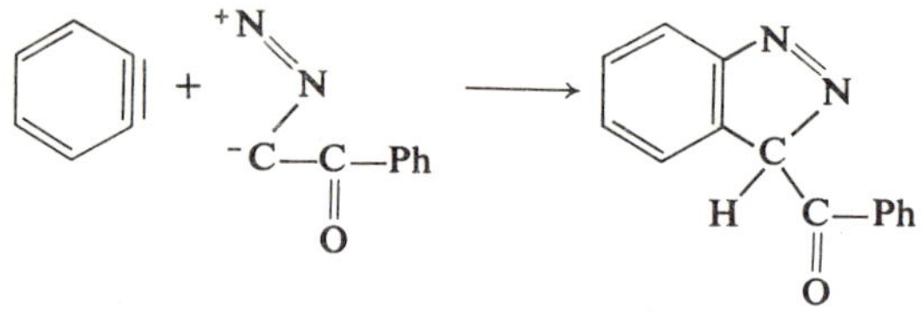

$$\text{benzene} + {}^+\text{N}{=}\text{N}{-}\text{C}^-{-}\underset{\text{O}}{\text{C}}{-}\text{Ph} \longrightarrow \text{indazole with C(H)(C(=O)Ph)} \quad (63)$$

(104) 88%

oxide yielded 3-phenylbenzisoxazole (105) (Minisci and Quilico, 1964).

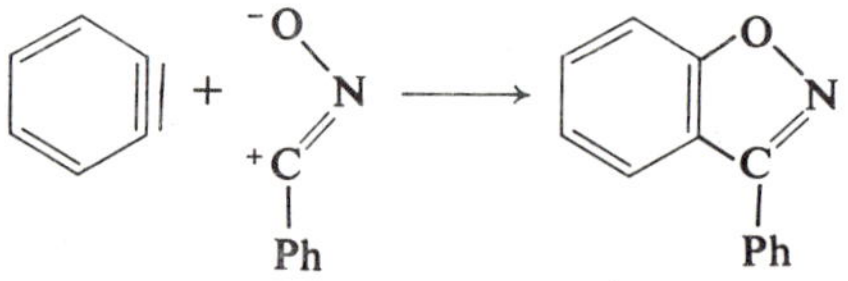

$$\text{benzene} + {}^-\text{O}{-}\text{N}{=}\text{C}^+{-}\text{Ph} \longrightarrow \text{benzisoxazole-Ph} \quad (64)$$

(105) 55%

The benzocyclobutanols were formed from benzyne and enolates of some cycloalkanones (Caubere et al., 1972). The high yields of products from 2 + 3 cycloaddition were further exemplified by the reaction of benzyne with benzylidene methylamine-N-oxide (Huisgen, 1969) to give the

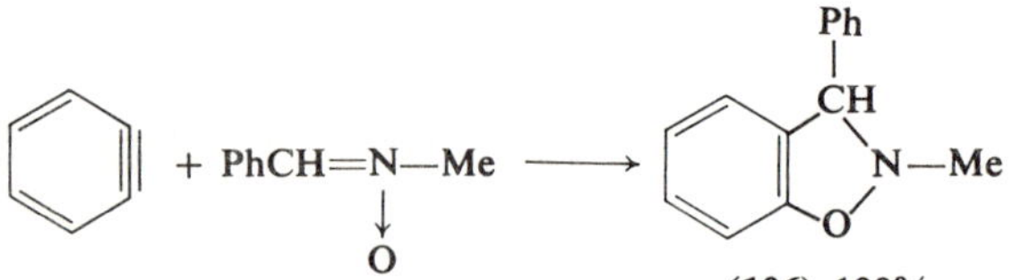

$$\text{benzene} + \text{PhCH}{=}\underset{\downarrow \text{O}}{\text{N}}{-}\text{Me} \longrightarrow \text{fused ring with CH(Ph)-N(Me)-O} \quad (65)$$

(106) 100%

benzisoxazole (106). Carbon trisulfide, from carbon disulfide and sulfur at 700°C, gave benzo-1,2-dithiole-3-thione (107) and benzo-1,3-dithiole-2-thione (108) (Fields and Meyerson, 1969b). The same study uncovered

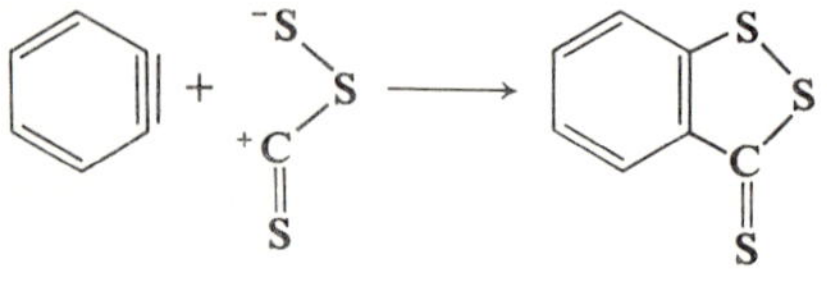

$$\text{benzene} + {}^-\text{S}{-}\text{S}{-}\text{C}^+{=}\text{S} \longrightarrow \text{benzo-S-S-C(=S)} \quad (66)$$

(107) 26.6%

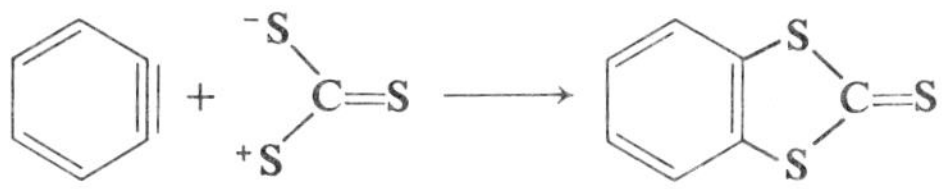

an apparent 1,3-dipolar addition of a carbenoid species (**109**) from carbon disulfide.

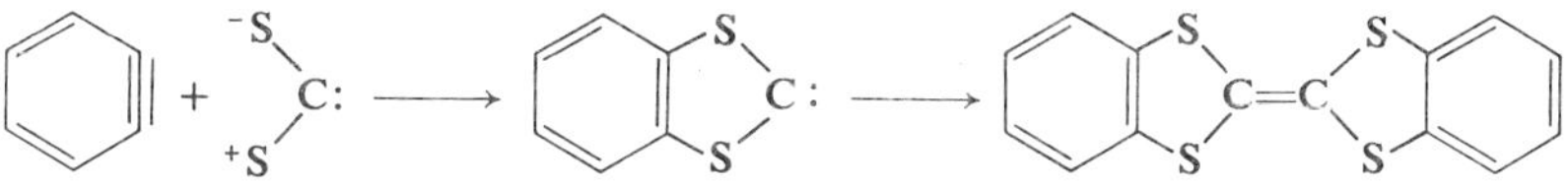

C. 2 + 4 CYCLOADDITION

The most familiar and widely used cycloaddition reaction of arynes is 2 + 4 cycloaddition, the Diels–Alder reaction. Many examples have already been cited in the discussion on trapping agents used to prove the formation of arynes. The reactions with tetracyclone, furan, diphenylisobenzofuran, cyclopentadiene, benzene, and pyridine are 2 + 4 cycloadditions. There are numerous other examples useful in the synthesis of compounds difficult to prepare by other means.

Thus, tetrachlorobenzyne from pentachlorophenylmagnesium chloride with mesitylene gave the product (**110**) (Heaney and Jablonski, 1966).

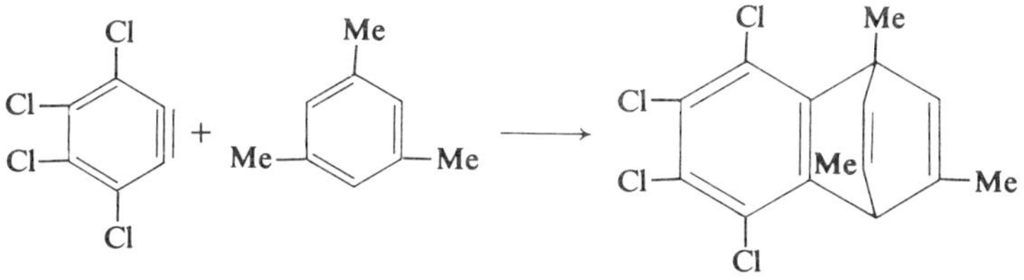

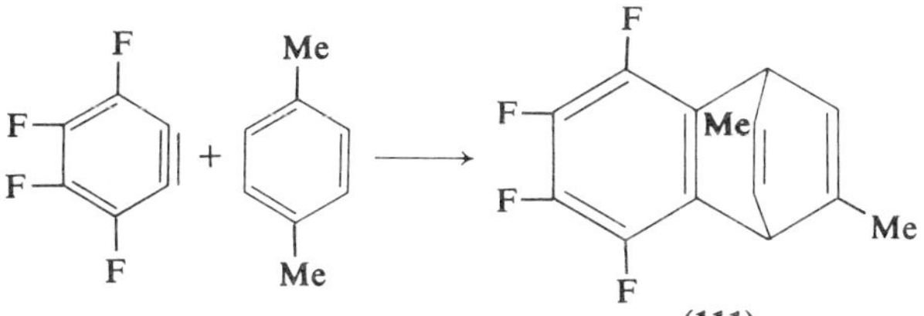

Tetrafluorobenzyne, generated from pentafluorobenzene with butyllithium or from pentafluorobromobenzene through the Grignard reagent (Callender et al., 1969), reacted with a series of aromatic hydrocarbons to give 1,4-adducts in about 25% yield; p-xylene gave the one product **111**. o- and m-xylenes gave two isomers each. Styrene gave tetrafluoro-9,10-dihydrophenanthrene (**112**) in about 30% yield; it was readily aromatized with Pd/C to the phen-

(112) (113) (71)

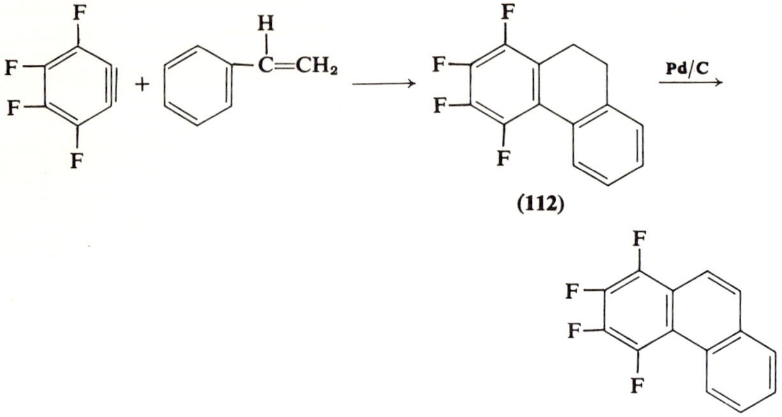

anthrene (**113**). Pyrrole yielded only intractable tars with tetrafluorobenzyne. 1-Methylpyrrole, probably because it lacked the acidic hydrogen attached to nitrogen, gave a high yield of 1,2,3,4-tetrafluoro-5,8-dihydro-5,8-*N*-methyl-

(114) (72)

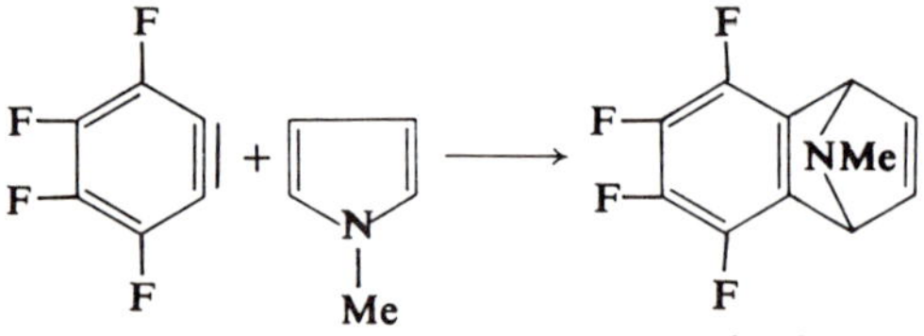

iminonaphthalene (**114**). Tetrafluorobenzyne with [2,2]-paracyclophane gave both a 1:1 (**115**) and 2:1 (**116**) adduct, depending on the mole ratios of the

(115) 44%

(116) 39%

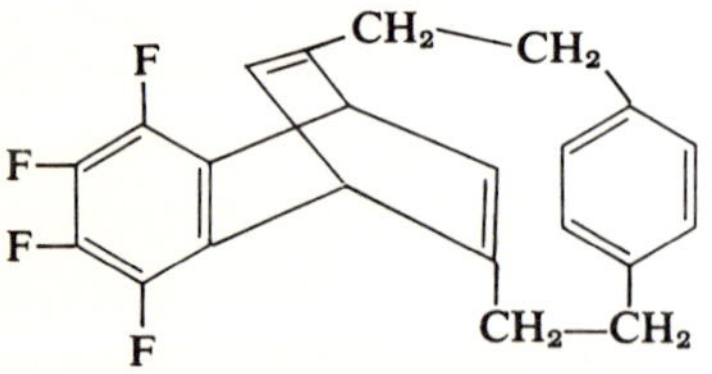

reagents (Brewer *et al.*, 1969). The reaction of benzyne with thionyl aniline gave a complex mixture from which some 6*H*-dibenzo[*c*,*e*][1,2]thiazine-5-dioxide (**118**), from oxidation of the intermediate sulfinamide (**117**), was isolated (Campbell and Rees, 1969):

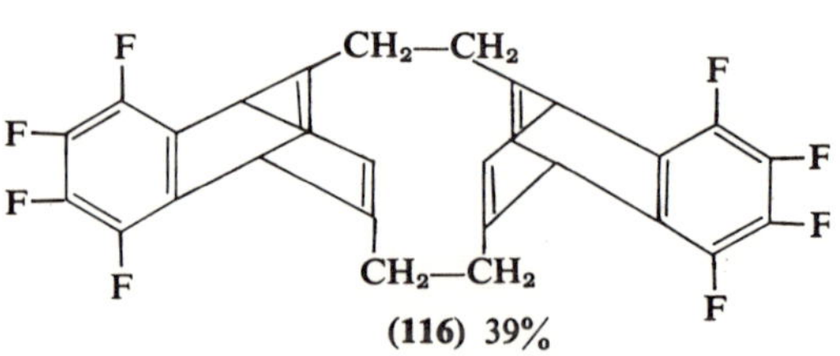

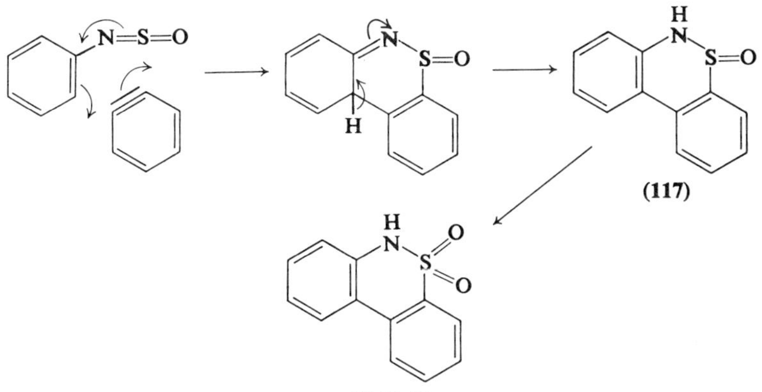

(73)

(117)

(118) 4%

A benzohomobarrelenone **(119)** was obtained by Ciabattoni and Crowley (1967) by 2 + 4 cycloaddition of benzyne to tropolone.

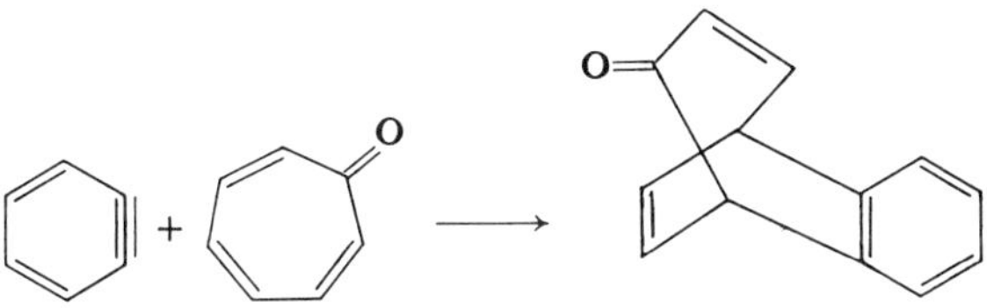

(74)

(119) 40%

Arynes in the gas phase from pyrolysis of aromatic anhydrides gave products with a wide variety of aromatic reagents derived via both insertion ("ene" reaction) and 2 + 4 cycloaddition (Fields and Meyerson, 1968b; Fields, 1970). Yields were in the range of 15–70%. The 2 + 4 adducts were not stable at 700°C; they split out acetylene or substituted acetylenes and aromatized to naphthalenes [Eqs. (75) and (76)].

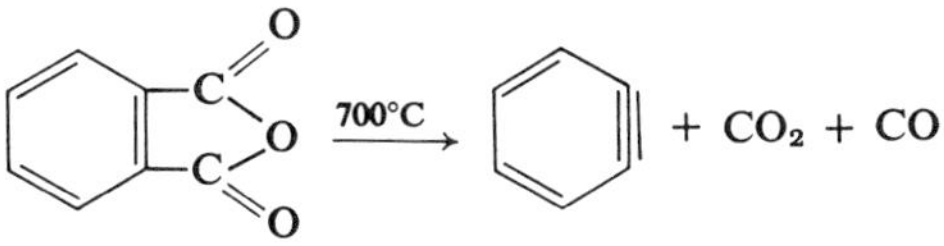

$$\xrightarrow{700°C} \quad + CO_2 + CO \quad (75)$$

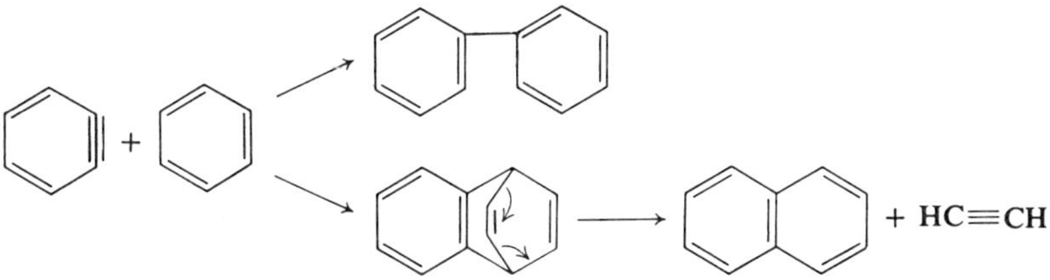

$$+ HC \equiv CH \quad (76)$$

Formation of biphenyl from benzene and benzene, the "insertion reaction," may be considered an "ene" reaction (see Section V,G on "ene" reaction) [Eq. (77)].

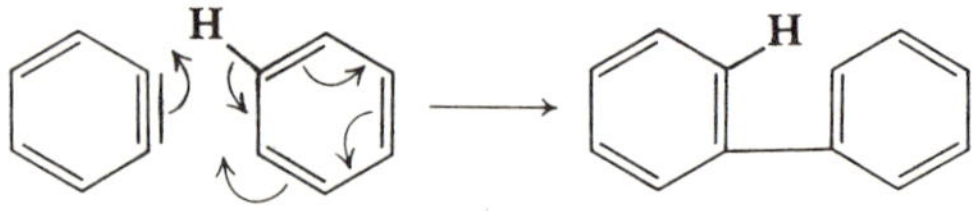

$$\text{(77)}$$

Table IV shows the ratio of insertion to addition products with a series of reagents (Fields, 1970). Although the ratio varied appreciably, many reagents gave ratios between 1 and 5. In the series benzene, toluene, and o- and p-xylene, the major factor involved was steric.

The 2 + 4 cyclo adduct of a halogenated benzyne with thiophene can extrude sulfur not only at 700°C in the vapor phase but also at 25°C. 3,5-Dichloro-2,4-difluorobenzyne from s-trichlorotrifluorobenzene with lithium reacted with thiophene to give dichlorodifluoronaphthalene (120) and dichlorotrifluorothiophenol (121) (Hayashi and Ishikawa, 1970). The thiophenol

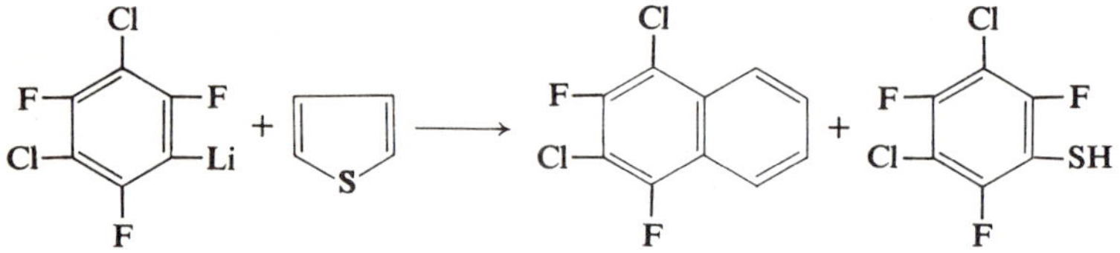

$$\text{(78)}$$

(120) 39% (121) 22%

resulted from the reaction of the lithium derivative with sulfur, either extruded from the aryne adduct as atomic sulfur or plucked off by the lithium compound.

With 1,4-disubstituted sym-tetrazines, benzyne from diazotized anthranilic acid in methyl chloride gave phthalazines (122) (Sauer and Heinriche, 1966).

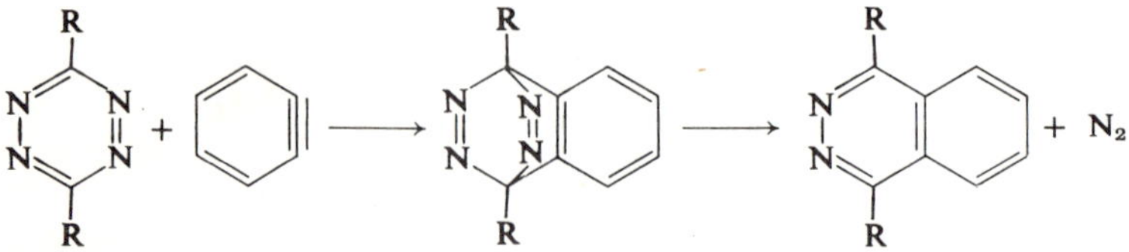

$$\text{(79)}$$

R = Ph, CO_2, Me   (122) 67–73%

Cadogan et al. (1971b) found that o-$tert$-butyl-N-nitrosoacetanilide gave 3-$tert$-butylbenzyne. Similarly, 2,5-di-$tert$-butyl-N-nitrosoacetanilide gave 3,6-di-$tert$-butylbenzyne [Eq. (80)]. 3,6-Di-$tert$-butylbenzyne gave 30 and

7. ARYNES

Table IV
Products from Arynes via Aromatic Anhydrides and Reagents

Aryne	Reagent	Products[a]	Ratio, insertion to addition
Benzyne	Benzene	Biphenyl, naphthalene	5
	Toluene	Methylbiphenyl, fluorene,[b] naphthalene, methylnaphthalene	3
	o-Xylene	Dimethylbiphenyl, naphthalene, 1-methylnaphthalene, 2,3-dimethylnaphthalene	0.14
	p-Xylene	2,5-Dimethylbiphenyl, 2-methylfluorene,[c] dimethylbiphenyl	0.5
	Hexafluorobenzene	Hexafluorobiphenyl, tetrafluoronaphthalene	5
	Methyl benzoate	Methyl biphenyl carboxylate, methyl naphthoate	1.5
	Furan	Phenylfuran, benzofuran, naphthalene	3
	Pyridine	Phenylpyridine, quinoline, naphthalene	3
	Quinoline	Phenylquinoline, acridine	5
	Thiophene	Phenylthiophene, benzothiophene, naphthalene	3
	3-Picoline	Phenylpicoline, azafluorene, methylquinoline, methylnaphthalene	0.5
	Phthalic anhydride	Biphenyl-2,3-dicarboxylic anhydride, naphthalene-2,3-dicarboxylic anhydride	1
	Chlorobenzene	Chlorobiphenyl, chloronaphthalene	1.2
Tetrachlorobenzyne	Pyridine	Tetrachlorophenylpyridine, tetrachloronaphthalene	8
Carboxybenzyne	Pyridine	Phenylpyridine, carboxyphenylpyridine, naphthoic acid	50
2,3-Naphthalyne	Pyridine	Naphthylpyridine, anthracene, benzoquinoline	1.6
1,8-Perinaphthalyne	Thiophene	Naphthylthiophene, anthracene, naphthothiophene	4.6

[a] Conditions: mole ratio, aromatic anhydride to reagent, 1:5; 700°C; N_2, 10 cm^3/min; contact time, 10–20 sec.
[b] By loss of hydrogen from 2-methylbiphenyl.
[c] By loss of hydrogen from 2,5-dimethylbiphenyl.

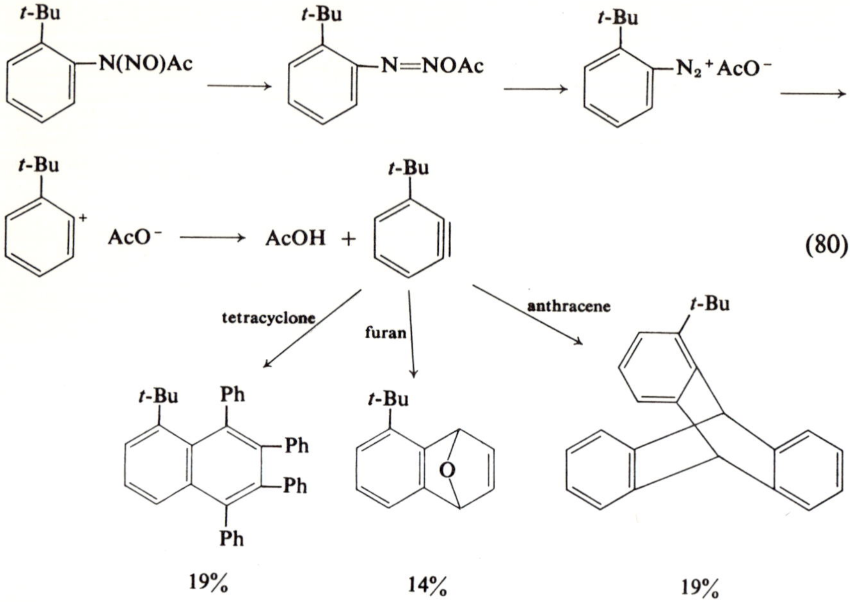

16% yields of adducts with furan and anthracene, respectively; no adduct was formed with tetraphenylcyclopentadienone. This was attributed to steric hindrance to the cycloaddition caused by the presence of two *p-tert*-butyl groups.

D. 2 + 6 Cycloaddition

Tabushi *et al.* (1971) reacted benzyne from benzenediazonium carboxylate with cycloheptatriene and obtained a 50% yield of the two products **123** and **124** in about equal amounts. These arise evidently by 2 + 6 cycloaddition and an "ene" reaction, respectively. The structure of **123** has been corrected by Lombardo and Wege (1971). In contrast to tropone, which gave mostly 2 + 4 cyclo adduct (**125**) (Ciabattoni and Crowley, 1967) and a little 2 + 6 cyclo adduct (**126**) (Miwa *et al.*, 1969), cycloheptatriene gave only the 2 + 6 and none of the 2 + 4 adduct.

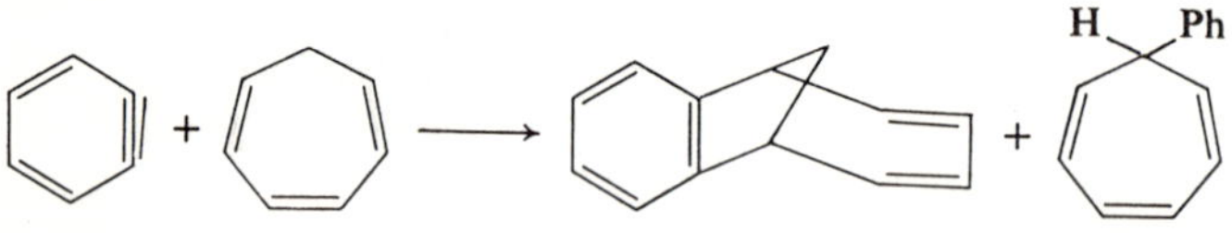

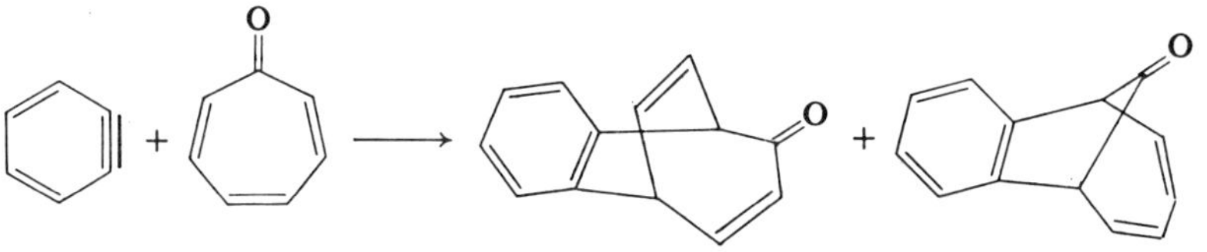

7. ARYNES

[Reaction scheme showing benzyne + tropone → products (125) and (126)] (82)

E. 2 + 8 Cycloaddition

An example of 2 + 8 cycloaddition was found by Oda and Kitahara (1970) in the reaction of benzyne with 8-cyanoheptafulvene (127) to give the adduct 128.

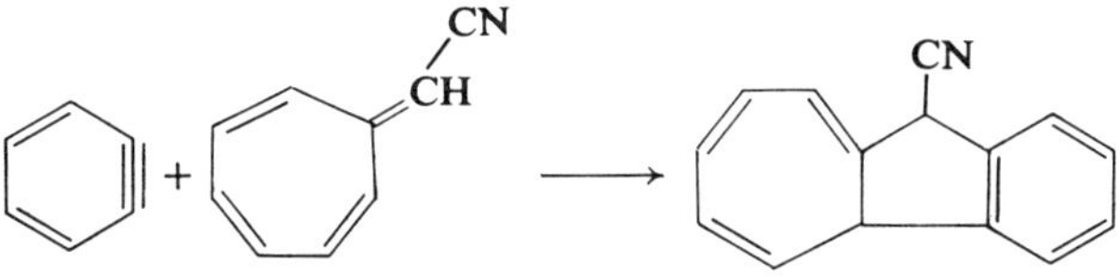

[Reaction scheme: benzyne + (127) → (128) 36%] (83)

F. Polymerization

In the absence of other reactive species, benzyne dimerizes to biphenylene (129) and trimerizes to triphenylene (130). The most convenient synthesis of

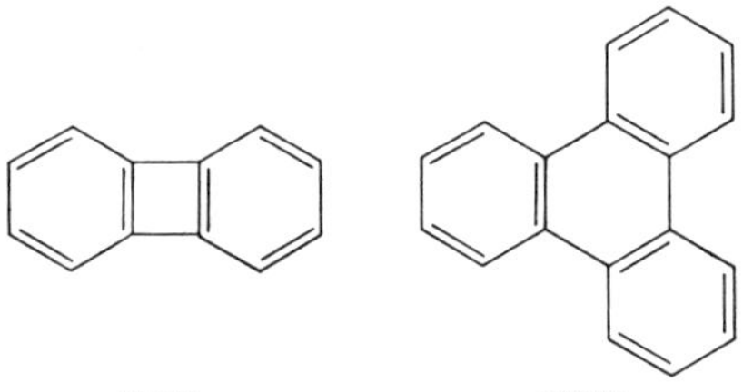

[Structures of biphenylene (129) and triphenylene (130)]

biphenylene is by decomposition of benzenediazonium o-carboxylate in dichloroethane (Friedman and Logullo, 1968) although some explosions have been reported with this procedure (Matuszak, 1971).

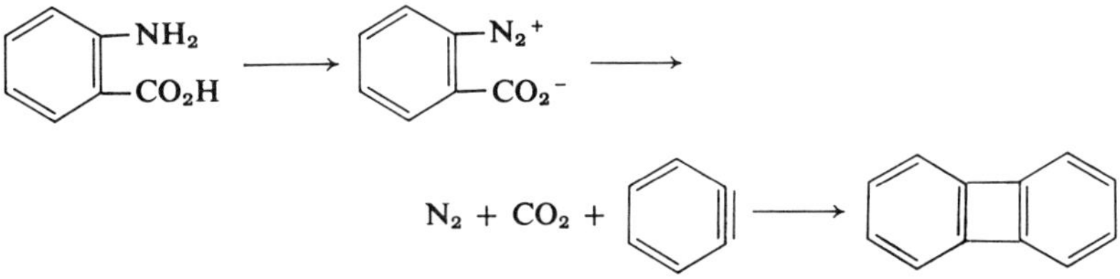

[Reaction scheme: o-aminobenzoic acid → benzenediazonium o-carboxylate → $N_2 + CO_2$ + benzyne → biphenylene] (84)

Another good synthesis of biphenylene is by oxidation of *N*-aminobenzotriazole with lead tetraacetate (Gilchrist *et al.*, 1968). Although the yield is high, the method is less attractive because of the unavailability of aminobenzotriazole relative to the abundant anthranilic acid.

Biphenylene and substituted biphenylenes are readily prepared by pyrolysis of phthalic anhydrides in the gas phase [Eq. (85)] (Fields and Meyerson, 1965; Brown *et al.*, 1966; Cava *et al.*, 1966). Because biphenylenes are not

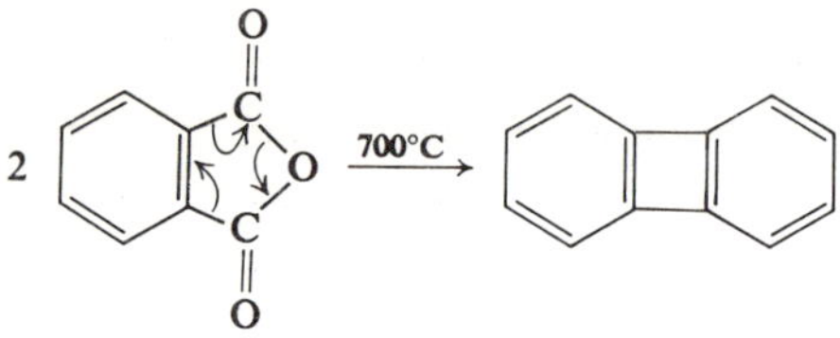

$$2 \; \text{[phthalic anhydride]} \xrightarrow{700°C} \text{[biphenylene]} + 2\,CO_2 + 2\,CO \qquad (85)$$

stable for long above 300°C, the products of pyrolysis are best condensed quickly on cold fingers.

Biphenylene is probably formed by concerted dimerization of benzyne; however, it is unlikely that triphenylene forms by a concerted reaction of three molecules of benzyne. In solution, triphenylene is most likely formed by the interaction of a metalloorganic biphenyl intermediate and benzyne [Eq. (86)].

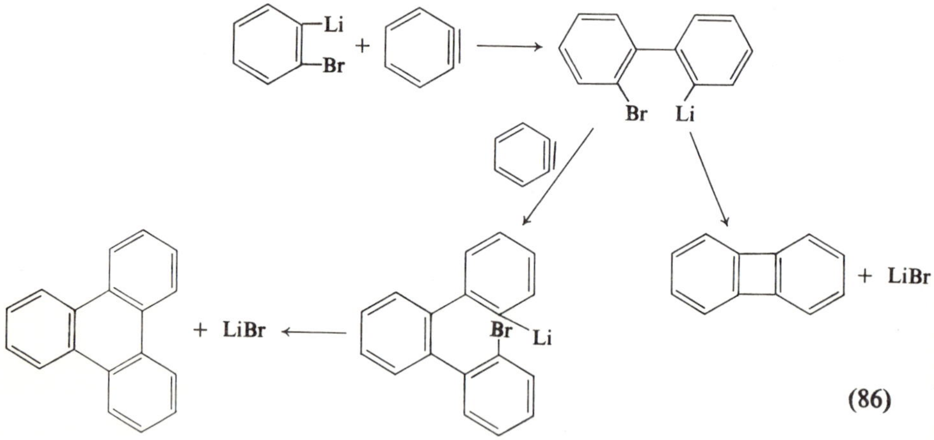

(86)

In the gas phase, triphenylene may result from opening of the biphenylene ring (131) and reaction with another mole of benzyne. A dimer of the bi-

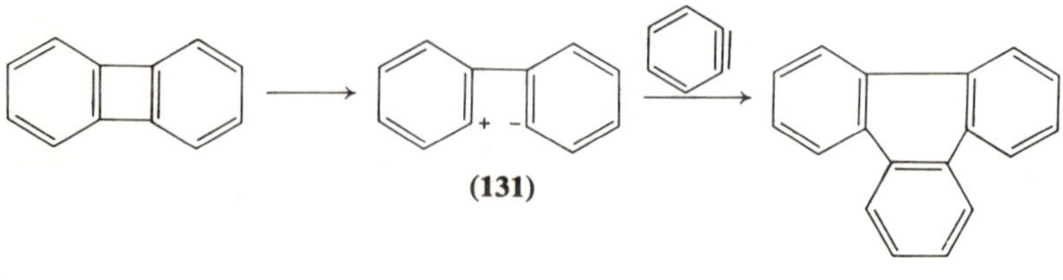

(87)

(131)

phenylene intermediate (131) would be anticipated; indeed, some product of molecular weight 304 was found by Fields and Meyerson (1971a) in the pyrolysis of phthalic anhydride that corresponded to tetrabenzocyclooctate-

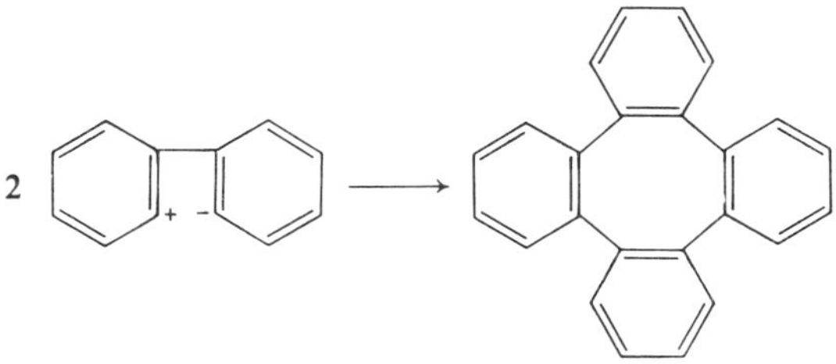

$$2 \; (131) \longrightarrow (132) \tag{88}$$

traene (132). Bi-2,3-naphthalyne (133) was also formed in the gas phase from naphthalene-2,3-dicarboxylic anhydride although it was not isolated (Fields and Meyerson, 1968b).

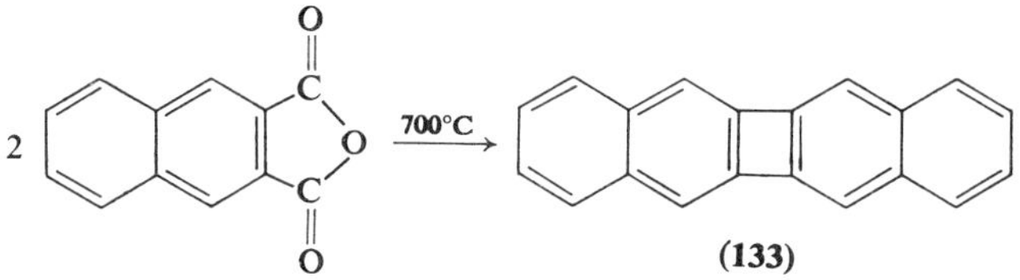

$$2 \xrightarrow{700°C} (133) \tag{89}$$

Barton and Grinham (1972) failed to obtain the biphenylene dimer of 9,10-phenanthryne, but succeeded in isolating unsymmetrical biphenylenes from 9,10-phenanthryne with benzyne and with 1,2-naphthalyne.

G. "Ene" Reaction

Activated olefins and acetylenes react with olefins containing an allylic hydrogen by a concerted cycloaddition, involving a six-membered transition state, known as the "ene" reaction.

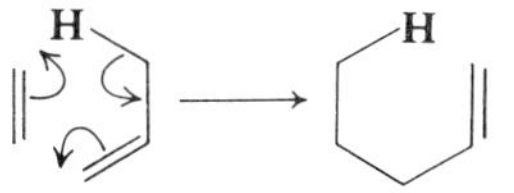

$$\tag{90}$$

Industrially, this reaction is of considerable importance. Long-chain alkenyl succinic acids as rust inhibitors for oils are made by the "ene" reaction of olefins with maleic anhydride [see Eq. (91), where R is 10 or more carbon atoms].

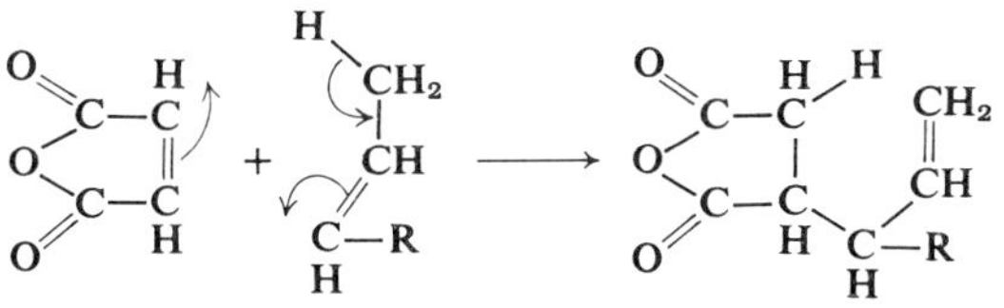

$$\tag{91}$$

Benzyne reacted with isobutylene to give isobutenylbenzene (**134**) (Wittig

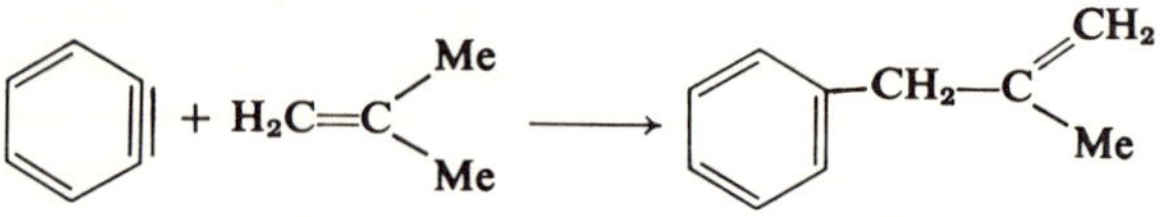

$$\text{[benzyne]} + H_2C=C\begin{smallmatrix}Me\\Me\end{smallmatrix} \longrightarrow \text{Ph}-CH_2-C\begin{smallmatrix}CH_2\\Me\end{smallmatrix} \quad (92)$$

(**134**) 4%

and Dürr, 1964). 1,2,3-Triphenylcyclopropene with benzyne gave a considerably higher yield of "ene" product, 1,2,3,3-tetraphenylcyclopropene (**135**)

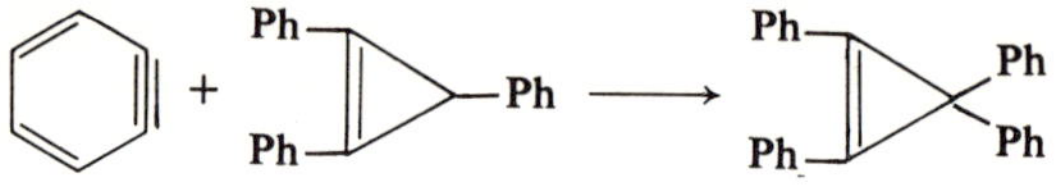

$$\text{[benzyne]} + \text{Ph-cyclopropene-Ph} \longrightarrow \text{tetraphenylcyclopropene} \quad (93)$$

(**135**) 60–70%

(Berson and Pomerantz, 1964). A good yield of "ene" product (**136**) was also obtained by Tabushi et al. (1969a,b) by the reaction of a cyclopropane derivative with benzyne.

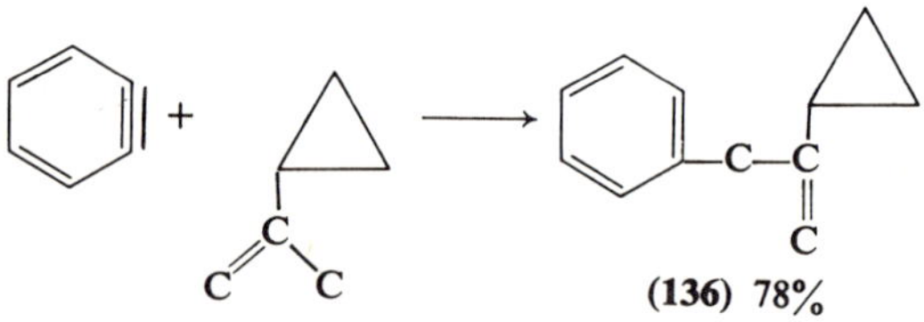

$$\text{[benzyne]} + \text{[cyclopropane-allene]} \longrightarrow \text{(product)} \quad (94)$$

(**136**) 78%

Often the "ene" reaction is accompanied by cycloaddition reactions, especially in the vapor phase.

With allenes, benzyne gave 2 + 2 cyclo adducts and "ene" products; the "ene" reaction yielded 1,3-dienes and acetylenes (Wasserman and Keller, 1970) [Eqs. (95)–(97)].

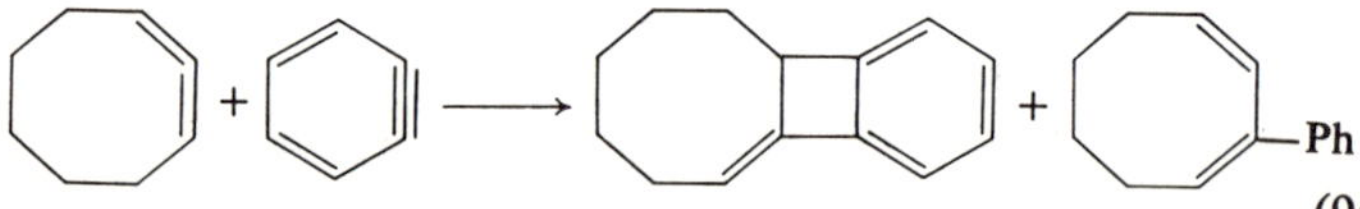

$$\text{[cyclooctatetraene]} + \text{[benzyne]} \longrightarrow \text{[fused adduct]} + \text{[cyclooctatetraene-Ph]} \quad (95)$$

$$Me_3C-CH=C=\overset{H}{C}OMe + \text{[benzyne]} \longrightarrow$$

$$\text{[benzocyclobutene with COMe, H, CMe}_3\text{]} + Me_3C-C\equiv C-\overset{H}{\underset{Ph}{C}}-OMe \quad (96)$$

$$Me_2C=C=CH_2 + \text{[benzyne]} \longrightarrow CH_2=C-\overset{Me}{\underset{Ph}{C}}=CH_2 \quad (97)$$

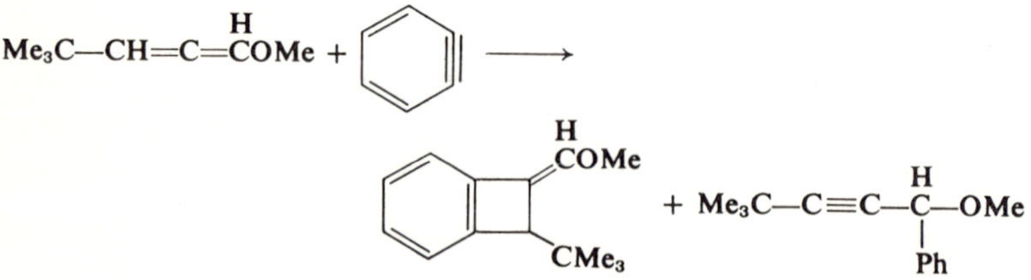

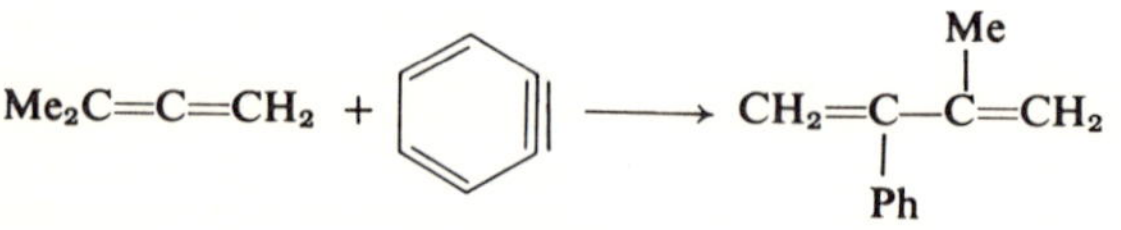

7. ARYNES

Products of both the "ene" reaction and the 2 + 2 cycloaddition were obtained by Vysochin et al. (1970). They reacted tetrafluorobenzyne with cis- and trans-hexene-3 and formed tetrafluorophenylhexene-2 (**137**) and the adduct (**138**).

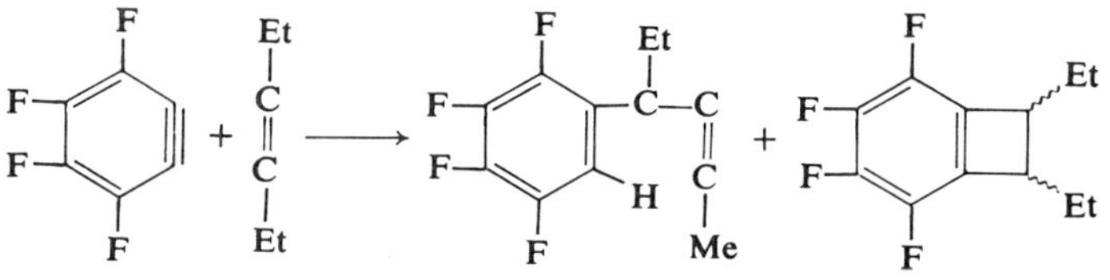

(98)

(**137**) 20%   (**138**) 18%

The first excited triplet level of benzyne is close to the ground state (Yonezawa et al., 1969). However, Ahlgren and Akermark (1970) found that benzyne shows no biradical character. They reacted benzyne with 1,2-dideuteriocyclohexene and obtained only the one "ene" product (**139**) formed by concerted, dipolar addition and none of the isomer (**140**) from a radical reaction.

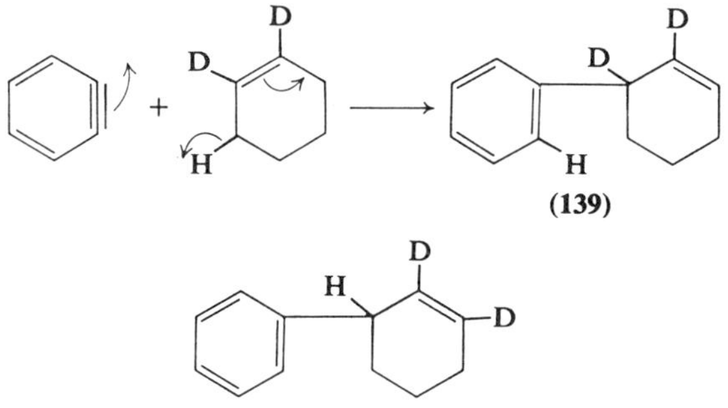

(99)

(**139**)

(**140**)

Mikhailova et al. (1970) found similar evidence with more highly substituted reagents. Tetrafluorobenzyne with 3,3,6,6-tetradeuteriocyclohexene gave only the "ene product" (**141**). The yield of adduct (**141**) was 86% when

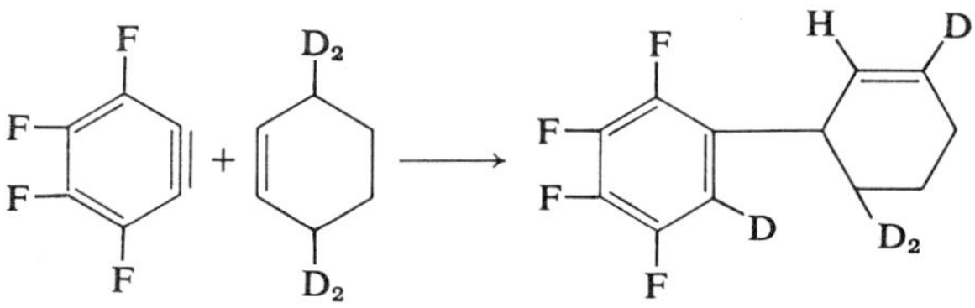

(100)

(**141**) 26%

tetrafluorobenzyne was formed from tetrafluoroanthranilic acid rather than from pentafluorophenylmagnesium chloride.

Although insertion of benzyne into a phenyl–H bond can be written as an "ene" reaction, a more direct insertion not amenable to this mechanism was found. Insertion of benzyne into an organometallic bond was described by Miller and Kuhlmann (1971). Decomposition of benzenediazonium 2-carboxylate in a solution of *trans*-(phenylethynyl) (trichlorovinyl)bis(triethylphosphine)nickel (II) (**142**) gave the compound **143**.

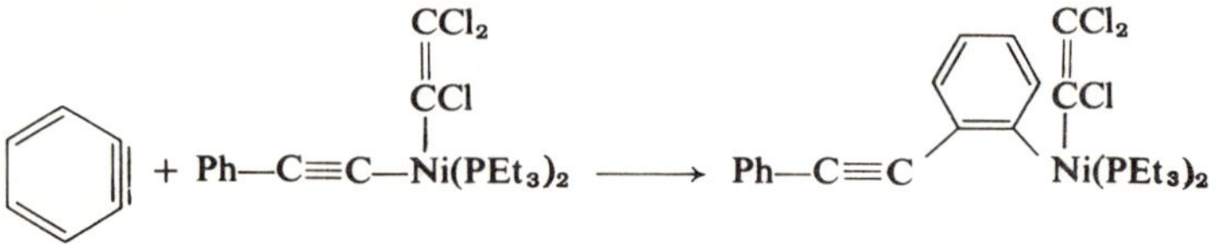

$$\text{benzyne} + \text{Ph—C≡C—Ni(PEt}_3)_2 \longrightarrow \text{Ph—C≡C—}\underset{(143)}{\text{Ar}}\text{—Ni(PEt}_3)_2 \quad (101)$$

(**142**)

H. Reaction with Nucleophiles

Nucleophilic addition usually occurs when the aryne is formed by the reaction of a base with an aromatic halide. As was described in earlier sections, action of sodium amide or potassium amide in liquid ammonia on chlorobenzene gives aniline with the NH_2 attached to any of three carbon atoms (shown by labeling).

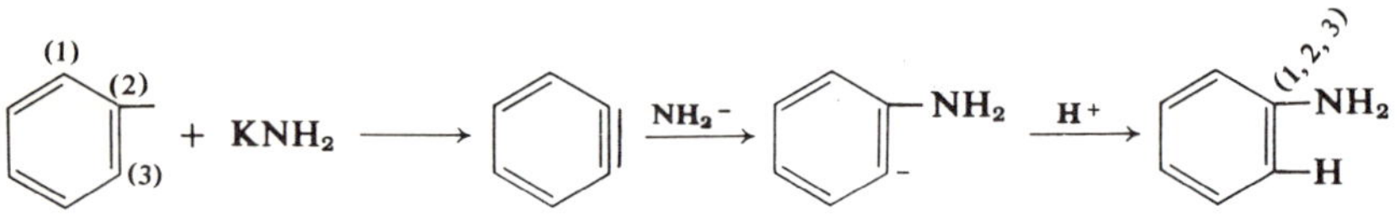

$$\text{(1,2,3-labeled benzyne)} + KNH_2 \longrightarrow \text{benzyne} \xrightarrow{NH_2^-} \text{Ar-NH}_2^- \xrightarrow{H^+} \text{Ar(1,2,3)-NH}_2,\text{H} \quad (102)$$

Arynes react rapidly with a wide variety of nucleophiles. Among these are alcohols, amines, thiols, water, carbanions, sulfides, and phosphines. The general scheme for the addition of nucleophiles is

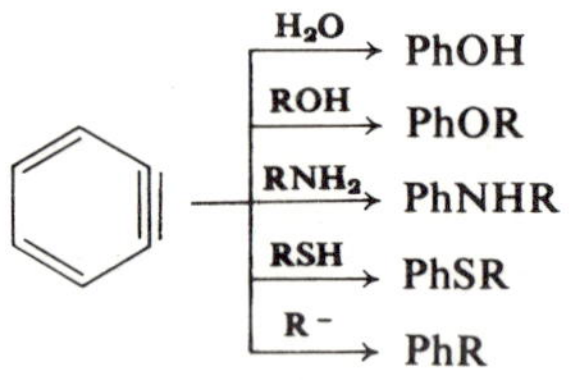

$$\text{benzyne} \begin{cases} \xrightarrow{H_2O} & PhOH \\ \xrightarrow{ROH} & PhOR \\ \xrightarrow{RNH_2} & PhNHR \\ \xrightarrow{RSH} & PhSR \\ \xrightarrow{R^-} & PhR \end{cases} \quad (103)$$

Sulfides, tertiary amines, and tertiary phosphines react with arynes to give zwitterions that eliminate olefin or rearrange. With a series of sulfides, Hellmann and Eberle (1963) found the following reactions (104)–(106):

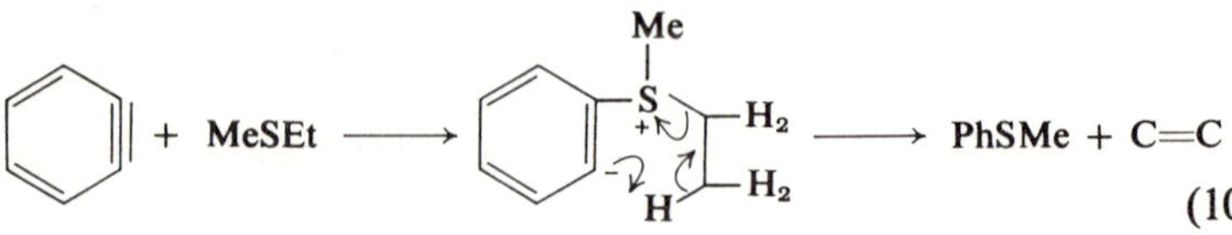

$$\text{benzyne} + \text{MeSEt} \longrightarrow \text{Ph-S}^+(Me)(CH_2-CH_2-H) \longrightarrow PhSMe + C=C \quad (104)$$

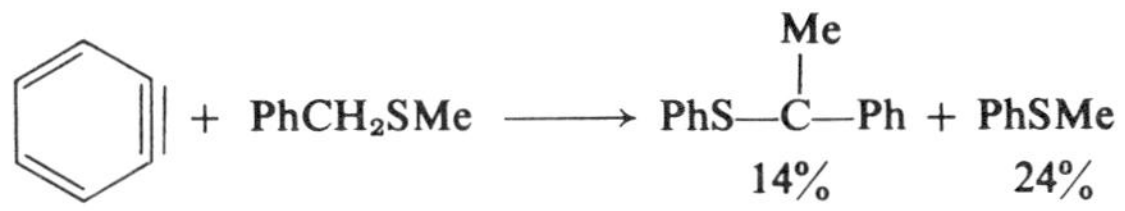

$$\text{benzene} + PhCH_2SMe \longrightarrow PhS\underset{\underset{Ph}{|}}{\overset{\overset{Me}{|}}{C}}H + PhSMe \qquad (105)$$
$$\qquad\qquad\qquad\qquad 14\% \qquad\quad 24\%$$

$$\text{benzene} + PhSCH_2CH=CH_2 \longrightarrow PhS-\underset{\underset{Ph}{|}}{CH}-CH=CH_2 + PhS-\overset{H}{C}=\overset{H}{C}-CH_2Ph$$
$$\qquad\qquad\qquad\qquad\qquad 49\% \qquad\qquad\qquad\qquad 7\%$$
$$(106)$$

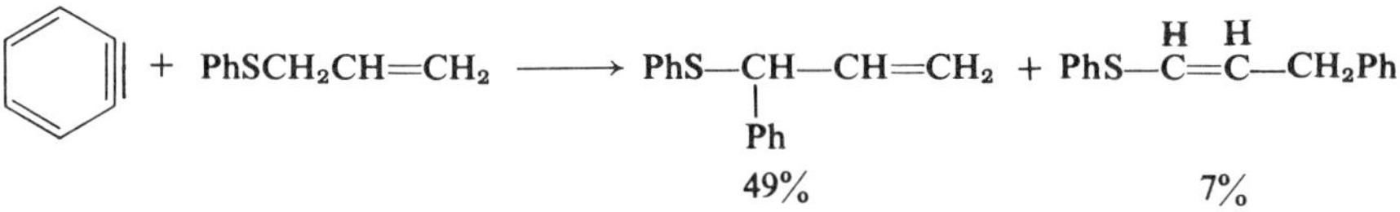

N-Benzylaziridine with benzyne gave N-benzylaniline by elimination of ethylene (Giumanini, 1972). Benzyne from decomposition of 1-(2-carboxyphenyl)-3,3-dimethyltetrazene at 190°C in dibenyl sulfide gave ESR signals denoting the free radical dissociation recombination of the first-formed sulfoniumylide (144) (Iwamura et al., 1971).

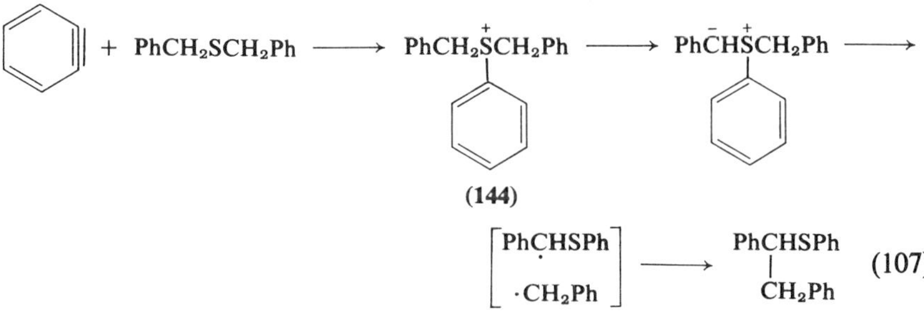

$$\text{benzene} + PhCH_2SCH_2Ph \longrightarrow PhCH_2\overset{+}{S}CH_2Ph \longrightarrow Ph\overset{-}{C}H\overset{+}{S}CH_2Ph \longrightarrow$$
$$\qquad\qquad\qquad\qquad\qquad\qquad\underset{Ph}{|} \qquad\qquad\underset{Ph}{|}$$
$$\qquad\qquad\qquad\qquad\qquad\qquad (144)$$

$$\begin{bmatrix} Ph\overset{\cdot}{C}HSPh \\ \cdot CH_2Ph \end{bmatrix} \longrightarrow \underset{CH_2Ph}{\overset{PhCHSPh}{|}} \qquad (107)$$

On the basis of kinetic studies, Bunnett and Pyun (1969) concluded that methoxide ion added to 4-chlorobenzyne 70 to 150 times as fast as methanol.

The reaction of sodamide with chloro- and bromobenzoic acids yielded isomeric aminobenzoic acids as well as iminodibenzoic acids (145) by further reaction with another mole of benzynecarboxylic acid (Biehl et al., 1969).

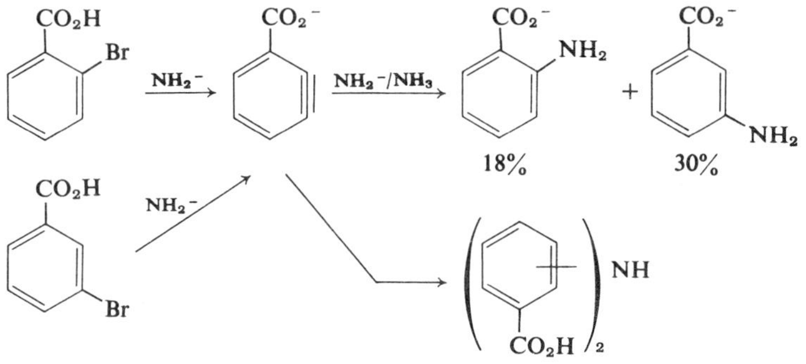

$$(108)$$
(145) 45%

Effect of solvent was shown by Haberfield and Seif (1969). In the reaction of bromobenzene with potassium anilide and aniline in benzene, diethyl ether, and 2,2-dimethoxydiethyl ether, they obtained 2-aminobiphenyl, as well as diphenylamine, in varying amounts. Highest yield of 2-aminobiphenyl (9%) was in dimethoxydiethyl ether. A six-membered ring transition state was invoked.

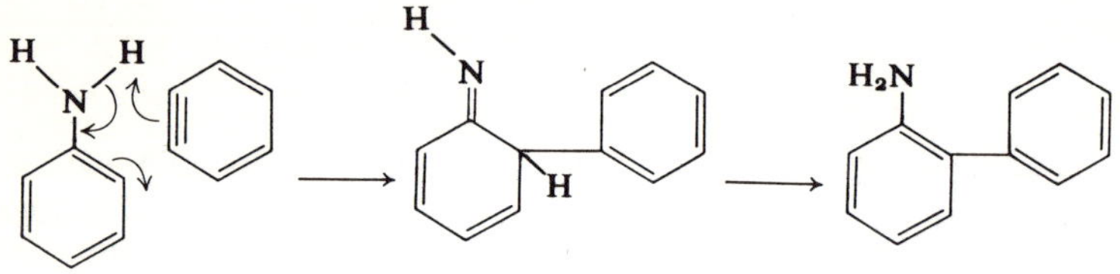

(109)

Direct solvent interaction with benzyne was shown by Wolthuis *et al.* (1970). Diazotization of anthranilic acid with amyl nitrite in tetrahydrofuran gave two products, **146** and **147**, derived from solvent.

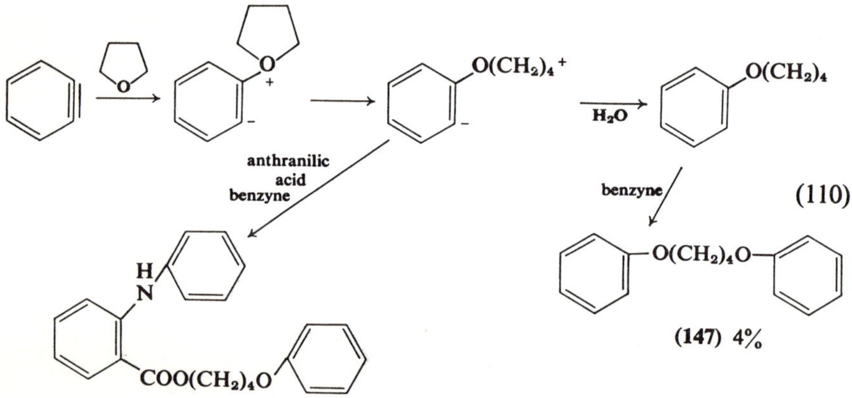

(110)

(147) 4%

(146) 17%

Intramolecular cyclizations of aromatic compounds involving the benzene ring are relatively easy via benzyne formation and anion addition. Bunnett and Hrutfiord (1961) evolved an ingenious synthesis of benzo compounds by the general reaction (111). Thiobenz-(o-bromo)anilide with potassium amide

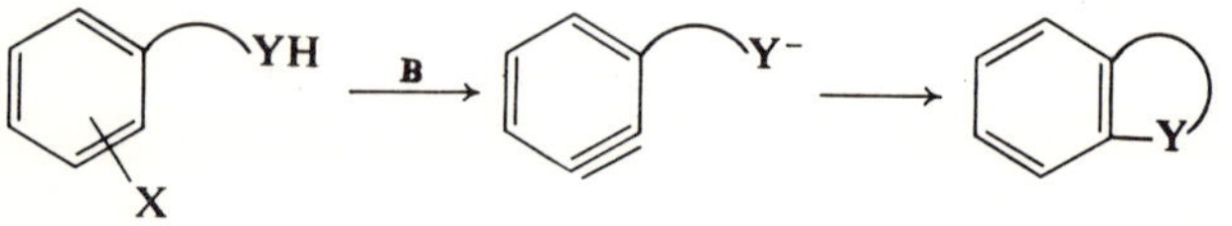

(111)

in liquid ammonia gave 2-phenylbenzothiazole (**148**). 2-Phenylbenzothiazole

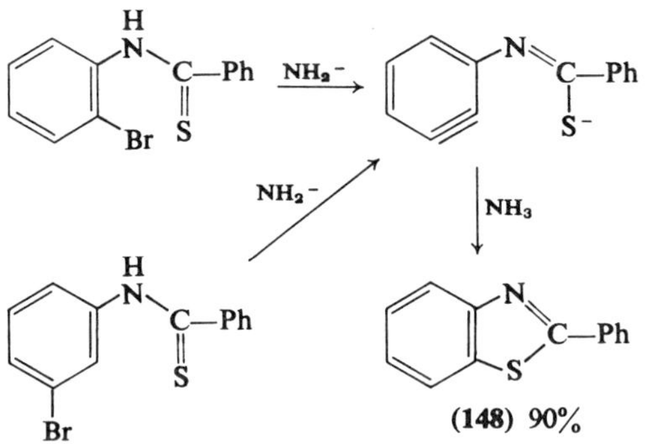

$$\text{(112)}$$

(**148**) 90%

also formed in 72% yield from the meta bromo isomer, providing strong evidence that the elimination–addition mechanism involves a benzyne intermediate. The corresponding chloro derivatives also gave phenylbenzothiazole in somewhat lower yields.

Benz (2-bromo)anilide was converted by potassium amide in liquid am-

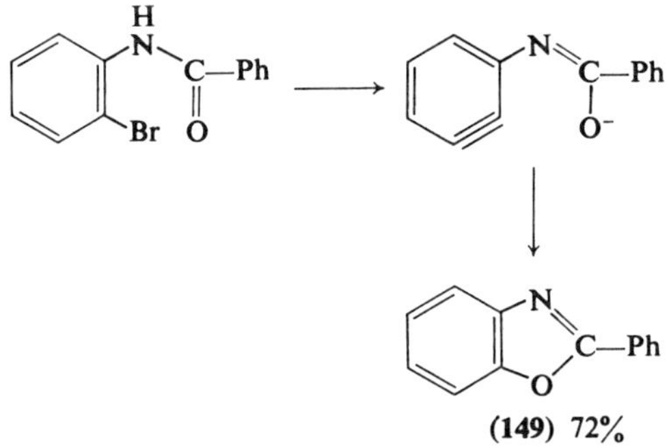

$$\text{(113)}$$

(**149**) 72%

monia to 2-phenylbenzoxazole (**149**). Carbocyclic rings can be synthesized by intramolecular aryne ring closures (Bunnett and Skorcz, 1962) [Eqs. (114)–

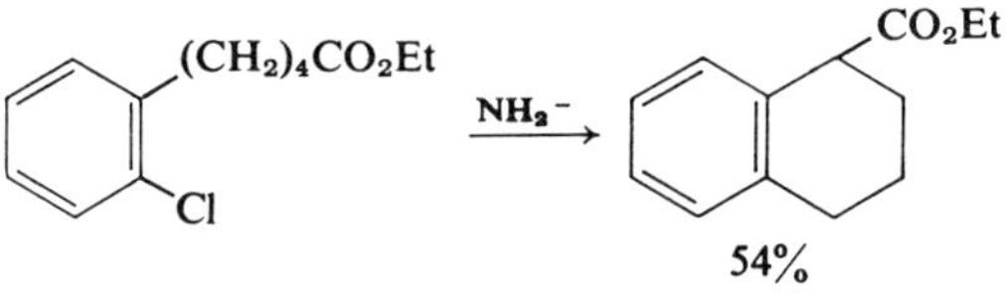

$$\text{(114)}$$

54%

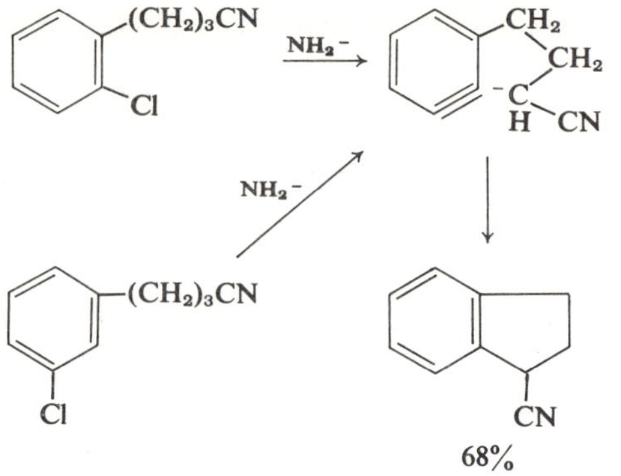

(115)

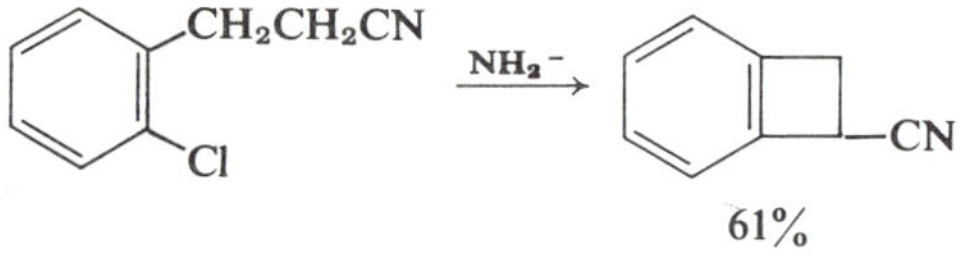

(116)

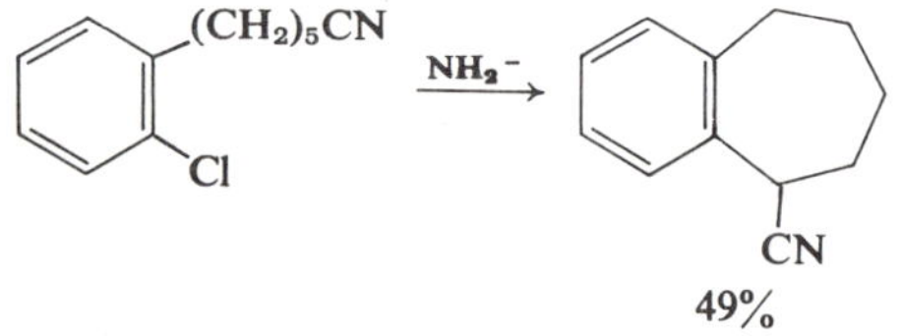

(117)

(117)]. Extensions and limitations of this general synthesis are discussed by Bunnett et al. (1963).

Aryne ring closures were applied by Kessar et al. (1971) to the synthesis of phenanthridines (**151**) from *o*-chloroanils (**150**). An unusual intramolecular

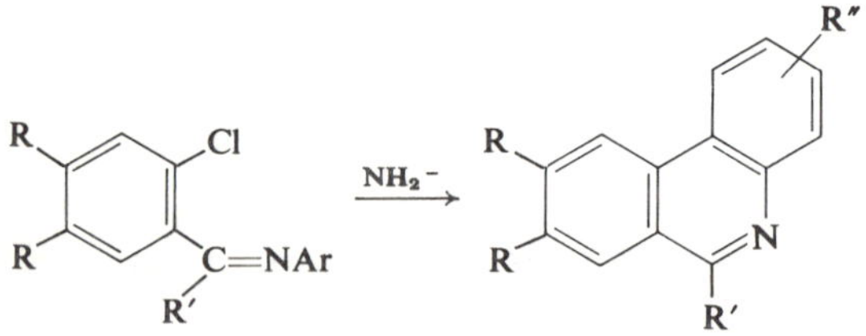

(118)

(**150**)   (**151**)

cyclization of arynes was found by Chambers and Spring (1969). Tetrafluorobenzyne reacted with pentafluorothiophenol to give octafluorodibenzothiophene (**152**). Evidently, the susceptibility of polyfluoroaromatic compounds to nucleophilic attack was enhanced by the sulfur substituent.

(119)

(152) 70%

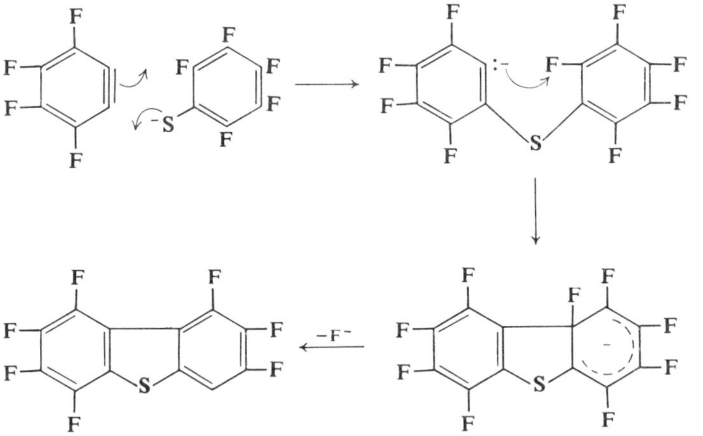

VI. Hetarynes

Heterocyclic rings containing a formal acetylenic bond are called hetarynes. The chemistry of hetarynes is somewhat confused and muddy, complicated because, unlike carbocyclic ring compounds, heterocyclics and especially *N*-heterocyclics can react by two mechanisms, elimination–addition (EA) or addition–elimination (AE). In many cases, products are formed by both mechanisms. This is well illustrated by the study of Kauffmann and Nürnberg (1967) on the reaction of lithium bases with 4-chloropyridine [Eqs. (120) and

52% 48% (120)

(EA)—100%

(121)

98.4%

(AE)—97%

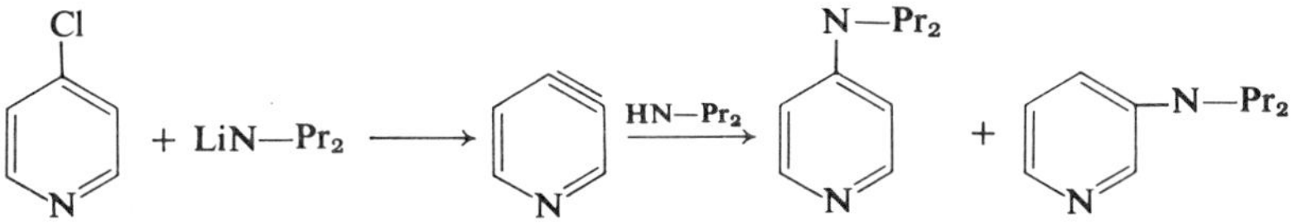

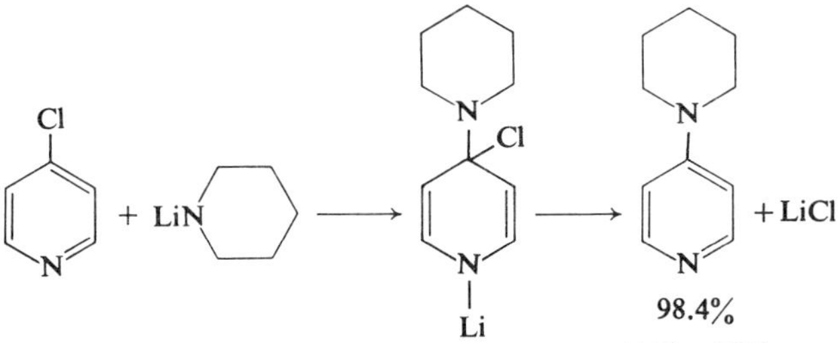

(121)]. The percentage of EA versus AE reactions in heterocyclic halide reactions with nucleophiles is estimated by competition reactions of series of halides with series of bases (Kauffmann and Wirthwein, 1971). If the concentration of products is essentially independent of the halogen substituent, the EA mechanism is considered strongly indicated.

More direct evidence for hetaryne formation was found by van der Plas et al. (1968) in the reaction of 5-bromo-6-*tert*-butyl-4-deuteropyrimidine (153) with potassium amide in liquid ammonia. The starting compound underwent no D/H exchange under their experimental conditions; the product 154 contained no deuterium.

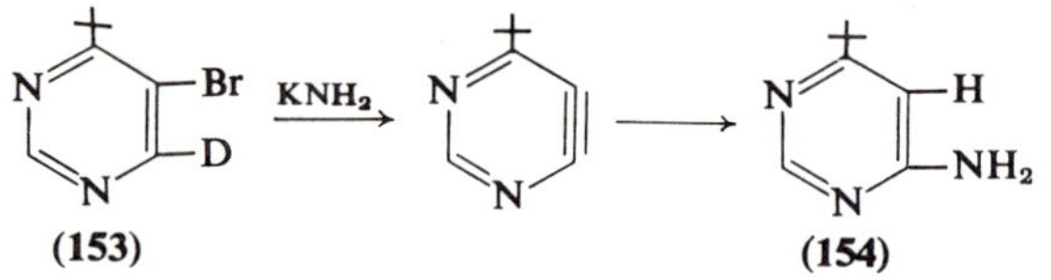

$$\text{(153)} \xrightarrow{KNH_2} \longrightarrow \text{(154)} \quad (122)$$

2,3-Pyridyne was formed in the gas phase by pyrolysis of quinolinic anhydride (Fields and Meyerson, 1968a). With benzene, it gave phenylpyridines; with pyridine, pyridylquinolines [Eq. (123)]. 3,4-Pyridyne was

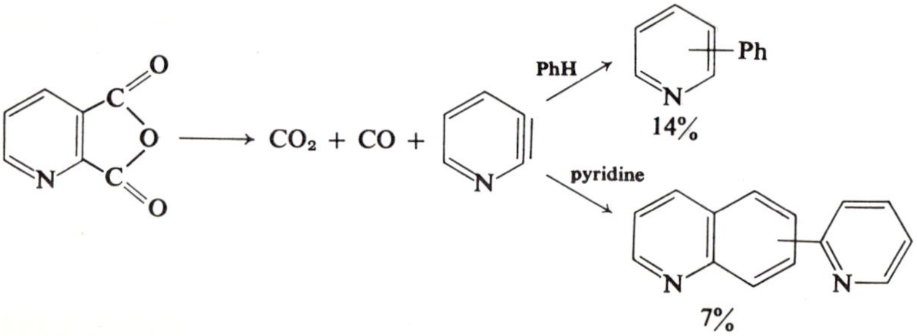

$$\longrightarrow CO_2 + CO + \cdots \quad (123)$$

obtained by Sasaki et al. (1971) by oxidation of 1-aminotriazolo[4,5-c]pyridine with lead tetraacetate. Although dimethylfulvene and norbornadiene gave no adducts, furan and 1,3-diphenylnitrilimine gave the adducts 155 and 156, respectively.

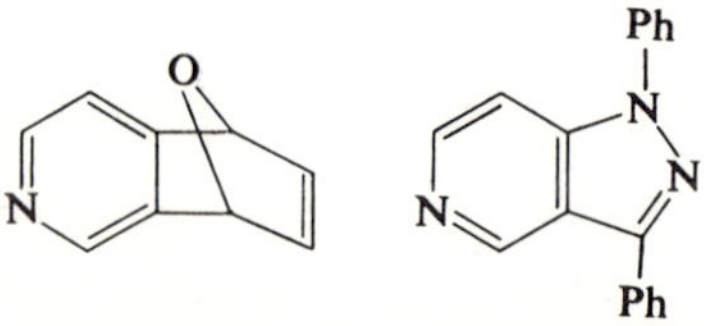

(155)   (156)

7. ARYNES

Intramolecular cyclization by addition of an anion to a hetaryne was described by Kauffmann and Wirthwein (1971). Phenyllithium and diethylamine reacted with 3-(β-aminoethyl)-5-bromopyridine to give 2,3-dihydro-1H-pyrrolo[3,2-c]pyridine (**157**). Other attempted ring closures gave much lower yields or none at all.

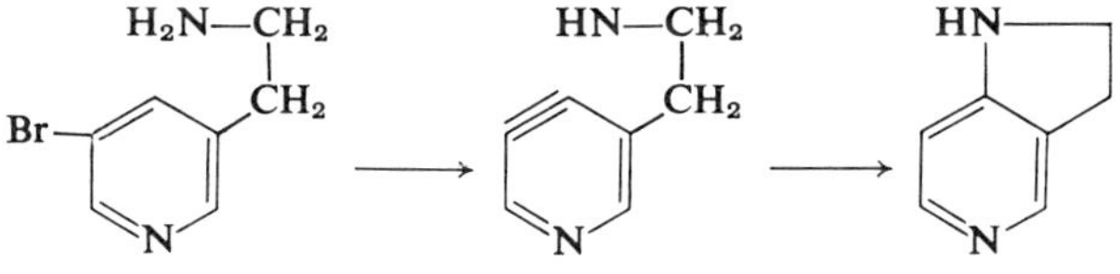

(124)

Five-membered ring hetarynes that have been postulated include thiophyne and dehydro-1-methyl imidazole. Formation of 2,3-thiophyne in the gas phase was proposed by Fields and Meyerson (1967a) to explain the products from the reaction of benzyne with thiophene among which were thiophthene (**158**), benzothiophthene (**159**), and naphthothiophene (**160**). The formation

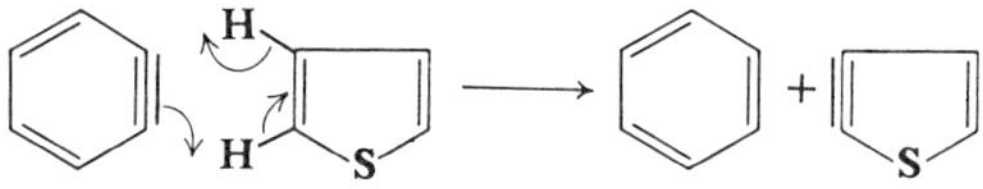

(125)

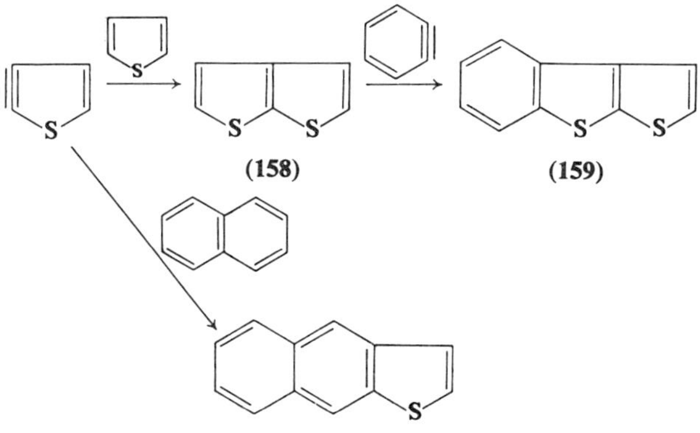

(126)

of thiophthene, benzothiophene, and phenylthiophene among the products of thiophene pyrolysis at 700°C were also rationalized by formation of thiophyne [Eqs. (127)–(129)].

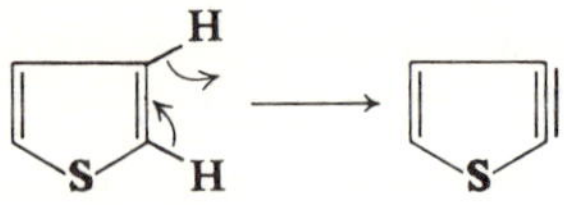

(127)

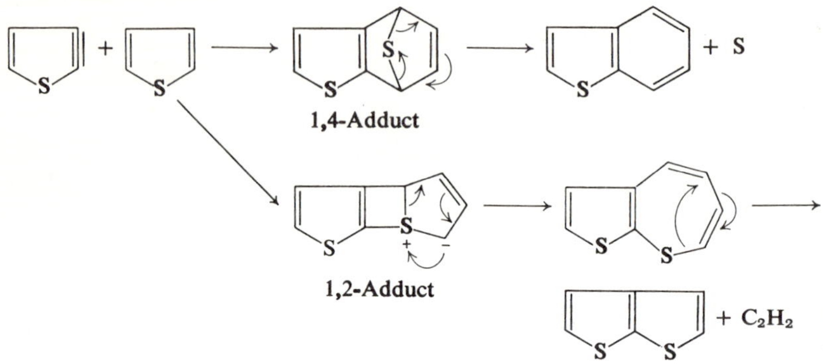

(128)

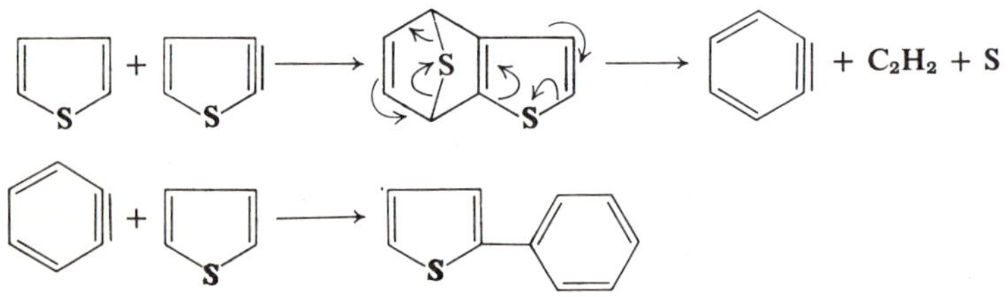

(129)

Kauffmann *et al.* (1967) reacted 5-chloro- and 5-bromo-1-methylimidazole with lithium piperidide (LiPip) and piperidine (Pip) and obtained a mixture of isomeric products from the intermediate dehydro-1-methyl imidazole [Eq. (130)].

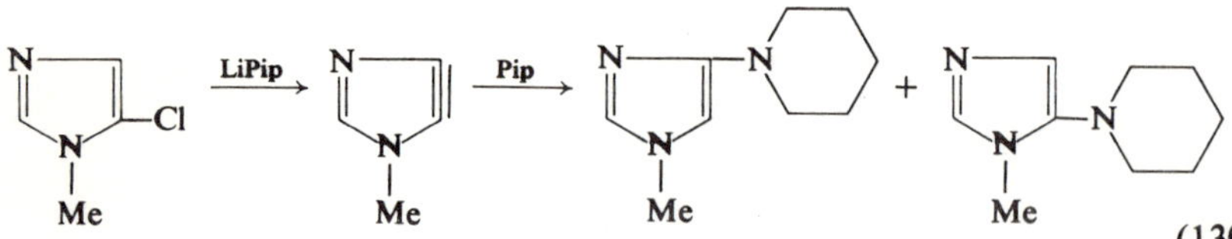

(130)

Hetarynes containing two different hetero atoms are known. A series of 2- and 3-aminophenothiazines were prepared by the reaction of chloro-10-methylphenothiazines with sodium amide and morpholine (D. H. Jones, 1971). Both 1- and 2-chloro-10-methylphenothiazine gave 10-methyl-2-morpholinophenothiazine (**161**); the 3- and 4-chloro derivatives both gave 10-methyl-3-morpholinophenothiazine (**162**) presumably by way of 1,2- and 3,4-phenothiazynes, respectively.

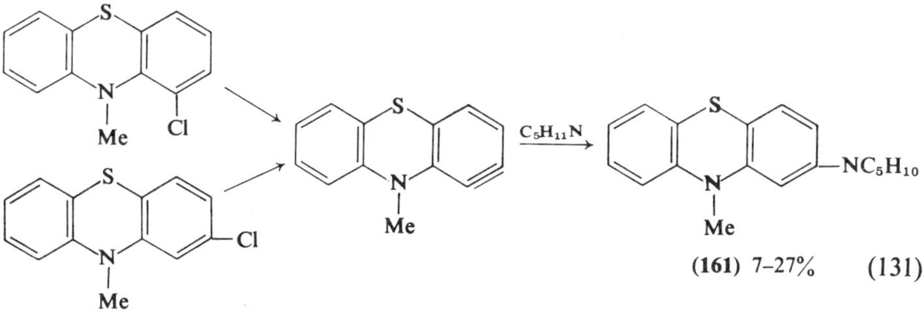

[Reaction schemes showing chlorophenothiazines converting to phenothiazine arynes and then to piperidinyl products (161) 7–27% (131) and (162) 5% (132)]

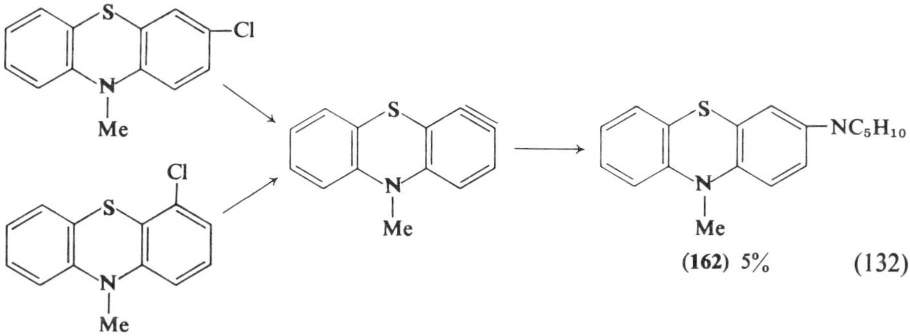

Researchers in the hetaryne field express reservations about some of the evidence for hetaryne formation (see den Hertog and van der Plas, 1969). Many reactions of heterocyclic compounds strongly indicate hetaryne intermediates; however, the hetaryne field is relatively new and offers many opportunities for productive investigation.

VII. Conclusion

Theory and mechanisms of aryne formation and reactions have been firmly established by Wittig, Roberts, and the many subsequent investigators. Use of aryne reactions in synthesis has become an accepted and almost routine procedure. The greatest effort at present appears to be in the discovery of new and easy methods of generating arynes from readily available chemicals and in trapping arynes as metal complexes, or perhaps in rigid glasses at low temperatures for spectroscopic investigations. The latter procedure could prove useful in photolysis of light-labile compounds such as phthaloyl peroxide (Jones and DeCamp, 1971).

By contrast to aryne chemistry, the field of hetarynes is in a somewhat confused state. This is due to several factors: the multiplicity of mechanisms available for nucleophilic substitutions, the relative ease of ring opening, and to some extent the reactivity of the heteroatoms. An attractive method of forming hetarynes that may become more widely exploited is the pyrolysis at relatively low temperatures of heterocyclic compounds substituted with good leaving groups such as N_2, SO_2, CO, and CO_2.

References

Abramovitch, R. A., Singer, G. M., and Vinutha, A. R. (1967). *Chem. Commun.* p. 55.
Adam, W., Grunson, A., and Hoffmann, R. (1969). *J. Amer. Chem. Soc.* **91**, 2590.
Ahlgren, G., and Akermark, B. (1970). *Tetrahedron Lett.* p. 3047.
Atkin, R. W., and Claxton, T. A. (1970). *Trans. Faraday Soc.* **66**, 257.
Badger, G. M., and Spotswood, T. M. (1960). *J. Chem. Soc., London* p. 4420.
Bailey, N. A., Churchill, M. R., Hunt, R., and Wilkinson, G. (1964). *Proc. Chem. Soc., London* p. 401.
Barton, J. W., and Grinham, A. R. (1972). *J. Chem. Soc. Perkin.* **1**, 634.
Benkeser, R. A., and Buting, W. E. (1952). *J. Amer. Chem. Soc.* **74**, 3011.
Berry, R. S., Spokes, G. N., and Stiles, M. (1962). *J. Amer. Chem. Soc.* **84**, 2738.
Berry, R. S., Clardy, J., and Schafer, M. E. (1964). *J. Amer. Chem. Soc.* **86**, 2738.
Berry, R. S., Clardy, J., and Schafer, M. E. (1965). *Tetrahedron Lett.* p. 1011.
Berson, J. A., and Pomerantz, M. (1964). *J. Amer. Chem. Soc.* **86**, 3896.
Berthelot, A. (1866). *C. R. Acad. Sci.* **63**, 790.
Berthelot, A. (1867a). *Ann. Chem.* **142**, 254.
Berthelot, A. (1867b). *Bull. Soc. Chim. Fr.* [2] **7**, 218.
Bertorello, H. E., Rossi, R. A., and de Rossi, P. H. (1970). *J. Org. Chem.* **35**, 3332.
Biehl, E. R., Nieh, E., Li, H-M., and Hong, C. (1969). *J. Org. Chem.* **34**, 500.
Biehl, E. R., Hsu, K. C., and Nieh, E. (1970). *J. Org. Chem.* **35**, 2454.
Blomquist, A. T., and Lin, L. H. (1953). *J. Amer. Chem. Soc.* **75**, 2153.
Bottini, A. T., and Roberts, J. D. (1957). *J. Amer. Chem. Soc.* **79**, 1458.
Bradford, C. W., Nyholm, R. S., Gainsford, G. D., and Guss, J. M. (1972). *Chem. Commun.*, 87.
Brewer, J. P. N., Heaney, H., and Marples, B. A. (1969). *Tetrahedron* **25**, 243.
Brown, R. F. C., and Solly, R. K. (1965). *Chem. Ind. (London)* p. 1462.
Brown, R. F. C., Crow, W. D., and Solly, R. K. (1966). *Chem. Ind. (London)* p. 343.
Brown, R. F. C., Gardner, D. V., McOmey, J. F. W., and Solly, R. K. (1967). *Aust. J. Chem.* **20**, 139.
Bunnett, J. F., and Hrutfiord, B. F. (1961). *J. Amer. Chem. Soc.* **83**, 1691.
Bunnett, J. F., and Pyun, C. (1969). *J. Org. Chem.* **34**, 2035.
Bunnett, J. F., and Skorcz, J. A. (1962). *J. Org. Chem.* **27**, 3836.
Bunnett, J. F., and Zahler, R. E. (1951). *Chem. Rev.* **49**, 273.
Bunnett, J. F., Kato, T., Flynn, R. R., and Skorcz, J. A. (1963). *J. Org. Chem.* **28**, 1.
Cadogan, J. I. G. (1971). *Accounts Chem. Res.* **4**, 186.
Cadogan, J. I. G., Mitchell, J. R., and Sharp, J. T. (1971a). *Chem. Commun.* p. 1.
Cadogan, J. I. G., Cook, J., Harger, M. J. P., Hibbert, P. G., and Sharp, J. T. (1971b). *J. Chem. Soc., B* p. 595.
Cadogan, J. I. G., Smith, D. M., and Thomson, J. B. (1972). *J. Chem. Soc. Perkin.* **1**, 1296.

7. ARYNES

Callender, D. D., Coe, P. L., Tatlow, J. C., and Uff, A. J. (1969). *Tetrahedron* **25**, 25.
Campbell, C. D., and Rees, C. W. (1969). *J. Chem. Soc.*, C p. 748.
Caubere, P., Derozier, N., and Loubinoux, B. (1971). *Bull. Soc. Chim. Fr.* [5] p. 302.
Caubere, P., Guillamet, G., and Mourad, M. S. (1972). *Tetrahedron*, **28**, 95.
Cava, M. P., Mitchell, M. S., DeJongh, D. C., and Van Fossen, R. Y. (1966). *Tetrahedron Lett.* p. 2947.
Chambers, R. D., and Spring, D. J. (1969). *Tetrahedron Lett.* p. 2481.
Ciabattoni, J., and Crowley, J. E. (1967). *J. Amer. Chem. Soc.* **89**, 2778.
Cook, C. D., and Jauhal, G. S. (1968). *J. Amer. Chem. Soc.* **90**, 1464.
Coulson, C. A. (1958). *Chem. Soc., Spec. Publ.* **12**, 85.
Crews, P., and Loffgren, M. (1971). *Tetrahedron Lett.* p. 4697.
Criegee, R., and Schroder, G. (1959). *Justus Liebigs Ann. Chem.* **623**, 1.
de Rossi, P. H., Bertorello, H. E., and Rossi, R. A. (1970). *J. Org. Chem.* **35**, 3328.
Deeming, A. J., Nyholm, R. S., and Underhill, M. (1972). *Chem. Commun.* p. 224.
den Hertog, H. J., and van der Plas, H. C. (1969). *In* "Chemistry of Acetylenes" (H. G. Viehe, ed.), p. 1149. Dekker, New York.
Dobson, J. E., Miller, R. G., and Wiggen, J. P. (1971). *J. Amer. Chem. Soc.* **93**, 554.
Domnin, N. A. (1938). *J. Gen. Chem. USSR* **8**, 851.
Ebel, H. F., and Hoffmann, R. W. (1964). *Justus Liebigs Ann. Chem.* **673**, 1.
Ellis, J. A., and Sargent, M. V. (1969). *J. Amer. Chem. Soc.* **91**, 4734.
Embree, H. D. (1971). *Chem. Eng. News* **49**, 3.
Erickson, K. I., and Wolinsky, J. (1965). *J. Amer. Chem. Soc.* **87**, 1142.
Fields, E. K. (1970). U.S. Patent 3,514,458 [to Standard Oil (Ind)].
Fields, E. K., and Meyerson, S. (1965). *Chem. Commun.* p. 474.
Fields, E. K., and Meyerson, S. (1966a). *J. Amer. Chem. Soc.* **88**, 21.
Fields, E. K., and Meyerson, S. (1966b). *J. Amer. Chem. Soc.* **88**, 3388.
Fields, E. K., and Meyerson, S. (1967a). *In* "Organosulfur Chemistry" (M. J. Janssen, ed.), p. 143. Wiley, New York.
Fields, E. K., and Meyerson, S. (1967b). *Tetrahedron Lett.* p. 571.
Fields, E. K., and Meyerson, S. (1968a). *Advan. Phys. Org. Chem.* **6**, 1.
Fields, E. K., and Meyerson, S. (1968b). *Advan. Phys. Org. Chem.* **6**, 15.
Fields, E. K., and Meyerson, S. (1969a). *Accounts Chem. Res.* **2**, 273.
Fields, E. K., and Meyerson, S. (1969b). *Tetrahedron Lett.* p. 629.
Fields, E. K., and Meyerson, S. (1971a). Unpublished results.
Fields, E. K., and Meyerson, S. (1971b). *Tetrahedron Lett.* p. 719.
Fieser, L., and Fieser, M. (1967). "Reagents for Organic Synthesis," p. 46. Wiley, New York.
Fisher, I. P., and Lossing, F. P. (1963). *J. Amer. Chem. Soc.* **85**, 1018.
Friedman, L. (1967). *J. Amer. Chem. Soc.* **89**, 3071.
Friedman, L., and Logullo, F. M. (1965). Private communication.
Friedman, L., and Logullo, F. M. (1967). Reported in Fieser and Fieser (1967).
Friedman, L., and Logullo, F. M. (1968). *Org. Syn.* **48**, 12.
Friedman, L., and Logullo, F. M. (1969). *J. Org. Chem.* **34**, 3089.
Gainsford, G. J., Guss, J. M., Ireland, P. R., and Mason, R. (1972). *J. Organometal. Chem.* **40**, C70.
Gilchrist, T. L., Graveling, F. J., and Rees, C. W. (1968). *Chem. Commun.* p. 821.
Gill, J. B. (1968). *Quart. Rev., Chem. Soc.* **22**, 338.
Giumanini, A. G. (1972). *J. Org. Chem.* **37**, 513.
Gowling, E. W., Kettle, S. F. A., and Sharples, G. M. (1968). *Chem. Commun.* p. 21.
Haberfield, P., and Seif, L. (1969). *J. Org. Chem.* **34**, 1508.

Hantzsch, A., and Davidson, W. B. (1896). *Chem. Ber.* **30**, 2548.
Hantzsch, A., and Ologauer, R. (1897). *Chem. Ber.* **30**, 2548.
Haeussermann, O. (1900). *Chem. Ber.* **33**, 939.
Haeussermann, O. (1901). *Chem. Ber.* **34**, 38.
Haselbach, E. (1971). *Helv. Chim. Acta* **54**, 1981.
Hayashi, S., and Ishikawa, N. (1970). *Yuki Gosei Kagaku Kyokai Shi* **28**, 533; *Chem. Abstr.* **73**, 45241c (1970).
Heaney, H. (1970). *Fortschr. Chem. Forsch.* **16**, 35.
Heaney, H., and Jablonski, J. M. (1966). *Tetrahedron Lett.* p. 4529.
Heaney, H., Mason, K. G., and Sketchley, J. M. (1971). *J. Chem. Soc.*, *C* p. 567.
Hellmann, H., and Eberle, D. (1963). *Justus Liebigs Ann. Chem.* **662**, 188.
Hess, B. A., and Schaad, L. S. (1971). *Tetrahedron Lett.* p. 17.
Hinton, R. C., Mann, F. G., and Millar, I. T. (1958). *J. Chem. Soc.*, *London* p. 4704.
Hoffmann, R. W. (1967). "Dehydrobenzene and Cycloalkynes." Academic Press, New York.
Hoffmann, R. W., Guhn, G., Preiss, M., and Dittrich, B. (1969). *J. Chem. Soc.*, *London* p. 769.
Huisgen, R. (1969). *Chem. Ber.* **102**, 904.
Huisgen, R., and Knorr, R. (1961). *Naturwissenschaften* **48**, 716.
Huisgen, R., and Rist, H. (1954). *Naturwissenschaften* **41**, 358.
Iwamura, H., Iwamura, M., Nishida, T., Yoshida, M., and Nakayama, U. (1971). *Tetrahedron Lett.* p. 63.
Jones, D. H. (1971). *J. Chem. Soc.*, *C* p. 132.
Jones, M., Jr., and DeCamp, M. R. (1971). *J. Org. Chem.* **36**, 1536.
Jones, M., Jr., and Levin, R. H. (1969). *J. Amer. Chem. Soc.* **91**, 6411.
Jones, R. R., and Bergman, R. G. (1972a). *J. Amer. Chem. Soc.* **97**, 660.
Jones, R. R., and Bergman, R. G. (1972b). *J. Amer. Chem. Soc.* **94**, 660.
Kauffmann, T., and Nürnberg, R. (1967). *Chem. Ber.* **100**, 3427.
Kauffmann, T., and Wirthwein, R. (1971). *Angew. Chem., Int. Ed. Engl.* **10**, 20.
Kauffmann, T., Nürnberg, R., Schulz, J., and Stabba, R. (1967). *Tetrahedron Lett.* p. 4273.
Kekule, A. (1864). *C. R. Acad. Sci.* **64**, 753.
Kessar, S. V., Singh, M., Jit, P., Singh, G., and Lumb, A. K. (1971). *Tetrahedron Lett.* p. 471.
Krebs, A. (1965). *Angew. Chem., Int. Ed. Engl.* **4**, 953.
Krebs, A. (1969). *In* "Chemistry of Acetylenes" (H. G. Viehe, ed.), p. 1006. Dekker, New York.
Krebs, A., and Kimling, H. (1971). *Angew. Chem. Int. Ed. Engl.* **10**, 509.
Kym, O. (1895). *J. Prakt. Chem.* [2] **51**, 325.
Limpricht, H. (1874). *Chem. Ber.* **7**, 1439.
Lombardo, L., and Wege, D. (1971). *Tetrahedron Lett.* 3981.
Lüttringhaus, A., and Saaf, G. (1930). *Justus Liebigs Ann. Chem.* **542**, 250.
McLafferty, F. W., and Gohlke, R. J. (1959). *Anal. Chem.* **31**, 2076.
Martin, J. C., and Bloch, D. R. (1971). *J. Amer. Chem. Soc.* **93**, 451.
Martinez de Bertorello, M., Bertorello, H. E., and Garcia-Martinez, N. (1970). *An. Asoc. Quim. Argent.* **58**, 291.
Matsuda, T., and Mitsuyasu, T. (1966). *Bull. Chem. Soc. Jap.* **39**, 1342.
Matuszak, C. A. (1971). *Chem. Eng. News* **49**, 39.
Meyer, K. H., and Bergius, F. (1914). *Chem. Ber.* **47**, 3159.
Meyerson, S. (1965). *Rec. Chem. Progr.* **26**, 257.
Meyerson, S., and Fields, E. K. (1966). *Chem. Commun.* p. 275.

Mikhailova, I., Sycheva, T. N., and Barkhash, V. A. (1970). *Zh. Org. Khim.* **6**, 1426.
Miller, R. G., and Kuhlmann, D. P. (1971). *J. Organometal. Chem.* **26**, 401.
Miller, R. G., and Stiles, M. (1963). *J. Amer. Chem. Soc.* **85**, 1798.
Millie, P., Praud, L., and Serre, J. (1971). *Int. J. Quantum Chem., Symp.* **4**, 187.
Minisci, F., and Quilico, A. (1964). *Chim. Ind. (Milan)* **46**, 428.
Miwa, T., Kato, M., and Tamano, T. (1969). *Tetrahedron Lett.* p. 1761.
Morton, A. A., Davidson, J. B., and Hakan, B. L. (1942). *J. Amer. Chem. Soc.* **64**, 2242.
Nieuland, J. A., and Vogt, R. R. (1945). "Chemistry of Acetylene," pp. 1–14, and references cited therein. Van Nostrand-Reinhold, Princeton, New Jersey.
Oda, M., and Kitahara, Y. (1970). *Bull. Chem. Soc. Jap.* **43**, 1920.
Paquette, L. A. (1971). *Chem. Commun.* p. 1076.
Petersen, C. E. L., and Berg, A. (1971). *Acta Chem. Scand.* **25**, 375.
Porter, G., and Steinfeld, J. I. (1968). *J. Chem. Soc.* p. 877.
Rees, C. W. (1971). "Hey Symposium." King's College, London.
Rees, C. W., and Storr, R. C. (1965). *Chem. Commun.* p. 193.
Rees, C. W., and Storr, R. C. (1969). *J. Chem. Soc., London* p. 765.
Rees, C. W., Gilchrist, T. L., and Graveling, F. J. (1971). *J. Chem. Soc., C* p. 977.
Reid, W., and Schön, M. (1965). *Justus Liebigs Ann. Chem.* **689**, 141.
Roberts, J. D., Simmons, H. E., Carlsmith, L. A., and Vaughan, C. W. (1953). *J. Amer. Chem. Soc.* **75**, 3290.
Roberts, J. D., Vaughan, C. W., Carlsmith, L. A., and Semenow, D. A. (1956). *J. Amer. Chem. Soc.* **78**, 611.
Roe, D. M., and Massey, A. G. (1970). *J. Organometal. Chem.* **23**, 547.
Rüchardt, C., and Tan, C. C. (1970). *Angew. Chem., Int. Ed. Engl.* **9**, 522.
Sasaki, T., Kanematsu, K., and Uchide, M. (1971). *Bull. Chem. Soc. Jap.* **44**, 858.
Sauer, J., and Heinriche, G. (1966). *Tetrahedron Lett.* p. 4979.
Schafer, M. E., and Berry, R. S. (1965). *J. Amer. Chem. Soc.* **87**, 4497.
Seibert, R. A., and Bergstrom, F. W. (1945). *J. Org. Chem.* **10**, 544.
Simmons, H. E. (1961). *J. Amer. Chem. Soc.* **83**, 1657.
Simmons, H. E. (1967). *In* "Dehydrobenzene and Cycloalkynes" (R. W. Hoffmann, ed.), pp. 264–272. Academic Press, New York.
Stiles, M., and Miller, R. G. (1960). *J. Amer. Chem. Soc.* **82**, 3802.
Stiles, M., Miller, R. G., and Burckhardt, U. (1963). *J. Amer. Chem. Soc.* **85**, 1798.
Stoermer, R., and Kahlert, B. (1902). *Chem. Ber.* **35**, 1633.
Suhr, H., and Szabo, A. (1971). *Justus Liebigs Ann. Chem.* **752**, 37.
Tabushi, I., Fujita, K., Okazaki, K., Yamada, H., and Oda, R. (1969a). *Kogyo Kagaku Zasshi* **72**, 1677; *Chem. Abstr.* **74**, 1286e (1971).
Tabushi, I., Fujita, K., Okazaki, K., Yamada, H., and Oda, R. (1969b). *Tetrahedron* **25**, 4401.
Tabushi, I., Yomada, H., Yoshida, Z., and Kuroda, H. (1971). *Tetrahedron Lett.* p. 1093.
van der Plas, H. C., Smit, P., and Koudijs, A. (1968). *Tetrahedron Lett.* p. 9.
Vedejs, E. (1968). *Tetrahedron Lett.* p. 2633.
Vedejs, E., and Shepherd, R. A. (1970). *Tetrahedron Lett.* p. 1863.
Viehe, H. G., ed. (1969). "Chemistry of Acetylenes." Dekker, New York.
Vollmer, J. J., and Servis, K. L. (1968). *J. Chem. Educ.* **45**, 214.
Vysochin, V. J., Mikhailova, I. F., and Barkhash, V. A. (1970). *Zh. Org. Khim.* **6**, 1341.
Warner, P. (1971). *Tetrahedron Lett.* p. 723.
Wasserman, H. H., and Keller, L. S. (1970). *J. Chem. Soc., D* p. 1483.
Wasserman, H. H., Solodar, A. J., and Keller, L. S. (1968). *Tetrahedron Lett.* p. 5593.
Wilhite, D. L., and Whitten, J. L. (1971). *J. Amer. Chem. Soc.* **93**, 2858.
Winkler, H. J. S., and Wittig, G. (1963). *J. Org. Chem.* **28**, 1733.

Wittig, G. (1961). *Justus Liebigs Ann. Chem.* **650**, 20.
Wittig, G. (1965). *Angew. Chem., Int. Ed. Engl.* **4**, 731.
Wittig, G., and Behnisch, W. (1958). *Chem. Ber.* **91**, 2358.
Wittig, G., and Dorsch, H. L. (1968). *Justus Liebigs Ann. Chem.* **711**, 46.
Wittig, G., and Dürr, H. (1964). *Justus Liebigs Ann. Chem.* **672**, 55.
Wittig, G., and Fritze, P. (1968). *Justus Liebigs Ann. Chem.* **712**, 79.
Wittig, G., and Fuhrmann, G. (1940). *Chem. Ber.* **73**, 1197.
Wittig, G., and Härle, H. (1959). *Justus Liebigs Ann. Chem.* **623**, 17.
Wittig, G., and Mayer, U. (1963). *Chem. Ber.* **96**, 342.
Wittig, G., and Meske-Schüler, J. (1968). *Justus Liebigs Ann. Chem.* **711**, 55.
Wittig, G., and Pohlke, R. (1961). *Chem. Ber.* **94**, 3276.
Wittig, G., and Pohmer, L. (1955). *Angew. Chem.* **67**, 348.
Wolthuis, E., Bouma, B., Modderman, J., and Sytsma, L. (1970). *Tetrahedron Lett.* p. 407.
Woodward, R. B., and Hoffman, R. (1968). *Accounts Chem. Res.* **1**, 17.
Yamatani, T., Yasunami, M., and Takase, K. (1970). *Tetrahedron Lett.* p. 1725.
Yonezawa, T., Konishi, H., and Kato, H. (1969). *Bull. Chem. Soc. Jap.* **42**, 933.

AUTHOR INDEX

Numbers in italics refer to the pages on which the complete references are listed.

A

Aaronson, A. M., 179, *191*
Abramovitch, R. A., 128, 129, 131, 132, 135, 137, 138, 139, 140, 142, 143, 144, 145, 148, 149, 150, 151, 158, 159, 160, 161, 166, 169, 171, 172, 173, 174, 176, 177, 178, 179, 180, 181, 182, 186, *188*, *504*
Achiwa, K., 163, *192*
Adam, W., *504*
Adamcik, J. A., 352, *415*
Adams, J. T., 357, *417*
Adams, K. A. H., 137, 179, *188*
Adams, M., 427, *444*
Adams, R., 304, *335*, *415*
Adams, R. N., 427, 429, 437, *446*, *447*
Adickes, H. W., 363, *419*
Adkins, H., 242, *330*
Ahlgren, G., 493, *504*
Ahmad, Y., 158, *188*
Ainsworth, C., 436, *446*
Akermark, B., 493, *504*
Alfrey, T., Jr., 48, *51*
Allen, C. F. H., 348, 364, *415*, *417*
Allen, L. C., 69, 70, *122*, 315, *335*
Allen, R. E., 346, *420*
Allen, R. G., 23, *59*
Allenstein, E., 150, *190*
Altschuld, J. W., 389, 408, *417*
Alukar, R. H., 414, *420*
Amrich, M. J., 83, *120*
Anastassiou, A. G., 105, *120*, 132, 142, 167, 168, *188*
Anderson, D. J., 156, *188*
Anderson, R. J., 354, *415*
Andes, B. A., Jr., 316, *335*
Ando, W., 92, 101, 106, 107, *120*, *122*
Andrews, E. B., 73, *123*
Andrews, G. C., 387, *420*
Andrews, L., 73, 74, *120*
Andrulis, P. J., 441, *444*
Angeleno, B., 436, *444*
Anschütz, L., 427, *444*
Anselme, J.-P., 83, *124*
Antheunis, D., 95, *123*
Appel, B., 205, *335*
Appenrodt, J., 425, *447*
Appl, M., 128, 163, *188*
Applequist, D. E., 46, *57*
Apsimon, J. W., 128, 163, *188*
Armbrecht, F. M., 88, *125*
Arndt, F., 200, *324*
Arnett, E. M., 259, 262, 265, 266, *324*
Arold, H., 100, 101, 115, *123*
Aryden, H. L., Jr., 349, *416*
Ashby, E., 346, *415*
Ashe, A. J., 316, *335*
Ashitaka, H., 74, 78, *123*
Atkin, R. W., 479, *504*
Atkinson, R. S., 156, *188*
Attridge, C. J., 94, *125*
Aue, D. H., 185, *191*
Ausloos, P., 256, *324*
Austen, D. E. G., 427, *444*
Aver, E. E., 51, *58*
Ayers, P. W., 435, *445*
Ayres, D. C., 338, *415*
Ayscough, P. B., 430, 432, *444*
Azogu, C. I., 142, 149, 159, 160, 161, 169, 177, 180, 181, 186, *188*

B

Bachman, G. L., 97, 99, 106, *122*
Bachman, W. E., *415*, 434, *444*
Backeberg, O. G., 3, *57*

Bäcklund, B., 397, *420*
Badger, G. M., 452, *504*
Baeyer, A., 196, 200, *324*
Bahary, W., 233, *325*
Bailey, N. A., 475, *504*
Bailie, J. C., 379, *417*
Baird, R. L., 308, 310, *324*, *327*
Bak, T. A., 136, *189*
Baker, A. D., 229, *325*
Baker, E. B., 210, 218, 221, 227, 228, *332*
Baker, R., 205, *335*
Balabanov, G. P., 166, *188*
Baldeschwieler, J. D., 181, *188*
Baldwin, J. E., 113, *120*, 385, *415*
Ballester, M., 348, *415*
Bamberger, E., *444*
Bamford, C. H., 97, *120*, 156, *188*
Bamford, D. A., 156, *188*
Barash, L., 66, 75, 76, *120*, *126*, 133, 134, *189*, *192*
Barbas, J. T., 436, *445*
Bank, S., 435, *444*, *445*
Barkhash, V. A., 493, *507*
Barnish, I. T., 372, 413, *415*
Baron, W. J., 110, 114, *120*, *123*
Barrow, R. F., 73, *123*
Bartlett, P. D., 12, *57*, *58*, 237, 255, 275, 276, 281, 284, 286, 295, 306, 307, 317, 321, *324*, *325*, *332*
Bartling, G. J., 347, 372, *418*
Bartolini, G., 387, *420*
Barton, B. L., 429, *444*
Barton, D. H. R., 161, *189*
Barton, F. E., 435, 436, *445*
Barton, J. W., 491, *504*
Bass, A. M., 119, *121*
Bastien, I. J., 210, 218, 221, *332*
Bateman, L. C., 206, *325*
Battenberg, E., 256, *331*
Battiste, M. A., 266, 320, *325*
Battisti, A., 435, *445*
Bauld, N. L., 249, *326*
Baumgarten, H. E., 402, *415*
Baun, C. E. H., 6, *57*
Bayes, K. D., *126*
Bayles, J. W., 226, *325*
Bayless, J. H., 80, *122*
Beach, W. F., 131, *189*
Beam, C. F., 375, 376, *415*, *417*
Beard, R. D., 373, 374, 377, *415*, *418*

Bebb, R. L., 403, *419*
Beck, B. H., 319, *329*
Becker, H.-D., 428, *447*
Becker, R. H., 381, 385, 386, *419*
Becker, R. S., 67, *120*
Beckmann, E., 425, *444*
Beckwith, A. L. J., 174, *189*
Behnisch, W., 465, *508*
Behr, F. E., 137, *190*
Bell, J. A., 83, *120*
Belloli, R., 129, 161, *189*
Bemis, A. G., 427, 431, 435, *446*
Bender, C. F., 70, *120*
Bender, M. L., 438, *445*
Benfey, O. T., 206, *325*
Benkeser, R. A., 404, 407, *415*, 454, *504*
Bennett, J. E., 427, *444*
Bennett, R. P., 179, *189*
Benson, S. W., 27, 28, *57*, 90, *121*
Bentley, M. D., 308, *325*
Bentrude, W. G., 262, *324*
Berg, A., 465, *507*
Berger, J. G., 111, *122*
Berguis, F., 453, *506*
Bergman, R. G., 452, *506*
Bergmann, E. D., *329*, 351, *416*
Berkey, R. A., 233, *334*
Berkheimer, H. E., *327*
Berliner, E., 307, *315*
Bernal, J. D., 259, *325*
Bernhard, R., 317, 319, *325*
Bernheim, R. A., 76, *120*
Bergstrom, F. W., 454, *507*
Bernal, I., 427, *444*, *446*
Bersohn, M., 255, *334*
Berson, J. A., 312, *325* 480, 492, *504*
Berry, R. S., 132, 143, *189*, 459, 468, *504*, *507*
Berthelot, A., 452, *504*
Bertini, F., 443, *446*
Bertoniere, N. R., 67, *120*
Bertorello, H. E., 471, *504*, *505*, *506*
Bethell, D., 81, 86, 87, 95, 101, 107, 109, 112, *120*, 196, 200, 286, 299, *324*, *325*
Betteridge, D., 229, *325*
Bevan, W. I., 88, *120*
Bevilacqua, E. B., 51, *58*
Bhide, B. H., 415, *420*
Bieber, J. B., 390, *419*
Biehl, E. R., 465, 495, *504*

AUTHOR INDEX 511

Billon, J. P., 429, *444*
Bingham, R. C., 278, *328*, 357, 419
Binsch, G., 113, *122*
Bird, C. W., *120*
Birkhimer, E. A., 136, *189*
Bissing, D. E., 350, *421*
Blanc-Guenee, J., 373, *419*
Blanchard, E. P., 66, 87, *120*, *125*
Blicke, F. F., 349, *416*
Bloch, D. R., 477, *506*
Blois, M. S., 427, *444*
Blomquist, A. T., 450, *504*
Bloomfield, J. J., 349, 357, *421*
Boatman, S., 367, 371, *416*, *417*
Bodenbanner, K., 251, *331*
Boer, F. P., 198, 227, *325*
Boggs, R. A., 340, 354, *416*
Bohl, W. R., 403, *420*
Boldt, P., *120*
Bollinger, J. M., 254, 267, *327*, *332*
Bolton, J. R., 427, 432, *444*
Bonner, W. A., 247, *325*, *326*
Book, G., 407, *421*
Booth, W. T., Jr., 356, *420*
Boozer, C. E., 13, *58*
Bordwell, F. G., 398, 399, *416*
Borner, P., 251, *331*
Bost, R. O., 67, *120*
Bottini, A. T., 453, *504*
Bouma, B., 496, *508*
Bourguel, M., 358, *419*
Bowers, K. W., 425, 432, *444*
Bowes, G., 130, 131, 132, *191*
Boyd, J. D., 298, *327*
Boyd, R. H., 265, *325*, 432, *444*
Boyd, S. D., 440, *445*, *446*
Boyer, J. H., 143, 152, 156, *189*
Boys, S. F., 70, *121*
Boyton, H. G., 256, *327*
Bradford, C. W., 461, *504*
Bradley, J. N., 81, *121*
Brandon, R. W., 75, *121*
Brasen, W. R., 388, *417*
Braun, W., 119, *121*
Braye, E. H., 379, *416*
Bredeweg, C. J., *58*
Brenneisen, P., 297, *325*
Brenner, M., 143, 152, 163, *191*
Brenner, S., 377, *419*
Brenninger, W., 431, *445*

Breslow, D. S., 131, 141, 159, 160, 161, 162, 186, *189*, *191*, 347, *417*
Breslow, R., 225, 233, 266, *325*, 394, *416*
Brewer, J. P. N., 484, *504*
Brewster, J. H., 382, *416*
Briody, R. G., *328*
Brizzolara, A., 352, 361, *421*
Broaddus, C. D., 403, *416*
Brockway, L. O., 199, *330*
Brockway, N. M., *123*
Broeker, K., 424, *444*
Brook, A. G., 324, *325*, 362, *416*
Brookhart, M., 313, *325*, *330*
Brouwer, D. M., 291, *325*
Brower, K. R., 141, *189*
Brown, B. B., 136, *191*
Brown, G. B., 361, *416*
Brown, H. C., 39, *57*, 221, 239, 241, 269, 273, 275, 276, 278, 289, 299, 306, 307, 308, 310, 313, 317, 319, *325*, *326*, *329*
Brown, K. C., 87, *120*
Brown, R. A., 142, 160, *188*
Brown, R. F. C., 468, 472, 490, *504*
Brunner, H., 406, *417*
Brunson, H. A., 351, 352, *416*
Bruylants, A., 35, *57*
Bryan, R. F., 227, *326*
Bryant, D. R., 370, 371, 372, *416*, *422*
Bryukhovestskaya, L. V., 441, *447*
Buchanan, G. L., 180, *189*
Buck, J. S., 349, *418*
Buddrus, J., *126*
Bunce, N. J., 44, *59*
Bung, W., 6, *58*
Bunnett, J. F., 363, *416*, 440, *445*, 454, 495, 496, 497, 498, *504*
Bunton, C. A., 286, 302, *326*
Bunyan, P. J., 128, *189*
Burckhardt, U., 464, *507*
Burgess, J. R., 436, *445*
Burgmaier, G. J., 47, *59*
Burke, J. J., 262, *324*
Burkett, A. R., 371, *419*
Burlitch, J. M., 66, 87, 88, 94, *125*
Burns, R. P., 116, *121*
Buschoff, M., 115, *123*
Buschow, K. H. J., 431, 432, *444*
Bush, D.C., 352, *419*
Bushick, R. D., 266, *324*

Buss, V., 315, *335*
Buting, W. E., 454, *504*
Butler, P. E., 150, *190*
Buttery, R. G., 96, 98, *121*
Bykhovskaya, E. G., 128, *190*

C

Cadogan, J. I. G., 128, 152, 163, 171, 172, 179, *189*, 466, 486, *504*
Callender, D. D., 483, *505*
Callister, J. D., 81, 95, *120*
Calvin, M., 156, *191*
Cameron-Wood, M., 179, *189*
Campbell, A., 382, *416*
Campbell, B. K., 402, *416*
Campbell, C. D., 150, 186, *189*, 484, *505*
Campbell, K. N., 402, *416*
Cannon, G. W., 361, *416*
Caplier, I., 379, *416*
Capon, B., 302, *326*
Carboni, R. A., 136, 137, *189*
Caress, E. A., 138, 139, *191*
Carey, F. A., 67, *121*
Carlsmith, L. A., 450, 451, 454, 478, *507*
Carney, J., 232, *326*
Carney, P. A., 377, *422*
Carr, R. W., 69, 98, *121*, *122*
Carrington, A., 5, 7, *57*, 430, *444*
Carroll, J. T., 295, *330*
Carter, M. K., 430, *444*
Casanova, J., 249, *326*
Casey, C. P., 340, 354, *416*
Cassimatis, D., 221, *334*
Casson, J. E., 97, *120*
Cast, J., 390, *416*
Castle, J. E., 136, 137, *189*
Castro, C. E., 66, *121*
Catlin, W. E., 346, *417*
Caubere, P., 481, 482, *505*
Cauquis, G., 429, *444*
Cava, M. P., 490, *505*
Cayen, C., 357, *419*
Chadwell, H. M., *419*
Challand, S. R., 173, 178, *188*
Chambers, R. D., 498, *505*
Chan, L. L., 339, *416*
Chan, T., 383, *418*
Chang, H. W., 266, *325*
Chang, K.-Y., 432, *447*
Chang, S., 207, *328*

Chaput, E. P., 402, *416*
Chattaway, F. D., 3, *57*
Chaudhuri, N., 96, 98, *121*
Chaykovsky, M., 350, *416*
Chen, F., 436, *446*
Chesick, J. P., 100, *121*
Chiang, Y., 267, *330*
Childs, R. F., 301, *326*
Chloupek, F. J., 269, 280, 307, 308, *325*, *326*
Cholod, M. S., 84, 108, 109, 110, 111, *125*
Chow, B. F., 231, *326*
Christie, J. B., 247, *326*
Christmann, A., 128, 141, 148, *190*
Christmann, F., 249, *328*
Christova, T., 369, *418*
Chu, T. L., 427, *448*
Chu, W., 233, *325*
Churchill, M. R., 475, *504*
Ciabattoni, J., 485, 488, *505*
Ciappenelli, D. J., *420*
Ciganek, E., 99, 106, 107, 113, *121*
Clardy, J., 459, 468, *504*
Clark, H. C., 66, *121*
Clark, M. T., 391, *416*
Claxton, T. A., 479, *504*
Clifford, P. R., 320, *331*
Clippinger, E., 205, *335*
Clopton, J. C., 81, *125*
Closs, G. L., 68, 75, 85, 87, 90, 95, 99, 102, 103, 106, 107, 109, *121*, *122*
Closs, L. E., 85, 87, 95, 106, *121*, *122*
Closson, W. D., 145, *189*, 435, *444*, 445
Coates, G. E., 339, *416*
Cocivera, M., 290, *326*
Cockerill, A. F., 86, *120*
Coe, P. L., 483, *505*
Coffey, R. S., 97, 99, 106, *122*
Coffin, B., 171, *189*
Coke, J. L., 308, *326*, *329*
Cole, T. M., 300, 302, *334*
Coleman, G. H., 355, *416*
Collins, C. J., 246, 247, 248, 293, 315, *325*, *326*
Colonge, J., 355, *416*
Colson, J. G., 182, *190*
Combrisson, J., 429, *444*
Comeford, J. J., 132, *189*
Comisarow, M. B., 200, 215, 282, 293, 310, 318, 319, *331*, *332*, *333*
Commeyras, A., 221, 222, 319, *332*
Commeyras, J. R., 219, *332*

AUTHOR INDEX

Conant, J. B., 231, *326*, 345, *416*
Condon, F. E., 295, *325*
Conradi, J. J., 430, *448*
Conrow, K., 297, *326*
Cook, A. G., 113, *121*
Cook, C. D., 6, *57*, 461, 462, *505*
Cook, D., 221, 228, *326*
Cook, J., 486, *504*
Cooke, B. J. A., 313, *334*
Cooke, I., 221, *326*
Cooper, A., 152, *189*
Cooper, G. D., 398, *416*
Cope, A. C., *326*, 345, 349, 359, *416*
Coppinger, G., 6, *57*
Corey, E. J., 67, *121*, 249, *326*, 350, 362, 371, 372, 399, *416*
Corfield, P. W. R., 399, *416*
Corse, J., 239, 319, *335*
Cotter, A. K., 212, *333*
Cotter, R. J., 131, *189*
Cotton, W. D., 387, *420*
Couch, M. M., 106, *121*
Coulson, C. A., 451, *505*
Cowan, D. O., 106, *121*
Cowell, G. W., 66, 79, 81, 82, *121*
Cox, L. E., 229, *326*
Cox, R. H., 436, *445*
Coyle, J. J., 99, 109, *121*
Craig, D., 355, *416*
Cram, D. J., 293, 307, 308, *326*, *334*, 337, 338, 340, 342, 402, *416*
Crandall, J. K., 67, *121*
Creger, P. C., 370, *416*
Crews, P., 402, *505*
Criegee, R., 459, *505*
Cristol, S. J., 314, *326*
Crow, W. D., 114, *121*, 164, 165, *189*, 490, *504*
Crowley, J. E., 485, 488, *505*
Crumbliss, A. L., 84, *122*
Cryberg, R. L., 184, *190*
Cue, B. W., 186, *188*
Curnutte, B., 218, *327*
Currie, C. L., 143, 170, *189*
Curtin, D. Y., 284, 285, 290, *326*, *335*, 407, *420*
Curtius, T., 128, 180, *189*

D

Damrauer, R., 88, 94, 110, *125*
Danby, R., 35, *57*

Danen, W. C., 86, *124*, 432, 438, 439, *446*, *447*
Darragh, K. V., 88, *125*
Darwent, B. de B., 143, 170, *189*
Das, B. P., 163, 164, *192*
Das, M. R., 427, *444*
Daub, G. H., 348, *418*
Dauben, H. J., Jr., 255, 300, *327*
Dauben, W. G., 351, *416*
Davidson, J. B., 453, *507*
Davidson, W. B., 463, *506*
Davies, T. M., 439, *445*, *446*
Davis, B. A., 128, 131, 132, 137, 138, 139, 142, 143, 160, 171, 182, *188*
Davis, N. R., 362, *416*
Davoust, C. E., 75, *121*
Day, A. R., 380, *417*
Deadman, W. G., 308, *331*
DeBoer, E., 430, *448*
DeCamp, M. R., 503, *506*
Deeming, A. J., 461, *505*
Dehl, R., 427, *444*
Dehmlow, E. V., 112, *121*
DeJongh, D. C., 490, *505*
Dekker, M., 150, *189*
DeMaria, C., 116, *121*
DeMember, J. R., 219, 221, 222, 319, *332*
DeMore, W. B., 90, *121*
Denault, G. C., 165, *192*
den Hertog, H. J., 503, *505*
den Hollander, J. A., 95, *123*
Denney, D. B., 100, *121*
Deno, N. C., 212, 214, 221, 223, 224, 226, 243, 255, 264, 266, 267, 274, 286, 287, 297, 298, *327*, 443, *444*
Derqunov, Y. I., 166, *188*
de Rossi, P. H., *504*, *505*
Derozier, N., 481, *505*
Dessau, R. M., 430, 437, 441, 442, *444*, *445*
Dewar, M. J. S., 195, 255, 308, *325*, *327*, 441, *444*
Deyrup, A. J., 264, *329*
Deyrup, C. L., 320, *325*
Diamanti, J., 357, *419*
Diaz, A., 205, 308, *327*, *335*
Dickinson, R. A., 3, *57*
Dieleman, J., 431, 432, *444*
Diesen, R. W., 132, *189*
Dietz, R., 441, *444*
Dieu, C., 35, *57*

Dilthey, W., 308, *327*
Dimroth, K., 322, *327*
Dimroth, O., 430, *444*
Dinklage, R., 200, *327*
Dirks, J. E., 402, *415*
Dittrich, B., 451, *506*
Doak, K. W., 48, *57*
Dobson, J. E., 460, *505*
Doering, W. von E., 63, 93, 94, 96, 97, 98, 99, 106, *121*, 143, 163, 185, *189*, 218, 226, 256, 266, 280, 281, *327*, 365, 391, *416*, *417*
Dolbier, W. J., 85, 100, *122*
DoMinh, T., 115, 119, *121*, *126*
Domnin, N. A., 472, *505*
Donaldson, M. M., 278, *326*, *333*
Donath, W. E., 284, *332*
Doran, M. A., 223, 226, *332*, *334*
Dorsch, H. L., 473, 475, *508*
Dotz, K. H., 89, *121*
Dowell, A. M., 91, *122*
Doyle, M. P., 254, *327*
Dravnieks, F., 430, *444*
Drowart, J., 116, *121*
Druey, J., 384, *418*
Dubois, J. E., *329*
Duff, J. M., 362, *416*
Duggleby, P. M., 262, *324*
Duncan, J. H., 81, *125*
Dunn, J. L., 381, *417*
Dunnavant, W. R., 348, 379, *417*, *419*
Durett, L. R., 67, *124*
Durr, H., 480, 492, *508*
Durst, T., 350, 371, 372, *416*
Dvoretzky, I., 67, 82, 96, *123*, *124*
Dyas, C., 200, 317, *330*, *332*
Dyckes, D., 308, *327*
Dye, J. L., 431, *445*
Dyer, M. C. D., 376, *415*

E

Earhart, H. W., 256, *327*
Earl, G. W., 439, *445*, *446*
Easthan, J. F., 390, 391, *421*
Ebel, H. F., 468, *505*
Eberle, D., 494, *506*
Eberson, L., 308, *327*
Edens, R., 93, 94, 100, *121*
Eder, T. W., 69, *121*
Edinger, J. M., 380, *417*

Edmison, M. T., 128, *190*
Edwards, O. E., 140, 185, *188*, *189*
Edwards, W. R., 256, *327*
Eiben, K., 428, *444*
Eisert, M. A., 88, *125*
Eiszner, J. R., 3, *59*, 97, *126*
Eliel, E. L., 285, *327*
Ellis, J. A., 477, *505*
Ellis, R. C., 361, *416*
Elphimoff-Felkin, I., 67, *121*
Embree, H. D., 464, *505*
Emmons, W. D., 350, *421*
Engel, R. R., 116, 117, 118, *125*
Engleman, C., 317, *332*
Engelman, K., 200, *330*
Englemann, H., 2, *58*
Englemann, T. R., 405, 407, 408, *421*
Engels, J., 130, 131, *191*
Ennis, C. L., 72, *123*
Erickson, B. W., 362, *421*
Erickson, K. I., 473, *505*
Erickson, W. F., 385, *415*
Eriks, K., 227, *327*, *328*
Ernst, C. E., 405, 407, *421*
Etter, R. M., 109, *125*
Etzemüller, J., *120*
Evans, A. G., 226, 259, *325*, *327*
Evans, J. C., 209, 218, 221, *327*
Evans, J. C. W., 364, *417*
Evans, T. C., 210, 218, *332*
Evans, W. L., *327*

F

Fagley, T. F., 136, 137, *189*
Fagone, F. A., 117, *125*
Fainburg, A. H., 205, 261, 263, *327*, *335*
Fairbourne, A., 364, *417*
Fanta, G. F., 46, *57*
Fanta, P. E., 295, *330*
Farge, D., 172, *191*
Fargher, J. M., 158, 161, *190*
Farnum, D. G., 209, *328*
Farrington, J. A., 430, *444*
Fateley, W. G., 218, *327*
Favorski, A. E., 427, *444*
Fawson, H. R., 364, *417*
Feldman, M., 232, 233, *327*, *328*
Felletschin, G., 381, *422*
Feng, R., 44, *58*
Fenton, S. W., *326*

Ferrington, T. E., *57*
Ferris, J. P., 443, *444*
Fessenden, R. W., 428, *444*
Feuer, B. I., 160, 161, 163, *191*
Fianu, P., 6, *57*
Field, F. H., 203, 272, 273, *328*
Fields, E. K., 113, *121*, 450, 452, 463, 466, 467, 468, 470, 472, 479, 482, 485, 486, 490, 491, 500, 501, *505*, *506*
Fieser, L., *505*
Fieser, M., *505*
Finelli, A. F., 352, *418*
Finger, C., *126*
Finkelstein, M., 280, *328*
Finnegan, R. A., 389, 408, *417*
Fischer, E., 425, *445*
Fischer, E. O., 89, *121*, 406, *417*, *420*
Fishbein, R., 443, *444*
Fisher, I. P., 77, *121*, 468, *505*
Fitzgerald, W. P., 407, *415*
Fleet, G. W. J., 187, *191*
Fleming, J. S., 431, *445*
Fletcher, R. S., 275, 289, *325*
Flynn, R. R., 498, *504*
Flythe, W. C., 232, 233, *327*, *328*
Folk, T. L., 377, *420*
Font, J., 119, *126*
Foote, C. S., 277, *328*
Foote, R. S., 375, *415*, *417*
Forbes, W. F., 430, 431, *445*
Forrester, A. R., 432, *445*
Fort, A. W., 394, *417*
Fort, R. C., Jr., 280, 282, 284, 318, 319, *328*, *333*
Foster, J. M., 70, *121*
Fowler, R. H., 259, *325*
Fraenkel, G. K., 209, *328*, 427, 429, 430, *444*, *445*, *446*
Frankel, M., 436, *448*
Frankham, D. B., 86, *120*
Franklin, J. L., 203, 259, 272, 273, *328*
Frantz, A. M., 227, *328*
Franzen, V., 66, 93, 94, 100, 114, *121*, *122*
Frater, C., 114, *122*
Freamo, M., 136, *191*
Freedman, H. H., 227, *328*
Freidenreich, P., 109, *124*
Freidlina, R. Kh., 23, *57*
Frey, H. M., 98, 105, *122*
Fricke, H., 113, *124*

Fridinger, T. L., 403, *417*
Friedman, L., 80, 111, *122*, *123*, 253, *328*, 461, 462, 464, 489, *505*
Friedman, N., 221, 254, 255, *327*, *331*, *332*
Friedolsheim, A., 401, *420*
Friedrich, E. C., 111, *124*, 317, *329*
Friedrich, H. J., 431, *445*
Fries, J. A., *418*
Frister, F., 430, *444*
Fritsch, J. M., 429, 437, *447*
Fritze, P., 475, *508*
Fry, A., 141, 155, *189*
Fry, J. L., 278, 279, 290, 291, 314, 315, *328*, *333*
Fuhrmann, G., 454, 465, *508*
Fujita, K., 461, 492, *507*
Fujita, S., 67, *124*
Fujito, T., 113, *124*
Fuller, A. E., 34, *57*
Fuller, G., 20, *57*
Funakubo, E., *122*
Furukawa, J., 87, 110, 113, *124*

G

Gadecki, F. A., 255, *327*
Gagosian, R. B., 395, *420*
Gainsford, G. D., 461, *504*, *505*
Galinao, F. R., 435, *446*
Gal'perin, V. A., 166, *188*
Gander, R., 346, *420*
Ganellin, C. R., 255, *327*
Gapski, G. R., 443, *444*
Garcia, Z., 3, *58*
Garcia-Martinez, N., 471, *506*
Gardini, G. P., 443, *446*
Gardlund, S. L., 380, *420*
Gardlund, Z. G., 380, *420*
Gardner, D. M., 430, *445*
Gardner, D. V., 472, *504*
Garner, A. Y., 102, 108, 109, *125*
Garst, J. F., 435, 436, *445*
Gaspar, P. P., 69, 90, 114, *120*, *122*
Gassman, P. G., 182, 184, 186, *190*, 322, *328*
Gaulhier, R., 35, *57*
Gay, R. L., 371, 372, 407, 413, *415*, *417*, *421*
Geels, E. J., 436, *447*
Geissler, G., 391, *420*
Genge, C. A., 160, *189*
George, R. S., 346, *422*

Gerlock, J. L., 427, *445*
Gerstl, R., 109, *124*
Gerwe, R. D., 443, *444*
Geske, D. H., 427, 431, *446*
Geuther, A., 62, *122*
Ghosez, L., 113, *122*
Gibbons, W. A., 74, *122*
Gierer, P., 389, 407, 409, *421*
Gilbert, B. C., 428, *446*
Gilchrist, T. L., 129, 156, *188*, *190*, 462, 475, 490, *505*, *507*
Gill, J. B., 451, *505*
Gillespie, R. J., 209, 265, *328*
Gilman, H., 341, 342, 346, 347, 379, 403, 407, *417*
Gilmore, W. F., 396, *418*
Gilmour, N. D., 23, *59*
Ginsburg, D., 351, *416*
Giumanini, A. G., 381, 385, *419*, 495, *505*
Given, P. H., 427, *444*
Glarum, S. H., 427, *445*
Gleave, J. L., 197, 198, *328*
Gleiter, R., 70, *122*
Goering, H. L., 207, *328*
Goh, S. H., 85, 87, *122*
Gohlke, R. J., 466, *506*
Gold, E. H., 223, *329*
Gold, H., 256, *331*
Gold, V., 196, 200, 226, 264, 289, 299, 324, *325*, *328*
Goldberg, S. I., 404, *417*
Goldman, L., 302, *328*
Goldschmidt, S., 249, *328*
Goldstein, M. J., 85, 100, *122*, 312, *328*
Golstein, J. P., 116, 118, *125*
Gomburg, M., 2, *57*, 196, 200, 229, *328*
Gomes de Mesquita, A. H., 227, *328*
Gordon, M. E., 66, 88, 94, *125*
Gornowicz, G. A., 377, 378, *417*, *422*
Gortler, L. B., 435, *445*
Gosselink, E. P., 67, *123*
Gotshal, Y., 136, 137, *191*
Goubeau, J., 150, *190*
Gould, E. S., 240, *328*
Gowling, E. W., 459, *505*
Gramas, J. V., 76, *120*
Granick, S., 426, 429, *446*
Grant, D. M., 214, 215, *328*, *329*
Graveling, F. J., 462, 475, 490, *505*, *507*
Gray, H. B., 145, *189*

Greeley, R., 300, 302, *334*
Green, G. S., 439, *446*
Green, S. A., 66, *122*
Greenburg, G. Y., 185, *192*
Greenberg, S., 48, *57*
Greenstreet, C. H., 286, *326*
Grelecki, C. J., 136, *191*
Griffin, G. W., 67, *120*, *124*, 156, *191*
Griffiths, J. S., 376, *417*
Grigat, E., 255, *332*
Grignard, V., 365, *417*
Grinham, A. R., 491, *504*
Grinter, R., 286, *328*
Grob, C. A., 297, 298, *325*, *328*
Gross, M. H., 228, *332*
Grovenstein, E., Jr., 392, 393, *417*
Groves, P. J., 226, 267, *327*
Grubert, H., 406, *420*
Grummitt, O., 345, *418*
Grunson, A., *504*
Grunwald, E., 239, 262, 263, 302, 319, *328*, *335*
Guhn, G., 451, *506*
Guillamet, G., 482, *505*
Gumby, W. L., 342, *422*
Gunning, H. E., 115, 119, *121*, *126*
Günter, F., 427, *446*
Guss, J. M., 461, *504*, *505*
Gustavson, G., 299, *328*
Gutsche, C. D., 97, 99, 106, *122*

H

Haaf, W., 255, *330*
Haberfield, P., 342, *416*, 496, *505*
Haberstadt, M. L., 69, *122*
Hackler, R. E., 385, *415*
Haeussermann, O., 453, *506*
Hafer, K., 267, *327*
Hafner, K., 174, *190*, 225, *329*
Hagen, E. L., 217, 292, *333*
Hakan, B. L., 453, *507*
Hall, J. H., 128, 130, 137, 158, 161, *190*, *191*
Hall, R. E., 278, *328*
Haller, J. F., 136, *190*
Hammar, W. J., 278, *325*
Hammel, O., 197, *331*
Hammett, L. P., 204, 240, 264, *329*
Hammond, G. S., 12, 13, 14, 31, *57*, *58*, *59*, 69, 90, 106, 107, *121*, *122*, *123*, 310, *329*

AUTHOR INDEX

Hampton, K. G., 367, 368, 375, 376, 377, *417*, *418*
Hamrick, P. J., Jr., 370, 378, *417*
Hancock, E. M., 345, *416*
Hanna, S. B., 86, *122*
Hansen, R. L., 250, *329*
Hansley, V. L., 427, *447*
Hantzsch, A., 62, *122*, 196, *329*, 463, 464, *506*
Harbordt, C. M., 324, *331*
Hardy, W. B., 179, *189*
Harger, M. J. P., 486, *504*
Härle, H., 465, *508*
Harmon, K. M., 255, *327*
Harned, H. S., 230, *329*
Harper, J. J., 278, 280, 308, *326*, *330*
Harris, C. M., 366, 367, *417*
Harris, J. M., 278, 308, 309, *328*, *329*
Harris, R. F., 118, 119, *122*, *125*
Harris, T. M., 366, 367, 368, 369, 371, 373, *416*, *417*, *419*, *422*
Harrison, A. G., 271, *329*
Harrison, A. M., 72, *123*
Harrison, J. F., 69, 70, *122*
Hart, E. J., 51, *58*
Hart, H., 284, *329*, 431, *445*
Hartwell, J. L., 185, *192*
Haselbach, E., 452, *506*
Hass, H. B., 438, 439, *445*, *447*
Hassall, C. H., 128, *190*
Haszeldene, R. N., 88, *120*
Hatch, M. J., 402, *416*
Hauser, C. F., 351, 353, *418*
Hauser, C. R., 346, 347, 348, 351, 353, 356, 357, 358, 360, 363, 366, 367, 368, 369, 370, 371, 372, 373, 374, 375, 376, 377, 379, 383, 385, 386, 387, 388, 398, 407, 411, 412, 413, 414, *415*, *416*, *417*, *418*, *419*, *420*, *421*, *422*
Hausser, K. H., 7, *58*, 432, *445*
Hayashi, S., 486, *506*
Hayashi, Y., 388, *417*
Haygood-Farmer, J., 320, *325*
Heacock, J. F., 128, *190*
Heaney, H., 478, 483, 484, *504*, *506*
Heck, R., 205, *335*
Hederich, V., 235, *331*
Heene, R., 430, *444*
Hegerty, A. F., *329*
Hehre, W. J., 315, *335*
Heiba, E. I., 430, 437, 441, 442, *444*, *445*

Heicklen, J., 66, *122*
Heilbronner, E., 256, 266, *331*, *332*
Heinriche, G., 486, *507*
Heinz, G., 111, *125*
Hellmann, H., 494, *506*
Henderson, A. T., 6, *58*
Henderson, G. G., 282, *329*
Hendley, E. G., 391, *416*
Hendrickson, B. W., 46, *57*
Hendrix, J. P., 407, *421*
Hendry, D. G., 90, *124*
Henery-Logan, K. R., 403, *417*
Henne, A. L., 35, *58*
Henoch, F. E., 375, 376, 377, *417*, *418*
Henrick, C. A., 354, *415*
Herbst, R. M., 348, *418*
Hercules, D. M., 229, *326*, *329*
Herr, R. W., 359, *418*
Herschmann, C., 221, *326*
Herzberg, G., 72, 73, 117, *122*
Herzog, B. M., 98, *122*
Hess, B. A., 459, *506*
Hess, C. G., 323, *329*
Hetzner, H. P., *415*
Hey, D. H., 2, *58*
Heyrovský, J., 431, *445*
Hiatt. R. R., 12, *57*
Hibbert, P. G., 486, *504*
Hickinbottom, W. J., 34, *57*
Hill, J. W., 158, 161, *190*
Hill, R. K., 383, *418*
Hine, J., 62, 84, 91, *122*
Hinkamp, F. B., 35, *58*
Hinton, R. C., 465, *506*
Hirota, N., 427, 432, 436, *445*
Hirschhorn, H. H., 282, *333*
Hisshon, J. M., 430, *445*
Hite, G., 396, *421*
Ho, S.-Y., 66, 96, 98, *122*
Hoberg, H., 88, *122*
Hodge, J. D., 212, 221, 255, 267, 298, *327*
Hoeg, D. F., 84, *122*
Hoergerlee, K., 150, *190*
Hofeditz, W., 2, *58*
Hofer, O., 409, *418*
Hoffman, R., 451, *504*, *508*
Hoffmann, A. K., 63, *121*
Hoffmann, A. V., 6, *58*
Hoffmann, R., 70, 73, 76, 81, 103, 105, *122*, *126*, 312, *328*

Hoffmann, R. W., 450, 451, 468, *505, 506*
Hofstra, A., 267, *330*
Hogen-Esch, T. E., 432, *445*
Hoijtink, G. J., 226, *334*, 431, 432, *444*
Holcomb, W. D., 142, 166, *188*
Hollander, J. M., 229, *329*
Holmes, H. L., 359, *416*
Holmes, J., 390, *416*
Holness, N. J., 274, *335*
Holy, N. L., 439, *445*, *446*
Homer, J. B., 77, *121*
Homeyer, J., 425, *446*
Hong, C., 495, *504*
Honour, R. J., 111, *122*
Hooz, J., 404, *415*
Hopper, S. P., 88, *125*
Horn, D. E., 280, *330*
Horne, R. J., 130, 131, 132, *191*
Horner, L., 16, *58*, 128, 141, 148, *190*
Horning, E. C., 352, *418*
Horwell, D. C., 156, *188*
Houlihan, W. J., 344, *420*
House, H. O., 354, 359, 396, *416*, *418*
Houser, J. J., 212, 267, *327*
Houser, R. W., 400, *420*
Houston, A. H. J., 382, *416*
Hover, H., 266, *325*
Howard, J. A., 51, *58*
Howard, R. D., 101, *120*, 286, *325*
Howe, R., 317, *329*
Howell, C. F., 349, *416*
Howes, B. W. V., 226, *328*
Hruby, V. J., 66, *122*
Hrutfiord, B. F., 496, *504*
Hsu, K. C., 465, *504*
Hubel, W., 379, *416*
Hudson, B. E., Jr., 356, *417*
Hughes, A. N., 97, *120*
Hughes, E. D., 197, 198, 206, 286, 289, *325*, *326*, *328*, *329*
Huisgen, R., 66, 79, 113, *122*, 128, 159, 163, 169, 173, *188*, *190*, *192*, 454, 481, 482, *506*
Hummel, K. F., 106, 107, *122*
Humphlett, W. J., 370, *420*
Hung, Y.-Y., 179, *190*
Hünig, S., 235, 302, *329*, 431, *445*
Hunt, R., 475, *504*
Hunt, R. L., 441, *444*
Hutchison, C. A., 74, 75, 76, *121*, *122*, 431, *445*

Hutton, R. S., 76, *126*
Huyser, E. S., 3, 10, 13, 20, 22, 23, 39, 44, 48, *58*, *59*
Hyman, H. H., 267, *330*

I

Ichibori, K., 101, *120*
Ichikawa, K., 269, *325*
Ide, W. S., 349, *418*
Impasto, F. J., 340, *418*
Indictor, N., 16, *59*
Ingold, K. U., 46, 51, *58*
Inghram, M. G., 116, *121*
Ikeler, T. I., 170, *192*
Ingalls, R. B., 136, *191*
Ingold, C. K., 197, 198, 204, 206, 260, 286, 289, 304, *325*, *326*, *328*, *329*
Ingold, K. U., 443, *447*
Ingram, D. J. E., 427, *444*
Ipatieff, V. N., 298, *329*
Ireland, P. R., 461, *505*
Irie, T., 320, *334*
Ishikawa, N., 486, *506*
Iskander, Y., 86, *122*
Ito, Y., 92, *124*
Itoh, K., 75, 76, *123*
Ivanoff, D., 369, *418*
Ivanova, T. M., 441, *447*
Iwamura, H., 495, *506*
Iwamura, M., 495, *506*

J

Jablonski, J. M., 483, *506*
Jablonski, R. J., 308, *329*
Jack, J. J., 229, *326*
Jackman, L. M., 209, *329*, 340, *418*
Jacknow, B. B., 36, *59*
Jackson, R. A., 297, *325*
Jacobs, T. L., 358, *418*
Jacox, M. E., 73, *123*
Jacqueman, W., 391, *420*
Jaffe, H., 240, *329*
Jagar-Grodzinski, J., 258, *332*
Jäger, G., 150, *190*
Jander, J., 128, *190*
Janke, W., 426, *446*
Jaruzelski, J. J., 226, 264, *327*
Jarvis, B. B., 399, *416*
Jauhal, G. S., 461, 462, *505*
Jayawant, M., 156, 158, *192*

Janzen, E. G., 424, 427, 428, 429, 439, *445*, *447*
Jefferson, G. D., 357, *418*
Jemison, R. W., 384, *418*
Jenkins, C. L., 249, 250, *329*
Jennen, J. J., 198, 200, *329*
Jennings, C. A., 405, 407, 408, 410, 412, *418*, *421*
Jenny, E. F., 384, *418*
Jensen, E. D., 231, *329*
Jensen, E. V., 11, *58*
Jensen, F. R., 255, 319, *329*
Jensen, L. H., 228, *334*
Jernow, J. L., 71, *124*
Jeuell, C. L., 316, *332*
Jewett, J. G., 314, 315, *332*, *333*
Ji, S., 435, *445*
Jit, P., 498, *506*
Johns, J. W. C., 72, 73, *122*
Johnson, A. W., 66, *122*, 388, *418*
Johnson, C. R., 359, *418*
Johnson, J. R., 345, 348, *418*
Johnson, J. R., Jr., 407, *421*
Johnson, W. S., 348, *418*
Joines, R. C., 114, *122*, *124*
Jolly, W. L., 229, *329*
Jones, D. A. K., 218, *330*
Jones, D. H., 502, *506*
Jones, F. N., 411, *418*
Jones, G., 345, *418*
Jones, J. R., 337, *418*
Jones, M., 67, 72, 106, 107, 110, 114, *120*, *122*, *123*
Jones, M., Jr., 451, 503, *506*
Jones, M. G., 308, *326*, *329*
Jones, M. T., *445*
Jones, P. C., 377, 378, 404, *422*
Jones, P. F., 362, *416*
Jones, P. R., 427, *445*
Jones, R. G., 341, *417*
Jones, R. R., 452, *506*
Jones, W., 405, *421*
Jones, W. M., 72, 114, *122*, *123*, *124*, 251, *329*
Jordan, R. W., 435, *446*
Jores, G. G., 2, *58*
Jorgenson, M. J., 356, *418*
Joschek, H. I., 66, *122*
Jugelt, W., 114, *123*
Julian, P. L., 357, *418*
Jungk, H., 299, *329*
Jurewicz, A. T., 80, *122*, *123*

K

Kabitzke, K. H., 67, *125*
Kahlert, B., 454, *507*
Kahn, A. A., 152, 159, 169, *190*
Kaiser, E. M., 346, 347, 351, 353, 360, 371, 372, 373, 374, 375, 376, 377, *418*
Kaiser, E. T., 249, *326*, 432, *447*
Kaiser, W., 174, *190*
Kampmeier, J. A., 290, *326*
Kanematsu, K., 500, *507*
Kanō, H., 163, *191*
Kantor, S. W., 385, 388, *417*, *418*
Kaplan, F., 114, *123*
Kapps, M., 101, *123*
Kaptein, R., 95, *123*
Karabatsos, G. J., 290, 291, 314, 315, *328*, *329*
Karplus, M., 214, *329*
Kato, H., 493, *508*
Kato, M., 488, *507*
Kato, T., 498, *504*
Katz, T. J., 223, *329*
Kauer, J. C., 137, *189*
Kauffmann, T., 499, 500, 501, 502, *506*
Kaufman, R. J., 376, *418*
Kaupp, G., 432, *447*
Kawabata, N., 87, 110, 113, *124*
Kazakova, V. M., 427, *447*
Keating, J. T., 203, 244, 246, *329*
Keeffe, J. R., 270, *333*
Kehrmann, F., 196, *330*
Keith, L. H., 404, *417*
Kekule, A., 453, *506*
Keller, L. S., 481, 492, *507*
Kelly, D. P., 316, *332*
Kelly, W. J., 308, *331*
Kempf, R. J., 76, *120*
Kenarle, P., 271, *327*
Kende, A. S., 393, *418*
Kennedy, J., 396, *419*
Kenyon, J., 382, *416*
Kenyon, W. G., 360, *418*
Kerber, R. C., 438, 439, *445*, *446*
Kerr, C. A., 282, *329*
Kerr, J. A., 27, 28, *58*, 77, *123*
Kessar, S. V., 498, *506*
Kessick, M. A., 289, *328*

Kessler, H., 66, 82, 113, *124*
Kettle, S. F. A., 459, *505*
Kevan, L., 5, *58*
Keys, R. T., 13, *58*
Kharasch, M. S., 2, 11, *58*, 341, 346, 347, 353, 354, 359, *419*
Khmelinskaya, A. D., 441, *447*
Khotsyanova, T. L., 218, 228, *330*
Kiesel, R. J., 93, *125*
Kilmurry, L., 67, *126*
Kilpatrick, M., 267, *330*
Kim, C. J., 308, 310, *326*, *332*
Kim, J. K., 440, *445*
Kim, L., 48, *58*
Kimball, G. E., 302, *333*
Kimball, R. H., 357, *418*
Kimling, H., 475, *506*
Kimura, K., 74, 78, *123*
King, G. J., 148, *191*
King, P. S., 270, *331*
Kingsbury, C. A., 342, *416*
Kiovsky, T. E., 253, *331*
Kirby, R. H., 347, *417*
Kirk, A. G., 115, *123*
Kirmse, W., 80, 100, 101, 106, 115, 116, *123*, 128, *190*
Kispert, L. D., 200, 317, *329*, *332*
Kitahara, Y., 489, *507*
Kitaigorodskii, A. I., 218, 228, *330*
Kitayama, M., 110, *124*
Klamann, D., *126*
Klanderman, B. H., 285, *326*
Klein, J., 377, *419*
Klemann, L. P., 380, *420*
Klemchuk, P. P., 100, *121*
Kliegman, J. M., 134, 135, *191*
Kline, M. W., 382, *416*
Kloosterziel, H., *126*
Knaus, E. E., 172, *188*
Knaus, G. N., 142, 150, 151, *188*
Knight, V., 66, *122*
Knorr, R., 481, *506*
Knowles, J. R., 187, *191*
Knox, G. R., 150, *189*
Knox, L. H., 93, 99, *121*, 218, 226, 266, 281, 284, *325*, *327*
Knunyants, I. L., 128, *190*
Knutsson, L., 395, *420*
Köbrich, G., 66, 115, *123*
Koch, E., 143, *190*

Koch, H., 255, *330*
Kochi, J. K., 21, *58*, 249, 250, *329*, *330*
Koehl, W. J., 249, *330*, 441, 442, *445*
Kofron, W. G., 363, 379, *419*
Koh, L. L., 227, *327*
Kohler, B. E., 75, 76, *121*, *122*
Kohler, E. P., *419*
Kolc, J., 67, *120*
Kolker, P. L., 427, 428, *445*
Konaka, R., 429, 432, *447*
Kondo, K., 95, *124*
Kondo, S., 92, 101, *120*
Konishi, H., 82, *124*, 493, *508*
Konizer, G., 340, *422*
Koonsvitsky, B. P., 389, 405, 407, 408, *421*
Kopecky, K. R., 106, 107, *121*, *123*
Kornblum, N., 438, 439, 440, *445*, *446*
Kornblum, R. B., 276, *326*
Korzan, D. G., 436, *446*
Koudijs, A., 500, *507*
Kovar, R. F., 406, *419*
Krapcho, A. P., 91, *123*, 280, *330*, 357, *419*
Kray, W. C., 69, *121*
Krbechek, L., 178, *190*
Krbechek, L. O., 128, *191*
Krebs, A., 394, *416*, 473, 475, 477, *506*
Kreher, R., 150, *190*
Kresge, A. J., 231, 267, *330*
Krubsack, A. J., 67, *126*
Krueger, W. E., 156, *189*
Ku, A. T., 252, *332*
Ku, T., 433, 435, *447*
Kuck, V. J., 76, *126*
Kuhlmann, D. P., 494, *507*
Kuhn, D. A., 6, *57*
Kuhn, S. J., 221, 227, 228, *332*
Kulczycki, A., 67, 106, 107, *122*
Kunert, F., 251, *331*
Kupfer, O., 62, *126*
Kurkov, V. P., 51, *59*
Kurland, R. J., 212, *333*
Kuroda, H., 488, *507*
Kursanov, D. N., 218, 228, *330*
Kůta, J., 431, *445*
Kuwata, K., 431, *446*
Kwart, H., 152, 159, 169, *190*
Kwiatkowski, G. T., 350, *416*
Kyba, E. P., 138, 139, 140, 143, 144, 146, 174, *188*
Kym, O., 453, *506*

L

L'abbé, G., 129, 137, 150, *190*
Laber, G., 256, *327*
LaCount, R. B., 388, *418*
LaFlamme, P., 106, *121*
Lagercrantz, C., 428, *446*
Laird, R. K., 73, *123*
LaLonde, R. T., 249, *326*
Lam, L. K. M., 278, 279, *328, 333*
Lamb, R. C., 435, *445*
Lambert, J. L., 344, *420*
Lamm, B., 270, *333*
Lampe, F. W., 271, 323, *329, 334*
Lancelot, C. J., 278, 279, 308, 309, *326, 328, 330, 333*
Lancelot, J. J., 308, 309, *329*
Landesman, H., 352, 361, *421*
Landgrebe, J. A., 94, 115, *123*
Lane, C. A., 270, *331*
Langer, A. W., Jr., 341, *419*
Lansbury, P. T., 182, *190*, 390, *419*
Lapore, R., 352, *419*
Lappert, M. F., 82, *123*
Larson, J. W., 199, 235, 259, 295, 296, 302, 320, *324, 332*
Latham, W. A., 315, *335*
Latimer, W. M., 259, *330*
Laughlin, K. C., 298, *334*
Laughlin, R. G., 96, 98, *121*
Laurent, A., 425, *446*
Lauterbur, P. C., 213, *330*
Lawson, D. F., 427, *447*
Lazdins, I., 308, *327*
Leal, J. R., 361, *416*
Ledwith, A., 66, 79, 81, 82, *121*, 430, *444*
Lee, C. C., 307, 314, 315, 316, 318, *330, 333*
Leedy, D. W., 429, 437, *447*
Leermakers, P. A., 3, *57*, 106, 107, *123*
Leffler, J. E., 141, *190*
Leftin, H. P., *330*
Lehmann, M., 62, *122*
Lemal, D. M., 67, *123*, 156, 170, *190*
Leone, R. E., 314, 315, *330*
Lepley, A. R., 381, 385, 386, *419*
Leroi, G. E., 73, *125*
Lespieau, R., 358, *419*
Lester, C. T., 247, *326*
Letsinger, R., 9, *58*
Levene, P. A., 357, *419*
Levin, R. H., 451, *506*
Levine, R. M., 180, *189*
Levitz, M., 280, 281, *327*
Levy, D. H., 427, *446*
Levý, H. A., 199, *330*
Levy, J. F., 207, *328*
Lewis, E. S., 281, *325*
Lewis, F. D., 138, 139, 144, *190*
Lewis, F. M., 49, *58*
Lewis, I. C., 95, *125*, 430, *446*
Lewis, J., 405, *421*
Ley, K., *58*
Li, H.-M., 495, *504*
Lias, S. G., 256, *324*
Libby, W. F., 118, *126*
Lichtin, N. N., 209, 231, *330*
Lide, D. R., 73, *124*
Liebermann, C., 425, *446*
Liggero, S. H., 280, *330*
Light, R. J., 368, *419*
Liler, M., 264, *330*
Limpricht, H., 453, *506*
Lin, L. H., 450, *504*
Lin, L.-H. C., 67, *121*
Lindert, A., 347, *420*
Lincoln, D. N., 267, *327*
Lindsay, J. K., 383, *419*
Linke, S., 140, *190*
Linn, C. B., 298, *329*
Linsay, E. C., 207, *328*
Lipkin, D., 435, *446*
Lippincott, E. R., 218, *327*
Lippman, A. E., 128, *190*
Liston, T. V., 404, *415*
Litchman, W. M., 214, 215, *328*
Littler, J. S., 16, *58*
Liu, C. Y., 221, 222, 319, *332*
Liu, J. S., 212, 267, *327*
Loffgren, M., 462, *505*
Loftfield, R. B., 394, *419*
Logan, A. V., 352, *419*
Logullo, F. M., 464, 489, *505*
Löhman, L., 389, *422*
Lokengard, J., 433, 435, *447*
Lomas, J. S., *329*
Lombardo, L., 488, *506*
Lomker, F., 391, *420*
Long, F. A., 265, 267, *330, 332*
Lorenz, L., 260, *324*
Lossing, F. P., 77, *121, 125*, 271, *329*, 468, *505*

Lotsch, W., 233, *334*
Loubinoux, B., 481, *505*
Lowry, T. H., 399, *416*
Lucas, H. J., 302, 304, *335*
LucBribes, J., 219, *332*
Luckenbach, T. A., 128, *191*
Ludt, R. E., 407, *419*
Ludwig, P., 427, *446*
Ludwig, R., 430, *448*
Lukas, J., 284, *331*, *332*
Lumb, A. K., 498, *506*
Lusk, D. I., 84, *122*
Lustgarten, R. K., 313, *325*, *330*
Lüttringhaus, A., 128, *190*, 454, *506*
Lux, G. A., 435, *447*
Lwowski, W., 128, 129, 130, 131, 132, 140, 141, 148, 149, 150, 153, 154, 155, 160, 162, 163, 166, 167, 181, *190*, *191*, 302, *330*
Lynch, B. M., 179, *190*
Lyssy, T., 436, *446*

M

McBride, W. R., 430, *447*
McCollum, J. D., 286, *325*
McCorkindale, R. A., 396, *419*
McCubbin, R. J., 242, *330*
McDonald, R. S., 320, *335*
McFarland, J. W., 351, *416*
McFarlane, F. W., 308, *326*
MacGillavry, C. H., 227, *328*
McGregor, S. D., 67, *123*
McIntyre, J. S., 210, 218, *332*
Mackay, C., 116, *124*
MacKay, F. F., 221, 255, *327*
McKelvy, D. R., *324*
McKenna, J. F., 402, *416*
Mackie, R. K., 179, *189*
McKillop, A., 358, 361, *421*
Mackor, E. L., 267, 291, *325*, *330*
McLachlan, A. D., 5, 7, *57*
McLafferty, F. W., 203, *330*, 466, *506*
MacLean, C., 291, *325*
McManus, S. P., 199, 235, 295, 296, 302, 320, *330*, *332*
McMaster, I. T., 142, 159, 160, 161, 180, 186, *188*
McMillan, F. L., 21, *58*
McMillan, G R., 51, *58*
McNesby, J. R., 69, 119, *121*, *122*

McNinch, H. A., 342, *410*
McOmey, J. F. W., 472, *504*
Maercker, A., 350, *419*
Magerlein, B. J., 348, *420*
Mageswaran, S., 384, *418*
Mahler, J. E., 218, *330*
Mahler, W., 67, *123*
Mai, V. A., 94, *125*
Majerski, Z., 316, *330*, 359, *420*
Maki, A. H., 427, *446*
Mallan, J. M., 403, *419*
Malte, A. M., 363, 400, *419*
Mamantov, A., 109, 110, 111, *124*
Mancuso, N. R., 182, *190*
Maness, D. D., 251, *329*
Mango, F. D., 82, *123*
Mangold, R., 381, *422*
Mann, D. E., 73, *123*, 132, *189*
Mann, F. G., 465, *506*
Mansoor, A. M., 97, *123*
Manthey, J. W., 439, *446*
Mao, C. L., 351, 353, 369, 371, 372, 413, 415, *418*, *419*, *421*, *422*
Marcantonio, A., 160, *189*
Marchand, A. P., *123*
Marcoux, L. S., 429, 437, *447*
Marey, R., 355, *416*
Margrave, J. L., 77, *126*
Maricich, T. J., 131, 153, 154, 162, *190*
Marino, J. P., 71, *124*
Marley, R., 143, *191*
Marples, B. A., 484, *504*
Marsh, F. D., 132, 167, 186, *188*, *190*
Martin, J. C., 11, 39, *58*, *59*, 477, *506*
Martin, J. M., 67, *124*
Martin, R., 9, *58*
Martin, R. A., 284, *329*
Martin, R. H., 271, *334*
Martinez-de Bertorello, M., 471, *506*
Marvel, C. S., 355, 360, *415*, *419*
Marvell, E. N., 352, *419*
Masamune, S., 185, *190*
Mason, K. G., *506*
Mason, R., 461, *505*
Mason, S. F., 286, *328*
Massey, A. G., 460, *507*
Mataga, N., 75, 76, *123*
Mateescu, G. D., 278, *332*
Matheson, M. S., 51, *58*
Mathews, C. W., 73, *123*

Matsuda, T., 481, *506*
Matsumoto, H., 163, *191*
Matthews, W. S., 363, 400, *419*
Mattingly, T. W., Jr., 131, *190*
Matuszak, C. A., 489, *506*
Matuszesky, F., 298, *334*
Maxwell, R. J., 203, *333*
May, J. M., 432, *445*
Mayer, J., 282, *334*
Mayer, K. K., 157, *191*
Mayer, R. P., 276, *334*
Mayer, U., 473, *508*
Mayo, F. R., 2, 49, *58*
Mazur, R. H., 316, *330, 333*
Medvetskaya, I. M., *333*
Meerwein, H., 196, 197, 235, 251, 256, *330, 331*
Meier, G. F., 276, *333*
Meinwald, J., 185, *191*
Meloy, G. K., 114, *123*
Melzer, A., 384, *418*
Melzer, M. S., 407, *415*
Merer, A. J., 73, *123*
Meske-Schüler, J., 474, *508*
Messer, M., 172, *191*
Metcalfe, A. R., 434, *446*
Meth-Cohn, O., 160, *191*
Meyer, E., 67, *124*, 156, *191*
Meyer, G. M., 357, *419*
Meyer, K. H., 453, *506*
Meyer, M. W., 226, *331*
Meyer, R. B., 370, 372, *419*
Meyers, A. I., 363, *419*
Meyers, C. Y., 363, 397, 400, *419, 421*
Meyers, R. J., 427, *446*
Meyerson, S., 291, *329*, 450, 452, 463, 466, 467, 468, 470, 472, 479, 482, 485, 490, 491, 500, 501, *505, 506*
Michael, A., 425, *447*
Michaelis, L., 4, *58*, 425, 426, 429, 431, *446*
Michel, R. E., 438, *445*
Miescher, K., 128, *192*
Migita, T., 92, 101, *120*
Mihova, M., 369, *418*
Mikhailova, I. F., 493, *507*
Miklasiewicz, E. J., 352, *415*
Mikol, G. J., 143, 156, *189*
Mile, B., 427, *444*
Miles, M. L., 358, 366, 369, *419*
Millar, I. T., 465, *506*

Miller, R. G., 460, 464, 481, 494, *505, 507*
Miller, S. F., 6, *57*
Miller, W., 243, *332*
Miller, W. T., 84, *123*
Millie, P., 453, *507*
Milligan, D. E., 73, *123*
Mineo, I. C., 377, *422*
Minisci, F., 443, *446*, 482, *507*
Miocque, M., 373, *419*
Mishra, A., 169, *191*
Mitchell, J. R., 466, *504*
Mitchell, M. S., 490, *505*
Mitsch, R. A., 73, 109, *123*
Mitsuyasu, T., 481, *506*
Miwa, T., 488, *507*
Mixon, L. W., 298, *334*
Miyashita, I., 430, *448*
Mo, Y. K., 201, 258, 324, *331*
Modderman, J., 496, *508*
Moersch, G. W., 371, *419*
Moffett, M. E., 221, *332*
Moffett, R. B., 351, *419*
Mole, T., 106, *121*
Morantz, D. J., 436, *446*
Moreland, C. G., 366, *419*
More O'Ferrall, R. A., 79, *123*
Morgan, A. F., 128, *191*
Morgan, K. J., 307, 308, *326*
Morgan, L. R., Jr., 161, *189*
Moriarity, R. M., 134, 135, 143, 145, 148, 161, *191*
Moritani, I., 74, 75, 76, 78, 82, 114, 115, *122, 123, 124, 126*
Moriuti, S., 82, *124*
Morrill, T. C., 314, *326*
Morris, J. I., 436, *445*
Morrison, R. C., 436, *445*
Morschel, H., 235, *331*
Morse, B. K., 239, 319, *335*
Morton, A. A., 454, *507*
Morton, J. W., Jr., 403, 407, *417*
Moser, W. R., 82, 111, *124*
Mosher, M. W., 44, *59*
Moskowitz, H., 373, *419*
Moss, R. A., 85, 99, 102, 109, 110, 111, *121, 124*, 253, 254, *331*
Mount, R. A., 291, *329*
Mourad, M. S., 482, *505*
Mourning, H. C., 308, *326*
Moyer, W. W., 355, *419*

Mui, J. Y.-P., 66, 87, 88, 94, 110, *125*
Mulay, L. N., 4, *58*
Müller, E., 6, *58*, 66, 82, 113, *124*, 426, 427, 431, *446*
Muller, N., 214, 215, *331*
Muller-Rodloff, I., 6, *58*
Mulliken, R. S., 224, *331*
Mulvaney, J. E., 377, 380, *420*
Munemo, E. M., 387, *420*
Murahashi, S.-I., 74, 75, 76, 78, 114, *122*, *123*
Murray, R. W., 75, 76, *126*, 134, *192*
Murrell, J. N., 432, *445*
Musker, W. K., 381, 385, *420*
Musser, M. T., 439, *445*, *446*

N

Nabeya, A., 363, *419*
Nagai, T., *122*
Nair, R. M. G., 67, *124*
Nakagawa, M., 289, *326*
Nakai, T., 322, *328*
Nakano, M., 95, *124*
Nakayama, K., 101, *120*
Nakayama, U., 495, *506*
Namanworth, E., 293, *332*
Narasimhan, N. S., 414, 415, *420*
Naville, G., 266, *331*
Nazarow, I. N., 427, *444*
Neber, P. W., 401, *420*
Neckers, D. C., 20, *58*
Necsoiu, I., 255, *331*
Nef, J. U., 62, *124*
Nelson, R. F., 429, 437, *447*
Nelson, S. F., 429, *446*
Nenitzescu, C. D., 196, 255, *331*
Nerdel, F., *126*
Nesmeyanov, A. N., 389, *421*
Neta, P., 436, *448*
Neureiter, N. P., 398, *420*
Neville, O. K., 391, *416*
Newall, A. R., 81, 101, *120*
Newburg, N. R., 131, 141, 159, 160, 161, 162, *189*, *191*
Newman, D., 158, *188*
Newman, M. S., 348, 356, *420*
Newton, D. J., 377, 380, *420*
Nicholas, J. E., 116, *124*
Nicholas, R. D., 277, 282, *333*
Nickon, A., 148, 149, *191*, 344, *420*

Nich, E., 465, 495, *504*
Nielsen, A. T., 344, *420*
Nieuland, J. A., 450, *507*
Nishida, S., *122*
Nishida, T., 495, *506*
Nishimura, J., 87, 110, 113, *124*
Nishino, M., 75, 76, *123*
Nordlander, J. E., 308, *331*
Norman, R. O. C., 428, 443, *446*, *447*
Norris, J. F., 196, *331*
Norris, K. K., 440, *447*
Norup, B., 136, *189*
Noyce, D. S., 270, *331*
Noyes, W. A., 66, 96, 98, *122*
Noyori, R., 67, 82, *124*
Nozaki, H., 67, 82, 95, *124*
Nozaki, K., *58*
Nürnberg, R., 499, 502, *506*
Nyholm, R. S., 461, *504*, *505*

O

Oae, S., 303, *331*
O'Brien, C., 401, *420*
O'Brien, D. H., 251, 324, *331*, *332*
Ochrymowycz, L. A., 427, *447*
O'Connor, B. R., 290, *326*
Oda, M., 489, *507*
Oda, R., 92, *124*, 388, *417*, 461, 492, *507*
Odum, R. A., 143, 152, 163, 179, 185, *189*, *191*
Offer, R. R., *331*
Ogata, M., 163, *191*
Oglukian, R. L., 136, 137, *189*
Ohta, M., 297, *325*
Okahara, M., 154, *191*
Okamoto, Y., 39, *57*, 241, 269, *326*
Okano, M., 92, *124*
Okazaki, K., 461, 492, *507*
Okhlobystina, L. V., 427, *447*
Olah, G. A., 196, 200, 201, 204, 209, 210, 211, 212, 213, 214, 215, 216, 217, 218, 219, 221, 222, 223, 226, 227, 228, 251, 252, 253, 254, 255, 256, 258, 282, 291, 292, 293, 298, 299, 310, 316, 318, 319, 320, 324, *331*, *332*, *333*
Olah, J. A., 204, 209, 251, 252, 256, 299, *331*, *332*
Oliver, J. J., 357, *418*
Ollis, W. D., 384, *418*
Olofson, R. A., 71, *124*

Ologauer, R., 464, *506*
Oosterhoff, L. J., 95, *123*
Osawa, E., 359, *420*
O'Sullivan, W. I., 370, *420*
Othneiser, A., 424, *444*
Oulevey, G., 221, *332*
Overberger, C. G., 83, *124*, 346, *420*
Owen, B. B., 230, *329*

P

Pake, G. E., 427, *448*
Panek, E. J., 440, *447*
Paneth, F., 2, *58*
Panfilov, V. N., 441, *447*
Pannell, K. H., 324, *325*
Pappo, R., 351, *416*
Paquette, L. A., 364, 365, 399, 400, *420*, 462, *507*
Paradka, M. V., 414, *420*
Parcell, R. F., 402, *420*
Parris, G., 346, *415*
Partridge, C. W. H., 361, *416*
Pasternak, V. Z., 365, *416*
Pasto, D. J., 303, *332*
Patai, S., 136, 137, *191*
Patterson, J. M., 258, *332*
Patterson, S., 9, *58*
Pattison, V. A., 390, *419*
Paul, D. E., 427, *448*
Paul, M. A., 265, *332*
Paul, R., *420*
Paul, T., 425, *444*
Pearson, D. L., 255, *327*
Pearson, R. E., 39, *58*
Pearson, R. G., 90, *124*
Pedersen, C. J., 340, *420*
Pelster, H., 225, *329*
Penczek, S., 258, *332*
Peover, M. E., 427, *444*
Pepper, D. C., 242, *332*
Perevalova, E. G., 383, 389, *421*
Perst, H., 198, 256, *332*
Petersen, C. E. L., 465, *507*
Peterson, H. J., *327*
Peterson, J. M., 402, *415*
Peterson, P. E., 320, *332*
Petrovich, J. P., 308, *327*
Pettit, R., 218, *330*
Pfeiffer, G. V., 315, *332*
Pfeiffer, P., 226, *332*

Pfeil, E., 256, *331*, 391, *420*
Philip, H., 23, *59*
Phillips, T. R., 387, *420*
Phillips, W. D., 432, *444*
Piccard, J., 426, *448*
Piette, L. H., 427, *446*
Pike, A. B., 357, *418*
Pimentel, G. C., 181, *188*
Pincock, R. E., 255, 320, *325*, *332*
Pine, S. H., 381, 385, 387, *420*
Pink, P., 438, *445*
Pinkney, P. S., 357, *420*
Pinnick, H. W., 439, *446*
Pischugin, F. V., 441, *447*
Pittman, C. U., Jr., 199, 200, 209, 210, 212, 213, 221, 223, 224, 226, 232, 243, 251, 274, 291, 293, 295, 296, 298, 302, 320, *326*, *327*, *330*, 331, *332*
Pitts, J. N., Jr., 9, *58*
Pitzer, K. S., 259, 284, *330*, *332*
Plattner, P. A., 256, *332*
Plonka, J. H., 67, 117, 118, 119, *125*
Pocker, Y., 292, *333*
Pohlke, R., 473, *508*
Pohmer, L., 455, 465, *508*
Poland, J. S., 82, *123*
Politzer, I. R., 363, *419*
Pomerantz, M., 480, 492, *504*
Pople, J. A., 315, *335*
Porter, G., 472, *507*
Porter, R. D., 310, 316, 320, *331*, *332*
Porter, R. R., 187, *191*
Posner, J., 394, *416*
Potter, S. E., 384, *418*
Poulter, C. P., 111, *124*
Poutsma, M., 34, 46, *59*
Powell, F. X., 73, *124*
Praud, L., 452, *507*
Preiss, M., 451, *506*
Prelog, V., 241, 274, *333*
Pretty, A. J., 384, *418*
Prinzbach, H., 94, 121
Pritchard, D. E., 214, *331*
Pritzkow, W., 140, *191*
Prokopov, T. S., 404, *417*
Prosser, T. J., 131, 160, *189*, *191*
Przybyla, J. R., 109, *124*
Puterbaugh, W. H., 371, 372, 387, 407, 412, 413, *420*
Putnam, W. E., 67, *124*

Putney, R. K., 136, *191*
Pyun, C., 495, *504*

Q

Quilico, A., 482, *507*

R

Raaen, V. F., 247, *326*
Raber, R. J., 278, *328*
Rabinovitch, B. S., 100, *125*
Rabinowitz, R., 8, 20, *59*
Rahman, M., 143, 148, 161, *191*
Raley, J. H., 10, 51, *59*
Ramberg, L., 397, *420*
Ranade, A. C., 414, *420*
Rando, R. R., 90, *124*
Rao, K. N., 231, *330*
Rappe, C., 395, *420*
Rathke, M. W., 347, *420*
Ratts, K. W., 350, *421*
Rausch, M. D., 380, 406, *419*, *420*
Razenberg, E., *126*
Readio, P. D., 45, *59*
Reagan, M. T., 148, 149, *191*
Reardon, R. C., 145, *191*
Rectenwald, G., 9, *58*
Redmond, J. W., 174, *189*
Reed, W. L., 270, *331*
Rees, C. W., 129, 150, 156, 186, *188*, *189*, *190*, 462, 472, 475, 484, 490, *505*, *507*
Reichle, W. T., 166, *191*
Reid, W., 482, *507*
Reimlinger, H., 92, *124*
Reinisch, R. F., 136, *191*
Reinmuth, O., 341, 346, 347, 353, 354, 359, *419*
Reinmuth, W., 233, *325*
Reinmuth, W. H., 427, *446*
Reiser, A., 130, 131, 132, 143, *191*
Renaud, D. J., *126*
Renfrow, N. B., 141, 159, 160, 161, 162, *189*
Renfrow, W. B., 403, *420*
Resemann, W., 128, *191*
Respess, W. L., 354, *418*
Rettig, K. R., 72, 106, 107, *122*, *123*
Reynolds, J., *447*
Riad, Y., 86, *122*
Rice, F. O., 128, 136, *191*
Rice, S. N., 167, *191*

Richards, J. H., 82, *126*
Richardson, D. B., 67, 96, *124*
Richey, H. G., Jr., 212, 267, *327*
Rickborn, B., 255. *329*, 342, *416*
Rickter, D. O., 291, *329*
Rieger, P. H., 427, *444*, *446*
Rieker, A., 427, *446*
Rinkler, H. A., 80, *123*
Risson, J., 180, *189*
Rist, H., 454, *506*
Ritchie, C. D., 343, *420*
Ritter, A., 94, *125*
Robbins, R. F., 171, *189*
Roberts, B. P., 46, *58*
Roberts, I., 302, *333*, 391, *420*
Roberts, J. D., 307, 314, 316, 318, *330*, *333*, 407, *420*, 450, 451, 453, 454, 478, *504*, *507*
Roberts, R. D., 436, *445*
Robertson, J. M., 282, *329*
Robinson, E. A., 209, *328*
Robinson, G. C., 205, *335*
Rochester, C. H., 265, *333*
Rockett, B. W., 407, *421*
Rodgers, A. S., 109, *123*
Roe, D. M., 460, *507*
Rosenberg, H., 406, *419*
Ross, J. R., 226, *325*
Rossi, R. A., *504*, *505*
Roth, H. D., 97, *124*
Rothberg, I., 86, *124*, 278, *326*
Rothstein, E., 197, *329*
Rowland, F. S., 66, 109, *126*
Rowley, A. G., 443, *447*
Roy, J., 159, 176, *188*
Royals, E. E., 344, 345, 353, *420*
Rüchardt, C., 466, *507*
Russell, G. A., 34, 36, *59*, 86, 90, *124*, 424, 427, 428, 429, 431, 432, 433, 435, 436, 437, 438, 439, 440, *446*, *447*
Russell, K. E., 136, *191*
Rust, F. F., 10, 20, 35, 51, *57*, *59*
Ryabova, R. S., *333*
Ryang, M., 88, *124*

S

Saaf, G., 454, *506*
Sacher, E., 297, *327*
Sachs, W. H., 67, *122*
Sadler, I. H., 110, *124*

Sado, A., 228, *334*
Saines, G., 221, 226, 255, 267, 286, 287, *327*
St. John, N. B., 347, *417*
Sanchez, R. A., 314, *326*
Sandrock, G., *328*
Sands, R. H., 427, *444*
Sarda, P., 67, *121*
Sargeant, P. B., 115, *124*
Sargent, G. D., 221, 317, *333*, 435, *447*
Sargent, M. V., 477, *505*
Sasaki, T., 500, *507*
Sauer, J., 130, 131, 156, 157, *191*, 486, *507*
Sauers, R. R., 93, *125*
Saunders, J. H., 346, *420*
Saunders, M., 217, 256, 292, 319, *327, 333*
Saunders, W. H., Jr., 138, 139, 144, *190, 191*, 314, 318, *333*
Sayigh, A., 280, 281, *327*
Schaad, L. S., 459, *506*
Schaad, R. E., 298, *329*
Schadt, F. L., 308, 309, *329*
Schaefer, H. F., 70, *120*
Schaefer, J. P., 349, 357, *421*
Schafer, M. E., 459, 468, *524, 507*
Schäffer, H., 390, *421*
Scheffler, K., 427, 432, *446, 447*
Scheiffele, E., 148, 149, 154, 166, *190*
Scheutzow, D., 431, *445*
Schiess, P. W., 298, *328*
Schlenk, W., 249, *333*, 425, 434, 436, *447*
Schleyer, P. von R., 196, 200, 211, 277, 278, 279, 280, 282, 283, 284, 308, 309, 314, 315, 316, 318, 319, *326, 328, 329, 330, 331, 333, 335*, 359, *420*
Schlögl, K., 409, *418*
Schlosser, M., 92, 111, *125*, 126, 403, *421*
Schmidt, A., 150, *190*
Schmidt, D., 114, *123*
Schmidt, U., 67, *125*
Schmitz, E., 67, 83, *125*
Schneider, A., 295, *325*
Schneider, R., 128, *190*
Schoenewaldt, E. F., 280, 281, *327*
Scholl, R., 425, *447*
Schölltopf, U., 389, 390, *421*
Schön, M., 482, *507*
Schreiber, K. C., 239, 319 *335*
Schriesheim, A., 214, 226, 264, 266, *327*
Schroder, G., 459, *505*
Schubert, M. P., 425, 426, 429, 431, *446*

Schubert, W. M., 270, *333*
Schulz, J., 502, *506*
Schulz, L., *120*
Schulze, F., 347, *417*
Schulze, J., 267, *330*
Schumacher, H., 300, 302, *334*
Schuster, I. I., 212, *333*
Schwartz, W., 297, *328*
Schwarz, M., 282, *333, 334*
Schwarz, R., 374, 377, *418*
Schwarz, R. A., 376, 414, *415, 422*
Schwarzenbach, G., 426, 431, *446*
Schwechten, H. W., 426, 437, *448*
Scott, C. B., 299, *334*
Scott, N. D., 427, *447*
Scott, R. B., Jr., 397, *421*
Scott, R. M., 385, *415*
Scott, W. T., 396, *419*
Scouten, G. G., 436, *445*
Scriven, E. F. V., 172, 173, 178, *188*
Searle, R. J. G., 179, *189*
Seebach, D., 362, *416, 421*
Seel, F., 231, *333*
Seibert, R. A., 454, *507*
Seif, L., 496, *505*
Selman, S., 390, 391, *421*
Selwood, P. W., 4, *59*
Semanov, D. A., 316, *330*
Semenow, D. A., 451, 478, *507*
Sen, J. N., 13, *58*
Seo, E. T., 429, 437, *447*
Serini, A., 197, *331*
Serre, J., 452, *507*
Serve, M. P., 303, *332*
Servis, K. L., 451, *507*
Seubold, F. H., 20, *59*
Seyferth, D., 66, 87, 88, 94, 110, *125*, 342, *421*
Shanley, W. B., 298, *329*
Shapiro, B. I., 427, *447*
Shapiro, J. S., 77, *125*
Shapiro, R. H., 81, *125*
Sharp, D. W. A., 227, *333*
Sharp, J. T., 466, 486, *504*
Sharples, G. M., 459, *505*
Shechter, H., 115, *124*
Shein, S., 441, *447*
Shemin, D., 348, *418*
Shen, Y. H., 110, *123*
Shenkin, P., 405, *421*

Shepelavy, J. N., 132, 167, *188*
Shepherd, R. A., 462, *507*
Sheppard, N., 227, *333*
Sherf, K., 16, *58*
Sherman, P. D., 303, *334*
Shih, S., 430, 437, 442, *444*
Shiner, V. J., 276, 289, 314, *329*, *333*
Shirley, D. A., 356, 407, *421*
Shoosmith, J., 72, *122*
Shoppee, C. W., 304, *333*
Short, W. T., 20, 48, *58*
Shriner, R. L., *331*, 347, *421*
Siddall, J. B., 354, *415*
Sidler, J. D., 390, *419*
Siegle, L. W., 439, *447*
Silbey, R., 75, *121*
Silver, M. S., 243, 289, 316, *330*, *333*
Simmons, H. E., 66, 87, *120*, *125*, 132, 137, 142, 167, 186, *188*, *189*, *190*, 450, 452, 454, *507*
Simmons, M. C., 96, *124*
Simon, J. W., 100, *125*
Simons, J. P., 74, *125*
Simonsen, W. J., *418*
Simson, J., 162, *190*
Singer, G. M., *504*
Singer, L. S., 95, *125*, 430, *446*
Singh, G., 362, *421*, 498, *506*
Singh, M., 498, *506*
Skell, P. S., 23, 45, *59*, 67, 76, 84, 91, 102, 106, 108, 109, 110, 111, 117, 118, 119, *120*, *122*, *125*, 203, 244, 246, *329*, *333*
Sketchley, J. M., *506*
Skorcz, J. A., 497, 498, *504*
Slansky, C. M., 259, *330*
Slates, R. V., 431, *447*
Slaymaker, S. C., 67, *124*
Sloan, M. F., 131, 141, *189*, *191*
Slocum, D. W., 386, 389, 405, 407, 408, 409, 410, 412, *421*
Small, L. F., 231, *326*
Smalley, R. K., 160, *191*
Smentowski, F. J., 392, *422*, 428, *447*
Smid, J., 339, 340, *416*, *422*, 432, *445*
Smissman, E. E., 396, *421*
Smit, P., 500, *507*
Smith, D. M., 466, *504*
Smith, H., 342, 372, *421*
Smith, J. G., 378, *421*
Smith, J. R. L., 443, *447*

Smith, L. I., 359, *421*
Smith, P. A. S., 128, 130, 136, 143, *191*, 322, *333*
Smith, R. A., 113, *120*
Smith, R. D., 66, 87, *125*
Smith, R. H., Jr., 163, 164, *192*
Smith, R. L., 67, *120*
Smith, W. H., 73, *125*
Smolinsky, G., 128, 133, 134, 148, 158, 160, 161, 163, 170, 171, *191*, *192*
Smoot, C. R., 299, *329*
Sneen, R. A., 246, *334*
Snow, D. H., 439, *445*, *446*
Snyder, E. I., 308, *329*
Snyder, L. C., 427, *445*
Soddy, T. S., 407, *417*
Soffer, L. M., 12, *58*
Solly, R. K., 468, 472, 490, *504*
Solodar, A. J., 481, *507*
Solter, L. E., 374, 377, *418*
Sommelet, M., 385, *421*
Sommer, L. H., 94, *125*, 323, *329*, *333*
Songstad, J., 90, *124*
Sorensen, T. S., 224, 225, 274, *333*
Spangler, M. S., 255, 286, 287, *327*
Spanwick, J., 443, *447*
Sparks, W. J., 295, *334*
Spencer, C. F., *326*
Speziale, A. J., 350, *421*
Splitter, J. S., 156, *191*
Spokes, G. N., 468, *504*
Spotswood, T. M., 452, *504*
Spring, D. J., 498, *505*
Sprecher, M., 280, 281, *327*
Sprung, J. L., 118, *126*
Stabba, R., 502, *506*
Stagner, B. A., 128, *191*
Stam, M. F., 430, *444*
Stang, P. J., 299, *334*
Starer, I., 91, *125*
Starichenko, V. F., 441, *447*
Starratt, A. N., 161, *189*
Staudinger, H., 62, *126*, 128, *192*
Stegmann, H. B., 432, *447*
Stehlik, D., 7, *58*
Steinfeld, J. I., 472, *507*
Sternhell, S., 209, *329*, 340, *418*
Stetter, H., 282, *334*
Stevens, G., 81, 101, 107, *120*
Stevens, T. S., 381, 390, *416*, *417*, *421*

AUTHOR INDEX

Stevens, I. D. R., 97, *123*
Stevenson, G. R., 427, *447*
Stewart, R., 286, *334*
Stieglitz, J., 128, *192*
Stiles, M., 464, 468, 481, *504*, *507*
Stiles, R. M., 237, 275, 276, *325*, *334*
Stoermer, R., 453, *507*
Stork, G., 755, *334*, 352, 361, *421*
Storr, R. C., 451, *507*
Story, P. R., 313, *334*, 436, *445*
Stothers, J. B., 214, *334*
Strauss, H., 266, *331*
Strausz, O. P., 114, 115, 119, *121*, *122*, *126*
Streitwieser, A., Jr., 238, 246, 260, 284, 299, *334*, 338, *421*
Strom, E. T., 424, 428, 429, 432, 439, *447*
Struchkov, V. T., 218, 228, *330*
Struchal, F. W., 439, 440, *445*, *446*
Suaz, B. P., 221, *326*
Sugden, S., 426, *447*
Suhr, H., 472, *507*
Sullivan, P. D., 430, 431, *445*
Sundaralingam, M., 228, *334*
Sundberg, R. J., 152, 160, 163, 164, *192*
Surmatis, J. D., 298, *334*
Suschitzky, H., 160, *191*
Susz, B. P., 221, 228, *332*, *334*
Sutherland, I. O., 384, *418*
Sutherland, R. G., 129, 142, 149, 159, 161, 169, 174, 177, 181, 186, *188*, 407, 409, *421*
Sutter, J. R., 136, 137, *189*
Swain, C. G., 86, *126*, 299, *334*
Swain, M. S., 237, 275, *325*
Swamer, F. W., 356, 357, 370, *417*, *420*, *421*
Swanwick, M. G., 428, *447*
Swenson, J. J., *126*
Swenton, J. S., 67, *126*, 156, 170, *192*
Swern, Y., 154, *191*
Swiger, R. T., 439, *446*
Sycheva, T. N., 493, *507*
Symons, M. C. R., 209, 223, *332*, 430, *444*
Sytsma, L., 496, *508*
Szabo, A., 472, *507*
Szmuszkovicz, J., 352, 361, *421*
Szwarc, M., 258, *332*, 431, 436, *447*

T

Tabushi, I., 461, 488, 492, *507*
Taft, R. W., 231, 240, 271, *329*, *334*
Taguchi, V., 301, *326*

Tai, W. T., 397, *421*
Takahashi, J., 243, *335*
Takase, K., 477, *508*
Takaya, H., 82, *124*
Takaya, T., 180, *188*
Takimoto, H., 178, *190*
Takimoto, H. H., 165, *192*
Tamano, T., 488, *507*
Tan, C. C., 466, *507*
Tang, Y.-N., 66, 109, *126*
Tanida, H., 320, *334*
Tanner, D. D., 44, *59*
Tatlow, J. C., 483, *505*
Tavares, D. F., 285, *326*
Taylor, B. S., 231, *326*
Taylor, E. C., 358, 361, *421*
Taylor, J. W., 11, *59*
Taylor, R., 9, *58*
Tchelitcheff, S., *420*
Terashima, S., 163, *192*
ter Borg, A. P., *126*
Terrell, R., 352, 361, *421*
Terry, G. C., 132, *191*
Tetenbaum, M. T., 372, *417*
Thal, A., 425, *447*
Thaler, W. A., 44, *59*
Thayer, F. K., 348, *421*
Thebtaranonth, Y., 384, *418*
Thomas, A., 427, *444*
Thomas, H., 322, *327*
Thomas, R. M., 295, *334*
Thompson, J. A., 308, *334*
Thompson, T., 381, *421*
Thomson, D., 430, *444*
Thomson, D. T., 233, *334*
Thomson, J. B., 466, *504*
Thomson, R. H., 432, *445*
Thornton, E. R., 86, *124*, *126*
Thun, W. E., 430, *447*
Thurman, D. E., 94, *123*
Tickle, P., 107, *120*
Tidwell, T. T., 276, *325*
Tiemann, F., 128, *192*
Timm, D., 140, *191*
Tisue, G. T., 140, *190*
Tits, M., 35, *57*
Tobey, S. W., 228, *334*
Todd, L. J., 88, *125*
Todd, M. J., 163, *189*
Tolgyesi, W. S., 210, 218, 221, 227, 228, *332*

Tolles, W. M., 430, *447*
Topor, M. G., 69, *121*
Torssell, K., 428, *446*
Townsend, J., 427, *448*
Trapp, O. D., 13, 14, *58, 59*
Travis, D. N., 73, *123*
Traynham, J. G., 241, *333*
Trevilyan, A. E., 404, *415*
Trifan, D. S., 221, 307, 317, 318, 321, *335*
Trifunac, A. D., 95, *121*
Trost, B. M., 66, *126*
Trozzolo, A. M., 74, 75, 76, 77, *122, 126,* 134, *192*
Tsubomura, H., 74, 78, *123*
Tsuji, T., 320, *334*
Tsuno, Y., 141, *190*
Tuleen, D. L., 45, *59*
Turkel, R. M., 88, *125*
Turner, A. B., 114, *122*
Turner, J. O., 267, 298, *327*
Turro, N. J., 395, *420*
Tuttle, N., 345, *416*
Tuttle, T. R., 427, *447*
Tye, F. L., 226, *328*

U

Uchide, M., 500, *507*
Uff, A. J., 483, *505*
Ulland, L. A., 94, *125*
Uma, V., 142, 148, 159, 176, 177, *188*
Umbach, W., 322, *327*
Underhill, M., 461, *505*
Unni, A., 407, 409, *421*
Urban, R. S., 391, *417*
Urberg, M. M., 432, *447*
Urey, H. C., 391, *420*
Uri, N., 16, *59*
Urry, R. W., 438, *445*
Urry, W. H., 3, 11, *58, 59,* 97, *126*
Uschold, R. E., 343, *420*
Ustynyuk, Y. A., 383, 389, *421*
Uy, O. M., 77, *126*

V

Van Allan, J., 348, *415*
van der Plas, H. C., 500, 503, *505, 507*
van der Veen, J. M., 91, *122*
van der Waals, J. H., 267, *330*
Van Dine, G. W., 70, 73, 76, *122*
Van Eenam, D. N., 386, *417*
Van Emster, K., 196, *330*
Van Fossen, R. Y., 490, *505*
Van Scoy, R. M., 10, *58*
van Tamelen, E. E., 300, 302, *334*
Vaughan, C. W., 450, 451, 454, 478, *507*
Vaughan, W. E., 10, 20, 35, 51, *59*
Vaulx, R. L., 371, 411, *418*
Veach, C. D., 378, *421*
Vedejs, E., 462, *507*
Velthorst, N. V., 226, *334*
Vernon, C. A., 269, *334*
Viehe, H. G., 450, *507*
Vignale, M. J., 231, *330*
Villaume, J. E., 117, *125*
Villiger, V., 196, *324*
Vincow, G., 425, 430, 432, *444, 447*
Vinnik, M. I., *333*
Vinutha, A. R., *504*
Vivian, D. L., 185, *192*
Voevodkii, V. V., 441, *447*
Vogt, R. R., 450, *507*
Vollmer, J. J., 451, *507*
Volpin, M. E., 218, 228, *330*
Volz, H., 233, *334*
von Fraunberg, K., 159, 169, 173, *190, 192*
von Schriltz, D. M., 366, *419*
von Stolz, H.-D., 100, 115, *123*
Vorster, J., 197, *331*
Vosburgh, I., 128, *192*
Vysochin, V. J., 493, *507*

W

Waack, R., 223, 226, *332, 334*
Wade, K., 339, *416*
Wadley, E. F., 256, *327*
Wadsworth, W. S., Jr., 350, *421*
Wagner, G., 196, *334*
Wagner, P., 43, *59*
Waiss, A. C., 246, *334*
Walborsky, H. M., 340, *418*
Walden, P., 196, 229, *334*
Walinsky, S. W., 71, *124*
Walker, J. F., 427, *447*
Walker, P., 130, 138, *192*
Waller, F. J., 67, *122*
Walling, C., 3, 8, 16, 20, 36, 43, 51, 57, *59*
Walsh, A. D., 68, *126,* 224, *334*
Wang, R. H. S., 20, 48, *58*
Ward, A. M., 62, *126,* 197, *334*
Ward, H. R., 303, *334*

Wardenten, J., 13, *58*
Ware, J. C., 138, *191*
Warhurst, E., 436, *446*
Waring, A., 435, *445*
Warner, C. D., 372, *418*
Warner, P., 462, *507*
Warnhoff, E. W., 397, *421*
Washburne, S. S., 94, *125*
Wasserman, E., 66, 75, 76, 77, *120*, *126*, 133, 134, 148, *189*, *191*, *192*
Wasserman, H. H., 481, 492, *507*
Watanabe, H., 371, 372, 407, 414, *421*, *422*
Waters, W. A., 2, 16, *58*, 130, 138, *192*, 427, 428, 431, 434, *445*, *446*, *447*
Watts, W. E., 278, 282, 318, 319, *333*
Wawzonek, S., 233, *334*
Weber, S., 256, *332*
Wecker, E., 425, *448*
Wege, D., 488, *506*
Weickel, T., 425, 434, 436, *447*
Weiner, H., 246, *334*
Weiner, M. D., 342, *421*
Weissberger, A., 428, *447*
Weissman, S. I., 427, 430, 432, 436, *445*, *447*, *448*
Weitz, E., 426, 430, 431, 437, *448*
Wentrup, C., 114, *121*, 164, 165, *189*, *192*
Wentworth, G., 392, 393, *417*
Wentzel, F., 196, *330*
Werner, H., 82, *126*
Wescott, L. D., 116, 118, *125*
West, R., 228, *334*, 377, 378, 404, *417*, *422*, 427, *445*
Westheimer, F. H., 284, *334*, 391, *422*
Weston, R. E., Jr., 231, *330*
Weyerstahl, P., *126*
Whalen, D. M., 84, *123*
Whelan, W. P., 280, 281, 284, *327*, *334*
White, A. M., 214, 215, 217, 221, 222, 292, 319, *331*, *332*
White, E. H., 253, *334*
White, W. N., 316, *330*
Whitesides, G. M., 354, *418*
Whitmore, F. C., 197, 242, 298, *334*, 346, *422*
Whittaker, D., 81, 95, 101, *120*
Whitten, J. L., 452, *507*
Whittle, J. R., 109, *124*
Wiberg, K. B., 47, *59*, 246, 316, *334*, *335*

Wideman, L. G., 86, *122*
Wieland, D. M., 359, *418*
Wieland, H., 425, *448*
Wierenga, W., 254, *327*
Wiesemann, W., 426, 431, *446*
Wiggen, J. P., 460, *505*
Wilhelm, M., 284, *335*
Wilhite, D. L., 452, *507*
Wilkinson, G., 425, *504*
Willcott, M. R., 100, *121*
Willets, F. W., 132, *191*
Willfang, G., 256, *331*
Williams, B. H., 170, *192*
Williams, J. E., 315, *335*
Williamson, R. C., Jr., 36, *59*
Willis, C., *126*
Willis, C. J., 66, *121*
Willstätter, R., 426, *448*
Wilson, L. A., 228, *332*
Wilt, J. W., 23, *59*
Wingler, F., 87, *126*
Winkler, H. J. S., 461, *507*
Winstein, S., 111, 118, *124*, *126*, 205, 221, 239, 243, 261, 262, 263, 274, 290, 302, 304, 307, 308, 310, 313, 317, 318, 319, 321, *324*, *325*, *326*, *327*, *328*, *329*, *330*, *335*
Winter, R. A. E., 67, *121*
Wirthwein, R., 500, 501, *506*
Wisotsky, M. J., 212, *327*
Witte, J. F., 403, *420*
Wittenbrook, L. S., 399, 400, *420*
Wittig, G., 87, 92, *126*, 381, 389, *422*, 450, 454, 455, 461, 465, 467, 473, 474, 475, 480, 492, *507*, *508*
Woerner, F. P., 163, *190*
Wold, S., 428, *446*
Wolf, A. P., 116, *126*
Wolf, D. C., 402, *415*
Wolf, R. A., 403, *420*
Wolfe, J. F., 367, 371, 372, *415*, *422*
Wolff, M. E., 443, *448*
Wolfgang, R., 116, *124*
Wolinsky, J., 473, *505*
Wolthius, E., 496, *508*
Wong, C. M., 397, *421*
Wong, D. Y., *120*
Wong, K. H., 340, *422*
Wood, L. S., 117, *125*
Woodcock, D. J., 253, *334*

Woodward, R. B., *126*, 451, *508*
Woodworth, R. C., 102, 106, *125*
Wooster, C. B., 431, 434, 436, *448*
Work, S. D., 370, 371, 372, *422*
Worka, K. V., 438, *445*
Wriede, P., 435, *444*, *445*
Wright, J. C., 141, 155, *189*
Wright, W. V., *329*
Wu, S. W., 13, *58*
Wuhrmann, J. J., 221, 228, *334*
Wulff, K., 282, *334*
Wunderlich, K., 235, 251, *331*
Wyckoff, J. C., 443, *444*

Y

Yager, W. A., 66, 75, 76, 77, *120*, *126*, 133, 134, 148, *189*, *191*, *192*
Yamada, H., 461, 488, 492, *507*
Yamada, S. I., 163, *192*
Yamamoto, K., 94, *125*
Yamamoto, Y., 75, 76, 82, 114, 115, *123*, *124*, *126*
Yamatani, T., 477, *508*
Yasunami, M., 477, *508*

Yates, K., 320, *335*
Yates, P., 67, 82, *126*
Yilmaz, H., 117, *126*
Yokozawa, Y., 430, *448*
Yonezawa, T., 493, *508*
Yoshida, M., 495, *506*
Yoshida, Z., 488, *507*
Young, J. C., 88, *120*
Young, M. C., 427, *447*
Yuan, C., 225, *325*

Z

Zahler, R. E., 363, *416*, 454, *504*
Zech, B., 66, 82, *124*
Zeiss, G. D., 70, 73, 76, *122*
Ziegler, G. R., 338, *421*
Ziemek, P., 427, *446*
Zilkha, A., 436, *448*
Zimmer, H., 156, 158, *192*
Zimmerman, H. E., 392, *422*
Zmbov, K. F., 77, *126*
Zollinger, H., 66, 79, *126*
Zook, H. D., 342, *422*
Zwanenburg, B., 396, *419*

SUBJECT INDEX

A

Acetyl peroxide, decomposition of, 10–11, 14–15
Acidity of hydrocarbons, 337–339
Aldol condensation, 344–345, 373
Appearance potentials of carbonium ions, 271–273
Aryl-bridged ions, 307–311
Aryl radicals, 12
Arynes
  factors affecting the stability of, 472–480
    ring size, 472–477
    substituents, 478–480
  methods of formation of, 463–472
    gas phase methods, 466–472
      by pyrolysis, 467–472
      under electron impact, 466–467
    solution methods, 463–466
      from addition reactions, 463
      from amines, 466
      from aryl dihalides, 465–466
      from aryl halides, 464–465
      from benzenediazonium 2-carboxylate, 463–464
  methods of investigation of, 455–462
    metal complexation, 459–462
    spectroscopic, 459
    trapping, 456
  nomenclature of, 450
  reactions of, 480–499
    2 + 2 cycloaddition, 480–481
    2 + 3 cycloaddition, 481–483
    2 + 4 cycloaddition, 483–488
    2 + 6 cycloaddition, 488–489
    2 + 8 cycloaddition, 489
    ene reaction, 491–494
    with nucleophiles, 494–499
    polymerization, 489–491
  structure of, 451–452
Atomic carbon, 116–119
Azobisisobutyronitrile (AIBN), decomposition of, 13–15

B

Benzene radical anion, 424–425
Benzene radical cation, 424–425
Benzil semidione, 425
Benzilic acid rearrangement, 390–391, 394, 397
Benzoin condensation, 349
Benzophenone ketyl, 425, see also anion radicals, reactions of
Benzyne, see also Arynes
  analogy with cycloalkynes, 449–450
  historical background, 452–455
  structure of, 451–452
m-Benzyne, 459
p-Benzyne, 452, 459
Binding energy measurements, see ESCA studies
Bridgehead carbonium ions, 281–285
Bromotrichloromethane, photolysis of, 14–15
t-Butyl cation, 219
t-Butyl peracetate, decomposition of, 12

C

Carbanions
  factors affecting the stability of, 342–344
    electronic, 343
    homoconjugative effects, 344
    medium effects, 342
    nucleophilicity, 342
  formation of, 341–342
    M–H exchange, 341
    M–X exchange, 341
    metal addition, 342
    reductive metalation of alkyl and aryl halides, 341
    reductive metalation of ethers, 342
    transmetalation, 342

534 SUBJECT INDEX

Carbanions—*Cont.*
  methods of investigation of, 339–341
    contact and ion pair nature, 339–340
    isotopic exchange, 340–341
    spectroscopic, 340
    use of chelating ethers, 340
  reactions of, 344–403
    multiple anion, 365–369
      1,3-dianions, 366–377
        from β-diketones and related compounds, 366–369
        from carboxamides and related compounds, 371–372
        from carboxylic acids, 369–371
        from nitrogen compounds, 374–377
        from sulfur-containing compounds, 372–374
      1,1- 1,2-, and 1,4-dianions, 377–380
    nucleophilic acyl substitutions, 354–358
    nucleophilic additions, 344–354
      to azo linkages, 347
      to carbonyl groups, 344–350, 353–354
      to imines, 347, 351
      to nitriles, 347, 349
      to thioketones, 347, 351
    nucleophilic substitutions, 358–365
      in aliphatic systems, 358–363
      in aromatic systems, 363–365
      benzyne mechanism, 364
      direct displacement, 364
      with epoxides, 359
      with esters, 359
      Grignard coupling, 358
      with halides, 359–363
    rearrangements, 381–403
      1,2-rearrangements, 381–393
      three-membered ring intermediates, 393–403
Carbenoid species, 40, 78–79, 82–83, 86–89
  reactions of, *see* Carbenes, reactions of
Carbenes
  electronic states and structure of, 65
  spectroscopic observations, 72–77
    electron paramagnetic resonance (EPR), 74–77, 78, 95
    electronic and infrared, 72–74, 78
    mass, 77
    nuclear magnetic resonance (NMR), *see* CIDNP
  formation of, 64–67
    in addition processes, 64
    from base-induced elimination, 83–86
    copper-catalyzed decompositions, 82–83
    from diazirines, 83
    from diazoalkenes, 79–83
    in α-elimination processes, 64, 79–89
    gas-phase heats of formation of, 77
    geometrical isomerism in, 76–77
    history of, 62–64
    in insertion processes, 64
    from ketones, 83
    methylene iodide and Zn–Cu couple, 87
    nomenclature of, 61–62
    organomercury halides, 87–88
    from organometallic reagents, 86–89
    photolytic decompositions, 79–82
    reactions of, 63, *see also* Carbene and carbenoid species
    in rearrangement processes, 64
    thermal decompositions, 79–82
  "free" carbene, 78–79
  reactions of, 89–115
    addition to multiple bonds, 101–113
      to acetylenes, 112
      affect of carbene structure, 110–111
      to aromatic compounds, 113
      to C—N multiple bonds, 113
      general reaction characteristics, 101–102
      olefin reactivity, 107–110
      polar and steric effects, 111
      solvent dependence, 111
      stereospecificity, 105–107
      theoretical considerations, 102–105
    effects of electronic state, 89–91
    insertion, 92–101
      mechanisms of, 92–95
      structure and reactivity in, 95–101
    reactions with nucleophiles having unbonded electron pairs, 91–92
    rearrangements, 113–115
      of 1,1-diaryl-2-haloethylenes, 115
      of diazoketones (Wolff rearrangement), 114
      migratory aptitudes in, 115
Carbenium ions, *see* Carbonium ions, nomenclature of

SUBJECT INDEX 535

Carbocations, *see* Carbonium ions, nomenclature of
Carbonium ions
  analogy with boranes, 199–200
  bonding and hybridization in, 194–196, 198–200
  factors affecting the stability of, 258–266
    inductive and resonance effects, 266–273
    medium effects, 259–268
      heats of formation, 259, 272–273
      heats of ionization, 259
      heats of solution, 261–262
      steric hindrance to solvation, 260, 279–281
    steric effects, 273–287
  formation of, 248–259
    by electron loss, 248–250
    by electrophilic addition, 256–258
    by heterolytic fission (ionization), 250–255
      of alcohols, ethers, and thiols, 251–252
      of diazonium salts and related compounds, 252–253
      of halides and esters, 250–252, 255
    by interaction of neutral molecules, 258
  historical development of, 196–198
  methods of investigation of, 202–203
    concentration and lifetime effects, 203–209
    electrolytic methods, 229–233
    gas-phase versus solution phase in, 202–203
    kinetic methods, 233–241
    product analysis, 241–248
    spectroscopic methods, 209–229
      infrared and Raman spectroscopy, 218–223, 228
      ion cyclotron resonance, 202
      mass spectrometry, 202–203, 248–249
      nuclear magnetic resonance methods, 209–218
      photoelectron spectroscopy, 228–229
      ultraviolet spectroscopy, 223–226
      X-ray spectroscopy, 227–228
  nomenclature of, 200–202
  reactions of, 287–302
    addition to alkenes, 287, 295
    addition to aromatic compounds, 287, 298–299
    anchimeric assistance in, 238–241, 302–321
    bridging, 288, 302–321
    elimination of a proton, 288–290
    fragmentation, 288, 297–298
    hydride abstraction, 287, 295
    internal alkylation, 288
    ion pairs in, 204–208
    isotope effects in, 241
    kinetics of, 205–207
    with nucleophiles, 287, 299–300
    photochemical reactions, 300–302
    solvent effects in, 243–246, 259–268
    stereochemistry of, *see* S_N1 and S_N2 reactions
    steric strain in, 237–239
    substituent effects in, 239–241, 266–271
    thermodynamic versus kinetic control in, 234–235
Carbynes, 119
Chemically induced dynamic nuclear polarization (CIDNP)
  in carbene reaction studies, 95–97
  in free radical reaction studies, 7, 436
  in the Stevens rearrangement, 383
i-Cholesterol rearrangement, 304–311
Claisen condensation, 356–357
Claisen-Schmidt condensation, 345
Common ion effect, 207
Conductivity measurements, 229–231
Conjugate addition, 350–354
Crossed-Claisen condensation, 357
Curtin-Hammett principle, 285–286
Curtius rearrangement, *see* Nitrenes, formation of, from azides
Cyanonitrene, 132, 136, 142–143
Cyclic voltametry, *see* Polarographic techniques
Cycloalkynes, 450, 472–477
Cycloheptyne, 474
Cyclohexyne, 473
Cyclooctyne, 473, 475
Cyclopentyne, 473–474
Cyclopropylcarbinyl cations, 315–317

D

Darzens glycidic ester synthesis, 348
Dehydrobenzene, *see* Benzyne
Dehydrocyclopentadienyl anion, 476–477
Dieckman condensation, 357

Diels-Alder reaction, 394
Dynamic nuclear magnetic resonance spectroscopy, 216–218

E
ESCA studies, 228–229

F
Favorskii-Zimmerman rearrangement, 392–393
Free radicals
  detection of, 3–7
    by use of electron spin resonance (ESR), 3–7
    by use of magnetic susceptibility, 3–4
    by use of nuclear magnetic resonance (NMR), 7
  formation of, 10–17
    in addition to aldehydes, 20
    in addition of alkyl radicals to alkynes, 19
    in addition to aromatics, 19
    in addition of halogen to alkenes, 19
    in addition to imines, 20
    in addition of polyhalomethanes to alkenes, 2, 19
    in autooxidation of alkyl aromatics, 2
    in bimolecular redox reactions, 16–17
    in fragmentation of tertiary alkyl hypohalites, 3
    in halogenation of alkanes, 2, 18
    in oxidation of alkenes, 3
    in photolysis reactions, 14–16
    in polymerization of olefins, 2
    in reduction of alkyl halides, 2
    in reaction of diazomethane with carbon tetrachloride, 3
    in thermolysis of covalent bonds, 10–14
      azo compounds, 13–14
      peroxides, 10–13
  gas-phase studies of, 2, 271
  reactions of, 17–26
    addition of carbon tetrachloride to an alkene, 7–8
    bimolecular radical-propagating reactions, 17–21
      addition, 19–20
      displacement (abstraction), 17–19
      reduction, 20–21
    chain reactions, 7–9, 50–57
      kinetic aspects, 50–57
      mechanisms of, 7–9
    fragmentation, 21–22, 24
      α-elimination, 22
      β-elimination, 21–22, 24
    free radical halogenation of alkanes, 7
    nonchain reactions, 9–10
    oxidative coupling of esters, 9
    photochemical dimerization of benzophenone, 9
    reaction of trialkyl phosphites with mercaptans, 8
    rearrangement, 22–24
    termination, 25–26
    vinyl polymerization, 10
  reactivities of, 26–50
    bridging effects, 40–43
    polar effects, 32–39
      complexing, 40–46
      donor–acceptor, 36–39
      inductive, 33–36
    resonance factors, 26–32
    solvent, 40–43
    steric, 46–50
  resonance energies, 27
  stable, 6–7
Foote-Schleyer relationship, 277–279

G
Grignard reaction, 346–347, 353, 355
Grignard reagents, 341, 346–347, 353, 355, 358, 359, 365, 391
Grovenstein-Zimmerman rearrangement, 392–393
Grunwald-Winstein relationship, 262–263

H
H_O Acidity function, 264–265
H_R Acidity function, 264–265
Hammond postulate, 31, 97, 236, 284, 310–311
Hammett acidity function, 264–265
Hammett equation, 240, 269–270
Hetarynes, 499–503
Heteronuclear bridged cations, 320
Hofmann rearrangement, see Nitrenes, formation by α-elimination
Hofmann-Löffler reaction, 443
Hofmann rule, 289
Homoallylic cations, 311–314

SUBJECT INDEX

Homoaromaticity, 319–320
Hydrazinium radical cation, 425–426
Hyperconjugation
  in carbonium ions, 195
  in free radicals, 26

I

Internal return, 205
Ivanoff reagents, 369

K

Kinetic acidity, 337–339
Knoevenagel condensation, 345–346
Kolbe reaction, 249

L

Lossen rearrangement, *see* Nitrene, formation by α-elimination

M

Mannich reaction, 349
Markownikoff rule, 200
Mass-law effect, *see* Common ion effect
McEwen-Streitweiser-Applequist-Dessy (MSAD) acidity scale, 338–339
Metal-carbene complexes, *see* carbenoid species
Metalation of aromatic compounds, 403–415
  directed, 406–415
  general methods, 403–406
Methylene, 65, 68–70
Michael addition, 352
Migratory aptitude, 293–294
Müller-Pritchard relationship, 214–215

N

Naphthalene anion radical, *see* Radical anion, reactions of
Neber rearrangement, 401–403
Neighboring group participation, *see also* Anchimeric assistance, 247, 276, 302–321
Nitrenes
  applications of, 184–187
  formation of, 135–157
    from 3-aryl-1,4,2-dioxazolidin-5-ones, 157
    by deoxygenation of nitro and nitroso compounds, 152–153
    by α-elimination, 153–155
    by irradiation of oxaziridines, 156
    from isocyanates, 156–157
    from phosphinamines, 156
    by oxidative routes, 155–156
    by thermolysis, 135–142
      Curtius rearrangement, 140–142
      migratory aptitudes in rearrangements, 138–139
      nature of the intermediate, 136–139
      neighboring group participation, 136–139
    from azides, 135–152
      by metal catalysts, 150–152
      by photolysis, 143–150
      by thermolysis, 135–147
  "free" nitrene, 131
  historical background of, 128–129
  nomenclature of, 128
  reactions of, 157–181
    addition to olefins, 167–169
    aromatic substitution, 171–178
    dimerization, 169–170
    hydrogen abstraction, 157–160
    insertion into aliphatic C–H bonds, 160–163
    rearrangements, 163–167
    trapping by nucleophiles, 179–181
Nitrenium ions, 181–184, 322
Nitrenoid species, 131, 150–153
Nonclassical carbonium ions, *see* Carbonium ions, reactions of, bridging
Norbornyl cation, 221–223, 317–319
Nucleophilicity constants, 299–300

O

Organocadmium reagents, 356
Organocopper reagents, 359
Organothallium reagents, 358, 361
Organozinc reagents, 356
Oxenium ions, 198, 199, 322
Oxinium ions, 198
Oxocarbonium ions, *see* Oxonium ions
Oxonium ions, 198–199, 221

P

Pentacoordinated carbocations, 257–258
1,8-Perinapthalyne, 451, 463, 479
Perkin reaction, 348
Phenonium ion, *see* Aryl-bridged ions

Phenothiazynes, 502–503
Pinacol rearrangement, 292–294
pK_a Values, see Acidity of hydrocarbons
pK_{R+} Values, 266–268
Polarographic techniques, 231–233
Protonated cyclopropanes, 216–218, 314–315
2,3-Pyridyne, 500
3,4-Pyridyne, 499–500

R

Radical anions
  classification of, 424–425
  definition of, 424
  formation of, 426–429
    by dismutation (disproportion) processes, 424, 428, 429
    by electron transfer processes, 428–429
    by oxidation processes, 428
    by reduction processes, 425–428
  historical development of, 425–426
  methods of investigation of, 431–433
    electrochemical methods, 431
    electron spin resonance method, 426, 432, 433
    magnetic susceptibility method, 426–431
    spectrophotometric methods, 432
  reactions of, 434–444
    by fragmentation, 440–441
    dimerization, 434
    electron transfer, 435–439
      by reaction with alkyl halides, 435, 438–439
      by reaction with ethers and esters, 435–436
      by reaction with oxygen, 435, 437–438
    in polymerization processes, 436
    hydrolysis, 434
Radical cations
  classification of, 424–425
  definition of, 424
  formation of, 429–431
    by dismutation (comproportionation) processes, 424, 430–431
    by oxidation processes, 429–430
    by reduction processes, 430
  historical development of, 425–426
  methods of investigation of, 431–433
    electrochemical methods, 431
    electron spin resonance spectroscopy, 426, 432–433
    magnetic susceptibility method, 426, 431
    spectrophotometric methods, 432
  reactions of, 437, 441–444
    by disproportionation, 437
    by hydrogen abstraction, 443
    by redox processes, 437
    with transition metals, 437, 441–444
Ramberg-Bäcklund reaction, 397–401
Reformatsky reaction, 347, 371
Robinson annelation procedure, 352

S

S_N1 Reactions, 198, 202–209, 233–241, 250–252, 260–262, 289, 304, 358
S_N2 Reactions, 198, 202–209, 304, 358
S_Ni Mechanism, 382, 390
Saytzeff rule, 289
Silicenium ions, 201, 323–324
Silicocations, see Silicenium ions
Siliconium ions, see Silicenium ions
Simmons-Smith reaction, 87
Solvolysis Reactions, see S_N1 Reactions and S_N2 Reactions
Sommelet-Hauser rearrangement, 381, 385–389
Steric hindrance to ionization, 277–279
Stevens rearrangement, 381–389
Stobbe condensation, 348
Sulfenium ions, 199
Sulfonium ions, 199
Sulfonium ylides, 350–351
Sulfoxonium ylides, 350–351
Super acids, 365–366

T

Tetracoordinated carbocations, 257–258
Thermodynamic acidity, 337–339
2,3-Thiophyne, 501–502
Three-center two-electron bonding, 216
Thorpe-Ziegler reaction, 349
Transannular effects, 241, 292
Triarylaminium radical cation, 425–426

U

Urey-Bradley force field calculations, 219

SUBJECT INDEX

V
Vinyl cations, 250–251, 256, 273
Vinyl radicals, 19

W
Wagner-Meerwein shift, 292
Walden inversion, 304
Winstein equation, 308–309
Wittig reaction, 350
Wittig rearrangement, 389–390

Wolff-Kishner reduction, 375
Wolff rearrangement, *see* Carbenes reactions of
Wurster's red, 425

Z
Zero-field splitting parameters, D and E
  of carbenes, 74–76
  of nitrenes, 133–135

ORGANIC CHEMISTRY
A SERIES OF MONOGRAPHS

EDITORS

ALFRED T. BLOMQUIST
Department of Chemistry
Cornell University
Ithaca, New York

HARRY WASSERMAN
Department of Chemistry
Yale University
New Haven, Connecticut

1. Wolfgang Kirmse. CARBENE CHEMISTRY, 1964; 2nd Edition, 1971
2. Brandes H. Smith. BRIDGED AROMATIC COMPOUNDS, 1964
3. Michael Hanack. CONFORMATION THEORY, 1965
4. Donald J. Cram. FUNDAMENTALS OF CARBANION CHEMISTRY, 1965
5. Kenneth B. Wiberg (Editor). OXIDATION IN ORGANIC CHEMISTRY, PART A, 1965; Walter S. Trahanovsky (Editor). OXIDATION IN ORGANIC CHEMISTRY, PART B, 1973
6. R. F. Hudson. STRUCTURE AND MECHANISM IN ORGANO-PHOSPHORUS CHEMISTRY, 1965
7. A. William Johnson. YLID CHEMISTRY, 1966
8. Jan Hamer (Editor). 1,4-CYCLOADDITION REACTIONS, 1967
9. Henri Ulrich. CYCLOADDITION REACTIONS OF HETEROCUMULENES, 1967
10. M. P. Cava and M. J. Mitchell. CYCLOBUTADIENE AND RELATED COMPOUNDS, 1967
11. Reinhard W. Hoffman. DEHYDROBENZENE AND CYCLOALKYNES, 1967
12. Stanley R. Sandler and Wolf Karo. ORGANIC FUNCTIONAL GROUP PREPARATIONS, VOLUME I, 1968; VOLUME II, 1971; VOLUME III, 1972
13. Robert J. Cotter and Markus Matzner. RING-FORMING POLYMERIZATIONS, PART A, 1969; PART B, 1; B, 2, 1972
14. R. H. DeWolfe. CARBOXYLIC ORTHO ACID DERIVATIVES, 1970
15. R. Foster. ORGANIC CHARGE-TRANSFER COMPLEXES, 1969
16. James P. Snyder (Editor). NONBENZENOID AROMATICS, VOLUME I, 1969; VOLUME II, 1971

17. C. H. Rochester. ACIDITY FUNCTIONS, 1970
18. Richard J. Sundberg. THE CHEMISTRY OF INDOLES, 1970
19. A. R. Katritzky and J. M. Lagowski. CHEMISTRY OF THE HETEROCYCLIC N-OXIDES, 1970
20. Ivar Ugi (Editor). ISONITRILE CHEMISTRY, 1971
21. G. Chiurdoglu (Editor). CONFORMATIONAL ANALYSIS, 1971
22. Gottfried Schill. CATENANES, ROTAXANES, AND KNOTS, 1971
23. M. Liler. REACTION MECHANISMS IN SULPHURIC ACID AND OTHER STRONG ACID SOLUTIONS, 1971
24. J. B. Stothers. CARBON-13 NMR SPECTROSCOPY, 1972
25. Maurice Shamma. THE ISOQUINOLINE ALKALOIDS: CHEMISTRY AND PHARMACOLOGY, 1972
26. Samuel P. McManus (Editor). ORGANIC REACTIVE INTERMEDIATES, 1973
27. H.C. Van der Plas. RING TRANSFORMATIONS OF HETEROCYCLES, Volumes 1 and 2, 1973

In preparation

Stanley R. Sandler and Wolf Karo. POLYMER SYNTHESIS